To Satyan,

    I compliment you on [...] that Phase Transfer Catalysis can greatly improve your personal performance and corporate performance while providing a basis for expressing scientific creativity.

Marc Halpern
Aug 29, 2002

PHASE-TRANSFER CATALYSIS

Charles L. Liotta
Department of Chemistry
Georgia Institute of Technology
Atlanta, GA 30332

Charles M. Starks
Cimmaron Technical Associates
Tulsa, OK 74119

Marc E. Halpern
Sybron Chemicals
Cherry Hill, NJ 08002

PHASE-TRANSFER CATALYSIS
Fundamentals, Applications, and Industrial Perspectives

CHARLES M. STARKS
CHARLES L. LIOTTA
MARC HALPERN

CHAPMAN & HALL
New York • London

First published in 1994 by
Chapman & Hall
One Penn Plaza
New York, NY 10119

Published in Great Britain by
Chapman & Hall
2-6 Boundary Row
London SE1 8HN

© 1994 Chapman & Hall, Inc.

Printed in the United States of America

All rights reserved. No part of this book may be reprinted or reproduced or utilized in any form or by any electronic, mechanical or other means, now known or hereafter invented, including photocopying and recording, or by an information storage or retrieval system, without permission in writing from the publishers.

Library of Congress Cataloging in Publication Data

Starks, Charles M., 1934–
    Phase-transfer catalysis : fundamentals, applications & industrial perspectives / Charles Starks, Charles Liotta, and Marc Halpern.
      p.   cm.
    Includes bibliographical references and index.
    ISBN 0-412-04071-9 (acid-free)
    1. Phase-transfer catalysis.   I. Liotta, Charles L., 1937–
II. Halpern, Marc, 1954– .   III. Title.
QD505.L56   1993
661'.8—dc20   93-19659
                                                            CIP

British Library Cataloguing in Publication Data available

Please send your order for this or any **Chapman & Hall book to Chapman & Hall, 29 West 35th Street, New York, NY 10001, Attn: Customer Service Department.** You may also call our Order Department at 1-212-244-3336 or fax your purchase order to 1-800-248-4724.

For a complete listing of Chapman & Hall's titles, send your requests to **Chapman & Hall, Dept. BC, One Penn Plaza, New York, NY 10119.**

Contents

Preface	xiii
Chapter 1: Basic Concepts in Phase-Transfer Catalysis	1
A. Phase-Transfer-Catalyzed Reactions	1
B. Basic Steps of Phase-Transfer Catalysis	2
C. The PTC Reaction Rate Matrix	5
D. Anion Transfer and Anion Activation	6
1. Transfer	6
2. Anion Activation	11
E. Effect of Reaction Variables on Transfer and Intrinsic Rates	12
1. Catalyst Structure	13
2. Agitation	14
3. Kind and Concentration of Inorganic Reagent; Amount of Water Added	15
4. Amount and Kind of Organic Solvent Used, If Any	16
5. Temperature and Microwave Heating	17
6. Cocatalysts	17
F. Outline of Compounds Used as Phase-Transfer Catalysts	18
1. Soluble Catalysts	18
2. Insoluble Catalysts	20
3. Catalysts for Vapor-Phase Reactions	21
Chapter 2: Phase-Transfer Catalysts: Fundamentals I	23
A. Introduction	23
B. Structural Factors Affecting the Distribution of Anions Between Aqueous and Organic Phases	24
C. Structural Factors Affecting the Distribution of Phase-Transfer Catalyst Cations Between the Aqueous and Organic Phases	26
D. Effects of the Organic Phase Polarity on the Distribution of Phase-Transfer Cation–Anion Pairs	29

E.	Effects of Changes in Organic Phase Polarity During Reaction	31
F.	Factors Affecting the Distribution of Phase-Transfer Catalyst Cation–Anion Pairs Between an Organic Phase and an Aqueous Phase Containing Hydroxide Ion	32
G.	Effect of Hydration of the Transferred Anion and the Effect of Inorganic Salt and/or Hydroxide Concentration in the Aqueous Phase	40

Chapter 3: Phase-Transfer Catalysis: Fundamentals II — 48
A. Introduction — 48
B. Liquid–Liquid PTC — 49
   1. Simple Displacement Reactions — 49
   2. Hydroxide-Promoted Reactions of Organic Acids — 89
   3. Alternative PTC Mechanisms Involving Hydroxide Ion — 106
C. Solid–Liquid PTC — 108
   1. Complexation and Solubilization of Potassium Salts with 18-Crown-6 — 111
   2. Simple Displacement Reactions — 113

Chapter 4: Phase-Transfer Catalysts — 123
A. Introduction — 123
B. Use of Quaternary Salts as Phase-Transfer Catalysts — 125
   1. Simple Tetraalkyl-Onium Salts as Phase-Transfer Catalysts — 125
   2. Special Quaternary Salts as Phase-Transfer Catalysts — 142
C. Macrocyclic and Macrobicyclic Ligands — 153
   1. Simple Crown Ethers — 153
   2. Cryptands — 155
   3. Special Crowns — 156
D. PEGs, Tris (3,6-dioxaheptyl)amine (TDA-1), and Related Ethoxylated Compounds as Phase-Transfer Catalysts — 158
   1. Synthesis of PEGs — 158
   2. Phase Distribution Behavior of PEGs — 159
   3. PEGs and Ethers as Phase-Transfer Catalysts — 162
   4. Special Ethoxylate Structures: Ethoxylate Derivatives as PTC Catalysts — 165
E. Other Soluble Polymers and Related Multifunctional Compounds as Phase-Transfer Catalysts — 171
F. Use of Dual PTC Catalysts or Use of Cocatalysts in Phase-Transfer Systems — 175
   1. Use of Dual PTC Catalysts — 176
   2. Use of Alcohols and Other Weak Acids as Cocatalysts in Hydroxide Transfer Reactions — 177
   3. Use of Metal Compounds and Salts as PTC Cocatalysts — 178
   4. Use of Iodide as a Cocatalyst — 179
G. Catalysts for Transfer of Species Other Than Anions — 179
   1. Inverse PTC: Transfer of Organic Reagents into Aqueous Solutions — 179
   2. Transfer of Acids — 183
   3. Transfer of Water — 184
   4. Transfer of Metals and Metal Hydrides — 185

	5. Transfer of Anhydrous Aluminum Chloride	185
	6. Transfer of Formaldehyde	186
	7. Cation Transfer	186
	8. Transfer of Radical Anions	186
	9. Transfer of Ammonia	187
	10. Transfer of Oxygen	187
H.	Separation and Recovery of Phase-Transfer Catalysts	188
	1. Extraction Methods	188
	2. Distillation Methods	189

Chapter 5: Insoluble Phase-Transfer Catalysts 207
A. Introduction 207
B. PTC Catalysts Bound to Insoluble Resins 208
 1. Basic Differences Between Soluble and Insoluble PTC Catalysts: The Importance of Diffusion Processes 208
 2. Some Examples of Use of Resin-Bound PTC Catalysts and Comparisons with Soluble Catalysts 210
 3. Preparation of Resin-Bound PTC Catalysts 210
 4. Effects of Reaction and Catalyst Parameters on Triphase Catalyst Effectiveness 221
 5. Kinetics of Reactions Catalyzed by Resin-Bound PTC Groups ... 247
C. Phase-Transfer Catalysts Bound to Inorganic Solid Supports 248
 1. PTC Catalysts Adsorbed on Inorganic Supports 249
 2. Catalysts with PTC Function Chemically Bonded to Inorganic Supports 250
D. PTC Catalysts Contained in a Separate Liquid Phase (Third-Liquid-Phase Catalyst) 252

Chapter 6: Variables in Reaction Design for Laboratory and Industrial Applications of Phase-Transfer Catalysis ... 266
A. Choice of Catalyst 266
 1. Structure–Activity Relationships of Quaternary Ammonium Catalysts .. 267
 2. Structure–Activity Relationships—Other Catalysts 286
 3. Catalyst Stability 288
 4. Catalyst Separation and Recycle 291
 5. Commercial Catalyst Reference 303
B. Choice of Solvent 303
 1. Choice of Solvent and the Nature of the Chemical Reaction 305
 2. Stabilization of the Transition State and Solvation of the Anion .. 306
 3. Solubility of the Catalyst–Anion Pair/Complex in the Organic Phase .. 307
 4. Rate of Transfer 307
 5. Solvent and the Nature of the Two Phases 310
 6. Examples of Effect of Solvent 311
 7. "Solvent-Free" PTC 314
 8. Choice of Solvent and Process Aspects 315
C. Presence of Water 318
D. Agitation 319

E.	Choice of Anion, Leaving Group, and Counteranion	322
F.	Choice of Base	325
G.	Guidelines for Exploring New PTC Applications	326

Chapter 7: Phase-Transfer Catalysis Displacement Reactions with Simple Anions ... 339

- A. General Considerations ... 339
  - 1. Important Factors in PTC Displacement Reactions ... 340
  - 2. Characteristics of Various Anions for Simple PTC Displacement Reactions ... 342
  - 3. PTC Catalysts for Simple Displacement Reactions ... 342
- B. Behavior of Various Anions in PTC Displacement Reactions ... 343
  - 1. Cyanide Displacements ... 343
  - 2. Halide Displacement and Exchange Reactions ... 347
  - 3. Displacements with Carboxylate Anions ... 355
  - 4. Azide Displacements ... 358
  - 5. Sulfide and Disulfide Displacements ... 362
  - 6. Thiocyanate Displacement ... 364
  - 7. Sulfite Displacement ... 366
  - 8. Nitrite Displacement ... 366
  - 9. Hydroxide Anion Displacements ... 367
  - 10. Carbonate and Bicarbonate Anion Displacement ... 368
  - 11. Displacement with Peroxide and Superoxide Anions ... 369
  - 12. Phosphide and Phosphinite Anion Displacements ... 370
  - 13. Cyanate Anion Displacements ... 370

Chapter 8: Phase-Transfer Catalysis Reaction with Strong Bases ... 383

- A. C-Alkylation ... 384
  - 1. Ketones ... 384
  - 2. Aldehydes ... 391
  - 3. Esters and Carboxylic Acids ... 392
  - 4. Imines ... 392
  - 5. Nitriles ... 395
  - 6. Sulfones ... 397
  - 7. Hydrocarbons ... 398
- B. N-Alkylation ... 400
  - 1. Nitrogen-Containing Heterocycles ... 400
  - 2. Amides ... 406
  - 3. Amines ... 408
- C. O-Alkylation—Etherification ... 410
  - 1. Etherification of Alkoxides ... 410
  - 2. Etherification of Phenoxides ... 413
- D. S-Alkylation—Thioetherification ... 418
- E. Dehydrohalogenation ... 420
- F. Carbene Reactions ... 424
  - 1. Dichlorocarbene Addition ... 424
  - 2. Dibromocarbene Addition ... 426

	3. Mixed Dihalocarbene Addition	427
	4. Other Reactions of Chloroform or Alternate Methods of Generating Carbenes Under PTC/OH Conditions	427
G.	Condensation Reactions	430
	1. Michael Addition	430
	2. Aldol Condensation	431
	3. Wittig	433
	4. Darzens	435
	5. Other Condensations	437
H.	Deuterium Exchange, Isomerization, and Oxidation	438

Chapter 9: Phase-Transfer Catalysis: Polymerization and Polymer Modification — 452

A.	Introduction	452
B.	Polymer Synthesis	452
	1. Condensation Polymerization	452
	2. Anionic Polymerizations	479
	3. Radical-Initiated Polymerizations	481
C.	Chemical Modification of Polymers	484
	1. Chemical Modification of Polymer Backbone	484
	2. Chemical Modification of Polymer Terminal Positions	489
	3. Chemical Modification of Pendant Groups Attached to Polymer Backbone	490

Chapter 10: Phase-Transfer-Catalyzed Oxidations — 500

A.	Introduction	500
B.	Permanganate Oxidations	500
	1. General Comments	500
	2. Transfer of Permanganate into Organic Phases	501
	3. PTC Permanganate Oxidations	503
C.	Oxidations with Hypochlorite and Hypobromite	508
	1. Hypochlorite Compositions in Aqueous Solutions	508
	2. Oxidation of Alcohols and Carbonyl Compounds	508
	3. Oxidation of Amines, Amides, Thioamides, and Related Compounds	512
	4. Oxidation of Sulfides and Related Compounds	514
	5. Oxidation of Olefins	514
	6. Oxidation of Nonolefinic Hydrocarbons	518
D.	PTC Oxidations with Hydrogen Peroxide	521
	1. Hydrogen Peroxide Transfer into Organic Solutions	521
	2. Hydrogen Peroxide Oxidations	522
E.	PTC Air or Oxygen Oxidations	534
	1. Carbanion Oxidations	534
	2. PTC Involvement in Free-Radical Oxidations	538
	3. PTC Oxidation with "Activated" Oxygen Carriers	538
	4. PTC with Singlet Oxygen Generation	539
	5. Transition-Metal-Mediated Oxidations Involving PTC	540
F.	Oxidations by Persulfates	540

x / Contents

	2. Electrochemical Regeneration of Chromium Oxidants in Combination with PTC Systems	547
H.	PTC Oxidations with Nitric Acid	547
I.	PTC Carbon Tetrachloride/Sodium Hydroxide Oxidations	548
J.	PTC Oxidations with Periodate and Related Oxidizing Anions	549
	1. Osmium as Cocatalyst	549
	2. Ruthenium as Cocatalyst	550
K.	PTC Oxidations with Perborate	550
L.	PTC Oxidations with Ferrate and Ferricyanide	550
M.	PTC Oxidations with Superoxide	551
N.	PTC Electrochemical Oxidations	551
O.	PTC Oxidations with Other Oxidants	552

Chapter 11: Phase-Transfer-Catalyzed Reductions — 565

A.	Sodium Borohydride Reductions	565
	1. Reduction of Carbonyl Compounds	565
	2. Azide Reductions	566
	3. Other PTC Borohydride Reductions	567
B.	Lithium Aluminum Hydride Reductions	568
C.	Reductions with Sodium Formate	568
D.	Reductions with Sulfur-Containing Anions	569
E.	Hydrogenation	570
F.	Reductions with Formaldehyde	571
G.	Electrochemical Reduction	571
H.	Photochemical Reduction	572
I.	Wolff–Kishner Reduction	572
J.	Reduction by Dodecarbonyltriiron and Related Species	572

Chapter 12: Phase-Transfer Catalysis: Chiral Phase-Transfer-Catalyzed Formation of Carbon–Carbon Bonds — 576

A.	Introduction	576
B.	Alkylation Reactions	577
	1. Methylation of 6,7-Dichloro-5-methoxy-2-phenyl-1-indanone	577
	2. Alkylation of 2,3-Dichloro-5-methoxy-2-n-propyl-1-indanone with 1,3-Dichloro-2-butene in Toluene–50% Aqueous Sodium Hydroxide	584
	3. Asymmetric Alkylation of Oxindoles	586
	4. Synthesis of Chiral Amino Acids	587
	5. Michael Addition Reactions	589

Chapter 13: Phase Transfer Catalysis–Transition Metal Cocatalyzed Reactions — 594

A.	Introduction	594
B.	Carbonylation and Reactions with Carbon Monoxide	595
	1. Formation of Metal Carbonyl Anions	595
	2. Carbonylation of Alkyl Halides and Aryl Halides	595
	3. Carbonylation of Olefins	600
	4. Carbonylation of Acetylenes	602

	5. Carbonylation of Aziridines and Azobenzenes	604
	6. Carbonylation of Thiiranes	605
	7. Carbonylation Reaction with Phenol	605
B.	PTC Reduction and Hydrogenation with Metal Cocatalysts	605
	1. Hydrogenolysis of Aryl and Alkyl Halides	605
	2. Reduction of Acid Chloride Groups to Aldehydes	609
	3. Hydrogenation of Arenes, Olefins, and Carbonyl Compounds	609
	4. Reduction of Nitrogen Compounds	610
	5. Other Reductions and Hydrogenations	613
C.	Coupling Reactions of Alkenes, Alkynes, and Alkyl Halides	613
	1. Acetylene and Olefin Reactions with Halo-compounds	613
	2. Acetylene and Olefins Coupling Reactions	616
D.	Other Reactions	617

Chapter 14:	Phase-Transfer Catalysis in Analytical Chemistry	622
A.	The PTC/Analytical Chemistry Match	622
B.	Esterification, Etherification, and Other Nucleophilic Derivatizations	622
C.	Non-Nucleophilic PTC Reactions Used in Analytical Chemistry	623

Chapter 15:	Phase-Transfer Catalysis: Industrial Perspectives	626
A.	Industrial Background	626
B.	Evaluation of PTC as a Commercial Manufacturing Process Technology	626
	1. General Considerations	626
	2. Advantages of PTC—Industrial Viewpoint	627
	3. Limitations of PTC and Barriers to Commercialization	631
	4. Identifying Future Opportunities for Making Economic Impact Using PTC	635
	5. Conclusion	637

Index   639

Preface

Since 1971 when useful working concepts for the technique of *phase-transfer catalysis* (PTC) were introduced, the understanding, development, and applications of this method for conducting organic reactions has expanded exponentially. PTC has brought vast new dimensions and options to chemists and chemical engineers. From its use in less than ten commercial processes in 1975, PTC use has increased so that in the early 1990s it is involved in more than 600 industrial applications to manufacture products valued at between 10 and 20 billion U.S. dollars. PTC is widely used for simple organic reactions, steps in synthesis of pharmaceuticals, agricultural chemicals, perfumes, flavorants, and dyes; for specialty polymerization reactions, polymer modifications, and monomer synthesis; for pollution and environmental control processes; for analysis of trace organic and inorganic compounds; and for many other applications. Often, PTC offers the best (and sometimes only) practical technique to obtain certain products.

The authors experience in teaching a short course on phase-transfer catalysis has shown to us that a newcomer to PTC can easily be frustrated and confused by the large amount of information available in the literature and in patents. The purpose of this book, therefore, was to bring this information together in a logical and user-friendly way, without sacrificing matters of scholarly and fundamental importance.

Concurrently with the proliferation of PTC applications, many advances in understanding of the mechanistic aspects of PTC have been made. These are extremely important to understand if one is to devise optimal reactions based on phase-transfer catalysis. Of particular value for understanding the fundamental steps within the PTC technique has been the recent elaboration of how the detailed kinetics of the various PTC steps affect the overall rates of conversion within PTC reactions, a topic fully described for the first time in this book. Although these kinetics can be very complex, one of the authors (CLL) has devised simple computer routines that easily handle the most complicated of PTC processes for chemists who have not had extensive experience in using kinetic data.

Selection of catalyst and other reaction conditions in PTC reactions is often found to be the most confusing aspect in the use of the PTC technique. This confusion results from the wide array of different catalysts that have been reported in the chemical literature, further complicated by the fact that sometimes the choice of catalyst is of extreme importance while in other situations the reaction proceeds very well with almost any PTC catalyst. Use of insoluble PTC catalysts, either solid or liquid, greatly expands the scope of PTC reactions. While they simplify procedures because they are easy to remove and reuse, their use usually demands great care in development and selection of catalysts and reaction conditions. Aspects of catalyst selection, including not only activity but toxicity, cost, recovery, reuse, and stability appear in almost every chapter of this book. Considerable attention has been devoted in Chapters 4, 5, and 6 toward reviewing and comparison of various materials used as phase transfer catalysts, and toward selection criteria for catalysts.

Other variables are also of concern in phase-transfer catalysis, and these are outlined in Chapter 6 along with some guidelines about exploration of new PTC reactions. One of the most beneficial features of phase-transfer catalysis is the ability greatly to expand the list of solvents that can be used in a given reaction, or even in many cases to omit a solvent altogether. Chapter 6 discusses aspects of solvent choice on PTC reactions, along with other practical concerns, such as the presence of water, agitation, choice of bases in base reactions, and commercial sources and relative costs of PTC catalysts.

The last half of this book is devoted to the chemistry and application of the PTC technique toward specific kinds of commonly used PTC reactions and processes. These chapters are not intended to review exaustively all the PTC chemistry reported in the chemical literature, but rather to illustrate with specific examples how the technique has been used with basic reactions as well as with subtle and advanced ideas on how to use PTC in special ways.

It is our belief that phase-transfer catalysis will continue to grow and be used extensively by organic chemists and engineers in both new developments and in improving old processes. Over the last 15 years, no decrease has been found in the frequency which *phase-transfer catalysis* is indexed in *Chemical Abstracts*. PTCs ability to be used beneficially in conjunction with polymers, electrochemistry, photochemistry, and various co-catalysts is still accelerating, and its substantial cost-reductions and avoidance of noxious and expensive reagents and solvents continues to find increasing use in a wide variety of processes. Although design and development of PTC-based processes is more difficult than use of classical chemical procedures, profitable results usually repay the extra development effort many times over.

1

Basic Concepts in Phase-Transfer Catalysis

A. Phase-Transfer-Catalyzed Reactions

Phase-transfer catalysis (PTC) is a powerful tool in many areas of chemistry. It is a technique for conducting reactions between two or more reagents in two or more phases, when reaction is inhibited because the reactants cannot easily come together. A *"phase-transfer agent"* is added to transfer one of the reagents to a location where it can conveniently and rapidly react with another reagent. It is also necessary that the transferred species be in a highly active state when transferred; otherwise large amounts of phase-transfer agent will be required. This activation function, plus the transfer function, allows *phase-transfer catalysis* to occur with only a catalytic amount of phase-transfer agent.

A simple and classic example of PTC is illustrated in Figure 1-1 for the reaction of 1-chlorooctane (no solvent) and aqueous sodium cyanide [1,2]. Without catalyst, heating of this two-phase mixture under reflux and with vigorous stirring for 1 or 2 days gives no apparent reaction, except possibly some hydrolysis of the sodium cyanide to ammonia and sodium formate. However, if 1 wt% of an appropriate quaternary ammonium salt is added, say $(C_6H_{13})_4N^+Cl^-$, then displacement reaction occurs rapidly producing 1-cyanooctane in near 100% selectivity and 100% conversion in 2–3 h. The quaternary ammonium cation (1) transfers cyanide anion into the organic phase, (2) activates the transferred cyanide for reaction with 1-chlorooctane, and allows displacement to occur rapidly, producing 1-cyanooctane and QCl, then (3) transfers displaced chloride anion back to the aqueous phase to start a new catalytic cycle.

PTC is a general technique for organic chemistry. It has been used for hundreds of reactions, including especially those anion transfer reactions represented in Table 1-1 [3]. Most publications and patents in PTC deal with reactions involving transfer of anions from an aqueous or solid phase into an organic phase, largely because

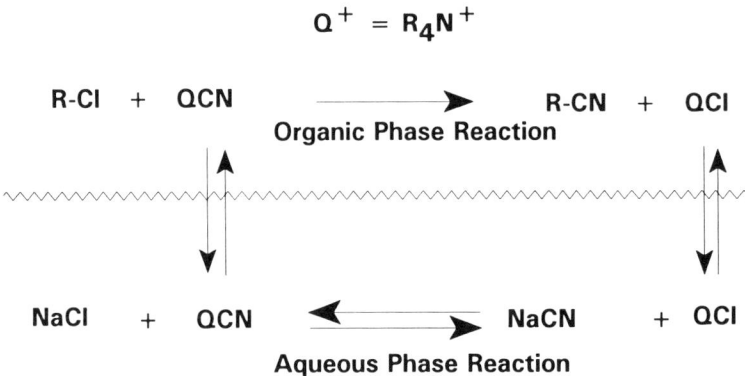

Figure 1-1. Schematic representation of phase-transfer-catalyzed cyanide displacement on 1-chlorooctane.

most practical PTC organic reactions require this sequence. However, the idea of PTC is quite general, extending to transfer of whole molecules, cations, free radicals, or other species, as described further in Chapter 4, Section G.

B. Basic Steps of Phase-Transfer Catalysis

Looking back at the cyanide displacement reaction outlined in Figure 1-1, it is important to recognize that at least two general steps are involved in the catalytic sequence:

Step 1: Anion transfer or delivery of anion from aqueous to organic phase:

The sequence of reactions that cause cyanide to be transferred into the organic phase, called the *transfer step*, is represented by three equilibria that (1) transfer quaternary ammonium chloride from the organic to the aqueous phase, (2) exchange chloride for cyanide anion in the aqueous phase, and (3) transfer quaternary ammonium cyanide from the aqueous to the organic phase. It is important to remember here that the transfer rate of interest is the *net rate of delivery* of cyanide to the organic phase. It is not simply the rate of the physical process of taking the cyanide across from the aqueous to the organic phase.

Step 2: The *intrinsic reaction* or *organic-phase* displacement reaction step:

The sequence of reactions in the organic phase, starting with transferred anion, which results in formation of product, is called the *intrinsic reaction* or more commonly the *organic phase reaction*. In Figure 1-1, this step consists only of the displacement reaction between quaternary ammonium cyanide and 1-chlorooctane to produce 1-cyanooctane. In other reactions,

Table 1-1. *Some types of phase-transfer reactions.*

SN2 displacement reactions with alkyl halides and anions:
   CN^-, SCN^-, CNO^-, F^-, CL^-, Br^-, I^-, N_3^-
   SH^-, HSO_3^-, RS^-, RO^-, ArO^-, RCO_2^-, etc

Alkylation reactions using NaOH(aq) and alkyl halides with:
   for C-Alkylations
      Activated nitriles, ketones, esters, nitro-compounds, cyclopentadienes, other acidic C–H compounds
   for N-Alkylations
      Imides, amides, sulfonamides, heterocyclic N-compounds
   for S-Alkylations

Elimination reactions to produce olefins and acetylenes

Oxidation reactions of many compounds using inorganic oxidants:
   MnO_4^-, OCl^-, H_2O_2, O_2, IO_4^-, HNO_3, etc.

Reduction reactions of many compounds with:
   BH_4^-, AlH_4^- HCO_2^-, etc.

Polymerization reactions and polymer modifications

Reactions containing transition metals as cocatalysts:
   Carbonylation, carboxylation, hydrogenation.

Deuterium exchange reactions.

the intrinsic organic-phase reaction may consist of more than one chemical reaction.

Why is it important to distinguish and understand the nature of these two steps?

Several variables may strongly affect PTC reactions. These variables include the type and amount of catalyst, agitation, amount of water in the aqueous phase, temperature, solvent (if one is used), and others. Some variables are much more important to one step than to another. For example, agitation is critically important to the transfer step, but of little importance to the rate of the intrinsic displacement reaction step. Therefore, to achieve maximal overall results we need to understand and design the whole phase-transfer system so that the reactivity factors will provide the most help where it is most needed. Thus, in the above cyanide displacement example, the intrinsic displacement reaction step (the reaction in the organic phase) is the slow step of the process, provided good agitation is used; therefore, to improve the overall rate, one needs to work most on those reaction variables that increase the rate of the cyanide displacement reaction, rather than on the transfer step.

Recognizing that high rates of both steps of the phase transfer process are necessary to good PTC processes, and that the kinetics of both steps are closely interrelated through the effect of catalyst on both steps, one can logically deduce (and later show through kinetics; see Chapter 3) that overall reactivity reaches a steady-state level when the rates of the transfer step and the intrinsic reaction step are equal. Neither step can be significantly faster than the other except for transient conditions at the start of reaction before a steady-state condition is reached. This equality of rates is represented by the diagonal in Figure 1-2.

In Figure 1-2, a point on the diagonal line represents a PTC reaction where the rate of delivery of cyanide to the organic phase is equal to the rate of its consumption in the intrinsic displacement reaction leading to product. Realistically, however, we expect few reactions to fall on the diagonal; most reactions will be above or below it. If transfer is faster than intrinsic reaction, the point representing the reaction rates will be below the diagonal, and the overall rate for the whole process will be the position on the diagonal horizontal to the point, as illustrated for point A, denoting that overall rate is limited by the intrinsic organic phase reaction rate. On the other hand, if the intrinsic rate is faster than transfer, as represented by point B, the overall rate will equal the position on the diagonal vertically below the point. The point A is generally located at a point on the matrix expected for the cyanide displacement on 1-chlorooctane. However, if benzyl chloride were to be substituted for 1-chlorooctane in the cyanide displacement reaction, then point B becomes more representative, because benzyl chloride undergoes displacement reactions 10 or 20 times faster than 1-chlorooctane. Then, if other reaction conditions stay the same, the overall reaction tends to become transfer limited.

Hydroxide and certain other anions are difficult to transfer, for reasons that will be discussed later, and it is common to find PTC hydroxide reactions have a slow transfer step. With hydroxide, the optimal catalyst type is often $C_{16}H_{33}N^+Me_3Cl^-$ or $PhCH_2N^+Et_3Cl^-$, which give a higher rate of anion transfer than the bulky catalysts that were most useful for cyanide displacement reaction on 1-chlorooctane. Thus to improve rates for PTC involving hydroxide transfer (or transfer of equivalents to hydroxide) we will look for ways to speed up

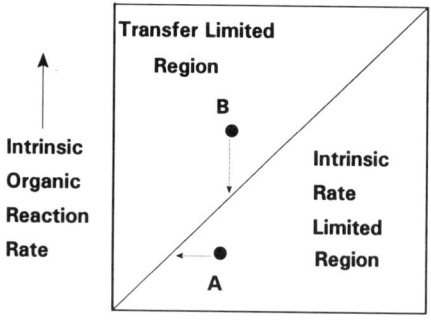

Figure 1-2. Linked effect of intrinsic reaction rate and transfer rate for PTC systems.

hydroxide transfer from the aqueous to the organic phase or to the organic–aqueous interface.

C. The PTC Reaction Rate Matrix

For more general understanding and design of PTC systems the differing features and factors surrounding PTC reactions suggest use of a plot like that shown in Figure 1-3, termed the *PTC reaction rate matrix*.

Transfer rate, plotted on the *x*-axis, represents the net rate at which the transferred species is delivered to the organic phase. Intrinsic rate, plotted on the *y*-axis is the net rate at which the organic-phase reaction(s) proceeds. No time units are specified here, although notice that the scales are logarithmic, and represent four orders of magnitude. For qualitative discussion no units are needed.

Midpoint lines divide the graph into quarters, which although arbitrary, serve to emphasize the predominant features of phase-transfer systems that fall into each of the quarters.

1. *Intrinsic reaction rate limited region.* In the lower-right quarter, transfer is fast but intrinsic reaction is slow, such as the cyanide displacement on 1-chlorooctane, indicating that if we wish to increase the overall rate, we need to increase the rate of the organic-phase intrinsic reaction step.

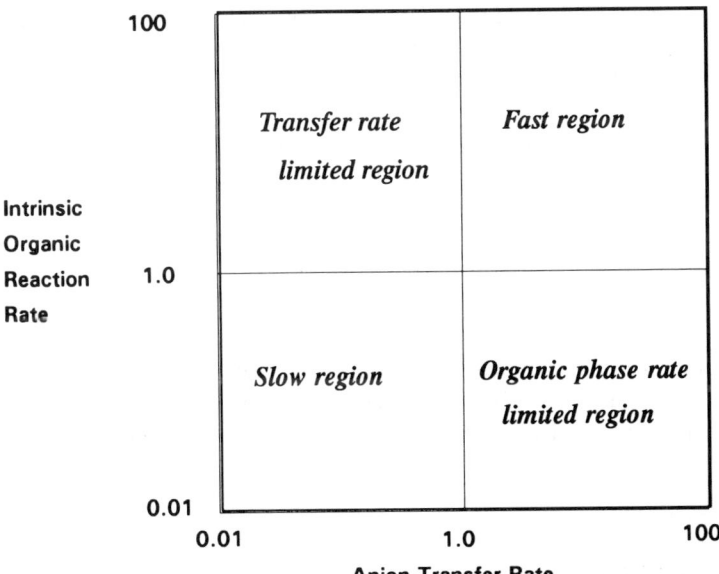

Figure 1-3. The PTC reaction rate matrix.

2. *Transfer rate limited region.* In the upper-left quarter, transfer is slow, but intrinsic rate is high, such as butylation of phenylacetonitrile with NaOH(aq) and 1-bromobutane, and indicating that to increase the overall rate we need to do something to increase the rate at which anion is delivered to the organic phase.
3. *Fast region.* In the upper-right quarter, both transfer and intrinsic rates are fast. Here, almost any catalyst and set of reaction conditions will yield satisfactory rates; in fact, our concern may be how to keep the reaction under control.
4. *Slow Region.* In the slow–slow quarter, we may need to use all our skills and planning to achieve a reaction that occurs within a reasonable time. Careful catalyst selection, maybe even use of two catalysts (one to assist transfer and one to speed the intrinsic reaction); and operation at the highest possible temperature (requiring a heat-stable catalyst), with the least amount of water, with a high degree of agitation, and perhaps with a polar solvent, are all options to increase rates. For certain reactions we may also be able to use a special catalyst, or cocatalyst, to increase rates.

Figure 1-4 shows the PTC reaction matrix with some reactions added as examples.

Permanganate oxidations are an excellent example for the "fast" quadrant. The MnO_4^- anion is easily transferred with almost any phase-transfer catalyst, including trialkyl amines (probably by intermediate formation of an amine oxide). Permanganate oxidations of many organic compounds are so fast that this phase transfer system can even be used for analytical titrations, for example, of olefins, alcohols, aldehydes, phenols, and others [4].

The fluoride displacement on 2-chlorooctane is also an excellent example for the "slow" quadrant. Fluoride anion, like hydroxide, is difficult to transfer from water, and, once transferred, is a poor nucleophile, especially in displacement of relatively inactive secondary chloro-groups.

D. Anion Transfer and Anion Activation

At this point let us consider some of the simple concepts of how quaternary ammonium salts transfer anions into organic phases, and how the quaternary cation activates anions. This discussion outlines several concepts that are important to an understanding of how catalyst structure and reaction variables affect PTC reactions.

1. Transfer

a. Simple Transfer

In the cyanide displacement reaction example represented by Figure 1-1, it was indicated that transfer of the cyanide occurred by movement of the quaternary

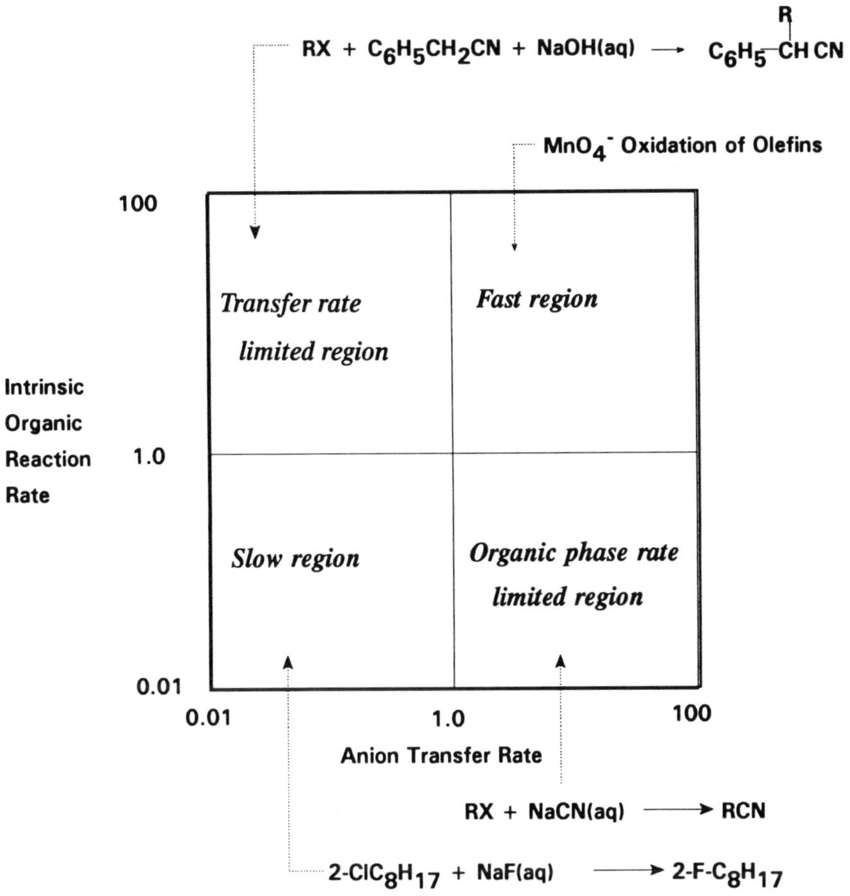

Basic Concepts in Phase-Transfer Catalysis / 7

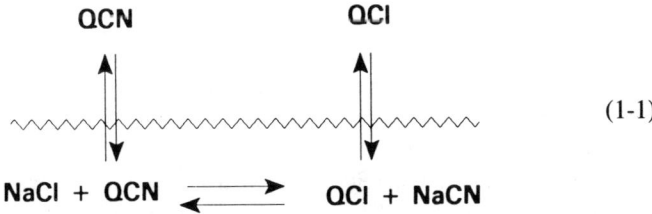

Figure 1-4. Reaction examples in the various quadrants of the PTC reaction matrix.

salt back and forth across the aqueous–organic interface, that is, as represented here:

$$\text{NaCl} + \text{QCN} \xrightleftharpoons{} \text{QCl} + \text{NaCN} \quad \text{(with QCN} \uparrow\downarrow \text{ and QCl} \uparrow\downarrow \text{ across interface)} \tag{1-1}$$

This behavior is certainly likely when the quaternary salts are small. If the quaternary cation is too small, for example, $Me_4N^+CN^-$, most of it will be partitioned into the aqueous phase and so little cyanide ion will be present in the organic phase that formation of 1-cyanooctane will be too slow to measure.

As larger quaternary salts are used, for example, tetraethyl, tetrapropyl, or tetrabutyl, more of the quaternary ammonium cyanide is partitioned into the organic phase where it can drive the slow organic reaction. Increased quaternary cation size, to tetrapentyl, tetrahexyl, or tetraoctyl causes most of the quaternary salt, 95% or more, to partition into the organic phase, giving maximal overall conversion rates. (More polar organic solvents allow smaller quaternary salts to be successfully partitioned into the organic phase.) Continued increase in the chain length of tetraalkyl quaternary salt does not lead to continued increase in catalyst activity. Rather, at some point (ca. tetradecyl), catalytic activity levels off, then slowly decreases with larger and larger quaternary salt. This loss of activity is due to inadequate interaction of the quaternary salt with the aqueous phase, causing the rate of anion transfer to slow, essentially resulting in a transfer-limited process due to restricted rates of anion transfer across the interface.

b. Simple Transfer Across the Interface

A second mechanism for anion transfer is simple anion interchange *across* the interface, as represented in the following diagram:

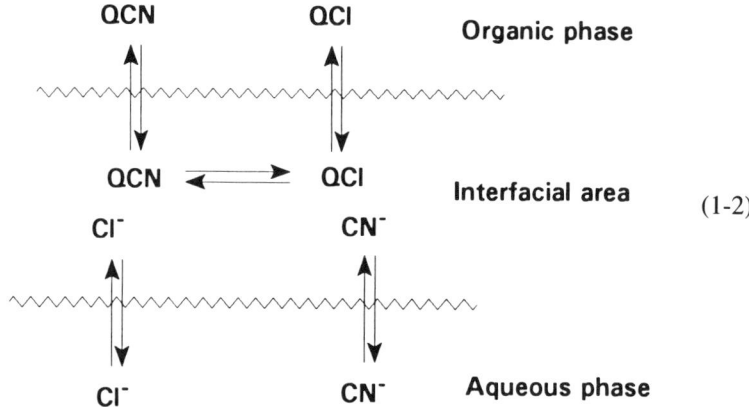

(1-2)

This mechanism does not require catalyst to actually leave the organic phase, to the extent that the interfacial area is still considered a part of the organic phase, but it does require the quaternary salt to have a dominant presence at the interface. Large water-insoluble quaternary salts, such as tetrahexyl- or trioctyl-methylammonium salts, which are predominantly (>99%) partitioned into the organic phase, function as catalysts mostly by anion transfer across the interface. However, if the quaternary cation is extremely large, the cation will not be frequently in the interfacial region, and then the reaction rate will become transfer-limited. On the other hand, quaternary salts that are good surfactants, such as $C_{16}H_{33}NMe_3^+Cl^-$, are dominantly positioned at or in the interfacial layer. These *accessible* quaternary salts provide for higher rates of anion transfer across the interface.

c. Transfer by Inverse Micelles

A third mechanism for anion transfer, at least with highly hydrated monovalent anions and most divalent ions, is by formation of inverse micelles [5]. For example, dilute (e.g., 10%) solutions of sodium hydroxide, fluoride, sulfate, and others may exist with the anions hydrated to the extent of 10 or more water molecules of hydration per anion. This high degree of hydration strongly binds these anions to the aqueous phase. One way this barrier to transfer can be lowered is to form inverse micelles having an aqueous interior and an organic exterior, as represented in Figure 1-5.

The quaternary cations provide the necessary organophilic exterior for inverse micelle formation and transfer of the micelle to the organic phase. The long alkyl groups attached to quaternary cations provide an organophilic pathway for organic reactant to closely approach hydroxide at the micelle center. Evidence for this type of transfer comes from the known aggregation behavior of quaternary salts in organic solvents [6]. Likewise, the organic phase may also be dispersed into micelles, which can be transported to the aqueous phase for reactions [7].

d. Anion Formation at the Interface Without Catalyst

A fourth transfer mechanism, acting with certain acidic organic phase reactants in the presence of sodium hydroxide, known as the *interfacial* mechanism, can

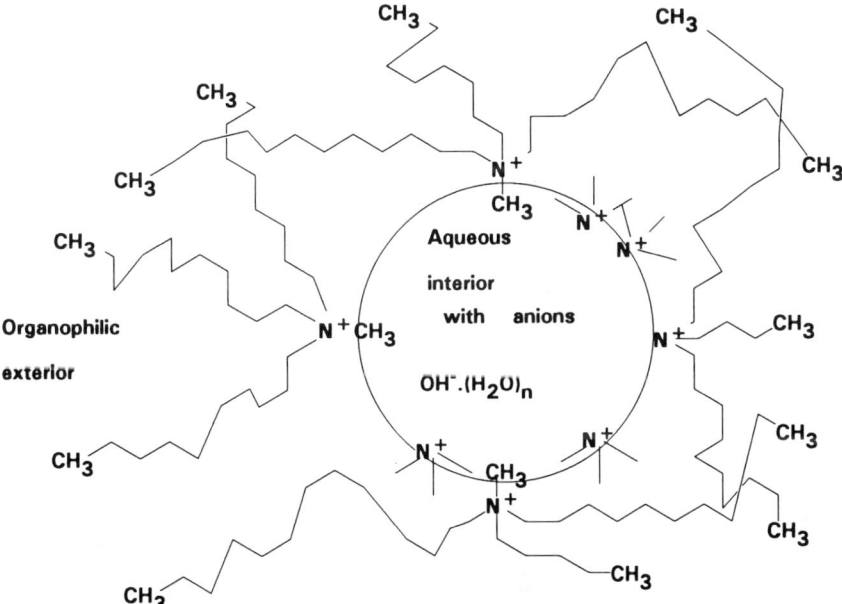

Figure 1-5. Representation of inverse micelles containing hydroxide solution, transferred to an organic phase.

be postulated in which organic anions are formed at the interface without the assistance of PTC. Once formed, the organic-anion-at-the-interface couples with an available quaternary cation to form an ion pair that can transfer away from the interface and into the bulk organic phase for further reaction [8]. For example, in the alkylation of phenylacetonitrile with 1-bromobutane in the presence of aqueous sodium hydroxide, this mechanism assumes that the phenylacetonitrile migrates to the aqueous interface where it reacts (without assistance of catalyst) with OH^- on the other side of the interface to form a $PhCHCN^-$ anion. Once formed, this anion combines with Q^+ (which may either be at the interface or in the aqueous phase), and now the $Q^+PhCHCN^-$ ion pair migrates away from the interface into the organic phase where displacement reaction occurs. This mechanism is believed to be correct only for a few special cases.

e. Formation of Third Phase

A fifth type of transfer mechanism takes place when the quaternary salt catalyst is not highly soluble in either the aqueous or the organic phase, but instead forms a third phase. For example, tetrabutylammonium salts in the presence of highly concentrated aqueous solutions, and with toluene as solvent for the organic phase, forms a third (quaternary compound) phase [9]. In such situations, most of the reaction actually occurs in the third phase with both aqueous and organic reagent transferring to this phase for conversion. Third-phase reactions of this type may be faster than simple PTC reactions. Because formation of the third phase offers simplified catalyst removal and recovery procedures, third-phase catalysis is highly attractive for commercial operations. Third-phase catalysis is discussed further in Chapter 5.

f. Transfer of Whole Molecules

A sixth type of transfer mechanism occurs when crown ethers, cryptands, polyethylene glycol (PEG) and derivatives, and other nonionic phase-transfer agents are used as catalysts. Here, the phase-transfer agent complexes with inorganic cation such that it, along with the anion, can be transferred to the organic phase. For example, with 18-crown-6 as a phase-transfer agent for sodium cyanide:

18-crown-6

$$\text{18-crown-6 (organic phase)} + \text{NaCN (solid)} \longrightarrow [\text{18-crown-6} \cdot Na^+] \; CN^- \quad (1\text{-}3)$$

(transferrable to organic phase)

Further discussion relating to this kind of transfer is given in Chapters 2 and 4.

g. Reactive PTC and reactive inverse PTC

In some PTC reactions, most notably those catalyzed by dialkyl or diaryl sulfides and having a highly reactive alkyl halide as one of the reactants, another mode of PTC occurs [10,11]. Here, for example, alkyl halide reacts first with the sulfide to produce a sulfonium salt.

$$R-S-R + R'X \rightarrow R_2S-R'^+ \ X^- \qquad (1\text{-}4)$$

The sulfonium salt exchanges anions at the interface, for example with cyanide,

$$R_2S-R'^+ \ X^-(\text{org}) + CN^-(\text{aq}) \rightarrow R_2S-R^+ \ CN^- \qquad (1\text{-}5)$$

and then the resulting sulfonium cyanide undergoes thermal decomposition, producing the cyanoalkane and regenerating the sulfide.

$$R_2SR'^+ \ CN^- \rightarrow R_2S + R'CN \qquad (1\text{-}6)$$

This process may be termed *reactive* PTC because it involves continuous formation and decomposition of an active catalyst–reactant component. If a small dialkyl sulfide is used, as, for example, diethyl sulfide, the sulfonium salt can be partitioned mostly into the aqueous phase where anion exchange and sulfonium salt decomposition occur. In this case the process can be termed *reactive inverse PTC*.

2. Anion Activation

For successful PTC, the anion must not only be transferred to the organic phase, but once there the anion must be in a highly reactive form. Some organic-phase reactions are so fast that the transferred anion requires little or no activation beyond just being delivered to the organic phase. Other reactions require substantial anion activation before useful and practical reaction rates can be achieved. Anion activation with quaternary ammonium salts is achieved mostly because these salts, transferred as ion pairs in organic phases, have greater distances separating cation from anion than in sodium or potassium salts. For example, compare sodium bromide and tetrabutylammonium bromide ion pairs:

$Na^+\ldots Br^-$ $(C_4H_9)_4N^+\ldots\ldots\ldots Br^-$
2.85Å   6.28Å

Although the difference in cation–anion interionic distances for the two ion pairs is only about 4.4 Å, it can, when translated into differences in energies of activation, be sufficient to cause anions from the quaternary salt to react up to four orders of magnitude faster than anions from the sodium salt, as discussed in Chapter 2.

A second mechanism by which transferred anions are activated for reaction in PTC systems stems from the ability to transfer anions with reduced levels of water of hydration [12]. In most organic-phase reactions involving anions, the less water of hydration around an anion the more reactive the anion. Thus, best PTC results are frequently obtained with very small amounts of water, perhaps just enough to allow the transfer step to occur rapidly, and frequently with substantial amounts of solids in the reactor. This complex subject is also discussed further in Chapter 2.

E. Effect of Reaction Variables on Transfer and Intrinsic Rates

As indicated in the preceding section, fundamental understanding and prediction of PTC reaction systems requires some knowledge of both the rates of transfer and the rates of the intrinsic organic-phase reaction. Such information is frequently unavailable to the chemist for unexplored reactions, at least in quantitatively useful form. For new systems it is necessary to make some guesses and extrapolations from known systems concerning the relative rates of organic-phase reactions and comparative ease of anion transfer.

Ease or difficulty of transfer of most anions into organic systems can be estimated by examination of similar PTC reactions using the same or a similar anion. For untested anions, rough estimates of ease of transfer can be made using theory, as discussed in Chapter 2.

Rates of intrinsic organic-phase reactions are best estimated by examination of the literature on examples of the reaction where all reactants are in solution. If reaction proceeds rapidly in solution in organic solvents, especially at room temperature (e.g., permanganate oxidation of olefins) then likely it will also be rapid in PTC systems. Conversely, if the reaction is slow and difficult in solution, even at elevated temperatures (e.g., fluoride displacement on 2-chlorooctane), then the intrinsic organic reaction is likely to be very slow. It should be emphasized that water of hydration of reactant anions could result in differences between homogeneous reaction in organic solvent and the same reaction under liquid–liquid PTC conditions.

Assuming literature data or preliminary experiments indicate PTC will be successful for a given reaction, then what can one do with the reaction variables to maximize rates, minimize cost, minimize process difficulties, and generally optimize the process for the use needed? Because these variables do not usually affect both the rates of transfer and intrinsic reaction equally it is best to sort out how each step is affected by each variable. Some general qualitative comparisons are outlined in Table 1-2. These variables are also discussed in detail in Chapters 2 and 3. Some points to note follow.

1. Catalyst Structure

Choosing a phase-transfer catalyst is usually the most important step in design and development of a phase transfer process. Two complete chapters of this book, Chapters 4 and 5 are devoted to describing and comparing catalyst features and performance, and substantial parts of Chapters 2 and 3 on PTC fundamentals also relate to explaining the effects of catalyst structure. Further discussion is provided in Chapter 6 regarding choice of PTC catalyst.

Some simple generalizations regarding quaternary salts are given here for illustration:

1. The catalyst cation and the anion to be reacted must be easily partitioned into the organic phase for the PTC sequence to be effective. This means that lower tetraalkylammonium salts such as tetramethyl- and tetraethylammonium cations with small highly hydrated anions may not be organophilic enough to be effectively transferred. Large quaternary salts, such as tetrahexyl- or trioctylmethylammonium salts, easily transfer most monovalent anions into most organic phases.

2. Large bulky quaternary cations such as tetrabutyl-, tetrapentyl-, or tetrahexylammonium salts provide the most "activation" for anions, and are best for reactions that tend to have slow intrinsic organic-phase reactions and require a highly activated anion. However, use of a highly polar organic solvent also tends to loosen the cation–anion binding, and such solvents may allow less bulky quaternary salts to be successful.

3. Quaternary cations that are relatively open-faced, or *accessible*, such as hexadecyltrimethyl-, tetraethyl-, or benzyltriethylammonium, readily occupy interfacial positions. These quaternary salts may also lower interfacial tension,

Table 1-2. General effect of reaction variables on PTC reactions.

Variable	Effect on transfer step	Effect on intrinsic reaction
Catalyst structure	+++	+++
Agitation	++++	0
Type of inorganic anion	++++	++++
Water concentration	++	+
Organic solvent	+	++
Temperature	+	+++
Cocatalysts	++	+++
Organic reactant structure	0	++++

increase interfacial area, and thereby increase the transfer rate of the anion to the organic phase [13]. Accessible quaternary salts are best for reactions where rates are limited due to slow anion transfer, such as normally encountered in reactions with hydroxide, fluoride, hypochlorite, or divalent anions. Tetrabutylammonium salts tend to occupy an intermediate position in their ability to activate anions and to transfer anions, and are recommended for exploratory PTC experiments.

4. When both the transfer step and the intrinsic organic reaction step are slow, it may be advisable to use two phase-transfer catalysts, one to increase transfer rates and the other to activate transferred anions. When both steps are fast, almost any catalyst will perform satisfactorily.

These relationships are illustrated in Figure 1-6.

2. *Agitation*

Transfer of anions from the aqueous to the organic phase requires agitation. Without agitation, PTC reactions are almost always too slow to be useful. Agitation increases interfacial area between organic and aqueous phases, speeding transfer of reactive species, and therefore accelerates transfer rates. Reactions limited by a slow transfer rate become faster as the level of agitation is increased. When the transfer rate substantially exceeds intrinsic reaction rate, then the PTC process will become independent of agitation rate. This topic is discussed further in Chapter III.

The use of ultrasound as an extraordinarily effective means of agitation has

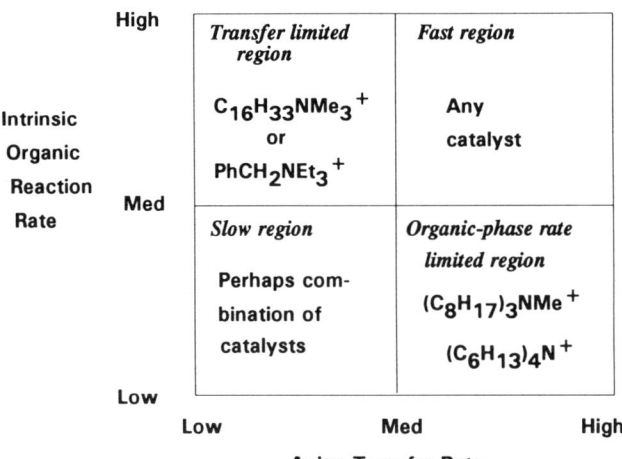

Figure 1-6. Some phase-transfer catalyst choices depending on the position of a PTC reaction in the matrix.

frequently been reported in the PTC literature; it usually causes a decrease in the required reaction time and often increases yields [14,15].

3. Kind and Concentration of Inorganic Reagent; Amount of Water Added

The ease of transfer of anions from an aqueous environment to an organic environment depends markedly on the kind of anion to be transferred. For example, as shown in Figure 1-7, hydroxide, fluoride, and hypochlorite anions are difficult to transfer; iodide, perchlorate, permanganate, and tosylate, are easy to transfer; chloride, cyanide, bromide, and nitrate occupy intermediate position, as indicated in Figure 1-7. (See also Chapter 2 for detailed discussion regarding the ease of transfer of different anions.) The concentration of the inorganic reagent in the aqueous phase is also important. For example, in the cyanide displacement reaction,

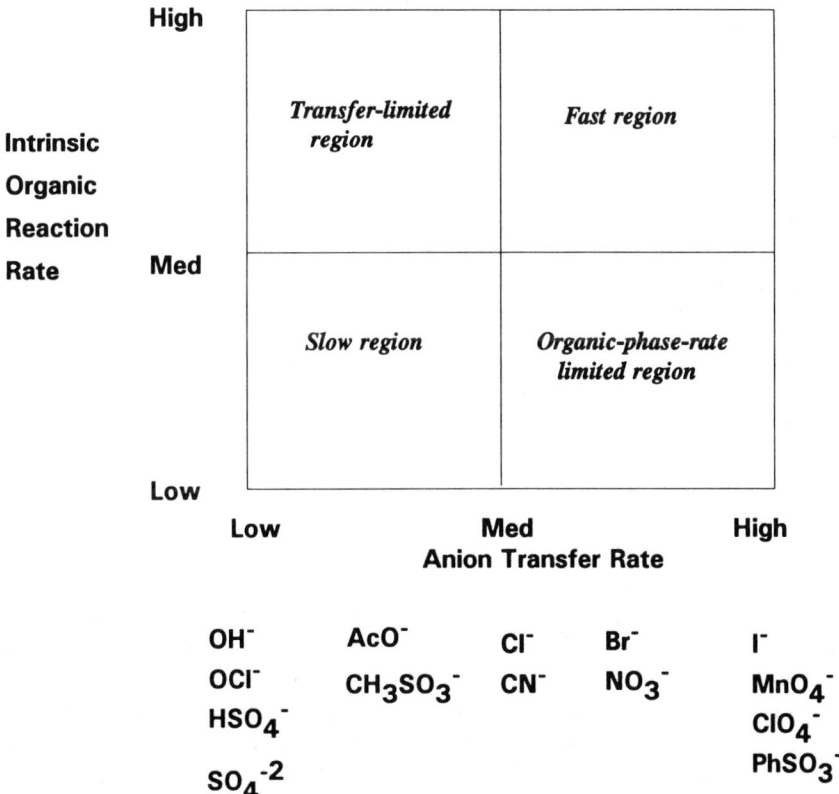

Figure 1-7. Relative ease of transfer for several common anions.

$$\text{CH}_3(\text{CH}_2)_6\text{CH}_2\text{Cl} + \text{NaCN} \xrightarrow{Q^+} \text{CH}_3(\text{CH}_2)_6\text{CH}_2\text{CN} + \text{NaCl}$$
$$\text{(org)} \qquad\qquad \text{(Aq.)} \qquad \text{(org)} \qquad\qquad\qquad \text{(Aq.)}$$

it is important to keep the aqueous phase concentration of cyanide ion high. This minimizes the extent to which quaternary salt exists as Q^+Cl^-, which species does not contribute to furthering the reaction. It is desirable to saturate the aqueous phase with NaCN, and even have solid NaCN present in the reaction mixture, to keep the amount of added water to the minimum necessary to obtain an adequate rate of anion transfer. Minimal use of water also reduces the amount of water available for hydration of ions. (See Chapter 2).

The cation associated with the inorganic salt can sometimes be important. If the aqueous-phase concentration of the anion to be reacted can be increased by changing to another cation, say potassium instead of sodium, then overall anion transfer and reactivity can often be increased by use of the more soluble potassium salt. For example, formate displacement reactions are better when potassium formate rather than sodium formate is used [16].

Sometimes it may be desirable to add no water or just minute amounts of water, especially with phase-transfer agents such as crown ethers and PEGs, which are capable of transferring inorganic compounds from their solid form. Although some tiny amount of water is required for transfer to occur rapidly [17], this amount of water may already be present on the inorganic solid. The presence of too much water inhibits formation of complexes with the polyethers.

On the other hand, for very fast PTC reactions it may be desirable to add water to moderate and control the reaction.

4. Amount and Kind of Organic Solvent Used, If Any

One outstanding feature of PTC reactions is the frequent opportunity to conduct reactions without organic solvent. This may lead to improved yields, rates, product purity, and avoidance of solvent environmental and recovery problems [18].

Sometimes a solvent is necessary, as, for example, a polar solvent may be necessary to obtain an appropriate rate of anion transfer to the organic phase; solvent may be necessary to increase the rate of the organic phase reaction; or solvent may be necessary simply because the organic reagent is an unreactive solid at reaction temperature. Methylene chloride has been extensively used as a polar solvent in the PTC work because it readily dissolves most quaternary salts and other phase-transfer agents, and because it is polar it speeds both the transfer step and the organic reaction step. More environmentally acceptable toluene and light hydrocarbons such as hexane or heptane have also been extensively used as solvents for PTC systems. Although the hydrocarbons suffer from lack of polarity, they are reasonably safe, inexpensive, and easy to recover in high purity.

Other solvent strategies may be followed: for example, a high-boiling solvent may be selected for reactions in which is product is a low boiler. A solvent might be chosen to minimize solubility of phase-transfer agent in the organic phase to force formation of third (catalyst) phase, or from which the phase-transfer catalyst may be more easily separated by extraction.

5. Temperature and Microwave Heating

Most organic reactions of interest are strongly accelerated by increasing temperature. Therefore, increased temperature is likely to be one of the first experimental variables considered for slow reactions, particularly for PTC systems that have slow organic-phase rates.

If polyethers are used as catalyst, then raising the temperature may decrease the stability of complex formation with salts [19], and increasing temperature may actually cause the catalyst to lose activity.

Quaternary ammonium and other -onium salts usually decompose at high temperatures, for example, greater than about 120–150°C for neutral salts, but at lower temperatures, for example, 50–70°C, for systems containing concentrated sodium or potassium hydroxide. This behavior is strongly dependent on catalyst structure, the reaction being catalyzed, the presence of cocatalysts, and other physical parameters of the system, and is more fully discussed in Chapter 4. Other catalysts such as polyethylene glycols and crown ethers are more resistant to thermal decomposition under basic conditions, but are sensitive to acidic conditions.

When microwave irradiation is used for heating of a reaction of o- or p-chloronitrochlorobenzene with ethanol in the presence of aqueous sodium hydroxide and a phase-transfer catalyst, the reaction is accelerated by a factor of 144- to 240-fold [20]. This remarkable acceleration beyond mere heating is probably due to the specific ability of microwave-range radiation to transfer energy specifically to water, presumably including water of hydration of hydroxide or ethoxide anions associated with the phase-transfer catalyst in the organic phase. Such specific transfer of energy probably causes the dissociation of water away from these anions, resulting in the enormous increase in anion reactivity. Use of microwave heating has been reviewed [32].

6. Cocatalysts

Cocatalysts may be added to increase either the transfer rate or the intrinsic organic reaction rate. For example, addition of alcohols, particularly diols, significantly increases the ease of hydroxide anion transfer (or its equivalent as alkoxide) [21]. Addition of iodide anion increases the rate of certain PTC benzyl chloride reactions [22], although in most PTC reactions iodide behaves as a catalyst poison. Added tungstate ion behaves as a cocatalyst with quaternary salts for hydrogen peroxide epoxidation of olefins [23].

Cocatalysts may also be added to enhance sequential or concurrent reactions other than the PTC reaction. For example, carbonylation of certain alkyl halides with carbon monoxide in the presence of NaOH to yield carboxylic acid salts is cocatalyzed by quaternary salts and transition metal compounds [24,25]. Many other examples of the use of cocatalysts are to be found throughout this book.

F. Outline of Compounds Used as Phase-Transfer Catalysts

The diagram in Figure 1-8 presents a general outline of several PTC catalyst categories.

1. Soluble Catalysts

Many compound types have been shown to have at least some activity as soluble phase-transfer catalysts, as reviewed in Chapter 4. Most practical work has centered on the types listed in Table 1-3. This group of compounds includes numerous materials that are readily available commercially, usually at reasonable cost.

Sometimes it is desirable for a catalyst to have a special property such as shown in Table 1-4. For example, the high-temperature catalyst is stable at >150°C, and is useful for displacement reactions on activated aromatic compounds, such as *p*-nitrochlorobenzene [26]. Chiral catalysts are able to catalyze formation of chiral products from optically inactive starting materials, as more fully discussed in Chapter 12 [27]. Quaternary salts containing β-oxyethyl groups transfer hydroxide and borohydride anions from aqueous to organic phases more

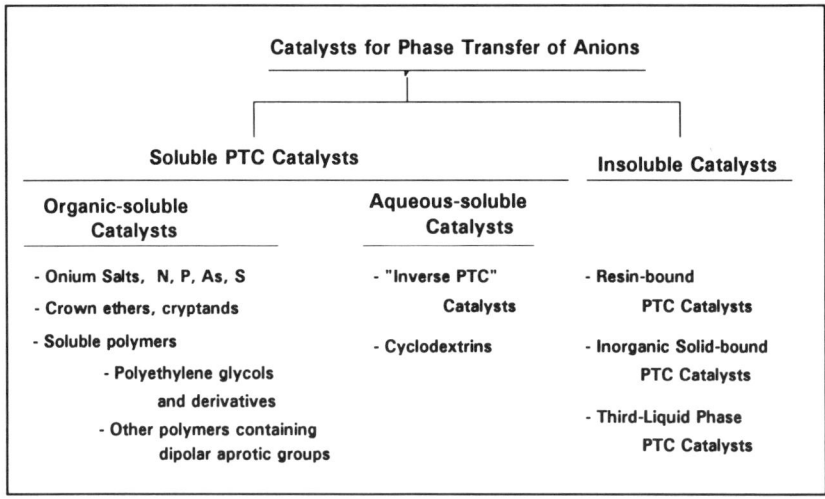

Figure 1-8. General outline of PTC catalyst types.

Table 1-3. Common types of soluble phase-transfer catalysts.

Quaternary Ammonium and Phosphonium Salts

- $(C_4H_9)_4N^+$ Br^-, $(C_8H_{17})_3NMe^+$ Cl^-
- $C_6H_5CH_2NEt_3^+$ Cl^-, $C_{16}H_{33}NMe_3^+$ Cl^-
- In-situ generation of quaternary salts

Polyethylene Glycol and Derivatives

- $HO(CH_2CH_2O)_nH$ n = 4 - 600
- $RO(CH_2CH_2O)_nH$, R = alkyl groups
- $N(CH_2CH_2OCH_2CH_2OCH_3)_3$

Crown Ethers and Cryptands

- 18-crown-6, 15-crown-5
- dibenzo-18-crown-6, dicyclohexano-18-crown-6

Polymeric "dipolar aprotic solvents"

for example:

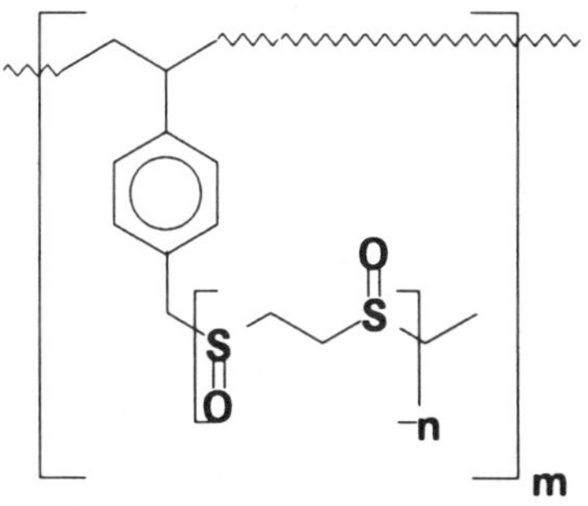

(or other dipolar aprotic functional groups bound to soluble polymers)

---

easily than unsubstituted quaternary salts, and are the catalysts of choice, for example, for dehydrohalogenation of 1,2-dichloro-3-butene to chloroprene [28]. Most divalent anions are difficult to transfer from aqueous to organic phases, but transfer is easier if a bis(quaternary salt) phase-transfer catalyst is employed [29,30]. The vast majority of work in PTC has been concerned with transfer of

Table 1-4. Examples of special quaternary salts for PTC.

High temperature catalysts for phenoxide substitution on p-chloronitrobenzene.	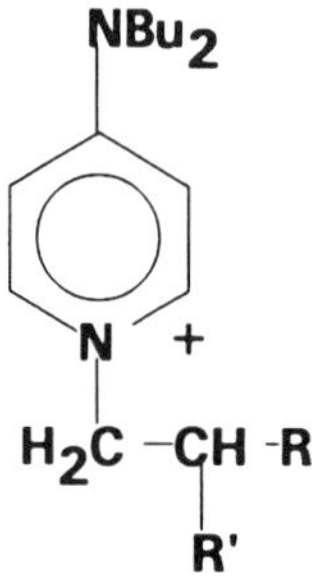 4-(NBu$_2$)-pyridinium with N-CH$_2$-CHR-R' substituent
High activity hydroxide transfer catalysts for elimination reactions.	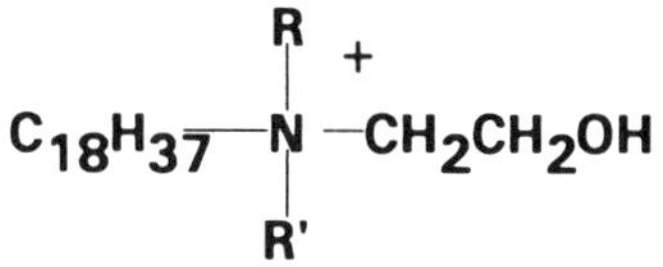 $C_{18}H_{37}$–N$^+$(R)(R')–CH$_2$CH$_2$OH
Catalyst for transfer of divalent anions.	R_3N^+-(CH$_2$)$_n$-N$^+$R$_3$
Catalyst for chiral induction in alkylation reactions.	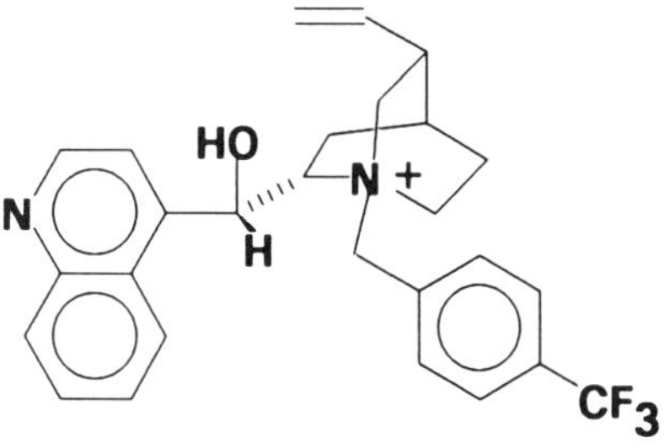 cinchona-derived quaternary ammonium with p-CF$_3$-benzyl group

anions from aqueous to organic phases for reaction. However, the primary concept of PTC is that any species can, in principle, be transferred to its non-normal phase and be activated for appropriate reactions. Some examples of non-anion transfer PTC are listed in Table 1-5. These special transfer situations are described further in Chapter 4.

2. Insoluble Catalysts

Insoluble phase-transfer catalysts, also called *triphase catalysts*, have the outstandingly useful feature of being easily removed from the reaction mixture and reused in subsequent reactions. The term *insoluble catalysts* includes both solid (where the PTC function is bound to an insoluble polymeric resin or an inorganic solid)

Table 1-5. Examples of species other than anions for phase-transfer reactions.

Water
Hydrogen peroxide
Oxygen
Hydrogen chloride and hydrogen bromide
Cations
Anhydrous $AlCl_3$
Transfer of organic compounds to aqueous phase for reaction ("inverse phase-transfer catalysis").

and liquid catalysts (where the PTC catalyst is predominately located in a third liquid phase). Much work has been published in this area, particularly with phase-transfer functions bound to insoluble resins or to inorganic solids. More recent publications clearly show that with certain systems a third phase, which contains most of the catalyst, can be formed. These catalysts are discussed in Chapter 5.

3. Catalysts for Vapor-Phase Reactions

Appreciable work has been done by Tundo and co-workers using vapor-phase flow-through reactors, by feeding reactants over supported PTC catalysts [31]. Where applicable, these reactions can be extremely useful for industrial-scale operations.

References

1. C.M. Starks, J. Am. Chem. Soc., **93**, 195 (1971).
2. C.M. Starks and D.R. Napier, to Continental Oil Co., U.S. Patent 3,992,432 (1976); Australian Patent 439,286 (1968); British Patent 1,227,144 (1971); Netherlands Patent 6,804,687 (1968).
3. C.M. Starks and C.L. Liotta, Phase Transfer Catalysts, Principles and Techniques, Academic Press, New York (1978).
4. W.A. Gibson and R.A. White, Anal. Chim. Acta, **12**, 413 (1955).
5. T.D. Shaffer and M.C. Kramer, Makromol. Chem., **191**(12), 3155 (1990).
6. C.M. Starks and R.M. Owens, J. Am. Chem. Soc., **95**, 2204 (1969).
7. F. Nome, A.F. Rubira, C. Franko, and L.G. Ionescu, J. Phys. Chem., **86**, 1881 (1982).
8. M. Makosza, Pure Appl. Chem., **43**, 439 (1975).
9. Der-Her Wang and Hung-Shan Weng, Chem. Eng. Sci., **43**, 2019 (1988).
10. T.D. Shaffer and M.C. Kramer, Makromol. Chem., **191**, 3155 (1990).
11. M. Takeishi, K. Se, N. Umeta, and R. Sato, Nippon Kagaku Kaishi, 824 (1992) [CA: 117.131693].
12. D. Landini and A. Maia, J. Chem. Soc., Chem. Commun., 1041 (1984).

13. D. Mason, S. Magdassi, and Y. Sasson, J. Org. Chem., **55**, 2714 (1990).
14. V. Ragaini, G. Colombo, P. Barzaghi, E. Chiellini, and S. D'Antone, Ind. Eng. Chem., Res., **27**, 1382 (1988).
15. Q. Lin, Y. Zhang, S. Chengru, and Q. Wenzhen, Chin. Chem. Lett. **2**, 517 (1991) [CA: 116: 193753].
16. O. Arrad and Y. Sasson, J. Am. Chem. Soc., **110**, 185 (1988).
17. C.L. Liotta, E.M. Burgess, C.C. Ray, E.D. Black, and B.E. Fair, ACS Symp. Ser., **326**, 15 (1985).
18. G. Bram, A. Loupy, and J. Sansoulet, Israel J. Chem., **26**, 291 (1985).
19. N. Kahana, A. Deshe, and A. Warshawsky, J. Polym. Sci., Chem. Ed., **23**, 231 (1985).
20. Y. Yuan, D. Gao, and Y. Jiang, Synth. Commun., **22**, 2117 (1992).
21. E.V. Dehmlow, R. Thieser, Y. Sasson, and E. Pross, Tetrahedron, **41**, 2927 (1985).
22. H.E. Hennis, J.P. Easterly, Jr., and L.R. Thompson, Ind. Eng. Chem. Prod. Res. Dev., **6**, 193 (1967); H.E. Hennis, L.R. Thompson, and J.P. Long, Ind. Eng. Chem., Prod. Res. Dev., **7**, 96 (1968).
23. O. Bortolini, F. DiFuria, G. Modena, and R. Seraglia, J. Org. Chem., **50**, 2688 (1985); O. Bortolini, V. Conte, F. DiFuria and G. Modena, J. Org. Chem., **51**, 2661 (1986).
24. H. Alper and H. des Abbayes, J. Organometal. Chem., (134) C11(1977).
25. L. Cassar and M. Foa, J. Organometal. Chem., (134), C15(1977).
26. D. Brunelle, to General Electric Co., U.S. Patent 4,595,760 (1986).
27. U. Dolling, P. Davis, and E. Grabowski, J. Am. Chem. Soc., **106**, 446 (1984).
28. L. Maurin, to Dupont Co., U.S. Patent 4,418,232 (1983).
29. M. Lissel, D. Feldman, M. Nir, and M. Rabinovitz, Tetrahedron Lett., **30**, 1683 (1989).
30. J.P. Idoux and J.T. Gupton, ACS Symp. Ser., **326**, 169 (1985).
31. P. Tundo, F. Trotta, G. Moraglio, and F. Ligorati, Ind. Eng. Chem. Res., **27**, 1565 (1988).
32. A. Loupy, G. Bram, J. Sansoulet, New J. Chem., **16**, 233 (1992).

2

Phase-Transfer Catalysis: Fundamentals I

A. Introduction

Critical to the success of phase-transfer catalytic (PTC) processes are (1) the maximization of the rate of transfer of reactant anions from the aqueous or solid phase to the organic phase, (2) the maximization of the rate of transfer of product anions from the organic phase to the aqueous or solid phase, and (3) the related equilibrium partitioning of the reactant and product anions between the organic and aqueous or solid phases. The common organic solvents employed in phase-transfer processes are usually relatively nonpolar and usually aprotic. Because anions do not have a great affinity for such solvents and prefer to reside in an aqueous environment, the desired transfer is not a particularly favorable process. The transfer of anions from an aqueous to an organic phase, however, may be achieved by choosing a phase-transfer cation that is not strongly solvated by water and that has organic-like characteristics and is thus compatible with the organic phase. For instance, the volume-to-charge ratio (as well as the organic-like nature) of quaternary ammonium and phosphonium salts can be adjusted over a wide range of values by simply changing the length of the alkyl (or aryl) substituents bonded to the quaternary heteroatom. Tetramethylammonium salts are highly soluble in aqueous media and only slightly soluble in most organic solvents, whereas tetradoecylammonium salts are soluble in most organic media but only slightly soluble in water. The former salt represents a quaternary ammonium ion with a small organic volume-to-charge ratio whereas the latter salt has a large organic volume-to-charge ratio. In a similar manner, macrocyclic multidentate ligands (crown ethers, cryptands, polyethylene oxides, etc.) may be employed to complex metal cations and carry them, along with their anions, from the aqueous or solid phase into the organic phase.

The factors that affect the mass transfer and distribution of the phase-transfer catalyst cation–anion pair between the organic and aqueous phases include (1) the combination of the charge-to-volume ratio, the polarizability, and the organic structure associated with the anion; (2) the hydrophilic–organophilic balance of the associated cation; (3) the polarity of the organic phase; (4) the hydration of the anion; and (5) the presence of aqueous salts and/or aqueous hydroxide ions. The following sections address each of these factors.

B. Structural Factors Affecting the Distribution of Anions Between Aqueous and Organic Phases

Table 2-1 summarizes the free energies of transfer of halide ions from water to acetonitrile ($\epsilon = 39$) and to dimethyl sulfoxide ($\epsilon = 45$) along with the ionic radius of each of the halide ions [1,2]. The positive free energies of transfer from the aqueous phase to the organic phase clearly indicate that halide ions prefer to reside in the aqueous phase. The transfer, however, becomes less unfavorable as one proceeds from chloride to bromide to iodide. This trend may be understood in terms of the change in charge-to-volume ratios of the halide ions. Because chloride has the largest charge-to-volume ratio it is the least polarizable and the most strongly hydrated. In contrast, iodide has a relatively diffuse charge and is less strongly hydrated.

The distributions of various tetraphenylphosphonium and -arsonium salts between water and chloroform ($\epsilon = 4.7$) are listed in Table 2-2 [3–7]. Comparison of the counteranions Cl^-, Br^-, and I^- shows that as the polarizability of the negatively charged species increases, the distribution ratio increases. These observations are consistent with the free energies of transfer listed in Table 2-1.

Table 2-2 contains a wide variety of simple and complex inorganic anions. Qualitatively, the trends within the two series of -onium salts are essentially identical. Some general trends with regard to the affinity of related anions for the chloroform phase are:

$$I^- > Br^- > Cl^-$$
$$ClO_3^- > BrO_3^- > IO_3^-$$
$$NO_3^- > NO_2^-$$
$$ClO_4^- > IO_4^-$$

The divalent anions listed in Table 2-2 have comparatively little affinity for the organic phase.

An important measure of the competitive partitioning of anions between the aqueous and organic phases in the presence of a quaternary cation is the selectivity constant, $K_{X/Y}^{sel}$, which is defined by Eq. (2-1) and (2-2).

$$[Q^+Y^-]_{org} + [X^-]_{aq} \rightleftarrows [Q^+X^-]_{org} + [Y^-]_{aq} \qquad (2-1)$$

Table 2-1. Free energies of transfer of halide ions from water to acetonitrile and dimethyl sulfoxide (DMSO).

Anion	Ionic radius (Å)	ΔG (kcal/mol) $H_2O \rightarrow CH_3CN$	ΔG (kcal/mol) $H_2O \rightarrow DMSO$
Cl^-	1.81	+11.6	+10.0
Br^-	1.95	+ 8.1	+ 6.9
I^-	2.16	+ 4.8	+ 2.9

[a] L. Pauling, The Nature of the Chemical Bond, Appendix 3.1. Cornell University Press, Ithaca, New York, 1940.

[b] R. Alexander and A.J. Parker, J. Am. Chem. Soc., 89, 5549 (1967).

Table 2-2. Distribution of various tetraphenylphosphonium and -arsonium salts between water and chloroform.[d]

$$\alpha = \frac{(anion)_{\text{organic phase}}}{(anion)_{\text{aqueous phase}}}$$

Anion, X^-	$(C_6H_5)_4P^+X^{-a}$	$(C_6H_5)_4As^+X^{-b}$
Cl^-	0.18	0.19
Br^-	3.4	4.8
I^-	54.	>300.
NO_3^-	4.6 → 5.6[c]	20.3 → 76.[c]
NO_2^-	0.11	0.2
ClO_3^-	100.	>150.
BrO_3^-	0.5	0.9
IO_3^-	<0.005	0.004
ClO_4^-	>200.	>200.
MnO_4^-	>300.	>300.
ReO_4^-	>600.	>200.
IO_4^-	0.017	0.02
CrO_4^{-2}	24.7 → 0.005[c]	71. → 0.04[c]
Molybdate, tungstate, vanadate, borate	<0.003	<0.005
SO_4^{-2}	<0.001	<0.001
S^{-2}	<0.002	<0.05
$S_2O_3^{-2}$	<0.003	<0.02
Arsenite	0.002	0.006 → 0.047[c]
Arsenate	0.013	<0.01
Selenite	<0.002	<0.01 → 0.09[c]
Tellurite	<0.002	<0.01 → 0.09[c]

[a] R. Bock and G.M. Beilstein, Z. Anal. Chem., 192, 45 (1963); R. M. Diamond and D.G. Tuck, Prog. Inorg. Chem. 2, 109 (1960).

[b] R. Bock and J. Jainz, Z. Anal. Chem., 198, 315 (1963).

[c] Dependent on pH, acidic → basic.

[d] Values of α for $(C_6H_5)_4Sb^+X^-$ have also been measured: R. Bock and E. Gallath, Z. Anal. Chem., 222, 283 (1966).

$$K_{X/Y}^{sel} = \frac{[Q^+X^-]_{org}[Y^-]_{aq}}{[Q^+Y^-]_{org}[X^-]_{aq}} \qquad (2\text{-}2)$$

Table 2-3 summarizes values of $K_{Cl^-\to X}$ for a series of inorganic and organic anions [8–11]. The organic solvents employed in the two-phase systems include 1-decanol, toluene, chloroform, and methylene chloride, and the quaternary salts varied widely in structure. In all cases the Cl^- is the reference anion. Those anions that are weakly hydrated and those that have considerable organic structure are more favorably partitioned into the organic phase. In general, trivalent anions have a less favorable distribution compared to divalent anions which have a less favorable distribution compared to monovalent anions. The following series illustrates this point:

$$H_2PO_4^- > HPO_4^{-2} > PO_4^{-3}$$
$$HS^- > S^{-2}$$

C. Structural Factors Affecting the Distribution of Phase-Transfer Catalyst Cations Between the Aqueous and Organic Phases

Table 2-4 [12] shows the effect of chain length on the distribution ratios, α, of alkyltriphenylphosphonium salts distributed between chloroform and water for a wide variety of counteranions [Eq. (2-3)].

$$\alpha = \frac{[Q^+X^-]_{organic\ phase}}{[Q^+X^-]_{aqueous\ phase}} \qquad (2\text{-}3)$$

As the length of the alkyl chain increases the distribution ratio increases. The trend is consistent for all of the counteranions listed in Tables 2-2 and 2-3.

A series of extraction constants, e_{QX}, defined by Eq. 2-4,

$$e_{QX} = \frac{[QX]_{org}}{[Q^+]_{aq}[X^-]_{aq}} \qquad (2\text{-}4)$$

are listed in Tables 2-5 [13,14], 2-6 [15], and 2-7 [15,16]. Extraction constants are tabulated (1) for several quaternary ammonium, phosphonium, and arsonium salts distributed between water and 1, 2-dichloroethane (Table 2-5), (2) for quaternary ammonium picrates distributed between water and a series of organic solvents (methylene chloride, chloroform, benzene, and carbon tetrachloride) (Table 2-6), and (3) values for a series of quaternary ammonium bromides distributed between water and 1, 2-dichloroethane (Table 2-7).

Competitive anion-exchange equilibria between halides and monovalent anions, halides and hydrogensulfate, and halides and divalent anions in methylene

Table 2-3. Selectivity constants $K_{Cl \to X}$ for anion extraction into organic solutions of quaternary salts.

Anion	0.2 M $(C_{10}H_{21})_3N^+ CH_3$ in 1-decanol[a]	Various Q^+ in toluene[b]	10^{-5} M $C_{16}H_{33}{}^+NC_5H_5$ in $CHCl_3$[c]	$(n-C_4H_9)_4N^+$ in $HCCl_3$ (and CH_2Cl_2)
Chloride	(1.0)	(1.0)[d,e,f,g,h]	(1.0)	(1.0)
Bromide	6	16.5[f]	18.8	25 (48.6)
Iodide	100	~5000[f]	338	1312 (6251)
Flouride	—	0.02[e]	—	—
Cyanide	—	1.0 ± 0.2[f,g,h]	—	—
Nitrate	4.4	18.4[f]	12.5	31.4 (226)
Hydroxide	—	0.01	—	—
Thiocyanate	73,200	—	670	—
Perchlorate	120	~500	—	3872 (1.3 × 10^5)
Chlorate	—	—	—	—
Bromate	—	0.22[h]	—	—
Permanganate	—	>100[f]	—	—
Formate	0.76	0.082[h]	—	—
Oxylate	0.1	—	—	—
Acetate	0.51	0.12[f]	—	9.7 × 10^{-3}
Propionate	0.69	—	—	—
Benzoate	2.0	—	—	3.14
m-Toluate	10.0	—	—	0.038
p-Toluate	25.0	—	—	—
Salicylate	>1000	—	—	337
Methane sulfonate	—	0.25[f]	—	—
p-Toluene sulfonate	320	>500	—	274
Sulfate	0.02	0.013[h]	—	—
Bisufate	—	0.22[h]	—	—
Carbonate	—	0.005[h]	—	—
Bicarbonate	—	0.05[h]	—	—
Phosphate	—	0.02[h]	—	—
Dihydrogen phosphate	—	0.02[h]	—	—
$C_6H_5CH_2CO_2^-$	—	—	—	2.38
Phenoxide	—	—	—	1.19
Picrate	—	—	—	1.04 × 10^6 (1.37 × 10^7)
Naphthaene-2-sulfonate	—	—	—	3613
Anthracene-2-sulfonate	—	—	—	1.67 × 10^5
Trinitro benzene sulfonate	—	—	—	3.7 × 10^4
2,4-Dinitro-1-napthoxide	—	—	—	3.6 × 10^6

[a]C.J. Cotzee and H. Freisee, Ana. Chem., **41**, 1128 (1969).

[b]C.M. Starks, unpublished results.

[c]H.K. Bisivas and B.M. Mandal, Anal. Chem., **44**, 1636 (1972).

[d]R.B. Grieves, W. Charewicz, and P.J.W. The, Sep. Sci. **10**, 77 (1975).

[e]$(C_{10}H_{37})_3(C_3H_7)N^+$.

[f]$(C_{18}H_{21})_3CH_3N^+$.

[g]$C_{16}H_{33}(n\text{-}Bu)_3P^+$.

[h]$(C_{18}H_{37})_2(CH_3)_2N^+$.

Table 2-4. *Effect of chain length on distribution ratios α of triphenylalkylphosphonium salts between chloroform and water.*

$\alpha = [Q^+X^-]$ in the organic phase / $[Q^+X^-]$ in the aqueous phase

Cation	Cl$^-$	Br$^-$	I$^-$	NO$_2^-$	SCN$^-$	ClO$_4^-$	ClO$_3^-$	BrO$_3^-$
Ph$_3$PCH$_3^+$	0.01	0.10	1.7	0.11	1.5	2.7	0.18	0.03
Ph$_3$PC$_2$H$_5^+$	0.02	0.17	3.6	0.25	3.1	6.5	0.40	0.05
Ph$_3$P–n–C$_3$H$_7^+$	0.04	0.52	6.9	0.53	6.3	12.0	0.84	0.11
Ph$_3$P–n–C$_5$H$_{11}^+$	0.18	1.7	9.5	2.2	16.0	25.0	3.0	0.51

Temperature, 25°C; ionic strength, 0.1; quaternary sat concentration initially in the aqueous phase, 0.0005 *M*.

From N.A. Gibson and D.C. Weatherburn, Anal. Chim. Acta, **58**, 160 (1972).

Table 2-5. *Extraction constants E_{QX} of various salts distributed between water and 1,2-dichloroethane.*

Cations	Cl$^-$	Br$^-$	I$^-$	CH$_3$OSO$_3^-$	Tos^a	pNP^b
NBu$_4^+$	0.13	6.15	830	18.1	180	247
	0.24^c	6.58^c	528^c			
PBu$_4^+$	0.57	28.4	3730	—	—	610
NPent$_4^+$	38.2	970	—	—	—	6.4 × 10^4
N-iso-Pent$_4^+$	9.95	—	—	—	—	1.6 × 10^4
NHex$_4^+$	—	1.3 × 10^5	—	—	—	~10^8
AsPh$_4^+$	7.56	349	4.3 × 10^4	820	1.1 × 10^4	1.9 × 10^4
PPh$_4^+$	7.2	—	—	—	—	1.9 × 10^4
C$_{12}$H$_{25}$NEt$_3^+$	13.2	357	—	—	—	—

^{a}*p*-Toluenesulfonate.

^{b}p-Nitrophenolate.

^{c}Values from Dehmlow, E.V.; Dehmlow, S.S. Phase Transfer Catalysis, 2nd ed.; Verlag Chemie: Weinheim (1983).

Table 2-6. *Extraction constants $E_{NR4pic} = [NR_{4pic}]_{org}/\{[NR_4^+]_{aq}[pic^+]_{aq}\}$ of quaternary ammonium picrates, extraction from water into an organic solvent.*

Solvent	N(CH$_3$)$_4^+$	N(C$_2$H$_5$)$_4^+$	N(n–C$_3$H$_7$)$_4^+$	N(n–C$_4$H$_9$)$_4^+$	N(n–C$_5$H$_{11}$)$_4^+$
CH$_2$Cl$_2$	1.5	220	2.9 × 10^4	4.8 × 10^6	2.45 × 10^8
ChCl$_3$	0.22	21	4.4 × 10^3	8.1 × 10^5	—
C$_6$H$_6$	—	0.22	35	3.9 × 10^3	7.9 × 10^5
CCl$_4$	—	—	—	87	2.9 × 10^4

Table 2-7. Apparent extraction constants for quaternary ammonium bromides from water to 1,2-dichloroethane.

Quaternary ammonium salt	e_{QX}
$R-N(CH_3)_3^+$	7590
$R = n-C_{16}H_{33}$	1860
$n-C_{15}H_{31}$	468
$n-C_{14}H_{29}$	33.1
$n-C_{12}H_{25}$	
$R-N(C_2H_5)_3^+$	5750
$R = n-C_{14}H_{29}$	355
$n-C_{12}H_{25}$	22.9
$n-C_{10}H_{21}$	
$R-N(C_3H_9)_3^+$	741
$R = n-C_{10}H_{21}$	190
$n-C_9H_{19}$	
$R-N(C_4H_9)_3^+$	30,903
$R = n-C_{10}H_{21}$	6.6
$n-C_4H_9$	.7
C_2H_5	.4
CH_3	
$N(n-C_5H_{11})_4$	1072
$N(n-C_6H_{13})_4$	165,959

chloride and 1, 2-dichloroethane are summarized in Tables 2-8 [17], 2-9 [17], and 2-10 [17], respectively. In this series, highly organophilic quaternary ammonium cations were employed and the results are reported in terms of percent exchange in the organic phase. For the simple anions (Table 2-8), the percent exchange is independent of the structure of the quaternary cation. In contrast, the results of the halide–hydrogen sulfate, halide–sulfate, and halide–chromate experiments show that the percent exchange (Tables 2-9 and 2-10) increases with increasing steric availability of the cationic nitrogen.

D. Effects of the Organic Phase Polarity on the Distribution of Phase-Transfer Cation–Anion Pairs

It is expected that factors such as polarity of the organic phase in conjunction with the structures of the anion and the catalyst cation affect selectivity of phase-transfer catalyst partitioning into the organic phase. For instance, increasing the polarity and hydrogen-bonding ability of the organic phase would have a strong favorable effect on the extraction of small ions (large charge-to-volume ratios) from an aqueous phase, but less effect on larger anions (small charge-to-volume ratios) and on anions with substantial organic structure, so that the whole range of selectivities would be compressed (a leveling effect would be observed).

Table 2-8. Competitive anion exchange of monovaent anions.

$$[Q^+X^-]_{org} + [Y^-]_{aq} \rightleftarrows [Q^+Y^-]_{org} + [X^-]_{aq}$$

	Percent exchange found in organic phase (solvent: A = CH_2Cl_2; B = $ClCH_2CH_2Cl$)					
Starting salt in organic phase	Br→ CN (B)	Br→ NO_3 (A)	Cl→ CN (B)	Cl→ CN (A)	I→ ClO_3 (B)	I→ ClO_4 (B)
$(n\text{-Hexyl})_4NX$	22	70	46	90	23.5	81
$Me(n\text{-Octyl})_3NX$	20		46.5		25	80
$Me_2(n\text{-Dodecyl})_2NX$	22.5		44	92	24	80
$Me_2(Benzyl)(n\text{-Dodecyl})NX$	22	61	49	87	24	80.5

	Percent exchange found in organic phase (solvent: A = CH_2Cl_2, B = $ClCH_2CH_2Cl$)	
Starting salt in organic phase	Cl→ Salicylate (A)	Cl→ Mandelate (A)
$(n\text{-Hexyl})_4NX$	90	
$(n\text{-Octyl})_4NX$		84
$ME_2(n\text{-Dodecyl})_2NX$	92	76
$Me_2 (Benzyl) (n\text{-Dodecyl})NX$	87	

E.V. Dehmlow and B. Vehre, J. Chem. Res. (S), 350 (1987).

Comparison of the data in columns 1 (1-decanol solution) and 2 (toluene solution) of Table 2-3 supports this expectation for every anion where data are listed in both columns. Additional data that support this qualitative generalization are shown in Table 2-11 [18]. Hydrogen-bonding solvents are compared to nonpolar and dipolar, aprotic solvents of varying polarities. The selectivity constants for the water-3-methyl-1-butanol and the water–chloroform two-phase systems are greatly compressed compared to the other organic solvents listed.

The effect of solvent polarity is more subtly illustrated in the increasing

Tabe 2-9. Hailde–hydrogen sulfate exchange in various solvents.

	Percent change in organic phase.					
	CH_2Cl_2		$ClCH_2CH_2Cl$		$HCCl_3$	CCl_4
Starting salt in organic phase	Cl→ HSO_4	Br→ HSO_4	Cl→ HSO_4	Br→ HSO_4	Cl→ HSO_4	Cl→ HSO_4
$(n\text{-Hexyl})_4NX$	24	5.6	37	9.6	12	
$(n\text{-Octyl})_4NX$	25		34			25.6
$Me(n\text{-Octyl})_3 NX$	30	7	43.5	11	17	
$Et(n\text{-Octyl})_3NX$	26					55
$Me_2(n\text{-Dodecyl})_2NX$	42.5	10	54	19	31	
$Me_2(Benzyl)(n\text{-Dodecyl})NX$	45.5	10	insol	19		

E.V. Dehmlow and B. Vehre, J. Chem. Res. (S), 350 (1987).

Table 2-10. Competitive extraction of divalent anions.

Starting salt in CCl$_2$H$_2$ phase	Percent exchange in methylene chloride phase				
	Cl→ SO$_4$	Cl→ CrO$_4$	Cl→ Cr$_2$O$_7$	Br→ Cr$_2$O$_7$	Cl→ S$_2$O$_3$
(n-Hexyl)$_4$NX	9	21	94	84	0
Me(n-Octyl)$_3$NX	15.5	34	97		2
Me$_2$(n-Dodecyl)$_2$NX	33	54	94.5		
Me$_2$(Benzyl)(n-Dodecyl)NX	37	51	98	90	11

Starting salt in CCl$_2$H$_2$ phase	Percent exchange in methylene chloride phase			
	Br→ Naph (1, 5)- (SO$_3$H)$_2$	Br→ Benz (1, 3)- (SO$_3$H)$_2$	Cl→ Phthalate	Cl→ Terephthalate
(n-Hexyl)NX	82.5	51	32	30.5
Me(n-Octyl)$_3$NX			40	
Me$_2$(n-Dodecyl)$_2$NX			54	57
Me$_2$(Benzyl)(n-Dodecyl)NX	85	62	53	52

E.V. Dehmlow and B. Vehre, J. Chem. Res. (S), 350 (1987).

selectivity of more polar solvents for CN$^-$ (more highly hydrated and more basic) relative to Cl$^-$ as shown in Table 2-12 [3]. These data demonstrate how, during the course of a reaction between aqueous sodium cyanide and an organic phase of neat 1-chlorooctane, the selectivity for transfer of the reactant anion CN$^-$ may increase as the reaction proceeds and the polarity of the organic phase increases. There are important kinetic consequences associated with increasing polarity of the organic phase during the course of a PTC process.

E. Effects of Changes in Organic Phase Polarity During Reaction

In many PTC reactions, particularly where no organic phase solvent is employed (neat organic reactant is used as the organic phase), a substantial change in the

Table 2-11. Selectivity constants, $K_{x \to Cl}$, for methyltrioctylammonium salts as a function of solvent.

Solvent	Br	I$^-$	ClO$_4^-$
Hexane	8.1	525	17,000
Benzene	24.5	3800	250,000
Methylene chloride	39.8	3090	60,250
Carbon tetrachloride	43.7	8510	43,650
Nitrobenzene	44.7	7585	1,150,000
3-Methyl-1-butanol	3.0	15.8	28
Chloroform	23.4	891	1738

Table 2-12. Selectivity constants $K_{Cl \to CN}$ for cyanide extraction by $(C_{10}H_{21})_3N^+(CH_3)$ relative to chloride extraction into various organic solvents[a]

Solvent	$K_{Cl \to CN}$ (0.2 M 25°C)	Dielectric (constant 20°C)
Toluene	1.2	2.4
1-Octanol	2.3	3.4
Chlorobenzene	2.2	5.9
1-Chlorooctane	1.3	~6
1-Cyanooctane	2.4	~12
Benzonitrile	3.1	25
cis-1,2-Dichloroethylene	0.9	—
Dioctylether	1.2	—

[a]Starks, C.M.; Liotta, C.L. Phase Transfer Catalysis: Principles and Techniques; Academic Press, New York (1978).

polarity of the organic phase may occur as the reaction proceeds. This may have some effect on the rate of the organic-phase reaction (may raise or lower the value of the rate constant k_2) although data will be presented that suggests that the magnitude of such changes will be rather small. A much more profound effect can be observed in some instances where changes in the polarity of the organic phase may increase or decrease the amount of catalyst cation–anion pair partitioned into the organic phase. This behavior is evident in the cyanide displacement on 1-bromooctane, catalyzed by tetra-*n*-butylphosphonium bromide [3,19]. The catalyst is only sparingly soluble in 1-bromoctane, but is substantially more soluble in aqueous sodium cyanide solution, so that initially little Q^+CN^- is in the organic phase and the displacement reaction is slow. However, tetra-*n*-butylphosphonium salts are more soluble in the product 1-cyanooctane, so that as conversion of the alkyl bromide to the alkyl cyanide continues, increasing quantities of the catalyst will be taken into the organic phase, and the reaction rate will accelerate. This behavior is clearly evident in the autocatalytic character of the upper curve in Figure 2-1, particularly when compared with the nicely linear first-order plot observed when using a catalyst that is essentially all in the organic phase from the start (lower part of Figure 2-1). The one-to-one relationship between rate and organic phase catalyst concentration is clearly evident in Figure 2-2 where the organic phase concentration of tetra-*n*-butylphosphonium ion was determined using ^{14}C-labeled phosphonium salt.

F. Factors Affecting the Distribution of Phase-Transfer Catalyst Cation–Anion Pairs Between an Organic Phase and an Aqueous Phase Containing Hydroxide Ion

PTC reactions involving hydroxide ion have been shown to have great synthetic utility. The ability of the phase-transfer cation to interact with hydroxide ion and

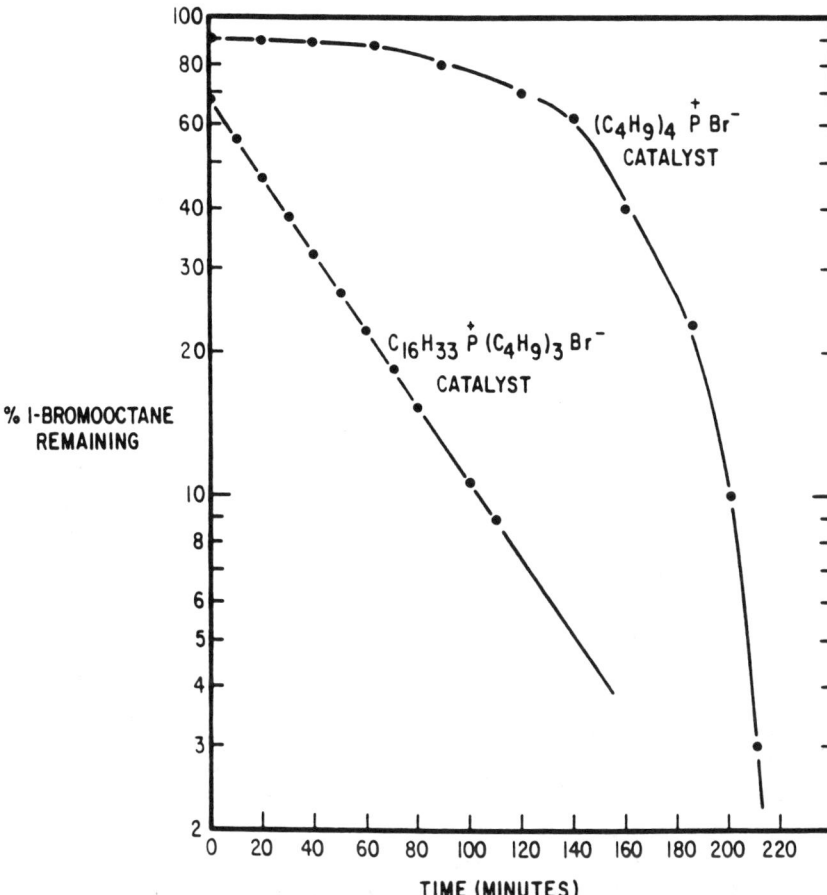

Figure 2-1. Comparison of cyanide displacement on 1-bromooctane catalyzed by tetra-*n*-butylphosphonium bromide and hexadecyltri-*n*-butylphosphonium bromide.

carry it to the region of the multiphase system where reaction can take place is critical to the success of the process. It is clear from the single entry in Table 2-3 that the selectivity constant relating hydroxide ion to chloride is comparatively small. Indeed, it is comparable in magnitude to the selectivity constants for fluoride ion and all the divalent anions listed in Table 2-3.

The quantity of hydroxide ion extracted into the organic phase is a function of the structure of the phase transfer cation. Table 2-13 shows the effect of quaternary ammonium salt structure on the extraction of OH^- into benzene [20]. It must be recognized that OH^- is an anion of relatively low polarizability (a hard anion) and, as such, difficult to extract into a nonpolar medium. Nevertheless, as the hydrocarbon nature of the quaternary salt increases the distribution of OH^-

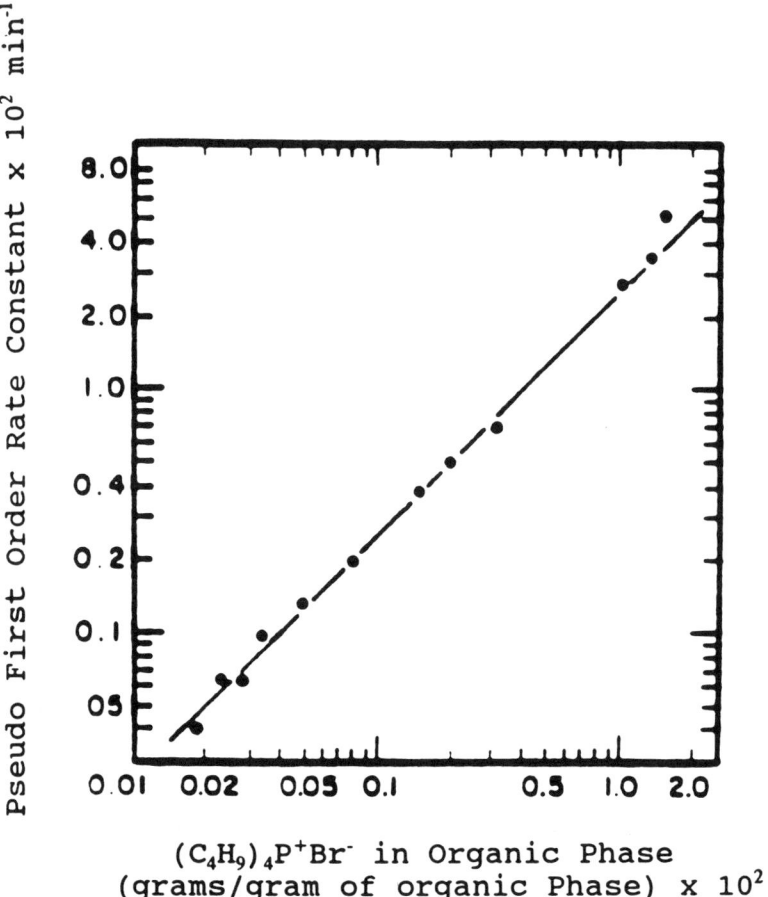

Figure 2-2. Relationship between cyanide displacement rate on 1-bromooctane and the concentration of tetra-*n*-butylphosphonium bromide in the organic phase.

into benzene increases. A qualitative relationship exists between the relative distribution constants and the number of carbon atoms associated with the quaternary ammonium cations. In PTC reactions involving hydroxide ion (alkylation, carbene additions and insertions, isomerization, deuterium exchange, etc.) the hydroxide ion competes with the other anions for the phase-transfer cation. $K_{OH/X}^{sel}$ values, defined by Eqs. (2-5) and (2-6), quantitatively

$$[Q^+X^-]_{org} + [OH^-]_{aq} \rightleftarrows [Q^+OH^-]_{org} + [X^-]_{aq} \qquad (2\text{-}5)$$

$$K_{OH/X}^{sel} = \frac{[Q^+OH^-]_{org}[X^-]_{aq}}{[Q^+X^-]_{org}[OH^-]_{aq}} \qquad (2\text{-}6)$$

Table 2-13. Effect of quaternary ammonium salt structure on the extraction of Q^+OH^- into benzene.

Cation	Number of C atoms	Distribution constant $(OH^-)_{Ph-H}/(OH^-)_{HOH}$
$(CH_3)_4N^+$	4	0.027
$C_6H_5CH_2N(C_2H_5)_3^+$	13	0.041
$(C_3H_7)_4N^+$	12	0.11
$C_{16}H_{33}N(CH_3)_3^+$	19	0.15
$C_5H_5NC_{12}H_{25}^+$	17	0.18
$C_{10}H_{21}N(C_2H_5)_3^+$	16	0.26
$C_{12}H_{25}N(C_2H_5)_3^+$	18	0.54
$(C_4H_9)_4N^+$	16	0.68

A.W. Herriott and D. Picker, J. Am. Chem. Soc., **97**, 2345 (1975).

describing the competition between a variety of monovalent or divalent anions and hydroxide ion for extraction into the organic phase by tetra-n-butylammonium cation, are summarized in Table 2-14 [21]. Almost all of the monovalent anions dominate in the competition with hydroxide for the quaternary cation. The competition is much less favorable for the monovalent anions with large charge-to-volume ratios and for the divalent anions.

The results of studies dealing with the extractability of OH^- into the organic phase as tetra-n-hexylammonium hydroxide with a variety of tetra-n-hexylam-

Table 2-14. Selectivity values for X in chlorobenzene–aqueous sodium hydroxide.

$$[Q^+OH^-]_{org} + [X^-]_{aq} \rightleftarrows [Q^+X^-]_{org} + [OH^-]_{aq}$$

$$K_{X,OH}^{sel} = \frac{[Q^+X^-]_{org}[OH^-]_{aq}}{[Q^+OH^-]_{org}[X^-]_{aq}}$$

Anion, X^-	$K_{X/OH}^{sel}$
$(COO^-)_2$	10
CO_3^{-2}	20
SO_4^{-2}	30
F^-	50
CH_3COO^-	120
Cl^-	950
$C_6H_5COO^-$	2×10^3
Br^-	3×10^3
NO_3^-	8×10^3
ClO_3^-	9×10^3
I^-	1×10^4
SCN^-	5×10^4
ClO_4^-	1×10^5
MnO_4^-	$> 1 \times 10^5$

J. de la Zerda and Y. Sasson, J. Chem Soc. Perkin Trans. II, 1147 (1987).

Table 2-15. Percent extractability of OH^- in the organic phase as $(C_6H_{13})_4N^+OH^-$ with tetrahexylammonium salts in chlorobenzene–aqueous sodium hydroxide two-phase systems at 25° and 60°C.

X^-	T (°C)	15% NaOH	30% NaOH	40% NaOH	50% NaOH
Cl^-	25	19	13	10	8
	60	28	22	17	10
$MeSO_3^-$	25	14	5	3	3
	60	18	12	6	4
Br^-	25	2.5	1.5		<0.5
	60	5	4	3	<1

[a] 40 mL of a chlorobenzene solution of Q^+X^- (4×10^{-2} M) and 40 mL of an aqueous solution of NaOH (15–50% w/w).

[b] Evaluated via acid–base titration of the organic phase.

D. Landini, A. Maia, and A. Rampoldi, J. Am. Chem. Soc., **51**, 5476 (1986).

monium salts in chlorobenzene–aqueous sodium hydroxide two-phase systems at 25 and 60°C have been reported. The data are summarized in Table 2-15 [22]. At both temperatures studied, as the concentration of aqueous sodium hydroxide increases from 15% to 50% the extractability of OH^- decreases. Table 2-16 summarizes the effect of the aqueous base concentration on both the selectivity constants ($K_{OH/X}^{sel}$) for the chloride, methanesulfonate, and bromide salts and the decomposition percentage via the Hofmann elimination process of the quaternary ammonium salts in chlorobenzene–aqueous NaOH two-phase systems at 60°C [23]. At a particular aqueous base concentration, the selectivity constant decreases in proceeding from chloride to methanesulfonate to bromide salts. This trend is not unexpected since the polarizability and organic solubility of these anions increase from chloride to methanesulfonate to bromide. As a consequence, the much less polarizable hydroxide ion has greater difficulty in competing for the quaternary cation for residency in the organic phase. For a particular anion, the selectivity constant decreases as the concentration of aqueous base increases. This decrease is quite dramatic with the chloride and methanesulfonate salts and much less so with the corresponding bromide salt.

Transfer of oxybases other than hydroxide can be important in promoting base-initiated reactions. In a formal sense, the following equilibria should be considered [Eqs. 2-7 and 2-8]: [22]

$$[R-OH]_{aq} + [OH^-]_{aq} \overset{K_a}{\rightleftarrows} [R-O^-]_{aq} + [H_2O]_{aq} \qquad (2\text{-}7)$$

$$[Q^+OH^-]_{org} + [R-O^-]_{aq} \overset{K_{RO/OH}^{sel}}{\rightleftarrows} [Q^+R-O^-]_{org} + [OH^-]_{aq} \qquad (2\text{-}8)$$

The first equation describes the acid–base equilibrium between an organic alcohol and hydroxide ion; the second equation describes the distribution equilibrium of

Table 2-16. *Effect of the aqueous base concentration on the selectivity constants and percent decomposition of tetrahexylammonium salts in chlorobenzene–aqueous sodium hydroxide two-phase systems, at 60°C.*

$$[Q^+X^-]_{org} + [OH^-]_{aq} \rightleftarrows [Q^+OH^-]_{org} + [X^-]_{aq}$$

$$K^{sel}_{OH/X} = \frac{[Q^+OH^-]_{org}[X^-]_{aq}}{[Q^+X^-]_{org}[OH^-]_{aq}}$$

	Cl$^-$		MeSO$_4^-$		Br$^-$	
NaOH, %	$K \times 10^5$	% dec. 2 h	$K \times 10^5$	% dec. 7 h	$K \times 10^5$	% dec. 7 h
15	91	2	33	7	2.2	1
30	25	17	6.5	36	0.67	15
40	9.3	76	1.0	65	0.25	
50	2.2	100	0.33	94	0.02	40

^{a}40 mL of a chlorobenzene solution of tetrahexylammonium salts (4 × 10^{-2} M) and 40 mL of an aqueous solution of NaOH (15–50% w/w).

D. Landini, A. Maia, and A. Rampoldi, *J. Am. Chem. Soc.*, **54**, 5476 (1986).

hydroxide and alkoxide ions between the aqueous and organic phases promoted by the quaternary ammonium cation. An acidity-selectivity constant may now be defined [Eq. 2-9] that combines the information contained in the above

$$K^S_{RO/OH} = K_a K^{sel}_{ROOH} \tag{2-9}$$

two equilibria. The K^{sel}_{ROOH} values for a series of representative primary alcohols and diols are shown in Table 2-17 [22]. As the organophilic nature of the alcohol increases, the acidity-selectivity value increases. Evidence exists that suggests that the extracted alkoxide is solvated by unionized alcohol molecules by means of intermolecular hydrogen bonding [3]. This solvation accomplishes at least two things that are not mutually exclusive. First, it makes the alkoxide more organophilic, and second, the negative charge is more delocalized resulting in a more polarizable species that can be more readily accommodated in the organic phase. With this in mind, it is interesting to compare the monoalcohols with the diols. From purely structural considerations alone, if the ratio of hydroxide to hydrocarbon is considered, it would *not* be anticipated that the diols would have acidity-selectivity values as large as those shown in Table 2-17. Unlike the monoalcohols, however, the diols can form monoalkoxides that can be stabilized by delocalization of the negative charge via intramolecular hydrogen bonding. The fact that their values are larger than expected suggests that these interactions may be operating and could account for the observations.

The amounts of base extracted into chlorobenzene from 50% aqueous sodium hydroxide in the presence of tetra-*n*-octylammonium bromide and a series of aliphatic primary, secondary, and tertiary alcohols, as well as a series of aliphatic

Table 2-17. *Acidity-selectivity values for RO⁻ formed in situ in chlorobenzene–aqueous sodium hydroxide.*

$$[R-OH]_{aq} + [OH^-]_{aq} \overset{K_a}{\rightleftharpoons} [R-O^-]_{aq} + [HOH]_{aq}$$

$$K^{sel}_{RO/OH}$$

$$[Q^+OH^-]_{org} + [R-O^-]_{aq} \rightleftharpoons [Q^+R-O^-]_{org} + [OH^-]_{aq}$$

$$K^s_{RO/OH} = K_a K^{sel}_{RO/OH}$$

Alcohol, R-OH	$K^s_{RO/OH}$[a]
CH_3OH	70
EtOH	420
Pr^nOH	710
Bu^nOH	960
Bu^sOH	970
Bu^iOH	980
$C_5H_{11}OH$	1150
$C_6H_{13}OH$	1300
$C_7H_{15}OH$	1380
$C_8H_{17}OH$	1380
$PhCH_2OH$	1900
Ethane-1,2-diol	210
Propane-1,2-diol	450
Pentane-1,5-diol	4100
2,5-Dimethylhexane-2,5-diol	4900

[a]Temperature = 41°C.

J. de la Zerda and Y. Sasson, J. Chem. Soc. Perkin II, 1147 (1987).

diols and diol monoethers, are listed in Table 2-18 [23]. The amounts are expressed as the percent of the maximum possible basicity. The data may be understood in terms of the equilibria described in Eqs. (2-4) and (2-5). It is well known that the order of acidity of alcohols in water is: primary > secondary > tertiary. It can be seen that the maximum possible basicity follows this order. In fact, the glycol monoethers, which are stronger acids than primary alcohols, are even more readily extracted into the organic phase. In general, the extraction of the diols is greater than that of all the monoalcohols. Again, intramolecular hydrogen bonding could be operating to enhance the extractability of the monoanion into the organic phase.

A systematic study has been reported for the extraction of the pinacol monoanion into the organic phase by tetra-*n*-octylammonium halides in a chlorobenzene–50% aqueous sodium hydroxide two-phase system. The data are shown in Table 2-19 [24]. Since the quaternary ammonium cation is totally soluble in the chlorobenzene phase, its concentration may be used as the reference for the maximum amount of base that can reside in the organic phase. The base content

Table 2-18. *Extraction by chlorobenzene solution of tetra-n-octylammonium bromide and alcohols from an equal volume of 50% NaOH (% of the maximum possible basicity).*

	Percentage
Primary alcohols	
Ethanol	4.5
1-Propanol	5.0
2-Methyl-1-propanol	4.4
1-Pentanol	4.3
1-Hexanol	4.3
1-Heptanol	4.8
1-Octanol	2.0
1-Dodecanol	0.8
Secondary alcohols	
2-Propanol	1.9
2-Pentanol	1.2
2-Hexanol	1.1
2-Octanol	0.7
Cyclohexanol	0.5
4-*tert*-Butylcyclohexanol	1.5
2-*tert*-Butylcyclohexanol	2.0
4-Methylcyclohexanol	2.0
Tertiary alcohols	
tert-Butyl	0.3
2-Methyl-2-butanol	0.2
Diols	
1,5-Pentanediol	<0.02
1,6-Hexanediol	(emulgation)
2,5-Hexanediol	5.2
2,2-Dimethyl-1,3-propanediol	18.4
2-Methyl-2,4-pentanediol	28.0
2,3-Dimethyl-2,3-butanediol	25.8
2,5-Dimethyl-2,5-hexanediol	32.0
Diol monoethers	
Ethyleneglycol monoethylether	8.9
Diethyleneglycol	8.7
Glycerol isopropylideneacetal	13.0

E.V. Dehmlow, R. Thieser, Y. Sasson, and E. Pross, *Tetrahedron,* **41**, 2927 (1985).

of the chlorobenzene phase is expressed in percent maximum possible base. The effect of the structure of the quaternary ammonium cation on the efficiency of extraction of the pinacol anion into the organic phase by -onium salts in a chlorobenzene–50% aqueous sodium hydroxide two-phase system is shown in Table 2-20 [24]. Both the halide ion and the structure of the quaternary cation have a pronounced effect of the extraction of the pinacol monoanion into the

Table 2-19. Extraction of the pinacol anion by tetraoctylammonium halides in chlorobenzene–50% aqueous sodium hydroxide.

Catalyst anion	Extracted base (% max. possible)		
	Molar ratio R_4NX/pinacol	R_4NX reference	Pinacol reference
Br	1:1	26	26
	1:2	50	25
	1:10	60	6
	1:100	50	0.5
Cl	1:1	40	40
	1:2	78	39
	1:10	88	8.8
	1:100	90	0.9

E.V. Dehmlow, R. Thieser, Y. Sasson, and E. Pross, *Tetrahedron*, **41**, 2927 (1985).

chlorobenzene phase. In addition, it may be seen that the phosponium salt is not as effective as the corresponding ammonium salt.

G. Effect of Hydration of the Transferred Anion and the Effect of Inorganic Salt and/or Hydroxide Concentration in the Aqueous Phase

Anions are hydrated to different extents depending, to a large degree, on the charge-to-volume ratio of the anion. The more strongly the anion is hydrated, the more strongly it will be attracted to the aqueous phase, and the more difficult it will be to transfer to the organic phase. This water of hydration may or may not accompany the anion when it is transferred into the organic phase, although most measurements indicate that water is transferred. The hydration data for some common monovalent anions in water are listed in Table 2-21 [3,20,24,25]. In addition, the hydration of quaternary cation–anion pairs in toluene and 1-cyanooctane are also listed for several quaternary ammonium and phosphonium

Table 2-20. Base extraction by chlorobenzene solutions of -onium salts and pinacol from an equal volume of 50% sodium hydroxide

	Bu_4N^+	Pen_4N^+	Hex_4N^+	Hep_4N^+	Oct_4N^+	Oct_4P^+
Cl^-	7	31	35	34	40	26
Br^-	8	20	23	22	26	
I^-	6	5	8	7	9	

^{a}0.1 M each of -onium salt and pinacol.

E.V. Dehmlow, R. Thieser, Y. Sasson, and E. Pross, *Tetrahedron*, **41**, 2927 (1985).

Table 2-21. Degree of hydration n of anions.

		n with cation in solvent		
Anion, X⁻	n in H$_2$O	(n-Octyl)$_4$N$^+$ X$^-$ in toluene	(n-Octyl)$_3$(n-Propyl)N$^+$ X$^-$ in toluene	(Hexadecyl)(n-Butyl)$_3$P$^+$ X$^-$ in 1-octylCN
NO$_2^-$	(2.0)	1.5	1.1	0.4
Cl$^-$	2.3	3.2	2.5	4
Br$^-$	1.7	2.4	1.6	
SO$_4^{-2}$		18		
CN$^-$				5
OH$^-$	5.2			

[a] D.S. Allam and W.H. Lee, J. Chem. Soc. A, 426 (1966).

[b] V.L. Kheifets, N.A. Yakovleva, and B. Ya Krasil'shchik, Zh. Prikl. Khim. (Lenningrad), **46**, 549 (1973).

[c] C.M. Starks and R.M. Owens, J. Am. Chem. Soc., **95**, 3613 (1973).

structures. It is assumed that the waters of hydration are primarily associated with the anion.

The hydration number for simple nucleophilic anions has been determined for the chlorobenzene–water two-phase system in the presence of the organophilic tetra-n-octylammonium cation [26]. The data are listed in Table 2-22. In general the larger the charge-to-volume ratio of the anion the greater the number of waters of hydration.

Table 2-22. Second-order rate constants for nucleophilic substitution reactions of anions with n-octyl methanesulfonate in the presence of $(C_8H_{17})_4N^+Y^-$ at 35°C.

Y$^-$	PhCl–H$_2$O		PhCl– aq. NaOH	Anhyd. PhCl
	$10^3\ k^a$ $M^{-1}\ \sec^{-1}$	Hydration, n of Q$^+$Y$^-$	$10^3\ k^b$ $M^{-1}\ \sec^{-1}$	$10^3\ k^c$ $M^{-1}\ \sec^{-1}$
N$_3^-$	2.9	3.0 ± 0.2	11.6	11.7
Cl$^-$	0.26	3.4 ± 0.2	3.4	3.4
Br$^-$	0.42	2.1 ± 0.2	1.1	1.1
I$^-$	0.29	1.0 ± 0.15	0.42	0.45

Reaction conditions:

[a] Chlorobenzene solution (20 mL) of substrate (0.6 M), catalyst (0.02–0.04 M), and an aqueous solution (15 mL) of K$^+$ salt (4.2 M).

[b] Same as (a) —15 mL 50% NaOH and 63 mmol KY.

[c] Substrate (0.02–0.06 M), catalyst (0.3–0.4 M).

D. Landini, A. Maia, and G. Podda, J. Org. Chem. 2264 (1982).

Table 2-23. Effect of inorganic salt concentration on $K_{Cl \to CN}$ values using $C_{16}H_{33}P^+Bu_3$ [a]

$(NaCN)_{aq}$ (moles/L)	$(NaCl)_{aq}$ (moles/L)	Total	Moles H_2O per mole ion present	$K_{Cl \to CN}$ Toluene	$K_{Cl \to CN}$ 1-Cyano octane
4.08	1.72	5.80	4.8	0.32	0.46
4.08	2.56	6.64	4.2	0.38	0.59
4.08	3.42	7.50	3.7	0.42	0.66
4.08	5.14	9.20	3.0	0.54	0.98
4.08	6.84[b]	10.92	2.6	0.73	1.29
4.08	1.72	5.80	4.8	0.31	0.46
8.16	1.72	9.88	2.8	0.40	0.53
12.24	1.72	13.96	2.0	0.40	0.55
20.40	1.72[c]	22.12	1.3	0.71	1.25

[a]C.M. Starks and R.M. Owens, J. Am. Chem. Soc., **95**, 3613 (1973).
[b]Solution is saturated with NaCl; not all NaCl is soluble.
[c]Solution is saturated with NaCN; not all NaCN is soluble.

Increasing the concentration of inorganic salts in the aqueous phase may have one or several of the following effects:

1. The distribution equilibrium represented by Eqs. (2-1) and (2-2) may be driven in one direction or the other depending on the anion added. This is nothing more than a restatement of the law of mass action.

2. Organic salts may be "salted-out" of the aqueous phase into the organic phase, thus affecting the value of $K_{Cl \to X}$.

3. The amount of water available for hydration of the anions may be reduced, thus affecting the distribution equilibrium represented by Eqs. (2-1) and (2-2) and the reactivity of the anion in the organic phase.

In general, as the salt concentration in the aqueous phase increases the hydration levels of all ions present in the system tend to decrease; the added salt ties up water molecules and has a dehydrating effect on the ions present in the system. As a consequence, on addition of salt to the aqueous phase containing two anions, it would be anticipated that their relative extractabilities into the organic phase would change. Table 2-23 shows the results of the effect of a systematic change in salt concentration on the value of $K_{Cl \to CN}$ for both toluene–water and 1-cyanooctane–water two-phase systems using the organophilic hexadecyltri-n-butylphosphonium cation [3,20]. As each of these anions becomes dehydrated with the addition of salt, the selectivity of the quaternary cation for cyanide increases relative to chloride.

The effect of aqueous salt concentrations on the second-order rate coefficients for the reaction of cyanide ion with 1-chlorooctane is summarized in Table 2-

Table 2-24. Effect of water concentration on the rate of NaCN reaction with 1-chlorooctane at 90°C catalyzed by $C_{16}H_{33}P^+Bu_3Br^a$

Moles H$_2$O added	Second-order rate constant ($\times 10^2$) (L mol-sec^{-1})
0	~0.001
0.56	8.8
0.83	9.9
1.67	7.6
b	b
3.34	7.6
5.56	5.4
8.35	4.4
27.8	3.7

[a] One mole of 1-chlorooctane, 4 mol of NaCN, 0.02 mol of $(C_{16}H_{33})(n-Bu)_3P^+Br^-$ at 90°C; Starks, C.M.; Liotta, C.L. Phase Transfer Catalysis: Principles and Techniques, Academic Press, New York (1978).

[b] Above this line NaCN is not completely soluble in H$_2$O phase; below this line NaCN is completely soluble in H$_2$O phase.

24 [3]. If it is assumed that the selectivity of Q^+ for CN^- relative to Cl^- is reflected in the rate of cyanide displacement on 1-chlorooctane, then as the concentration of salt in the aqueous phase is increased, the second-order rate constant would increase. It should be noted that very little reaction occurs when no water is present. This particular point will be addressed again when the mechanism of solid–liquid phase transfer catalysis and the role of the omega phase are discussed.

Table 2-25 summarizes the second-order rate coefficient for the effect of added sodium bromide on the cetylpyridinium cation-catalyzed reaction of sodium thiophenoxide with 1-bromooctane [21]. This quaternary cation is particularly sensitive to salt concentration since it also has a strong tendency to form micelles in the aqueous phase rather than to be dissolved in the organic phase. Usually, the best PTC conditions are realized when the aqueous phase is saturated with the inorganic salt.

Table 2-25. Effect of added NaBr on the rate of reaction of 1-bromobutane with sodium thiophenoxide in a benzene water two-phase system.

Added NaBr (M)	$k \times 10^3\ M^{-1}\ \text{sec}^{-1}$
0	0.48
0.2	1.7
0.5	3.7

Catalyst: Cetylpyridinium bromide.
A.W. Herriott and D. Picker, J. Am. Chem. Soc., **97**, 2345 (1975).

Table 2-26. *Effect of salt concentration of the aqueous phase on reaction rates for nucleophilic substitution reaction of* n-*octyl methanesulfonate with bromide ion in a chlorobenzene–water two-phase system at 60°C.*

Catalyst	KBr, M	$10^5 k_{obs}$ sec^{-1}	10^2 (Cat) M	$10^3 k$ $M^{-1} sec^{-1}$
$(C_{16}H_{33})$ $(n-Bu)_3P^+Br^-$	1.1	11.6	4.0	2.9
	2.0	11.2	4.0	2.8
	4.0	12.8	4.0	3.2
	6.0	11.6	4.0	2.9
	>6.0	12.4	4.0	3.1
$(n-Bu)_4P^+Br^-$	2.0	7.3	2.6	2.8
	3.0	8.9	3.3	2.7
	4.0	10.0	3.7	2.7
	6.0	10.2	3.9	2.6
$(n-Bu)_4N^+Br^-$	1.0	0.60	0.16	3.7
	2.1	1.59	0.43	3.7
	3.0	3.42	0.90	3.8
	4.0	5.97	1.6	3.7
	5.0	9.57	2.5	3.8
	6.0	12.07	3.2	3.8
	>6.0	12.58	3.4	3.7

Reaction conditions: Chlorobenzene solution (20 mL) of substrate (0.6 M) and a volume of aqueous KBr to produce a 5:1 molar ratio of salt/substrate.

D. Landini, A. Maia, and G. Podda, J. Org. Chem., **47**, 2264 (1982).

The effect of the salt concentration in the aqueous phase on reaction rates has been reported for the nucleophilic substitution reaction of *n*-octyl methanesulfonate with bromide ion in a chlorobenzene–water system [27]. The reaction rates were studied under pseudo-first-order conditions as a function of potassium bromide concentration in the aqueous phase. Pseudo-first-order rate constants, the concentrations of phase-transfer catalyst in the organic phase, and the corresponding second-order rate coefficients are summarized in Table 2-26. When the organophilic hexadecyltri-*n*-butylammonium bromide is used as the phase-transfer catalyst, the pseudo-first-order rate constants are independent of the aqueous salt concentration. However, when tetra-*n*-butylphosphonium bromide or tetra-*n*-butylammonium bromide is used, the pseudo-first-order rate constants increase with increasing aqueous salt concentration. Concomitant with this increase is an increase in the phase-transfer catalyst concentration in the organic phase. These facts are reflected in the constancy of the tabulated second-order rate coefficients—values derived by dividing the pseudo-first-order rate constants by the actual concentration of quaternary bromide in the organic phase. When chloride ion is used as the nucleophile and tetra-*n*-octylammonium as the phase-transfer cation, it is found that the percent conversion decreased and asymptotically approached 24% as the aqueous chloride ion concentration increased.

Table 2-27. *First-order rate constants for the Hofmann elimination of tetrahexylammonium hydroxide in chlorobenzene–aqueous sodium hydroxide at 25°C.*

$$(C_6H_{13})_4N^+OH^- \xrightarrow[\text{aq.NaOH}]{Ph-Cl} (C_6H_{13})_3N + 1-\text{hexene} + H_2O$$

NaOH, M	Hydration state n of (n-Hexyl)$_4$N$^+$OH$^-$	$k \times 10^6$
4.8	11.0 ± 1.0	0.019
6.5	9.0 ± 1.0	0.14
10.0	5.0 ± 0.5	0.74
15.0	4.0 ± 0.5	26
20.0	3.5 ± 0.5	210

D. Landini and A. Maia, J. Chem. Soc., Chem. Commun., 1041 (1984).

The aqueous salt and/or aqueous base concentration not only affects the selectivity constant but also the reactivity of the transferred anion. For instance, Table 2-15 [23] shows that as the hydration of hydroxide ion decreases its organic solubility also decreases, and Table 2-16 [23] shows that as the concentration of aqueous base increases the rate of decomposition of the quaternary ammonium salt increases. Thus, while the concentration of hydroxide ion in the organic phase decreases with increasing aqueous base concentration, the reactivity of the hydroxide ion increases due to its decreased solvation.

The effect of aqueous base concentration of the hydration state of hydroxide ion in the organic phase and its corresponding reactivity is shown in Table 2-27 [27]. A chlorobenzene–aqueous sodium hydroxide two-phase system containing tetra-n-hexyl ammonium cation as the phase-transfer catalyst was used. The

Table 2-28. *Effect of aqueous base concentration on the reaction of* n-*octyl methanesulfonate with bromide ion under PTC conditions.*

Base	Percent base	Hydration n of Q$^+$Y$^-$	$10^3\,k$ $M^{-1}\,\text{sec}^{-1}$
NaOH	30	2.0 ± 0.2	0.45
	40	1.0 ± 0.15	0.80
	50	<0.2	1.13
KOH	30	1.9 ± 0.2	0.49
	40	1.3 ± 0.15	0.75
	50	0.8 ± 0.10	0.83
	60	<0.2	1.18

[a]Bromide ion derived from tetra-n-octylammonium bromide.

[b]Chlorobenzene-aqueous base two-phase system.

D. Landini, A. Maia, and G. Podda, J. Org. Chem. **47**, 2264 (1982).

hydration of the transferred hydroxide as well as the rate constant associated with the decomposition of the quaternary ammonium hydroxide by Hofmann elimination has been reported. As the concentration of aqueous sodium hydroxide is varied from 4.8 to 20.0 M, the hydration of the transferred hydroxide ion changes from approximately 11 to 3.5 water molecules. The corresponding rates of decomposition of the transferred quaternary ammonium hydroxide are also listed. As anticipated, as the hydration number decreases, the reactivity increases.

The effect of aqueous base concentration on the reaction of n-octyl methanesulfonate with bromide ion in the presence of tetra-n-octylammonium bromide in chlorobenzene–water has been reported. The results are shown in Table 2-28 [27]. As the concentration of aqueous base is increased the hydration of the quaternary ammonium bromide in the organic phase decreases and the nucleophilic reactivity of bromide increases.

References

1. L. Pauling, The Naure of the Chemical Bond, Cornell University Press, Ithaca, New York, Appendix 3.1 (1940).
2. R. Alexander and A.J. Parker, J. Am. Chem. Soc., **89**, 5549 (1967).
3. C.M. Starks and C.L. Liotta, Phase Transfer Catalysis: Principles and Techniques; Academic Press, New York (1978).
4. R. Bock and G.M. Beilstein, Z. Anal. Chem., **192**, 45 (1963).
5. R.M. Diamond and D.G. Tuck, Prog. Inorg. Chem., **2**, 109 (1960).
6. R. Bock and J. Jainz, Z. Anal. Chem, **198**, 315 (1963).
7. Values of α for $(C_6H_5)_4Sb^+X^-$ have also been measured: Bock, R.; Gallath, E. Z. Anal. Chem., **222**, 283 (1966).
8. C.J. Cotzee and H. Freisee, Anal. Chem., **41**, 1128 (1969).
9. H.K. Bisivas and B.M. Mandal, Anal. Chem., **44**, 1636 (1972).
10. R.B. Grieves, W. Charewicz, and P.J.W. The, Sep. Sci., **10**, 77 (1975).
11. E.V. Dehmlow and S.S. Dehmlow, Phase Transfer Catalysis, 2nd ed; Verlag Chemie, Weinheim (1983).
12. N.A. Gibson and D.C. Weatherburn, Anal. Chim. Acta, **58**, 159 (1972).
13. J.P. Antoine, I. de Aquirre, F. Janssens, and F. Thyrion, Bull. Soc. Chim. Fr., **207** (1980).
14. J. Czapkiewica, T. Czapkiewica, and D. Struk, Pol. J. Chem., **52**, 2203 (1978).
15. E.V. Dehmlow and S.S. Dehmlow, Phase Transfer Catalysis, 2nd ed.; Verlag Chemie: Weinheim, p. 13 (1983).
16. B. Czapkiewicz-Tutaj and J. Czapkiewica, Rocz. Chem., **49**, 1353 (1975).
17. E.V. Dehmlow and B. Vehre, J. Chem. Res. (S), 350 (1987).
18. Y. Inoue and O. Tochiyama, Bull. Chem. Soc. Jpn., **53**, 1618 (1980).

19. C.M. Starks and R.M. Owens, J. Am. Chem. Soc., **95**, 3613 (1973).
20. A.W. Herriott and D. Picker, J. Am. Chem. Soc., **97**, 2345 (1975).
21. J. de la Zerda and Y. Sasson, J. Chem. Soc. Perkin Trans. II, 1147 (1987).
22. D. Landini, A. Maia, and A. Rampoldi, J. Am. Chem. Soc., **51**, 5476 (1986).
23. E.V. Dehmlow, R. Thieser, Y. Sasson, and E. Pross, Tetrahedron, **41**, 2927 (1985).
24. D.S. Allam and W.H. Lee, J. Chem. Soc. A, **426** (1966).
25. V.L. Kheifets, N.A. Yakovlova, and B. Ya Krasil'shchik, Khim. (Lenningrad), **46**, 549 (1973).
26. D. Landini, A. Maia, and G. Podda, J. Org. Chem. **47**, 2264 (1982).
27. D. Landini and A. Maia, J. Chem. Soc. Chem. Commun., 1041 (1984).

3
Phase-Transfer Catalysis: Fundamentals II

A. Introduction

Phase-transfer catalytic (PTC) systems are characterized by the presence of at least two phases and at least one interfacial region separating the phases [1–8]. Reactions taking place in such systems usually involve (1) transfer of an anionic reactant from its "normal" phase into the reaction phase or interfacial region, (2) reaction of the transferred anion with the nontransferred reactant located in the reaction phase or interfacial region, and (3) transfer of the anionic product from the reaction phase or interfacial region into its "normal" phase. Thus, PTC reactions involve several steps occurring in series and/or in parallel. As a consequence, a quantitative description of the entire process necessitates defining the reactions and corresponding rates taking place within each of the phases and within the interfacial region, and the rates of mass transfer of the reactant and product anions into and through the interfacial region.

PTC reactions usually take place between an aqueous or a solid phase and an organic phase. The former process is termed liquid–liquid phase-transfer catalysis (ll-PTC) whereas the latter process is termed solid–liquid phase-transfer catalysis (sl-PTC). It is important to realize that the ability of a phase-transfer catalyst to successfully transfer the reactant anion into the interfacial region and/or the organic phase is *not* sufficient to achieve a successful process. Once in the interfacial and/or organic phase the anion must be in a reactive state. If not, the process will either be very slow or stop altogether. If the organic phase reaction does successfully take place after the transfer, it is then necessary that the phase-transfer catalyst transport the product anion back to the aqueous phase so that the catalyst will be available for transport of another reactant anion. If this latter process is slow or does not take place at all, the catalytic process will come to a halt. This is sometimes referred to as "catalyst poisoning." Under such circumstances stoichiometric quantities of phase-transfer catalyst would be necessary; the phase-transfer catalyst is no longer a catalyst.

In the following discussion, the PTC process will be presented and analyzed in terms of the differential rate expressions describing the various kinetic models and the thermodynamic equilibria (discussed in Chapter 2) describing certain steps in the process. The factors that affect (1) the transfer of the anion from the aqueous or solid phase to the interfacial region and through the interfacial region to the organic phase, as well as the reverse process; and (2) the reactivity of the anion in the organic phase will be discussed. For convenience, liquid–liquid PTC and solid–liquid PTC are discussed separately. The relationship between these two processes will then be addressed.

B. Liquid–Liquid PTC

Over the last several decades two mechanistic models describing ll-PTC involving *simple displacement reactions* and *hydroxide-promoted reactions of organic acids* (alkylations, carbene reactions, etc.) have evolved. The former has usually been represented by the Starks *Extraction Mechanism* [9–11] and/or the Brandstrom–Montanari modification [5,8,12] whereas the latter has been represented by either the Starks *Extraction Mechanism* or the Makosza *Interfacial Mechanism* [13–15]. Much controversy exists in the literature concerning under what circumstances each of the mechanistic models is operating. At this juncture, however, the experimental facts suggest that there is a spectrum of mechanisms that fall within these limiting mechanistic models. The *Extraction Mechanism* and the *Interfacial Mechanism,* as well as some pertinent variations of these models, will be presented and discussed. The experimental facts reported in the literature in support of each of the models will be subsequently discussed and critically evaluated.

1. Simple Displacement Reactions

a. The Extraction Mechanism: Mechanistic Descriptions

The Starks Extraction Mechanism [2,6,9–11] describing liquid–liquid PTC is illustrated in Figure 3-1 and the Brandstrom [5,8]–Montanari [12] modification is illustrated in Figure 3-2. In the Starks mechanism the phase-transfer catalyst has both organophilic and hydrophilic characteristics and is distributed between the aqueous and the organic phases. The reactant and product anions are formally transferred across the interfacial region into the organic phase as an intact phase-transfer cation–anion pair. In the Brandstrom–Montanari modification the catalyst is highly organophilic and exists exclusively in the organic phase. In this case the transfer of the reactant and product anions involves the initial exchange of these species in the presence of the cationic phase-transfer agent at the interfacial

50 / *Phase-Transfer Catalysis*

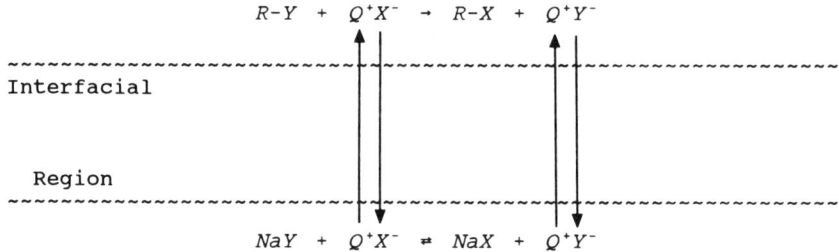

Figure 3-1. Liquid–liquid phase transfer catalysis: Starks extraction mechanism.

region of the system followed by transfer into the organic phase. In both the Starks mechanism and the Brandstrom–Montanari modification the phase–transfer cation–anion pair is the reactive species in the organic phase.

Figure 3-3 represents an additional variation of the above extraction mechanisms. In the mechanism illustrated in Figure 3-3 the metal salt has sparing organophilicity and transfers itself from the aqueous phase through the interfacial region into the organic phase where it undergoes ion-pair exchange with the phase-transfer catalyst. In this case the phase-transfer catalyst is not involved in the anion transfer step. As a consequence, the process described by Figure 3-3 is *not* strictly PTC. Nevertheless, the coordination of the phase-transfer cation with the transferred anion could result in increased anion reactivity. The origins and factors affecting anion activation are discussed in a later section. The reactive species in the organic phase in the mechanism shown in Figure 3-3 is the phase-transfer cation–anion pair.

Each of the mechanisms illustrated in Figures 3-1, 3-2, and 3-3 show the

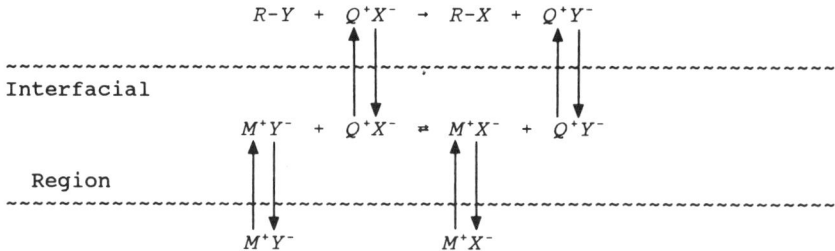

Figure 3-2. Liquid–liquid phase transfer catalysis: Brandstrom–Montanari modification of Starks extraction mechanism.

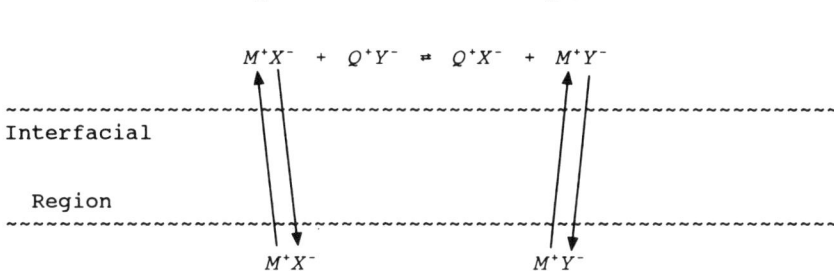

Figure 3-3. Liquid–liquid PTC: organophilic salt mechanism.

reactive organic species as the phase-transfer cation–anion pair. Depending on the polarity of the organic phase and the structure of the anion and cation, the reactive species could be the free anion, the ion pair (as shown in Figures 3-1, 3-2, and 3-3), or some complex aggregate. The structures of the reactive species in the organic phase and the related kinetic consequences are discussed in a later section.

b. The Extraction Mechanism: Kinetic Descriptions

The Starks Extraction Mechanism and the Brandstrom–Montanari modification can both be described by the reaction sequence illustrated in Scheme 3-1. The first step describes the distribution of the reactant and product anions between the aqueous and organic phases. Q^+ is the phase-transfer cation, X^- is the reactant anion (the nucleophile), Y^- is the product anion (the leaving group) derived from the organic substrate $R-Y$, and the subscripts aq and org represent the aqueous and organic phases, respectively. This step is used as the mechanistic model for introducing the reactant anion into the organic phase from the aqueous phase and returning the product anion from the organic phase into the aqueous phase. In other treatments, this step has been described by an equilibrium constant. In the treatment that follows, two rate constants (k_1 and k_{-1}) are used in order to address the effect of mass transfer across the interfacial region of the two-phase reaction system. The mass transfer rate constants k_1 and k_{-1} are each functions of the interfacial area. Specifically, k_1 and k_{-1} are each equal to an intrinsic mass transfer rate constant (k_1^o and k_1^o) each multiplied by some function of the interfacial area. Such definitions will allow the rate of phase-transfer processes to be related to stirring rates. The second step (k_2) describes an irreversible substitution reaction taking place within the organic phase. The result of the second step is to produce product $(R-X)_{org}$ and product anion paired with the phase-transfer cation $(Q^+Y^-)_{org}$ which is subsequently redistributed with $(Q^+X^-)_{aq}$ by repeating the first step. It should be emphasized that the organic phase reaction need not be irreversible. The following treatment may be easily modified to include reversibility. At this juncture, however, for simplicity the organic phase reaction will be considered irreversible. None of the generalizations will be lost in making such an assumption.

$$(Q^+Y^-)_o + (X^-)_{aq} \underset{k_{-1}}{\overset{k_1}{\rightleftarrows}} (Q^+X^-)_o + (Y^-)_{aq}$$

$$(Q^+X^-)_o + (R-Y)_o \overset{k_2}{\rightarrow} (Q^+Y^-)_o + (R-X)_o$$

Scheme 3-1. Kinetic description: Starks–Brandstrom–Montanari mechanism.

The rate of disappearance of reactant and the rate of formation of product is

$$-\frac{d[R-Y]_{org}}{dt} = \frac{d[R-X]_{org}}{dt} = k_2[Q^+X^-]_{org}[R-Y]_{org} \qquad (3\text{-}1)$$

where $[R-Y]_{org}$ is the concentration of reactant at a particular time during the course of reaction. The differential rate expression describing the formation of $(Q^+X^-)_{org}$ is

$$\frac{d[Q^+X^-]_{org}}{dt} = k_1[Q^+Y^-]_{org}[X^-]_{aq} \qquad (3\text{-}2)$$
$$- k_{-1}[Q^+X^-]_{org}[Y^-]_{aq} - k_2[Q^+X^-]_{org}[R-Y]_{org}$$

The operation of differential Eqs. (3-1) and (3-2) dictates that during the course of reaction (a) the concentration of phase-transfer cation in the organic phase remains constant, and (b) the relative concentrations of $(Q^+X^-)_{org}$ and $(Q^+Y^-)_{org}$ can change. These conditions are described by Eq. (3-3).

$$\frac{d[Q^+X^-]_{org}}{dt} + \frac{d[Q^+Y^-]_{org}}{dt} = 0 \qquad (3\text{-}3)$$

If, however, there is *no* change in the individual concentrations of $[Q^+X^-]_{org}$ and $[Q^+Y^-]_{org}$ then the steady-state approximation may be made for $[Q^+X^-]_{org}$ and Eq. (3-2) may be set equal to zero.

$$\frac{d[Q^+X^-]_{org}}{dt} = 0 \qquad (3\text{-}4)$$

If the total concentration of phase-transfer salts in organic phase $(Q)_{org}$ is

$$Q_{org} = [Q^+X^-]_{org} + [Q^+Y^-]_{org} \tag{3-5}$$

then applying the steady-state approximation and solving for $[Q^+X^-]_{org}$ gives Eq. (3-6).

$$[Q^+X^-]_{org} = \frac{k_1[X^-]_{aq} Q_{org}}{k_1[X^-]_{aq} + k_{-1}[Y^-]_{aq} + k_2[R-Y]_{org}} \tag{3-6}$$

Substitution into Eq. (3-1) gives the following differential rate expression describing the rate of formation of product:

$$\frac{d[R-X]_{org}}{dt} = \frac{k_1 k_2 [Q_{org}] [X^-]_{aq} [R-Y]_{org}}{k_1[X^-]_{aq} + k_{-1}[Y^-]_{aq} + k_2[R-Y]_{org}} \tag{3-7}$$

Kinetic studies dealing with PTC reactions usually involve following the rate of disappearance of $(R-Y)_{org}$ or the related rate of appearance of $(R-X)_{org}$.

$$-\frac{d[R-Y]_{org}}{dt} = \frac{d[R-X]_{org}}{dt} \tag{3-8}$$

It is instructive to analyze Eq. (3-7) in terms of its limiting cases and to relate each of these cases to the expected overall kinetic behavior of the phase-transfer reaction and, in particular, to the observation of pseudo-first-order kinetics with respect to $[R-Y]_{org}$. Special conditions within each of the limiting cases will also be discussed. The limiting cases are listed in Table 3-1.

The application of the steady-state approximation to $(Q^+X^-)_{org}$ simplifies the description of a rather complex process and allows a tractable analysis of a variety of limiting cases. Two pertinent questions, however, must be asked. Is this approximation reasonable? Can this approximation be justified? In order to address these questions, the rate constants and initial concentrations used in each of the limiting cases described in Table 3-1 were also applied to the differential Eqs. (3-1) and (3-2) which do not invoke the steady-state treatment. Using these equations, $[R-Y]_{org}$ can be calculated as a function of time. It should be emphasized again that the application of Eqs. (3-1) and (3-2) dictates that while the concentrations of $[Q^+X^-]_{org}$ and $[Q^+Y^-]_{org}$ are allowed to vary, the total concentration of phase-transfer salt in the organic phase is constant. (The kinetic description of a process in which the concentrations of the phase-transfer salts in the organic phase are not constant are discussed in a later section.) Plots of $[R-Y]_{org}$ vs. time using Eqs. (3-1) and (3-2) (indicated by 0) are superimposed on the corresponding plots derived from the steady-state treatment (indicated by *). In addition, the change in concentration of Q^+X^- in the organic phase (indicated by +) is also shown. These latter plots are additional indications of

Table 3-1. Starks–Brandstrom–Montanari extraction mechanism: limiting differential rate expressions.

	Condition	$d(R-X)/dt$
1	$k_1 [X^-]_{aq} = k_{-1} [Y^-]_{aq} = k_2 [R-Y]_{org}$	$\dfrac{k_1 k_2 Q_{org} [X^-]_{aq} [R-Y]_{org}}{k_1 [X^-]_{aq} + k_{-1} [Y^-]_{aq} + k_2 [R-Y]_{org}}$
2	$k_{-1} [Y^-]_{aq} \gg k_1 [X^-]_{aq} + k_2 [R-Y]$	$\dfrac{k_1 k_2 Q_{org} [X^-]_{aq} [R-Y]_{org}}{k_{-1} [Y^-]_{aq}}$
3	$k_{-1} [Y^-]_{aq} + k_1 [X^-]_{aq} \gg k_2 [R-Y]$	$\dfrac{k_1 k_2 Q_{org} [X^-]_{aq} [R-Y]_{org}}{k_{-1} [Y^-]_{aq} + k_1 [X^-]_{aq}}$
4	$k_1 [X^-]_{aq} + k_2 [R-Y]_{org} \gg k_{-1} [Y^-]_{aq}$	$\dfrac{k_1 k_2 Q_{org} [X^-]_{aq} [R-Y]_{org}}{k_1 [X^-]_{aq} + k_2 [R-Y]_{org}}$
5	$k_2 [R-Y]_{org} \gg k_{-1} [Y^-]_{aq} + k_1 [X^-]_{aq}$	$k_1 Q_{org} [X^-]_{aq}$
6	$k_{-1} [Y^-]_{aq} + k_2 [R-Y]_{org} \gg k_1 [X^-]_{aq}$	$\dfrac{k_1 k_2 Q_{org} [X^-]_{aq} [R-Y]_{org}}{k_{-1} [Y^-]_{aq} + k_2 [R-Y]_{org}}$
7	$k_1 [X^-]_{aq} \gg k_{-1} [Y^-]_{aq} + k_2 [R-Y]$	$k_2 Q_{org} [R-Y]_{org}$

the applicability of the steady-state approximation to $[Q^+ X^-]_{org}$. Finally, plots of $\ln[R-Y]_{org}$ vs. time (first-order behavior) and $\ln([R-Y]_{org}/[X^-]_{aq})$ vs. time (second-order behavior) for each of the limiting cases are also included. The rate constants and initial concentrations of species in the reaction system are summarized in Table 3-2. It will be seen that in each of the figures showing the change in concentration of $(R-Y)_{org}$ and $(Q^+ X^-)_{org}$ with respect to time and the corresponding figures showing the first- and second-order plots, there is little difference between the steady-state treatment and the non-steady-state treatment with the exception of Case 4. Nevertheless, even in Case 4 the deviations are small and do not seriously detract from the generalizations concerning the reaction process under these circumstances.

c. Examination of the Limiting Cases

In each of the limiting cases listed in Table 3-1 the rate terms that are being compared are those involving the transfer of X^- and Y^- between the aqueous and organic phases and the reaction of $Q^+ X^-$ with $R-Y$ in the organic phase. This is conveniently accomplished because each of these processes are represented by individual terms in the denominator of Eq. (3-7): $k_1(X^-)_{aq}$, $k_1(Y^-)_{aq}$, and $k_2(R-Y)_{org}$. When each of these terms is comparable in magnitude Case 1 best describes the rate of formation of product. It is clear that Case 1 represents a situation in which both the mass transfer of the anions and the organic phase reaction rate are important contributors to the overall rate. Clearly, simple integral-order kinetics would not be expected to operate under such circumstances.

Table 3-2. Rate constants and initial concentrations for each of the limiting cases.

Case	k_1	k_{-1}	k_2	$[R-Y]_{org}$	$[X^-]_{aq}$	$[Q^+Y^-]_{org}$	$[Y^-]_{aq}$
1a	0.1	0.1	0.2	0.1	0.2	0.01	0.2
1b	0.1	0.2	0.2	0.1	0.2	0.01	0.1
2a	0.1	2.0	0.2	0.1	0.2	0.01	0.5
3a	2.0	2.0	0.2	0.1	0.2	0.01	0.2
3b	2.0	2.0	0.2	0.1	0.4	0.01	0.4
4a	2.0	0.1	4.0	0.1	0.2	0.01	0.1
4b	2.0	0.1	4.0	0.2	0.4	0.01	0.1
4c	4.0	0.1	8.0	0.1	0.2	0.01	0.1
4d	4.0	0.01	8.0	0.1	0.2	0.01	0.1
4e	4.0	0.01	2.0	0.1	0.2	0.01	0.1
5a	0.1	0.1	4.0	0.1	0.2	0.01	0.1
5b	0.1	0.1	4.0	0.5	0.2	0.01	0.1
6a	0.1	4.0	4.0	0.1	0.2	0.01	0.1
7a	4.0	0.1	0.1	0.1	0.2	0.01	0.1

Figs. 3-4a and 3-4b show plots of $(R-Y)_{org.}$ vs. time for Case 1 conditions. The corresponding first- and second-order plots show slight curvature, as expected. In this situation it would not be surprising for the experimentalist to unhesitatingly draw a straight line between the points in either the first- or second-order plots, rationalizing that the slight deviations from straight lines are due to experimental error. Nevertheless, the plots of $\ln(R-Y)_{org}$ vs. time and $\ln([R-Y]_{org}/(X^-]_{aq})$ vs. time do not represent acceptable first- or second-order plots, respectively, and to interpret them as such is incorrect. Case 1 represents a set of conditions in which the concentration of (Q^+X^-) in the organic phase rapidly reaches a steady state. It is not surprising, therefore, that the plots based on the steady-state and non-steady-state treatments behave virtually the same. If the rate experiment was designed with a large initial excess of $(X^-)_{aq}$ and $(Y^-)_{aq}$ where the concentrations of these species are not changing over the course of the reaction, only the first two terms in the denominator would be constant. Under these circumstances if the magnitudes of the three terms in the denominator were still comparable then, again, simple integral-order kinetics and, in particular, pseudo-first-order kinetics, would not be observed. If, however, the presence of a large initial concentration of $(X^-)_{aq}$ and $(Y^-)_{aq}$ made the first two terms large compared to the third then pseudo-first-order kinetics would be observed. These conditions are exactly Case 3 which will be discussed shortly. Thus, in designing rate experiments in the hope of simplifying the rate law, nonintegral order kinetics could still result. Clearly, Case 1 represents a process in which both the mass-transfer rates and the organic-phase reaction rate are contributing to the overall rate of the process.

When the term describing the rate of transfer of Y^- from the aqueous phase to the organic phase is large compared to the other two terms, Case 2 is operating. Under these circumstances, the rate-controlling step is k_2 and the mass-transfer

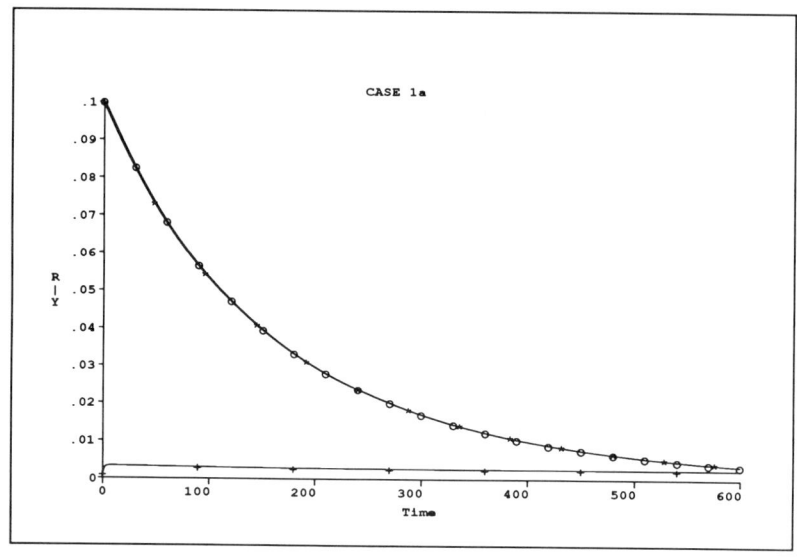

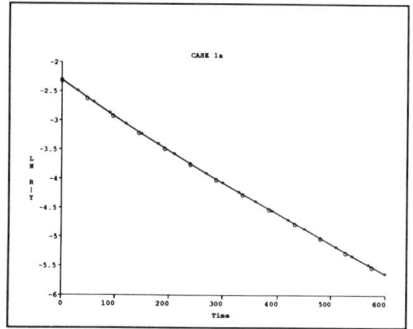
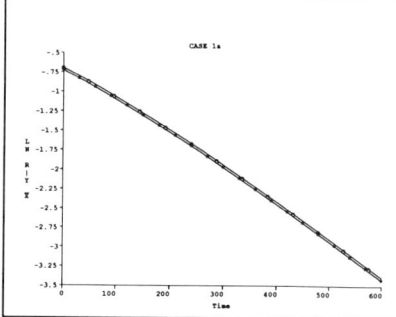

Figure 3-4a

steps may be treated as an equilibrium constant ($k_1/k_{\text{-}1} = K$). The overall rate of product formation is determined by both the mass-transfer equilibrium and the rate of reaction in the organic phase. Under conditions in which the rate of mass transfer of $(Y^-)_{\text{aq}}$ is very much greater than $(X^-)_{\text{aq}}$ (K is a very small number) there will be very little $(Q^+X^-)_{\text{org}}$ in the organic phase. As a consequence, the overall rate of the entire process would be very slow. If slow enough the situation would result in "catalyst poisoning." Nonintegral order kinetics should be observed under these circumstances since the concentrations of $(X^-)_{\text{aq}}$ and $(Y^-)_{\text{aq}}$ are changing along with the concentration of $(R-Y)_{\text{org}}$. This is clearly *not* the case as illustrated in Figure 3-5a. The first- and second-order plots both appear to be linear. The reason is that the rate was followed over only a few percent reaction. Following the rate over a greater percentage reaction would indeed

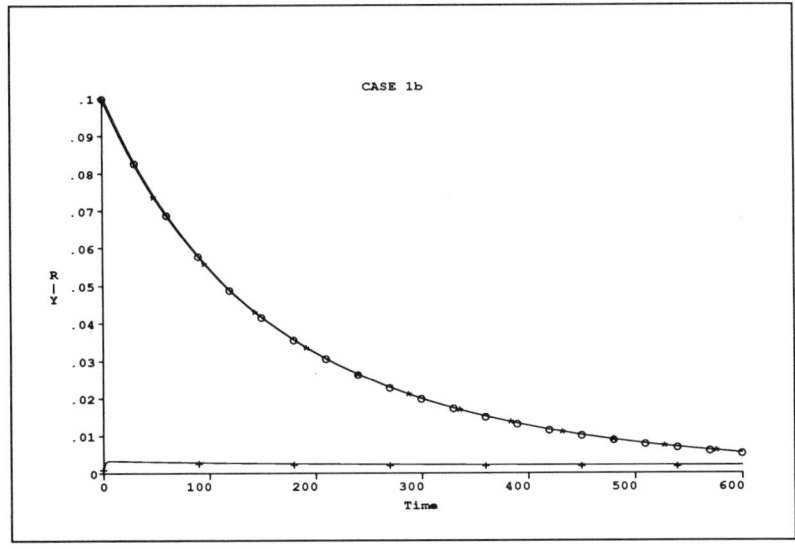
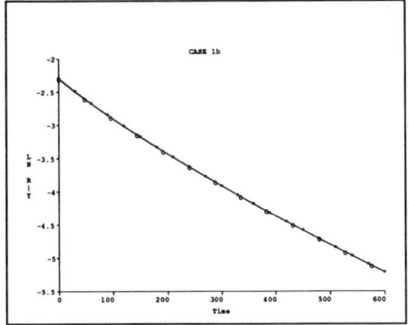
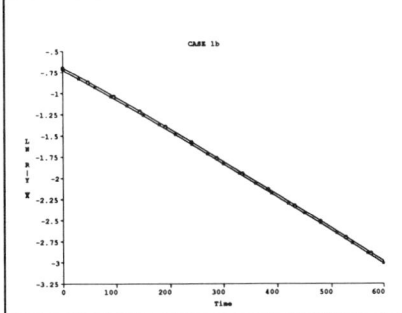

Figure 3-4b

show curvature in the plot. Assigning first- or second-order behavior to the plots in Figure 3-5a would be incorrect. It should be noted that the concentration of (Q^+X^-) in the organic phase is very small during the course of reaction. The steady-state approximation applied to $[Q^+X^-]_{org}$ appears to be reasonable. In the presence of excess initial $[X^-]_{aq}$ and $[Y^-]_{aq}$ simple pseudo-first-order kinetics should be observed. It should be emphasized that Case 2 is *not* desirable for successful PTC.

Case 3 is the limiting case in which the rate of reaction in the organic phase is slow compared to the mass transfer steps. Several examples of second-order kinetics under PTC conditions have been reported. These reports include the alkylation of tetra-*n*-pentylammonium nitrophenolates in methylene chloride–water [16] and in the reaction of thiophenoxide ion with 1-bromooctane in

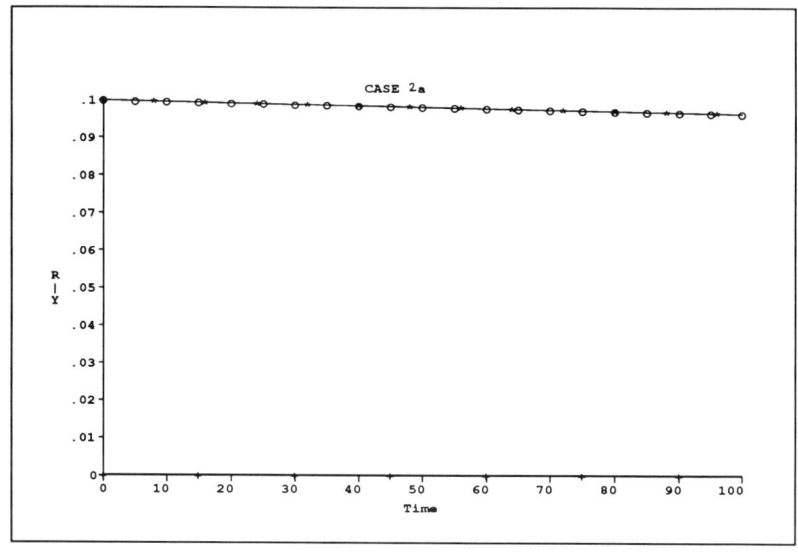
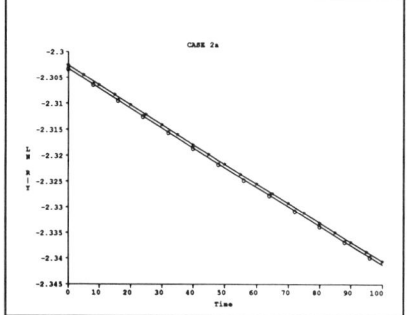
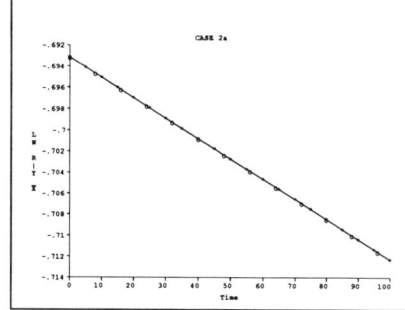

Figure 3-5a

benzene–water using a variety of phase-transfer catalysts [17]. The observation of second-order behavior suggests that mass transfer is rapid compared to the rate of the organic phase reaction. In general, reasonably good second-order kinetics should be observed (Figures 3-6a and b) since the denominator in the Case 3 differential equation remains nearly "constant" during the course of reaction; as $[X^-]_{aq}$ decreases, $[Y^-]_{aq}$ increases. Under Case 3 conditions, the concentration of (Q^+X^-) in the organic phase rapidly reaches a steady state, again suggesting that the steady-state approximation applied to $[Q^+X^-]_{org}$ is reasonable. In the presence of excess initial $[X^-]_{aq}$ and $[Y^-]_{aq}$ pseudo-first-order kinetics should be observed. Indeed, if the numerator and the denominator are divided by (k_1) $[X^-]_{aq}$, then Case 3 is exactly the form of the differential rate expression derived by Starks [11].

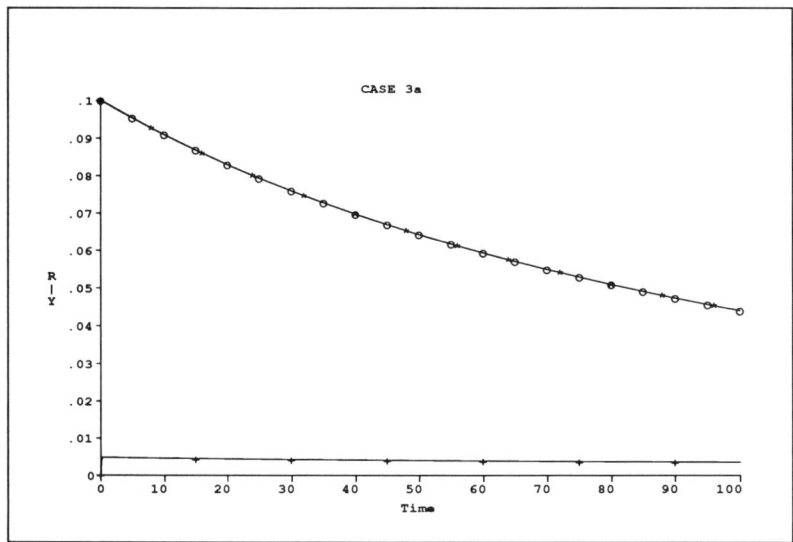
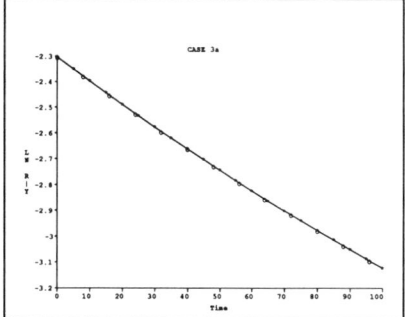
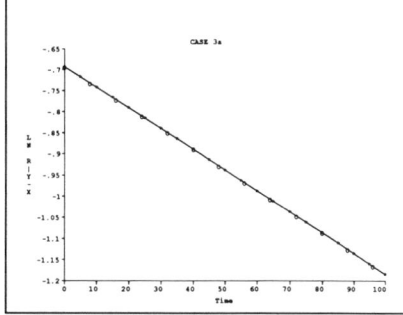

Figure 3-6a

$$\frac{d[R-X]_{org}}{dt} = \frac{Kk_2 Q_{org}[R-Y]_{org}}{[Y^-]_{aq}/[X^-]_{aq} + K} \quad (3\text{-}9)$$

In the PTC reaction of sodium cyanide with 1-chlorooctane to form 1-cyanooctane and sodium chloride the concentration ratio of aqueous anions in Eq. (3-9) was held constant by using a small volume of water (compared to the volume of the organic phase) and saturating the aqueous phase with sufficient sodium chloride and sodium cyanide so that excess solid is present throughout the reaction [2]. In this way, the thermodynamic activities of Na^+, Cl^-, and CN^- in the aqueous phase are maintained constant. This technique is satisfactory in this and certain other reactions, since essentially little or no anion transfer occurs from solid sodium chloride and sodium cyanide [2,18]. (This particular point will be ad-

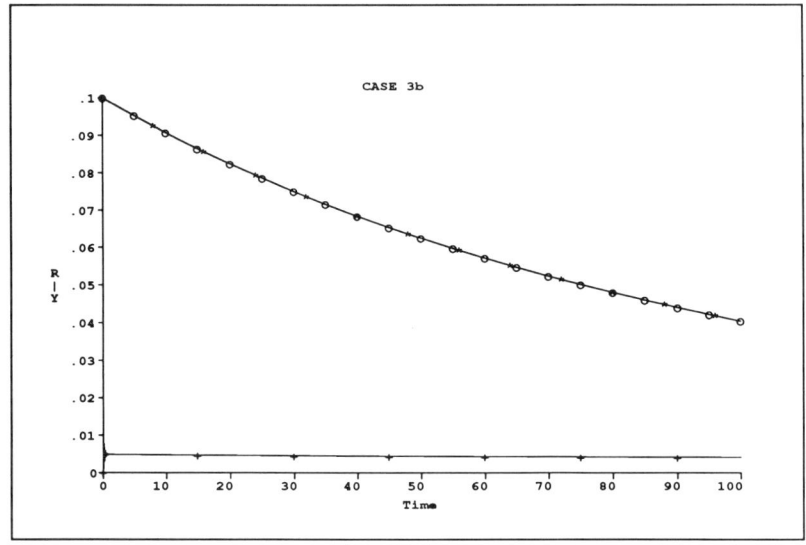

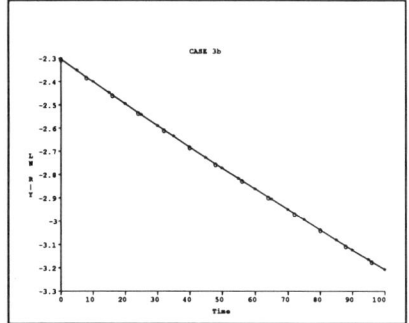
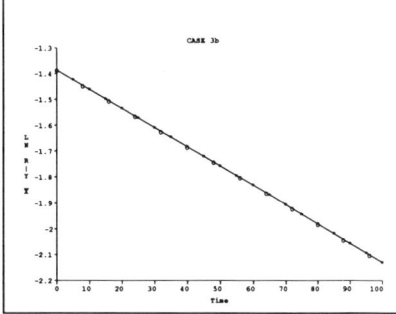

Figure 3-6b

dressed during the discussion of solid–liquid PTC). Employing these conditions, excellent pseudo-first-order kinetics were realized for a number of quaternary salts (Figure 3-7) [2]. Pseudo-first-order behavior has also been observed for the reaction of n-octyl methanesulfonate with a large excess of aqueous chloride, bromide, or iodide in a chlorobenzene–water two-phase system using tri-n-butylhexadecylphosphonium halides as phase-transfer catalysts [19,20].

The experimental procedure using saturated solutions of sodium chloride and sodium cyanide has been extensively used to ascertain many of the general kinetic features of PTC cyanide displacements. Thus, the value Q_{org}, although usually relatively constant for any given run, is an experimentally adjustable parameter, so that according to Eq. (3-9) one should observe the rate to be proportional to the concentration of added catalyst. This expected behavior is illustrated by

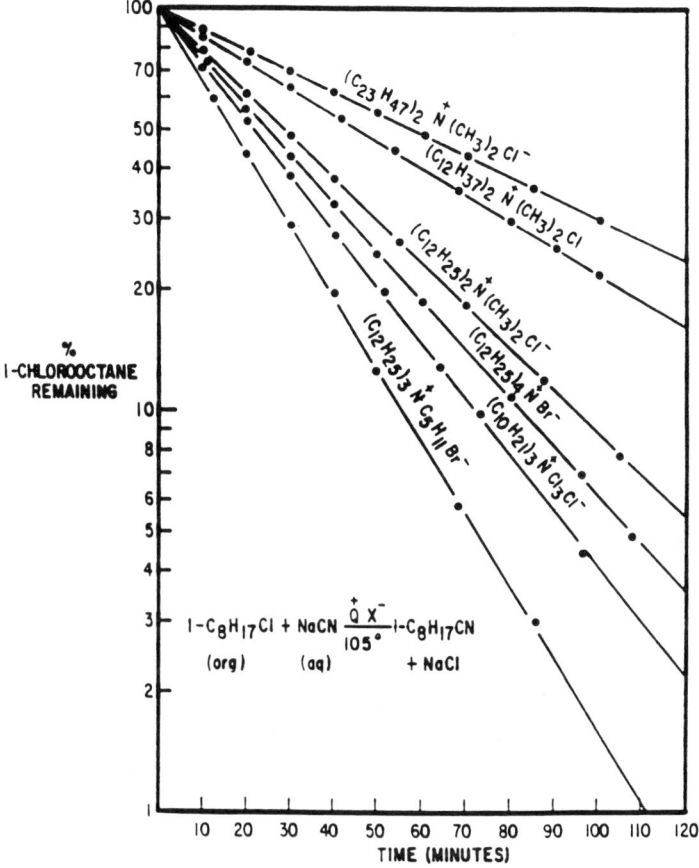

Figure 3-7

experimental data plotted in Fig. 3-8 [11]. This behavior has also been found in displacement reactions involving thiophenoxide [17] and alkoxide [21]. In the former case the rate constant was linearly dependent on the catalyst concentration over a 20-fold change in concentration.

Consider again the reaction of 1-chlorooctane with sodium cyanide. Under conditions of constant Cl^- and CN^- in the aqueous phase and in a well-stirred mixture, the catalyst cation will exist in the organic phase as $[Q^+CN^-]$ and $[Q^+Cl^-]$ in constant proportion. Since

$$\frac{[Q^+CN^-]_{org}}{[Q^+Cl^-]_{org}} = \Phi \tag{3-10}$$

it follows that

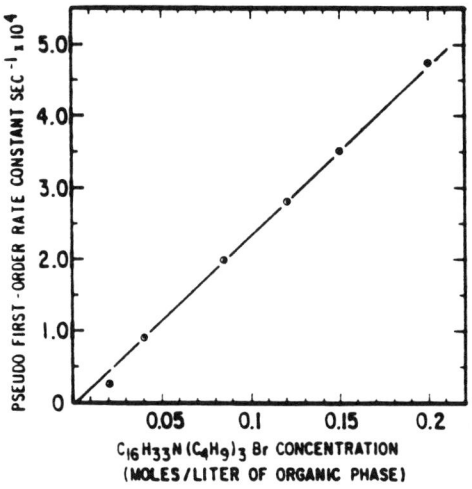

Figure 3-8

$$[Q^+CN^-]_{org} = \Phi Q_{org}(1 + \Phi) \quad ((3\text{-}11)$$

where Q_{org} is the total concentration or moles of catalyst present in the organic phase. For an experiment where a highly organophilic catalyst is used such that essentially all of it stays in the organic phase throughout the reaction, the rate of the organic phase displacement (represented by the second equation in Scheme 3-1) will be given by

$$-\frac{d[\text{R}-\text{Cl}])}{dt} = k_2[\text{R}-\text{Cl}][Q^+CN^-] = [k_2 \Phi Q_{org}(1+\Phi)][\text{R}-\text{Cl}] \quad (3\text{-}12)$$

Since ɸ is a constant, it may be independently measured by analysis of the organic phase for chloride and cyanide ions. ɸ was found to have a value of 2.0 at 90°C. Thus, from knowledge of k, ɸ, and Q_o the value of k_o is found to be 0.08 M^{-1} sec^{-1} at 90°C, using hexadecyltri-n-butylphosphonium cation as the phase-transfer ion.[2]

Apparent first-order kinetics can also be observed with many PTC reactions if an excess of the inorganic reagent is used. Thus, if excess cyanide is used, then most of the catalyst will be in the Q^+CN^- form, so that

$$[Q^+CN^-] \simeq Q_{org} \quad (3\text{-}13)$$

and

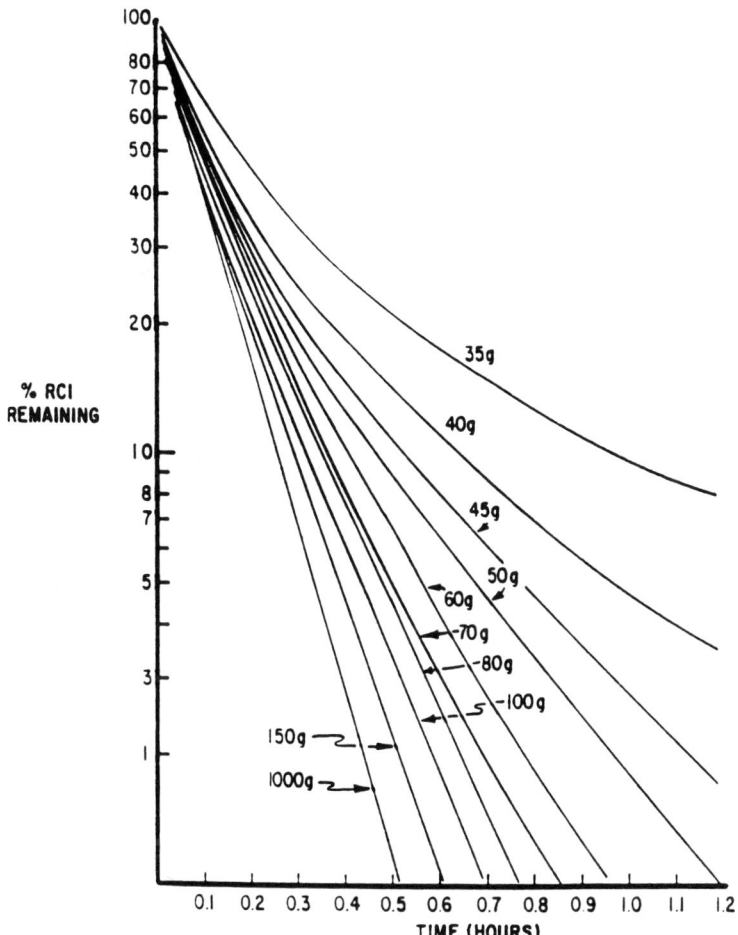

Figure 3-9. Effect of quantity PTC of NaCN (in grams) on the reaction profile of PTC cyanide displacements. Reaction mixture: 100 g 1-chloropentane, 1 g hexadecyltri-*n*-butylphosphonium bromide, 20 g water; 90°C; stoichiometric NaCN = 33 g.

$$-\frac{d[R-Cl]}{dt} \simeq k_2 Q_{\text{org}}[R-Cl] \qquad (3\text{-}14)$$

This technique works reasonably well when the anion partitioning equilibrium is not especially unfavorable (i.e., $K_{Cl \rightarrow CN}$ is equal to or greater than unity), but the numerical results obtained will depend on the ratio of reactants used (e.g., NaCN/R−Cl) as illustrated in Figure 3-9 [2].

In the extraction mechanism for the PTC process, reactant anions are transferred from the aqueous phase into the organic phase, where reaction with an organic substrate takes place. The organic product remains in the organic phase while the product anions are transferred from the organic phase into the aqueous phase. Since the phase-transfer catalyst is usually present in small concentrations both the reactant and the product anions compete for the limited quantity of catalyst. During the initial stages of the reaction, the catalyst is primarily associated with the reactant anions. However, as the reaction progresses, the competition between the two anions becomes a major factor in determining the success of the process. If the product anion has little affinity for the phase-transfer catalyst in the organic phase then, other factors omitted, the reaction will proceed to a reasonable completion. On the other hand, if the product anions have a greater affinity for the catalyst than the reactant anions, the reaction could come to a halt after a few percent of the product is formed. This latter situation is what was referred to as "catalyst poisoning" (see Case 2).

Thus, during the PTC process reactant and product anions compete for the phase-transfer cation. As discussed above this competition could be a significant factor in determining the success of a particular phase-transfer process. The greater facility with which some anions are transferred into organic media compared to other anions is directly related to the proportion of the reactant anion that will be in the active form (paired with Q^+ in the organic phase). An excellent measure of the competition between anions is the selectivity constant, $K_{X/Y}^{sel}$, which is defined by the following general reversible reaction and corresponding equilibrium expression [1–5]:

$$[Q^+Y^-]_{org} + [X^-]_{aq} \rightleftarrows [Q^+X^-]_{org} + [Y^-]_{aq} \qquad (3\text{-}15)$$

$$K_{X/Y}^{sel} = \frac{[Q^+X^-]_{org}[Y^-]_{aq}}{[Q^+Y^-]_{org}[X^-]_{aq}} \qquad (3\text{-}16)$$

Consider the following reaction of cyanide ion with alkyl halides [2,11]:

$$[Q^+X^-]_{org} + [CN^-]_{aq} \underset{\rightleftarrows}{K_{X-CN}} [Q^+CN^-]_{org} + [X^-]_{aq} \qquad (3\text{-}17)$$

$$[R-X]_{org} + [Q^+CN^-]_{org} \underset{\rightarrow}{k} [R-CN]_{org} + [Q^+X^-]_{org} \qquad (3\text{-}18)$$

The equilibrium constants, K_{X-CN}, for chloride, bromide, and iodide are 1.2, 0.05, and 0.0002, respectively. If it is assumed that the phase-transfer cation is highly organophilic so that it is essentially all in the organic phase, then the concentration of the reactant anion in the organic phase, $[Q^+CN^-]_{org}$ may be calculated from the following expression derived from the above equilibrium reaction [also see limiting Case 3 of Eq. (3-9)]:

$$[Q^+CN^-]_{org} = \frac{K_{X \to CN} Q_{org}}{(X^-/CN^-) + K_{X \to CN}} \quad (3\text{-}19)$$

A plot of percent Q^+ in the Q^+CN^- form vs. X^-/CN^- mole ratio is shown in Figure 3-10. If it is assumed that at $t = 0$ there were equal molar quantities of NaCN and R−X, then at approximately 17% reaction, the ratio of halide to cyanide will equal 0.2. At this point in the reaction, 85% of the quaternary cation will be in the Q^+CN^- form when $X^- = Cl^-$, 20% when $X = Br^-$, and 0.1% when $X^- = I^-$. In homogeneous media alkyl bromides undergo displacement reactions approximately 50 times faster than alkyl chlorides. This relative rate more than offsets the fourfold reduction in Q^+CN^- resulting when the alkyl bromide is used instead of the alkyl chloride. In contrast alkyl iodides undergo displacement in homogeneous media approximately 100 times faster than alkyl chlorides. This rate factor unfortunately does not compensate for the approximately 850-fold reduction in active catalyst form Q^+CN^- when using an alkyl iodide instead of an alkyl chloride. Experimentally, in the reaction of aqueous sodium cyanide with 1-iodooctane virtually no reaction occurs after 15–25% conversion. "Catalyst poisoning" has occurred. Kinetically, this situation was described by Eq. (3-7) under limiting Case 2. In order to overcome these difficulties a stoichiometric quantity of quaternary salt may be necessary.

There are several reports in which iodide ion, paired with quaternary ammonium cations, acts as a cocatalyst instead of a "catalyst poison." The examples include the benzylation reactions of isobutyraldehyde [22], of the sodium salt of phthalic acid mono-n-butyl ester [23], and of sodium benzoate [24], in a toluene–aqueous base two-phase system. The cocatalytic effect of iodide ion is rationalized by Scheme 3-2. The quaternary ammonium iodide, located in the organic phase, reacts with benzyl chloride to form the more reactive benzyl iodide. The quaternary ammonium chloride formed in this reaction is partitioned into the aqueous phase where an exchange reaction ensues in which the quaternary cation pairs with the organic anion. This is followed by the partitioning of the ion pair into the organic phase, where it reacts with benzyl iodide to form product and the quaternary ammonium iodide. It was reported that as the concentration of aqueous iodide is increased, the rate of reaction first increases and then decreases. The initial rate increase was attributed to an increase in the concentration of the quaternary ammonium iodide in the organic phase. The subsequent decrease in rate with an increasing concentration of aqueous iodide was attributed to competition of iodide with the organic anion for partitioning into the organic phase.

Case 4 represents a process in which the rate of mass transfer of $(X^-)_{aq}$ and the rate of reaction of the transferred anion with R−Y are fast compared to the rate of mass transfer of $(Y^-)_{aq}$ into the organic phase. Nonintegral order kinetics should be observed (Figures 3-11a–e). The first- and second-order plots clearly show curvature. In fact, the reactions appear to be accelerating with time. This may be easily understood by examining the plot of $[Q^+X^-]_{org}$ with respect

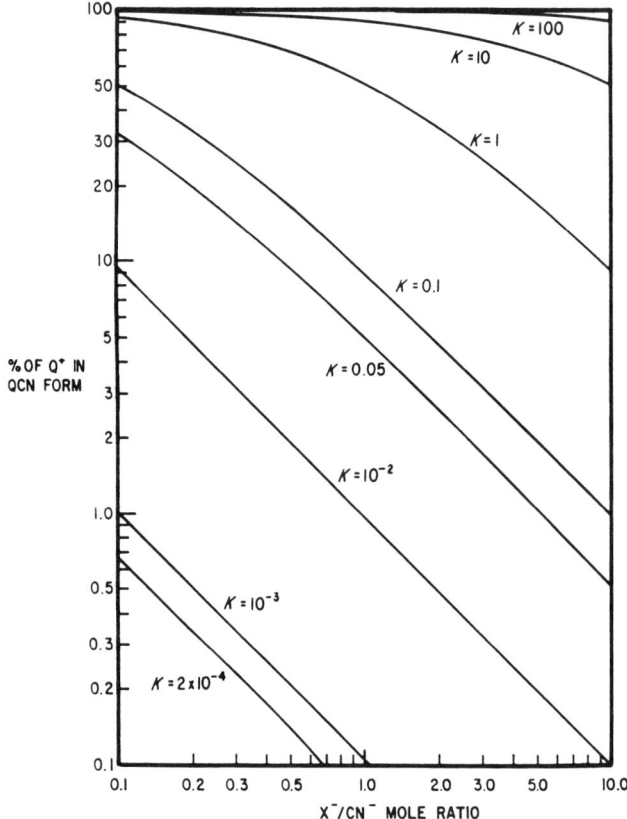

Figure 3-10. Percentage of catalyst cation $Q^=$ in active QCN form as calculated for various k values.

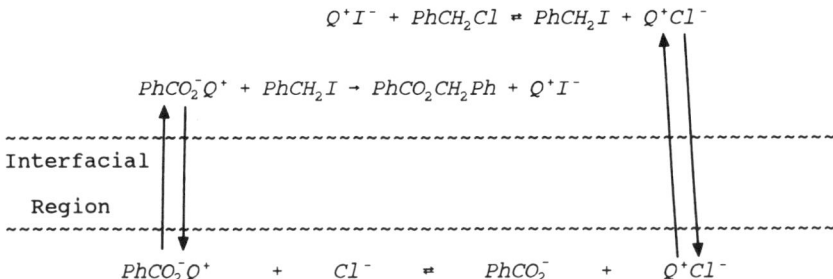

Scheme 3-2. Cocatalyst effect of iodide ion paired with quaternary cations.

to time. The concentration of this species continually increases during the course of reaction. Nevertheless, the plots with and without the application of the steady state for $[Q^+X^-]_{org}$ are amazingly similar. Clearly, the steady-state treatment in Case 4 is a rough, but instructive, approximation. In the presence of excess initial $[X^-]_{aq}$ nonintegral order kinetics will still be observed unless $k_1[X^-]_{aq} \gg k_2[R-Y]_{org}$. Then the differential rate expression becomes first order in $[R-Y]_{org}$ (Case 7).

$$\frac{d[R-X]_{org}}{dt} = k_2 Q_{org}[R-Y]_{org} \qquad (3-20)$$

In the presence of excess $[R-Y]_{org}$, if $k_2[R-Y]_{org} \gg k_1[X^-]_{aq}$ then the rate will be zero-order in $[R-Y]_{org}$ and first-order in $[X^-]_{aq}$ (Case 5).

$$\frac{d[R-X]_{org}}{dt} = k_1 Q_{org}[X^-]_{aq} \qquad (3-21)$$

It should be emphasized that Case 4 represents a desirable set of conditions since both of the rapid steps lead in the direction of product.

The kinetics associated with the sequential substitution reactions of hexachlorocyclotriphosphazene with 2,2,2-trifluoroethanol using a wide variety of quaternary ammonium salts in a chlorobenzene–aqueous sodium hydroxide two-phase system have been reported [25–27]. Pseudo-first-order rate constants for the first substitution reaction were reported as a function of concentration of reactants, structure of the phase-transfer catalyst, temperature, etc. For organophilic quaternary ammonium salts, as the concentration of the phosphazene increased the pseudo-first-order rate constant decreased. It was concluded that for these catalysts the reaction rate is controlled by both the organic phase reaction and the mass transfer of 2,2,2-trifluoroethoxide. Case 4 may be a reasonable representation of this reaction process. According to the equation describing Case 4, pseudo-first-order behavior should *not* be observed. Nevertheless, from an operational point of view, the slight curvature of a few of the plots of ln $[R-Y]_{org}$ vs. time (first-order behavior), as exemplified in Figure 3-11e, might be interpreted as arising from experimental error associated with the rate measurements and may understandably go undetected. In addition, slight deviations from linearity for the hexachlorophosphazene kinetics may be attributed to the operation of other competing reactions since the overall reaction process is a series of competing consecutive rate steps. Since the phosphazene concentration term is in the denominator, it is easy to understand why the observed "pseudo-first-order rate constants" decrease with increasing concentrations of phosphazene. For less organophilic quaternary salts such as tetraethylammonium chloride and benzyltrimethylammonium chloride, the pseudo-first-order rate constant did not change with changing concentrations of phosphazene.

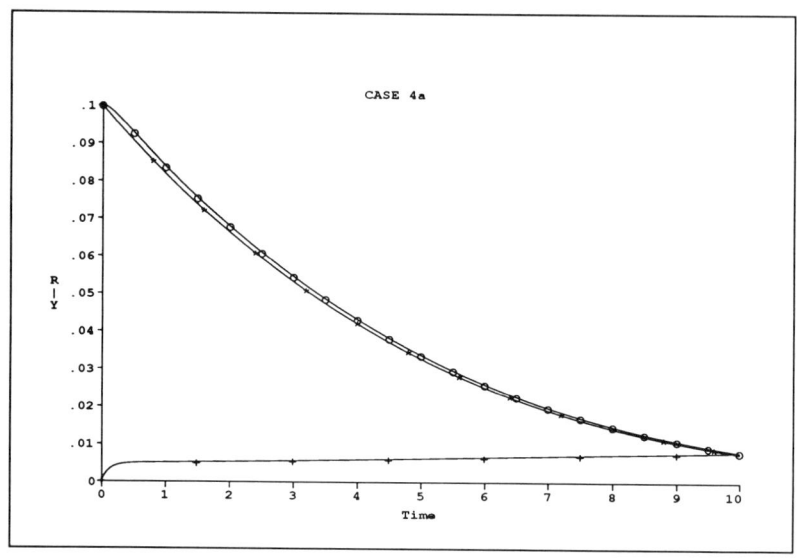
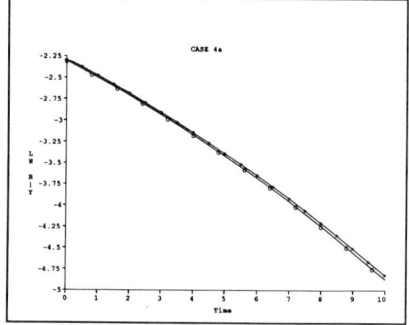
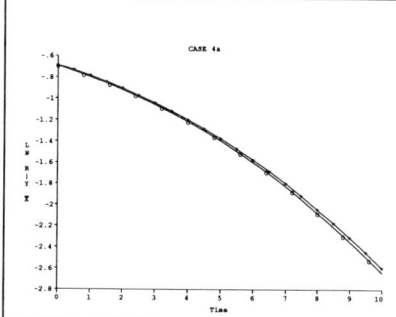

Figure 3-11a

Case 5 represents a process in which the reaction in the organic phase is very fast compared to the mass-transfer steps. The rate of reaction is dependent only on the concentration of $(X^-)_{aq}$ and is independent of the concentration of $(R-Y)_{org.}$. The rate-controlling step is the mass transfer of $(X^-)_{aq}$. This condition is exactly Eq. (3-21) as described in Case 4. Zero-order kinetics in $[R-Y]_{org}$ should be observed if the steady-state approximation is used for $[Q^+X^-]_{org}$. Figures 3-12a and b show two different Case 5 conditions. It is clear that as $k_2[R-Y]_{org}$ becomes larger compared to $k_1[X^-]_{aq} + k_1[Y^-]_{aq}$, the plots of $[R-Y]_{org}$ vs. time approach zero-order behavior. Under Case 5a conditions, the concentration of (Q^+X^-) in the organic phase is continually increasing during the course of reaction. A kinetic investigation dealing with the hypochloride oxidation of di-*n*-butyl sulfide to the corresponding sulfone in a methylene chloride–water

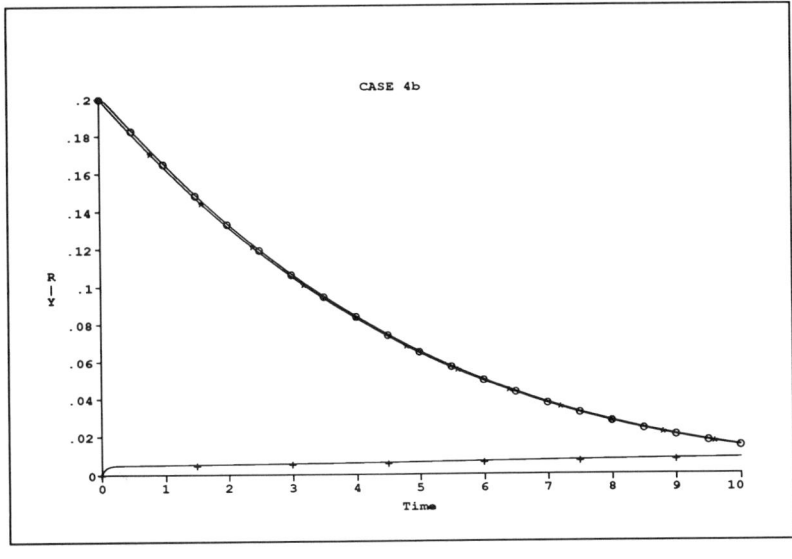
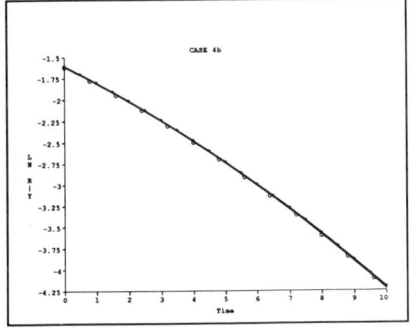
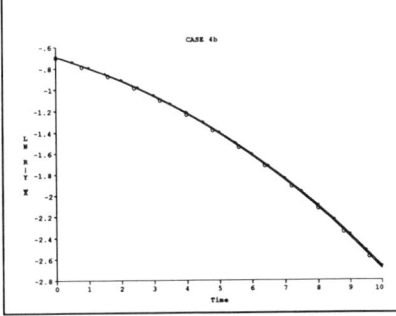

Figure 3-11b

system containing Aliquat 336 was reported [39]. The results were consistent with the oxidation reaction taking place in the organic phase. It was observed that when the amount of phase-transfer catalyst was varied from zero to 2.9 mmol the order with respect to the sulfide changed from a value close to zero to a value of unity. It was concluded that, at the low end of the catalyst concentration range, the transfer of hypochlorite from the aqueous to the methylene chloride phase was rate-limiting (Case 5). As the concentration of the catalyst increased the oxidation reaction in the organic phase became rate-limiting.

Case 6 represents a process in which the rate of mass transfer of $(Y^-)_{aq}$ and the rate of the reaction in the organic phase are fast. Nonintegral order kinetics might be expected to be observed (Figures 3-13a). In fact reasonably good second-order kinetics is observed. The origin of this observation is that the

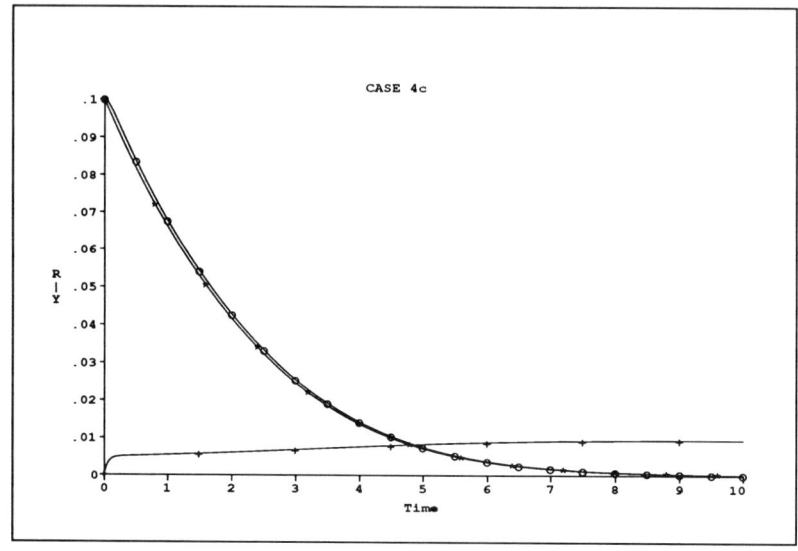
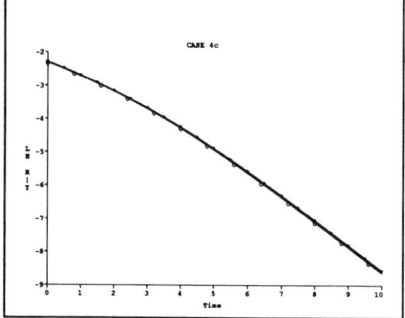
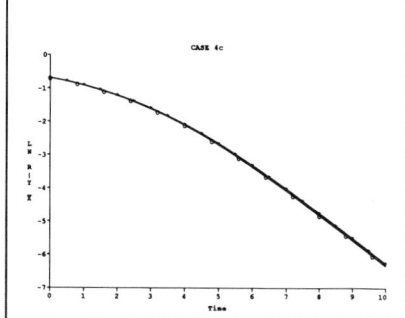

Figure 3-11c

denominator in the differential rate expression remains essentially "constant" during the reaction; as $[R-Y]_{org}$ decreases, $[Y^-]_{aq}$ increases. Under Case 6 conditions the concentration of $[Q^+X^-]$ in the organic phase is constant and negligibly small during the course of reaction. In the presence of excess $[Y^-]_{aq}$ and $[X^-]_{aq}$ if $k_{-1}[Y^-]_{aq} \gg k_2[R-Y]_{org}$ then pseudo-first-order kinetics will be observed (Case 2). In the presence of excess $[R-Y]_{org}$, if $k_2[R-Y]_{org} \gg k_{-1}[Y^-]_{org}$ zero-order kinetics will be observed.

Case 7 represents a process in which the rate of mass transfer of $(X^-)_{aq}$ is fast. Under these circumstances, $(Q^+X^-)_{org}$ rapidly accumulates in the organic phase and maintains a steady-state concentration; the phase-transfer cation in the organic phase becomes saturated with the anion X^-. Pseudo-first-order kinetics will be observed (Figures 3-14a). Again, it must be emphasized that the phase-transfer catalyst is usually present in very small quantities compared to the other reagents.

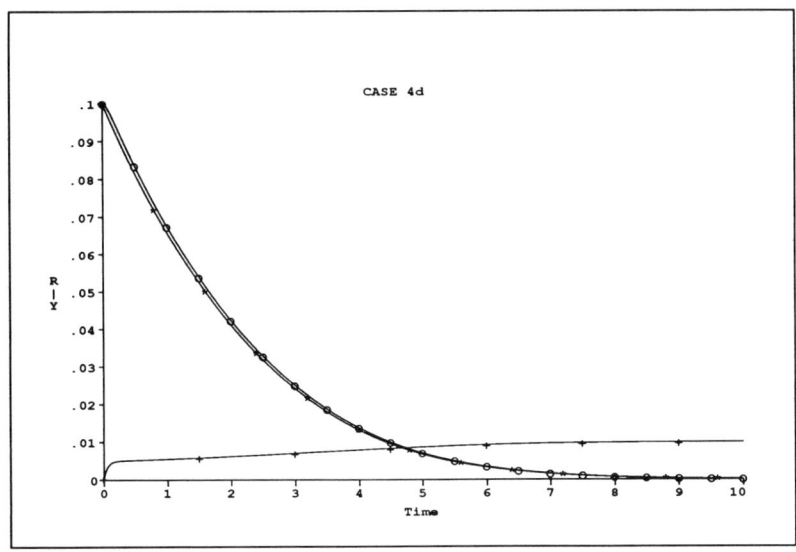

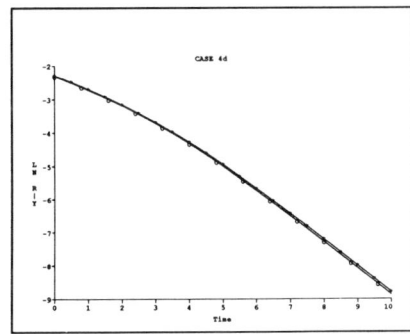
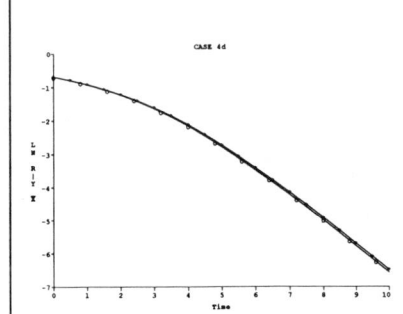

Figure 3-11d

d. Graphical Overview of Extraction Mechanism Limiting Cases

Employing the above analysis, a quantitative kinetic profile may be developed for substitution reactions involving the phase transfer catalytic processes proceeding via the extraction mechanism. It is clear from the mechanistic sequence shown in Scheme 3-1 that the three possible limiting components in the extraction mechanism are (1) the rate associated with the mass transfer of $(X^-)_{aq}$, the rate associated with the mass transfer of $(Y^-)_{aq}$, and (2) the rate of reaction of the transferred anion $(Q^+X^-)_{org}$ with the organic substrate $(R-Y)_{org}$. Figure 3-15 represents a quantitative three-dimensional representation of the two transfer rates and the reaction rate. Proceeding from left to right along the x-direction represents an increasing mass transfer rate for $(X^-)_{aq}$; proceeding from front to back along the y-direction represents an increasing mass transfer rate for $(Y^-)_{aq}$;

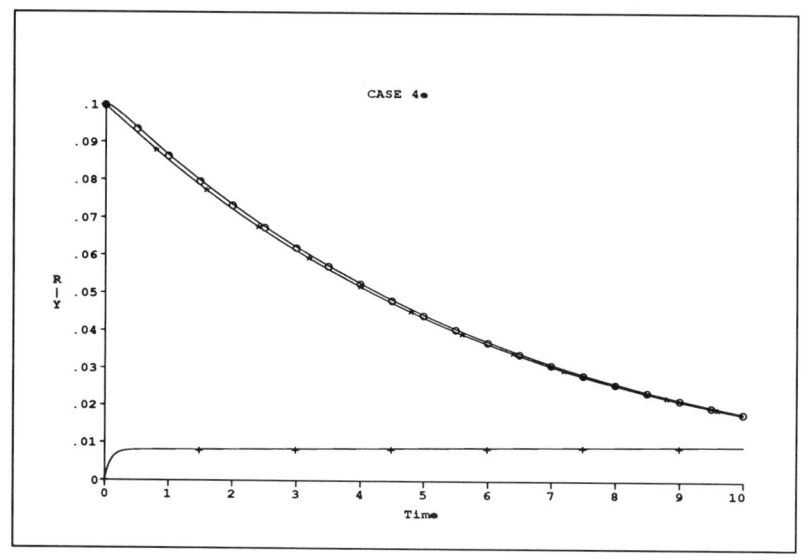
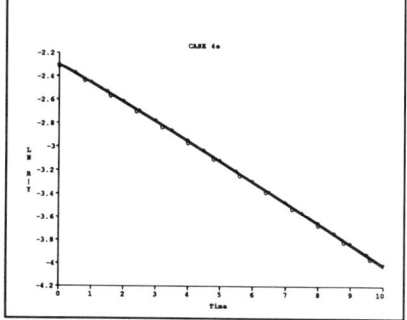
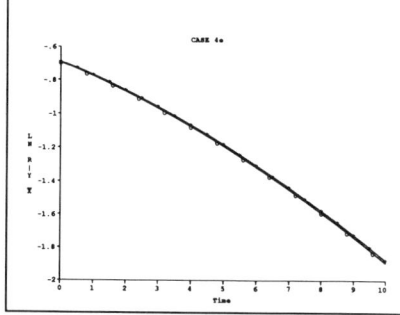

Figure 3-11e

and proceeding from bottom to top along the z-direction represents an increasing rate for the reaction in the organic phase. For instance, moving along line OA the rate of mass transfer of $(X^-)_{aq}$ is increasing while the rate of mass transfer of $(Y^-)_{aq}$ and the reaction rate in the organic phase are negligibly slow. Moving along line OB the rate of mass transfer of $(Y^-)_{aq}$ is increasing while the mass transfer rate of $(X^-)_{aq}$ and the reaction rate in the organic phase are negligibly slow. It is instructive to view Figure 3-15 in terms of the differential rate expression derived for the disappearance of $(R-Y)_{org}$ based on the steady-state treatment of $(Q^+X^-)_{org}$. The rate processes along line OA and OB may be best represented by the differential rate expressions described by Case 7 and Case 2, respectively. Moving along line OC the reaction rate in the organic phase is increasing while the mass transfer rates are negligibly slow. The differential rate expression

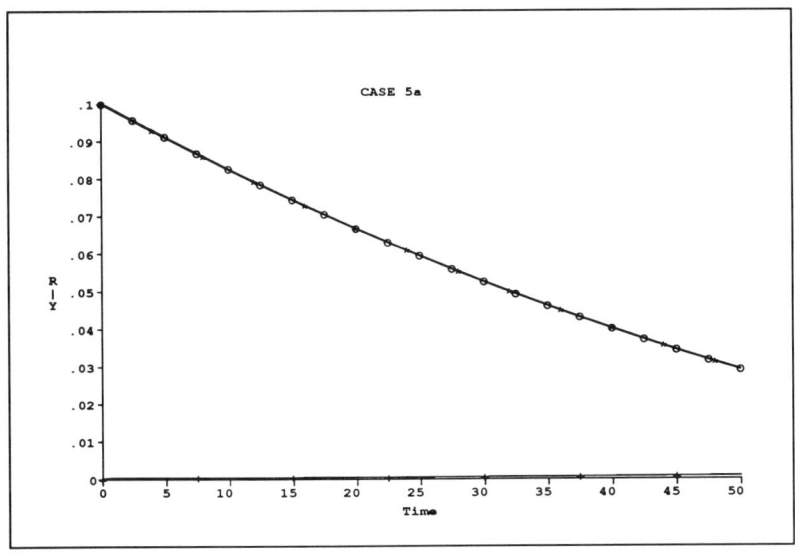
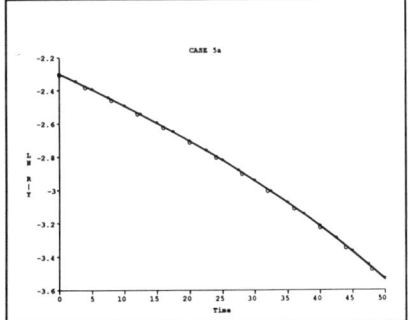
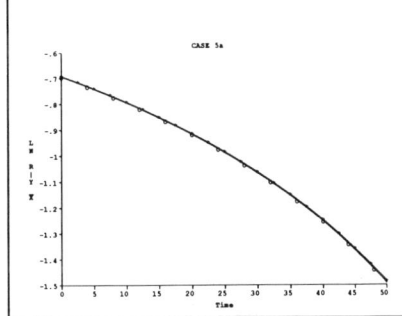

Figure 3-12a

represented by Case 5 best describes this rate process. The rate processes along lines OD, OE, OF, and OG are represented by the rate expressions described by Cases 3, 4, 6, and 1, respectively.

Considering the plane described by OAEC and proceeding in a circular counter clockwise manner from rapid transfer rate of $(X^-)_{aq}$ to rapid reaction rate of $(R-Y)_{org}$ (quarter circle), $k_1[X^-]_{aq}$ decreases and $k_2[R-Y]_{org}$ increases. Of particular importance is the line OE. Along the line OE, $k_1[X^-]_{aq}$ is equal to $k_2[R-Y]_{org}$. Both of these steps represent a forward process in the formation of product. When the rate of mass transfer of $(X^-)_{aq}$ and the rate of the organic phase reaction are equal the maximum rate of PTC should be observed. The further along the line OE the faster the overall process should be; an ideal balance of rates has been achieved. Of course the rate of mass transfer of $(Y^-)_{aq}$ is negligibly slow

74 / *Phase-Transfer Catalysis*

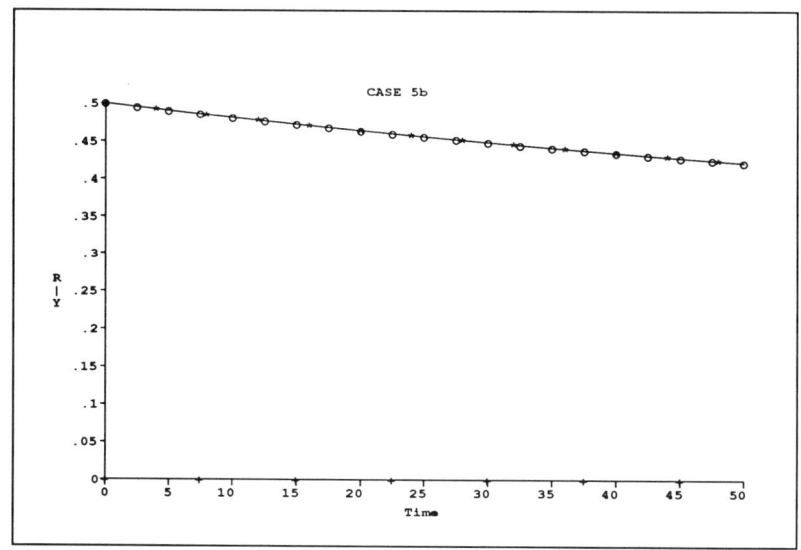

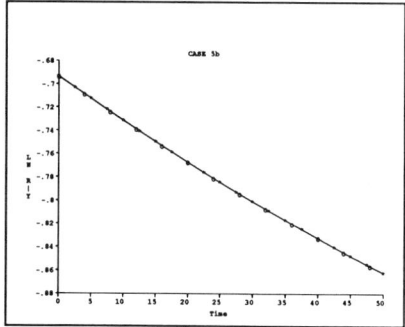
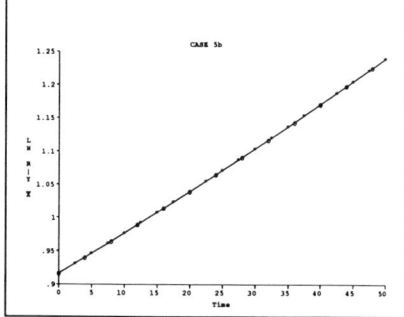

Figure 3-12b

in the OAEC plane. However, an infinite number of parallel planes exist between the OAEC and BDGF planes. Progressing along the OB axis between these two planes leads to rate processes in which the mass transfer of $(Y^-)_{aq}$ becomes more and more important. Clearly, in proceeding in this direction, the conditions for successful PTC are becoming less favorable. In principle, the most favorable conditions for a successful phase-transfer process would be in the OAEC plane along line OE.

e. Extraction Mechanism: Non-Steady-State Kinetic Description

The kinetics associated with a constant concentration of phase-transfer cation in the organic phase (application of the steady-state approximation to $[Q^+X^-]_{org}$ has already been discussed. The results of relaxation of this condition will now

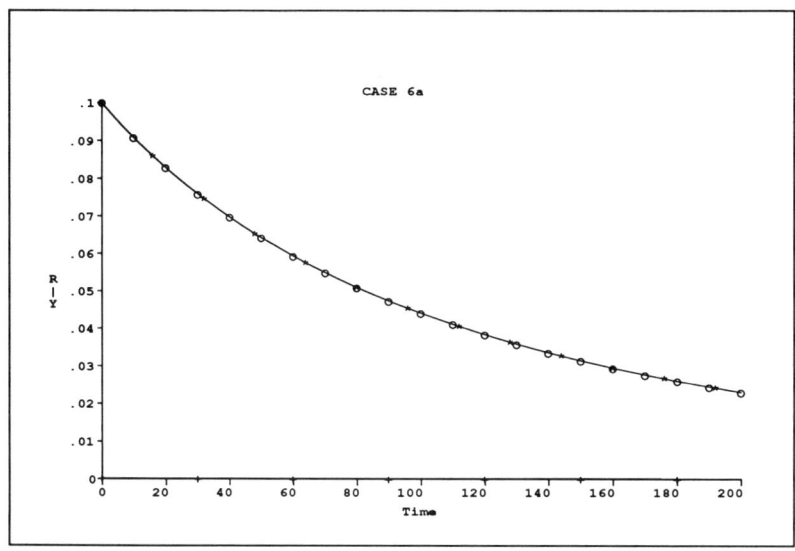
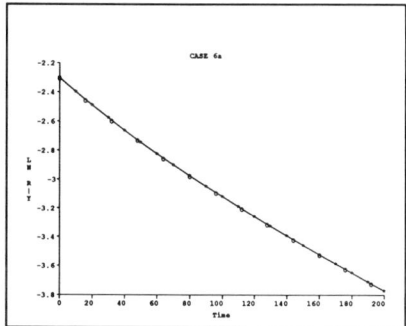
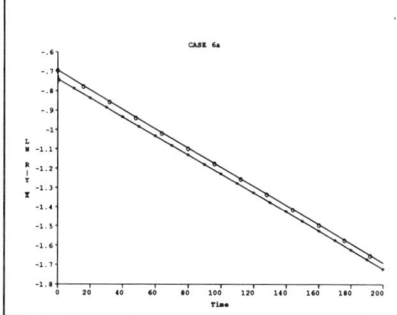

Figure 3-13a

be addressed. Under the circumstances in which the concentration of the phase-transfer cation in the organic phase is changing, the steady-state approximation for $[Q^+X^-]_{org}$ [Eq. 3-6] is no longer valid; a more complex, non-steady-state equation must be employed to describe the kinetic process.

In the derivation of the differential rate expression for the formation of product via the Starks extraction mechanism shown in Scheme 3-1 it was assumed that the distribution of phase-transfer salt between the aqueous and the organic phase remained constant throughout the course of reaction. There are circumstances where this assumption may not be correct. Consider the case in which the organic phase is neat organic reactant. If the organic reactant is relatively nonpolar but the organic product is relatively polar, as the reaction progresses, the distribution of the phase-transfer salt between the phases may be changing. This is exactly what was observed for the reaction of aqueous sodium cyanide with neat 1-

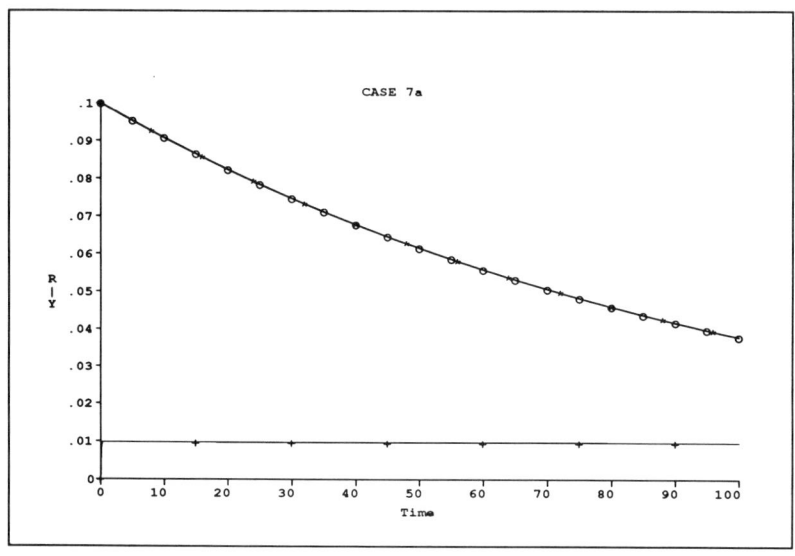
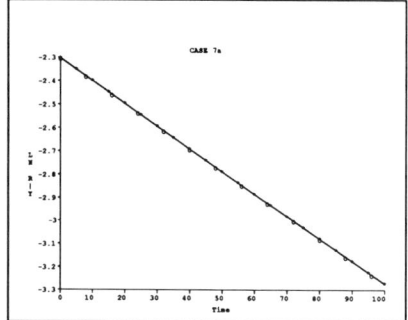
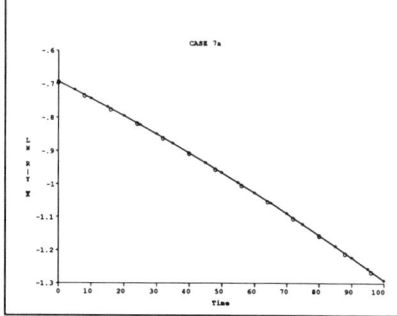

Figure 3-14a

bromooctane using tetra-*n*-butylphosphonium bromide as the phase-transfer catalyst. The steady-state approximation made for the change in concentration of the reactant phase-transfer salt in the organic phase would not be valid under such circumstances. Thus, in order to kinetically describe the reaction process the change in phase-transfer salt distribution must be specifically taken into account. In kinetic terms, the rate of mass transfer of $(X^-)_{aq}$ and $(Y^-)_{aq}$ are changing.

From Scheme 3-1, the rate of product formation and the rate of change of $(Q^+X^-)_{org}$ concentration are given by Eqs. (3-1) and (3-2). The concentration of $(Q^+X^-)_{org}$ at any time t is given by Eq. (3-22).

$$[Q^+X^-]_{org} = \frac{k_1[X^-]_{aq} Q_{org} - \dfrac{d[Q^+X^-]_{org}}{dt}}{k_1[X^-]_{aq} + k_{-1}[Y^-]_{aq} + k_2[R-Y]_{org}} \tag{3-22}$$

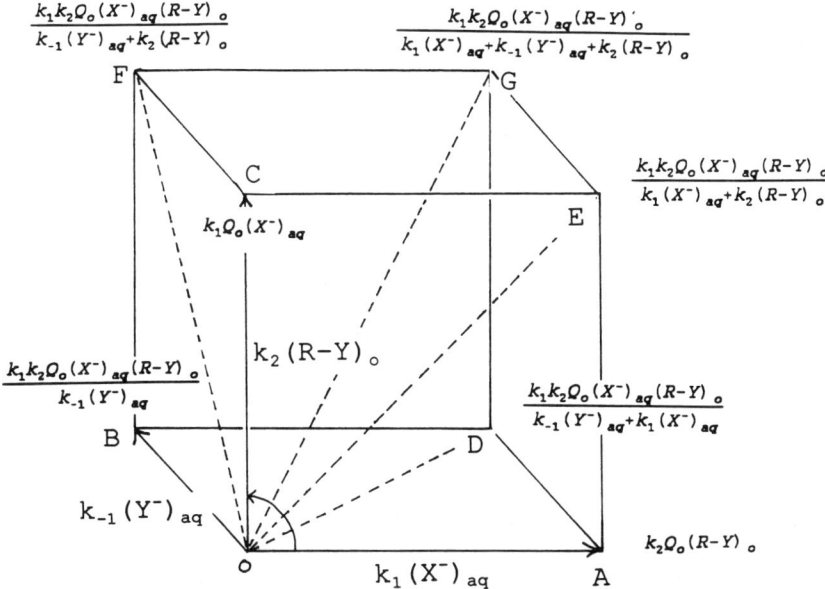

Figure 3-15. Graphical overview of extraction mechanism limiting cases.

where Q_{org} is the concentration of phase-transfer salt in the organic phase. In order to estimate the derivative term in the above equation, it is assumed that the concentration of $(Q^+X^-)_{org}$ varied linearly with the mole fraction of product, n_p, formed at any time, t, during the course of reaction. [Of course, it is also possible that a quadratic or cubic equation may better describe the concentration of $(Q^+X^-)_{org}$ with respect to formation of product.] Thus,

$$[Q^+X^-]_{org} = [Q^+X^-]'_{org} + an_p \tag{3-23}$$

where $[Q^+X^-]'_{org}$ is the concentration of the quarternary salt at $t = 0$, n_p is the mole fraction of product formed at time t, and a is a constant whose units are moles/liter. Differentiating this expression with respect to t gives

$$\frac{d[Q^+X^-]_{org}}{dt} = a\frac{dn_p}{dt} \tag{3-24}$$

From the definition of mole fraction, it follows that

$$n_p = \frac{m_P}{n_T} = \frac{V_{org} m_P}{V_{org} n_T} = \frac{V_{org}}{n_T}[R-X]_{org} \tag{3-25}$$

where m_P is the moles of product, n_T is the total moles of reactant plus product, V_{org} is the volume of the organic phase, and $[R-X]_{org}$ is the molarity of the product in the organic phase. Differentiating with respect to time and substituting in Eq. (3-24) the following equation is obtained:

$$\frac{d[Q^+X^-]_{org}}{dt} = \frac{aV_{org}}{n_T}\frac{d[R-X]_{org}}{dt} \tag{3-26}$$

Since the concentration of $(Q^+X^-)_{org}$ is changing, it follows that the total concentration of quaternary salt in the organic phase, Q_{org}, is also changing. Thus

$$Q_{org} = Q'_{org} + bn_p \tag{3-27}$$

where b is a constant whose units are moles/liter. Again, it is assumed that Q_{org} varies linearly with the mole fraction of product. As the polarity of the organic phase changes, concurrent with the change in the distribution of $(Q^+X^-)_{org}$, there will be a change in the rate constants k_1, k_{-1}, and k_2. In order to address this problem, it is assumed that the rate constants varied linearly with the formation of product. In particular, it is assumed that the rate constants varied linearly with the mole fraction of product formed during the course of reaction. Again, the variation in rate constant with respect to product formation could also obey a quadratic or cubic relationship. Thus,

$$k_1 = k'_1 + k''_1 n_p \tag{3-28}$$

$$k_{-1} = k''_{-1} + k''_{-1} n_p \tag{3-29}$$

$$k_2 = k'_2 + k''_2 n_p \tag{3-30}$$

Substitution of Eqs. (3-26)–(3-30) into Eq. (3-1) gives

$$\frac{d[R-X]_{org}}{dt} = \tag{3-31}$$

$$\frac{(k'_1+k''_1\left(\frac{V}{n_T}\right)[R-X]_{org})(k'_2+k''_2[R-X]_{org})[X^-]_{aq}[R-Y]_{org}\left(Q'_{org}+\frac{bV}{n_T}[R-X]_{org}\right)}{(k'_1+{}''_1[R-X]_{org})[X^-]_{aq}+(k'_{-1}+k''_{-1}[R-X]_{org})[Y^-]_{aq}+(k'_2+k''_2[R-X]_{org})[R-Y]_{org}\left(1+\frac{aV}{n_T}\right)}$$

Application of Eq. (3-31) in a kinetic rate simulation is shown in Figure 3-16. Using the initial concentrations and constants listed in Figure 3-16, an induction period is observed. A plot of $\ln[R-Y]_{org}$ vs. time (Figure 3-16) shows a dramatic deviation from first-order behavior reminiscent of Figure 2-1 in Chapter 2.

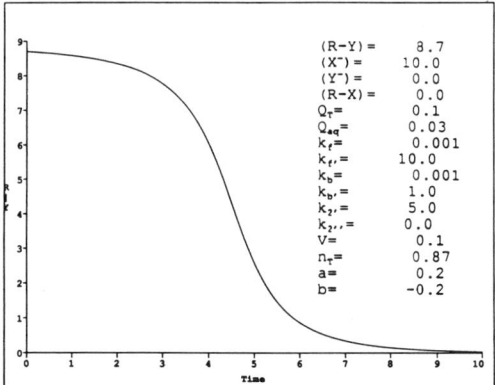

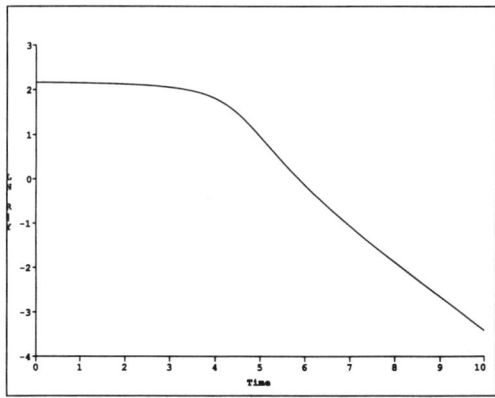

Figure 3-16.

f. Structure of the Reactant Anion in the Organic Phase

Once the anion is transferred into the organic phase, knowledge of the structure of the reactive nucleophilic species in the organic phase is critical to our understanding of the PTC process. In principle, the reactive species could be the free anion, the ion pair, some simple or complex aggregate, or perhaps a combination of all of these each reacting at a particular rate. Some of the important variables that will affect the nature of the quaternary salt in the organic solvent include (1) the structure of the quaternary ion, (2) the structure of the anion (including the number of waters of hydration), and (3) the polarity of the solvent. When the organic medium is one of higher dielectric constant, the free anion may be the reactive species. Nevertheless, the solvents used in most PTC processes are of relatively low dielectric constants and ion pairs and other aggregates are to be expected.

According to the Starks extraction mechanism, as well as the Brandstrom–Montanari modification for PTC processes described in Scheme 3-1, it is assumed that the quaternary salt (Q^+X^-) is a monomeric ion pair in the organic phase and reacts as the monomeric ion pair with the organic electrophile in the organic phase to yield product. The assumption that the reactive form of the quaternary salt is the monomeric ion pair is, however, too simplistic. Since the polarity of the organic phase in PTC reactions can vary over a wide range of dielectric constants, it would be anticipated that the aggregation state of the quaternary salt in the organic phase will also vary. Thus, in principle, the quaternary salt may exist as free ions, as ion pairs, or as more complex aggregates such as triple ions, quadrapoles, or higher aggregates [28,29].

$$Q^+ + X^- \rightleftarrows Q^+X^- \rightleftarrows (Q_2X)^+X^- / Q^+(QX_2)^- \rightleftarrows (Q^+X^-)_2 \rightleftarrows \text{etc.} \quad (3\text{-}32)$$

free ions   ion pair   triple ions   quadrapole

The states of aggregation of quaternary salts have been explored using electrochemical conductance measurements. Figure 3-17 shows a log–log plot of equivalent conductance, Λ, of a series of quaternary salts vs. concentration in the solvent benzene [29]. At concentrations below $10^{-4.5}$ the slopes approach $-1/2$. This may be compared with the dashed line in the lower left of Figure 3-17 with a slope of exactly $-1/2$. A slope of $-1/2$ represents the theoretical value for simple ionization behavior. Minima in the curves occur between the concentrations of 10^{-4} and $10^{-4.5}$. The current-carrying species within this range are believed to be simple and triple ions. However, the low value of the equivalent conductance indicates that at these minima more than 99.9% of the quaternary salts are in the ion pair form. At concentrations $> 10^{-4}$ the slopes become positive, but their magnitude is much less than $+1/2$, the theoretical value for triple ions. This may be compared with the dashed line in the lower right of Figure 3-17 with a slope of exactly $+1/2$. It is postulated that within this concentration range triple ions as well as nonconducting quadrapoles are present. At concentrations $>10^{-2.5}$ the slopes are greater than $+1/2$, indicating that charged higher aggregates are the major current-carrying species. At a concentration of 10^{-2} tetra-n-butylammonium nitrate has a slope of approximately unity. While most of the salt is probably in the quadrapole form, the slope corresponds to an average aggregation number of 2 for the current carrying species. It should be emphasized that the concentration range 10^{-2}–10^{-3} represents the lower range of catalyst concentrations ordinarily used in PTC. In contrast to benzene as a solvent, electrical conductance measurements and vapor pressure studies indicate that in solvents of higher polarity, such as acetonitrile or methanol, the salts are completely ionized within the 10^{-2}–10^{-3} concentration range.

Figure 3-18 shows a plot of stoichiometric molality vs. degree of association for a variety of quaternary ammonium and phosphonium salts in benzene and

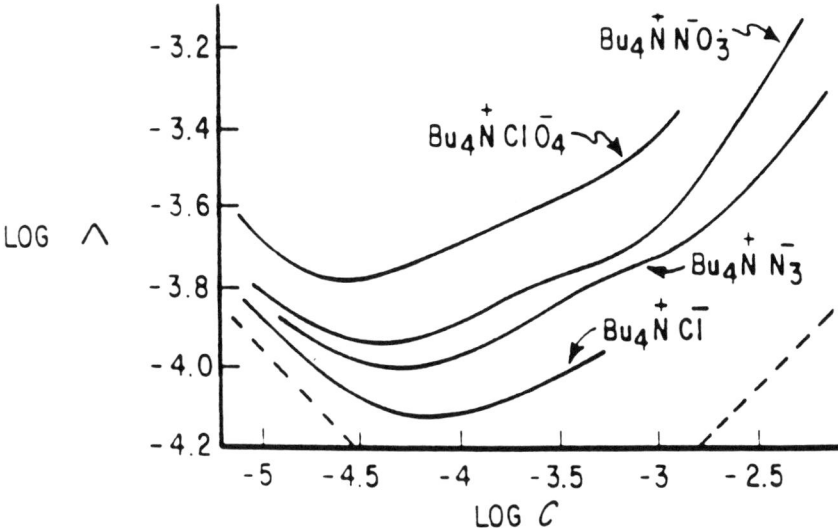

Figure 3-17. Log–log plot of equivalent conductance vs. concentration of quaternary salts.

1-bromopropane [2]. In these solvents the degree of association is substantial and can approach reasonably large values at the higher concentrations.

Any one of the quaternary salt aggregates in Eq. (3-32) or for that matter any combination of aggregates could be the reactive species in a PTC reaction. It is instructive to explore the kinetic consequences associated with several limiting cases for the reactive form of the anion. These include the reactive species being (1) very small quantities of the free anion in equilibrium with the monomeric ion pair and (2) very small quantities of the monomeric ion pair in equilibrium with free ions, quadrapoles, or other ion aggregates.

g. Ion Pair–Free Ion Equilibria

Consider the case in which the quaternary salt is primarily in the ion pair form with a very small equilibrium concentration of free ions.

$$(Q^+X^-) \rightleftarrows Q^+ + X^- \quad (3\text{-}33)$$

Since the total amount of quaternary salt $[Q_{org}]$ is approximately equal to the concentration of the ion pair

$$Q_{org} = [Q^+X^-] \quad (3\text{-}34)$$

it follows that

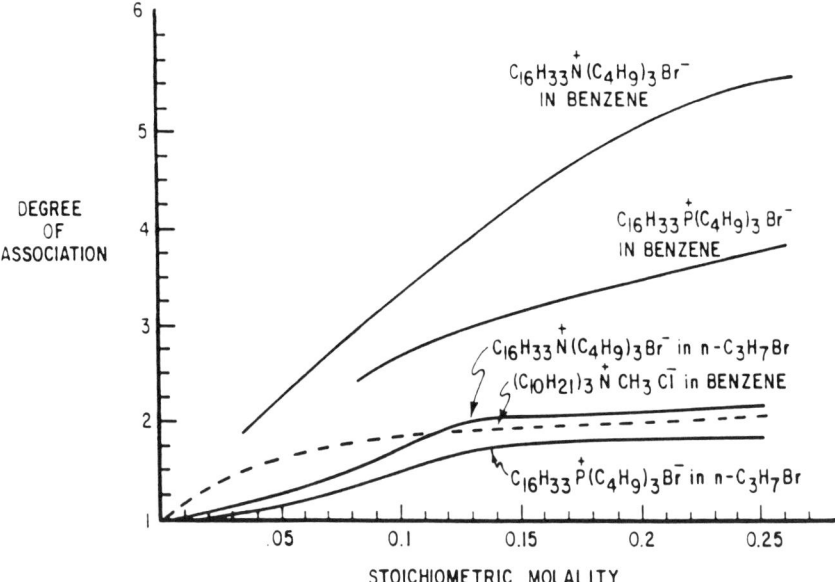

Figure 3-18. Aggregation of quaternary salts in benzene and 1-bromopropane.

$$[Q^+] = \sqrt{KQ_{\text{org}}} \qquad (3\text{-}35)$$

Thus, if the free anion were the reactive specie, the kinetic order with respect to the quaternary salt would be expected to be 1/2. In contrast, if the quaternary salt were almost completely ionized then the concentration of the free anion would be equal to the concentration of the quaternary salt. The kinetic order with respect to the quaternary salt under these circumstances would be expected to be unity. Table 3-3 lists the equilibrium constants, K_i, for several quaternary salts in solvents of varying dielectric constants [30]. In the range of dielectric constants most often used in PTC reactions (i.e., $\epsilon = 2\text{–}20$) the K_i values are uniformly less than 10^{-2}. It is clear that in these solvents and under the concentration conditions usually employed in PTC the quaternary ammonium salts are *not* primarily free ions; they exist as ion pairs or perhaps some higher aggregate.

The rate coefficients for the reaction of 1-chlorooctane with aqueous sodium cyanide at 90°C with various concentrations of hexadecyltributylphosphonium bromide have been reported [11]. Excess initial chloride and cyanide ion concentrations were employed. Since the rate-controlling step in this reaction is the displacement reaction in the organic phase, pseudo-first-order kinetics were observed (Case 3). A linear relationship between the pseudo-first-order rate constants and the concentration of the quaternary salt (Figure 3-19) was observed, indicating that the reactive form of the reactant anion is the monomeric ion pair. Halide exchange reactions have been investigated under PTC conditions [31].

Table 3-3. Ionization equilibrium constants for quaternary salts in various solvents.

	$K_i \times 10^4$ at 25°C			
Quaternary salts	Ethylene dichloride	Pyridine	Acetone	Nitro-benzene
NH_4 picrate	—	2.8	11.1	—
$(CH_3)_4N$ picrate	0.32	—	112.0	400
$(CH_3)_3(C_2H_5)N$ picrate	—	8.2	—	440
$(C_2H_5)_4N$ picrate	1.59	—	—	1400
$(C_2H_5)_4N$ chloride	0.510	—	—	—
$(C_2H_5)_4N$ bromide	0.697	—	—	—
$(C_2H_5)_4N$ nitrate	0.74	—	—	—
$(C_3H_7)_4N$ picrate	1.94	—	—	a
$(CH_3)(C_4H_9)_3N$ picrate	1.20	—	—	—
$(C_3H_7)(C_4H_9)_3N$ picrate	2.03	—	—	—
$(C_4H_9)_4N$ picrate	2.28	12.8	—	a
$(C_4H_9)_4N$ nitrate	1.18	3.7	—	250
$(C_4H_9)_4N$ bromide	—	2.5	—	162
$(C_4H_9)_4N$ acetate	1.34	1.7	—	67
$(C_5H_{11})_4N$ picrate	2.38	—	—	a
$(C_5H_{11})_4N$ nitrate	1.29	—	—	—
$(iso\text{-}C_5H_{11})_4N$ picrate	2.39	—	—	—
$(C_{18}H_{37})(CH_3)_3N$ picrate	—	7.66	—	—
$(C_{18}H_{37})(CH_3)_3N$ chloride	—	0.358	—	—
$(C_{18}H_{37})(CH_3)_3N$ acetate	0.062	—	—	—
$(C_{18}H_{37})(CH_3)_3N$ acetate octadecylsulfate	0.118	—	—	—
$(C_{18}H_{37})(C_4H_9)_3N$ acetate	1.33	3.88	—	—
$(C_{18}H_{37})(C_4H_9)_3N$ iodide	—	—	59.9	—
$(C_{18}H_{37})_2(C_4H_9)_2N$ iodide	—	4.21	67.7	—

^aCompletely dissociated.

The rates were found to be directly proportional to the concentration of the catalyst, again indicating that the ion pair is the reactive specie. In the PTC alkylation reaction of β-diketones using tetra-*n*-butylammonium cation, it was concluded that both ion pairs and free ions were taking part in the reaction [32]. In the reaction of thiophenoxide with 1-bromooctane in benzene–water using a variety of catalysts, it was found that the rate constant was linearly dependent on the catalyst concentration to the first power over a concentration change of 20-fold [17]. The pseudo-first-order rate constants obtained from the reaction of excess aqueous chloride, bromide, and iodide with *n*-octyl methanesulfonate (Case 3) showed a linear dependence on the concentrations of the corresponding tri-*n*-butylhexadecylphosphonium halide catalysts [19,20].

Thus, in the above experimental studies the rates of PTC reactions have been found to be directly proportional to the concentration of catalyst cation in the organic phase. These results are in the agreement with the observations reported by Uglestad, who demonstrated that ion pairs can react with alkyl halides with

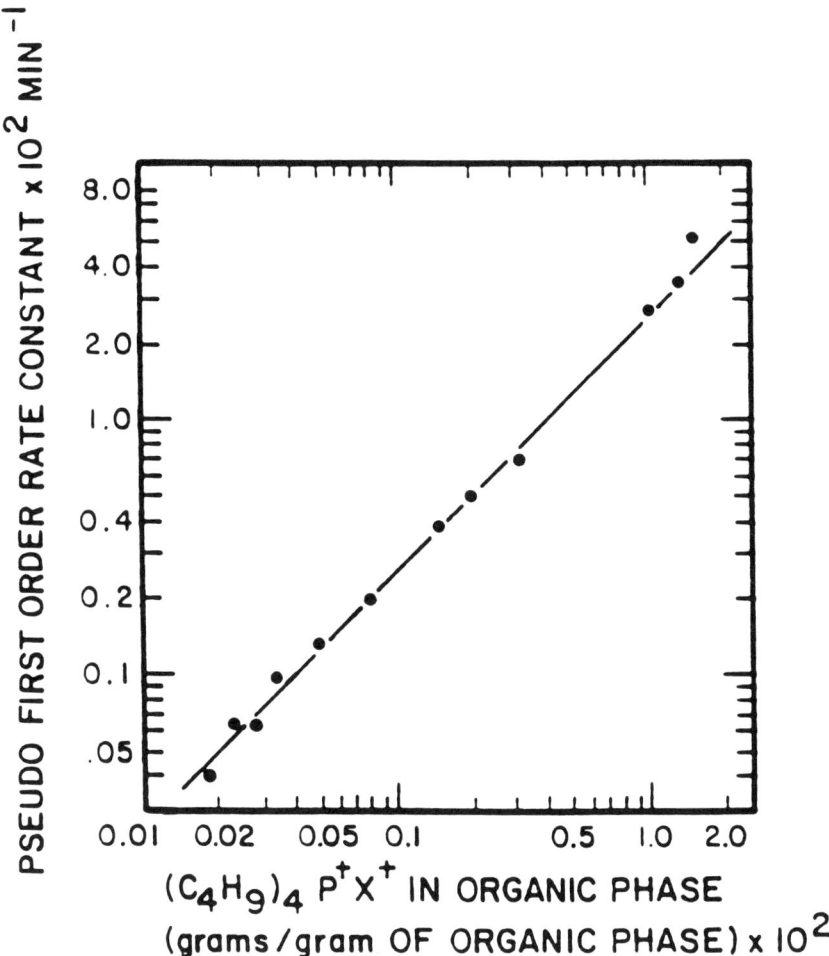

Figure 3-19. Relation between cyanide displacement rate on 1-bromooctane and the concentration of tetra-*n*-butylphosphonium halide present in the organic phase.

high reaction rates and postulated that the reactivity of the anionic component depends on cation–anion interaction energy [33].

The rates of reaction of the tetra-*n*-butylammonium and potassium salts of phenoxide ion with 1-chloro- and 1-bromobutane in pure solvents and solvent mixtures varying in dielectric constant from 2.2 to 39 have been reported [33]. The results are summarized in Table 3-4. The rates of reaction of potassium phenoxide vary over three orders of magnitude with the changes in the dielectric constant of the solvent, whereas the corresponding rates with the tetra-*n*-butylammonium salt vary by approximately a factor of 6. In addition, in all cases except for the reactions conducted in *N,N*-dimethylformamide, the rates of the quaternary salt are much higher than those of the potassium salt. Conductance measurements on the two salts indicate that both exist as ion pairs in dioxane and as free ions in acetonitrile. It was postulated that the higher reactivity of the quaternary salt resulted from reduced cation–anion electrostatic interaction energy compared to the potassium salt. This was primarily attributed to the greater distance of separation of the cationic and anionic centers of the quaternary salt compared to the corresponding distance of the potassium salt. This may be easily understood by invoking Coulomb's law as described by the following equation:

$$E_{\pm} = \frac{e^2 N}{\epsilon r} = \frac{33.18}{\epsilon r} \tag{3-36}$$

where E_r is the cation–anion interaction energy in kcal/mol, e is the electronic charge, ϵ is the dielectric constant of the medium separating the charges, r is the effective distance separating the cationic and anionic charge centers expressed in Angstroms, and N is Avogadro's number [2]. The interaction energies between the alkali metal bromide and quaternary ammonium bromide ion pairs have been calculated for dielectric constants varying from 1 to 39 and are summarized in Table 3-5 [2]. In addition, the ionic radii of the alkali metal cations [34] and the "effective" ionic radii of the quaternary ammonium cations [35] are also tabulated. Table 3-5 shows the following:

1. As the distance between the charged centers increases the electrostatic interaction energy decreases, approaching some limiting value.
2. As the dielectric constant of the medium increases not only does the electrostatic interaction energy decrease but the differences in energy between the various cations decrease.

If it is assumed that a decrease in cation–anion interaction energy is accompanied by a corresponding decrease in the free energy of activation for a displacement reaction, then the results of the rate experiments dealing with phenoxide ion as the nucleophile become easy to rationalize. Thus, since the distance separating the negative charge on the oxygen atom of phenoxide from the positive charge on the nitrogen atom of the quaternary ammonium cation compared to the corresponding separation in the potassium salt, the electrostatic model suggests that the former should be more reactive than the latter.

Studies related to the structures of ion pairs in solution using interionic nuclear Overhauser effects have been reported for tetra-*n*-butylammonium borohydride in deuterochloroform solution [36]. The results indicated that a close equilibrium association exists between the borohydride anion and the quaternary nitrogen. It was postulated that "each of three BH_4^- hydrogens are in van der Waals contact with two α-CH_2 protons and one β-CH_2 proton in a pyramidal site created by

Table 3-4. Comparison of potassium and tetrabutylammonium phenoxide rates of reaction with halobutanes.[a]

Solvent	Dielectric constant of solvent	Rate constant (L mol-sec^{-1} × 10^5) 1-Chlorobutane K$^+$ $^-$OC$_6$H$_5$	Rate constant (L mol-sec^{-1} × 10^5) 1-Chlorobutane Bu$_4$N$^+$ $^-$OC$_6$H$_5$	Rate constant (L mol-sec^{-1} × 10^5) 1-Bromobutane K$^+$ $^-$OC$_6$H$_5$	Rate constant (L mol-sec^{-1} × 10^5) 1-Bromobutane Bu$_4$N$^+$ $^-$OC$_6$H$_5$
Dioxane	2.2	—	—	0.01	330
10% Acetonitrile 90% Dioxane	6	0.0025	2.8	0.22	400
50% Acetonitrile 90% Dioxane	20	0.084	4.0	12	600
Acetonitrile	39	0.33	2.2	40	300
Tetrahydrofuran	—	0.0023	4.9	0.16	750
Dimethylformamide	38	12	17	—	—

[a] Data from Uglestad et al.; temperature, 25°C; phenoxide concentration, 0.2 M; halobutane concentration, 0.05 or 0.1 M.

three n-butyl side chains and the quaternary nitrogen." In addition, "the α-CH$_2$ and β-CH$_2$ groups of each n-butyl group are staggered (anti) conformation, and a C_3 symmetry axis through the nitrogen and the C-α carbon of the fourth n-butyl side chain is present." In another study, the solution structure of 8α-9(R)-hydroxy-1-(phenylmethyl) cinchonanium borohydride ion pair in deuterochloroform was investigated [37]. It was concluded that the NOEs "cannot result from a single ion binding site, but arise from the approach of the BH$_4^-$ to the cation

Table 3-5. Comparison of calculated cation–anion interaction energies for various bromide salts[a]

Cation	Cation radius (Å)	Interaction energy (kcal/mol) $\epsilon = 1.0$	Interaction energy (kcal/mol) $\epsilon = 2.2$	Interaction energy (kcal/mol) $\epsilon = 39$
Li$^+$	0.60	12.8	5.8	0.33
Na$^+$	0.90	11.4	5.2	0.30
K$^+$	1.33	9.9	4.5	0.26
Rb$^+$	1.48	9.5	4.3	0.25
Cs$^+$	1.69	9.0	4.1	0.23
(CH$_3$)$_4$N$^+$	2.85	6.8	3.1	0.18
(C$_2$H$_5$)$_4$N$^+$	3.48	6.2	2.8	0.16
(n-C$_3$H$_7$)$_4$N$^+$	3.98	5.5	2.5	0.14
(n-C$_4$H$_9$)$_4$N$^+$	4.37	5.3	2.4	0.13

[a] Anion radius for Br = 1.95.

at two of the four trigonal–pyramidal sites provided by the quaternary nitrogen" [37]. It was not clear from the experimental results if the two sites were occupied simultaneously as part of an aggregate, or if the sites are occupied only nonsimultaneously from time averaging of different solution structures.

Under the conditions of most PTC reactions kinetic studies, conductance measurements, and aggregation studies suggest that free anions are probably not the dominant reactant species. The data appear to be consistent with the reactive species being the ion pair. The nuclear Overhauser experiments suggest that the anion may be more closely associated with the quaternary nitrogen than previously thought. This suggests that the origins of relative reactivity of quaternary cation–anion pairs may be more complex than the simple electrostatic interaction model based upon Coulomb's law.

h. Ion Pair–Ion Aggregate Equilibria

As discussed previously, if the quaternary salt is present as the monomeric ion pair and the monomeric ion pair is the reactive species then the kinetic order with respect to the salt would be unity [Eq. (3-7)]. If, however, a higher aggregate is the dominant specie present but a small equilibrium concentration of monomeric ion pair is the reactive specie, then the kinetic order with respect to the quaternary salt could be different from unity. Consider the example of a quadrapole ion aggregate in equilibrium with the reactive monomeric ion pair [Eq. (3-27)].

$$[Q^+X^-]_2 \rightleftarrows 2\,[Q^+X^-] \quad (3\text{-}37)$$

where

$$K = \frac{[Q^+X^-]^2}{[Q^+X^-]_2} \quad (3\text{-}38)$$

When K is a very small number the stoichiometric quaternary salt (Q_{org}) is approximately equal to twice the concentration of the quadrapole specie ($Q_{org} = 2\,[Q^+X^-]_2$). It follows that

$$K\!\left(\frac{Q_{org}}{2}\right) = [Q^+X^-]^2 \quad (3\text{-}39)$$

and

$$[Q^+X^-] = \left(\frac{K}{2}\right)^{1/2} Q_{org}^{1/2} \quad (3\text{-}40)$$

Table 3-6. Aggregate–monomeric ion pair equilibria and the corresponding order with respect to the quaternary salt.

Equilibrium	$[Q^+X^-]_{org}$
$[Q^+X^-]_2 \rightleftarrows 2[Q^+X^-]$	$\sqrt{\dfrac{K}{2}} Q_{org}^{1/2}$
$[Q_2X]^+ [X]^- \rightleftarrows 2[Q^+X^-]$	$\sqrt{\dfrac{K}{2}} Q_{org}^{1/2}$
$[Q_2X]^+ + [X]^- \rightleftarrows 2[Q^+X^-]$	$\sqrt{\dfrac{K}{2}} Q_{org}$
$[Q_2X]^+ + [QK_2]^- \rightleftarrows 3[Q^+X^-]$	$\sqrt[3]{\dfrac{K}{4}} Q_{org}^{2/3}$
$[Q_2X]^+ [QX_2]^- \rightleftarrows 3[Q^+X^-]$	$\sqrt[3]{\dfrac{K}{3}} Q_{org}^{1/3}$

Thus, the kinetic order with respect to the quaternary salt would be equal to 1/2.

Since the kinetic order with respect to the stoichiometric quantity of quaternary salt may vary depending on its state of aggregation it is useful to tabulate the expected order for a given state of aggregation. Table 3-6 summarizes the results for a few selected cases. It must be emphasized that the monomeric ion pair is considered the reactive specie in each of cases in Table 3-6 and that almost all of the quaternary salt is present in the aggregated form with only a minute quantity of monomeric ion pair present in equilibrium.

For situations in which both the aggregate and the monomeric ion pair are present in comparable quantities, the kinetic order with respect to the quaternary salt becomes a function of the magnitude of the equilibrium constant and the concentrations of quaternary salt. Consider, for example, the equilibrium shown in Eq. (3-37) and the corresponding expression in Eq. (3-38). If

$$Q_T = [Q^+X^-] + 2[Q^+X^-]_2 \qquad (3-41)$$

then

$$K = \dfrac{[Q^+X^-]^2}{\left[\dfrac{Q_T - [Q^+X^-]}{2}\right]} \qquad (3-42)$$

and

$$[Q^+X^-] = \dfrac{-K/2 + \sqrt{(K/2)^2 + 4KQ_T/2}}{2} \qquad (3-43)$$

The stoichiometric quantity of quaternary salt may be empirically related to the concentration of the monomeric ion pair by the following equation

$$[Q^+X^-] = Q_T^n \tag{3-44}$$

where the exponent n is related to the kinetic order with respect to Q_T. Since

$$\log[Q^+X^-] = n \log [Q_T] \tag{3-45}$$

then a plot of log $[Q^+X^-]$ vs. log $[Q_T]$ *over a small concentration range* should produce a straight line with a slope equal to *n* (the order with respect to Q_T). Calculations have been carried out for a very large number of quaternary salt concentration increments ranging from 0.1 M to 0.0001 M and equilibrium constants varying over 14 orders of magnitude. Figure 3-20 summarizes the results of these calculations. Because the equilibrium constants cover such a wide range of values, log K (instead of K) was plotted against n for four different concentrations of quaternary salt. The following conclusions may be drawn from Figure 3-20:

1. When the concentration of quaternary salt is 0.1 M the kinetic order with respect to the catalyst may range anywhere between 0.5 and 1.0 depending on the value of K. When K is very large n approaches unity, whereas when K is very small n approaches 0.5.

2. When the concentration of quaternary salt is 0.0001 M the profile is somewhat different. The value of n is equal to or very close to unity for equilibrium constants ranging from 10^{-4} to 10^4. Between K values of 10^{-4} and 10^{-10}, n may assume values between 0.62 and unity.

2. Hydroxide-Promoted Reactions of Organic Acids

PTC reactions involving hydroxide ion are among the most useful applications of this reaction technique. Nevertheless, much controversy exists as to the mechanism of this process. Two separate and distinct mechanistic pathways have been proposed to account for the experimental observations—the Starks Extraction Mechanism and the Makosza Interfacial Mechanism. In the following sections a third mechanism—the Modified Interfacial Mechanism—is proposed and discussed. In the following sections, each of these mechanisms is discussed along with each of the associated differential rate expressions. A comparison between mechanisms is then presented.

a. Extraction Mechanism: Mechanistic Descriptions

The Starks Extraction Mechanism describing the reaction of a carbon acid (R—H) with an alkylating agent (R'—X) promoted by hydroxide ion is illustrated in Figure 3-21. In this mechanism, the phase-transfer cation (Q^+) is distributed between the aqueous and organic phases. The phase-transfer cation–hydroxide

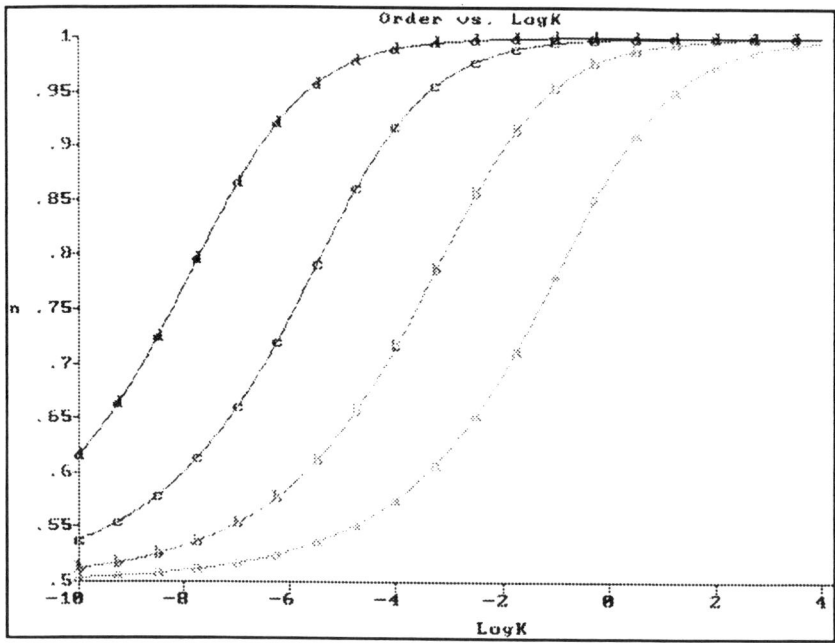

Figure 3-20. Relationship between kinetic order and equilibrium constant. (a) 0.1 M; (b) 0.01 M; (c) 0.001 M; (d) 0.001 M.

ion pair is transferred from the aqueous phase through the interfacial region into the organic phase where it is available for reaction with an organic acid R−H to form the corresponding carbanion. Reaction of the resulting carbanion with the electrophilic reagent R′−Y in the organic phase produces the substitution product R−R′ and the phase-transfer salt Q^+Y^-. The leaving group Y^- is then transported across the interfacial region into the aqueous phase, where it reequilibrates when OH^- to prepare more base to enter the organic phase. Fundamental to the extraction mechanism is that the phase-transfer salt must have some solubility in the organic phase in order to distribute hydroxide ion and leaving group anions between phases.

The Brandstrom–Montanari modification is illustrated in Figure 3-22. In this case, the phase-transfer catalyst is located exclusively in the organic phase. Exchange of X^- for OH^- takes place at the interfacial region followed by transfer of OH^- into the organic phase, where reaction with the carbon acid and then the alkylating agent takes place. Hydroxide-promoted counterparts of the mechanistic description illustrated in Figure 3-3 are also possible. Since no experimental evidence is presently available to illustrate this mechanism, it will not be the subject of further discussion.

Organic Phase

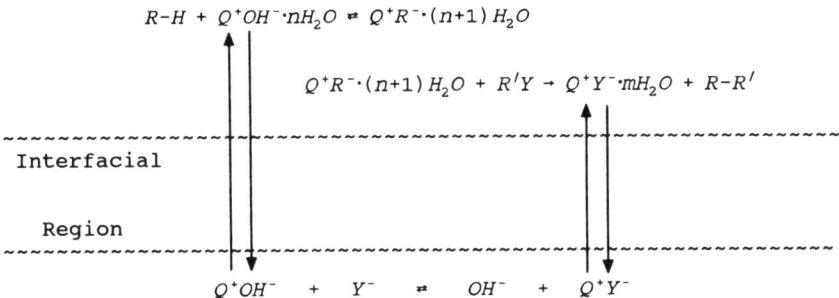

Aqueous Phase

Figure 3-21. PTC/OH reactions: Starks Extraction Mechanism.

Organic Phase

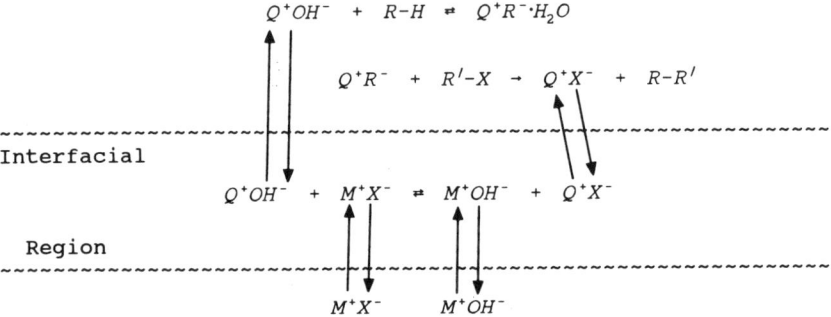

Aqueous Phase

Figure 3-22. PTC/OH extraction mechanism: Brandstrom–Montanari modification.

b. Extraction Mechanism: Kinetic Descriptions

The Starks Extraction Mechanism and the Brandstrom–Montanari modification can both be described by the reaction sequence illustrated in Scheme 3-3. The rate of product formation is

$$\frac{d[\text{R}-\text{R}']_{\text{org}}}{dt} = k_3 [Q^+R^-]_{\text{org}} [\text{R}-'\text{X}]_{\text{org}} \qquad (3\text{-}46)$$

$$(Q^+X^- \cdot mH_2O)_o + (OH^-)_{aq} \underset{k_{-1}}{\overset{k_1}{\rightleftarrows}} (Q^+OH^- \cdot nH_2O)_o + (X^-)_{aq}$$

$$(Q^+OH^- \cdot nH_2O)_o + (R-H)_o \underset{k_{-2}}{\overset{k_2}{\rightleftarrows}} (Q^+R^- \cdot (n+1)H_2O)_o$$

$$(Q^+R^- \cdot (n+1)H_2O)_o + (R'-X)_o \overset{k_3}{\rightarrow} (Q^+X^- \cdot mH_2O)_o + (R-R')_o$$

Scheme 3-3. PTC/OHH reactions: Starks–Brandstrom–Montanari Extraction Mechanism.

The differential rate expressions for the formation of $(Q^+R^-)_{org}$ and $(Q^+OH^-)_{org}$ are

$$\frac{d[Q^+R^-]_{org}}{dt} = k_2[Q^+OH^-]_{org}[R-H]_{org} \qquad (3\text{-}47)$$
$$- k_{-2}[Q^+R^-]_{org} - k_3[Q^+R^-]_{org}[R'-X]_{org}$$

$$\frac{d'[Q^+OH^-]_{org}}{dt} = k_{-2}[Q^+R^-]_{org} - k_2[Q^+OH^-]_{org} \qquad (3\text{-}48)$$
$$+ k_1[Q^+X^-]_{org}[OH^-]_{aq} - k_{-1}[Q^+OH^-]_{org}[X^-]_{aq}$$

If the steady-state approximation is made for the species $(Q^+R^- \cdot (n+1)H_2O)_{org}$ and $(Q^+OH^- \cdot nH_2O)_{org}$ then the rate of formation of product may be expressed by the following differential rate expression:

$$\frac{d[R-R']_{org}}{dt} = \frac{k_1 k_2 k_3 [Q^+X^-]_{org}[OH^-]_{aq}[R-H]_{org}[R'-X]_{org}}{k_{-1}k_{-2}[X^-]_{aq} + k_{-1}k_3[R'-X]_{org}[X^-]_{aq} + k_2 k_3[R'-X]_{org}[R-H]_{org}} \qquad (3\text{-}49)$$

c. Interfacial Mechanism: Mechanistic Descriptions

It is well known that the transfer of an anion from an aqueous phase to an organic phase is accompanied by a positive free energy change. For spherically symmetrical anions (I^-, Br^-, and Cl^-) the process becomes less favorable as the charge-to-volume ratio of the anion increases [34,38]. As a consequence, the corresponding transfer of OH^- is expected to be energetically more difficult. Nevertheless, the use of organic-like cations (quaternary ions) has allowed small but measurable transfer of OH^- into a variety of organic media. In most hydroxide-promoted reactions of organic acids, OH^- is in competition with other anions (usually halides) for the quaternary cation. Extraction experiments reveal that in a phenylacetonitrile/50% NaOH two-phase system, more than 99% of the benzyltriethylammonium ion is associated with chloride. It is natural to assume that when employing catalytic quantities of quaternary salts the accompanying decrease in organic-phase hydroxide ion might result in extremely slow reactions. This is not observed. These factors have prompted Makosza to suggest the Interfacial Mechanism.

Makosza claims to have demonstrated the importance of interfacial phenomena by successfully conducting the alkylation of phenylacetonitrile with n-butyl iodide and 50% aqueous sodium hydroxide in the absence of a phase-transfer catalyst [15]. A 70% yield of alkylated product was achieved in 5 h at 80°C. Using colorimetry and flame photometry, respectively, Makosza reports that the solubility of phenylacetonitrile in the aqueous phase was below 2 ppm and the solubility of the sodium salt of phenylacetonitrile in the organic phase was approximately 5 ppm. His conclusion is that the only region of the system where reaction could take place is at the interface. Makosza also reports that n-propyl bromide does not react under identical conditions and that only minute yields of alkylated product are obtained with benzyl chloride, but in the presence of n-butyl iodide reaction does take place with both of these electrophiles. It was suggested that n-propyl iodide and benzyl iodide are formed in situ during the course of reaction. In the presence of added sodium iodide and in the absence of an alkyl iodide no increase in yield is observed with benzyl chloride. In order to rationalize the experimental facts Makosza postulates a mechanism in which carbanion formation, alkylation, and iodide displacement all take place at the aqueous–organic interface. An important detail of the suggested mechanism, and one that was not addressed by Makosza, is that the rate of displacement of iodide ion on n-propyl bromide and benzyl chloride must be faster than the rate of diffusion of iodide ion from the organic side of the interface to the aqueous side.

Several other pertinent two-phase reactions have been carried out in the absence of a phase-transfer catalyst and have been postulated to take place at the interfacial region of the reaction system. A kinetic study of the di-n-butyl sulfide oxidation by sodium hypochlorite to di-n-butyl sulfoxide in a methylene chloride–water system has been reported [39]. The oxidation was found to take place at very slow rates in the absence of a phase-transfer catalyst. Since (1) the organic reactant had a negligible solubility in the aqueous phase and the hypochloride had essentially no solubility in the organic phase, and (2) as the stirring speed decreased the rate of reaction also decreased, it was concluded that the reaction took place at the interfacial region of the system.

Hexachlorophosphazene has been reported to react very slowly with 2,2,2-trifluoroethanol in a chlorobenzene–aqueous sodium hydroxide two-phase system in the absence of a phase-transfer catalyst [25–27]. Since sodium 2,2,2-trifluoroethoxide is not soluble in chlorobenzene the process probably proceeds at the interfacial region of the system.

The kinetics of benzylation of isobutyraldehyde in the presence of tetra-n-

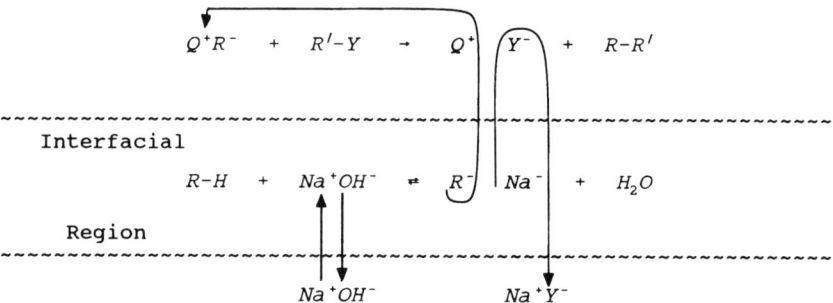

Figure 3-23. PTC/OH reactions: The Makosza Interfacial Mechanism.

butylammonium iodide in a toluene–aqueous sodium hydroxide two-phase system has been reported [22]. The Makosza Interfacial Mechanism was employed to rationalize the experimental results.

The Makosza Interfacial Mechanism is illustrated in Figure 3-23 [3,5,6,8,13–15]. The first step involves the reaction of OH$^-$ with the organic acid at the interfacial region to produce the corresponding solvated carbanion within the interfacial region of the system. The second step involves the transfer of the carbanion from the interfacial region into the bulk organic phase by coordination with the phase-transfer cation. The final step is the alkylation reaction within the organic phase to produce product. It is important to distinguish between the Extraction Mechanism and the Interfacial Mechanism. In the Interfacial Mechanism deprotonation of the organic acid takes place at the aqueous–organic interface—a process *not* assisted by the phase-transfer cation. The role of the phase-transfer cation is to complex with the already formed carbanion to carry it into the organic phase for subsequent reaction with the organic electrophile. In the Extraction Mechanism the phase-transfer cation carries the hydroxide ion into the organic phase where it subsequently reacts with the organic acid.

d. Interfacial Mechanism: Kinetic Descriptions

The Makosza Interfacial Mechanism can be described by the reaction sequence illustrated in Scheme 3-4. The rate of product formation is

$$\frac{d[\text{R}-\text{R}']_{\text{org}}}{dt} = k_3 [\text{Q}^+\text{R}^-]_{\text{org}} [\text{R}'-\text{X}]_{\text{org}} \qquad (3\text{-}50)$$

$$(OH^-)_{aq} + (R-H)_o \underset{k_{-1}}{\overset{k_1}{\rightleftarrows}} (R^-\cdot H_2O)_i$$

$$(R^-\cdot H_2O)_i + (Q^+X^-)_o \underset{k_{-2}}{\overset{k_2}{\rightleftarrows}} (Q^+R^-\cdot H_2O)_o + (X^-)_{aq}$$

$$(Q^+R^-\cdot H_2O)_o + (R-X)_o \overset{k_3}{\rightarrow} (R-R')_o + (Q^+X^-)_o$$

Scheme 3-4. PTC/OH reactions: The Makosza Interfacial Mechanism.

The differential rate expressions for the formation of $(Q^+R^-)_{org}$ and $(R^-)_i$ are

$$\frac{d[Q^+R^-]_{org}}{dt} = k_2[R^-]_i[Q^+X^-]_{org} \qquad (3\text{-}51)$$
$$- K_{-2}[Q^+R^-]_{org}[X^-]_{aq} - k_3[Q^+R^-]_{org}[R'-X]_{org}$$

$$\frac{d[R^-]_i}{dt} = k_1[OH^-]_{aq}[R-H]_{org} - k_{-1}[R^-]_i \qquad (3\text{-}52)$$
$$- k_2[R^-]_i[Q^+X^-]_{org} + k_{-2}[Q^+R^-]_{org}[X^-]_{aq}$$

If the steady-state approximation is made for the species $(R^-)_i$ and $(Q^+R^-)_{org}$ then the rate of formation of product may be expressed by the following differential rate expression:

$$\frac{d[R-R']_{org}}{dt} = \qquad (3\text{-}53)$$
$$\frac{k_1 k_2 k_3 [Q^+X^-]_{org}[OH^-]_{aq}[R-H]_{org}[R'-X]_{org}}{k_{-1}k_{-2}[X^-]_{aq} + k_{-1}k_{-3}[R'-X]_{org} + k_2 k_3 [R'-X]_{org}[Q^+X^-]_{aq}}$$

e. Modified Interfacial Mechanism: Mechanistic Descriptions

The Modified Interfacial Mechanism is illustrated in Figure 3-24. In this mechanism the phase-transfer catalyst is distributed between the organic and interfacial regions of the reaction system. The first step involves the interfacial reaction of aqueous hydroxide ion with the phase-transfer cation halide salt located at the interfacial region to produce the phase-transfer cation hydroxide pair. The second step involves the reaction of the phase-transfer cation hydroxide pair located at the interfacial region with the carbon acid to produce the phase-transfer cation carbanion pair in the organic phase. The final step is the reaction of the carbanion with the alkyl halide within the organic phase to produce product. In this mechanism, as in the Extraction Mechanism, the phase-transfer cation is involved in the formation of the carbanion. In this respect, both the Modified

96 / Phase-Transfer Catalysis

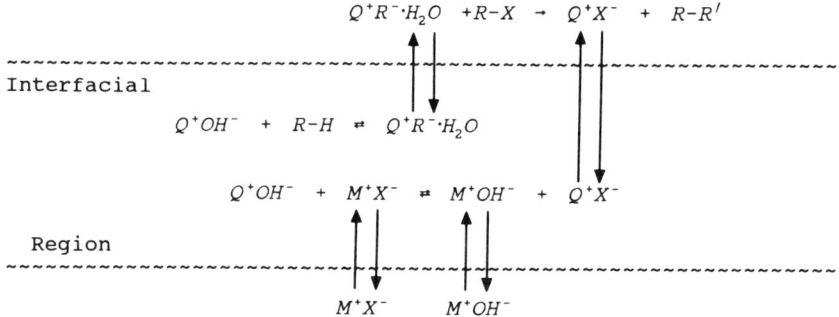

Figure 3-24. PTC/OH reactions: The Modified Interfacial Mechanism.

Interfacial Mechanism and the Extraction Mechanism are clearly different from the Interfacial Mechanism.

f. Modified Interfacial Mechanism: Kinetic Descriptions

The Modified Interfacial Mechanism can be described by the reaction sequence illustrated in Scheme 3-5. The rate of product formation is

$$\frac{d[R-R']_{org}}{dt} = k_3 [Q^+R^-]_{org} [R'-X]_{org} \tag{3-54}$$

The differential rate expression for the formation of $(Q^+R^-)_{org}$ and $(Q^-OH^-)_i$ are

$$\frac{d[Q^+R^-]_{org}}{dt} = \tag{3-55}$$
$$k_2 [Q^+OH^-]_i [R-H]_{org} - k_{-2} [Q^+R^-]_{org} - k_3 [Q^+R^-]_{org} [R'-X]_{org}$$

$$\frac{d[Q^+OH^-]_i}{dt} = \tag{3-56}$$
$$k_1 [OH^-]_{aq} [Q^+X^-]_{org} - k_{-1} [Q^+OH^-]_i [X^-]_{aq} - k_2 [Q^+OH^-]_i [R-H]_{org} + k_{-2}(Q$$

If the steady-state approximation is made for the species $(Q^+OH^-)_i$ and $(Q^+R^-)_{org}$ then the rate of formation of product may be expressed by the following differential rate expression:

$$\frac{d[R-R']_{org}}{dt} = \tag{3-57}$$
$$\frac{k_1 k_2 k_3 [Q^+X^-]_{org} [OH^-]_{aq} [R-H]_{org} [R'-X]_{org}}{k_{-1} k_{-2} [X^-]_{aq} + k_{-1} k_3 [R'-X]_{org} [X^-]_{aq} + k_2 k_3 [R'-X]_{org} [R-H]_{org}}$$

$$(Q^+X^- \cdot mH_2O)_o + (OH^-)_{aq} \underset{k_{-1}}{\overset{k_1}{\rightleftharpoons}} (Q^+OH^- \cdot nH_2O)_i + (X^-)_{aq}$$

$$(Q^+OH^- \cdot nH_2O)_i + (R-H)_o \underset{k_{-2}}{\overset{k_2}{\rightleftharpoons}} (Q^+R^- \cdot (n+1)H_2O)_o$$

$$(Q^+R^- \cdot (n+1)H_2O)_o + (R'-Y)_o \overset{k_3}{\rightarrow} (Q^+Y^- \cdot mH_2O)_o + (R-R')_o$$

Scheme 3-5. PTC/OH reactions: Modified Interfacial Mechanism.

9. Kinetic Criteria to Distinguish Between Mechanisms

A summary of characteristic kinetic criteria to distinguish between the operation of the Extraction and Interfacial Mechanisms has been suggested [40]. The Extraction Mechanism is characterized by (1) increased rates with increased organophilicity of Q^+; (2) reaction rates that are independent of stirring speed above a certain value; (3) first-order or fractional dependence of reaction rate on catalyst concentration; and (4) pseudo-first- or second-order kinetics if reaction in the organic phase reaction is rate controlling or zero-order kinetics if diffusion across the interface is rate controlling.

In contrast, the Interfacial Mechanism is characterized by (1) increased rates with increased hydrophilicity of Q^+; (2) reaction rates that are dependent on stirring speed; (3) fractional kinetic order with respect to the catalyst concentration; and (4) in general, "complex kinetics."

Although these characteristics of the reaction kinetics appear to present well-defined packages, it is not clear that they represent operational orthogonal sets. In the following sections each of the above characteristics is explored and critically evaluated.

Hydrophilicity–Organophilicity of the Quaternary Cation

In PTC reactions the success of hydroxide-promoted reactions of organic acids depends to a large extent on the interaction of the organic acid with the hydroxide ion to form the intermediate carbanion. Usually, the organic acid, as well as the alkylating agent, are located in the organic phase. In order for the deprotonation step in the reaction to take place the hydroxide ion must be brought proximate to the organic acid. In principle, this can be accomplished by hydroxide ion located at the interfacial region or within the organic phase. The Interfacial Mechanism does *not* invoke the phase-transfer cation in the deprotonation step whereas the Extraction Mechanism intimately involves the phase-transfer cation in this step. In the alkylation step both mechanisms invoke the phase-transfer cation–carbanion pair. As a consequence, *it is important to have a knowledge of the location of the catalyst in the multiphase reaction system in order to understand its role in the mechanism of reaction.*

In principle, a phase-transfer salt (ion pair or ion aggregate) can be distributed between the aqueous phase, the interfacial region, and the organic phase.

$$[Q^+X^-]_{aq} \underset{}{\overset{K^i_{aq}}{\rightleftarrows}} [Q^+X^-]_i \underset{}{\overset{K^i_{org}}{\rightleftarrows}} [Q^+X^-]_{org} \qquad (3\text{-}58)$$

For a given anion and a given organic phase polarity, the magnitude of the distribution constants depends on the organic structure associated with the quaternary cation. For symmetrical quaternary cations, as the length of the alkyl chains increases, the equilibria shown in Eq. (3-58) shift to the right; more and more of the quaternary salt will reside in the organic phase [5,8,41–44].

The PTC ethylation of dexoybenzoin in methylene chloride, benzene, and *p*-xylene using symmetrical quaternary salts has been reported [45].

$$\text{PhCH}_2\text{CHPh} + \text{EtBr} \xrightarrow{50\%\text{NaOH},5\%\text{QBr}} \text{PhCH(Et)COPh} + \text{PhCH}=\text{C(OEt)Ph} \qquad (3\text{-}59)$$

In all cases the reaction rate passed through a maximum as the size of the catalyst increased. The catalysts were also found to reduce the interfacial tension between the phases, this effect also passing through a maximum as a function of the catalyst's alkyl chain length. A correlation was found between the surface pressures and the rates of reaction. It was observed that those catalysts reducing interfacial tension most markedly were also the best catalysts. It was suggested that interfacial tension can be a guide to catalytic activity in certain PTC reactions. It was suggested that the Makosza Interfacial Mechanism best accounts for the observations. If, indeed, the most effective catalysis is observed when the quaternary cation is located at the interfacial region of the system, then it follows that the hydroxide ion located within the same region of the reaction system should be paired with the quaternary cation and it is this ion pair that is responsible for the deprotonation of the organic acid, a conclusion consistent with the Modified Interfacial Mechanism.

It is well known that when electrolyte is added to an aqueous solution of surfactant in large quantities a phenomenon called "coascervation" sometimes occurs. This is the term used to describe the formation of an additional phase that is rich in surfactant [46,47]. It has been reported that tetra-*n*-butylammonium bromide forms a third phase when used as a catalyst in the dehydrobromination of β-phenylethyl bromide to styrene with 40% sodium hydroxide [48].

$$(\text{PhCH}_2\text{CH}_2\text{Br})_{org} + (\text{NaOH})_{aq} \xrightarrow{\text{QBr,toluene},60°\text{C}} (\text{PhCH}=\text{CH}_2)_{org} + (\text{NaBr})_{aq} + \text{H}_2\text{O} \qquad (3\text{-}60)$$

These results correlate well with the reported maximum interfacial tension measurements for tetra-n-butylammonium bromide in xylene measured under more dilute conditions [45]. The results clearly indicate that the concentration of quaternary salt at the interfacial region is high. Again, if the quaternary salt is primarily localized at the interfacial region and if the hydroxide ion deprotonates the organic acid at the interfacial region, it follows that the quaternary salt must be the primary countercation associated with hydroxide. The Modified Interfacial Mechanism best describes these results.

In the conversion of benzyl chloride to benzyl bromide using tetra-n-butylammonium bromide as the catalyst, a third liquid phase was also observed [49]. More rapid reaction rates were obtained in the presence of this additional phase.

The kinetics associated with the base-catalyzed isomerization of p-allylanisole in the presence of a variety of polyethylene glycols (PEGs) has been reported [50]. The reaction followed first-order kinetics in p-allylanisole and the reaction system was characterized by three phases consisting of an organic solvent phase, an aqueous base phase, and a complex liquid phase consisting of PEG and potassium hydroxide. It was suggested that the isomerization reaction took place in the complex third phase.

Studies have been reported concerning the rate of deuterium exchange on fluorene in benzene using 16.1 M NaOD/D_2O in the presence of triethyl-n-butylammonium chloride or bromide [51]. No exchange took place in the absence of the quaternary salt. The Makosza Interfacial Mechanism was invoked to explain the results. It was concluded that deprotonation takes place at the interfacial region of the reaction system and that this process is independent of the quaternary cation. It was also concluded that the role of the catalyst was to shift the equilibrium in the first step to the right by removing the fluorenyl anion into the organic phase, thereby increasing the rate of deuteration. If the Interfacial Mechanism were important, isotopic exchange should have taken place in the absence of the catalyst since the mechanism dictates that a steady-state concentration of fluorenyl anion must be established prior to ion pairing with the quaternary ion. It would be a strange chemistry, indeed, if once this anion is formed it could be reprotonated only by the proton of the HOD molecule formed in the deprotonation step and not the deuterium. This would require an unbelievably high k_H/k_D and a rigidly oriented water molecule with respect to the carbanionic center. Since it is well known that the OD$^-$ ion has approximately three to four D_2O molecules of solvation, exclusive reprotonation without accompanying deuteration would be highly unlikely statistically. The experimental results of these deuterium exchange studies demand the inclusion of the quaternary ion in the mechanistic description of the deprotonation step. This investigation, along with the studies of the correlation of interfacial tension with reaction rate, provide evidence for the operation of the Modified Interfacial Mechanism.

The initial rates of n-butylation of phenylacetonitrile and α, α-d_2-phenylacetonitrile using 50% aqueous sodium hydroxide in the presence of benzyltriethylammonium bromide have been reported [51]. The initial rates for the protio- and deutero-phenylacetonitrile were identical—2.9 × 10^{-2} M/min. Although these results provided an enigmatic basis for a variety of conclusions concerning the operation of the Makosza Interfacial Mechanism, it was later determined that the rate of exchange of the deuteriums for protiums was faster than the rate of alkylation, therefore negating any discussion concerning isotope effects in the alkylation process. Nevertheless, the use of deuterium exchange rates should provide useful and sensitive probes for determining the role of the quaternary cation in the deprotonation of the organic acid.

Studies related to the rates of isotope exchange have been conducted on chloroform-d and phenylacetonitrile with varying concentrations of aqueous sodium hydroxide in the presence and absence of tetra-n-butylammonium bromide [52]. The results are summarized in Tables 3-7 and 3-8. The following experimental observations are pertinent:

1. Deuterium exchange takes place in the absence of tetra-n-butylammonium bromide. At a constant stirring speed (1860 ± 10 rpm) with a mechanical stirrer half-life increases with increasing concentrations of aqueous NaOH. In experiments involving chloroform-d a comparison of different stirring rates and devices was made. When employing a 9.1% aqueous NaOH (w/w) at constant stirring speed (1560 ± 10 rpm) with a magnetic stirrer the half-life is substantially greater than the corresponding half-life with 10% aqueous NaOH (w/w) using a mechanical stirrer.

2. At all concentrations of aqueous NaOH studied deuterium exchange takes place faster in the presence of tetra-n-butylammonium bromide than in its absence. At constant stirring speed with the mechanical stirrer (1860 ± 10 rpm) the half-life of the quaternary ammonium catalyzed reaction decreases sharply with increasing concentrations of aqueous NaOH. When a magnetic stirrer was employed for exchange experiments on chloroform-d using 9.1% (w/w) aqueous sodium hydroxide the half-life was substantially greater than the corresponding half-life using 10% aqueous NaOH (w/w) with a mechanical stirrer.

3. For both the catalyzed and the uncatalyzed reactions, the half-life decreases with simultaneously increasing stirring speeds and changing from the more efficient paddle of the mechanical stirrer to the less efficient magnetic stirring bar. For both catalyzed and uncatalyzed processes, the half-life approaches infinity when the stirring is stopped.

4. A cyclohexane organic phase containing tetra-n-butylammonium bromide was equilibrated with aqueous NaOH solutions of varying concentrations, separated from the aqueous phase, and $DCCl_3$ added to the organic phase. No deuterium exchange could be detected over a period of 48 h. In fact, organic phase samples taken from each of the kinetic experiments were analyzed over a period of days with no detectable increase in deuterium exchange.

Table 3-7. Reaction of $DCCl_3$ in cyclohexene–Cyclohexane[a] with aqueous $NaOH$[b] at $23 \pm 1°$ C in the absence and presence of tetra-n-butylammonium bromide (0.05 mol%).

Aq. NaOH, % (w/w)	With quat. salt $t_{1/2}$ (min)	Without quat. salt $t_{1/2}$ (min)
9.1[c]	58.8 ± 3.3[e]	78.0 ± 2.0[f]
10.0[d]	13.5 ± 0.6[e]	30.0 ± 0.1[g]
15.0[d]	7.0 ± 0.1[f]	44.5 ± 2.3[f]
20.0[d]	<0.33[g]	74.0 ± 1.0[g]
25.0[d]	<0.25[g]	88.3 ± 2.8[g]
50.0[d]	<0.25[g]	—[h]

[a] Volume of organic phase = 50 mL (3.01 g $DCCl_3$, 2.05 g cyclohexene, and 45.5 mL cyclohexane).

[b] Volume of aqueous phase = 25 mL.

[c] Magnetic stirrer at 1540 rpm.

[d] Mechanical stirrer at 1860 rpm.

[e] Average of four experiments.

[f] Average of three experiments.

[g] Average of two experiments.

[h] The silicone grease used to lubricate the reaction vessel joints reacts with the concentrated aqueous base, resulting in poor reproducibility; $t_{1/2}$ appeared to be greater than 113 min.

In the absence of quaternary salt, the experimental results cited above appear to be consistent with the operation of some sort of interfacial process. According to the Interfacial Mechanism carbanion formation is independent of the quaternary salt. The data presented above clearly show that the presence of the quaternary salt substantially enhances the exchange rate. Therefore, it is reasonable to assume that the quaternary salt is actively involved in carbanion formation. According to the Extraction Mechanism the quaternary salt is involved in carbanion formation but this process takes place in the organic phase after the hydroxide ion is extracted into this phase by the quaternary ammonium ion. Since it is observed that the separated organic phase, after equilibration with the aqueous

Table 3-8. Reaction of phenylacetonitrile with aqueous $NaOH$ at $23 \pm 1°C$ in the absence and presence of tetra-n-butyl-ammonium bromide (0.50 mol%).

Aq. NaOH, % (w/w)	With quat. salt $t_{1/2}$[b]	Without quat. salt $t_{1/2}$[b]
10.0[a]	0.5	3
20.0[a]	<0.25	10
50.0[a]	<0.25	33

[a] 2.93 g Phenylacetonitrile in 47 mL cyclohexane; 25 mL aqueous phase.

[b] Average of two experiments.

base phase in the presence of quaternary salt, does not promote deuterium exchange and that the rate of exchange is dependent on the degree of contact between phases, it may be assumed that the major portion of this process does not take place in the organic phase. The experimental results indicate that neither the Interfacial Mechanism nor the Extraction Mechanism is adequate to describe the deuterium exchange process. While it is clear that deuterium exchange does take place in the absence of tetra-n-butylammonium bromide, the process is enhanced in the presence of the quaternary ammonium salt. This is especially true at the higher aqueous base concentrations. While a small percentage of the exchange process may be taking place by the Makosza Interfacial Mechanism the major protion of the process is clearly in agreement with the Modified Interfacial Mechanism.

Effect of Stirring and Agitation Rates

If it is assumed that (1) the rates of mass transfer of reactant and product anions are rapid compared to the rate of reaction in the organic phase, and (2) $k_1[X^-]_{aq}$ and $k_{-1}[Y^-]_{aq}$ are comparable in magnitude, the differential rate expression described by Case 3 is an excellent representation of the rate process. This assumption may not be correct if the interfacial area is small. Consider the conditions of "slow" stirring or agitation. Under these circumstances, the mass transfer of the anionic species between phases could become an important component of the overall rate process; the mass transfer of the anions could become rate limiting. Figure 3-25 represents the effect of stirring rate on the pseudo-first-order rate constant for the reaction of aqueous sodium cyanide and 1-chlorooctane [11]. Below stirring speeds of 200 rpm, anion transfer is rate limiting. At stirring speeds greater than about 250 rpm the overall reaction rate remained essentially constant. This observation could mean that between stirring speeds of 200 and 250 rpm a transition exists from anion transfer limited to organic phase reaction limited kinetics. Constant overall reaction rates with stirring speeds ranging from 200 to 350 rpm have been reported for a number of reactions not involving hydroxide ion [14,17,19,20].

For PTC reaction involving hydroxide ion it has been found that the stirring speeds necessary to achieve constant rates vary widely. In the synthesis of ethers from alkyl chlorides and alcohols in the presence of phase-transfer catalyst and concentrated sodium hydroxide stirring speeds of only 80 rpm were needed [21]. If dimethyl sulfate is substituted for the alkyl chloride much more rapid stirring is required [53,54]. Stirring speeds of 750–800 rpm are required for reaction generating dihalocarbenes using haloform, phase-transfer catalyst, and 50% sodium hydroxide [65]. In the ethylation of phenylacetonitrile in the presence of tetra-n-butylammonium bromide and 50% sodium hydroxide stirring speeds from 50 to 2000 rpm were reported [55]. The rates continued to increase with increasing stirring speeds; no levelling of reaction rate was observed. This observation led

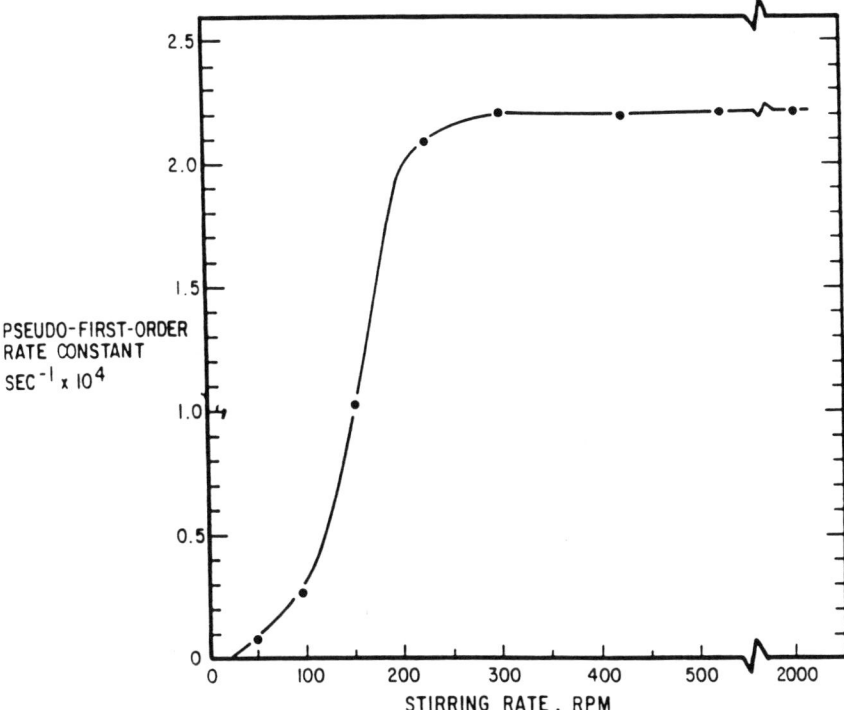

Figure 3-25. Effect of stirring rate on the PTC cyanide displacement on 1-chlorooctane.

the workers to postulate that the reaction was taking place by means of the Makosza Interfacial Mechanism. In the reaction of hexachlorophosphazene with 2,2,2-trifluoroethanol with a variety of quaternary ammonium salts in chlorobenzene–aqueous hydroxide it was reported that the stirring rate did not affect the reaction rate when the stirring rate exceeded 800 rpm [25–27].

In the benzylation of isobutyraldehyde, stirring speeds were varied from 44 to 111 rpm [22]. The rate of reaction was not affected by this change and it was concluded that diffusion resistance between phases is negligible and the process may be treated as chemical reaction controlled.

Kinetic data have been reported for the isomerization of allylbenzene to *trans*- and *cis*-β-methylstyrene with aqueous sodium hydroxide in a toluene–water system using tetra-*n*-butylammonium salts as catalysts [56]. The reaction rate increased with increasing stirring speed. A constant reaction rate was reached above a stirring speed of 300 rpm. It was also reported that an induction period was observed at the initial stages of the reaction. The length of time of the induction period depended on the counteranion associated with the quaternary cation and on the stirring speed. Above stirring speeds of 550 rpm, the induction period became constant.

The effect of stirring rate on the percent decomposition of hexadecyltri-n-butylphosphonium bromide in a chlorobenzene–50% aqueous sodium hydroxide two-phase system has been reported [57]. It was observed that the rate of decomposition of the quaternary salts became independent of stirring rate above 900 rpm.

The observation of a rate maximum with increased stirring speeds has prompted several researchers to suggest that at the maximum rate the phase-transfer process is reaction rate limited. This may not necessarily be a valid conclusion [58,90]. In experiments dealing with the alkaline hydrolysis of n-octylformate dispersed in aqueous sodium hydroxide, it has been demonstrated that the rate of mass transfer of ester from the organic phase into the aqueous phase is much faster than the rate of the hydrolysis reaction and that "the interfacial area per unit volume of dispersion increased linearly with increasing stirring speed till a stage is reached where there is no significant increase in interfacial area per unit volume of dispersion with the corresponding increase in the speed. It appears that a limiting size of droplets is attained." Thus, in this case increasing the stirring speed changes the particle size of the dispersed phase. Above certain stirring speeds, the particle size does not change appreciably. At these speeds the measured rate of reaction does not change. The constancy of the reaction rate may be observed not because the process is necessarily reaction rate limited but because the mass transfer rate has reached constant value. At this juncture it is not known how general the above observations may be. Nevertheless, it may be dangerous to use the effect of stirring speed on rates of reaction to determine which step is rate-limiting in the process without knowledge of the rates of the individual steps.

Kinetic Order with Respect to Catalyst Concentration

In a previous section the kinetic order with respect to catalyst concentration was discussed. It was shown that because the quaternary salt can exist as free ion, ion pairs, quadrapoles, triple ions, or more complex aggregates, the kinetic order with respect to the catalyst concentration could vary over a wide range of values. If the reactive form of the quaternary salt is the ion pair, the kinetic order can assume values between 0.33 and 1.0 depending on the structure of the aggregate (Table 3-6 and Figure 3-20). Eqs. (3-49) and (3-53) represent the differential rate expression for product formation for the Extraction and Interfacial Mechanism. In the former mechanism the catalyst concentration appears *only* in the numerator but in the latter mechanism the catalyst concentration appears both in the numerator and denominator. This subtle difference could, in principle, affect the observed kinetic order with respect to catalyst. This, coupled with the aggregation phenomenon, could easily result in fractional order behavior.

Studies dealing with the PTC reaction of phenylacetonitrile with ethyl bromide using 50% sodium hydroxide in the presence of tetra-n-butylammonium bromide have been reported [55]. No diluent organic solvent was used. From a bilogarithmic plot of initial rate vs. catalyst concentration a straight line with slope equal to 0.6 was obtained. Because the reaction rate was found to depend on the catalyst concentration with a fractional order it was concluded that the Interfacial Mechanism was operating. Since the first step in the alkylation process is the deprotonation of phenylacetonitrile and since it has been demonstrated from deuterium exchange studies that the quaternary cation is involved in the deprotonation step, it is reasonable to postulate that the overall alkylation process may best be described by the Modified Interfacial Mechanism. It should be noted that although deuterium exchange does take place with phenylacetonitrile in the absence of quaternary salt, the process is slower.

A kinetic order of 0.67 in quaternary ammonium iodide was reported for the benzylation of isobutyraldehyde in a toluene–aqueous sodium hydroxide two-phase system. Again, ion pair aggregates may be the origin of the observed kinetic order [22].

It was reported that in the base-catalyzed isomerization of allylbenzene to *trans-* and *cis-*β-methylstyrene in a toluene–water two-phase system, a bilogarithmic plot of first-order rate constant vs. concentration of tetra-*n*-butylammonium catalyst produced a straight line with a slope equal to unity [56]. In this case, it was suggested that the Extraction Mechanism was operating. The observation of a kinetic order of unity does not necessarily exclude the Interfacial and Modified Interfacial Mechanisms. It should be reiterated that the latter mechanism is kinetically indistinguishable from the Extraction Mechanism.

Kinetic Order of the Reaction

The kinetic studies reported for most PTC reactions involve following the disappearance of the organic reactant or the appearance of the organic product. It has been postulated that if the Extraction Mechanism is operating pseudo-first-order or second-order kinetics will be observed if the organic phase reaction is rate determining or zero-order kinetics if diffusion across the interface is rate determining. It has also been suggested that in general "complex kinetics" will be observed if the Interfacial Mechanism is operating. Analysis of the differential rate expression [Eq. 3-7] for simple displacement reactions proceeding via the Extraction Mechanism reveals that depending on the relative magnitudes of the rate constants and the initial concentration conditions, a variety of kinetic behaviors may be observed. Although it is true that pseudo-first-order, second-order, and zero-order are among these, it is also true that there are many conditions under which nonintegral order kinetics ("complex kinetics") will be observed.

The differential rate expressions for the hydroxide-promoted PTC alkylation reactions of organic acids have been derived for the Extraction Mechanism [Eq. 3-49], the Interfacial Mechanism [Eq. 3-53], and the Modified Interfacial Mechanism [Eq. 3-57]. It has been pointed out that the Starks Extraction Mechanism and the Modified Interfacial Mechanism are kinetically indistinguishable, and the Makosza Interfacial Mechanism differs from the first two only by some rather subtle features associated with the denominators of the differential equations. Indeed, in the limiting case in which $k_{-1}k_{-2}(X^-)_{aq}$ is greater than the other terms in the denominators all three mechanisms become kinetically indistinguishable.

It may be concluded that the simple criteria suggested in the literature to distinguish between various mechanisms are equivocal.

Some Additional Comments

In the study dealing with the hydroxide-promoted reaction of phenylacetonitrile and ethyl bromide using tetra-*n*-butylammonium bromide as catalyst [55] it was reported that (1) a bilogarithmic plot of initial rate vs. aqueous sodium hydroxide concentration fit a straight line with a slope of 5.3; and (2) reaction rates decreased with increasing concentrations of aqueous sodium bromide. Both observations were proported to be consistent with the operation of the Makosza Interfacial Mechanism. It was suggested that the superlinear dependence of the reaction rate on sodium hydroxide concentration is far beyond the expected trend for classical phase-transfer controlled reactions. The concentration of hydroxide was varied from 11 to 19 M. Over this range of concentrations it has been demonstrated that the hydration of transferred hydroxide may vary from an average of approximately five to three water molecules [59,60]. Accompanying this change in hydration number is a reactivity increase of greater than two orders of magnitude. Consequently, not only is the structure of the transferred hydroxide changing with change in aqueous base concentration but the corresponding reactivity is also changing. The observed slope of 5.3 is probably not indicative of the operation of a particular mechanism but a reflection of changes in hydration and reactivity of the hydroxide ion. In the kinetic study dealing with the isomerization of allylbenzene a similar bilogarithmic plot of aqueous hydroxide concentration vs. first-order rate constant produced a straight line with a slope of 5.0 [56]. An identical rationalization can be put forth to explain these results. In addition, it must be emphasized that at the high stoichiometric concentrations of hydroxide ion used in many phase-transfer reactions, it is more appropriate to use *activities* rather than concentrations. Inspection of Eqs. (3-49), (3-53), and (3-57) indicates that in all mechanistic cases increasing the concentration of aqueous bromide ion should be accompanied by a decrease in a reaction rate.

It may be concluded that neither of the above experimental observations demonstrates the operation of any one of the three mechanisms.

3. Alternative PTC Mechanisms Involving Hydroxide Ion

a. Reverse PTC

There are many preparative examples of two-phase dehydrohalogenation reactions of alkyl halides in the presence of hydroxide ion and quaternary salts to form alkenes and alkynes [61]. (2-Bromoethyl)benzene undergoes dehydrobromination with 50% aqueous sodium hydroxide with a wide variety of organophilic quaternary salts. In contrast, the dehydrobromination of dibromostilbene to diphenylacetylene requires three equivalents of tetra-n-butylammonium hydrogen sulfate. If the corresponding bromide or iodide salts are used no reaction takes place [62]. If the elimination reactions are conducted at higher temperatures, the quaternary salt alone will promote reaction; no aqueous hydroxide is necessary [63,64]. In this case gaseous HCl or HBr is produced. Two distinct mechanisms have been proposed to describe these elimination processes.

1. In those cases where the active base in the organic phase is the quaternary hydroxide the Brandstrom–Montanari modification of the Extraction Mechanism is invoked; the dehydrohalogenation is promoted by hydroxide ion. It has been shown that hydroxide ion may be extracted into organic media only in systems where highly lipophilic ammonium cations are paired with hydrophilic anions [65–68]. In the reaction of (2-bromoethyl)benzene with 50% aqueous sodium hydroxide in the presence of tetra-n-octylammonium bromide zero-order kinetics was observed along with an activation energy of 8 kcal/mol [69]. It was rationalized that the organophilic quaternary cation extracts the hydroxide into the organic phase where it reacts rapidly with the halide to form alkene product. The rate-controlling step in the reaction sequence is the mass transfer of hydroxide from the aqueous to the organic phase.

2. In those cases where the active base in the organic phase is the quaternary halide the Reverse Transfer Mechanism is invoked [70]. The reaction sequence associated with this mechanism is shown in Figure 3-26. Step 1 involves an organic phase dehydrohalogenation reaction promoted by the halide ion associated with the quaternary cation. The products of this first step are the alkene and the hydrogen halide associated with the quaternary halide by a hydrogen bond. Step 2 involves the migration of the complexed hydrogen halide to the aqueous–organic interfacial region where it rapidly becomes neutralized by hydroxide ion. A kinetic study dealing with the dehydrohalogenation of a series of substituted (1-haloethyl)benzenes with 50% aqueous sodium hydroxide in the presence of tetra-n-butylammonium chloride, bromide, iodide, or hydrogen sulfate has been reported [70]. It was found that the reactions followed first-order kinetics and that the catalytic activity of the quaternary salts decreased in the order Cl^-, Br^-, I^- in accordance with the decreasing basicity and nucleophilicity of these ions. The hydrogen sulfate salt showed an initial rapid reaction due to the extraction of the very active quaternary ammonium hydroxide. As halide formed from the elimination reaction it was preferentially extracted into the organic phase and assumed the role of base for the remainder of the reaction. Thus, when the

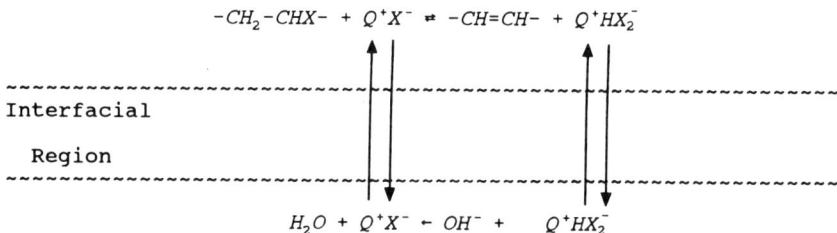

Figure 3-26. Mechanism of reverse PTC.

extraction of hydroxide ion is not possible, the Reverse Transfer Mechanism may becomes probable.

b. Inverse PTC

The literature contains several examples of two-phase reactions in which an organic soluble reagent is converted to a reactive ionic intermediate and transported into the aqueous phase where reaction takes place [71–75]. These processes have been termed Inverse Phase-Transfer Catalysis [76]. The reaction of benzoyl chloride and sodium benzoate in the presence of a catalytic quantity of pyridine-1-oxide in a methylene chloride–water two-phase system to form benzoic anhydride has been reported [77,78]. Figure 3-27 describes the mechanistic process. From competition experiments using different carboxylates, it was concluded that anhydride formation takes place in both the aqueous phase and at the interfacial region of the reaction system [77]. Evidence, based on detailed kinetic investigations, suggests that a component of the reaction may be taking place in the organic phase by the Extraction Mechanism [11].

C. Solid–Liquid PTC

Solid–liquid PTC (sl-PTC) using quaternary salts and cyclic and linear polyethers has been employed as a useful vehicle for conducting a wide variety of organic transformations [1–4,6,79–88]. While liquid–liquid PTC involves heterogeneous reaction processes between two liquid phases (usually an aqueous phase and an organic phase), solid–liquid PTC usually involves reaction of an anionic reagent in a solid phase (usually a salt) with a reactant located in a contiguous organic phase [89,90]. Although the analogy is real the two methods of carrying out PTC reactions are by no means equivalent. Several reports have appeared in which ll-PTC fails while sl-PTC is successful [91]. In particular, the generation

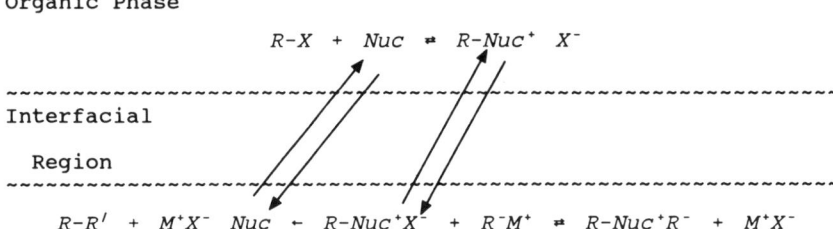

Figure 3-27. Mechanism of inverse PTC.

of dichlorocarbene from sodium trichloroacetate under solid–liquid PTC conditions is superior to the liquid–liquid counterpart. In addition, the "elimination" of waters of hydration often increases the nucleophilicity of anionic species in nonpolar, aprotic organic solvents [79–83].

The usual mechanistic descriptions of sl-PTC found in the literature are not too unlike the ll-PTC counterpart [90]. The first step involves the transport of a reactant anion from the solid phase to the organic phase by a phase-transfer cation. This could be an organophilic quaternary cation or an organophilic cation derived from the complexation of a metal cation with a multidentate ligand (crown ether, cryptand, polyethylene oxide, etc.). Several examples of macrocyclic and macrobicyclic multidentate ligands employed as phase-transfer catalysts are illustrated in Figure 3-28. The second step involves the reaction of the transferred anion with the reactant located in the organic phase. As in the discussion concerning ll-PTC, in order for this step to be successful, the reactant anion must be in a reactive form. Finally, the third step involves the transport of the product anion by the phase-transfer cation to the solid phase and the transport of another reactant anion into the organic phase. If any of the above steps does not occur in an efficient manner the catalytic process will not be successful. There are cases in which the reactant salt has a small solubility in the organic phase in the absence of the phase-transfer catalyst. Under these circumstances, the phase-transfer cation can exchange with the metal cation in the organic phase and promote anion activation [89].

Important questions have been asked regarding the mechanistic details of sl-PTC. These include the following:

1. What are the mechanisms of transport of anions from the solid phase to the organic phase?
2. What are the mechanisms of formation of reactive ion pairs?
3. What are the mechanisms of exchange of product anions located in the organic phase with reactant anions located in the solid phase?

110 / Phase-Transfer Catalysis

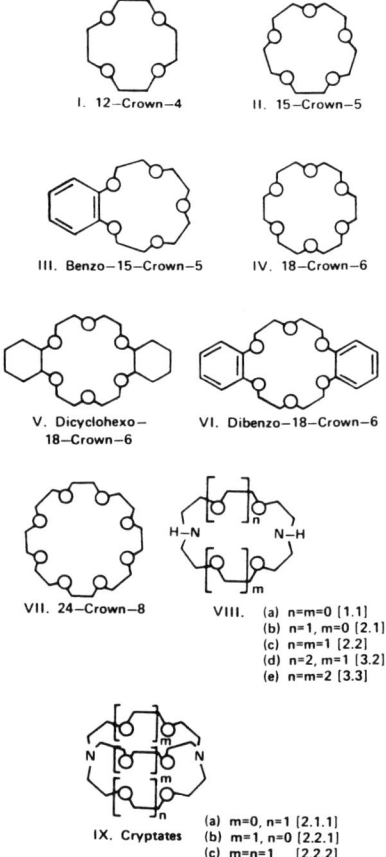

Figure 3-28. Representative macrocyclic multidentate ligands.

4. What are the effects of particle size on the rates of reaction?
5. What are the mechanistic differences between quaternary cation and crown ethers as phase-transfer catalysts?

It must be emphasized that each of the above questions is not separate and distinct from one another. Inherent in the above sl-PTC mechanistic descriptions is the implication that the process is carried out under anhydrous conditions, that both the solid and liquid phases are "dry." Indeed, this may or may not be the case since, unless strict precautions are taken, adventecious water is usually associated with inorganic as well as organic salts. Despite the fact that mechanistic models describing most sl-PTC processes do not invoke a role for water, numerous reports have appeared in the literature indicating that small amounts of water

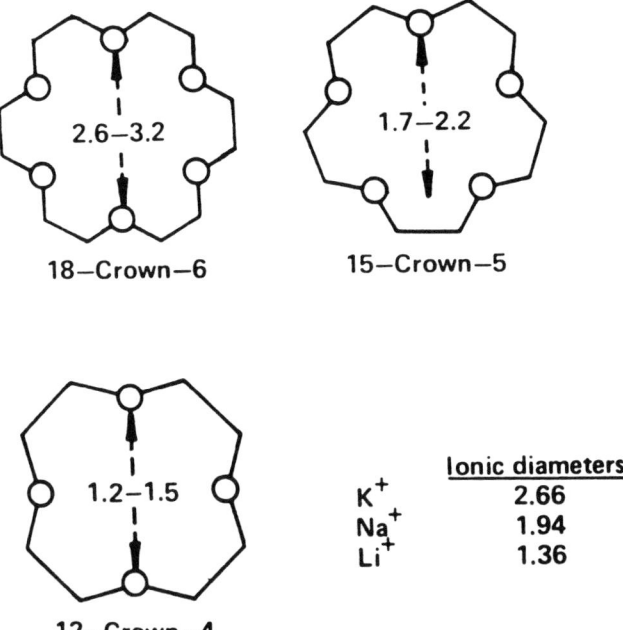

Figure 3-29. Cavity diameters of cyclic polyether and ionic diameters of metal cations.

dramatically affect the overall rate of reaction [11,18,19,92–99]. Therefore, another important question must be asked.

6. What is the mechanistic role of small quantities of water in sl-PTC?

In the following sections, the complexation and solubilization of potassium salts with 18-crown-6 in acetonitrile, the nucleophilicities of anions in acetonitrile, and studies related to the effect of water on the rates of simple displacement reaction under sl-PTC conditions are presented. A mechanistic model, consistent with the kinetics and thermodynamics of the process, are then described in an attempt to answer some of the above questions.

1. Complexation and Solubilization of Potassium Salts with 18-Crown-6

It is well known that crown ethers of different cavity diameters have different selectivities toward metal ions. Figure 3-29 shows the relationship between the size of the cavities of 12-crown-4, 15-crown-5, and 18-crown-6 and the ionic diameters of potassium, sodium, and lithium cations. Crown ethers have been shown to be useful vehicles for the complexation and solubilization of metal salts. Solubilities of a series of potassium salts in acetonitrile at 25°C in the presence and absence of 18-crown-6 were reported and are summarized in Table

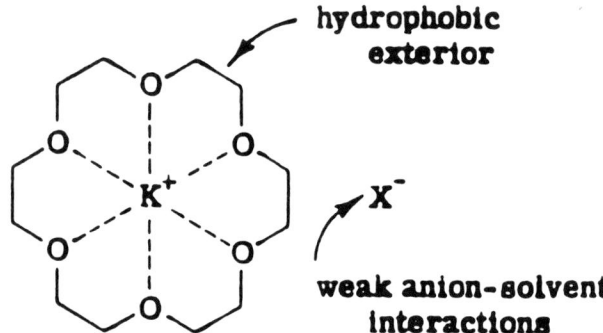

Figure 3-30. Generalized complex of 18-crown-6 with potassium salts.

3-9 [100]. It was believed that an important driving force for the solubilization of these salts was the organophilic nature of the complex which results in an increased affinity for the organic solvent. This is pictorially represented in Figure 3-30. The magnitude of the solubilization of a particular salt was conjectured to be a complex function of the crystal lattice free energy of the solid, the polarity of the organic solvent, and the organophilic nature of the macrocyclic or macrobicyclic multidentate ligand.

It has been pointed out that the successful transfer of an anion from the solid to the organic phase does not necessarily result in a successful PTC process. The anion must be in a reactive form. The second-order rate coefficients for the reaction of benzyl tosylate with a series of anionic nucleophiles dissolved in acetonitrile using 18-crown-6 were reported and are summarized in Table 3-10 [100]. It was pointed out that there appeared to be a levelling effect in nucleophilic reactivity. An anion such as fluoride, which is considered to be a poor nucleophile in protic media, was as reactive as iodide ion and more reactive than thiocyanate.

Table 3-9. Solubilities of potassium salts in acetonitrile at 25°C in the presence and absence of 18-crown-6[a].

	Solubility		
Potassium salt	In 0.15 M crown in acetonitrile	In acetonitrile	Solubility enhancement
KF	4.3×10^{-3}	3.18×10^{-4}	4×10^{-3}
KCl	5.55×10^{-2}	2.43×10^{-4}	5.52×10^{-2}
KBr	1.35×10^{-1}	2.08×10^{-3}	1.33×10^{-1}
KI	2.02×10^{-1}	1.05×10^{-1}	0.97×10^{-1}
KCN	1.29×10^{-1}	1.19×10^{-3}	1.28×10^{-1}
KOAc	1.02×10^{-1}	5.00×10^{-4}	1.02×10^{-1}
KN$_3$	1.38×10^{-1}	2.41×10^{-3}	1.36×10^{-1}
KSCN	8.50×10^{-1}	7.55×10^{-1}	0.95×10^{-1}

[a]Solubility was measured using Coleman Model-21 Flame Photometer.

Table 3-10. Second-order rate constants at 30°C in acetonitrile for the reaction with benzyl tosylate[a, b]

Nucleophile	k (Liters/Mol-sec^{-})	Relative rates	
		This study	Swain and Scott
N_3^-	1.02	10.0	20.0
CH_3COO^-	0.95	9.6	1.0
CN^-	0.23	2.4	250
Br^-	0.12	1.3	16.0
Cl^-	0.12	1.3	2.0
I^-	0.09	1.0	200.0
F^-	0.14	1.4	0.2
SCN^-	0.02	0.3	125

[a]The second-order rate coefficients of Cl^-, Br^-, and SCN^- toward n-Bu-I have previously been determined in CH_3CN, taking into account ion-pairing effects to be 4.0, 3.4, and 0.58 ($\times\ 10^{-3}$) ($M^{-1}\ sec^{-1}$), respectively, which gives approximately the same order of nucleophilicity reported here.

[b]The second-order rate coefficients of SCN^-, Br^-, N_3^-, Cl^-, and AcO^- toward CH_3-OTs have been determined in CH_3CN by spectrophotometric procedures to be 2.3, 50, 25, 230, and 2600 ($\times\ 10^{-4}$) ($M^{-1}\ sec^{-1}$), respectively. The relative nucleophilicities determined from these rate constants are considerably different from the values reported here.

The data indicated that once the anion was placed in the acetonitrile phase it became highly reactive—a condition that is necessary for a successful PTC process.

2. Simple Displacement Reactions

The kinetics associated with sl-PTC nucleophilic substitution processes involving the reaction of bromide and iodide with 1-bromooctane and the mesylate of 1-octanol in the presence of tetra-n-butylammonium bromide, 18-crown-6, and dicyclohexyl 18-crown-6 were reported [88]. From the observed "pseudo-first-order rate constants" it was concluded that 18-crown-6 was a more effective catalyst than tetra-n-butylammonium bromide. It was conjectured that the ineffectiveness of the quaternary salt in the overall reaction process was in part due to its inefficiency in transporting the nucleophilic anion from the solid to the organic phase. The kinetics associated with the sl-PTC reaction of 1-bromohexane dissolved in toluene and solid metal chloride (lithium, sodium, potassium, rubidium, and cesium) in the presence of tetra-n-butylammonium halides was reported [101–104]. The data were rationalized in terms of the formation of a ternary absorption complex between the substrate, catalyst, and the solid salt.

Early in the development of mechanism of PTC, a kinetic study dealing with the reaction of cyanide ion and 1-halooctanes suggested that the quaternary salt cannot exchange halide ions with solid sodium cyanide [2,11]. It was further reported that primary and secondary halides will not react with solid sodium salts under anhydrous conditions; trace quantities of water were essential if the substitution reactions were to be successfully catalyzed by quaternary salts [92]. It has been reported that in the displacement of chloride from 1-chlorooctane by fluoride ion catalyzed by Aliquat 336, no reaction took place in the absence of water. Indeed, 100% conversion was attained when the water/KF ratio was 1:4 [93]. In the reaction of sodium formate with a series of alkyl chlorides, conversions were observed to be higher when small amounts of water were present [94]. The nucleophilicity of anions in anhydrous systems compared to systems in which trace quantities of water were added have been reported [19,95]. The observations with regard to the necessity of water in sl-PTC are not without conflicting viewpoints. Some reports have shown that sl-PTC reactions are not sensitive to trace quantities of water [105], whereas other reports suggest that small amounts of water substantially increase the rates of reaction compared to anhydrous conditions and that the rate goes through a maximum with increasing quantities of added water [2,18,92]. This particular phenomenon will now be explored in greater detail.

The rates of reaction of benzyl bromide and benzyl chloride with potassium cyanide were studied as a function of added water in the presence and absence of crown ether in toluene at 85°C and 25°C [18]. In order to ensure tractable kinetics, the rate experiments were conducted under conditions in which the reactant and product salts (KCN and KBr or KCl, respectively) were initially present in large excess. The logic is not too unlike the special conditions involving excess salts discussed in Case 3 for simple displacement reaction using 11-PTC conditions. In addition, in order to achieve reproducible kinetic data, the salts and crown ether were stirred together in toluene for a period of approximately 1 h prior to the addition of the benzyl halide [18]. This procedure allowed the salts to be continually solubilized and precipitated until a constant particle size was achieved. Without employing these procedures reproducible data were difficult to achieve. The rate constants obtained in the presence of 18-crown-6 are summarized in Tables 3-11–3-13. **It is important to note that in the absence of added water the rates followed zero-order kinetics while in the presence of added water rates followed first-order kinetics.**

The data in Tables 3-11–3-13 clearly indicate that in the presence of crown ether addition of water caused dramatic changes in rate. The first-order rate constant sharply increased and reached a maximum with the addition of minute quantities of water. As the quantity of water was increased further, the rate constant decreased. In the absence of crown no reaction took place during the same time period. The rate constants for the reaction of potassium cyanide with benzyl chloride in the presence of the more lipophilic dicyclohexo-18-crown-6 were also investigated. The response to the addition of small quantities of water was similar to that of 18-crown-6.

In another series of experiments it was shown that 92% of the 18-crown-6 added to a system composed of toluene and salts (KCN and KCl) resided in the

Table 3-11. Rates of reaction of benzyl bromide with potassium cyanide at 85°C in the presence and absence of 18-crown-6 as a function of added water.

Water (mL)	$k \times 10^4$ sec^{-1} (crown)	$k \times 10^4$ sec^{-1} (no crown)	$k \times 10^4$ sec^{-1} (cat)
0	0.02[a]	0.0	0.02[a]
1	13.9	0.0	13.9
10	11.9	0.0	11.9
20	6.8	0.6	6.2
30	4.4	1.0	3.4
40	3.2	1.3	1.9
50	2.8	2.1	0.7

Reaction conditions: 0.05 mol benzyl bromide, 0.0025 mol 18-crown-6, 0.15 mol KBr, 0.15 mol KCN.

[a]Zero-order kinetics (M sec^{-1}).

C.L. Liotta, E.M. Burgess, C.C. Ray, E.D. Black, and B.E. Fair, in Phase Transfer Catalysis: New Chemistry, Catalyst and Applications, ed. C.M. Starks, A.C.S. Symposium Series No. 326, Am. Chem. Soc., Washington, 1987, ch. 3.

Table 3-12. Rates of reaction of benzyl chloride with potassium cyanide at 85°C in the presence and absence of 18-crown-6 as a function of added water.

Water (mL)	$k \times 10^5$ sec^{-1} (crown)	$k \times 10^5$ sec^{-1} (no crown)	$k \times 10^5$ sec^{-1} (cat)
0.0	3.2[a]	0.0	3.2[a]
0.36	9.2	0.0	9.2
0.50	9.4	0.0	9.4
1.0	11.6	0.0	11.6
2.0	14.7	0.0	14.7
10.0	10.2	0.0	10.2
15.0	6.9	0.7	6.2
20.0	5.8	1.3	4.5
30.0	5.0	1.7	3.3
40.0	3.9	1.9	2.0
50.0	4.2	2.5	1.7
75.0	4.8	3.2	1.6

Reaction conditions: 0.05 mol benzyl chloride, 0.01 mol 18-crown-6, 0.15 mol KBr, 0.15 mol KCN.

[a] Zero-order kinetics (M sec^{-1}).

C.L. Liotta, E.M. Burgess, C.C. Ray, E.D. Black, and B.E. Fair, in Phase Transfer Catalysis: New Chemistry, Catalyst and Applications, ed. C.M. Starks, A.C.S. Symposium Series No. 326, Am. Chem. Soc., Washington, 1987, ch. 3.

Table 3-13. Rates of reaction of benzyl bromide with potassium cyanide at 25°C in the presence of 18-crown-6 as a function of added water.

Water (mL)	$k \times 10^5 \text{ sec}^{-1}$
0.0	2.06[a]
0.5	1.71
1.2	2.54
1.4	2.84
1.5	9.40
1.6	5.10
1.75	5.34
2.0	5.12
2.5	3.36
3.0	3.50
3.5	3.26
4.0	2.20
6.0	1.49

Reaction conditions: 0.05 mol benzyl bromide, 0.01 mol 18-crown-6, 0.15 mol KBr, 0.15 mol KCN.

[a]Zero-order kinetics ($M \text{ sec}^{-1}$).

C.L. Liotta, E.M. Burgess, C.C. Ray, E.D. Black, and B.E. Fair, in Phase Transfer Catalysis: New Chemistry, Catalyst and Applications, ed. C.M. Starks, A.C.S. Symposium Series No. 326, Am. Chem. Soc., Washington, 1987, ch. 3.

organic phase. Upon addition of small quantities of water, however, all but approximately 1–2% of the crown was translocated onto the surface of the salt. That the crown was located on the surface of the salt was experimentally determined by filtering the salt, drying it, and extracting with methylene chloride; all the crown was recovered. The results are summarized in Table 3-14. It was conjectured that the initial water added to the system coats the surface of the salt particles and it was this aqueous salt coating that extracted the crown from the organic phase. This new region of the reaction system, *the omega phase*, appeared to be intimately involved in the catalytic reaction process since the quantity of added water that corresponds to maximum crown on the surface of the salt particles approximated the quantity of added water needed to achieve maximum catalysis. The 8% of the crown located on the surface of the salt particles prior to the addition of water was probably due to the presence of adventitious water already present on the salt. In another series of experiments it was determined that dicyclohexo-18-crown-6 could be translocated onto the surface of the salt particles to the extent of 75%. This was not surprising since this crown is substantially more lipophilic than 18-crown-6. The observation that dicyclohexo-18-crown-6 produced a similar rate profile as the 18-crown-6 with the addition of small quantities of water suggested that the displacement reaction was taking place in the organic phase.

The distribution of 18-crown-6 between the organic phase and the omega

Table 3-14. The effect of added water on the concentration of 18-crown-6 in toluene at ambient temperatures.

Water (μL)	Equiv. of water[a]	Percent crown in toluene
0	0.00	91.5
10	0.14	81.4
15	0.21	77.3
21	0.29	72.7
22	0.31	50.0
23	0.32	40.0
25	0.35	34.6
30	0.42	17.7
36	0.50	5.8
45	1.25	2.5
50	1.39	2.0
80	2.22	1.0

Conditions: 0.0040 mol 18-crown-6, 0.027 mol KCl, 0.027 mol KCN, 10 mL toluene.

[a]Moles of H_2O/moles of crown.

C.L. Liotta, E.M. Burgess, C.C. Ray, E.D. Black, and B.E. Fair, in Phase Transfer Catalysis: New Chemistry, Catalyst and Applications, ed. C.M. Starks, A.C.S. Symposium Series No. 326, Am. Chem. Soc., Washington, 1987, ch. 3.

phase was investigated as a function of added crown ether. The results are summarized in Table 3-15 [106]. The results indicated that although the millimoles of added 18-crown-6 were varied from 3.50 to 11.45 the number of millimoles of crown in the organic phase remained low and relatively constant. The omega phase acted like a sponge and adsorbed the added crown. As a companion series of experiments the kinetics of reaction of benzyl bromide with potassium cyanide was studied as a function of added 18-crown-6. The rate experiments were conducted as described above. The pseudo-first-order rate constants are summarized in Table 3-16 [106]. The results indicated that there was a slight increase in rate as the number of millimoles of 18-crown-6 was increased from 3.0 to 5.0 but the rate remained essentially constant with increases of crown up to 12.0 mmol. Since both the quantity of crown in the organic phase and the pseudo-first-order rate constants remained constant between 3.0 and 12.0 mmol of 18-crown-6 it was conjectured that the displacement reaction probably took place in the organic phase and *not* in the omega phase.

The experimental results presented above suggested that small quantities of water facilitated the distribution of crown ether between the organic and solid phases via the omega phase and that the displacement reaction took place in the organic phase. In particular, it has been demonstrated that, in the absence of added water, pseudo-zero-order kinetics were observed while in the presence of water the kinetics became pseudo-first-order. The following mechanism was proposed to account for these experimental observations:

Table 3-15. Distribution of crown between organic and omega phases as a function of added 18-crown-6.

Total millimoles of 18-crown-6	Millimoles of 18-crown-6 in organic phase
3.50	0.06
6.87	0.06
10.33	0.07
11.45	0.07

Conditions: 1.0 mL H$_2$O, 0.15 mol KBr, 0.15 mol KCN, 50 mL toluene, 25°C.

$$(K^+-C-CN^- \cdot nH_2O)_\omega + (K^+-C-X^- \cdot mH_2O)_{org} \quad (3\text{-}61)$$

$$k_{-1} \uparrow \downarrow k_1$$

$$(K^+-C-CN^- \cdot nH_2O)_{org} + (K^+-C-X^- \cdot mH_2O)_\omega \quad (3\text{-}62)$$

$$(K^+-C-CN^- \cdot nH_2O)_{org} + (PhCH_2X)_{org} \xrightarrow{k_2} \text{products}$$

where C represents the crown ether and the subscripts ω and org designate the omega phase and the organic phase, respectively. Step 1 represents the distribution of crown potassium salt complexes between the omega and the organic phases, and Step 2 represents the nucleophilic substitution reaction in the organic phase. In the absence of added water n and m are zero whereas in the presence of added water n and m are the degree of hydration of the potassium salt complexes. Assuming the steady-state approximation for $(K^+-C-CN^- \cdot nH_2O)_{org}$ the following differential expression for the rate of formation of products was derived:

$$\frac{d[(R-Y)_{org}]}{dt} = \frac{k_1 k_2 [C_T - C_F - C_\omega][K^+-C-CN^- \cdot nH_2O]_\omega [PhCH_2X]_{org}}{[K^+-C-CN^- \cdot nH_2O]_\omega [k_1 - k_{-1}] + k_{-1} C_{\omega} + k_2 [PhCH_2X]_{org}} \quad (3\text{-}63)$$

where C_T is the total concentration of crown ether (complexed and uncomplexed) in the organic phase, C_F is the concentration of uncomplexed crown in the organic

Table 3-16. 18-Crown-6 catalyzed reactions of benzyl bromide with KCN as a function of added crown.

mmoles 18-C-6	$k \times 10^5$ sec^{-1}
3.0	2.16, 2.47
5.0	3.97, 3.63
7.0	3.86, 3.99
10.0	3.75, 4.00
12.0	3.80, 3.60

Reaction conditions: 1.0 mL H$_2$O, 0.15 mol KBr, 0.15 mol KCN, 50 mL. toluene, 25°C.

phase, C_ω is the total crown in the omega phase, and n and m can have values of zero or any integer. It was experimentally observed that in the absence of added water the crown ether was primarily located in the organic phase. Under these conditions $k_2 [PhCH_2X]_{org} >> [K^+-C-CN^- \cdot nH_2O]_\omega (k_1 - k_{-1}) + k_{-1}C_\omega$ and Eq. (3-3) takes the form of a zero-order kinetic expression. Experimentally, zero-order kinetics was observed. In the presence of small quantities of added water, it was experimentally observed that the crown ether was primarily on the surface of the salt in the omega phase. Under these conditions $[K^+-C-CN^- \cdot nH_2O]_\omega (k_1 - k_{-1}) + k_{-1}C_\omega >> k_2 [PhCH_2X]_{org}$ and Eq. (3-3) takes the form of a first-order kinetic expression. Experimentally, first-order kinetics is observed.

Finally, it must be cautioned that it would be dangerous to generalize the mechanistic description of the specific reactions discussed above to all solid–liquid PTC reactions. The discussion was only meant to provide a deeper insight to the complexities of these multiphase reaction processes.

References

1. W.P. Weber and G.W. Gokel, Phase Transfer Catalysis in Organic Synthesis; Springer-Verlag: Berlin, Heidelberg, New York (1977).
2. C.M. Starks and C.L. Liotta, Phase Transfer Catalysis: Principles and Techniques; Academic Press: New York (1978).
3. E.V. Dehmlow and S.S. Dehmlow, Phase Transfer Catalysis, 2nd ed; Verlag Chemie: Weinheim (1983).
4. F. Montanari, D. Landini and F. Rolla, Top. Curr. Chem., **101**, 147 (1982).
5. A. Brandstrom, Principles of Phase Transfer Catalysis by Quaternary Ammonium Salts, in Advances in Physical Organic Chemistry, Academic Press, London and New York (1977).
6. E.V. Dehmlow, Angew, Chem., Int. Ed. Engl., **16**, 493 (1977).
7. M. Makosza, In Survey of Progress in Chemistry, A.F. Scott, ed., Academic Press, New York, 1 (1980).
8. H.H. Freedman, Pure Appl. Chem., **58**, 857 (1986).
9. C.M. Starks, J. Am. Chem. Soc., **93**, 195 (1971).
10. C.M. Starks and D.R. Napier, U.S. Patent 3,992,432 (1976); British Patent 1,227,144 (1971); French Patent 1,573,164 (1969); Australian Patent 439,286 (1968); Netherlands Patent 6,804,687, (1968).
11. C.M. Starks and R.M. Owens, J. Am. Chem. Soc., **95**, 3613 (1973).
12. D. Landini, A. Maia, and F. Montanari, J. Chem. Soc. Chem. Comm., 112 (1977).
13. M. Makosza, Pure Appl. Chem., **43**, 439 (1975).
14. M. Makosza, Russ. Chem. Rev., **46**, 1151 (1977).
15. M. Makosza and E. Bialecka, Tetrahedron Lett., **2**, 1983 (1977).
16. A.M. Tivert and K. Gustavil, Acta Pharm. Suec., **16**, 1 (1979).

17. A.W. Herriott and D.H. Picker, J. Am. Chem. Soc., **97** (1975).
18. C.L. Liotta, E.M. Burgess, C.C. Ray, E.D. Black, and B.E. Fair, ACS Symp. Ser., No. 326. 15 (1987).
19. D. Landini, A.M. Maia, F. Montanari, and F.M. Pirisi, J. Chem. Soc. Chem. Comm., 950 (1975).
20. D. Landini, A. Maia, and F. Montanari, J. Am. Chem. Soc., **100**, 2796 (1978).
21. H.H. Freedman and R.A. Dubois, Tetrahedron Lett., 3251 (1975).
22. T-C. Huang and S-C Lin, J. Chin, Ch. E., **19**, 193 (1988).
23. M-Y. Yeh, H-H. Guo, L-C. Chen, and Y-P. Shih, Ind. Eng. Chem. Res., **27**, 1582 (1988).
24. J-R. Chang, M-Y. Yeh, and Y-P. Shih, J. Chim Chem. Soc., 31, 185 (1984); J. Chin. Ch. E., **14**, 457 (1983).
25. M-L. Wang and H-S. Wu, J. Org. Chem., **55**, 2344 (1990).
26. M-L. Wang and H-S Wu, J. Chem. Soc. Perkin Trans. **2**, 841 (1991).
27. M-L. Wang and H-S. Wu, Chemical Engineering Science, **46**, 509 (1991).
28. W.F. Lunder, P.B. Kraus, and R.M. Fuoss, J. Am. Chem. Soc., **58**, 225 (1936).
29. E.D. Hughes, C.K. Ingold, S. Patai and Y. Pocker, J. Chem. Soc., 1209 (1957).
30. M.J. McDowell and C.A. Draus, Tetrahedron Lett., 3251 (1975).
31. A. Brandstrom and H. Kolind-Anderson, Acta Chem Scand., **29B**, 201 (1975).
32. A. Brandstrom, Acta Chem. Scand., **30B**, 203 (1976).
33. J. Uglestad, T. Ellingsen, and A. Beige, Acta Chem. Scand., **20**, 1593 (1966).
34. L. Pauling, The Nature of the Chemical Bond, Appendix 3.1, Cornell Univ. Press, Ithaca, New York (1940).
35. B.E. Conway, R.E. Verall, and J.E. Desnoyers, Trans, Faraday Soc., **62**, 2738 (1966); and references contained therein.
36. T.C. Pochapsky and P.M. Stone, J. Am. Chem. Soc., **112**, 6714 (1990).
37. T.C. Pochapsky, P.M. Stone, and S.S. Pochapsky, J. Am. Chem. Soc., **113**, 1460 (1991).
38. R. Alexander and A.J. Parker, J. Am. Chem. Soc., **89**, 5549 (1967).
39. J.H. Ramsden, R.S. Drago, and R. Riley, J. Am. Chem. Soc., **111**, 3958 (1989).
40. M. Rabinovitz, Y. Cohen, and M. Halpern, Angew, Chem., Int. Ed., Engl., **25**, 960 (1986); and references contained therein.
41. N.A. Gibson and D.C. Weatherburn, Anal. Cheim. Acta, **58**, 160 (1972).
42. J.P. Antoine, I.F. de Aguirre, F. Janssons, and F. Thyrion, Bull. Soc. Chim. Fr., 207 (1980).
43. K. Gustavii and G. Shill, Acta Pharm. Suec., **3**, 241 (1966).
44. K. Gustavii, Acta. Pharm. Suec., **4**, 233 (1967).
45. D. Mason, S. Magdassi, and Y. Sasson, J. Org. Chem., **55**, 2714 (1990).

46. A.E. Vassiliades, In Surfactant Science; Marcel Dekker, New York, 1970; Series Vol. 4, Chapter 12.
47. Y. Sela and S. Magdassi, Tenside, **27**, 179 (1990).
48. D. Mason, S. Magdassi, and Y. Sasson, J. Org. Chem., **56**, 7229 (1991).
49. D.H. Wang and H.S. Weng, Chem. Eng. Sci., **43**, 2019 (1988).
50. R. Neumann and Y. Sasson, J. Org. Chem., **49**, 3448 (1984).
51. M. Halpern, Y. Cohen, Y. Sasson, and M. Rabinovitz, Nouv. J. Chim., **8**, 443 (1984).
52. C.L. Liotta and J. Hurley, unpublished results.
53. A. Merz, Angew, Chem., **85**, 868 (1973).
54. A. Merz, Angew, Chem. Int. Ed. Engl., **11**, 846 (1973).
55. R. Solaro, S. D'Antone, and E. Chielline, J. Org. Chem., **45**, 4179 (1980).
56. M. Halpern, Y. Sasson, and M. Rabinovitz, J. Org. Chem., **47**, 1022 (1983).
57. D. Landini, A. Maia, and A. Rampoli, J. Org. Chem., **51**, 3187 (1986).
58. J. Fitzjohn, private communication.
59. D. Landini and A. Maia, J. Chem. Soc., Chem. Commun., 1041 (1984).
60. D.K. Bohme and G.I. Mackay, J. Am. Chem. Soc., **103**, 978 (1981).
61. J. Dockx, Synthesis, 441 (1973).
62. A. Gorgoes, A. Le Coq, Tetrahedron Lett., 4521 (1976).
63. K. Seenewald, K. Gehrmann, and G. Viertel, German Patent 1,271,107 (1968).
64. E.F. Lutz, J.T. Kelly, and D.W. Hall, British Patent 1,112,068 (1968).
65. E.V. Dehmlow and M. Lissel, Tetrahedron, **37**, 1653 (1981).
66. D. Landini and A. Maia, J. Chem. Soc., Chem. Comm., 1041 (1984).
67. E.V. Dehmlow, M. Slopianka, and J. Heider, Tetrahedron Lett., 2361 (1977).
68. E.V. Dehmlow, R. Thieser, Y. Sasson, and E. Pross, Tetrahedron, **41**, 2972 (1985).
69. M. Halpern, Y. Sasson, and M. Rabinovitz, J. Org. Chem., **49**, 2011 (1984).
70. M. Halpern, H.A. Zahalka, Y. Sasson, and M. Rabinovitz, J. Org. Chem., **50**, 5088 (1985).
71. M. Ellwood, J. Griffiths, and P. Gregory, Chem. Commun., 181 (1980).
72. H. Iwamoto, T. Sonoda, and H. Kobayashi, Tetrahedron Lett., 4703 (1983).
73. H. Kobayashi, T. Sonoda, and H. Iwamoto, Chem. Lett., 1185 (1982).
74. R.K. Smalley and H. Suschitzky, J. Chem. Soc., 775 (1964).
75. M. Yamada, Y. Watabe, T. Sakakibara, and R. Sudoh, J. Chem. Soc., Chem. Commun., 179 (1979).
76. L.J. Mathias and R.A. Vaidya, J. Am. Chem. Soc., **108**, 1093 (1986).
77. W.K. Fife and Y. Xin, J. Am. Chem. Soc., **109**, 1278 (1987); and references contained therein.
78. C-S. Kuo and J-J. Jwo, J. Org. Chem., **57**, 1991 (1992).

79. D.J. Sam and H.F. Simmons, J. Am. Chem. Soc., **94**, 4024 (1972).
80. C.L. Liotta and H.P. Harris, J. Am. Chem. Soc., **96**, 2250 (1974).
81. D.J. Sam and H.F. Simmons, J. Am. Chem. Soc., **96**, 2252 (1974).
82. F.L. Cook, C.W. Bowers and C.L. Liotta, J. Org. Chem., **39**, 3416 (1974).
83. C.L. Liotta, et al. Tetrahedron Lett., **28**, 2417 (1974).
84. J. Barry, et al. Tetrahedron Lett., **23**, 5407 (1982); J. Barry, et al. J. Org. Chem., **49**, 1138 (1984); J. Barry, et al., Synthesis 40 (1985).
85. G. Bram, et al. Synth. Commun., **49**, 1138 (1984); G. Bram, and G. Decodts, Synthesis, 543 (1985); G. Bram, A. Loupy, and J. Sansoulet, Israel J. Chem., **26**, 291 (1985).
86. P. Singh, M.S. Batra, and H. Singh, J. Chem. Res. (S), 204 (1985).
87. M.C. Vander Zwan and F.W. Hartner, J. Org. Chem., **43**, 2655 (1978).
88. H.A. Yee, J.J. Palmer, and S.H. Chen, Chem. Engin. Prog., 33 (1987).
89. J.B. Melville, and J.D. Goddard, Ind. Eng. Chem. Res., **27**, 551 (1988); J.B. Melville, and J.D. Goddard, Chem. Engin. Sci., **40**, 2207 (1985).
90. K. Wong and A.P.W. Wai, J. Chem. Soc. Perkin Trans., **11**, 317 (1983).
91. E.V. Dehmlow, Tetrahedon Lett., 91 (1976); E.V. Dehmlow and T. Remmler, J. Chem. Res. (S), 72 (1977).
92. H.A. Zahalka, and Y. Sasson, J. Chem. Soc. Chem. Commun., 1652 (1984).
93. S. Deremik and Y. Sasson, J. Org. Chem., **47**, 2264 (1985).
94. Y. Sasson, and H.A. Zahalka, J. Chem. Soc., Chem. Commun., 1347 (1983).
95. D. Landini, A.M. Maia, and G. Podda, J. Org. Chem., **47**, 2264 (1982).
96. O.I. Danilova, I.A. Esikova, and S.S. Yufit, Izv. Akad. Nauk SSSR, Ser. Khim., 314 (1988); Chem. Abstr., **109**, 169773j (1988).
97. S.S. Yufit, I.A. Esikova, and O.I. Danilova, Dokl. Akad. Nauk SSSR., **295**, 621 (1987).
98. O. Arad and Y. Sasson, J. Am. Chem. Soc., **110**, 185 (1988).
99. E.V. Dehmlow and H.J. Rarhs, J. Chem. Res. (S), 384 (1988).
100. C.L. Liotta, Application of Macrocyclic Polydentate Ligands to Synthetic Transformations, in Synthetic Multidentate Macrocyclic Compounds, R.M. Izatt, J.J. Christensen, eds. Academic Press, New York, p. 111 (1978).
101. S.S. Yufit and I.A. Esikova, J. Phys. Org. Chem., **4**, 336 (1991).
102. I.A. Esikova and S.S. Yufit, Izv. Akad, Nauk SSSR, Ser, Khim. 1524 (1988); Chem. Abstr., **110**, 94247a (1989).
103. I.A. Esikova and S.S. Yufit, J. Phys. Org. Chem., **4**, 149 (1991).
104. I.A. Esikova and S.S. Yufit, J. Phys. Org. Chem., **4**, 341 (1991).
105. J.W. Zubrick, B.I. Dunbar, and H.D. Durst, Tetrahedron Lett., **1**, 71 (1975).
106. C.L. Liotta, E.M. Burgess, and E.D. Black, Polymer Preprints, **31**, 65 (1990).

4

Phase-Transfer Catalysts

A. Introduction

In previous chapters we learned that a phase-transfer catalyst must have two particular chemical functions to be successful, that is, it must rapidly transfer one of the reactant species into the normal phase of the other reactant, and second it must make the transferred species available in a highly reactive form. The need for one or the other of these two functions to be strongly catalyzed often plays an important role in the selection of the best transfer catalyst for a particular reaction, as, for example, to pick a phase-transfer catalysis (PTC) catalyst that is especially useful for activating an anion, or for a catalyst that is especially good for facilitating anion transfer to the organic phase. Catalyst structure may also be of great significance when more than one product can be formed, to cause one product to be favored over another. Sometimes catalyst structure is not important at all, and almost any PTC catalyst will perform satisfactorily for some reactions, whereas for others it may be desirable to have two (or more) catalysts to facilitate each of the various steps in a reaction sequence. Some examples of the effect of catalyst structure are discussed in this chapter to illustrate these points.

Selection or development of a catalyst for a PTC reaction is usually the most important action taken in development of a PTC reaction system. As outlined in Chapter 1 a variety of different genera and species of compounds have been used as phase-transfer catalysts, mostly for anion transfer to organic phases, but also for transfer of other species. The soluble variety of PTC catalysts, predominantly including quaternary ammonium and phosphonium salts, polyethylene glycols (PEGs) and related compounds, crown ethers and cryptands, and soluble polymers containing dipolar aprotic functional groups are discussed in this chapter. Also included in this chapter are related topics such as the use of dual- and cocatalysts, catalysts for the transfer of species other than anions, and methods for separation or removal of PTC catalysts from reaction products. Additional discussion on some practical points in choice of a phase-transfer catalyst are given in Chapter 6. Coverage of insoluble ("triphase") catalysts, both liquid and solid, is given in Chapter 5; and the use of chiral phase-transfer catalysts to yield optically active products is reviewed in Chapter 12.

In addition to having a sufficiently active or efficient catalyst, the chemist or engineer developing a phase-transfer process is also concerned about other catalyst features:

1. *Stability under reaction conditions.* No phase-transfer catalyst is stable under all reaction conditions. Some, such as quaternary ammonium salts, are sensitive to strong bases; others, such as crown ethers and PEGs, are sensitive to acids. Some are sensitive to high temperatures, whereas others have considerable high-temperature stability. Some are sensitive to water, and can be used only under adhydrous liquid–solid PTC conditions, whereas others perform best when an aqueous phase is present. Usually, the chemist can find suitable catalysts and reaction conditions to obtain an acceptable level of catalyst stability.

2. *Catalyst cost and availability.* Many commonly used phase-transfer catalysts are commercially available at moderate cost; for example, many quaternary ammonium salts and PEGs fit these conditions, and sometimes a low-cost amine can be used, forming a quaternary ammonium catalyst in situ. Anion-exchange resins, when they can be used as catalysts, are also widely available in large quantity and at reasonable cost, particularly when they can be reused for many reaction cycles. Specially produced compounds for use as phase-transfer catalysts, such as crown ethers, cryptands, specialty quaternary salts and other compounds, and specially designed resin-bound phase-transfer catalysts are quite expensive, and used only when their advantages justify their high cost.

Cost and availability may not be of great significance for laboratory-scale applications and for some premium-value commercial applications, but for most commercial applications cost and availability are usually of considerable importance. The cost factor is less significant if the catalyst can be easily removed and reused.

3. *Catalyst removal, recovery, and recycle or disposal.* This aspect of PTC may be one of the most important for development of commercial PTC processes. A variety of processes can and are being used to remove PTC catalysts from reaction products, frequently in such a way that the catalyst can be reused. However, the separation step adds another process unit, and consequently more cost to the process. These issues are discussed in section H. Sometimes a resin-bound catalyst, or a *third-phase* catalyst, can be used for facile removal and reuse of catalysts, as discussed in Chapter 5.

4. *Low-toxicity catalysts.* In addition to easy recovery and recycle, commercial processes are far easier to handle when the ingredients are nontoxic and easily

biodegradable. Fortunately, most of the commonly used quaternary ammonium salts and PEGs fit these conditions, although both can damage biopond operations under some circumstances. More exotic catalysts may require specialized handling, as when a metal cocatalyst is used in conjunction with phase-transfer catalysts.

5. *Use of multiple catalysts and cocatalysts.* Use of two (or possibly more) phase-transfer catalysts, or use of cocatalysts, can sometimes significantly increase rates, alter product distributions, increase catalyst stability, or provide other benefits. For example, many oxidation reactions are highly facilitated by the presence of molybdate or tungstate ions, as described in Chapter 10. Many reactions involving transfer of hydroxide are facilitated by use of alcohols as cocatalyst.

This chapter deals primarily with efficiencies of various materials to catalyze PTC reactions. The reader is referred to Chapter 6 for additional information and guidance for choosing a soluble catalyst, and to Chapter 5 for discussion of the possibility of using an insoluble catalyst.

B. Use of Quaternary Salts as Phase-Transfer Catalysts

1. Simple Tetraalkyl-Onium Salts as Phase-Transfer Catalysts

The most commonly used phase-transfer catalysts are quaternary ammonium and phosphonium salts. Of these, tetrabutylammonium salts, trioctylmethylammonium chloride (or its commercial equivalent), benzyltriethylammonium chloride, and hexadecyltributylphosphonium bromide are the most frequently reported quaternary salts used as phase-transfer catalysts. A list of the quaternary compounds often used as catalysts, their costs, and other properties is given in Chapter 6. Reviews on general catalyst aspects have been published [1–5]. Further detailed discussions on specific PTC catalysts commonly used are provided in each chapter of this book as appropriate to the reactions being catalyzed.

When considering use of quaternary onium salts as phase-transfer catalysts, structural aspects of the catalyst and features of the reaction system to be catalyzed become important. These include distribution of the catalyst and reactive species into the organic phase; consideration of use of an *anion-activating* or an *accessible* quaternary salt; the stability of the quaternary salt under the reaction conditions; the effect (if any) of the anion associated with the quaternary cation; the effect of the catalyst structure (if any) on the product selectivity; the possibility for in situ generation of quaternary salts in the reaction mixture from low-cost trialkylamines; and consideration for use of onium anions other than nitrogen, for example, phosphonium, arsonium, and sulfonium salts. In addition to these general aspects it is sometimes worthwhile for the catalyst to contain some additional structural features or functional groups, grouped into the category of *special quaternary salts* and discussed in Section B.2.

a. Distribution of Q^+Y^- into the Organic Phase

In a general sense, a phase-transfer catalyst (for anions) can be successful only if it transfers enough of the anion from the aqueous phase to the organic phase for the reaction to proceed in a reasonable time span. This distribution of catalyst, plus the associated anion into the organic phase, depends strongly on the structure of the quaternary cation for small anions such as Cl^-, Br^-, CN^-, etc., but less strongly as the anion being transferred, such as $C_6H_5O^-$, $C_6H_5CHCN^-$, RCO_2^-, has considerable organophilic structure itself.

Thus, the catalyst cation, plus its associated anion, need to be sufficiently distributed into the phase where reaction is to occur. Assuming reaction occurs in the organic phase, the organic structure of the onium cation, Q^+, plus the organic structure of the anion to be transferred, Y^-, together need to be sufficient to overcome the normal tendencies of the salt Q^+Y^- to be distributed mostly into the aqueous phase. For example, among the simple tetraalkylammonium salts:

- Tetramethylammonium cation plus simple anions, $(CH_3)_4N^+Y^-$, ($Y = Cl^-$, Br^-, CN^-, AcO^-, etc.) are not easily distributed into most organic solutions and therefore tetramethylammonium salts with such anions are usually not good phase-transfer catalysts. However, if the tetramethylammonium cation is coupled with a large organic anion, such as stearate, or in high-temperature reactions (>150°C), such in some fluoride ion displacements, then tetramethylammonium salts may function as reasonable phase-transfer catalysts [6–11].

- Tetraethylammonium and tetrapropylammonium salts are also usually found to be poor, or even ineffective, catalysts for the transfer of small anions into most organic phases, but may be quite suitable for transfer of organic anions [12,13].

- Tetrabutylammonium salts show medium to high efficiencies as phase-transfer catalysts. Tetrabutylammonium salts, in reaction systems with low-polarity organic phases (such as with a toluene solvent) may form a third catalyst-rich phase, which can enhance reactivity (see Section D). Tetrabutylammonium salts are frequently used as phase-transfer catalysts, not necessarily because they offer the highest reactivity, but because they are readily available in high purity on a commercial scale, they are relatively inexpensive, and they can be removed from organic reaction products by extraction into water (see Section H).

- Quaternary ammonium cations, R_4N^+, having R = pentyl to decyl groups, easily extract anions into almost all organic phases, and usually demon-

Table 4-1. *Effect of catalyst structure on the rate of thiophenoxide displacement on 1-bromooctane in benzene/water [14].*

Catalyst	$k \times 10^3$ (L/M-sec^{-1})
$Me_4N^+Br^-$	< 0.0016
$Pr_4N^+Br^-$	0.0056
$Bu_4N^+Br^-$	5.2
$Bu_4P^+Cl^-$	37.
$(C_8H_{17})_3NMe^+Cl^-$	31.
$(C_8H_{17})_3PEt^+Br^-$	37.
$C_6H_{13}NEt_3^+Br^-$	0.015
$C_8H_{17}NEt_3^+Br^-$	0.16
$C_{10}H_{21}NEt_3^+Br^-$	0.24
$C_{12}H_{25}NEt_3^+Br^-$	0.28
$C_{16}H_{33}NEt_3^+Br^-$	0.48
$C_{16}H_{33}PEt_3^+Br^-$	1.8
$C_6H_5CH_2NEt_3^+Cl^-$	< 0.0016
$C_{16}H_{33}NMe_3^+Br^-$	0.15
$(C_6H_5)_4P^+Br^-$	2.5
$(C_6H_5)_4As^+Cl^-$	1.4
$(C_6H_5)_3PMe^+Br^-$	1.7

strate higher catalytic activities than other simple tetraalkylammonium salts. For testing of a reaction to determine whether it will work under PTC conditions, tetrahexyl- or triocylmethylammonium catalysts are recommended as the most likely to show some activity.

- Higher tetraalkylammonium salts, $R_4N^+X^-$; where R = dodecyl and higher groups, can easily extract any monovalent anion into an organic phase, but because these highly organophilic cations do not easily approach the organic–aqueous interface, the interchange of anions between organic and aqueous phase is slowed, and reaction rates decrease compared quaternary salts where R = C_5–C_{10} alkyl groups. That is, the higher quaternary ammonium salts cause the reaction to exhibit less than optimum reactivity because of restricted anion transfer rates.

The effects of several catalyst structures on rate of PTC reaction of thiophenoxide with 1-bromooctane are listed in Table 4-1. These data clearly show that even with an anion of substantial organic structure, $C_6H_5S^-$, good catalytic efficiencies are obtained only with cations large enough to transfer the anion, and bulky enough to activate the anion for nucleophilic displacement.

The beneficial effect of having a more organophilic structure on the anion to be extracted has been demonstrated in esterification with sodium carboxylates. Thus, phase-transfer esterification of epichlorohydrin with the sodium salts of C_4 and C_{9-17} carboxylic acids behaved differently with various simple tetraalkylammonium halides (R_4N^+ X^-; $R = C_1-C_4$]) [15]. With sodium butanoate the esterification rate depended significantly on the number of carbons attached to the quaternary nitrogen, being fastest with the tetrabutylammonium catalyst. But with salts of the higher C_{9-17} carboxylic acids the size of alkyl groups attached to the quaternary nitrogen catalyst had little effect on reactivity.

Data for the transfer capability of variously-sized quaternary salts shows that the size of the catalyst is of significant importance in liquid–liquid PTC reactions, but is of much less importance in solid–liquid PTC reactions [414]. That is, in the absence of water, lower quaternary salts can usually be used as PTC catalysts, but on addition of water they will be dissolved into the aqueous phase. This effect may be useful for removal of PTC catalysts from reaction mixtures.

b. Anion-Activation vs. Accessible Nature of -Onium Salts

i. Anion-Activating Quaternary Salts

Many PTC reactions require that the transferred anion not only be transferred from the aqueous phase to the organic phase, but also that it be delivered in an *activated form*. Bulky quaternary cations activate anions by increasing the distance separating cation from anion in the ion pair, Q^+-X^-. (See Chapter 2.) Simple displacement reactions, such as cyanide displacement on 1-chloroalkanes, require the higher activation afforded by quaternary ammonium cations having three or four large alkyl groups (butyl or larger) to achieve useful reaction rates. These catalysts are referred to as having an *anion-activating* structure. This effect can be seen in the data of Table 4-1 by comparison of the anion-activating quaternary compounds trioctylmethylammonium, trioctylmethylphosphonium, tetrabutylphosphonium, and to a lesser extent tetrabutylammonium salts with the series of open-faced or *accessible* alkyltriethylammonium salts.

In another example, reaction of sodium benzoate with benzyl chloride to produce benzyl benzoate, bulky anion-activating catalysts give the best results [16]. Thus:

$RN^+Me_3Cl^-$ (R = decyl, hexadecyl or octadecyl): low yield of ester.
$RR'N^+Me_2Cl^-$ (R = R' = dodecyl, octadecyl, R = tetradecyl,
   R' = benzyl): yield > 95%.
$Bu_4N^+I^-$: yields > 95%.

As a general rule, bulky anion-activating-onium salts are the best sort for displacement reactions, whereas accessible onium salts tend to be best for reactions involving strong bases.

ii. Accessible Quaternary Onium Compounds

Quaternary salts having one large group and three small groups, for example, hexadecyl trimethylammonium salts, $C_{16}H_{33}N(CH_3)_3^+X^-$, are considered to be highly *accessible* because the openness of the structure allows relatively close association of the anion, and its waters of hydration, with the cation. This accessibility is important for several reactions, particularly those involving the use of hydroxide anion and other anions that are slow to be transferred into organic media.

The accessible quaternary cations reside easily in the interface and are therefore able to lower the interfacial tension between the organic and the aqueous phase. This lowering of interfacial tension allows the organic phase to be dispersed as smaller droplets into the aqueous phase (or of aqueous droplets into the organic phase), thereby increasing the total interfacial surface area, and consequently increasing the net total number of anions transferred into the organic phase per unit of time. In the case of hexadecyltrimethylammonium salts, this tendency to lower interfacial tension is so strong that its use as a catalyst often leads to undesirable formation of emulsions. This tendency to emulsion formation can be moderated by use of slightly larger alkyl groups on the quaternary cation, as, for example, with benzyltriethylammonium chloride, which is extensively used as a catalyst for alkylation reactions involving use of hydroxide and concurrent formation of organic anions [17–19].

Quaternary ammonium cations having intermediate unsymmetrical structures between R_4N^+ and $RNMe_3^+$, for example, $R_3NCH_3^+$ and $R_2N(CH_3)_2^+$, R = hexyl, octyl, decyl, hexadecyl, etc., are intermediate between the symmetrical tetraalkylammonium salts and highly accessible alkyltrimethylammonium salts in their catalytic behavior. Thus, trioctylmethylammonium chloride and didodecyldimethylammonium salts are commonly used as phase-transfer catalysts, with excellent results.

An example of the differences in rates observed with differently sized and configured quaternary ammonium salts is shown in data of Table 4-2, which compares the rates of different catalysts for PTC NaOH(aq) dehydrochlorination of 1,1,2,2,3-pentachloropentane. These data illustrate the need to have both sufficient organic structure to transfer hydroxide anion into the organic phase, and the higher activity advantage of accessible-type quaternary ammonium salt structures. Similarly, the data in Table 4-3 show the beneficial effect of accessible catalyst structure on the yield of α-ethylphenylacetonitrile from alkylation of phenylacetonitrile with ethyl chloride in the presence of aqueous sodium hydroxide. [As a note of caution here, yields of products under a standard set of reaction conditions should not necessarily be regarded as reflective of reaction *rates*, particularly for reactions in the presence of hydroxide. It is possible that some of the quaternary salts are more active catalysts, but also decompose faster under the reaction conditions, and may give lower yields than a less active catalyst.]

Table 4-2. PTC Catalytic activity of various quaternary ammonium salts for dehydrochlorination of 1,1,2,2,3-pentachloropentane with aqueous sodium hydroxide [20].

Catalyst	Relative rate
$(CH_3)_4N^+Br^-$	<0.00002
$(HOCH_2CH_2)_3NBu^+Br^-$	<0.00002
$(HOCH_2CH_2)_3CH_2C_6H_5^+Cl^-$	0.00040
$O(CH_2CH_2)_2NBu^+Br^-$	0.00095
$Et_4N^+Br^-$	0.00268
$C_5H_5NBu^+Br^-$	0.00647
$BuNEt_3^+Br^-$	0.0200
$Bu_4N^+OH^-$	0.0240
$C_6H_5CH_2NEt_3^+Cl^-$	0.0240
$(C_8H_{17})_3NBu^+Cl^-$	0.0820
$(C_8H_{17})_4N^+Cl^-$	0.0907
$C_8H_{17}NBu_3^+Cl^-$	0.254
$C_{16}H_{33}NMe_3^+Br^-$	0.394
$C_8H_{17}NEt_3^+Cl^-$	3.45

Friedi et al. [22] compared several quaternary ammonium salts as catalysts for three different reactions with the following results:

- For acetate displacement on benzyl chloride with KOAc to produce esters, trioctylmethyl ammonium chloride was most efficient. (Shows the need to have sufficient anion activation.)
- For sodium hypochlorite oxidation of benzyl alcohol to benzaldehyde, the C_{12-14} trimethylammonium catalysts performed well. This suggests that anion activation is not important; but transfer rate of anion to the organic phase is important, favoring the use of an accessible catalyst.

Table 4-3. Effect of PTC catalyst structure on yield of α-phenylacetonitrile from alkylation of phenylacetonitrile with ethyl chloride using aqueous NaOH [21].

Catalyst	Percent yield $C_6H_5CH(Et)CN$
$C_6H_5CH_2NMe_3^-$	32
$C_6H_5CH_2NMe_2Et^+Cl^-$	40
$C_6H_5CH_2NMEt_2^+Cl^-$	45
$C_6H_5CH_2NEt_3^+Cl^-$	50
$C_6H_5CH_2NEt_3^+Br^-$	35
$p-MeOC_6H_5CH_2NEt_3^+Cl^-$	54
$C_6H_5CH_2NEt_2Pr^+Cl^-$	44
$C_6H_5CH_2NEt_2Bu_+Cl^-$	45
$C_6H_5CH_2NMeEtPr^+Cl^-$	43
$C_6H_5CH_2NPr_3^+Cl^-$	43
$(C_6H_5CH_2)_2NEt_2^+Cl^-$	15
$(C_2H_5)_4N^+Cl^-$	51

- For dichlorocarbene addition to cyclohexene with chloroform and aqueous sodium hydroxide, C_{16-18} trimethylammonium chloride was the catalyst of choice. (Also shows the desirability of having an accessible catalyst to increase transfer rate.)

Dehmlow and Vehre [23] found the extraction of monovalent anions by large lipophilic quaternary cations,

$$(n-C_6H_{13})_4 N^+$$
$$(n-C_8H_{17})_3 NMe^+$$
$$(n-C_{12}H_{25})_2 N(Me)_2^+$$
$$(n-C_{12}H_{25})(PhCH_2)N(Me)_2^+$$

to be generally independent of the structure of the quaternary cation. However, bisulfate and bivalent anions were extracted more effectively by sterically accessible cations than by bulky quaternary cations.

c. Solvent and Polarity of the Organic Phase

Use of higher polarity solvents, for example, methylene chloride instead of toluene, allows easier distribution of quaternary salts into organic solutions, and therefore allows use of smaller quaternary salts as catalysts. With methylene chloride, or a similarly polar solvent, it may be possible to use tetrapropylammonium or even tetramethylammonium salts as catalysts, depending on the nature of the anion. Table 4-4 shows the effect of solvent on the distribution of tetrabutylammonium bromide between aqueous and organic phases. In addition to increasing the ease of transfer of anions to an organic phase, more polar solvents are commonly recognized in physical organic chemistry to enhance the rate of organic-phase ionic reactions by providing a more ion-compatible reaction medium.

d. Stability of the Quaternary Salt Catalyst During PTC Reaction: [2]

Quaternary ammonium salts undergo decomposition mostly by two kinds of reactions: (1) internal displacement to yield trialkylamine and displacement product,

$$R_4N^+ \; Y^- \xrightarrow{100\text{-}200^\circ C} R_3N + R\text{-}Y \qquad (4\text{-}1)$$

and (2) in the presence of strong bases, decomposition by Hoffman degradation to trialkylamine and olefin [25–27],

$$RCH_2CH_2NR_3^+ \; OH^- \longrightarrow RCH=CH_2 + R_3N + H_2O \qquad (4\text{-}2)$$

Table 4-4. Effect of solvent on distribution of $(C_4H_9)_4N^+Br^-$ between organic and aqueous phases.

Solvent	Extraction constant, K $K = [Q^+Br^-]_{org}/[Q^+]_{aq}[Br^-]_{aq}$
Benzene	<0.1
Et_2O	<0.1
C_6H_5Cl	<0.1
$o-C_6H_4Cl_2$	<0.1
EtOAc	0.2
$n-C_4H_9Cl$	<0.1
$Cl(CH_2)_4Cl$	0.3
$Cl(CH_2)_3Cl$	2.9
$Cl(CH_2)_2Cl$	6.1
$Cl-CH_2-Cl$	35.
$CHCl_3$	47.
CCl_4	<0.1
EtCOEt	1.1
MeCOEt	14.
$n-C_4H_9OH$	69.
$n-C_3H_7NO_2$	9.0
CH_3NO_2	168.

Data from [24].

Tetraalkylphosphonium salts are much more resistant than equivalent quaternary ammonium salts to internal displacement, and the R_4P^+ are therefore more stable and long-lived catalysts at temperatures between 100 and 150°C under nonalkaline conditions. However, tetraalkylphosphonium salts in the presence of sodium hydroxide at low to moderate temperatures, and strongly basic fluoride solutions at high temperatures [28], undergo decomposition to trialkylphosphine oxide and alkane. (Sometimes trialkyl phosphine oxides and amine oxides behave as phase-transfer catalysts, but they are usually effective only for reactions that are in the fast–fast quadrant of the PTC matrix.)

$$R_4P^+ \; OH^- \longrightarrow R_3PO + RH \qquad (4-3)$$

Loss of catalytic activity through internal displacements is not usually a serious problem at temperatures below about 100°C. However, highly active R-groups, such as benzyl, methyl, and allyl, attached to the nitrogen of quaternary ammonium salts, undergo internal displacement more readily and can exhibit substantial loss of catalytic activity. For example, trioctylmethylammonium salts in the presence of aqueous sodium cyanide at 105°C decompose to trioctylamine and acetonitrile with a half life of 2–4 h due to the formation of trioctylamine and acetonitrile.

Table 4-5. Decomposition of various quaternary salts in a two-phase system with 50% aq. NaOH.

	Half-life (h) for Q^+X^{-a} in the presence of 50% NaOH					
	Cl^-		Br^-		I^-	
Q^+	25°	60°	25°	60°	25°	60°
Hex_4N^+	35	0.4	Days	13	Stable	Stable
Oct_4N^+	—	0.24	—	3.5	—	—
Bu_4N^+	18	—	—	—	—	—
$(PhCH_2)_4N^+$	6.7	—	240	—	—	—
$PhCH_2(Hex)_3N^+$	13	0.05	432	2.3	—	—
Oct_3N^+Me	32	0.7	—	—	—	—
Bu_4P^+	0.03	—	3.5	—	—	—
$C_{16}H_{33}P^+Bu_3$	0.03	—	2	0.25	—	—

[a] Reactions conducted with 40 mL of ca. 0.03 M Q^+X^- in PhCl with 40 mL 50% aq. NaOH.

Decomposition of quaternary ammonium and phosphonium salts in the presence of sodium hydroxide and other strong bases is a more severe problem. Landini et al. [29] studied the parameters that affect the stability of quaternary ammonium and phosphonium salts in chlorobenzene–aqueous NaOH systems. Some of the results are given in Table 4-5. The effect of sodium hydroxide concentration is shown by the data in Table 4-6. The effect of stirring rate on decomposition of hexadecyltributyl phosphonium bromide at 25°C is shown in Figure 4-1.

These data led to the following conclusions:

- For a given quaternary ammonium cation in chlorobenzene solution, mixing with aqueous sodium hydroxide, decomposition rate is a function of the associated cation initially present (or produced during reaction) in the following order:

$$Q^+ Cl^- > Q^+ Br^- > Q^+ I^-$$

Table 4-6. Effect of NaOH concentration in the aqueous phase on half-life of Q^+X^- decomposition rates in chlorobenzene–aq. NaOH system at 60°C.

	Half-life (h) for Q^+X^-		
Percent NaOH	$Hex_4N^+Cl^-$	$Hex_4N^+Br^-$	$C_{16}H_{33}P^+Bu_3Br^-$
10	78	—	—
15	29	312	20.
25	12	—	—
30	6	30	4.5
40	0.75	—	—
50	0.4	13	0.25

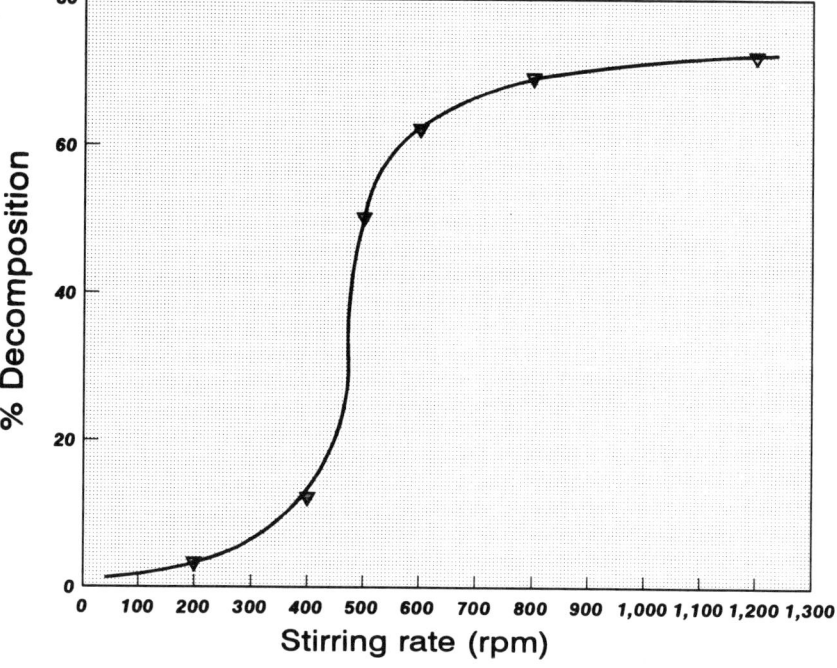

Figure 4-1. Agitation effect on decomposition of $C_{16}H_{33}PBu_3^+Br^-$ in PhCl–50% NaOH at 25°C, 4 h.

This order is due to the increasing reluctance of the halide anions to exchange with OH^-, producing Q^+OH^- in the organic phase where decomposition occurs.

- Decomposition rates dramatically increase as the concentration of NaOH in the aqueous phase is increased. This results from the fact that increasing NaOH concentration causes the transferred hydroxide anion to have less water of hydration, and consequently greater reactivity. In the presence of 50% NaOH or 53% KOH, OH^- transferred to the organic phase carries with it no water of hydration [2]. In fact, at elevated temperatures (greater than about 65°C) decomposition of Q^+OH^- is so fast (minutes) that no catalysis can occur unless the OH^- anion reacts immediately with some component of the organic phase. For example, in the alkylation of phenylacetonitrile with ethyl chloride and aqueous NaOH, reaction of transferred OH^- with the nitrile to form $C_6H_5CHCN^-$ and water is believed to occur almost instantaneously, forming $Q^+[C_6H_5CHCN]^-$, which then undergoes displacement with ethyl chloride to yield α-ethylphenylacetonitrile.

- Decomposition rates of quaternary salts increase with agitation rates up to a point, after which the rate remains constant with increased agitation. This behavior is consistent with slow OH$^-$ transfer rates at low agitation rates but slow intrinsic reaction rate control at high agitation levels.

If catalyst decomposition in systems containing strong base becomes too severe with common quaternary ammonium salts, then switching to other phase-transfer catalysts such as a PEG (see Section D) or a crown ether (see Section C) or use of special hydroxide-stable catalysts such as triphenylsulfonium salts [30] or tris(3,6-dioxaheptyl)amine (see Section D) is recommended.

Landini and co-workers also showed that addition of inorganic salt, NaX, to the aqueous NaOH solution slowed decomposition of Q^+X^-, clearly due to NaX reversing the equilibrium forming quaternary ammonium hydroxide.

$$Q^+ X^- + OH^- \rightleftharpoons Q^+ OH^- + X^- \qquad (4\text{-}4)$$

Zerda et al. also examined the effect of Hofmann degradation on quaternary ammonium salts in the presence of strong bases, showing that the presence of alcohols as cocatalysts can reduce the rate of catalyst loss by transfer of alkoxide, rather than hydroxide [31].

$$Q^+ OH^- + ROH \rightleftharpoons Q^+ {}^-OR + H_2O \qquad (4\text{-}5)$$

In many reactions requiring the transfer of hydroxide anion or an equivalent base to an organic phase, alcohols may be used as cocatalysts (see Section F.2). Thus, for example, addition of a small amount of methanol to the PTC dehydrobromination of 2-bromooctane with aqueous potassium hydroxide, catalyzed with trioctylmethylammonium chloride, resulted in increased reaction rates, due to the easier transfer of CH_3O^- compared to OH^-, and also reduced the loss of catalyst activity due to quaternary salt decomposition [32]. Tertiary diols, such as pinacol, have been demonstrated to be superior cocatalysts for transfer of strong base for aqueous hydroxide to organic phases [33].

e. Effect of the Anion Associated with Quaternary Cation

It is often possible to obtain quaternary ammonium salts in different anion forms, as for example, with tetrabutylammonium salts, it is possible to purchase $Bu_4N^+Cl^-$; $Bu_4N^+Br^-$; $Bu_4N^+I^-$; $Bu_4N^+HSO_4^-$. In general it is preferable to employ the chloride or bisulfate forms and less preferable to use the bromide, and least preferable to use the iodide form. Iodide anions, and other anions that tend to associate strongly with the quaternary cation in the organic phase, can poison the activity of the catalyst, especially with anions that are difficult to transfer

Table 4-7. *Effect of anion, X^-, on the yield of dichloronorcane under standard reaction conditions* [34].[a]

Catalyst $(C_4H_9)_4N^+X^-X^-$	Percent yield of dichloronorcarane
Cl^-	38.8
Br^-	29.
I^-	22.6
HSO_4^-	45.7
ClO_4^-	13.6
Picrate	36.6
Benzoate	24.5
p-Nitrobenzoate	16.3
β-Naphthalenesulfonate	8.6

[a]Standard conditions: 0.1 mol cyclohexene, 0.4 mol $CHHCl_3$, 0.2 mol 50% NaOH, 1 mmol catalyst; 4 h at 23°C, magnetically stirred.

such as hydroxide. (See Chapter 2.) For example, the data in Table 4-7 illustrate the effect of anion on the yield of dichloronorcarane from the reaction of cyclohexene with chloroform and sodium hydroxide.

This order of preference is not always followed, as for example, in the esterification of 1,4-dichlorobutane with sodium formate, the reactivity of the quaternary ammonium salts with regard to the counterion was $Cl^- > Br^- > I^- > HSO_4^-$ [35]. In some rare reactions, particularly those involving benzyl chloride, a small amount of iodide acts as a cocatalyst rather than a poison, although too much iodide does poison catalyst activity [36–38].

f. Effect of Catalyst Choice on Product Selectivity

Most phase-transfer reactions catalyzed by quaternary ammonium salts having sufficient stability, organic-phase solubility, and anion-activation ability or accessibility, and other qualities discussed previously, are not overly sensitive to the structure of the quaternary ammonium salt, even though some may be slightly better than others. However, for some reactions the choice of quaternary salt structure is of great importance in determination of which products are formed, and may require use of quaternary salts containing special structures or functional groups. (See Section B.2 for discussion of some quaternary catalysts having special structures.) Some reactions that have been shown to have a special sensitivity to the structure of the quaternary salt catalyst are shown in the following examples.

Reactions that may lead to several possible products by exchange of halide anions (as a competing reaction) provide a good example of selective behavior

Table 4-8. *Effect of PTC catalyst on product distributions in etherification of 2,5-dimethylphenol with 1-bromo-3-chloropropane.*[a]

Catalyst	Percent yield of monoethers (I + II)	Ratio of products I/II
$C_6H_5CH_2NEt_3^+Br^-$	71	24.6
$Et_4N^+Cl^-$	60	26.2
$C_6H_5CH_2NMe_3^+Cl^-$	58	25.6
18-Crown-6	64	33.7
TDA-1	62	30.6
$Bu_4N^+Br^-$	75	1.9
$C_6H_5CH_2NBu_3^+Cl^-$	76	2.5
$Bu_4N^+HSO_4^-$	72	2.1
$(C_8H_{17})_4N^+Br^-$	63	2.1
$(C_6H_{13})_4N^+Br^-$	56	2.3

[a]PTC catalyst (0.01 mol), 2M NaOH (50 mL), toluene (50 mL), 1-bromo-3-chloropropane (0.11 mol), and the phenol (0.1 mol) were heated under reflux with vigorous stirring for 24 h.

dependent on catalyst structure. For example, reaction of 2,5-dimethylphenol with 1-bromo-3-chloropropane can give two halopropylethers [39]:

(4-6)

Various catalysts tested produced two types of product distributions, as seen by the data in Table 4-8. The significant difference between the two groups of catalysts is that the first group is relatively accessible and does not activate anions, whereas the second catalyst group is rather bulky, and activates anions readily. Thus, although both catalyst groups satisfactorily transfer 2,5-dimethylphenolate anion into the organic phase and allow it to react by displacement with the bromo-group of 1-bromo-3-chloropropane, the second group also transfers and activates bromide anion so that it can apparently exchange bromide for chloride on the initially formed product. The substantial preference for distribution of bromide ion into the organic phase when associated with quaternary

Table 4-9. Effect of catalyst structure on the yield of various dihalocarbene addition products from cyclohexene and chlorodibromomethane in PTC reaction with NaOH [39].[a]

Catalyst[b]	Total yield (%)	Relative adduct yields (%)		
		CBrCl	CCl$_2$	CBr$_2$
Me$_4$N$^+$Cl$^-$	22	97.8	2.2	—
Dibenzo-18-crown-6	63	96.2	3.2	0.4
3,3′,5,5′-di-tert-butyl-18-C-6	46	97.5	2.5	—
3,5-di-tert-Bu-benzo-15-C-5	75	96.4	3.6	—
C$_6$H$_5$CH$_2$NEt$_3^+$Cl$^-$	56	50.8	40.8	8.2
(C$_6$H$_5$)$_4$As$^+$Br$^-$	71	45.5	44.2	7.2
18-Crown-6	77	74.2	18.7	7.1
Benzo-18-C-6	79	74.5	18.5	7.0
Dicyclohexano-18-C-6	81	75.8	18.7	5.5

[a]Reaction: 20 mmol cyclohexene with 20 mmol HCBr$_2$Cl, 100 mmol NaOH, and 0.4 mmol catalyst, 45°C, 4 h.

[b]The notation, e.g., 18-C-6, represents 18-crown-6.

ammonium phase-transfer catalysts drives this reaction to displace chloride by bromide:

$$\text{I} \xrightarrow[\text{Reflux, 24 hours, Toluene}]{\text{NaBr (aq), Bulky } R_4N^+Br^- \text{ catalyst}} \text{II} \quad (4\text{-}7)$$

where I is 2-(2-chloroethoxy)-1,4-dimethylbenzene (OCH$_2$CH$_2$-Cl) and II is the corresponding bromide (OCH$_2$CH$_2$-Br).

Another example of product selectivity due to catalyst choice includes the addition products of carbenes derived from CHBr$_2$Cl in PTC reaction with cyclohexene [40]. Three dihalocarbene addition products were observed due to bromo–chloro group exchange prior to dihalocarbene addition to cyclohexene:

$$\text{cyclohexene} + \text{CHBr}_2\text{Cl} \xrightarrow[45°\ \ 4\text{ Hr}]{\text{PTC, NaOH (aq)}} \text{[Cl,Br-bicyclic]} + \text{[Cl,Cl-bicyclic]} + \text{[Br,Br-bicyclic]} \quad (4\text{-}8)$$

The effect of catalyst structure on product distributions is shown by the data in Table 4-9.

Other examples where choice of catalyst can markedly affect product composition include dibromocarbene formation vs. CBr$_3^-$ substitution in reaction with allyl bromide [41],

(4-9)

$$\text{CH}_2\text{=CHCH}_2\text{Br} + \text{CHBr}_3 + \text{NaOH(aq)} \xrightarrow{\text{PTC}}$$

[structures: dibromocyclopropane with CH₂Br substituent; CH₂=CHCH₂CBr₃; dibromocyclopropane with CH₂CBr₃ substituent]

where the observed ratio of dibromocarbene addition to tribromomethyl substitution in the primary step varies between 92:1 and 1:91, depending on whether the catalyst is primarily accessible or primarily anion-activating, respectively. A similar situation exists in the reaction of methyl acrylate with chloroform and sodium hydroxide under PTC conditions, where initially formed CCl_3^- can either add by Michael addition to the double bond, forming $CCl_3CH_2CH_2CO_2R$, or lead to: CCl_2 which adds to the double bond forming dichlorocyclopropane adducts [42]. In this case accessible catalysts produced predominantly the dichlorocarbene addition products, whereas anion-activating catalysts produced mostly Michael addition products.

Also, in reactions with ambident anions, choice of catalyst can affect the distribution of products formed. For example, solid–liquid PTC reactions of KSCN with benzyl chloride at 180° give benzyl thiocyanate and benzyl isothiocyanate in ratios that depend on catalyst structure, although liquid–liquid PTC at 100°C yields benzyl thiocyanate exclusively [43]. PTC methylations of desoxybenzoin by dimethyl sulfate can be steered toward enol–ether formation by large, sterically shielded ammonium ions, or, more strongly, by large highly delocalized cations, while C-alkylation is favored by small, hard ammonium ions of the type $RNMe_3^+$ and by crown ethers, particularly benzo-crowns [44]. $O-C$-alkylation product ratios can be varied between 0.75 and 63.

Normal PTC alkylation of phenols by reactive alkylating agents and aqueous sodium hydroxide is almost completely selective for O-alkylation, showing that the reaction environment is nonpolar [45]. However, bulky quaternary ammonium or phosphonium salts immobilized on silica gel or alumina provide an aqueous or polar environment giving a high yield of C-alkylation products [46].

Phase-transfer catalyst structure and also cocatalyst (alcohol) structure affect the average molecular weight of polycarbonates formed from phosgene and diphenols [47]. The efficiency of quaternary ammonium catalyst decreased in the following order:

$$Bu_4N^+ HSO_4^- > C_6H_5CH_2NBu_3^+ Br^- > Bu_4N^+ Br^-$$
$$> Bu_4N^+I^- > C_6H_5CH_2NEt_3^+Cl^-$$

The cocatalyst effect of alcohols decreased in the following order: benzyl alcohol > methanol > ethanol > propanol > n-butanol; primary alcohol > secondary alcohol > tertiary alcohol > diols > monoalcohols > triols. For preparation of polyester–polyethers from the reaction of 2,2-bis(1,4-phenyleneoxy)propylidenediacetic acid and related diacids with diphenols best results were obtained with benzyltriethylammonium chloride which was a hydrophilic catalyst and did not promote hydrolysis of the polymeric chains [48,49]. That is, the benzyltriethylammonium catalyst, although efficient for forming and transferring $[Q^+R-CO_2^-]$, was not efficient at transferring the small and highly hydrated OH^- anion, thereby avoiding the ester hydrolysis associated with larger catalyst cations which transfer hydroxide more easily.

g. In-Situ Generation of Quaternary Salts and Use of Amines as Catalysts

Often it is possible, when conducting PTC reactions with alkylating reagents, to simply add trialkylamines or trialkylphosphines to a reaction mixture, allowing tetralkylammonium or phosphonium salts to form in situ. Tertiary amines are used in this way, for example, to form benzyl ethers from reaction of benzyl chloride and 1-butanol with NaOH(aq) [37], in the reaction of potassium acetate with 1-bromobutane [50], in the formation of butylbenzyl phthalate from reaction of benzyl chloride with sodium monobutyl phthalate [51]. The kinetics of in situ quaternary salt formation and decomposition in a reaction of benzyl chloride with sodium cyanide to produce phenylacetonitrile have been described [52]. Tributylamine and higher trialkyl amines react in situ with dimethyl sulfate to produce quaternary ammonium salts, which are good phase-transfer catalysts for the reaction of 2-phenylethanol or other alcohols with dimethyl sulfate and 50% aq. NaOH to produce ethers [53]. In the synthesis of aminoethers from *trans*-2-phenoxycyclohexanol a quaternary salt is formed in situ, catalyzing the two-phase reaction [54].

Substitution of poly(chloromethylstyrene) with nucleophilic reagents such as potassium acetate proceeds smoothly with addition of tributylphosphine, which forms catalytic quantities of the phosphonium salt of the polymer in situ [55]. This phosphonium salt has a higher catalytic activity than benzyltributylphosphonium chloride. Polymerization of α,α'-dihalo-*p*-xylene with bisphenol-A in the presence of sodium hydroxide proceeds when catalyzed by trialkyl amines. The quaternary ammonium salt, existing at the end of growing polymer chain, showed nearly the same catalytic activity as when benzyltriethylammonium chloride was the initial stage of the polymerization [56].

h. Use of Different -Onium Ions: Phosphonium, Arsonium, Sulfonium Selenonium

In addition to quaternary ammonium salts, other -onium salts having a different cationic center have been used as phase-transfer catalysts. Only a few isolated examples comparing use of these compounds with quaternary ammonium compounds have been published, such as the data listed in Table 4-10 for thiophenoxide on 1-bromobutane.

Phosphonium salts [57] have been used in a variety of reactions and are generally more thermally stable than quaternary ammonium compounds and

Table 4-10. Comparison of some quaternary salts with different central onium ions for the displacement reaction of 1-bromobutane with aqueous sodium thiophenoxide.[a]

Catalyst	Rate constant for PhS⁻ reaction with 1−$C_8H_{17}Br^-$, (×10³) (L mol-sec⁻¹)		
	X = R_4N	X = R_4P	R_4As
$(C_4H_9)_4X^+Cl^-$	—	37	—
$(C_4H_9)_4X^+Br^-$	5.2	—	—
$(C_4H_9)_4X^+I^-$	7.4	—	—
$(C_6H_5)_4X^+Cl^-$	—	2.7	1.4
$(C_6H_5)_4X^+Br^-$	—	2.5	—

[a]Data from Herriott and Picker [14].

somewhat more active catalysts under neutral or acidic conditions, but they are much less stable than quaternary ammonium salts under strongly alkaline conditions. The greater cost of phosphonium salts compared to ammonium salts works against their use for industrial reactions, although in some cases the expense may be justified. Tetraphenylphosphonium salts show a high level of stability and have been used with excellent results in fluoride displacement reactions, such as the fluoro-denitration of 1-chloro-2,5-dinitrobenzene with KF to give 2-chloro-1-fluoro-4-nitrobenzene in one step [58]. Clark [59] showed that PTC reactions with fluoride anion transfer can be appreciably enhanced by use of tetraphenylphosphonium bromide as a catalyst. Rate accelerations are especially large in nondipolar aprotic solvents. Cyclic phosphonium and arsonium salts have been used as phase-transfer catalysts for Finkelstein and Kolbe reactions [60]. Tricyclohexyl-n-dodecylphosphonium bromide has been used as a phase-transfer catalyst for the reaction of mercaptide anions with 1,2,4-trichlorobenzene [61].

Gibson [62] first showed the use of tetraphenylarsonium compounds to be useful in PTC analytical chemistry by using these compounds for PTC analytical titration of highly organophilic unsaturated compounds with aqueous potassium permanganate. Other uses of arsonium compounds include alkylation of aldehydes [63], and preparation of ferrocenes [64]. Triphenylalkylarsonium salts have been used in PTC Wittig reactions [65].

Triphenylsulfonium halides are effective phase-transfer catalysts for nucleophilic substitution reactions of 1-bromooctane with KSCN, NaCN, NaOPh, and KI and for cyclopropanation of cyclohexene with dichlorocarbene [30]. The triphenylsulfonium cation is stable in the presence of concentrated sodium hydroxide solution [66].

1,5-Dithiacyclooctane promotes reaction of aromatic aldehydes with dialkyl sulfides–alkyl halides in aqueous NaOH–benzene two-phase systems or solid KOH–acetonitrile (solid-liquid two-phase system), producing oxiranes in a single-step process [67]. The dialkyl sulfides themselves acted as mediators to transfer alkyl groups to the aldehydes whereas sulfonium salts acted as phase-transfer catalysts.

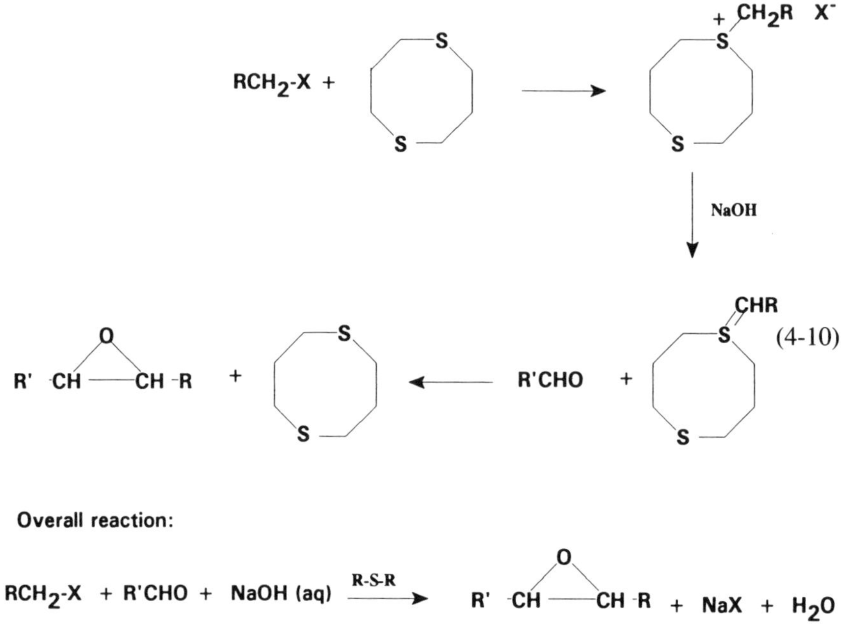

$$RCH_2\text{-}X + \underset{\text{(1,5-dithiocane)}}{\bigcirc} \longrightarrow \underset{\text{sulfonium salt}}{\bigcirc}^{+CH_2R\ X^-} \tag{4-10}$$

$$\downarrow NaOH$$

$$R'\overset{O}{\underset{CH-CH-R}{\triangle}} + \underset{}{\bigcirc} \longleftarrow R'CHO + \underset{}{\bigcirc}^{CHR}$$

Overall reaction:

$$RCH_2\text{-}X + R'CHO + NaOH\ (aq) \xrightarrow{R\text{-}S\text{-}R} R'\overset{O}{\underset{CH-CH-R}{\triangle}} + NaX + H_2O$$

Lauryl dimethylsulfonium and tributylsulfonium salts are highly efficient catalysts for the two-phase reaction of 1-bromooctane with KSCN in a benzene–water system, giving 88% yield of octyl thiocyanate [68]. Other PTC reactions using organosulfonium and sulfoxide compounds as phase-transfer catalysts have been reviewed [69].

Triphenylselenonium salts are said to be highly efficient phase-transfer catalysts for nucleophilic substitution reactions of octyl bromide and for addition reactions of dichlorocarbene to olefins [415].

2. Special Quaternary Salts as Phase-Transfer Catalysts

a. Heterocyclic Quaternary Salts

i. Derivatives of 4-Aminopyridinium Salts and Related Compounds; High-Temperature Catalysts

Several kinds of special quaternary salts have been found to have enhanced thermal stability and resistance to hydroxide-induced decomposition, including tetramethylammonium salts, tetraphenylphosphonium salts, and triphenylsulfonium salts, as well as nonquaternary salts, including PEGs, soluble polymers

Table 4-11. Catalyst stability in the presence of sodium phenoxide.

Catalyst	Solvent	Temp. (°C)	Half-life
Bu$_4$NBr	Toluene	110	7 min
I	Toluene	110	8 h
II	Toluene	110	11 h
II	PhCl	125	9 h
III	Toluene	110	12 h
IV	Toluene	110	2 h

containing dipolar aprotic groups, and crown ethers. However, for many commercial applications these catalysts have too little activity, are too expensive, or have other drawbacks. Brunelle and his co-workers at the General Electric Company demonstrated that quaternary salts based on 4-aminopyridine, and quaternized with an alkyl group having minimal β-hydrogens, are substantially more stable at high temperatures than conventional tetraalkylammonium salts, and are particularly useful for aromatic displacement reactions [70,71]. For example, starting with 4-hydroxypyridine, sequential reactions with (1) dibutyl amine, (2) neopentyl methanesulfonate, and (3) hydrogen bromide

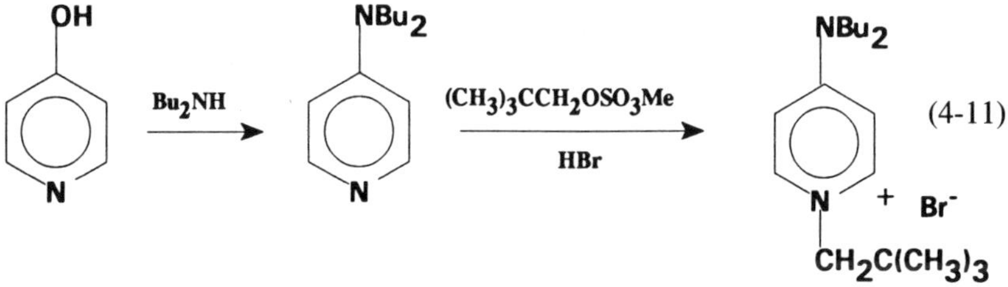

$$\text{4-hydroxypyridine} \xrightarrow{Bu_2NH} \text{4-(NBu}_2\text{)pyridine} \xrightarrow[HBr]{(CH_3)_3CCH_2OSO_3Me} \text{quaternary salt} \quad (4\text{-}11)$$

give a quaternary salt that is much more than 100 times as stable as tetrabutylammonium bromide when used as catalyst for the reaction of N-methyl-4-nitrophthalimide with the disodium salt of bisphenol-A [72]. The ethylhexyl derivative is somewhat less stable than the neopentyl salt, but is much less expensive to prepare, has higher organic-phase solubility, and overall is more cost-effective for commercial applications [73]. Saturated heterocyclic quaternary salts, such as 1,1-dibutylpiperidinium bromide, are also claimed to be more stable than conventional tetraalkylammonium quaternary salts [74]. The relative stability of the quaternary salts represented by these structures is illustrated in Table 4-11.

Some of Brunelle's results comparing these catalysts with other common PTC catalysts for preparation of 4-nitrophenylphenyl ether from sodium phenoxide and 4-chlorobenzene using 5 mol % catalyst at reflux is given in Figure 4-2 [71]. The Brunelle catalysts can be recovered from products by precipitation with diethyl ether, or they can be extracted into dilute aqueous solutions of mineral acids.

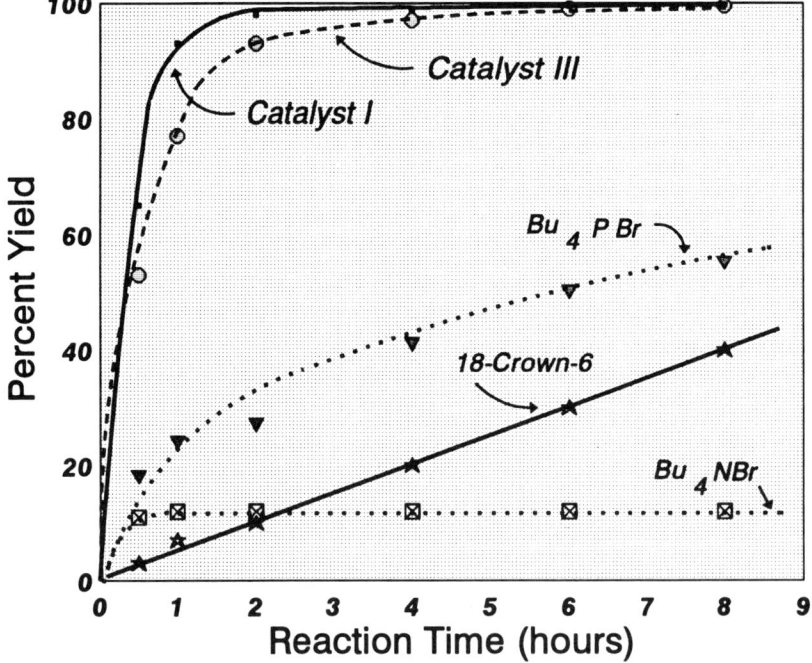

(4-12)

In addition to the 4-aminopyridinium derivatives, Brunelle and co-workers [75] have patented the use of hexaalkylguanidinium salts for use as high-activity phase-transfer catalysts for aromatic displacement reactions. Guanidinium salts have been used as higher-stability phase transfer catalysts for the preparation of diaryl ethers [423]. Hexaalkylguanadinium salts are made from tetraalkylguanid-

Figure 4-2. 4-Nitrophenylphenyl ether yields with various catalysts.

ine with alkyl bromide, potassium carbonate and a phase transfer catalyst [416]. Trialkylacylammonium salts, $Me(CH2)_nCONR3^+$, easily prepared by reaction of a trialkylamine with the corresponding acid chloride, were said to be superior catalysts for alkylation of phenylacetonitrile with alkyl halides in the presence of aqueous sodium hydroxide [417]. These materials were stable in strongly alkaline solutions at 40–120°C up to 3 hours.

Although the pyridinium catalysts are excellent for anhydrous reactions with solid sodium phenoxides, they have limited utility in the presence of aqueous sodium hydroxide or sodium sulfide, even at moderate temperatures. These catalysts decompose to pyridones or thiopyridones within hours at 100°C. Dehmlow and Knufinke [76] also compared the stability and catalytic activity of 11 different quaternary ammonium catalysts based on derivatives of 4-aminopyridine. Cantrell [77] patented the use of 4-aminopyridine derivatives similar to the GE quaternary salts for PTC catalysts for substitution of aryl halides by fluoride to produce fluoroaromatics.

Heterocyclic compounds containing no easily lost β-hydrogens and having the following structures,

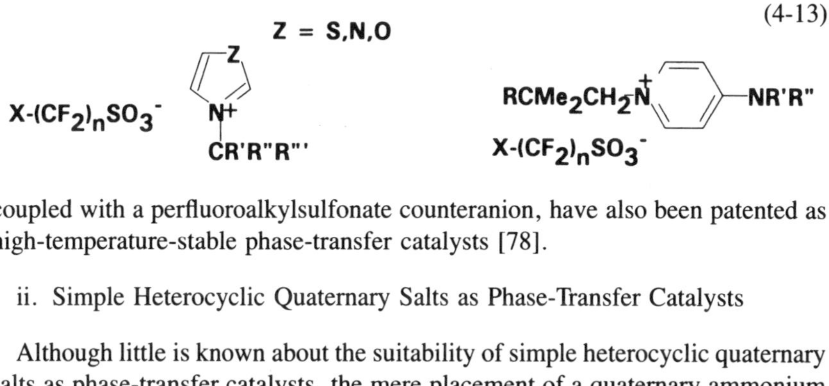

(4-13)

coupled with a perfluoroalkylsulfonate counteranion, have also been patented as high-temperature-stable phase-transfer catalysts [78].

ii. Simple Heterocyclic Quaternary Salts as Phase-Transfer Catalysts

Although little is known about the suitability of simple heterocyclic quaternary salts as phase-transfer catalysts, the mere placement of a quaternary ammonium center in a heterocyclic ring would not necessarily be expected to have a significant effect on their catalytic effectiveness. However, the more compact cyclic ring structures may affect physical properties, mainly solubility, and the cation–anion distances, which can affect catalytic efficiencies.

Heterocyclic quaternary ammonium salts based on morpholine and piperidine,

(4-14)

have been patented as phase-transfer catalysts for oxidation of *p*-xylene by oxygen, cocatalyzed by cobalt bromide [79], and Mitsubishi Petrochemical [80,81] patented heterocyclic quaternary ammonium salts for use as PTC etherification catalysts soluble in both organic and aqueous solvents.

Chiral azapropellanes having the structures

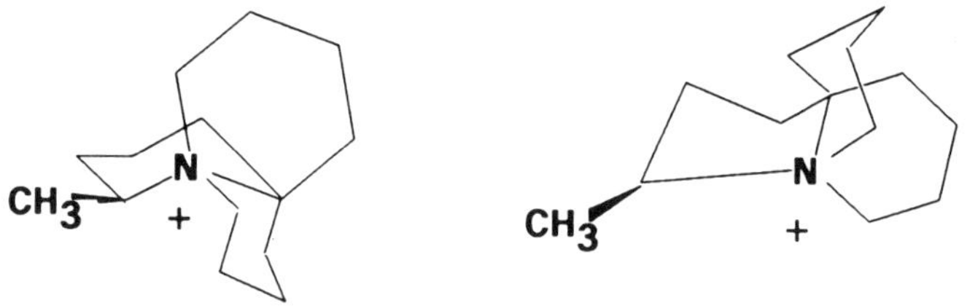

(4-15)

have PTC catalytic activities comparable to nonheterocyclic quaternary salts, and their use did not lead to significant asymmetric induction [82]. Similarly, chiral 3-hydroxy- and 3-(benzyloxy)-1-benzylquinuclidinium salts were tested as phase-transfer catalysts for a nucleophilic substitution and for two *C*-alkylation reactions [83]. These salts had activity comparable to that of benzyltriethylammonium chloride but lower than that of tetraalkylammonium bromides. Asymmetric induction with chiral 3-hydroxy-1-alkylquinuclidinium salts was lower than that with the conformationally mobile *N*-benzylquininium chloride. The chemically related cinchonidinium-based quaternary salts used by Dolling and associates are discussed in Chapter 12.

N-Methyl-*N'*-acylimidazole derivatives, easily prepared from *N*-methylimidazole and acid chlorides, were found to be superior catalysts for alkylation of phenylacetonitrile because of their high degree of stability [85].

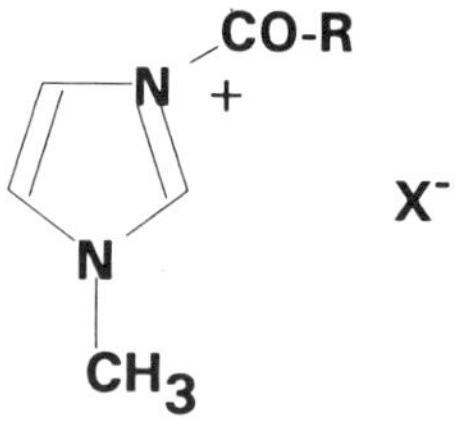

(4-16)

iii. Tetrazolium Salts as PTC Catalysts

Oxidation of alkylformazans, PhN = NCR:NNHPh, with potassium permanganate gives 2,3-diphenyl-5-alkyltetrazolium salts,

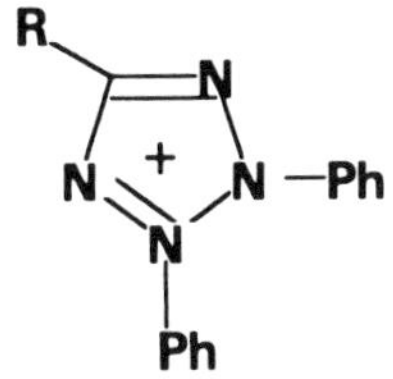

$$\underset{\underset{Ph}{N}}{\overset{R}{\underset{N^+}{\diagdown}}}\!\!\!\!\diagup\!\!\!\!\overset{N}{\diagdown}\!\!\!N-Ph \qquad (4\text{-}17)$$

which are themselves phase-transfer catalysts for this oxidation [86]. Similarly, 2,3,5-triphenyltetrazolium chloride was used as a phase-transfer catalyst for oxidation of toluenes and xylenes to toluic acids [87]. This tetrazolium salt, and hexadecyltrimethylammonium chloride, were the most effective catalysts tested for these reactions. 2,3,5-Triphenyltetrazolium chloride is as effective as tetrabutylammonium salts for oxidation of aldehydes to acids [88]. 2,3-Diphenyl-5-butyltetrazolium iodide [89] and 2,3,5-triaryltetrazolium [90] salts have been successfully used in azide reactions. Use of tetrazolium salts as phase-transfer catalysts has been reviewed [91].

b. Bis-(Quaternary Ammonium Salts) as Phase-Transfer Catalysts

Bis(quaternary ammonium salts) have a PTC activity similar to ordinary monoquaternary ammonium salts when used with monovalent anions, but the bis(quaternary salts) are clearly superior catalysts for use with divalent anions.

i. Monovalent Anion Reactions

Bis(quaternary ammonium salts) of the general structure,

$$R_3\overset{+}{N}\text{-}(CH_2)_n\text{-}O\text{-}(CH_2)_n\text{-}\overset{+}{N}R_3 \qquad (4\text{-}18)$$

are said to be useful as selective phase-transfer catalysts [92]. For example, alkylation of phenylacetonitrile with 1-bromobutane using aqueous NaOH, with bis-[2N-benzyl-N,N-diethylammonium)ethyl]ether dichloride, gave 70% monoalkylated product and 22% dialkylated product, compared to yields of 61% and 27%, respectively, when tetrabutylammonium bromide was used as catalyst under the same reaction conditions. Bis(pentaalkylguanidinium)alkane salts have been shown to function as phase-transfer catalysts useful for aromatic displacement reactions [93].

Table 4-12. Comparison of various quaternary salts for their ability to transfer multivalent anions into organic phase.

Quaternary Catalyst	Loading (moles of anion/mole of quaternary cation)					
	$Cr_2O_7^{-2}$	CrO_4^{-2}	MoO_4^{-2}	$Fe(CN)_6^{-3}$	Terephthalate	Phthalate
$Bu_3N^+(CH_2)_4N^+Bu_3Br$	0.88	<0.32	—	<0.1	0.05	0.1
$Hex_3N^+(CH_2)_4N^+Hex_3Br$	0.95	0.26	—	0.28	0.05	0.13
$Oct_3N^+(CH_2)_4N^+Oct_3Br$	0.81	<0.24	—	0.14	0.05	0.24
$Oct_3N^+(CH_2)_4N^+Oct_3HSO_4$	1.30	1.73	0.73	0.16	0.57	0.95
Bu_4N^+Br	0.47	0.075	<0.2	0.04	0.01	0.05
$Bu_4N^+HSO_4$	0.59	0.94	0.58	<0.03	0.02	0.08
Bu_4P^+Br	0.48	0.20	—	—	0.02	—
$Bu_3MeN^+HSO_4$	0.28	<0.19	—	<0.06	—	—
$Oct_3MeN^+HSO_4$	0.67	0.97	—	0.16	—	—
Oct_3MeN^+Cl	0.56	0.22	—	0.32	0.08	0.17
Hex_4N^+Br	0.50	0.14	—	0.18	—	—
Hex_4N^+Cl	0.56	0.25	—	0.28	0.10	0.12
$PhCH_2Et_3N^+Cl$	0.01	0.02	—	0.02	—	—
PEG 400	—	0.02	0.02	—	0.01	0.01
18-Crown-6	—	0.01	0.02	—	0.02	0.01

*a*X was the original anion associated with the quaternary salt.

Idoux [94] described the syntheses of a variety of multisite phase-transfer catalysts and the determination of their catalyst activity in some simple SN2 reactions and some weak nucleophile–weak electrophile SNAr reactions. In general, at the same molar ratio, the multisite phase-transfer catalysts are as effective or somewhat more effective than similar single-site phase-transfer catalysts.

ii. With Divalent Anions

Lissel and co-workers [95] found that bis(quaternary salts) transfer divalent anions into organic solutions more easily then monoquaternary salts. This comparison was done by equilibration of equal volumes of 0.001 M solution of various quaternary salts with 0.01 M aqueous solution of various inorganic salts, and measurement of the "loading" of the organic phase with polyvalent anion, that is, moles of anion per mole of quaternary salt, as typified by the data in Table 4-12.

Brunelle [71] demonstrated the use of bis(dialkylaminopyridinium salts) for PTC nucleophilic substitution by bisphenoxides to enhance reaction rates and reduce catalyst use. He found the dialkylaminopyridinium structure, when incorporated into bis(quaternary salts), functioned particularly well for aromatic displacement reactions of the dianion of bisphenol-A [96,97]. As illustrated, the compound where $n = 10$ methylene groups has a long enough chain to allow strain-free formation of 1:1 species,

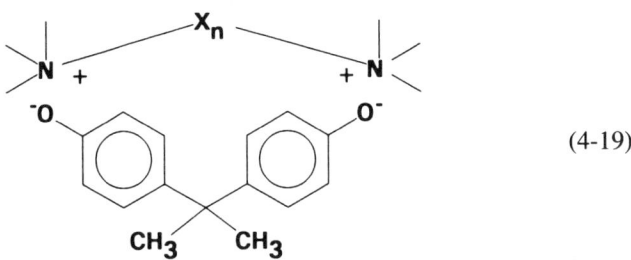

(4-19)

Formation of 1:1 ion pairs permits 1 mol of catalyst to transfer 1 mol of dianion, rather than requiring 2 mol of catalyst per mole of dianion. The kinetics of the reaction are first order in catalyst with the bis(quaternary salts), but are approximately second order in catalyst with monoquaternary salt catalyst. For example, with 4-chloronitrobenzene reacting with the dianion of bisphenol-A in a liquid–solid system, the time for complete reaction was reduced from > 6 h with 5 wt% monoquaternary salt to about 1 h with 2.5 wt% of bis(quaternary salt).

Unsaturated bis(quaternary ammonium salts) catalyze acetate displacement more effectively than saturated bis(quaternary salts) and monoquaternary salts. The bis(quaternary salt) 1,5-bis(*N*-benzyl-*N*,*N*-diethylammonio)-3-oxapentane dichloride produced higher yields at lower catalyst concentrations in several PTC Williamson reactions with diphenols than did more commonly used monoquaternary salt phase-transfer catalysts [98].

4,4′-Dialkylbipyridinium salts, for example, the dioctyl derivative, have been used as "electron carriers" between organic–aqueous phases in which nitroalkenes in the organic phase were reduced to the corresponding oximes and/or carbonyl compounds [99].

$$C_8H_{17}-\overset{+}{N}\overset{+}{N}-C_8H_{17} \qquad (4\text{-}20)$$

v. Cluster Quaternary Salts

Cluster quaternary salts are formed by clustering quaternary ammonium cations around a polyvalent anionic species, as, for example, in the structure around a polysulfonated triphenylphosphine [100]:

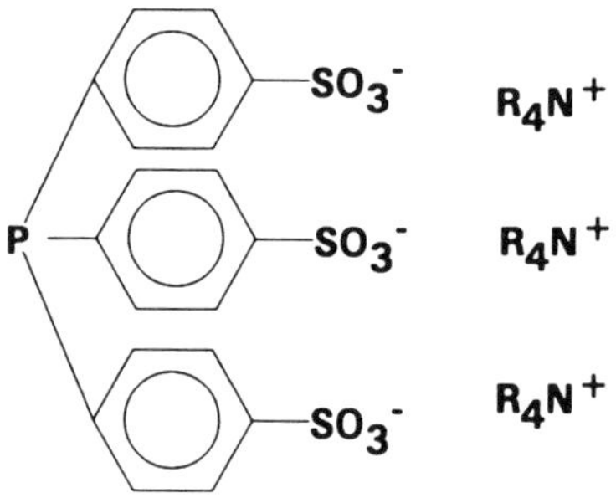

$$P \begin{pmatrix} \text{—}\bigcirc\text{—}SO_3^- & R_4N^+ \\ \text{—}\bigcirc\text{—}SO_3^- & R_4N^+ \\ \text{—}\bigcirc\text{—}SO_3^- & R_4N^+ \end{pmatrix} \qquad (4\text{-}21)$$

These clusters are claimed to be useful as phase-transfer catalysts, presumably by providing a highly polar ionic center which can easily accommodate additional ionic species as triplet or quadrupole species, soluble in the organic phase.

c. Ethoxylated Quaternary Salts (PEG–Quaternary Combination)

Because both quaternary ammonium salts and PEGs are good phase-transfer catalysts, it would be logical to expect that quaternary salts also having bound ethoxylate groups would be even better phase-transfer catalysts. Several experimental studies indicate that this is true.

Quaternary ammonium compounds having the structure $Me(OCH_2CH_2)_nN^+Me_2(CH_2)_{17}Me$ ($n = 1$–5) produced a higher level of PTC behavior than commonly used quaternary ammonium compounds and dibenzo-18-crown-6 for dichlorocarbene formation [101–103]. The same type of bis-ethoxylated quaternary salt produces high yields of mandelic acid from the reaction of benzaldehyde and chloroform with sodium hydroxide [104]. For dehydrohalogenation of alkyl halides the most effective PTC catalysts were quaternary salts having one alkyl group replaced by an ethoxylate group [105]. Urethane-functionalized polyoxyethylene quaternary ammonium chlorides were claimed also to be phase-transfer catalysts [106].

d. Quaternary Salts Containing Crown Ether Groups (Crown-Quaternary Salt)

Czech et al. [107], noting that crown ethers tend to do a better job as phase-transfer catalysts for solid–liquid reactions, whereas quaternary salts are better for liquid–liquid systems, prepared a crown ether–quaternary phosphonium salt as represented by structure I, and tested it as a phase-transfer catalyst, compared to similarly constructed monoquaternary, II, and crown ether, III, compounds.

Results from use of these catalysts, listed in Table 4-13, indicate that the presence of both a quaternary phosphonium group and a crown ether function do not significantly increase catalytic activity.

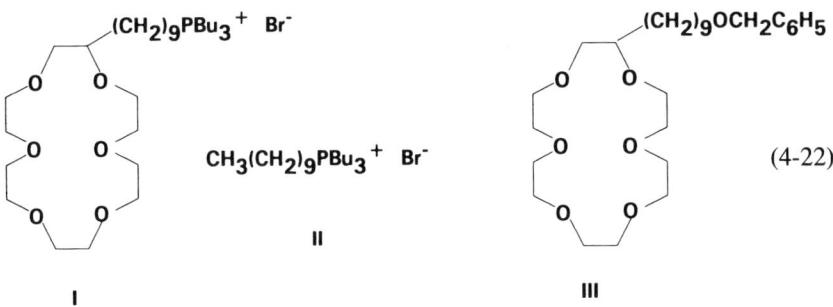

(4-22)

Quaternary salts having the quaternary cation center as a part of the crown polyether ring are said to be active phase-transfer catalysts [108]. A variety of crown ethers containing many possible functional groups have been patented as phase-transfer catalysts [109].

Table 4-13. *Comparison of crown–quaternary salts with crown and quaternary salt for catalytic efficiency.*

Catalyst	Catalyst conc. (mol%)	Liquid–solid rxn. $1-C_4H_9Cl + KCN$ $k \times 10^4$ (mol L$^-$sec^{-1})	Liquid–liquid rxn.^{a} $1-C_8H_{17}Br + KCN$ $k \times 10^6$ (sec^{-1})
Quaternary II	10	2.7	1030
Crown III	10	4.2	7.2
Crown-quaternary I	10	7.2	380
Q II + C III	10+10	6.8	990

^{a}Second-order rate constant for 1-chlorobutane with KCN in LS reaction in acetonitrile at 50°C.
^{b}Pseudo-first-order rate constant of 1-bromooctane with aqueous KCN at 90°C.

e. *Quaternary Salts Containing Other Functional Groups*

i. *Quaternary Salts Containing a β-Hydroxyl Group*

Hydroxyl groups attached to the β-carbon from a quaternary center can interact with the anion to increase its binding to the quaternary salt and displace some of the water of hydration around the anion. For most anions this is not too important, but for anions such as hydroxide and borohydride, which are highly hydrated, the added binding of the β-hydroxy group produces a substantial boost in PTC activity. This intramolecular binding may be represented as:

(4-23)

The stabilizing effect of β-hydroxyl groups has been utilized in anion-exchange resins to increase the selectivity of the resin for OH$^-$ anions many-fold [110].

Hiyama and co-workers [111,112] and Colonna and co-workers [113] have made particular use of ephedrine-based quaternary salts,

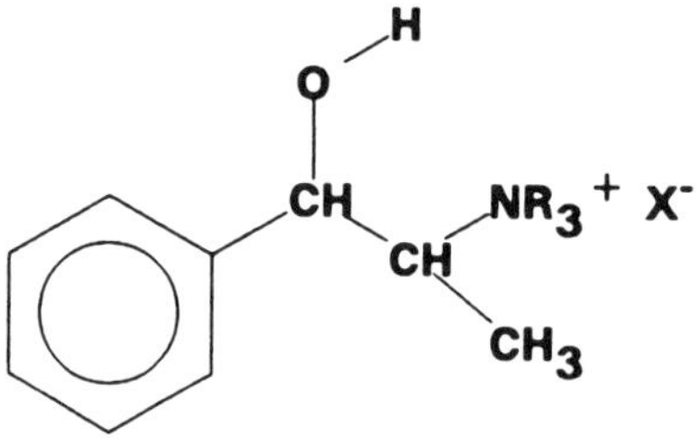

$$\text{PhCH(OH)CH(CH}_3\text{)NR}_3^+ \text{ X}^- \qquad (4\text{-}24)$$

which are 10–50 times more active than ordinary tetraalkylammonium salts as phase-transfer catalysts for borohydride reduction of 2-octanone.

Quaternary salts containing β-hydroxyethyl groups are also particularly useful for dehydrohalogenation reactions, and most of the chloroprene manufacturing processes throughout the world use quaternary salts containing one or more β-hydroxyethyl groups [114]. Similarly, 2,3-dichlorobutadiene is produced from dehydrochlorination of 2,3,4-trichloro-1-butene at −10° to +60°C in the presence of aqueous alkali hydroxides, inhibitor, and a catalyst of the structure PhCH$_2$(C$_{16}$H$_{33}$)N(CH$_2$CHMeOH)$_2^+$ Cl$^-$ [115].

Quaternary ammonium salts having β-hydroxyethyl groups appear to be more thermally stable than analogous quaternary ammonium salts without this substituent [116,117], probably due to formation of the zwitterionic (alkoxide) form [112]. Sulfhydryl groups in the β-position impart even greater thermal stability to quaternary cations in hot aqueous sodium hydroxide solution, presumably due to the greater acidity of the −SH group [118].

ii. Betaines as PTC Catalysts

Betaines, having the zwitterionic structure R$_3$N$^+$CH$_2$CO$_2^-$ are more effective catalysts, by 20- to 50-fold, than simple tetraalkylammonium salts for reaction of alkyl halides with aqueous sodium hydroxide solutions [119].

$$\text{R-Br (org)} + \text{NaOH (aq)} \xrightarrow{\text{PTC}} \text{ROH and R-O-R} \qquad (4\text{-}25)$$

The high rate for the betaine-catalyzed reaction is believed to be due to ability of the zwitterion to react with alkyl bromide to produce intermediate quaternary salt-ester, which rapidly hydrolyzes with aqueous NaOH, regenerating the betaine catalyst.

(4-26)

$$R_3N^+CH_2CO_2^- + R'Br \longrightarrow R_3N^+CH_2CO_2\text{-}R' \ Br^- \xrightarrow{OH^- \ (aq)} R_3N^+CH_2CO_2^- + R\text{-}OH$$

Betaines of the structure $Me_3N^+(CH_2)_nCO_2^-$, ($n = 1$–3) are effective phase-transfer catalysts for dichlorocarbene reactions with styrene, benzamide, and pyrrolidine [120].

β-Hydrazinobetaines of the structure $RN^+Me_2NHCH_2CH_2CO_2^-$, where R = alkyl, have been found to be excellent catalysts for dichlorocyclopropanation of olefins and for conversion of amides to nitriles by treatment with chloroform and aqueous sodium hydroxide [121]. Best results were obtained when R was n-hexyl, or n-decyl. For catalysis of etherification of n-propanol with benzyl chloride in methylene chloride–50% NaOH, these betaines were less effective than $Bu_4N^+HSO_4^-$.

iii. Quaternary Salts Containing Siloxane Groups

3-Siloxanylpropylammonium halides containing fluorocarbon groups have been used as phase-transfer catalysts for fluorocarbon syntheses [122]. α,α,ω-Trihydroperfluoralkyl allyl ethers of the structure $H(CF_2CF_2)_nCH_2OCH_2CH=CH_2$ ($n = 1$-3), useful in the preparation of fluorocarbon surfactants, were obtained in excellent yields with these quaternary salts in the PTC reaction of $H(CF_2CF_2)_nCH_2OH$ and allyl chloride.

C. Macrocyclic and Macrobicyclic Ligands [395–397]

The syntheses and cation complexing properties of macrocyclic and macrobicyclic ligands (crown ethers and cryptands, respectively) have been extensively reviewed. This section deals only with a discussion of the synthesis of those crowns and cryptands used most often as catalysts in phase-transfer processes. A "special" class of chiral crowns is also addressed.

1. Simple Crown Ethers

The synthesis of 12-crown-4, 15-crown-5, 18-crown-6, and their monobenzo-, dibenzo-, monocyclohexano-, and dicyclohexano-derivatives has been described by a number of researchers. Most of the procedures involve a modified Williamson ether synthesis (Figure 4-3). In many cases it has been found that certain metal cations increase the yield of cyclic product and it has been suggested that the metal cations act as templates by a mechanism similar to that shown in Figure 4-3.

Dibenzo-18-crown-6 was the first crown to be synthesized [398, 399]. Catechol was reacted with 2,2'-dichloroethyl ether in the presence of sodium hydroxide in refluxing n-butanol (Figure 4-4). A 40% yield of crown was realized. It is

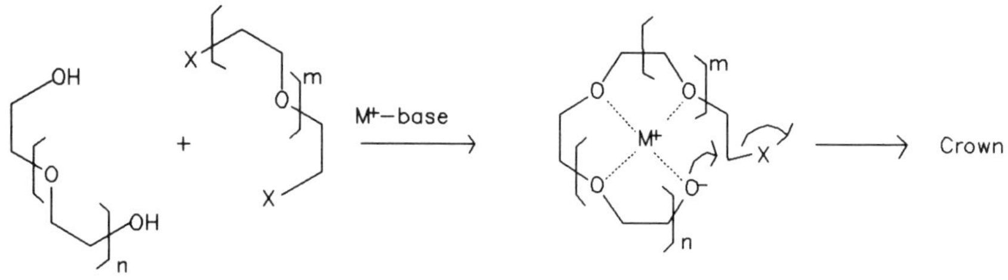

Figure 4-3

interesting to note that high dilution was not necessary. Catalytic hydrogenation using a ruthenium catalyst at high pressure produced an isomeric mixture of dicyclohexano-18-crown-6 products (Figure 4-5) [398–402].

18-Crown-6 has been synthesized by a variety of procedures [403–407]. For instance, reaction of triethylene glycol dichloride, triethylene glycol, and potassium hydroxide in tetrahydrofuran containing 10% water followed by precipitation of the crown–acetonitrile complex produced an overall yield of 25% (Figure 4-6) and represents a simple as well as an economical approach to this macrocyclic ligand.

The syntheses of 12-crown-4 [408] and 15-crown-5 [406,407] are outlined in Figures 4-7 and 4-8, respectively.

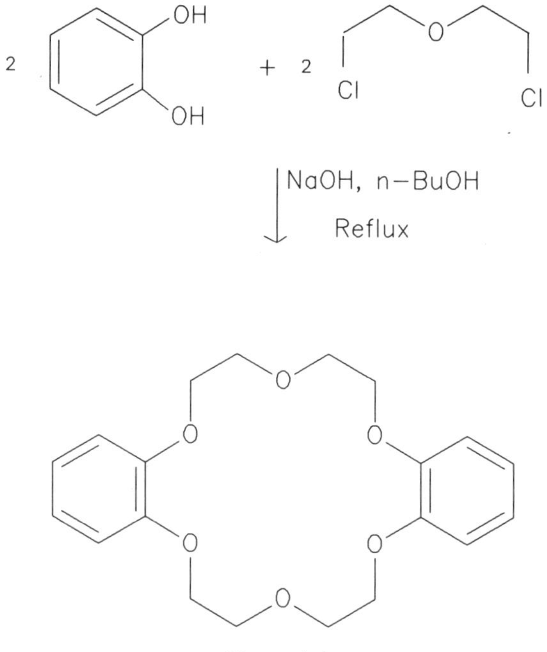

Figure 4-4

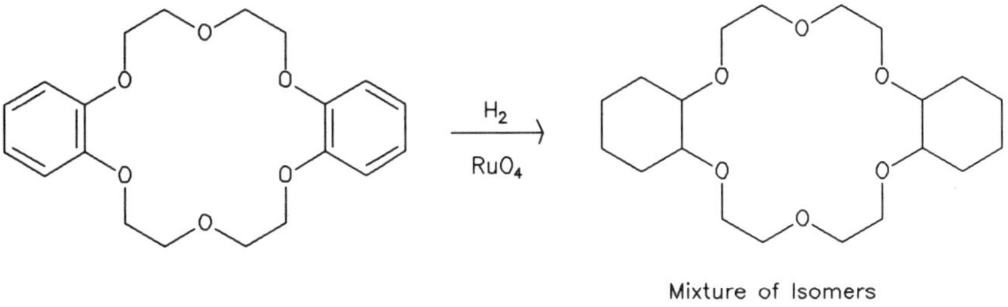

Figure 4-5

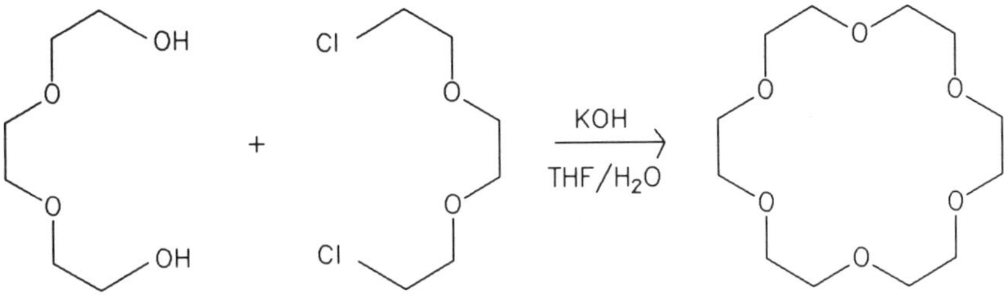

Figure 4-6

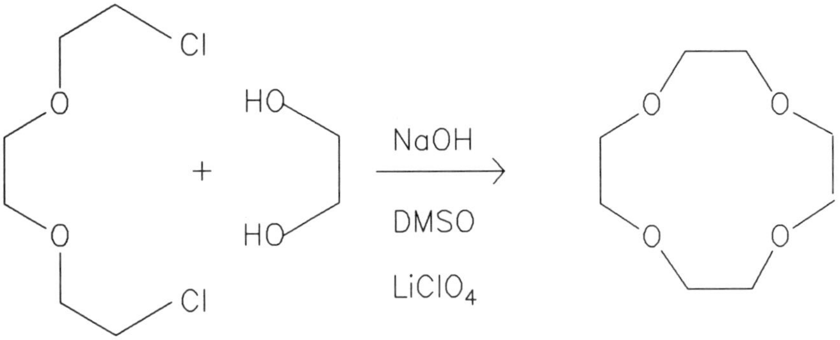

Figure 4-7

The effect of metal cation templates has been explored in the cyclic oligomerization of ethylene oxide using boron trifluoride as the acid catalyst in the presence of metal salts of BF_4^-, PF_6^-, and SbF_6^- [409–411]. The relative percentages of 12-crown-4, 15-crown-5, and 18-crown-6 are shown in Table 4-14.

2. Cryptands

Macrobicyclic multidentate ligands (cryptands) are usually superior to their macrocyclic counterparts in their ability to complex metal cations [395]. The syntheses

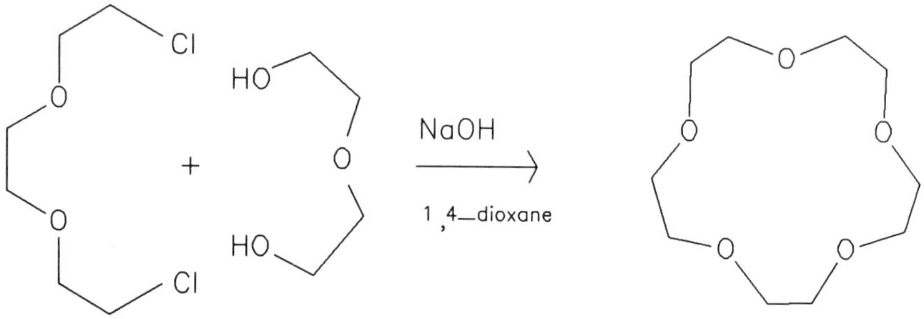

Figure 4-8

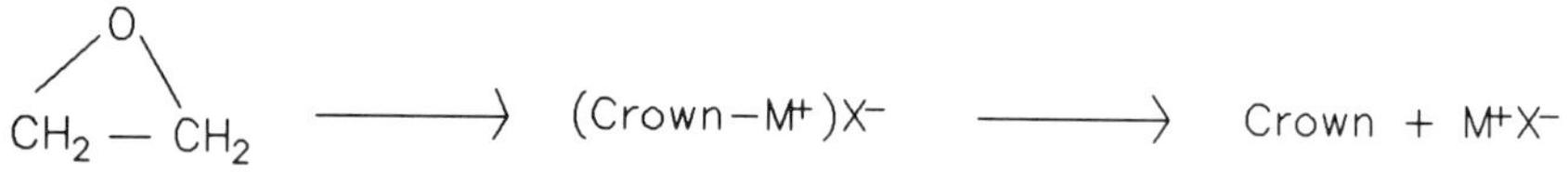

Figure 4-9

Table 4-14. *Percent product distributions for reactions in dioxane.*

M⁺X⁻	12-Crown-4	15-Crown-5	18-Crown-6
LiBF$_4$	30	70	—
NaBF$_4$	25	50	25
KBF$_4$	—	50	50
KPF$_6$	20	40	40
KSbF$_6$	40	20	40
RbBF$_4$	—	—	100
CsBF$_4$	—	—	100
Ca(BF$_4$)$_2$	50	50	—
Sr(BF$_4$)$_2$	10	45	45
Ba(BF$_4$)$_2$	10	30	60
AgBF$_4$	35	30	35
Hg(BF$_4$)$_2$	20	70	10
Ni(BF$_4$)$_2$	20	80	—
Cu(BF$_4$)$_2$	5	90	5
Zn(BF$_4$)$_2$	5	90	5

of these ligands, however, are far more intricate [395,396]. Figure 4-10 illustrates the scheme for the synthesis of the [2.2.2] cryptand [395,396,412]. The other members of the series have been synthesized by similar procedures [395,396].

3. Special Crowns

Bis-binaphthyl crowns have been effectively used as chiral complexing agents. A typical synthetic procedure is outlined in Figure 4-11 [413]. Optically pure binaphthol was reacted with potassium *t*-butoxide and the ditosylate of diethylene

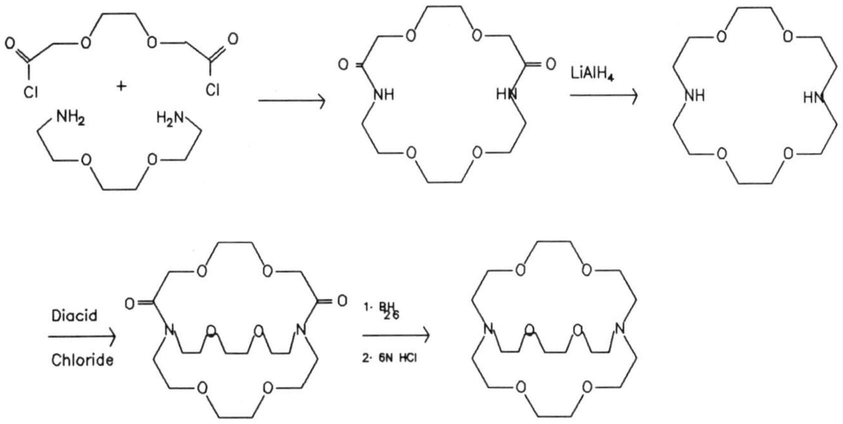

Figure 4-10

glycol in tetrahydrofuran at reflux for 15 h. The "1 + 1" and the "2 + 2" products were isolated in 5% and 31% yields, respectively.

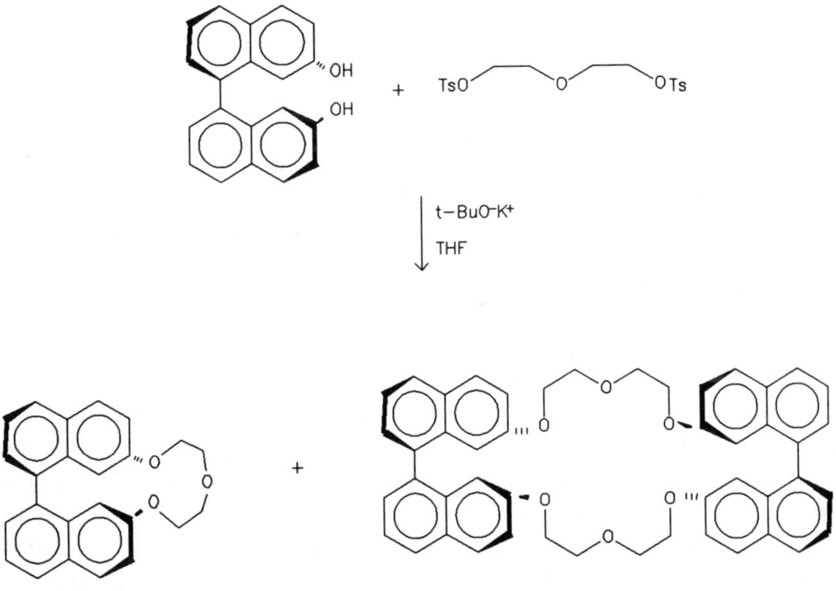

Figure 4-11

D. PEGs, Tris(3,6-dioxaheptyl)amine (TDA-1), and Related Ethoxylated Compounds as Phase-Transfer Catalysts

PEGs and their many derivatives have been extensively investigated as phase-transfer catalysts, and are used in many commercial processes for this purpose. These materials are inexpensive, thermally stable in the absence of strong acids, usually easy to remove and recover, nontoxic, easily biodegradable, and universally available. For some reactions such as with hydroxide transfer, PEGs are excellent catalysts, sometimes better than crown ethers, especially when used in liquid–solid PTC reactions with potassium salts, and with little or no added water, and with at least moderately polar organic solutions. For liquid–liquid reactions PEGs are usually mediocre or poor catalysts, or even noncatalytic. PEGs are water soluble and if the organic phase is not sufficiently polar the PEG will reside almost completely in the aqueous phase; or, with concentrated aqueous solutions of organic salts, the PEG may form a third catalyst-rich phase, a change that normally leads to a high level of catalytic activity.

1. Synthesis of PEGs

PEGs are made by reaction of ethylene oxide with a limited amount of water, using sodium or potassium hydroxide as a catalyst:

$$\text{(ethylene oxide)} + H_2O \xrightarrow{\text{NaOH}} HO(\text{-}CH_2CH_2O\text{-})_nH \quad (4\text{-}27)$$

The number n of repeating $-CH_2CH_2-O-$ethoxylate groups in PEGs is an important parameter for their use as phase-transfer catalysts. Average values of n equal to or greater than 8 are generally preferred for good catalyst activity. Polyethylene oxide oligomerization products tend to follow statistical distributions for higher products ($n_{\text{avg}} = 4$ or higher), and these mixtures are normally used for most commercial applications. If PEGs with exactly defined values of n are required, say for a kinetics study, then these products must be synthesized by appropriate synthetic procedures [123]. Ethoxylate chains bound to insoluble resins or adsorbed on certain inorganic solids can be used as insoluble phase-transfer catalysts, as discussed in Chapter 5.

PEGs having one hydroxyl group capped, for example, by an alkyl group, have been frequently examined as phase-transfer catalysts, because the organic-phase solubility of these can be easily adjusted by the size of the alkyl or aryl group attached. These compounds are easily and inexpensively manufactured from compounds containing an active hydrogen by similar reactions with ethylene oxide. Doubly substituted PEG derivatives, where one of the substituents is an alkyl group, are prepared by simple etherification with an alkyl halide or alkyl methanesulfonate (X = Cl, Br, I, O_3SCH_3), as represented in the last two equations of 4-28 [124,125].

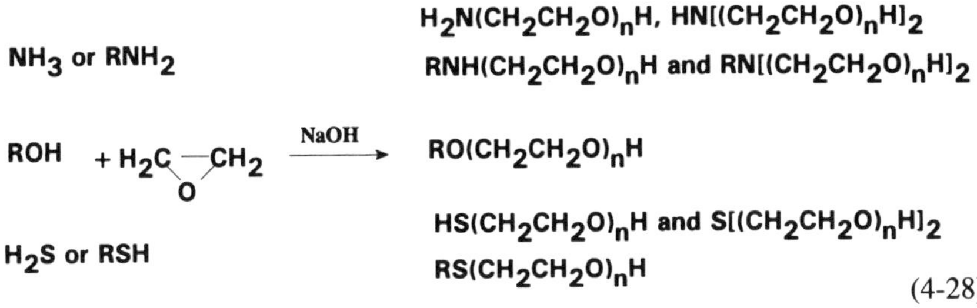

NH$_3$ or RNH$_2$

H$_2$N(CH$_2$CH$_2$O)$_n$H, HN[(CH$_2$CH$_2$O)$_n$H]$_2$
RNH(CH$_2$CH$_2$O)$_n$H and RN[(CH$_2$CH$_2$O)$_n$H]$_2$

ROH + H$_2$C—CH$_2$ $\xrightarrow{\text{NaOH}}$ RO(CH$_2$CH$_2$O)$_n$H
         \O/

H$_2$S or RSH

HS(CH$_2$CH$_2$O)$_n$H and S[(CH$_2$CH$_2$O)$_n$H]$_2$
RS(CH$_2$CH$_2$O)$_n$H

(4-28)

HO(CH$_2$CH$_2$O)$_n$H   RO(CH$_2$CH$_2$O)$_n$R
                         + RX $\xrightarrow{\text{NaOH}}$
Y-(CH$_2$CH$_2$O)$_n$H   Y-(CH$_2$CH$_2$O)$_n$R

Unsymmetrically substituted PEGs are readily prepared by combining two monosubstituted PEGs, after converting the hydroxyl group of one to a displaceable group, X, such as methanesulfonate:

(4-28A)

R-(CH$_2$CH$_2$O)$_n$H + Br-(CH$_2$CH$_2$O)$_m$-R' $\xrightarrow{\text{NaOH}}$ R-(CH$_2$CH$_2$O)$_{m+n}$-R'

2. Phase Distribution Behavior of PEGs

a. Ability of PEGs to Complex with Inorganic Compounds

Polyethylene oxide chains form complexes with cations, much like crown ethers, and these complexes cause the anion to be transferred into the organic phase and to be "activated" [126]. The data in Table 4-15 show that stability constants, K, for PEG complexes with sodium anion depend on both the value of n (i.e., average molecular weight of PEG) and on the end-group substituents:

Gokel and co-workers [130,131] determined the binding strength for Na$^+$ in anhydrous methanol solution with PEGs in the molecular weight range 200–14,000. The stability constants, K_s, ranged from 44 to 12,000 and when plotted as log K_s vs. log(mol wt), gave a straight line with a slope of about 1.4 [131]. They concluded that the strength of complexation is a function of the total number of binding sites present and not the number of polymer chains, suggesting that a long PEG chain may be involved in binding more than one cation.

PEGs are themselves soluble in water and in dilute aqueous solutions. To obtain partitioning of PEG into an organic solution may require use of a mono- or diether derivative. For example, Harris and Case [124] examined the relationship

Table 4-15. *Stability constants for complexation of PEGs and some symmetrical derivatives with sodium cation [127–129].*

$$Na^+ + PEG \overset{K}{=} [Na^+ \cdot PEG] \text{ complex}$$

Avg. PEG	Avg. n	Log K (stability constant) with various sodium salts				
		HO–	CH_3O-	C_2H_5O-	PhO–	$cyC_5H_{10}N-$
200	4.1	1.64	—	—	0.5	—
300	6.4	2.02	1.55	1.25	1.05	1.16
400	8.7	2.26	—	—	1.49	1.51
600	13.2	2.59	2.09	1.99	1.87	—
1000	22.3	2.88	2.55	2.48	2.37	2.46
1500	33.7	3.09	2.86	2.80	2.68	—
2000	45.0	3.28	3.08	3.05	2.81	3.08

between the ease of permanganate transfer into benzene and the rate of oxidation of 1-octene. These authors noted that with the exception of methylene chloride as an organic phase solvent, most PEGs are themselves partitioned into the aqueous phase, depending also on the concentration of dissolved salts in the aqueous phase [132]. ($KMnO_4$ is only soluble in water to the extent of about 5g/100 ml at room temperature.) To improve organic-phase solubility of PEG they prepared and tested several dialkyl ethers of PEGs as permanganate-PTC catalysts with partitioning results as shown by the data in Table 4-16.

The purple coloration of permanganate in benzene could be observed when small amounts of ethers were added to potassium permanganate–water–benzene systems, but not when unetherified PEGs were added.

Table 4-16. *PEG-dialkyl ethers, partitioning in benzene/water.*

PEG Ether [a]	Partitioning[b] (% in C_6H_6/% in Water)
C_4–PEG1500–C_4	14
C_6–PEG1500–C_6	84
C_{18}–PEG6000–C_{18}	Emulsion
C_{18}–PEG 750–Me	108
C_{18}–PEG1900–Me	39
C_{18}–PEG5000–Me	37
C_8–PEG5000–Me	12
C_4–PEG5000–Me	13
PEG6000	<1

[a] The designation "C_4–PEG1500–C_4" represents a PEG of mol wt 1500, capped at both ends by a n-butyl group.

[b] The partitioning ratio = [PEG in organic phase]/[PEG in aqueous phase] · 100.

Table 4-17. Transfer of various potassium and sodium salts from the solid phase to PEG 400.

Anion	K^+ transferred mol%	Na^+ transferred mol%
OH	38.7	22.2
F	26.1	<0.01
Cl	15.8	9.7
Br	38.3	37.0
I	30.5	33.0
SCN	43.6	39.2
HCO_3	10.2	0.7
HF_2	4.1	<0.05
HSO_4	9.4	—
NO_2	58.4	40.7
NO_3	38.4	28.8
CO_3^{-2}	9.6	0.08
SO_4^{-2}	0.1	0.01

Complexation of solid alkali metal salts by PEG is strongly dependent on both the anion and the cation of the salt, as shown by the data in Table 4-17 [133, 134].

From these studies Sasson and co-workers concluded:

- Potassium salts are more easily complexed than sodium salts.
- Anions capable of hydrogen bonding with the −OH groups of the polyglycol, such as OH^-, F^-, HSO_4^-, and HCO_3^- are transferred from the solid potassium salt to the glycol with relative ease. These anions were believed to be relatively free of hydration.
- Other anions such as Cl^-, Br^-, I^-, SCN^-, NO_3^-, NO_2^-, and HF_2^- are believed to exist as dissociated anions with significant solvation shells.

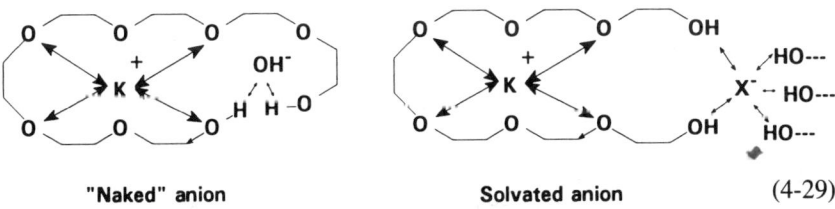

"Naked" anion   Solvated anion   (4-29)

Anion extraction data listed in Table 4-17 indicate that best results will be obtained with PEGs as phase-transfer catalysts (1) when the PEG molecular weight is high, (2) with potassium salts, (3) with a polar solvent, and (4) with very little added water. In general practice these expectations have been borne out, although in some reaction mixtures the decreasing organophilicity of PEG with increasing chain lengths causes the medium chain length ($n = 7-9$) PEGs to have the highest catalytic efficiencies, [135–137].

b. Solvent Effects

Neuman et al. [133] concluded that with PEGs as phase-transfer catalyst, it was important to select a good organic-phase solvent such as aromatic hydrocarbons, chlorinated hydrocarbons, or acetonitrile. Aliphatic hydrocarbons are immiscible with PEGs. In toluene, for benzyl chloride reaction with solid potassium acetate, PEGs were more effective catalysts than crown ethers; in butanol the effectiveness of polyethylene oxides and crown ethers as phase-transfer catalysts was the same [138].

c. Third-Phase Formation with PEG

PEG forms a third phase between toluene and aqueous KOH when some methanol is added [139]. Thus, dehydrobromination of 2-bromooctane in toluene, using PEG as catalyst, could be accomplished by removal and replacement of the organic and aqueous phases after completion of reaction, and recycle of the catalyst phase. After four cycles no catalyst was lost. Reaction rates for KOH dehydrohalogenation of 2-bromooctane in toluene with PEG catalysts were increased by a maximum factor of 126 by addition of methanol [140]. The base efficiency (moles base per mole of catalyst) PEGs with molecular weights 3000 and 20,000 exceeded unity and reached a maximum of 12 on addition of methanol.

3. PEGs and Ethers as Phase-Transfer Catalysts

a. General

Since Lehmkuhl and co-workers [141] introduced the use of PEG and its derivatives as phase-transfer catalysts, the popularity of these compounds as phase-transfer catalysts has grown dramatically as shown by the number of reviews [142–148]. PEG has been referred to as "a poor chemist's crown ether" [149].

b. Anion Activation by PEGs

The polyethylene oxide structure seems to be rather special in its ability to behave as an anion-activating phase-transfer catalyst, as Yamazaki et al. [150] have demonstrated in the homogeneous reaction of sodium phenoxide with 1-bromobutane in dioxane, Table 4-18. Likewise, Slaoui and co-workers [151] have shown that PEGs cause significant activation of hydroxide and acetate anions, in saponification reactions and in acetate displacement reactions in homogeneous solutions. Also, Szabo and co-workers [152] showed the catalytic effect of PEG chains.

Table 4-18. Catalytic activity of polyethers in reaction of sodium phenoxide with 1-bromobutane [150].

Polyether	Rate constant (10^6 mol L-min^{-1})[a]
None	0.35
HO(CH$_2$CH$_2$O)$_n$H	39.0
HO(CH$_2$CH$_2$CH$_2$O)$_n$H	0.4
HO(CH$_2$CH$_2$CH$_2$CH$_2$O)$_n$H	0.8

—(CH$_2$—CH)$_n$—
       |
    OCH$_2$CH$_3$
	0.4

[a]Rxn. 0.5 mmol PhONa, 5.4 mmol 1-BrC$_4$H$_9$, 2.7 mmol polyether, 20 mL dioxane.

Table 4-19. Rates for PTC-dehydrobromination of 2-bromoethylbenzene with KOH [153, 154].

Catalyst	$k \times 10^5$ sec^{-1}[a]
PhCH2N$^+$Et$_3$Cl$^-$	2.5
18-Crown-6	4.9
HO(CH$_2$CH$_2$O)$_1$H	<0.01
HO(CH$_2$CH$_2$O)$_2$H	<0.01
HO(CH$_2$CH$_2$O)$_3$H	1.0
HO(CH$_2$CH$_2$O)$_4$H	88.0
HO(CH$_2$CH$_2$O)$_5$H	208.0
HO(CH$_2$CH$_2$O)$_{13.5}$H	130.0
n-C$_4$H$_9$O(CH$_2$CH$_2$O)$_{13.2}$C$_4$H$_9$	<0.01

[a]Pseudo-first-order rate constants for 115 mL 0.2 M 2-bromoethyl-benzene in 0.5 mL of 60% aq. KOH with 0.1 equivalent catalyst at 35°C.

c. Behavior as Phase-Transfer Catalysts

PEGs, as would be expected from the complexation data given in Table 4-19, are especially good phase-transfer catalysts for reactions involving hydroxide ion. For example, Sasson and co-workers [153,154] found that in dehydrobromination of 2-bromoethylbenzene to styrene under mild conditions, the higher PEGs were superior to all other catalysts tested.

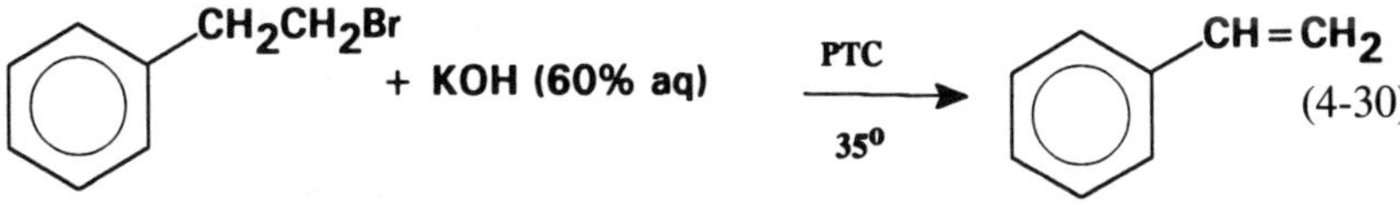

$$\text{PhCH}_2\text{CH}_2\text{Br} + \text{KOH (60\% aq)} \xrightarrow[35°]{\text{PTC}} \text{PhCH=CH}_2 \qquad (4\text{-}30)$$

The importance of complexation by terminal hydroxyl groups on the PEG for hydroxide ion transfer is shown by the lack of catalytic activity by the dibutyl ether of PEG. However, in PEG-catalyzed oxidation of picolines and diarylmeth-

Table 4-20. Comparison of catalysts for PTC cyanide displacement on 1-chlorooctane [131].[a]

Catalyst[b]	Amt. g (mol%)	Relative Rate[c]
Aliquat 336	0.404 (1.5)	269.
None	—	0
PEG-400	0.400 (1.5)	1.0
PEG-400	3.400 (12.25)	18.7
PEG-3400	3.400 (1.5)	1.5
PEG-MME-750	0.750 (1.5)	0.3
	3.400 (6.8)	2.5
18-Crown-6	0.264 (1.5)	0.8
18-Crown-6	3.400 (19.3)	27.

[a]Reaction at 105°C.
[b]MME = monomethyl ether, "Aliquat 336" is a commercially available quaternary ammonium salt with composition of approximately trioctylmethylammonium chloride.
[c]The relative rate of 1.0 corresponds to a rate constant of 1.34×10^{-6} $(M\text{-sec})^{-1}$.

anes with oxygen in the presence of strong bases, the dimethyl- and monomethyl ethers of PEG are more active than nonmethylated PEG [155].

PEG 400 strongly catalyzes elimination of hydrogen halides from 1,2-dihalides to yield 1- and 2-alkynes as well as 1,2-dienes. However, if quaternary salts are used as catalysts, along with bis-tertiary alcohols as cocatalysts, then 1-alkyne is the only product [156].

For reasons where considerable anion activation is necessary to achieve good PTC rates, PEGs may be relatively poor catalysts. For example, Gokel and coworkers [130] examined PTC activities of PEGs, crown ether, and "Aliquat 336" for liquid–liquid cyanide displacement on 1-chlorooctane with results as listed in Table 4-20.

Not all displacement reactions require so much activation as 1-chlorooctane, and for these PEG catalysts may be excellent phase-transfer catalysts. Thus, polyethylene oxide gels were highly effective catalysts for the phase-transfer substitution reaction of benzyl chloride with potassium acetate, although they were easily poisoned by the substrate and products, presumably by etherification of the PEG end-hydroxyl groups [135]. In reactions of 1-bromobutane and other alkyl halides with sodium acetate, PEGs were good phase-transfer catalysts, having increasing catalytic activity with increasing molecular weight [157]. Ester yields increased also with solvents of higher dielectric constant, but reactions were poisoned when iodide was present. Good catalysis was observed in the Williamson ether synthesis for reaction of 1-bromobutane with potassium phenoxide, where higher PEG oligomers had a high degree of activity, comparable to that of 18-crown-6 [158].

d. Tabular Survey of Use of PEG as Phase-Transfer Catalyst

Table 4-21 lists examples for the use of PEGs and their ethers as phase-transfer catalysts in a number of different types of reactions.

4. Special Ethoxylate Structures: Ethoxylate Derivatives as PTC Catalysts

a. Tris(3,6-dioxaheptyl)amine (TDA-1)

Tris(3,6-dioxaheptyl)amine, sold under the name TDA-1 by its manufacturer,

$$[CH_3\text{-}O\text{-}CH_2CH_2\text{-}O\text{-}CH_2CH_2]_3N \qquad (4\text{-}31)$$

is an especially effective phase-transfer agent for transfer of sodium and potassium salts from the solid into organic solutions and it is a highly effective phase-transfer catalyst, particularly for solid–liquid reactions [215]. A review on use of TDA-1 as a phase-transfer catalyst states that this material is more soluble than crown ethers, and has a wider application and lower toxicity than crown ethers [215]. This amine is also commercially available (although its use is covered by patents), and it is easily removed and recovered from reaction mixtures by extraction into dilute aqueous acid solutions. TDA-1 is prepared by reductive aminolysis of $MeO(CH_2CH_2O)_2H$ [216].

A particularly interesting application for TDA-1 is in dissolving metal reactions with sodium and/or potassium. This amine produces a deep blue solution from sodium–potassium alloy in tetrahydrofuran, and was found to be useful for deoxygenation of acetates, dehalogenation, and hydrolysis of tosylates and mesylates [217]. Hexaalkyldisilanes are prepared by reaction of trialkylsilyl halides with alkali metals in hydrocarbon solvents in the presence of alkali metal salts and TDA-1 phase-transfer catalysts [218]. In the Ullmann reaction there is synergistic catalytic activity between TDA-1 and CuCl [218].

TDA-1 has been reported as a good catalyst for several other phase-transfer reactions, including production of azo dyes for which TDA-1 was a better catalyst than quaternary salts or PEGs [219]. Also, esterification reactions [220], alkoxydechlorination of pyridines [221], aromatic nucleophilic displacements with potassium thioacetate [222], and production of N'-(tetrahydrofuran-2-yl)-uracil antitumor agents [223] are applications for TDA-1.

Derivatives related to TDA-1 have been made as linear, star-shaped, and comb-shaped graft copolymers by treating N,N'-dimethylethylenediamine (linear), diethyltriamine (5-branched star), triethylenetetramine (6-branched star), or polyethyleneimines (comb-shaped) with a preformed graft prepared from polyethylene glycol monomethyl ether and 1,6-diisocyanatohexane [224]. All these

Table 4-21. Examples of PEG-used as Phase-transfer catalysts.

Reaction	Products	References
Dehydrohalogenation, alkyl halides		[153, 154]
1,2-Dihalides	1- and 2-Alkynes	[156]
1,1,2-Trichloroethane	Chloroolefins	[159]
1-Bromoethylbenzene	Styrene	[134]
Williamson ether syntheses		[160–165]
Phenols/chloronitrobenzenes	Ethers (22–88%)	[166]
1-Naphthol/glycidyl chloride	Naphthylglycidyl ether (86%)	[167]
Methanol/chloronitrobenzene	Methyl nitrophenyl ether	[168]
Alcohols/halobenzenes	Alkylaryl ethers	[169]
Phenols/ClCH$_2$CO$_2$H	ArOCH$_2$CO$_2$H	[170]
Alcohols and polyols	Monoglycidyl ethers	[171]
Substitution reactions		[172–174]
Benzyl chloride/KOAc	Benzyl acetate	[135]
PhCOCH$_2$Br/K$_2$CO$_3$/ArCO$_2$H	ArCO$_2$CH$_2$COPh	[175]
1-Chlorooctane/NaCN	1-Cyanooctane	[130]
Acid chlorides/Na$_2$S	(RCO)$_2$S (91–98%)	[176]
Hydroxyprogesterone-21 + iodide/NaOAc	Hydroxyprogesterone-21 acetate (94%)	[177]
Desyl chloride/RCO$_2$Na	Desyl esters (excellent)	[178]
1−C$_4$H$_9$Br/NaOAc	Butyl acetate (52–92%)	[157]
Halohydrocarbons/NaOAc	Acetate esters	[178]
Benzyl bromide/KOAc	Benzyl acetate	[138, 179]
Acid chlorides/Na$_2$S	Diacyl sulfides (91–98%)	[176]
POCl$_3$/ArONa	(RO)$_3$P (90–96%)	[180]
Hydrolysis		
Beeswax + H$_2$O/NaOH	Fatty acids, glycerol	[181]

N-Alkylation

PEGs were as effective as crown ethers or tetraalkylammonium salts catalysts for *N*-alkylation of pyrrole, indole, and other nitrogen heterocycles using solid KOH [182, 183]. Diethyl or dibutyl ethers of PEG 400 or 1000 gave the best results. First-order kinetics were observed in alkylation of imidazole by alkyl halides [184].

Amides, imides, lactams	*N*-alkylated products	[185]
Caprolactam	*N*-alkylated products	[186]
Pyrrole, indole RX/KOH	*N*-Alkyl derivatives	[187]
Imidazoles + RX/base	*N*-Alkylimidiazoles	[183]
1,2,4-Triazole ClCH$_2$COMe$_3$	1-*N*-Alkylated products	[188]
Indoles + RX/KOH	*N*-Alkylindole (45%)	[189, 190]
Phthalimide/RBr/K$_2$CO$_3$	*N*-Alkylphthalimides (63–94%)	[191]
ε-Caprolactam /1−C$_{12}$H$_{25}$Br	"Azone" (72%)	[192]
Acylanilines + Alkoxymethyl halides	*N*-Alkoxymethylanilines	[193]
Aromatic amines (K-salt) RX	*N*-Alkyl derivatives	[189]

Table 4-21. (Continued).

Reaction	Products	References
C-Alkylations		
Cyclopentadienes/RX/KOH	Alkylcyclopentadienes	[194]
Phenylacetonitrile/RX/KOH	α-Alkylphenylacetonitriles	[195, 196]
α-Triazolylpinacolones + RX + KOH	C-alkylation products	[197]
Ethyl acetoacetate + BrCH$_2$CH$_2$Br/ K$_2$CO$_3$	Ethyl 1-acetylcyclopropane −1−carboxylate (79%)	[198]
Olefin isomerization		
4-Methoxyallylbenzene/KOH	Isomers	[199]
Allylbenzene (gas phase) K$_2$CO$_3$/PEG/Al$_2$O$_3$	Isomers	[200]
Dichlorocarbene reactions		
Phenol	Salicylaldehyde (Reimer Tiemann reaction)	[201]
ClCH = CHCH = CH$_2$	2-(2−Chlorovinyl)−1,1−dichlorocyclopropane (84%)	[202]
cis-1,4-Poly(butadiene) rubber	Dichlorocyclopropanated rubber	[203]
Olefins and others	Dichlorocarbene products	[204]
PhHgCl	PhHgCCl$_3$ (70–95%)	[205]
Condensations		
Darzens condensation PhCHO and PhCH = CHCHO/ClCH$_2$CO$_2$Et/K$_2$CO$_3$	Glycidic esters (50–77%)	[206]
Biltz synth:benzil and urea	Phenyltoin (87–93%)	[207]
Carbozate cyclization		[208]
Oxidations and reductions		[209–211]
PEGs and their dimethyl ethers are excellent phase transfer catalysts for borohydride reductions.		
Benzyl bromide + NaBH$_4$	Toluene	[212]
Steroid ketones + NaBH$_4$	Alcohols	[213]
Carbonyls + NaBH$_4$	Alcohols	[154]
Aryl methanes/O$_2$/strong base	Diaryl ketones	[155]
Benzyl alcohol/NaOCl	Benzaldehyde	[214]
Olefins/KMnO$_4$	Carboxylic acids	[150]

polymers were efficient complexing and phase-transfer agents for metal ions. The star-shaped block copolymers with two arms of polystyrene and two arms of polyethylene oxide were successful for the Williamson ether synthesis, with catalytic efficiency increasing as the molecular weight of the polymer increased [225,226].

b. Polypod Polyethers as Catalysts

Molecules having multiple polyethylene oxide *legs* attached are called *polypod ethers* or *octopus ethers* [227,228], and these have been demonstrated to be reasonably good phase-transfer catalysts.

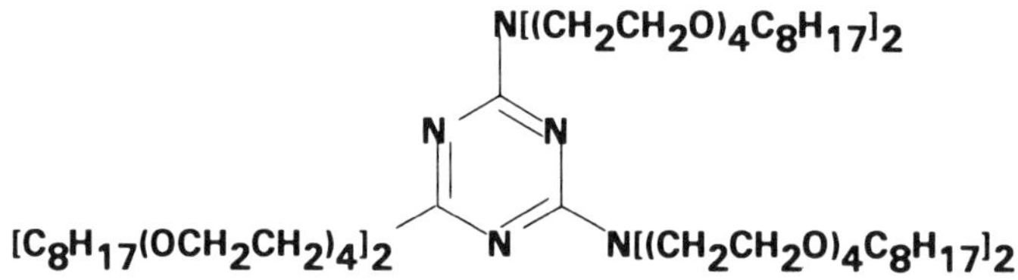

$$[C_8H_{17}(OCH_2CH_2)_4]_2 \underset{\text{"Polypode Ethers"}}{\overset{N[(CH_2CH_2O)_4C_8H_{17}]_2}{\text{triazine}}} N[(CH_2CH_2O)_4C_8H_{17}]_2$$

(4-32)

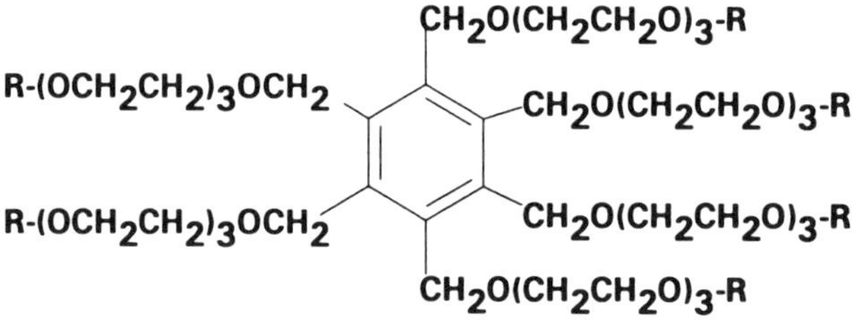

$$\text{"Octopus Ethers" — hexasubstituted benzene with six } -CH_2O(CH_2CH_2O)_3\text{-R arms}$$

Acryclic polypodans have also been made from sucrose [229], and the commercially available material, Polysorbate 80, also known commercially as TWEEN 80, has a complex polyethoxylate structure,

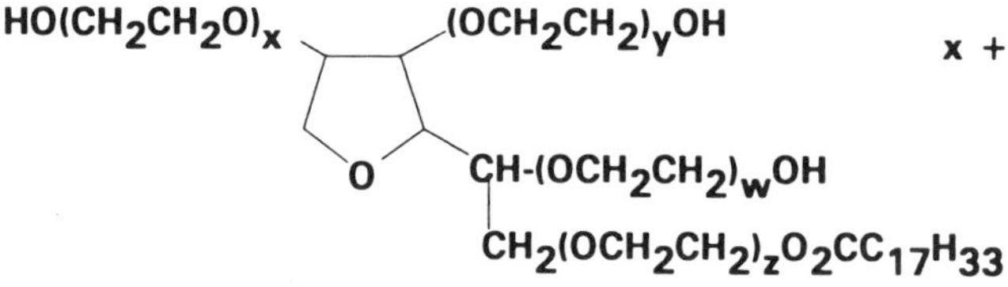

$$\text{HO(CH}_2\text{CH}_2\text{O)}_x \cdots (\text{OCH}_2\text{CH}_2)_y\text{OH} \qquad x + y + z + w = 20$$

$$\text{CH-(OCH}_2\text{CH}_2)_w\text{OH}$$
$$\text{CH}_2(\text{OCH}_2\text{CH}_2)_z\text{O}_2\text{CC}_{17}\text{H}_{33}$$

(4-33)

and has been used for cyanide displacement, Williamson ether synthesis, and dehydrohalogenation [230]. Catalytic activities were substantially lower than observed with, for example, 18-crown-6 as catalyst.

Particular attention has been given to making soluble polypodand copolymers. For example, comb-shaped polymers have been made by treating poly[(β-chloroethyl) glycidyl ether] with the monosodium salts of methanol, ethylene glycol monomethyl ether, and then diethylene glycol monomethyl ether [228]. These polyethers exhibited good complexing capacities for K^+ and Na^+, and PTC activities for nucleophilic substitution of 1-bromobutane with potassium phenoxide [231]. Linear, star-shaped, and comb-shaped graft copolymers have been shown to be excellent PTC catalysts for the Williamson ether synthesis [231–234].

Acyclic polypod polyethers from the reaction of poly[(β-chloroethyl) glycidyl ether] with monosodium salts of methanol, ethylene glycol monomethyl ether, and diethylene glycol monomethyl ether complex with sodium and potassium picrate and were found to be effective phase-transfer catalysts for substitutions such as iodide substitution on octyl bromide, for Wolf–Kishner reduction, and for addition of dichlorocarbene to styrene [235].

A polysiloxane-based polyethylene glycol was similarly found to be a useful catalyst for the same reactions [236].

s-Triazines having ethoxylated arms and capped with ether groups have been claimed by Au [237] as phase-transfer catalysts. These compounds are made according to the following:

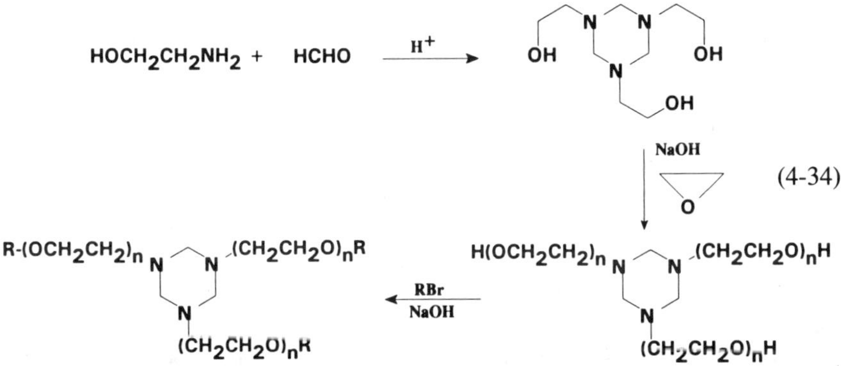

(4-34)

Examples for use of these materials as PTC catalysts are shown in Table 4-22.

Cyclophosphazenic polypodands, a class of compounds analogous to the above triazines, and represented by the following structures,

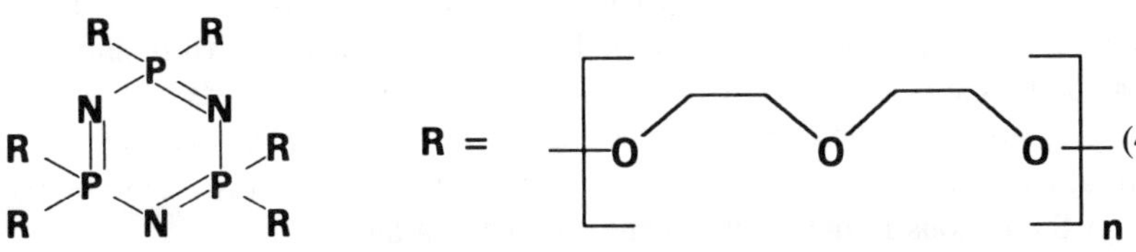

(4-35)

Table 4-22. Use of polyethoxylated-s-triazines as phase-transfer catalysts.

Catalyst	Yield at 1 h of product from reaction with $C_8H_{17}Br$					
	$NaOC_6H_5$	NaI	NaSCN	KI	KSCN	K-o-Phthalimide
PTP-EO$_7$-Oc	52	55	30	77	—	—
PTP-EO$_7$-H	39	—	—	—	—	—
PTP-EO$_{5.5}$-Oc	51	100	58	74	57	28
PTP-EO$_{5.5}$-Bu	43	—	—	—	46	59
PTP-EO$_{5.5}$-H	32	—	—	—	6	—
PTP-EO$_4$-Oc	25	—	—	—	—	—
PTP-EO$_3$-Oc	12	—	—	—	—	—
PTP-EO$_2$-Oc	10	—	—	—	—	—

^{a}The designation, for example, PTPEO$_{5.5}$Bu, indicates a triazine having an average $n = 5.5$ per leg, capped by a butyl group.

have been prepared by Landini and co-workers [238], and found to be powerful complexing agents of alkali metal salts and very efficient transfer catalysts in solid–liquid nucleophilic substitution, alkylation, reduction, and oxidation reactions. Cyclophosphazenic polypodands are excellent phase-transfer catalysts for nucleophilic substitution reactions [418]. The relative reactivities found ($I^- >$ $C_6H_5O^- > SCN^- > C_6H_5CH_2O^-$, $I^- > Br^- Cl^-$) indicate that anion activation levels are comparable with those found for cyclic crown ethers and cryptands. Unlike TDA-1, which is generally restricted to liquid–solid reactions, the more lipophilic of the polypodands, that is, $R' = $ n-dodecyl or p-octylphenyl, are also excellent catalysts for liquid–liquid PTC reactions.

In a further study [239], the catalytic activity of cyclophosphazenic polypodands was evaluated in typical reactions performed under solid-liquid and liquid–liquid PTC conditions. The activity was largely determined by the complexation power of the catalysts, which is, in turn, related to the number of binding sites of the ligand and to the nature of the inorganic salt M^+Y^- (Na > K >> Rb, and SCN = I > Br). Comparison with the open-chain analogs PEG and TRIDENT showed the cyclophosphazenes to be excellent catalysts, especially for solid–liquid systems.

Hexa-substituted cyclophosphazenes are formed by the reaction of hexachlorocyclophosphazenes with polyethylene glycol monoalkyl ethers and are said to be useful as phase-transfer catalysts [240]. Catalysts prepared by treating polydichlorophosphazenes with tetraethylene glycol monododecyl ether or pentaethylene glycol p-tert-octylphenoxy ether had less catalytic activity in phase-transfer reactions than cyclophosphazenes with the same ethyleneoxy chains and end groups due to a smaller cavity size effect [241]. The linear catalysts had similar complexing ability to the cyclophosphazenes with alkali iodides and bromides, but lithium and sodium iodide were complexed better, presumably due to better fitting of the cations in the cavity of the complexing agents.

E. Other Soluble Polymers and Related Multifunctional Compounds as Phase-Transfer Catalysts

We have seen in the previous section that ethylene oxide polymers or oligomers function quite well for a wide variety of PTC jobs. We would expect that many other polymers having dipolar functional groups, such as sulfoxide, sulfone, amides, phosphoramides, and others, would function similarly, particularly if the microenvironment around the polymer resembles the bulk phase of a dipolar aprotic solvent such as N,N-dimethylformamide or dimethyl sulfoxide, or others. Structural representations, such as

$$\sim\sim[CH_2CH_2CH_2S(O)]_n\sim\sim + KOH \longrightarrow$$

(4-36)

indicate both anion and cation of inorganic salts could be reasonably well complexed, transferred from solid or aqueous phase to organic phase, and activated for organic phase reaction. Experimental studies have shown that these soluble polymers can, in fact, perform excellently as phase-transfer catalysts.

Polymeric formamide products from polymerization of N-methyl-N-vinylformamide or N-methyl-N-(4-vinylbenzyl)formamide and copolymerization of these monomers with styrene were shown to be active phase-transfer catalysts for several liquid–liquid reaction systems [242]. For vinylformamide copolymers with styrene, the catalytic activity is strongly dependent on composition, and there are maxima at certain compositions. However, copolymers of the vinyl formamide monomers with acrylonitrile show little or no PTC catalytic activity, even though they can extract all alkali metal ions with high efficiency, and with extraction efficiency increasing with increasing density of active sites. Their lack of catalytic activity is evidently due to lack of organophilicity around the active sites of the monomer and the inability of organic reagents to approach the active sites. Similar results were reported for polymers of N-methyl-N-(p-vinylbenzyl) formamide [243], and the analogous polymer based on N-methylacetamide [244].

In general, it has been found that polymers, such as the above amides, can be good phase-transfer catalysts whereas the unpolymerized monomer itself is not catalytically active, even though present in the same concentration as polymer, suggesting that the functional groups on the polymer are cooperative [245]. Further, copolymers are frequently found to have higher activities than homopolymers, suggesting that the environment around the transferred anions must not be so polar as to exclude close diffusion of the organic reactant into the reactant zone. This behavior is also observed in insoluble phase-transfer catalysts (see

Chapter 5). Polymeric analogs of dimethyl sulfoxide, sulfonane, N,N-dimethylformamide (DMF), N,N-dimethylacetamide, N-methylpyrrolidone, and tetramethylurea are all effective phase-transfer catalysts. They extract all alkali-metal cations more effectively than their monomeric analogs, and the extraction ability (but not necessarily the PTC catalytic activity) increases with increasing density of active sites. Even partially alkylated poly-L-leucine and Nylon-66 exhibit PTC activity.

Polymeric analogs of tetramethylurea, prepared by polymerization of N,N,N'-trimethyl-N'-p-vinylbenzylurea and copolymers with styrene [246], are soluble in toluene and have the ability to extract alkali metal ions such as lithium, sodium, and potassium. These polymers function as phase-transfer catalysts for typical SN2 reactions, although the monomeric analogs tetramethylurea and N,N,N'-trimethyl-N'-benzylurea do not.

Copolyester-amides from polycondensation of $(CH_2)_6[NHCOC_6H_4CO_2Me-pl_2$, 1,6-hexanediol, and PEG have PTC catalytic activity in nucleophilic substitutions, eliminations, alkylations of activated methylene groups, dichlorocyclopropanation of double bonds, reduction of ketones, and oxidation of primary alcohols [247]. Oligo N-acetyliminoethylenes, both free and grafted on polystyrene, polymethacrylate, and cellulose, have phase-transfer catalytic activity in nucleophilic substitution reactions [248]. Regularly sequenced polyamides with sufficient oxyethylene units also function as phase-transfer catalysts [249].

Poly N-vinylpyrrolidone-costyrenes have a high activity as phase-transfer catalysts for substitution reactions in two-phase systems, although the monomeric analog is not a catalyst [250]. Catalytic activity is affected remarkably by the composition of copolymer, indicating the importance of organophilic environment around the active site. Poly-N-vinylpyrrolidone-costyrenes extract potassium cations into the organic phase, and this extraction ability increases with increasing density of pyrrolidone units in the polymer. When the polymers are crosslinked by inclusion of some divinylbenzene, the resulting insoluble polymers also exhibit high activity as phase-transfer catalysts. Copolymers of 1-vinyl-2-pyrrolidinone and diethylene glycol divinyl ether are phase-transfer catalysts for etherification of alcohols with alkyl halides in the presence of sodium hydroxide to gave 22–85% yields of ethers [251].

Pseudodipeptides, cyclized to 24- and 36-membered macrocyclic pseudopeptides, such as

Boc-Tyr-OHex
|
CO-Ala-OCH$_2$Ph $\longrightarrow$

(4-37)

were effective and selective phase-transfer agents for Li$^+$ ions, solubilizing up to 0.6 molar equivalents of Li$^+$ ions in chloroform. The uptake of Na$^+$ and K$^+$ ions was only 0.025 and 0.007 molar equivalents, respectively [252].

A number of 2- and 2,6-sulfoxy-substituted pyridines, including those represented by structures I–VI, were tested as phase-transfer catalysts [253–255]. These polyfunctional complexing agents are useful as catalysts for alkylation reactions and some nucleophilic substitution reactions, although they are not as active as commonly used quaternary salt or crown-ether phase-transfer catalysts, as seen by the data in Table 4-23. They are, however, quite

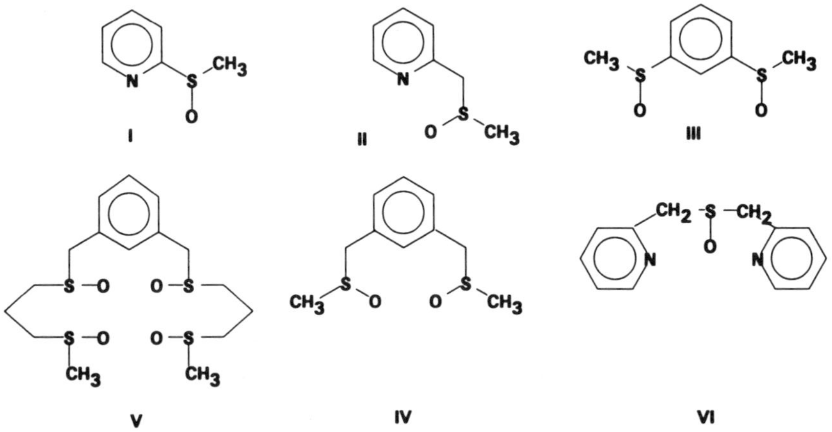

(4-37A)

selective for monoalkylation reactions, as illustrated by the data in Table 4-24. This selectivity also carries over for alkylations of phenylacetone, which with these catalysts give only C-alkylation products, also with very high selectivity to the mono-derivative, and no O-alkylation products.

The catalysts are easily removed from product mixtures by extraction in aqueous mineral acid solution.

Use of polymers having sulfur-containing functions as phase-transfer catalysts has been reviewed [256–258]. Soluble polymeric sulfoxides catalyze displacement and alkylation reactions, presumably by coordination of the metal cation with the sulfinyl oxygen atom, resulting in transfer of the cation together with its associated anion from the aqueous phase to the organic phase [259–262]. The catalytic activity of these sulfoxides depends on their composition and they were more effective than the monomeric analog both for liquid–liquid and solid–liquid reactions with NaOH. Conformation and shape factors of the polymer backbone are believed to play an important role in the catalytic activity, suggesting an analogy with enzymic catalysis.

Polymers prepared from 2,6-dichloropyridine and HS(CH$_2$)$_n$SH [$n = 6,10$]

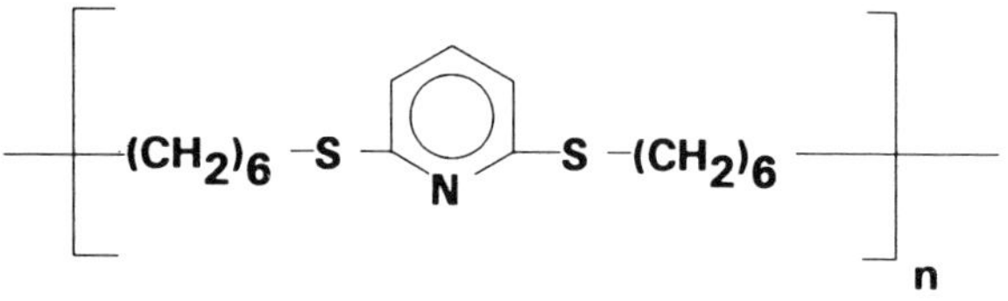

$$\left[-(CH_2)_6-S-\underset{N}{\underset{|}{C_5H_3N}}-S-(CH_2)_6-\right]_n \qquad (4\text{-}38)$$

are phase-transfer catalysts for the substitution reaction of 1-bromooctane with sodium thiocyanate and the sodium borohydride reduction of ketones [263]. 2-Pyridinethiol itself was not a catalyst. *p*-[(2)-Pyridylthio)methyl]styrene and *p*-[(2)-pyridylthio)ethyl]styrene give polymers that extract alkali metal ions such as sodium and potassium cations into organic media [264], having increasing extraction efficiency with increasing active site density. These polymers serve as phase-transfer catalysts for the reduction of carbonyl compounds by sodium borohydride.

Polymeric sulfones, prepared by radical copolymerization of styrene and

Table 4-23. *S-L PTC displacement reactions with sulfoxy-substituted pyridines as catalysts.*

$$1\text{-}C_8H_{17}\text{-}Br + M^+ Y^- (s) \xrightarrow[\text{Xylene}]{\text{Catalyst}} 1\text{-}C_8H_{17}\text{-}Y + M^+ Br^- (s)$$

M^+Nu^+	Catal.[a]	Temp. (°C)	Time (h)	Product yield (%)
NaCN	I	100	54	13
NaCN	II	100	46	87
NaCN	IV	100	32	84
NaCN	V	100	32	0
NaCN	VI	100	50	64
KSCN	I	100	40	96
KSCN	II	100	27	96
KSCN	III	100	40	20
KSCN	IV	100	40	23
KSCN	VI	100	17	75
KSPh	II	r.t.	0.5	91
KSPh	III	r.t.	12	93
KSPh	IV	r.t.	12	88
KSPh	VI	r.t.	0.5	90
KOPh	I	70	36	93
KOPh	II	70	36	93
KOPh	III	70	40	33
KOPh	IV	70	40	88
KOPh	VI	70	14	93

[a] 4.0 Molar equivalent of catalyst per mole of substrate.

Table 4-24. *Use of Sulfoxide-substituted pyridines as phase-transfer catalysts for alkylation.*

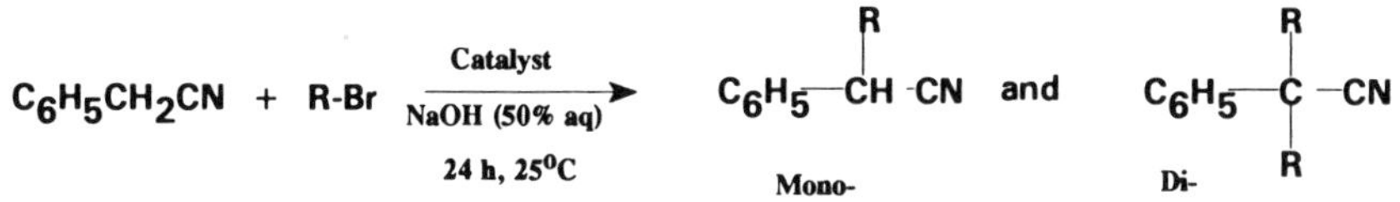

$$C_6H_5CH_2CN + R\text{-}Br \xrightarrow[\text{NaOH (50\% aq)}]{\text{Catalyst}} \underset{\text{Mono-}}{C_6H_5\text{-}CHR\text{-}CN} \text{ and } \underset{\text{Di-}}{C_6H_5\text{-}CR_2\text{-}CN}$$

24 h, 25°C

RX	Catalyst	Alkylation product yield	
		Percent mono	Percent Di-
MeI	I	73	5
MeI	II	79	13
MeI	IV	70	5
MeI	VI	83	15
EtI	I	85	0
EtI	II	93	0
EtI	IV	73	0
EtI	VI	90	0
BuI	I	54	0
BuI	II	88	0
BuI	IV	70	0
BuI	VI	89	0

methyl *p*-styryl sulfone, have been shown to extract metal salts from aqueous into organic phases, and to be relatively weak phase-transfer catalysts for iodide displacement on 1-bromooctane [265]. Maximal catalytic activity was observed for 1:1 copolymer composition.

Polydichlorophosphazenes with tetraethylene glycol monododecyl ether or pentaethylene glycol *p-tert*-octyl-phenoxy ether side chains had less catalytic activity in phase-transfer reactions than non-polymer cyclophosphazenes with the same chains and end groups [266]. This was due to a smaller cavity size effect.

Polysiloxane-based soluble polymers were useful as phase-transfer catalysts for iodide substitution reactions and for the Wolff–Kishner reaction [267].

F. Use of Dual PTC Catalysts or Use of Cocatalysts in Phase-Transfer Systems

Some PTC reaction systems perform much better when a cocatalyst is present. The kinds of cocatalysts used depend greatly on the needs of the reaction being conducted, generally to enhance reaction rate, improved product selectivity, or decrease the need for excess of an aqueous phase reagent such as NaOH. Some kinds of cocatalysts that have been used include:

- Use of two phase-transfer catalysts (dual PTC catalysts)
- Use of alcohols or related compounds as cocatalysts
- Use of metal compounds as cocatalysts
- Use of iodide to boost certain displacement reactions

1. Use of Dual PTC Catalysts

In that quadrant of the PTC rate matrix (Chapter 1) where both organic phase reaction and transfer are slow, one can increase reaction rates by using *two* phase-transfer agents, one designed to maximize the rate of the transfer step of anions to the organic phase, and the other to maximize the rate of the organic phase reaction. The increase in rates, or synergism, involves complex kinetic interactions. Thus, for example, for preparation of azoisoheptanonitrile [268], useful as a polymerization initiator similar to AIBN, by oxidative coupling of α-aminoisoheptanonitrile, use of three phase-transfer catalysts, $Me(CH_2)_{17}NMe_3Cl$, $Me(CH_2)_{11}N(CH_2Ph)Me_2Cl$, and a betaine gives high reaction rates at the low temperatures, 5–10°C, necessary to prevent thermal decomposition of the product. In the C-alkylation of esters reaction rates could not be accelerated above a limiting rate by increasing the quaternary ammonium catalyst concentration but the rate plateau could be exceeded by adding a second PTC catalyst such as crown ether or PEG cocatalyst [269].

Dolling and co-workers, in the preparation of (R)-fluorenyloxyacetic acid, useful in the treatment of brain edema, using a PTC chiral alkylation with sodium hydroxide as base, and using 3,4-dichlorobenzyl-cinchonidinium as catalyst, found the reaction to be accelerated by use of an achiral nonionic surfactant cocatalyst, Triton X-405 [270]. In this reaction mixture, the cocatalyst presumably functions by increasing the rate of transfer of hydroxide anion to the chiral catalyst. Similarly, preparation of 2,3-dicyano-1,4-dithiaanthraquinone, by reaction of 2,3-dichloro-1,4-naphthoquinone and alkali metal salt of dimercaptomaleonitrile, works best when the phase-transfer catalyst is aided by a surfactant cocatalyst (calcium dodecylbenzene sulfonate) to speed the transfer step [271].

Synergetic results have been shown in PTC dichlorocarbene addition to cyclohexene, wherein a mixture of triethylamine and benzyltriethylammonium chloride has substantially greater catalytic activity than either catalyst by itself [272]. Thus, under the reaction conditions used, triethylamine alone as PTC catalyst gave a 7% yield of 7,7-dichloronorcarane, benzyltriethyl ammonium chloride alone gave 16%, whereas an equimolar mixture of the two catalysts gave a 51% yield. Other additives, for example, phenol, tributylphosphine oxide, PEG-4000, and 15-crown-5 were also synergetic.

In the preparation of 2,3-dicyano-1,4-dithiaanthraquinone by treating 2,3-dichloro-1,4-naphthoquinone with salts of dimercaptomaleonitrile use of an anionic surfactant, for example, calcium dodecylbenzenesulfonate, as a dual or cocatalyst with trioctylmethylammonium chloride, accelerated the reaction to give high yields of product at room temperature [273].

In the alkylation of cyclopentadiene with t-alkyl halides, use of 4,4'-bipyridinium salts in addition to a common phase-transfer catalyst gives better results, apparently because the reaction involves a reaction with a single electron transfer step [274].

A technique named "bimechanistic phase-transfer catalysis" [275,276] is a process devised by Shaffer and Kramer to combine the effect of normal PTC with inverse PTC for polymerization reactions, such as between sodium sulfide and 1,8-dibromooctane. Two catalysts are used: an ammonium salt and a cyclic or acyclic sulfide. With hexadecyltrimethylammonium chloride and tetrahydrothiophene, the optimum catalyst concentrations were 10 mol% and 5 mol%, respectively. The activity of each was independent of the other.

2. Use of Alcohols and Other Weak Acids as Cocatalysts in Hydroxide Transfer Reactions

Hydroxide is one of the most difficult anions to transfer from aqueous to organic phases, yet it is one of the most valuable and most commonly used anions in PTC systems. Addition of small amounts of alcohols to PTC systems requiring hydroxide transfer causes a dramatic increase in rates. At least two processes are believed to contribute to the cocatalytic effect:

- Formation of alkoxide anions, RO^-, which are more readily transferred than the highly hydrated hydroxide anion, and which can serve as a strong base just as well as OH^-, and
- Solvation of the hydroxide with alcohol rather than with water, making the hydroxide anion more organophilic and more easily transferred [277].

1,2-Diols have been shown to be the best cocatalysts in competitive halide/alkoxide (hydroxide) extraction experiments [278]. For example, pinacol, better than other alcohols tested, increases the yields of dihalocyclopropanes made by PTC reactions of chloroform or bromoform with olefins under standardized conditions [279].

Not only does alcohol addition increase the rate of transfer of "hydroxide equivalents," but it also retards decomposition of quaternary ammonium salt catalysts, a serious catalyst-destroying side reaction commonly observed with quaternary salts in the presence of strong hydroxide solutions. For example, in dehydrohalogenation of 2-bromooctane with KOH, catalyzed by various quaternary ammonium salts, decomposition of the quaternary salt was prevented by addition of methanol [280]. The protective action of methanol is believed to be due to formation of its anion, MeO^-, replacing HO in association with the quaternary cation, and thereby retarding Hofmann degradation.

Reaction rates for KOH dehydrohalogenation of 2-bromooctane in toluene with PEG catalysts were increased by a maximum factor of 126 by addition of methanol [140]. PTC dehydrochlorination of 1,2-dichloroethane with aqueous sodium hydroxide, catalyzed by quaternary ammonium salts in the presence of alcohol cocatalysts, proceeds both in the organic phase and at the interface [277]. Makosza and co-workers have shown that both H-O or H-N weak acids are cocatalysts with quaternary ammonium salts for dehydrobromination of alkyl bromides with aqueous sodium hydroxide [281].

PEG 400 catalyzes the elimination of hydrogen halides from 1,2-dihalides to yield 1- and 2-alkynes as well as 1,2-dienes. However, if quaternary salts are used as catalyst, along with bis-tertiary alcohols as cocatalyst, then 1-alkyne is the only product [282].

Benzylation of sugar by benzyl chloride with sodium hydroxide and a phase-transfer catalyst are greatly facilitated by addition of a cocatalyst such as a tertiary alcohol and/or a cosolvent such as dimethyl sulfoxide (DMSO) [283]. By use of a cocatalyst efficient procedures were developed for benzylation of sugar derivatives having three to eight hydroxyl groups per molecule, in a two-phase system employing an almost stoichiometric amount of the alkylating agent.

Cyanide displacements catalyzed by quaternary ammonium salts usually do not proceed without the presence of some water to facilitate exchange and transfer of anions. However, in the presence of a catalytic amount of alcohol, PTC displacement reactions do proceed in the absence of added water [284]. The cocatalytic effect was dependent on alcohol structure, benzyl alcohol being about 1.5–2 times as effective as either methanol or ethanol.

Although alcohols and related compounds can be of exceeding great use as cocatalysts for PTC reactions, this may lead to byproduct formation, by formation of ethers, and thereby complicate the process.

3. Use of Metal Compounds and Salts as PTC Cocatalysts

Use of metal compounds as cocatalysts with PTC has developed extensively in several directions. Certain metal-containing anions, such as molybdate, tungstate, chromate, and others are highly useful cocatalysts in selective PTC oxidations using aqueous hydrogen peroxide or aqueous sodium hypochlorite, as the primary oxidant, and these are discussed in some detail in Chapter 10. Likewise, metalate anions may be useful cocatalysts in PTC reductions, as noted in Chapter 11. Additionally, much work has been done on use of transition metal compounds as cocatalysts with phase-transfer catalysts for reactions such as hydroformylation, carbonylation, and related reactions, and these are discussed in Chapter 13.

Cuprous salts, as cocatalysts with ordinary phase-transfer catalysts, can often accelerate two-phase reactions, as, for example, in allylic substitution of allyl halides by 1-alkynes at or near room temperature under solid–liquid PTC conditions [285,286], Tri-*n*-butyltin chloride as a cocatalyst with a polyether phase-transfer catalyst is highly effective for reduction of alkyl halides with sodium borohydride to produce hydrocarbons [287]. Chiral cupric complexes have been used as phase-transfer catalysts to promote asymmetric alkylation on achiral glycine equivalents [419]. Positively charged Ni(II) and Cu(II) complexes of Schiff bases of chiral diamines are novel phase-transfer catalysts for C-alkylation of achiral Schiff bases [420].

Cesium fluoride is said to activate anions [288], whereas anhydrous aluminum chloride can act as a cocatalyst for certain kinds of phase-transfer reaction, as in the reaction of aryl ketones with metal cyanides to produce α-arylacrylonitriles. Similarly, aluminum chloride cocatalyst with the phase-transfer catalyst TDA-1 catalyze the reaction of 5-fluorouracil and 2,3-dihydrofuran to give N-(tetrahydrofuran-2-yl) 5-fluorouracil, a low-toxicity antitumor agent [223].

4. Use of Iodide as a Cocatalyst

Although iodide anion is a "poison" for most PTC reactions with alkyl chlorides or bromides, some exceptions have been noted where it is actually a cocatalyst. For example, in displacement on benzyl chloride with sodium carboxylates, iodide accelerated the reaction when it was present within certain concentration ranges [289]. At too high a concentration iodide poisoned the catalytic effect. The optimum concentration varied with the reaction system, but the critical concentration was shown to depend on the distribution coefficients of the intermediates formed in the reaction. Small amounts of iodide function as cocatalysts in PTC systems for esterification of sodium monobutylphthalate with benzyl chloride [290], and for conversion of benzyl chloride or bromide to benzyl fluoride [291].

G. Catalysts for Transfer of Species Other Than Anions

1. Inverse PTC: Transfer of Organic Reagents into Aqueous Solutions

Although the overwhelming majority of PTC literature deals with transfer of anions into organic phases, it is clearly appropriate to consider also systems where organic reagents are transferred into aqueous solutions of a second reactant.

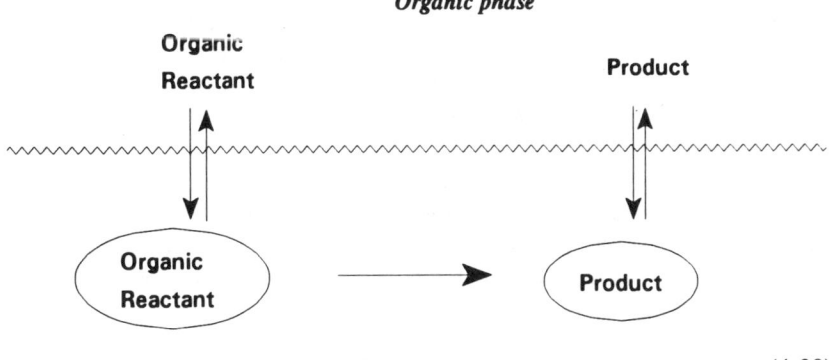

(4-39)

This kind of catalysis has been given the name *inverse phase transfer catalysis* (IPTC) by Mathias and Vaidya [292–294]. In principle, any number of different second-phase solvents other than water can be considered. As an example, formamide instead of water as solvent for KF in fluoride–chloride exchange of octyl chloride led to higher yields of alkyl fluoride [295].

a. Cyclodextrins and Derivatives as Inverse Phase-Transfer Catalysts

Cyclodextrins are carbohydrate structures that behave as general inverse phase-transfer catalysts in that they can solubilize many different kinds of organic compounds into aqueous solutions. These unusual compounds in aqueous solution form cylindrical-like structures having organophilic interiors and hydrophilic exteriors.

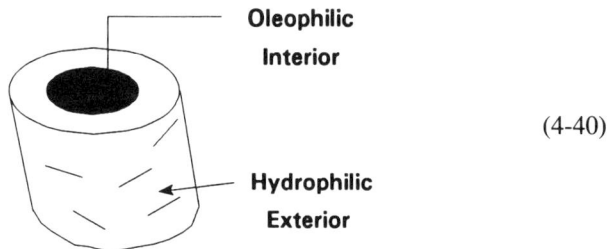

(4-40)

Solubilization of organic compounds occurs by formation of host–guest complexes within the interior of the cyclodextrin structure, and once in the aqueous phase the transferred organic reagent can react with other reagents dissolved in the aqueous phase.

β-Cyclodextrin has been used for inverse PTC with $IrCl_3$ for isomerization of 4-allylanisole [296]; for reduction of 2- 3- and 4-MeOC_6H_4Br to PhOMe with sodium formate in high yields under mild conditions with Pd/C cocatalyst [297]; and for epoxidation of alkenes such as norbornene, styrene, cis-cyclooctene with iodosylbenzene with a chromium(III) salen complex cocatalyst [298].

Chemical modification of the outer hydroxyl groups of β-cyclodextrins can improve the performance of these materials as inverse phase-transfer catalysts [299,300].

Soluble copolymers of β-cyclodextrin with epichlorohydrin produces a material that has much higher inverse PTC activity than does β-cyclodextrin itself for liquid–liquid two-phase nucleophilic displacement reactions [301]. The enhanced efficiency of the polymer catalysts was attributed to the cooperative effect of more than one β-cyclodextrin group in polymers in bringing a substrate from an organic phase to an aqueous phase.

b. Metal Salts and Complexes as Inverse Phase-Transfer Catalysts

Some metal compounds, particularly transition metals such as platinum, palladium, rhodium, etc., complex strongly with water-soluble ligands, such as the trisodium salt of triphenylphosphine trisulfonic acid:

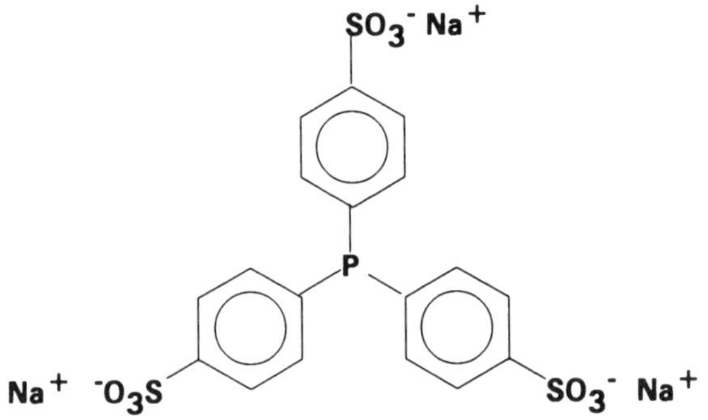

$$\text{[structure of trisodium triphenylphosphine trisulfonate with } SO_3^- Na^+ \text{ groups]} \tag{4-41}$$

Water-soluble rhodium complex

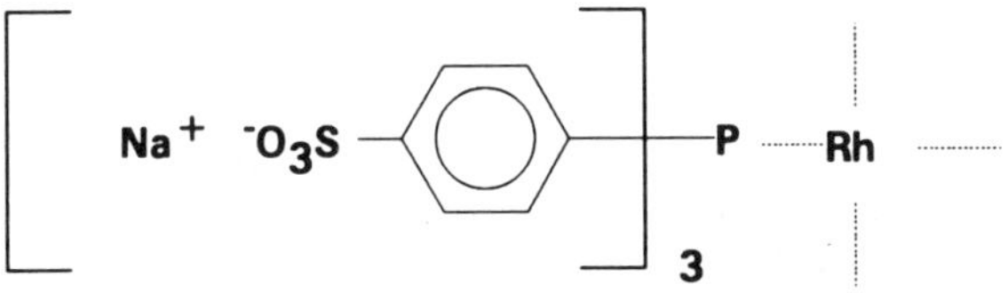

$$\left[Na^+ \ {}^-O_3S-C_6H_4- \right]_3 P\text{----}Rh\text{----}$$

These complexes are completely retained in the aqueous phase of aqueous–organic two-phase mixtures, but they have the ability to bring olefins, and possibly other organic reagents, into the aqueous phase a little at a time, and to catalyze reactions with other species present there. This approach has significant commercial appeal, because expensive metal catalysts can be maintained under complete control and without loss, while the organic products are extracted back into the organic phase. For example, diphenyl(2-potassium sulfonatobenzyl)phosphine and its oxide, complexed with palladium chloride, catalyze reductive dehalogenation of allyl and benzyl halides [302]. Surfactants and a water-soluble Rh catalyst (4-$Ph_2PC_6H_4CO_2H$ and quaternary ammonium surfactants) gave good normal-isoselectivity and conversion with a P ligand for hydroformylation of hexene, dodecene, and hexadecene in aqueous systems at 80°C and 5.5 atm [303].

An aqueous-phase soluble palladium triphenylphosphine catalyzes nucleophilic substitutions at allylic carbons in 93–100% yields [304]. For example, reaction of cinnamylethyl carbonate with nucleophiles such as ethyl acetoacetate or acetylacetone in an organic solvent gave regio- and stereospecific alkylation products. The aqueous catalyst solution is easily separated and recycled. No other solvent was required for these reactions.

Complexes of lower olefins with cuprous salts and some other metal compounds are soluble in water, and thus cuprous chloride may sometimes be used as an inverse phase-transfer catalyst [305].

c. Pyridine-N-oxide

Amines and amine oxides have been found to reside primarily in aqueous phases and to behave as inverse PTC catalysts for certain reactions. Mathias and Vaidya examined the reaction of alanine with decanoyl or *p*-chlorobenzoyl chloride, and found it to be greatly accelerated by 4-(dimethylamino)pyridine [294]. They showed this reaction to involve transfer of the acid chloride into the aqueous phase for reaction with alanine. Fife and Zin [306] similarly described a procedure using amine oxides to form acylpyridinium intermediates, which are transferred to an aqueous phase containing carboxylate anions, and where reaction takes place to yield anhydrides and regenerate amine oxide. Silicon-substituted pyridines and pyridine oxides have been patented as inverse phase-transfer catalysts [307].

d. Tetramethylammonium Salts and Related Compounds as Inverse Phase-Transfer Catalysts

Tetramethylammonium chloride, normally a poor phase-transfer catalyst for most reactions because of its lack of solubility in organic media, has been found to be highly suitable for certain other systems where reaction does not take place in the organic phase. For example, tetramethylammonium salts have been used as phase-transfer catalysts for fluorination of chlorobenzaldehydes with alkai metal fluorides at 190°C [308], to prepare 1,2,2,2-tetrafluoroethyldifluoromethyl ether, a known anesthetic [309] and for ethoxylation of chloronitrobenzenes to produce ethylnitrophenyl ethers [310, 311].

Carbohydrates are insoluble in both water and most organic solvents, but are wetted by water, and are difficult to transfer. Using low molecular weight quaternary salts, such as tetramethyl- or tetraethyammonium salts as inverse PTC catalysts, sometimes allows highly selective carbohydrate reactions to occur in the aqueous phase. For example, benzylation of chitin to give 6-*O*-monobenzyl- or 3,6-di-*O*-benzyl-2-acetamido-2-deoxyglucosan with benzyl chloride and NaOH was best accomplished with tetraethylammonium chloride as catalyst [312]. Acetalization of sorbitol with benzaldehyde proceeds readily and with high selectivity with inverse phase-transfer catalysis [313].

Ion-exchange resins, having a high concentration of trimethylammonium groups attached to the resin, retain a substantial amount of water around these active sites, and behave much like aqueous phases. Thus with anion-exchange resins as catalysts, oxidation of benzyl alcohol by NaOCl in water at 30°C to yield benzaldehyde occurred in the aqueous phase and on the solid phase [314]. No reaction occurred in the organic phase.

e. Reactive-Inverse PTC

Cyclic and acyclic sulfides such as tetrahydrothiophene, diethyl sulfide, etc. catalyze nucleophilic substitution of benzyl chloride in an organic phase with KSCN in an aqueous phase contacting the organic phase [315]. This mode of catalysis, termed *reactive inverse PTC*, proceeds by first reaction between benzyl chloride and the sulfide to form an intermediate sulfonium salt, which transfers to the aqueous phase to react with thiocyanate ion, and decomposing to regenerate the sulfide and benzyl thiocyanate, both of which return to the organic phase. This is closely related to the *bimechanistic phase transfer catalysis* noted in Section F.1.

2. Transfer of Acids

As early as 1973 it was noted that dry HCl could be transferred into organic solutions in the form of triethylammonium bichloride, $Et_3NH^+HCl_2^-$, when a highly polar solvent was present, and addition of HCl to acetylenic compounds could be achieved [316–319].

Landini and co-workers further explored the transfer/reaction of hydrogen halides from aqueous solutions, and used the system for conversion of alcohols to the corresponding alkyl chlorides in > 90% yields, employing tributylhexadecylphosphonium bromide as catalyst. For conversion of *n*-alkanols to the corresponding bromide, however, no catalyst was required, and addition of the quaternary phosphonium salt did not appreciably accelerate the reaction rate [318]. For cleavage of ethers using aqueous HBr, tributylhexadecylammonium bromide was a strong catalyst [320]. Other reactions using quaternary salts to transfer and catalyze hydrogen halide reactions include the transformation of 2-acetylbutyrolactone into 5-bromopentan-2-one with benzyltriethylammonium chloride at room temperature [321]; production of 1-bromododecane, used to manufacture benzyldodecyldimethylammonium chloride (the disinfectant "ajatin") from 1-dodecanol and aqueous HBr [322].

Landini also found that addition of HCl, HBr, or HI to olefins occurs by heating the olefin with concentrated aqueous HX and hexadecyltributylphosphonium halide [323]. Chlorination of olefins with partial asymmetric induction results from reaction of aqueous HCl and hydrogen peroxide is reported to occur in the presence of a chiral quaternary salt catalyst [324]. Conversion of 1,4-butanediol to 1,4-dichlorobutane results from treatment with aqueous HCl in the presence of tetraalkylammonium salts [325].

Dakka and Sasson have further studied the transfer of hydrogen halides, particularly with branched alcohols, finding reduced skeletal rearrangement when phase-transfer catalysts were used [326]. They found that, for example, 2-pentanol reacted with 48% HBr at 70°C for 24 h to give 59% conversion to a product mixture of 70% 2-bromopentane and 30% 3-bromopentane [327]. Replacement of the aqueous acid with anhydrous HBr improved the selectivity, yielding 13% 3-bromopentane and 87% 2-bromopentane. However, under PTC conditions, even aqueous hydrobromic acid led to practically pure 2-bromopentane. They suggested that under conditions of PTC the mechanism of hydrobromination shifted from partial SN1 to a pure SN2 reaction, free from rearrangements, as the adduct $R_4N^+XHBr^-$ is a stronger nucleophile and a far weaker acid than free hydrobromic acid.

Dehmlow examined the transfer of HCl from aqueous solutions into methylene chloride, showing that in the absence of quaternary salt only a small amount of acid is transferred into the organic phase, but in the presence of tetrabutylammonium chloride a little more than an equimolar amount of hydrochloric acid can be titrated in the methylene chloride layer [328]. He concluded that the hydrogen bond between chloride and hydrochloride is strong enough to permit extraction of $Bu_4N^+HCl_2^-$, noting that species such as $NR_3 \cdot 2HCl$ have been assumed to occur in benzene solution [329] and that $NEt_3H^+HCl_2^-$ is a known compound both in the solid state and in solution [330]. Dehmlow also demonstrated that with more lipophilic catalysts such as $[C_7H_{14}]_4NCl$, extraction of hydrochloric acid can be accomplished even into nonpolar solvents such as benzene. Hydrogen fluoride can also be extracted in the form of $Bu_4N^+HF_2^-$ from aqueous potassium hydrogen difluoride. Dehmlow noted that HBr was more difficult to transfer from aqueous solution than was HCl [328].

In other work [331] it was shown that quaternary salts transfer HCl from 32–37% aqueous solutions into alcohols at 50–60°C to produce alkyl halides in 85–95% yields and > 99.5% selectivities. In commercial use, these products may be used directly for production of amines or quaternary ammonium salts by reaction with ammonia or other amines. This PTC-based process, using aqueous HCl, is economically superior and provides for easier processing on a commercial scale than the classic process using anhydrous HCl, catalyzed by anhydrous zinc chloride. Symmetrical quaternary salts were more efficient catalysts than those containing only one long-chain substituent [332]. Tetrahexylammonium iodide was the most efficient of the salts tested. Addition of zinc chloride, calcium chloride, or sodium chloride increased the reaction rate even further.

Gas–liquid PTC reaction of (vaporized) aqueous hydrogen halides with vaporized alcohols produced the corresponding alkyl halides using a catalyst of $Bu_4P^+Br^-$ supported on silica gel at 170°C [333]. Byproduct dialkyl ethers can be avoided either by increasing the contact time, or by recycle.

3. Transfer of Water, Glycerol, Carbohydrates

Transfer of sufficient water into organic phases for reactions with water to occur rapidly usually does not require a phase-transfer catalyst if good agitation is used. However, for some applications it is desirable to control amount and rate of water more precisely, and one might choose a PTC method to accomplish this. For example, spherical Y-doped ZrO_2 powders were prepared from water in oil emulsions of acid stabilized aqueous solutions [334]. Precipitation of hydroxides inside the droplets of oil was achieved by an ion-exchange process between the aqueous phase and an anion-exchange resin using a phase-transfer catalyst as a carrier for the anions through the organic solvent.

C-Undecylcalix[4]reesorcinarene,

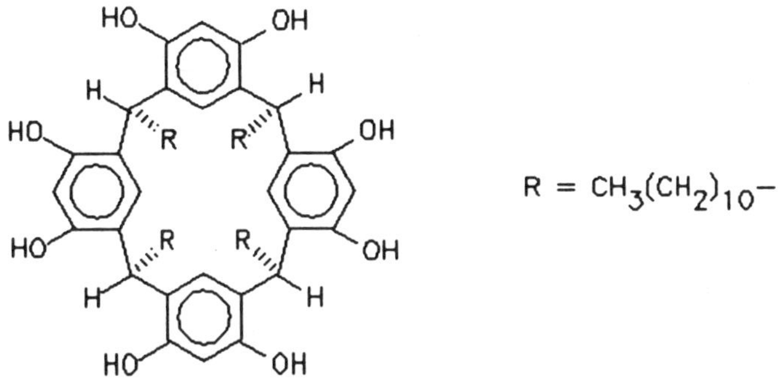

$$R = CH_3(CH_2)_{10}-$$

developed by Aoyama and co-workers is a bowl-shaped macrocyclic compound which complexes with such polar compounds as carbohydrates, water, glycerol, etc., and transfers these into non-polar organic solutions. The transferred species undergo appropriate reactions [422].

4. Transfer of Metals and Metal Hydrides

Tris(3,6-dioxaheptyl)amine (TDA-1) dissolves Na–K alloy to produce a deep blue solution in tetrahydrofuran, useful for deoxygenation of acetates, dehalogenation, and hydrolysis of tosylates and mesylates [335]. Likewise sodium can be solubilized by crown ethers in organic solutions that are extremely strong bases capable of deprotonation of triphenyl methane and ammonia [336]. The crown ethers are therefore useful PTC agents for sodium in reactions such as formation of polysilanes through Wurtz type coupling reactions [337,338]. Metallic sodium, complexed with crown ethers, has been used to purify cyclic acetals, prior to their use as monomers for producing polyformaldehyde [339].

Transfer of magnesium and magnesium hybride using anthracene as a phase-transfer agent has been reviewed. The active MgH_2–Mg system can be used for chemical synthesis, for example, for the preparation of silane and magnesium alkyls [340].

5. Transfer of Anhydrous Aluminum Chloride

Anhydrous aluminum chloride, widely used as a catalyst for the Friedel–Crafts reaction, frequently has limited solubility in aromatic hydrocarbons, requiring the use of solvents such as nitrobenzene which are sometimes expensive and troublesome to remove from the products. It has been shown that anhydrous aluminum chloride forms complexes with some agents that can transfer the $AlCl_3$ into organic media in an active form to catalyze Friedel–Crafts reactions [341]. In another process [342] quaternary salts are used along with anhydrous aluminum chloride to catalyze reaction of araalkyl ketones to produce unsaturated nitriles.

6. Transfer of Formaldehyde

p-Toluenesulfonic acid and alkylbenzenesulfonic acids of mol wt 300–550 catalyze phase-transfer cyclocondensation of styrene [343] or α-methylstyrene [344] with formaldehyde to give 4-phenyl-1,3-dioxane. The maximum activity of the alkylbenzene sulfonic acids occurred at mol wt of 440 (both water- and oil-soluble) in the presence of ethylbenzene. Reaction occurred mainly in the organic phase when alkylbenzenesulfonic acids were present.

7. Cation Transfer

Tetraphenylborate anion and its derivatives have been examined as phase-transfer catalysts for cation transfer in several reactions [345–347]. For example, tetrakis [3,5-bis(trifluoromethyl)phenyl]borate promoted the formation of oxiranes from carbonyl substrates and trimethylsulfonium chloride in a two-phase mixture of methylene chloride and concentrated aqueous sodium hydroxide via a sulfonium ylide [348].

Long-chain carboxylic acids, acting as phase-transfer agents or surfactants, catalyzed the metalation of mesotetratolylporphyrin by Zn^{2+}, Cu^{2+}, and Mn^{2+} in a water–toluene two-phase system [349]. The metalation rate was affected by both the chain length and the hydrophobicity of the carboxylic acid. These reactions are seen as models for metalation of porphyrins in geochemical sediments and in biological environments.

8. Transfer of Radical Anions

Radical anions, such as the one resulting from decomposition of peroxydisulfate anion, $\cdot OSO_3^-$, or anions such as $^-O_3S\text{-}O\text{-}O\text{-}SO_3^-$, which can decompose after transfer to produce free radicals, have been rather extensively studied as convenient and low-cost free-radical initiators for polymerizations. Inversely, organic-soluble free radicals, or their precursors, may be transferred from the organic phase into an aqueous (or other) phase. Thus cyclodextrin has been used to transfer organic-soluble free radical polymerization initiators into aqueous solutions to initiate polymerization [350,357,359].

Rasmussen and Smith [351] reported many examples of free-radical polymerizations under phase-transfer conditions with aqueous persulfates, and found this method to be more facile than when common organic-soluble initiators were used [352–360]. Extensive kinetic information has been developed for polymerization of butyl acrylate in an ethyl acetate–water system, using potassium persulfate with quaternary ammonium salts for initiation [354]. Symmetrical quaternary salts are more efficient than unsymmetrical ones, and Aliquat 336-catalyzed polymerization was about three times faster than when 18-crown-6 was used as PTC. Other studies on polymerization have been reported [361–363].

Shah [355] and Mandal [364, 365] and their co-workers have also studied the kinetics of PTC systems for transfer of radical anions or peroxy-anions for initiation of polymerizations. Choi and Lee [366] found that polymethyl methacrylate prepared by PTC had higher molecular weight and more uniform molecular weight distribution than that prepared by azoisobutyronitrile.

A redox catalyst system of sodium dithionate–carbon tetrachloride has been studied for polymerization of methyl mechacrylate in an aqueous–organic two-phase system, using tetrabutylammonium chloride as a phase-transfer catalyst [367].

Other polymerization studies on polymerization of butyl methacrylate with potassium peroxydisulfate–quaternary ammonium salt initiators use emulsifying quaternary salts such as cetyltrimethylammonium chloride [368]. Butyl acrylate polymerization using tetrabutylated calixarenes as PTC gave better results than when 15-crown-5-was used [369]. With tetrabutyl-ammonium salts, known to easily form third (catalyst-rich) layers, it was concluded that initiation of polymerization of acrylonitrile and methacrylic acid in aqueous alcohol media occurred in the interfacial layer (third phase) [370].

Oxygen radical-anions [371] and fluorenone ketyl anions could be generated and handled by phase-transfer agents [372]. An organic-soluble bis(quaternary salt), N,N'-dioctyl-4,4'-bipyridinium dibromide, behaves as an electron carrier between water–organic two-phase mixtures to cause reduction of nitroalkenes to the corresponding oximes and/or carbonyl compounds. Organic-soluble bipyridinium compounds are electron transfer agents [373,374].

9. Transfer of Ammonia

Reaction of aqueous ammonia with diethyl 1,1-cyclopropanedicarboxylate in the presence of PEG-400 as a phase-transfer catalyst gave the diamide in 83% yield without decarboxylation [376].

10. Transfer of Oxygen

Activated oxygen requires a special type of phase-transfer catalyst such as hemoglobin which can complex with gaseous oxygen, transfer it into the desired phase, and activate the oxygen for oxidation reactions. A simple chemistry for "activated" oxygen carriers is the reversible complex formation of cobalt–salcomine complexes with oxygen is described in Chapter 10.

H. Separation and Recovery of Phase-Transfer Catalysts

One of the more frequently encountered technical problems in use of PTC for industrial applications is the need to separate the product and the phase-transfer catalyst [377]. This problem may be largely avoided by use of insoluble catalysts, easily separated by filtration, centrifugation, or phase separation, if use of such catalyst is applicable to the reaction being conducted. (See Chapter 5.) Use of soluble phase-transfer catalysts requires a method for catalyst separation from the reaction mixture or product separation from the catalyst; and, for industrial use, it is also desirable that the catalyst be reusable or recyclable.

The most commonly used methods for separation of products and PTC catalysts on an industrial scale are extraction and distillation, as outlined below. Other methods include: separation of quaternary salts by sorption on ion-exchange resins [378–380], silica gel [381], or Florisil [382]; by use of catalysts containing conjugated diene groups that can react with an immobilized triazolinedione dienophile [383]; by use of chemically decomposable quaternary ammonium catalysts [384]; and use of PTC functions bound to a polyethylene chain 1000–3000 mol wt) that is soluble in most organic solutions at temperatures greater than about 90°C, but separates as a solid when cooled.

1. Extraction Methods

Several techniques for the separation of quaternary salts by extraction into water or aqueous solutions have been published, with particularly good results obtained with tetrabutylammonium salts as catalysts [385]. Thus, for example, tetrabutylammonium bromide is soluble to the extent of 27% in dilute (1% NaOH) aqueous solutions, but when the solution is made more concentrated (15% NaOH) the solubility of $Bu_4N^+Br^-$ decreases to 0.07%. Extractive separation has been patented for use in a process for production of aromatic ether imides, used for polyether-imide manufacture, prepared by aromatic substitution of halo- or nitrophthalimide with alkali metal bisphenolates in dichlorobenzene [386]. The tetrabutylammonium halide is removed from the product by extraction into an aqueous solution of sodium nitrite, from which it is then recovered by extraction into o-dichlorobenzene. Similarly, tetrabutylammonium catalyst is removed by aqueous-phase extraction from a process for manufacture of an ibuprofen intermediate [387], and also in a PTC process for production of benzoisothiazoline derivatives for pharmaceutical use [388]. In another patented method, quaternary ammonium salts are removed from organic solutions by extraction into and recovery from dilute sodium hydroxide solutions [389]. Use of quaternary salts containing β-hydroxyethyl groups, such as used in the production of chloroprene by dehydrochlorination of 1,2-dichlorobutene, allows their extractive recovery from the reaction product [390].

One patent for the separation and recovery of a phase-transfer catalyst, phenyl-triethylammonium chloride, involves extraction of the quaternary salt from the reaction product into water, concentration of the aqueous solution, then addition of an organic solvent to precipitate residual inorganic salts, then removal of solvent to precipitate crude quaternary salt, which could be used directly as catalyst [391].

PEGs, particularly the lower molecular weight homologs (mol wt < 1500), are easily extractable into water from relatively nonpolar organic solutions. Use of dilute aqueous acid solutions to extract quaternary ammonium salts also containing amino groups has been used to separate these catalysts from reaction products [392]. Likewise, TDA-1, cryptands, and soluble pyridine-containing sulfoxide polymers used as phase-transfer catalysts [393] are removable from organic reaction mixtures by extraction into dilute hydrochloric acid. Neutralization of the extracts permits recovery of the catalysts.

2. Distillation Methods

When lower-boiling compounds are being produced in a PTC reaction, they can usually be distilled away from phase-transfer catalysts, as, for example, in the production of chloroprene. PTC catalyst in the distillation residue may sometimes be reusable, although contamination from reaction byproducts and other residual material is likely to severely restrict this practice. With quaternary ammonium salts as catalysts, temperatures above ca. 100–120°C usually result in partial or total decomposition of the quaternary salt to trialkylamines and other products. If this occurs, then separation may be achieved by first thermally decomposing the quaternary salt and then removing it by extraction of the trialkylamine into dilute acid.

In the manufacture of 1,6-dicyanohexane by a PTC cyanide displacement reaction, the catalyst tetrabutylammonium chloride was partially removed by extraction into water, then the washed organic product was flash-distilled at a high temperature to decompose any remaining catalyst [394]. This treatment prevents continuous slow thermal decomposition of quaternary salt during fractional distillation, thereby permitting the production of very high purity product.

PEG, crown ethers, and many of the soluble polymers used as PTC catalysts are reasonably stable under nonacidic distillation conditions.

References

1. F. Montanari, D. Landini, A. Maia, S. Quici, and P.L. Anelli, Reactivity and application of soluble and polymer-supported phase-transfer catalysts, ACS Symp. Ser., **326**, 54–66 (1987).
2. D. Landini, A. Maia, and F. Montanari, Israel J. Chem., **26**, 263 (1985).
3. W. Ma, Yiyao Gongye, **18**, 428 (1987) [CA **108**.027455].
4. G. Pantini, Technol. Chim., **5**, 54 (1985) [CA: **103**:021855].

5. J. Bian, Phase transfer catalysts, Huaxue Shijie, **23**, 25 (1982).
6. Y. Sasson and S. Zbaida, (to Gadot Petrochemical Industries), Ger. Patent DE 3,307,164 (1983) [CA: 100.085389].
7. T. Kawai to Central Glass Co., British Patent 2,219,292 (1989) [CA: 112.234804].
8. C.R. White, to Mallinckrodt, Inc., U.S. Patent 4,642,399 (1987) [CA: 106.138076].
9. J. Wild and N. Goetz, to BASF A.-G., Ger. Pat. DE 3,820,979 (1989) [CA: 113.040150].
10. H. Namba, N. Takahashi, K. Abe, and M. Saito, to Mitsubishi Gas Chemical Co., Japanese Patent, Jpn. Kokai Tokkyo Koho 63/196,547 (1988) [CA: 110.023358].
11. W.H. Daly, J.D. Caldwell, V.P. Kien, and R. Tang, Polym. Prepr. Am. Chem. Soc., Div. Polym. Chem.) **23**, 145 (1982) [CA: 100.008801].
12. Kohjin Co., Ltd., Japanese Patent Jpn. Kokai Tokkyo Koho JP 58/147,402 (1983) [CA: 100.105396].
13. M.B. Smith, Synth. Commun., **16**, 85 (1986).
14. A.W. Herriott and D. Picker, J. Am. Chem. Soc., **97**, 2345 (1975).
15. M.F. Sorokin, L.G. Shode, and A.I. Kuz'min, Izv. Vyssh. Uchebn. Zaved., Khim. Khim. Tekhnol., **28**, 82 (1985) [CA 104.087902].
16. J.R. Chang, M.Y. Yeh, and Y.P. Shih, J. Chin. Inst. Chem. Eng., **14**, 457 (1983) [CA 100.191042].
17. C.M. Starks and C.L. Liotta, Phase Transfer Catalysis, Principles and Techniques, Academic Press, New York, 1978, p. 51.
18. D. Mason, S. Magdassi, and Y. Sasson, J. Org. Chem., **55**, 2714 (1990).
19. R.A. Moss and W.J. Sanders, J. Am. Chem. Soc., **100**, 5247 (1978).
20. F.S. Sirovskii, S.M. Velichko, Yu.A. Treger, A.L. Chimishkyan, and M.V. Panova, Kinet. Katal., **26**, 1478 (1985) [CA 105.005920].
21. M. Makosza and B. Serafinowa, Rocz. Chem., **39**, 1223 (1965).
22. F. Friedl, T.L. Vetter, and M.J. Bursik, J. Am. Oil Chem. Soc., **62**, 1058 (1985).
23. E.V. Dehmlow and B. Vehre, J. Chem. Res., Synop., 350 (1987) [CA 108.094061].
24. A. Brandstrom, Principles of phase transfer catalysis by quaternary ammonium salts, in Advances in Physical Organic Chemistry, Vol. 15, V. Gold, ed., Academic Press, New York, 1977, p. 276.
25. F. Montanari, D. Landini, and F. Rolla, Top. Curr. Chem., **101**, 147 (1982) and references contained therein.
26. E.V. Dehmlow, M. Slopianka, and J. Heider, Tetrahedron Lett., 2361 (1977).
27. F. Montanari and P. Tundo, J. Org. Chem., **47**, 1298 (1982).
28. S. Yonemori, Y. Hayashi, S. Kumai, and A. Wada, Nippon Kagaku Kaishi, (8) 1146–8 (1991) [CA: 115.159279].
29. D. Landini, A. Maia, and A. Rampoldi, J. Org. Chem., **51**, 3187 (1986).
30. S. Kondo, Y. Takeda, and K. Tsuda, Synthesis, **11**, 862 (1989).

31. J. de la Zerda, R. Neumann, and Y. Sasson, J. Chem. Soc., Perkin Trans. II, 823 (1986).
32. T. Ido, Y. Matsuura, and S. Goto, Kagaku Kogaku Ronbunshu, **14**, 174 (1988) [CA 109.072972].
33. E.V. Dehmlow and R. Thieser, Tetrahedron, **41**, 2927 (1985).
34. E.V. Dehmlow and M. Lissel, Tetrahedron Lett., 1783 (1976).
35. H.A. Zahalka and Y. Sasson, Can. J. Chem., **67**, 245 (1989).
36. M.Y. Yeh, D.H. Hwu, C. Hwang, T.K. Hwang, and Y.P. Shih, J. Chin. Chem. Soc. (Taipei) **38**, 221–30 (1991) [CA: 115.135239].
37. D.H.H. Hwu, S.Y.P. Chyi, M.Y. Yeh, and C.L. Chao, Ind. Eng. Chem. Res. **31**, 177 (1992).
38. Z. Liu and X. Chen, Huaxue Shijie, **23**, 104 (1982) [CA 100.087638].
39. E. Reinholz, A. Becker, B. Hagenbruch, S. Schaefer, and A. Schmitt, Synthesis, 1069 (1990).
40. E.V. Dehmlow and J. Stuetten, Liebigs Ann. Chem., 187 (1989).
41. E.V. Dehmlow and J. Wilkenloh, Liebigs Ann. Chem., 125 (1990).
42. E.V. Dehmlow and J. Wilkenlow, Chem. Ber. **123**, 583 (1990).
43. E.V. Dehmlow and G.O. Torossian, Z. Naturforsch. B: Chem. Sci., **45**, 1091 (1990) [CA: 114.005984].
44. E.V. Dehmlow and S. Schrader, Z. Naturforsch., B., Chem. Sci., **45**, 409 (1990) [CA 113.058227].
45. E. D'Incan, P. Viout, and R. Gallo, Israel J. Chem., **26**, 277 (1986).
46. P. Tundo and M. Badiali, React. Polym., **10**, 55 (1989).
47. L.H. Tagle and F.R. Diaz, Eur. Polym. J., **23**, 109 (1987) [CA 107.040445].
48. L.H. Tagle, F.R. Diaz, and M.A. Vargas, Acta Polym., **42**, 398 (1991) [CA: 116.042114].
49. L.H. Tagle, F.R. Diaz, and M.I. Arroyo, Bol. Soc. Chil. Quim., **35**, 367–75 (1990) [CA:115.029974].
50. K. Murai, C. Kimura, and H. Hazawa, Nippon Kagaku Kaishi (5) 805 (1982) [CA 97.126579].
51. Z. Liu and X. Chen, Huaxue Shijie, **23**, 104 (1982) [CA 100.087638].
52. V.L. Afanas'eva, M.R. Bagreeva, A.V. Lyubeshkin, V.M. Pechenina, N.A. Epshtein, and R.G. Glushkov, Khim.-Farm. Zh., **21**, 1114 (1987) [CA 108.149703].
43. T. Miyakoshi, M. Nakajima, H. Omichi, and S. Saito, Nippon Kagaku Kaishi, 986 (1985) [CA 104.109121].
54. P. Depreux and A. Marcinal-Lefebvre, Can. J. Chem., **64**, 626 (1986).
55. T. Iizawa, K. Hayasgi, Y. Endo, and T. Nishikubo, J. Polym. Sci., Polym. Lett. Ed., **23**, 623 (1985).
56. N. Yamazaki and Y. Imai, Kobunshi Ronbunshu, **43**, 105 (1986) [CA 104.168929].

57. G. Pantini, Tecnol. Chim., **5**, 54 (1985) [CA 103.021855].
58. J.H. Clark and N. Boechat, Chem. Ind. (Lond., 436 (1991).
59. J.H. Clark and D.J. Macquarrie, Tetrahedron Lett., **28**, 111 (1987).
60. S. Samaan and F. Rolla, Phosphorus Sulfur, **4**, 145 (1978) [CA 89.041853].
61. D.J. Brunelle, J. Org. Chem., **49**, 1309 (1984).
62. N.A. Gibson and J.W. Hosking, Aust. J. Chem., **18**, 123 (1965).
63. Y. Huang, L. Shi, J. Yang, W. Xiao, S. Li, and W. Wang, Hetroat, Chem. [Internat. Conf.] 2nd. 1989 (Pub. 1990). E. Block, ed., VCH: New York [CA:115.028319].
64. J. Huang and Y. Yuang, J. Organomet. Chem., **414**, 49 (1991).
65. W. Tao, X. Wang, and W. Zeng, Huaxue Shiji, **14**, 71 (1992) [CA: 117.111209].
66. S. Kondo, Y. Takeda, and K. Tsuda, Synthesis, **11**, 862 (1989).
67. N. Furukawa, K. Okano, and H. Fujihara, Nippon Kagaku Kaishi, 1353 (1987) [CA 108.021618].
68. S. Kondo, Y. Takeda, and K. Tsuda, Synthesis, 403 (1988).
69. H. Fujihara and N. Furukawa, Kagaku to Kogyo (Osaka), **62**, 134 (1988) [CA 109.116726].
70. D.J. Brunelle, Prepr. Am. Chem. Soc., Div. Pet. Chem., **30**, 378 (1985) [CA 105.114663].
71. D.J. Brunelle, ACS Symp. Ser. 326, 38 (1987).
72. D.A. Singleton, to General Electric Co., Eur. Patent 126950 (1984) [CA 102.151217].
73. D.J. Brunelle to General Electric Co., U.S. Patent 4,460,778 (1984).
74. J.W. Verbicky, Jr., and A.M. Colley, to General Electric Co., U.S. Patent 4,577,033 (1986).
75. D.J. Brunelle, D.A. Haitko, J.P. Barren, and S. Singh, to General Electric Co., Can. Pat. Appl. CA 2,034,435 (1991) [CA:117.026338].
76. E.V. Dehmlow and V. Knufinke, J. Chem. Res., Synop., 224 (1989).
77. G. Cantrell, to Mallinckrodt, Inc. U.S. Patent 4,973,772 (1990); U.S. Patent, 4,973,771 (1990).
78. H. Vorbrueggen, to Schering A.-G., Ger. Patent DE 3,733,471 (1989) [CA 112.035686].
79. M. Harustiak, M. Hronec, J. Ilavaky, and S. Witek, Stud. Org. Chem. (Amsterdam) **35** (Chem. Heterocycl. Compd.) 308 (1988) [CA 110.172760].
80. T. Kawahara, S. Yamamoto, K. Hasegawa, and K. Konno, to Mitsubishi Petrochemical Co., Ltd., Jpn. Kokai Tokkyo Koho JP 61/103,872 (1986) [CA 105.226552].
81. S. Mori, K. Ida, and M. Ue, to Mitsubishi Petrochemical Co., Ltd., Jpn. Kokai Tokkyo Koho JP 63/115,876 (1988) [CA 110.008219].
82. J.M. McIntosh and S.O. Acquaah, Can. J. Chem., **66**, 1752 (1988).
83. I.A. Esikova and E.P. Serebryakov, Izv. Nauk SSSR, Ser. Khim., 1836–43 (1989) [CA: 112.054595].

84. I.A. Esikova and E.P. Serebryankov, Izv. Akad. Nauk SSSR, Ser. Khim., 1836 (1989) [CA 112.054595].
85. U.T. Bhalerao, S.N. Mathur, and S.N. Rao, Synth. Commun., **22**, 1645 (1992).
86. I.V. Nikonova, G.I. Koldobskii, A.B. Zhivich, and V.A. Ostrovskii, Zh. Obshch. Khim., **61**, 2104 (1991).
87. J. Kulic, M. Adamek, A.B. Zhivich, G.I. Koldobskii, and Yu. E. Myznikov, Zh. Obsch. Khim., **60**, 2370 (1990).
88. A.B. Zhivich, Yu. E. Myznikov, G.I. Kolodobskii, and V.A. Ostrovskii, Zh. Obshch. Khim., **58**, 1906 (1988).
89. I.V. Nikonova, G.I. Koldobskii, A.B. Zhivich, and V.A. Ostrovskii, Zh. Obshch. Khim., **61**, 1483-4 (1991).
90. A.B. Zhivich, G.I. Koldobskii, and V.A. Ostrovskii, Zh. Org. Khim., **25**, 668 (1989).
91. G.I. Koldobaski, A.B. Zhivich, and V.A. Ostrovaskii, Zh. Obsch. Khim., **62**, 3 (1992).
92. C. Sunkel, et al., to Alter S.A., Eur. Pat. Appl. EP 267,136 (1988) [CA 109.157276].
93. D.J. Brunelle to General Electric Co., U.S. Patent 5,081,298 [CA: 116:128171].
94. J.P. Idoux and J.T. Gupton, ACS Symp. Ser., 326, 169 (1987).
95. M. Lissel, D. Feldman, M. Nir, and M. Rabinovitz, Tetrahedron Lett., **30**, 1683 (1989).
96. E.A. O'Neil, Jr., and J.W. Verbicky, Jr., to General Electric Co., Eur. Pat. Appl. EP 141692 (1985) [CA 103.195993].
97. D.J. Brunelle, to General Electric Co., Eur. Pat. Appl., EP 142,835 (1985) [CA 103.196007].
98. J. Alvarez-Builla, et al., Tetrahedron, **45**, 967 (1990).
99. H. Tomioka, K. Ueda, H. Ohi, and Y. Izawa, Chem. Lett., 1359 (1986).
100. H. Bahrmann, B. Cornils, W. Lipps, P. Lappe, and H. Springer, to Ruhrchemie A.-G., Can. Pat. 1,247,642 (1988) [CA 111.097503], Ger. Offen. DE 3,420,493 (1985) [CA 106.033308].
101. B. Wu and Y. Lu, Yingyong Huaxue 7(3) 64-5 (1990) [CA: 114.042203].
102. P.M. Chakrabarti, S.G. Desai, and L.S. Wood, Jr., to GAF Corp., U.S. Patent 4,694,104 (1987).
103. Y. Lu and B. Wu, Zhejiang Daxue Xuebao, Ziran Kexueban, **24**, 46 (1990) [CA: 114.163576].
104. B. Wu and Y. Lu, Zhejiang Yike Daxue Xuebao, **19**, 255 (1990) [CA: 114.228463].
105. S. Grinberg and E. Shaubi, Tetrahedron, **47**, 2895 (1991).
106. J. Beger, R. Jacobi, U. Rehbeil, T. Roethling, and H. Luthardt Horst (Bergakademie Freiberg) Ger. (East) DD 295,346 (1990) [CA: 116.151347].
107. B.P. Czech. M.J. Pugia, and R.A. Bartsch, Tetrahedron **41**, 5439 (1985).

108. S. Inokuma and T. Kuwamura, Nippon Kagaku Kaishi, (10), 1494 (1983). [CA 100.008973].

109. B. Mailard, M.J. Bourgeois, E. Montaudon, and R. Lalande, to Centre National de la Recherche Scientifique, FR 2,628,107 (1988) [CA 112.118868].

110. O. Samuelson, Ion Exchange Separation in Analytical Chemistry, p. 71, Wiley, New York, 1963.

111. T. Hiyama, H. Sawada, M. Tsukanaka, and H. Nozaki, Tetrahedron Lett. 3013 (1975).

112. T. Hiyama, T. Mishima, H. Sawada, and H. Nozaki, J. Am. Chem. Soc., **97**, 1626 (1975).

113. J. Balcells, S. Colonna, and R. Fornasier, Synthesis, **8**, 266 (1976).

114. L. Maurin, to E.I. Dupont Co., U.S. Patent 4,418,232 (1983).

115. J. Heinrich, R. Casper, and M. Beck, to Bayer A.-G., Ger. Offen. DE 3,208,796 (1983) [CA 100.008204].

116. A.C. Cope, E. Ciganek, and J. Lazor, J. Am. Chem. Soc., **84**, 259 (1962).

117. W.J. Traynelis and J.G. Dadura, J. Org. Chem., **26**, 686, 1813 (1961).

118. A. Ekambaram, R. Thoman, D. Buckley, J. Kampmeier, and D.S. Tarbell, J. Org. Chem., **32**, 2985 (1967).

119. C.M. Starks, unpublished results.

120. Yu. Sh. Gol'berg, E. Abele, I. Kalvins, P.T. Trapentsier, M.V. Shimanskaya, and E. Lukevics, Zh. Org. Khim., **23**, 1561 (1987).

121. Yu. Sh. Gol'dberg, E. Abele, I. Kalvins, G. Bremanis, M.V. Shimanskaya, and E. Lukevic, Dokl. Akad. Nauk SSSR, **294**, 1387 (1987).

122. G. Sonnek, C. Rabe, G. Schmaucks, R. Kaden, and I. Lehms, J. Organomet. Chem., **405**, 179 (1991).

123. R.A. Bartsch, C.V. Cason, and B.P. Czech, J. Org. Chem., **54**, 857 (1989).

124. J.M. Harris and M.G. Case, J. Org. Chem., **48**, 5390 (1983).

125. A.V. Bogatskii, S.A. Kotlyar, and E.I. Klimova, Ukr. Khim. Zh. (Russ. Ed.) **51**, 1206 (1985) [CA 106.032317].

126. For a review see: R. Vladea, T. Simandan, Rev. Chim. (Bucharest), **41**, 421–5 (1990) [CA: 114.229413].

127. L. Toke and G.T. Sazbo, Acta Chem. Acad. Sci., Hung, **93**, 421 (1977), as summarized by G.E. Totten and N.A. Clinton, Rev. Macromol. Chem. Phys., **C28**, 293 (1988).

128. L. Toke, G.T. Szabo, and K. Somogyi-Werner, Acta Chim. Acad. Sci. Hung., **101**, 47 (1979) [CA 92.065502].

129. G.T. Szabo, K. Aranyosi, and L. Toke, Acta Chim. Acad. Sci., Hung., **110**, 215 (1982) [CA 97.169893].

130. G.W. Gokel and co-workers, ACS Symp. Ser., **326**, 24 (1987).

131. G.W. Gokel, D.M. Goli, and R.A. Schultz, J. Org. Chem., **48**, 2837 (1983).

132. J.M. Harris, N.H. Hundley, T.G. Shannon, and E.C. Struck, J. Org. Chem. **47**, 4789 (1982).
133. R. Neumann, S. Dermeik, and Y. Sasson, Israel J. Chem., **26**, 239 (1985).
134. R. Neumann and Y. Sasson, J. Mol. Catal., **31**, 81 (1985).
135. O.E. Filippova, I.N. Topchieva, and V.P. Zubov, Vestn. Mosk. Univ., Ser. 2: Khim., **24**, 590 (1983) [CA 100.102832].
136. L. Toke, G.T. Szabo, K. Aranyosi, and K. Somogyi-Werner, Period. Polytech. Chem. Eng., **32**, 131 (1988) [CA 111.056941].
137. S. Jin and X. Huang, Fenzi Kexue Yu Huaxue Yanjiu, **5**, 159 (1985) [CA 105.023742].
138. O.E. Filippova, I.N. Topchieva, V.V. Lutsenko, and V.P. Zubov, Vysokomol. Soedin., Ser. A, **26**, 402 (1984) [CA 101.022686].
139. T. Ido, M. Saiki, and S. Goto, Kagaku, Kogaku Ronbunshu, **15**, 403 (1989) [CA 111.056994].
140. T. Ido, M. Saiki, and S. Goto, Kagaku Kogaku Ronbunshu, **14**, 539 (1988) [CA 110.038372].
141. H. Lehmkuhl, F. Rabet, and H. Hauschild, Synthesis, 184 (1977).
142. G.E. Totten and N.A. Clinton, J. Macromol. Sci., Rev. Macromol. Chem. Phys., **C28**, 293 (1988).
143. L. Xu, F. Tao, and S. Wu, Youji Huaxue (4) 265 (1984) [CA 101.151046].
144. X. Huang and Z. Huang, Huaxue Shiji, **7**, 20, 1985 [CA 102.226556].
145. F. Tao, L. Xu, G. Rong, and C. Wang, Huaxue Shijie, **23**, 324 (1982) [CA 100.070339].
146. G.T. Szabo, K. Gero, and L. Toke, Acta Chim. Hung, **113**, 11 (1983) [CA 99.139240].
147. L.J. Mathias and J.B. Canterberry, ACS Symp. Ser., **195**, 139 (1982).
148. Y. Kimura, P. Kirszensztejn, and S.L. Regen, J. Org. Chem., **48**, 385 (1983).
149. D. Balasubramanian and B. Chandani, J. Chem. Educ., **60**, 77 (1983).
150. N. Yamazaki, A. Hirav, and S. Nakahama, J. Macromol. Sci-Chem., **A13**, 321 (1979).
151. S. Slaoui, R. LeGoaller, J.L. Pierre, and J.L. Luche, Tetrahedron Lett. **23**, 1681 (1982).
152. G.T. Szabo, K. Aranyosi, and L. Toke, Acta Chim. Acad. Sci. Hung, **110**, 225 (1982) [CA 97. 197588].
153. Y. Kimura and S.L. Regen, J. Org. Chem., **48**, 195 (1983).
154. B.G. Zupancic and M. Kokaij, Synthesis Commun., **12**, 881 (1982).
155. R. Neuman and Y. Sasson, J. Org. Chem., **49**, 1282 (1984).
156. E.V. Dehmlow, R. Thieser, Y. Sasson, and R. Neumann, Tetrahedron, **42**, 3569 (1986).

157. N. Yang, S. Wang, and L. Pan, Jilin Daxue Ziran Kexue Xuebao, (3) 113 (1986) [CA 106.158256].
158. S. Itsuno, et al., Kobunshi Ronbunshu, **43**, 91 (1986) [CA 105.171740].
159. F.S. Sirovskii, Yu. A. Treger, M.V. Panova, and A.V. Voronkina, Zh. Vses. Khim. O-va. im. D.I. Mendeleeva, **30**, 580 (1985). [CA 105.152291].
160. A. Hirao, S. Nakahama, M. Takahashi, and N. Yamazuki, Makromol. Chem., **179**, 915 (1978).
161. J. Kelly, W.M. Mackenzie, D.C. Sherrington, and G. Reiss, Polym. Commun., **20**, 1048 (1979).
162. S. Slaoui, R. LeGoaller, J.L. Pierre, and J.L. Luche, Tetrahedron Lett., 1137 (1982).
163. E. Angeletti, P. Tundo, and P. Venturello, J. Chem. Soc., Perkin Trans. I, 1137 (1982).
164. K. Sukata, Yuki Gosei Kagaku Kyokaishi, **39**, 443 (1981) [CA 95.080037].
165. F. Tao, L. Xu, and G. Rong, Fudan Xuebao. Ziran Kexueban, **21**, 279 (1982) [CA 98.197695].
166. J. Chen, Y. Dai, L. Wang, J. Niu, and S. Chen, Gaodeng Xuexiao Huaxue Xuebao, **8**, 240 (1987) [CA 108.094148].
167. I.M. Gella, Khim.-Fram. Zh., **20**, 730 (1986) [CA 106.049898].
168. G. Qi and X. Wang, Huadong Huagong Xueyuan Xuebao, **11**, 317 (1985) [CA 106.017646].
169. R. Neumann and Y. Sasson, Tetrahedron, **39**, 3437 (1983).
170. Y. Li, J. Chen and M. Li, Xibei Shifan Xueyuan Xuebao, Ziran Kexuebanm (2) 47 (1986) [CA 106.175898].
171. M. Feng, X. Lu, and Y. Chen, Huaxue Shiji **12**, 135–8 (1990) [CA: 114.042408].
172. T. Kitazume and N. Tahikawa, Chem. Lett., 283 (1978).
173. M.M. Movsumzade, A.L. Shabanov, F.K. Agaev, and I.M. Abdulabekov, Zh. Org. Khim., **12**, 2477 (1976).
174. L. Tlke, G.T. Szabo, G. Szabo, L. Nacy, and I. Rusnak, British Pat. Appl. 2,008,098 (1979).
175. X. Huang and L. Xie, Hangzhou Daxue Xuebao, Ziran Kexueban, **13**, 332 (1986) [CA 106.156002].
176. J. Wang and L. Geng, Xibei Shifan Xueyuan Xuebao, Ziran Kexueban, (3) 36 (1988) [CA 111.057014].
177. N. Yang, S. Wang, and Y. Yan, Jilin Daxue Ziran Kexue Xuebao, (1) 120 (1986) [CA 107.217918].
178. G.R. Shenoy and D.W. Rangnekar, Dyes Pigm., **10**, 165 (1989) [CA 111.024952].
179. M.J. Harris, N.H. Hundley, T.G. Shannon, and E.C. Struck, Polym. Sci. Technol. (Plenum), 24 (Crown Ethers Phase Transfer Catalysis in Polymer Science), 371 (1984) [CA 100.176044].

180. V.K. Krishnakumar, Synth. Commun., **14**, 189 (1984).
181. X. Jian and X. Su, Huaxue Shijie, **31**, 252 (1990) [CA 114.084214].
182. K. Sukata, Bull. Chem. Soc., Japan, **56**, 280 (1983).
183. H. Liu and C. Yuan, Youji Huaxue, 123 (1983) [CA 99.087391].
184. X. Huaxue, Chem. Abstracts, **102**, 131300 (1985).
185. J. Cen and G. Geng, Zhongguo Yiyao Gongye Zazhi, **21**, 218 (1990) [CA:114.122016].
186. R.Z. Zhou and R. Hongkui, Xiamen Daxue Xuebao, Zuan Kexueban, **29**, 596 (1990) [CA: 116.151542].
187. K. Sukata, Bull. Chem. Soc. Japan, **56**, 280 (1983) [CA 98.179147].
188. I. Liao, G. Hongkuei, and Q. Guo, Faming Zhuanli Shenqing Gongkai Shomingshu CN 85102994 (1986) [CA 107.077805].
189. R. Davidson, Ali M. Patel, and A. Safdar, J. Chem. Res., Synop., (3) 88 (1984).
190. Y. Cai and R. Du, Jinan Daxue Xuebao, **11**, 52 (1990) [CA 114.081487].
191. Q. Zhong, J. Shao, J. Wang, and Z. Gong, Gaodeng Xuexiao Huxue Xuebao, **8**, 441 (1987) [CA 108.075172].
192. R. Zhou, Y. Chen, H. Zhao, and Q. Guo, Xiamen Daxue Xuebao, Ziran Kexueban, **27**, 671 (1988) [CA 111.194558].
193. B.G. Zupancic and M. Sopcic, Synthesis, **11**, 942 (1982).
194. G. Rong, F. Tao, and L. Xu, Gaodeng Xuexiao Huaxue Xuebao, **4**, 576 (1983) [CA 100.209196].
195. K. Sukata, Yuki Gosei Kagaku Kyokaishi, **39**, 1131 (1981) [CA 97.005926].
196. Y. Hu and Y. Tao, Zhejiang Yike Daxue Xuebao, **14**, 39 (1985) [CA 104.168075].
197. L. Liao and Q. Guo, Xiamen Daxue Xuebao, Ziran Kexueban, **27**, 590 (1988) [CA 110.212710].
198. G. Zeng and S. Yin, Xinan Shitan Daxue Xuebao, Ziran Kexueban, (2) 88 (1988) [CA 110.134733].
199. R. Neumann and Y. Sasson, J. Org. Chem., **49**, 3448 (1984).
200. R. Neuman and Y. Sasson, J. Mol. Catal., **33**, 201 (1985).
201. Y. Yang, Hunan Shifan Daxue Ziran Kexue Xuebao, **12**, 318–23 (1989) [CA: 116.083313].
202. L.A. Khachatryan, G.V. Mirzoyan, R.A. Kazaryan, A.Ts. Malkhasyan, and N.G. Martirosyan, Arm. Khim. Zh., **41**, 305 (1988). [CA 110.134729].
203. G.S. Huvard, P.P. Nicholas, and S.E. Horne, Jr., J. Polym. Sci., Polym. Chem., **23**, 2005 (1985).
204. F. Tao and J. Xu, Gaodeng Xuexiao Huaxue Xuebao, **2**, 460 (1981) [CA 97.038121].
205. A.K. Saxena, C.S. Bisaria, and L.M. Pande, Synth. React. Inorg. Met.-Org. Chem., **16**, 667 (1986) [CA 106.156597].
206. L. Xiong and L. Yu, Gaodeng, Xuexiao Huazue Xuebao, **7**, 426 (1986) [CA 107.058754].

207. J.H. Poupaert, J.L. de Keyser, D. Vandervorst, and P. Dumont, Bull. Soc. Chim. Belg., **93**, 493 (1984).

208. C. Sastry, K. Reddy, V.S.H. Krishnan, G.K.A.S.s. Narayan, and K. Vemana, Chem. Ind (Lond.), (7), 227 (1989).

209. B.G. Zupancic and M. Kokalj, Synth. Commun., **12**, 881 (1982).

210. E. Santaniello, A. Manzocchi, and P. Sozzani, Tetrahedron Lett., **47**, 4581 (1979).

211. D.G. Lee and V.C. Chang, J. Org. Chem., **43**, 1532 (1978).

212. W. Hou and X. Jiang, Hangzhou Daxue Xuebao, Ziran Kexueban, **12**, 275 (1985) [CA 104.148071].

213. Y. Jiang, Z. Xu, and Q. Guo, Xiamen Daxue Xuebao, Ziran Kexueban, **28**, 519–22 (1989) [CA: 114.024312].

214. D. Balasubramanian, P. Sukamar, and B. Chandoni, Tetrahedron Lett., 3543 (1979).

215. P. Lavelle, Spec. Chem., **6**, 16,18,20 (1986) [CA 105.208179].

216. G. Soula, J. Org. Chem., **50**, 3717 (1985).

217. A.K. Bose and P. Mangiaracina, Tetrahedron Lett., **28**, 2503 (1987).

218. J.S. Ferlut, to Rhone Poulenc Chimie de Base, Eur. Pat. Appl. EP 255,453 (1988) [CA 108.204834].

219. N.R. Ayyangar and K.V. Srinivassan, Colourage, **37**, 37-9 (1990) [CA: 114.166331].

220. R.A. Bartsch and J.B. Phillips, Synth. Commun., **16**, 1777 (1986).

221. P. Ballesteros, R.M. Claramunt, and J. Elguero, Tetrahedron, **43**, 2557 (1987).

222. D.M. Spyriounis, E. Rekka, and V.J. Demopoulos, Synth. Comm., **20**, 2417 (1990).

223. M. Iki and T. Hayashi, Jpn. Kokai Tokkyo Koho JP 62/22780 (1987) [CA 106.176418].

224. S. Huang and P.T. Trazasko, Polym. Sci. Technol. (Plenum) **25**, (New Monomers Polym.) 163 (1984) [CA 100.210583].

225. H. Xie and J. Xia, Makromol. Chem., **188**, 2543 (1987) [CA 107.237427].

226. H.Q. Xie and J. Xia, Polym. Prepr. (Am. Chem. Soc., Div. Polym. Chem.), **28**, 203 (1987) [CA 107.134740].

227. F. Vogtle and E. Weber, Angew. Chem., Int. Ed. Engl., **13**, 814 (1974).

228. Y. Xu, X. Zhang, F. Wang, and H. Wang, Wuhan Daxue Xuebao, Ziran Kexueban, (94) 71 (1985) [CA 105.098046].

229. H. Gruber and G. Greber, Monatsh. Chem., **112**, 1063 (1981) [CA 96.069323].

230. C.J. Thoman, T.D. Habeeb, M. Huhn, M. Korpusik, and D.F. Slish, J. Org. Chem., **54**, 4476 (1989).

231. M. Feng and Y. Chen, Yingyong Huaxue, **6**, 57 (1989) [CA 111.078723].

232. S. Huang and P.T. Trazasko, Polym. Sci. Technol. (Plenum) **25** (New Monomers Polym.), 163 (1984) [CA 100.210583].

233. H. Xie and J. Xia, Makromol. Chem., **188**, 2543 (1987) [CA 107.237427].
234. H.Q. Xie and J. Xia, Polym. Prepr. (Am. Chem. Soc., Div. Polym. Chem.), **28**, 203 (1987) [CA 107.134740].
235. Y. Xu, X. Zhang, F. Wang, and H. Wang, Wuhan Daxue Xuebao, Ziran Kexueban, **94**, 71 (1985) [CA 105.098046].
236. X. Zhang, T. Huang, and H. Duan, Wuhan Daxue Xuebao, Ziran Kexueban, (1) 70 (1986) [CA 106.102765].
237. A.T. Au, to Dow Chemical Co., U.S. Patent 4,308,031 (1981).
238. D. Landini, A. Maia, L. Corda, A. Maccioni, and G. Podda, Tetrahedron Lett., **30**, 5781 (1989).
239. D. Landini, A. Maia, L. Corda, A. Maccioni, and G. Podda, Tetrahedron, **47**, 7477 (1991).
240. L. Corda, C. Anchisi, G. Podda, P. Traldi, and M. Gleria, Heterocycles, **24**, 2821 (1986) [CA 106.176520].
241. L. Bonsignore, L. Corda, E. Maccioni, G. Podda, M. Gleria, and M. Gazz. Chim. Ital., **121**, 341 (1991) [CA:115.280649].
242. S. Kondo, Y. Inagaki, M. Ozeki, and K. Tsuda, J. Polym. Sci., Part A: Polym. Chem., **27**, 3383 (1989).
243. S. Kondo, Y. Inagaki and K. Tsuda, J. Polym. Sci., Polym. Lett. Ed., **22**, 249 (1984).
244. S. Kondo, M. Minafuji, Y. Inagaki, and K. Tsuda, Polym. Bull. (Berl.) **15**, 77 (1986) [CA 104.187665].
245. S. Kondo, et al., Pure Appl. Chem., **60**, 387 (1988).
246. S. Kondo, N. Nakashima, and K. Tsuda, J. Macromol. Sci., Chem., **a26**, 1425 (1989) [CA 112.177696].
247. F. Montanari, M. Penso, G. della Fortuna, and A. Re, Gazz. Chim. Ital. **115**, 427 (1985) [CA 104.147893].
248. J. Kahovec, M. Jelinkova, and V. Janout, Polym. Bull. (Berl.) **15**, 485 (1986) [CA 105.134440].
249. S. Iwabuchi, T. Nakahira, A. Tsuchiya, K. Kojima, and V. Boehmer, Makromol. Chem., **184**, 535 (1983).
250. S. Kondo, M. Ozeki, N. Nakashima, K. Suzuki, and K. Tsuda, Angew. Makromol. Chem., **163**, 139 (1988) [CA 111.114700].
251. A. Zicmanis, M. Karulis, and M.P. Kovaleva, Latv. PSR Zinat. Akad. Vestis, Kim. Ser., 353 (1986) [CA 106.175867].
252. Y. Wu and J. Kohn, J. Am. Chem. Soc., **113**, 687 (1991).
253. N. Furukawa, S. Ogawa, T. Kawai, and S. Oae, J. Chem. Soc., Perkin Trans. I, 1833 (1984).
254. N. Furukawa, S. Ogawa, T., Kawai, K. Kishimoto, H. Fujihara, and S. Oae, Heterocycles, 16, 1927 (1981) [CA 96.068769]; **19**, 2041 (1982) [CA 98.053635].

255. N. Furukaws, K. Kishimoto, S. Ogawa, T. Kawai, H. Fujihara, and S. Oae, Tetrahedron Lett., **22**, 4409 (1981).
256. Y. Wu and D. Yang, Huaxue Shiji, **12**, 27 (1990) [CA 113.058099].
257. K. Tsuda, Kobunshi Kako, **34**, 493 (1985) [CA 105.225438].
258. K. Tsuda, Kobunshi Kako, **34**, 545 (1985) [CA 104.149426].
259. S. Kondo, K. Ohta, R. Ojika, H. Yasui, and K. Tsuda, Makromol. Chem., **186**, 1 (1985).
260. N. Furukawa, K. Imaoka, H. Fujihara, and S. Oae, Chem. Lett., 1421 (1982).
261. S. Kondo, K. Ohta, and K. Tsuda, Makromol. Chem., Rapid Commun. **4**, 145 (1983).
262. E. Yaacoub and P. LePerchec, React. Polym., Ion Exch, Sorbents, **8**, 285 (1988) [CA 110.56805].
263. S. Kondo, M. Nakanishi, and K. Tsuda, J. Polym. Sci., Polym. Chem. Ed., **23**, 581 (1985).
264. S. Kondo, M. Nakanishi, K. Yamane, A. Horibe, and K. Tsuda, J. Appl. Polym. Sci., **32**, 4255 (1986).
265. S. Kondo, H. Yasui, K. Ohta, and K. Tsuda, J. Chem. Soc., Chem. Commun., 400 (1985).
266. L. Bonsignore, L. Corda, E. Maccioni, G. Podda, and M. Gleria, Gazz. Chim. Ital., **121**, 341 (1991) [CA: 115.280649].
267. X. Zhang, T. Huang, and H. Duan, Wuhan Daxue Xuebao, Ziran Kexueban, (1) 70 (1986) [CA 106.102765].
268. G. Qin, et al., Faming Zhuanli Shenqing Gongkai Shuomingshu CN 1,039,413 (1990) [CA: 113,230791].
269. G.T. Szabo, K. Aranyosi, M. Csiba, and L. Toke, Synthesis, 565 (1987).
270. U.H. Dolling, to Merck and Co., U.S. Patent, 4,605,761 (1986).
271. S. Kawada, H. Maruyama, H. Kobayashi, to Nippon Kayaku Co., Ltd., Jpn. Kokai Tokkyo Koho JP 03,240,782 (1991) [CA: 116.106303].
272. S.M. Shostakovskii, V.N. Mochalov, V.M. Shostakovskii, Yu. Z. Karasev, and O.M. Nefedov, Dokl. Akad. Nauk SSSR, **302**, 1122 (1988).
273. S. Kawada, H. Maruyama, and H. Kobayashi, to Nippon Kayaku Co., Ltd., Jpn. Kokai Tokkyo Koho JP 03,240,782 (1991) [CA: 116.106303].
274. E.V. Dehmlow and C. Bollmann, Tetrahedron Lett., **32**, 5773 (1991).
275. T.D. Shaffer and M.C. Kramer, Makromol. Chem., **191**, 3155 (1990).
276. T.D. Shaffer and M.C. Kramer, Polym. Prepr. (Am. Chem. Soc., Div., Polym. Chem.) **30**, 171 (1989).
277. S.S. Shavanov, G.A. Tolstikov, T.V. Shutenkova, and G.A. Viktorov, Zh. Obshch. Khim., **57**, 1587 (1987).
278. E.V. Dehmlow, R. Thieser, Y. Sasson, and E. Pross, Tetrahedron, **41**, 2927 (1985).
279. E.V. Dehmlow, H.C. Raths, and J. Soufi, J. Chem. Res., Synop., 334 (1988).

280. T. Ido, Y. Matsuura, and S. Goto, Kagaku Kogaku Ronbunshu, **14**, 174 (1988) [CA 109.072972].

281. M. Makosza and W. Lasek, Tetrahedron, **47**, 2843 (1991).

282. E.V. Dehmlow, R. Thieser, Y. Sasson, and R. Neumann, Tetrahedron, **42**, 3569 (1986).

283. W. Szeja, I. Fokt, and G. Grynkiewicz, Recl. Trav. Chim. Pays-Bas, **108**, 224 (1989).

284. S.S. Shavanov, G.A. Tolstikov, and G.A. Viktorov, Zh. Obsch. Khim., **59**, 1615 (1989).

285. T. Jeffery, Tetrahedron Lett., **30**, 2225 (1989).

286. V.V. Grushin and H. Alper, J. Org. Chem., **57**, 2188 (1992).

287. D.E. Bergbreiter and J.R. Blanton, J. Org. Chem., **52**, 472 (1987).

288. S. Sebti and A. Foucaud, J. Chem. Res., Synop., 72 (1987).

289. M.Y. Yeh, D.H. Hwu, C. Hwang, and T.K. Hwang, J. Chin. Chem. Soc. (Taipei) 38, 221–30 (1991) [CA: 115.135239].

290. M.Y. Yeh, H.H. Guo, L.C. Chen, and Y.P. Shih, Ind. Eng. Chem. Res., **27**, 1582 (1988).

291. T.J. Mason, J.P. Lorimer, A.T. Turner, and A.R. Harris, J. Chem. Res., Synop., 300 (1986).

292. R.K. Smalley and H. Suschitzky, J. Chem. Soc., 755 (1964).

293. M. Yamada, Y. Watabe, T. Sakakibara, and R. Sudoh, J. Chem. Soc., Chem. Commun., 179 (1979).

294. L.J. Mathias and R.A. Vaidya, J. Am. Chem. Soc., **108**, 1093 (1986).

295. B. Escoula, I. Rico, and A. Lattes, Bull. Soc. Chim. Fr., 256 (1989).

296. G. Barak and Y. Sasson, Bull. Soc. Chim. Fr., 584 (1988).

297. S. Shimizu, Y. Sasaki, and C. Hirai, Bull. Chem. Soc., Japan, **63**, 176 (1990).

298. P.A. Ganeshpure and S. Satish, J. Chem. Soc., Chem. Commun., 981 (1988).

299. A. Deratani, G. Lelievre, T. Maraldo, and B. Sebille, Carbohydr. Res., 192, 215 (1989) [CA 112.099040].

300. H. Ryoshi, N. Kunieda, and M. Kinoshita, Makromol. Chem., **187**, 263 (1986).

301. N. Tanaka, A. Yamaguchi, Y. Araki, and M. Araki, Chem. Lett., 715 (1987).

302. E. Paetzold, G. Oehme, and B. Costisella, Z. Chem., **29**, 447 (1989) [CA 112.244852].

303. M.J.H. Russell, Platinum Met. Rev., **32**, 179 (1988) [CA 110.117096].

304. M. Safi and D. Sinou, Tetrahedron Lett., **32**, 2025 (1991).

305. C.M. Starks and C.L. Liotta, Phase Transfer Catalysis, Principles and Techniques, Academic Press, New York, 1978.

306. W.K. Fife and Y. Xin, J. Am. Chem. Soc., **109**, 1278 (1987).

307. M. Zeldin, to Indiana University Foundation, U.S. Patent 4,855,433 (1988) [CA 112.099456].
308. J. Wild and N. Goetz, to BASF A.-G., Ger. Offen. DE 3820979 (1989) [CA: 113.040150].
309. T. Kawai, to Central Glass Co., Ltd., Brit. UK Pat. Appl., GB 2,219,292 (1989) [CA: 112.234804].
310. Y. Sasson and S. Zbaida, to Gadot, Petrochemical Industries Ltd., Ger. Offen. DE 3,307,164 (1983) [CA: 100.085389].
311. C.R. White, to Mallinckrodt, Inc., U.S. Patent 4,642,399 (1987) [CA: 106.138076].
312. T. Jiang, Kexue Tongbao, **29**, 1627 (1984) [CA 103.006649].
313. G. Fleche, Starch/Staerke, **42**, 31 (1990) [CA 112.120970].
314. T. Ido, N. Ohyama, S. Goto, and H. Teshima, Kagaku Kogaku Ronbunshu, **9**, 58 (1983) [CA 98.125171].
315. M. Takeishi, K. Se, N. Umeta, and R. Sato, Nippon Kagaku Kaishi, 824 (1992) [CA: 117.131693].
316. J. Cosseau, L. Gouin, L.V. Jones, G. Jugic, and J.A.S. Smith, J. Chem. Soc., Faraday Trans., **2**, 1821 (1973).
317. J. Cousseau and L. Gouin, J. Chem. Soc., Perkin Trans., **1**, 1797 (1977).
318. E.V. Dehmlov, Phase Transfer Catalysis, Verlag Chemie, Deerfield Beach FL, 1983, ref. 2, Chapter 3.
319. J. Cousseau, Synthesis, 805 (1980).
320. D. Landini, F. Montanari, and F. Rolla, Synthesis, 37 (1974).
321. J. Simon and R. Seguin, Synth. Commun., **10**, 897 (1980).
322. D. Berkes, R. Kacina, J.F. Gomory, N. Maria, and J. Niznansky, Czech. Patent CS 270,876 (1991) [CA: 116:128161].
323. D. Landini and F. Rolla, J. Org. Chem., **45**, 3524 (1980).
324. S. Julia and A. Ginebreda, Tetrahedron Lett., 2171 (1979).
325. W. Schoenleben, J. Datow, H. Hoffmann, and S. Winderl, Ger. Pat. OFfen. 2,149,822 (1973) [CA 79.052768].
326. G. Dakka and Y. Sasson, Tetrahedron Lett., **28**, 1223 (1987).
327. B.A. Howel and R.E. Kohman, J. Chem. Ed., **61**, 932 (1984).
328. E.V. Dehmlow and M. Slopianka, Chem. Ber., **112**, 2768 (1979).
329. M. Muhammed, J. Szabon, and E. Hogfeldt, Chem. Ser., **6**, 61 (1974).
330. J. Cosseau, L. Gouin, L.V. Jones, G. Jugic, and J.A.S. Smith, J. Chem. Soc., Faraday Trans. **2**, 1821 (1973).
331. R.V. Vladea, D.E. Oltean, T.L. Simandan, L.M. Rusnac, and C. Vladea, Rom. Patent, RO 89387 (1986) [CA 107.006763].
332. R. Vladea, T. Simandan, L. Rusnac, G. Pop, and G. Musca, Rev. Chim. (Bucharest), **37**, 109, (1986) [CA 106.049562].

333. P. Tundo, P. Venturello, and E. Angeletti, J. Chem. Soc., Perkin Trans., **1**, 2157 (1987).

334. G. Rinn and H. Schmidt, Ceram. Powder Process. Sci., Proc. Int. Conf., 2nd. 1988 (publ. 1989). 221–8. Edited by H. Hausner, G. Messing, and S. Hirano [CA:114.127544].

335. A.K. Bose and P. Mangiaracina, Tetrahedron Lett., **28**, 2503 (1987).

336. R.R. Dewald, S.R. Jones, and B.S. Schwartz, J. Chem. Soc., Chem. Commun., 272 (1980).

337. S. Gauthier and D.J. Worsfold, Macromolecules, **22**, 2213 (1989).

338. J.S. Ferlut, to Rhone Poulenc Chimie de Base, Eur. Pat. Appl. EP 255,453 (1988) [CA 108.204834].

339. J.T. Fenton, to Conoco Inc., U.S. Patent, 4,423,238 (1983).

340. B. Bogdanovic, Acc. Chem. Res., **21**, 261 (1988).

341. H. Kobayashi, T. Sonoda, and H. Iwamoto, Chem. Lett., 1185 (1982).

342. V. Ramachandran, to Ethyl Corp., U.S. Patent, 4,536,343 (1985).

343. V.Z. Sharf, E.F. Litvin, K.A. Kasymova, and V.A. Afans'ev, Izv. Akad. Nauk SSSR, Ser. Khim., 1463 (1985).

344. V.Z. Sharf, K.A. Kasymova, and E.F. Litvin, Izv. Akad. Nauk SSSR, Ser. Khim., 1013 (1986).

345. H. Kobayashi, T. Sonoda, M. Kashiwagi, and A. Ichikawa, to Central Glass Co., Ltd., Japan Kokai Tokkyo Koko JP 63238087 (1988) [CA 111.007610].

346. J. Ichikawa, H. Kobayashi, and T. Sonoda, Yuki Gosei Kagaku Kyokaishi, **46**, 943 (1988) [CA 109.237969].

347. H. Iwamoto, T. Sonoda, and H. Kobayashi, J. Fluorine Cjhem., **24**, 535 (1984) [CA 101.072348].

348. Y. Shiraki, K. Onitsuka, K. Takuma, T. Sonoda, and H. Kobayshi, Bull. Chem. Soc., Japan, **58**, 3041 (1985).

349. G. Lipiner, I. Wilner, and Z. Aizenshtat, Nouv. J. Chim., **10**, 91 (1986).

350. H. Taguchi, N. Kunieda, and M. Kinoshita, Makromol. Chem., **184**, 925 (1983).

351. J.K. Rasmussen and H.K. Smith, II, J. Am. Chem. Soc., **103**, 730 (1981).

352. J.K. Rasmussen and H.K. Smith II, in Crown Ethers and Phase Transfer Catalysts in Polymer Science, L.J. Mathias and C.E. Carraher Jr., eds., Plenum, New York, 1984, pp. 105–119.

353. M. Takeishi, H. Ohkawa, and S. Hayama, Makromol. Chem., Rapid Commun., **2**, 457 (1981).

354. J.K. Rasmussen and H.K. Smith, II, Makromol. Chem., **182**, 701 (1981).

355. A. Jayakrishnan and D.O. Shah, J. Polym. Sci., Polym. Chem. Ed., **21**, 3201 (1983).

356. A. Jayakrishnan and D.O. Shah, J. Appl. Polym. Sci., **29**, 2937 (1984).

357. H. Taguchi, N. Kunieda, and M. Kinoshita, Makromol. Chem., Rapid Commun., **3**, 495 (1982).
358. H. Taguchi, N. Kunieda, and M. Kinoshita, Makromol. Chem., **184**, 925 (1983).
359. H. Royshi, N. Kunieda, and M. Kinoshita, Makromol. Chem., **187**, 263 (1986).
360. N. Kunieda, S. Shiode, H. Royshi, H. Taguchi, and M. Kinoshita, Makromol. Chem., Rapid Commun., **5**, 137 (1984).
361. J.K. Rasmussen and H.K. Smith, II, Polym. Prepr. [Am. Chem. Soc., Div. Polym. Chem.] **23**, 152 (1982).
362. J.K. Rasmussen, S.M. Heilmann, L.R. Krepski, and H.K. Smith, II, Prepr. Am. Chem. Soc., Div. Pet. Chem., **30**, 402 (1985).
363. J.K. Rasmussen, S.M. Heilmann, L.R. Krepski, and H.K. Smith, II, ACS Symp. Ser., **326**, 116 (1987).
364. N.N. Ghosh and B.M. Mandal, Macromolecules, **19**, 19 (1986).
365. N.N. Ghosh, G.N. Gupta, B.M. Mandal, and S.C. Guhaniyogi, J. Polym. Mater. **5**, 227 (1988) [CA 110.136115].
366. K.Y. Choi and C.Y. Lee, Ind. Eng. Chem. Res., **26**, 2079 (1987).
367. K. Tabuchi, H. Suzuki, K. Hasegawa, K. Nakagawa, and M. Kidawara, Nihama Kogyo Koto Semmon Gakko Kiyo, Rigogaku-hen, **24**, 72 (1988) [CA 109.129726].
368. C. Simionescu, C. Mihailescu, and V. Bulacovschi, Acta Polym., **38**, 502 (1987).
369. B. Kneafsey, J.M. Rooney, and S.J. Harris, to Loctite Co., Ireland, Brit. Pat. Appl. 2,185,261 (1987) [CA 108.095105].
370. R. Yu. Makushka, A. Usaitis, G. Bajras, and M. Seno, Vysokomol. Soedin., Ser. A. **31**, 1419 (1989).
371. A.D. Grebenyuk and L.V. Kosareva, Khim. Prir. Soedin, 515 (1982) [CA 98.053344].
372. C.T. Cazianis and C.G. Screttas, Tetrahedron, **39**, 165 (1983).
373. H. Tomioka, K. Ueda, H. Ohi, and Y. Izawa, Chem. Lett., 1359 (1986).
374. T. Endo, Y. Saotome, and M. Okawara, J. Am. Chem. Soc., **106**, 1124 (1984).
375. R. Narayan and G.T. Tsao, Prepr. Pap-Am. Chem. Soc., Div. Fuel Chem, **28**, 261 (1983) [CA 100.123791].
376. S. Li, R. Cen, Z. Zeng, and Z. Liang, Huaxue Tongbao, (6) 28, (1987) [CA 108.037240].
377. B. Zaidman, Y. Sasson, and R. Neumann, Ind. Eng. Chem., Prod. Res. Dev., **24**, 390 (1985).
378. G.E. Boyd and Q.V. Larson, J. Am. Chem. Soc., **89**, 6038 (1967).
379. H.P. Gregor and J.I. Bergman, J. Colloid Sci., **6**, 323 (1951).
380. S.F. Belaya, O.D. Kurielenko, and E.F. Nekryach, Ukr. Khim. Zh., **41**, 277 (1975) [CA: 83.016158].
381. H. Ledon, Synthesis, 347 (1974).
382. G. Bram, A. Loupy, and J. Sansoulet, Israel J. Chem., **26**, 291 (1985).

383. J.F.W. Keana and D.D. Ward, Synth. Comm., **13**, 729 (1983).
384. D. Jaeger, M.D. Ward, and A.K. Dutta, J. Org. Chem., **53**, 1577 (1988).
385. K. Sjoberg, Aldrichimica Acta, **13**, 55 (1980).
386. T.L. Evans, to General Electric Co., U.S. Patent 4,520,204 (1985).
387. F. Hampl, et al., Czech. Patent CS 265,360 (1990) [CA: 114.206808].
388. F. Hampl, J. Hajek, V. Votava, J. Svoboda, J. Paleck, and J. Mostecky, Czech. Patent CS 265,560 (1989) [CA: 114.207024].
389. B.C. Berris, to Ethyl Corp., U.S. Patent 5,030,757 (1991).
390. L. Maurin, to DuPont Co., U.S. Patent 4,418,232 (1983).
391. M. Taracon Estrada, to Ercros, S.A., Span. ES 2,023,606 (1992) [CA: 117:048107].
392. D.J. Brunelle, ACS Symposium Ser., **326**, 38 (1987).
393. N. Furukawa, S. Ogawa, T. Kawai,and S. Oae, J. Chem. Soc., Perkin Trans., **I**, 1833 (1984).
394. H. Nanba, K. Abe, and M. Saito, to Mitsubishi Gas Chemical Co., Inc., Japanese Patent, Jpn. Kokai Tokkyo Koho JP 63/196550 (1988) [CA: 110.023361].
395. Gokel, G.W. and Korzeniowski, S.H., Macrocyclic Polyether Syntheses, Springer-Verlag, Berlin, Heidelberg, New York (1982).
396. Gokel, G.W., Crown Ethers and Cryptands, Monographs in Supramolecular Chemistry, No. 3, J.F. Stoddart, series ed., The Royal Society of Chemistry (1991).
397. Inoue, Y. and Gokel, G.W., Cation Binding by Macrocycles, Marcel Dekker, Inc., New York and Basel (1990).
398. Pedersen, C.J., J. Am. Chem. Soc., **89**, 7017 (1967).
399. Pedersen, C., J. Org. Synth. **52**, 66 (1972).
400. Izatt, R.M., Haymore, B.L., Bradshaw, J.S., Christensen, J.J., Inorg. Chem., 14, 3132 (1975).
401. Burden, I.J., Coxon, A.C., Stoddart, J.F., Wheatley, C.M., J. Chem. Soc., Perkin Trans. I, 220 (1977).
402. Stoddart, J.F., and Wheatley, C.M., J. Chem. Soc. Chem. Commun., 390 (1976).
403. Greene, R.N., Tetrahedron Lett., 1793 (1972).
404. Dale, J., Kristiansen, P.O., Acta Chem. Stand., **26**, 1471 (1972).
405. Gokel, G.W., Cram, D.J., Liotta, C.L., Harris, H.P., Cook, F.L., J. Org. Chem., **39**, 2445 (1974).
406. Gokel, G.W., Cram, D.J., Liotta, C.L., Harris, H.P., Cook, F.L., Org. Synth., **57**, 30 (1977).
407. Johns, G., Ransom, C.J., Reese, C.B., Synthesis, 515 (1976).
408. Cook, F.L., Caruso, T.C., Byrne, M.P., Bowers, C.W., Speck, D.H., Liotta, C.L., Tetrahedron Lett., 4029 (1974).
409. Dale, J., Borgen, G., Daasvatn, J., Acta Chem. Scand., **26**, 1471 (1974).
410. Dale, J., Daasvatn, J., J. Chem. Soc., Chem. Commun., 295 (1976).

411. Dale, J. and Daasvatn, K., U.S. Patent 3,997,563 (1976).
412. Lehn, J.M., Structure and Bonding, 16 (1973).
413. Kyba, E.P., Koga, K., Sousa, L.R., Siegel, M.G., Cram, D.J., J. Am. Chem. Soc., **95**, 2692 (1973).
414. Y. Zhao, L. Zheng, Y. Shen, Chem. Res. Chin. Univ., **8**, 5 (1992) [CA: 118:021895].
415. S. Kondo, A. Shibuta, H. Kunisada, Y. Yuki, Bull. Chem. Soc. Jpn. **65**, 2555 (1992).
416. D.J. Brunelle (to General Electric Co.) U.S. Patent 5,082,968 (1992).
417. U.T. Bhalerao, S.N. Mathur, S.N. Rao, Synth. Commun., **22** 1645 (1992).
418. D. Landini, A. Maia, G.-S. Podda, Y.M. Yan, Chem. Soc. Perkins Trans. II, 1721 (1992).
419. Y.N. Belokon, Pure. Appl. Chem., **64**, 1917 (1992).
420. Yu. N. Belokon, Akad. Nauk., Ser. Khim. 1106 (1992).
422. Y. Aoyama, Y. Tanaka, H. Ogoshi, J. Am. Chem. Soc., **110**, 634 (1988); Y. Tanaka, Y. Kato, Y. Aoyama, J. Am. Chem. Soc., **112**, 2807 (1990); H. Toi, Y. Aoyama, J. Am. Chem. Soc., **113**, 1349 (1991); Y. Aoyama, Y. Tanaka, H. Ogoshi, J. Am. Chem. Soc., **111**, 5397 (1989); Y. Tanaka, Y. Ubukata, Y. Aoyama, Chem. Lett. 1905 (1989); Y. Tanaka, C. Khare, M. Yonezawa, Y. Aoyama, Tetrahedron Lett. **31**, 6139 (1990); Y. Tanaka, Y. Aoyama, Bull. Chem. Soc. Japan **63**, 3343 (1990); Y. Kikuchi, K. Kobayashi, Y. Aoyama, J. Am. Chem. Soc., **114**, 1351 (1992).
423. D.J. Brunelle, D.A. Haitko, J.P. Barren, S. Sineh (to General Electric Co.) U.S. Patent 5,132,423 (1992).

5

Insoluble Phase-Transfer Catalysts

A. Introduction

An important problem facing the designer of industrial phase-transfer catalysis (PTC) processes using soluble PTC catalysts concerns the removal of the catalyst from the reaction mixture, and its economic recycle, as noted in Chapter 4. This problem affects costs and may also affect product purity, byproduct disposal, and environmental concerns. Although methods are available for removal of the traditionally used soluble phase-transfer catalysts, significant process simplification is likely to be realized if insoluble PTC catalysts can be used because of their easy separation and potential for recycle. This chapter is directed toward the possibility using insoluble solid catalysts where the PTC function is (1) bound to an insoluble resin, (2) the PTC function is bound to an insoluble inorganic solid, or (3) the PTC catalyst is maintained in an insoluble *third-liquid layer*.

In spite of the obvious advantages of using insoluble PTC catalysts, they are currently used only infrequently for industrial reactions. No more than an estimated 5–10% of industrial PTC reactions use insoluble catalysts. Those that do use insoluble PTC catalysts normally use a commercially available ion-exchange resin as catalyst. Why the lack of use of insoluble PTC catalysts? Several problems contribute to their lack of use:

1. Many reactions are much slower with insoluble (solid) PTC catalysts than with a corresponding amount of soluble analogous PTC catalysts. These reactions tend to be the ones with slow intrinsic (organic-phase) reaction rates. Reactions with fast intrinsic rates and slow transfer rates normally do not suffer as much rate loss when changing from a soluble PTC catalyst to a resin-bound PTC catalyst. (However, with both types of reaction, special insoluble solid PTC catalysts can usually be developed that do have high rates, and sometimes even exceed the activity of soluble PTC catalysts. The commercial use of these special

catalysts requires extensive development and special manufacturing that cause cost and time problems frequently judged to be too high to justify their industrial use.

2. The cost of resin-bound PTC catalysts is usually prohibitively high if a special catalyst composition is required. Commercially available ion-exchange resins sometimes show excellent behavior as resin-bound PTC catalysts, and these pose few barriers to commercial use. Insoluble PTC catalysts made by adsorbing simple quaternary ammonium cations on organophilic clays appear to provide the least expensive and most robust materials suitable for commercial use.

3. Solid insoluble PTC catalysts generally lack the robust physical and chemical stability necessary to survive repeated use for long periods in industrial-scale reactors. This barrier can often be overcome or significantly reduced by sufficient development work with the reaction system being studied.

4. Insoluble liquid PTC catalysts, although not particularly affected by the above problems, require much more time during the process development stage to obtain an industrially useful process. Such systems have not been extensively explored, although several examples of such reaction systems are described in Section D of this chapter. These systems, once developed, can be extremely useful.

As technology for insoluble PTC catalyst preparation and use expands, costs decrease, and recyclability increases, it is expected that insoluble PTC catalysts will become the dominant type for conducting industrial-scale PTC reactions [1].

Regen and co-workers [2] were the first to demonstrate that quaternary salts chemically bound to insoluble resins could function as phase-transfer catalysts, and coined the term *triphase catalysis* to describe this special branch of PTC. This name is now generally used for all types of PTC catalysts bound to, adsorbed on, or consisting of any insoluble solid whether organic or inorganic. Tomoi and Ford [3], however, suggest the more specific name, *polymer-supported phase-transfer catalysis,* to describe reactions that occur within the polymer gel phase. Both these terms, as well as *resin-bound PTC catalyst* or *clay-supported PTC catalyst,* are used in this chapter.

The remainder of this chapter is divided into three general sections: the first deals with resin-bound PTC catalysts, the second with PTC catalysts on inorganic supports, and the third with insoluble liquid PTC systems.

B. PTC Catalysts Bound to Insoluble Resins

1. Basic Differences Between Soluble and Insoluble PTC Catalysts: The Importance of Diffusion Processes

The basic steps involved in reactions with resin-supported PTC catalysts differ from ordinary two-phase PTC reactions in one important respect: ordinary PTC reactions require only one of the reagents to be transferred from their normal phase to the phase of the second reactant. Use of resin-supported catalysts requires that both reagents diffuse to active PTC sites on the catalyst surface, or for reactions with slow intrinsic rates, both reagents must also diffuse to active sites inside the resin bulk phase. The need for diffusion processes with solid catalysts also means that both reagents are required to diffuse to and penetrate the stagnant outer layer of liquid(s) (the Nernst layer) coating the catalyst particle.

As an example, consider the reaction of 1-bromooctane with aqueous sodium cyanide, known to have a slow intrinsic rate, catalyzed by tributylphosphonium groups bound into beads of an insoluble styrene–divinylbenzene resin, a reaction extensively explored by Ford and Tomoi [4]:

$$1\text{-}C_8H_{17}\text{-}Br + NaCN\ (aq) \xrightarrow{\text{Polymer}-\!\bigcirc\!-PBu_3^+\ Br^-} 1\text{-}C_8H_{17}\text{-}CN + NaBr \quad (5\text{-}1)$$

As schematically illustrated in Figure 5-1 several steps are required. These steps include:

1. Diffusion of aqueous sodium cyanide through the Nernst layer and through the resin bulk to active sites
2. Equilibrium exchange of CN^- for Br^- at the active sites
3. Diffusion of Br^- out of the catalyst particle and into the aqueous phase
4. Diffusion of R-Br through the Nernst layer and through the bulk resin to active sites. Some reaction may occur at sites on the catalyst surface, but since the number of surface sites is small compared to the number of sites within the bulk of the catalyst, and in this example intrinsic reaction is slow, most of the reaction will occur inside the catalyst bulk.
5. Chemical reaction (the intrinsic reaction) between R-Br and Resin-$PR_3^+CN^-$ at active sites to produce R-CN and Br^-.
6. Diffusion of RCN out of the catalyst particle and into the organic phase.

If the system had been one having a fast intrinsic reaction rate, such as butylation of phenylacetonitrile, more, maybe all, of the reaction would occur at surface sites.

$$C_6H_5\text{-}CH_2\text{-}CN + 1\text{-}C_4H_9\text{-}Br + KOH\ (aq) \xrightarrow{\text{Polymer}-\!\bigcirc\!-PBu_3^+\ Br^-} C_6H_5\text{-}CH(C_4H_9)\text{-}CN + KBr + H_2O \quad (5\text{-}2)$$

In this reaction less diffusion of reagents into the bulk phase is needed, as the intrinsic reaction, that is, reaction of butyl bromide with $Q^+[C_6H_5CHCN]^-$, is fast.

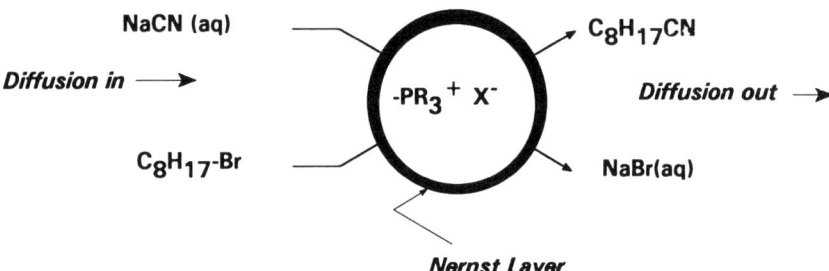

Figure 5-1. Schematic representation of the diffusion steps involved in use of resin-bound PTC catalysts for the displacement reaction of 1-bromooctane with NaCN.

2. Some Examples of Use of Resin-Bound PTC Catalysts and Comparisons with Soluble Catalysts

At this point it is useful to review some results obtained with resin-bound phase-transfer catalysts to obtain a clearer picture of what can be achieved with these systems. These results are summarized in Table 5-1. [Note that the abbreviation S/DVB represents a resin having a basic styrene–divinylbenzene structure.]

Aside from increasing the ease of removal and reuse of PTC catalyst, an important feature of triphase catalysis may be an increase in product selectivity as compared to soluble catalysts. This occurs, for example, in the alkylation of phenylacetonitrile with 1-bromobutane and 50% aqueous NaOH. Using a polymer-supported benzyltrimethylammonium catalyst the selectivity to monoalkylated product is about 98% at 85% phenylacetonitrile conversion, compared to about 80% selectivity to monoalkylated product at the same conversion level using a soluble catalyst [89].

A process equivalent to molar-scale polymer-bound molar phase transfer catalysis, that is, where 1 mole of phase-transfer agent is used per mole of reactant, has been demonstrated in use of insoluble polymer-bound poly(vinylbenzyltriphenylphosphonium chloride [37]. These reagents were used for the synthesis of carboxylic acid esters and sulfones and for alkylation of phenols by reaction with alkyl halides. In addition to the ease and simplicity of the method and regeneration of the polymeric byproduct, the polymeric reagent seems to increase the nucleophilicty of the anions. The products were obtained in higher yields than those for the corresponding polymeric phase-transfer catalyst which need longer reaction times.

3. Preparation of Resin-Bound PTC Catalysts

Resin-bound phase-transfer catalysts may be generically represented by a structure having three elements: the basic supporting crosslinked resin, an optional spacer chain, and the PTC functional group, as illustrated in Figure 5-2.

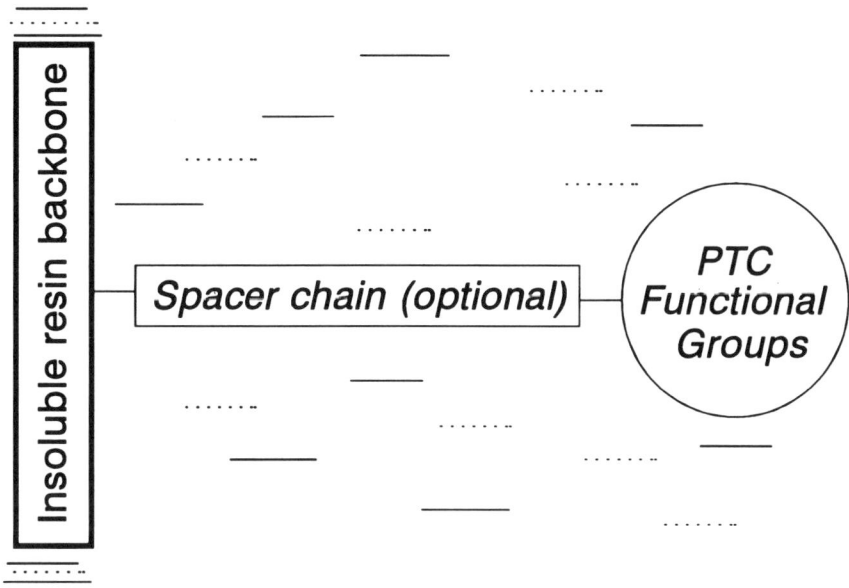

Figure 5-2. Schematic diagram of usual resin-bound PTC catalysts.

Preparation of PTC analyst groups bound to insoluble resins and their activity for catalyzing two-phase reactions has been extensively reviewed [38–43].

a. The Basic Resin Support

Much work has been done on the preparation and testing of phase-transfer catalysts supported on resins including extensive work by Montanari and co-workers [44,45] and by Ford and Tomoi [46] and their co-workers. Most published works on resin-bound phase-transfer catalysts use a styrene-divinylbenzene resin (SDV) and related resins, taking advantage of the huge amount of technology available on these resins due to their use as ion-exchange resin supports.

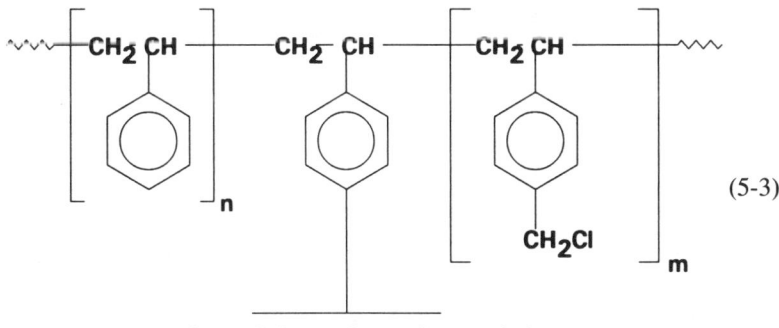

(5-3)

Cross-link to other polymer chains

Table 5-1. Summary of some results with Resin-Bound PTC Catalysts

PTC group type	Reaction and comments	References
Polymer–NR_3^+	$C_4H_9Br + C_6H_5CH_2CN + NaOH$ PS/DBV quaternized with Me_3N and Bu_3N (polymer-bound catalysts more selective for monoalkylation than soluble equivalents)	[5]
	$C_6H_5OH, C_6H_5CO_2H + EtBr + NaOH$, and $BuBr + KI$ Chloromethylated and Br-containing macroporous S/DVB copolymers reacted with amines to quaternize. Catalysts showed good activity and reproducibility.	[6]
	$MeSO_2Cl + NaN_3$ to $MeSO_2N_3$, then $RR'CH_2$ to $RR'CN_2$ On catalyst of tributylammonium groups bound to an insoluble polystyrene resin.	[7]
Polymer–PR_3^+	$R-X + NaCN$, mechanism kinetics	[4]
	$ArSO_2Cl + NaN_3$ on resin–$CH_2PBu_3^+ Cl^-$ catalysts	[8]
	$1-C_8H_{17}Br + NaCN$ on a catalyst of 2-Phenyl-1,2-poly-1,2-oxaphospholane quaternized with $1-C_{16}H_{32}I$ crosslinked polyquaternary salt coated on Al_2O_3	[9]
Polymer–PPh_3^+	Alkylation reactions Hydroxide-stable polymer-bound tetraphenylphosphonium catalysts salts as triphase catalysts—stable under strongly alkaline conditions	[10]
Polymer–SR_2^+	$R-X$ Substitutions and BH_4 reductions on Polymer-supported sulfonium salts from S-methylation of partially p-sulfomethylated polystyrene	[11]
Polymer–crown ethers	$C_8H_{17}Br + KOC_6H_5$ or KI Polystyrene-supported benzo-15-crown-5 and polystyrene-supported acyclic crown ethers	[12]
	$C_8H_{17}Br + KI$ on Polymer-supported crown ethers from grafting hydroxymethyl-crown ethers to 1–2% crosslinked chloromethylated polystyrene	[13]

	$C_8H_{17}Br$ or $-Cl + KI$ on	[14]
	Polymer-supported crown ethers from chloromethylated or bromoalkylated polystyrenes and hydroxymethylbenzo-18-crown-6 and 15- or 18-membered monoazacrown ethers	
	PTC Reactions on	[15]
	Urushiol + $(ClCH_2CH_2)_2O$ or $o\text{-}C_6H_4[OCH_2\text{-}CH_2OCH_2)_mCH_2Cl]_2$ ($n = 1$ or 2) to prepare the corresponding crown ethers, which were copolymerized with formaldehyde or formaldehyde and phenol.	
	Wolff–Kishner; $C_8H_{17}Br + KI$; Cl_2C: addition to styrene	[16]
	bis(benzo-15-crown-5-methyl)ether-formaldehyde-phenol resin and poly(vinyl alcohol)-15-crown-5-acetal resins; also: polymeric crown ethers from poly(β-chloroethylglycidyl ether) with disodium polyethylene glycols $[H(OCH_2CH_2)_nOH]$ ($n = 1\text{–}4$). Comparable in activity to dibenzo-18-crown-6 and 18-crown-6.	
	Polymeric crown ethers: effect of structure and temperature on anion binding properties[a]	[17]
	PTC reactions on	[18]
	Polymer with dibenzo-18-crown-6 as an anchor group	
	$C_8H_{17}Br + KOC_6H_5$ on	[19]
	Polymer-supported dibenzo-18-crown-6 and polymer-PEG C_{10-16} alkyl ethers	
	PTC reactions on	[20]
	Low-molecular weight polyether with benzo-15-crown-5 in the side chain from reaction of hydroxymethylbenzo-15-crown-5 with a copolymer of β-iodoethylglycidyl ether, then immobilized on silica	
Polymer–crowns and cryptands	PTC reactions on	[21, 22]
	Polymer-supported crown ethers and cryptands bonded by a long alkyl chain to a polystyrene matrix	
Polymer–aza-crowns	PTC nucleophilic reactions on	[23]
	Amino-functionalized crown ethers with microporous polystyrene bearing carboxyl groups gave macrocyclic polyethers bound to the matrix through an amido bond.	

continued

Table 5-1. (Continued).

PTC group type	Reaction and comments	References
Polymer–thia-crowns	BuCl + Na_2S on Thiacrown ether polymers from copolymerization of ethylene glycol diglycidyl ether or diethylene glycol diglycidyl ether with 2-alkoxymethylthiiranes	[24]
Polymer–calixarenes	PTC reactions on Calixarene polymer having a very high sodium and potassium complexing ability and good PTC activity	[25]
Polymer–polyethylene glycol	C_4H_9Br + $NaOC_6H_5$ on α,ω-bis(p-Vinylbenzyl)oligo(ethylene oxides) with oxyethylene repeating units of 1–21 copolymerized with styrene to give insoluble crosslinked polymer beads	[26]
	RBr dehydrobromination with NaOH on Polymer-supported polyethylene glycols with diol structures by reaction of polyethylene glycol with crosslinked polystyrene resins containing epoxide groups, in the presence of NaH	[27]
	Cyclopentadiene + $FeCl_2$ to ferrocene on Divinylbenzene–styrene copolymer-supported polyethyleneglycol; catalyst could be reused many times	[28]
	$C_8H_{17}Br$ + $NaOC_6H_5$; KOAc(s) + $C_6H_5CH_2Br$ on Copolymers of H_2C = $CHMeCO_2(CH_2CH_2O)_nMe$, kinetics studied with n = 4, 8, or 22, with styrene, methyl methacrylate, acrylonitrile, or 4-vinylpyridine, and crosslinking agents	[29, 30]
	KOAc + $C_6H_5CH_2Cl$ on Polymer-bound polyoxyethylene and crown ethers, kinetics studied	[31]
Polymer–CH_2SOCH_3	C-Alkylations on Polymer-bound-DMSO: styrene-(p-chloromethyl)–styrene copolymers	[32]
	R–X Displacements on Crosslinked polystyrene with pendant $MeSOCH_2$-groups (efficient phase-transfer catalysts). Direct introduction of the $MeSOCH_2$ groups into halogenated polystyrene with (methylsulfinyl)methyl sodium	[33]

Polymer–pyrrolidone	RX displacement reactions on Polymers of *N*-vinylpyrrolidone–styrene–divinyl benzene (high-activity insoluble PTC catalysts)	[34]
Polymer–sulfonamide	$C_6H_5S^- + n-C_8H_{17}Br$ on Copolymers from *N*-methyl-*N*-(*p*-vinylbenzyl)methanesulfonamide and *N,N*-dimethyl-1-*p*-styrene-sulfonamide with divinylbenzene. Catalysts are stable under strongly alkaline conditions; recovered catalysts reused without loss of activity.	[35]
Polymer–pyriazinyl thio- groups	Substitutions and BH_4^- reductions on Polymers containing pyriazinylthio groups	[36]

^aUsing distribution and column techniques, the equilibrium distribution values in methanol at 20–60°C for perchlorate, thiocyanate, and bromide salts were measured for these resin-bound materials. The order of binding constants was K > Cs > Na > Li for the polymeric benzocrown-6 and benzocrown-8 and K > Cs = Na > Li for polymeric benzocrown-4 and benzocrown-5. A temperature increase of 40° caused spontaneous elution with quantitative release of all the bound salt.

These resins are functionalized (1) by chloromethylation [47], (2) by chlorination of resins made by *p*-methylstyrene as a comonomer [48,49], or (3) most easily by including *p*-chloromethylstyrene as a comonomer [50,51].

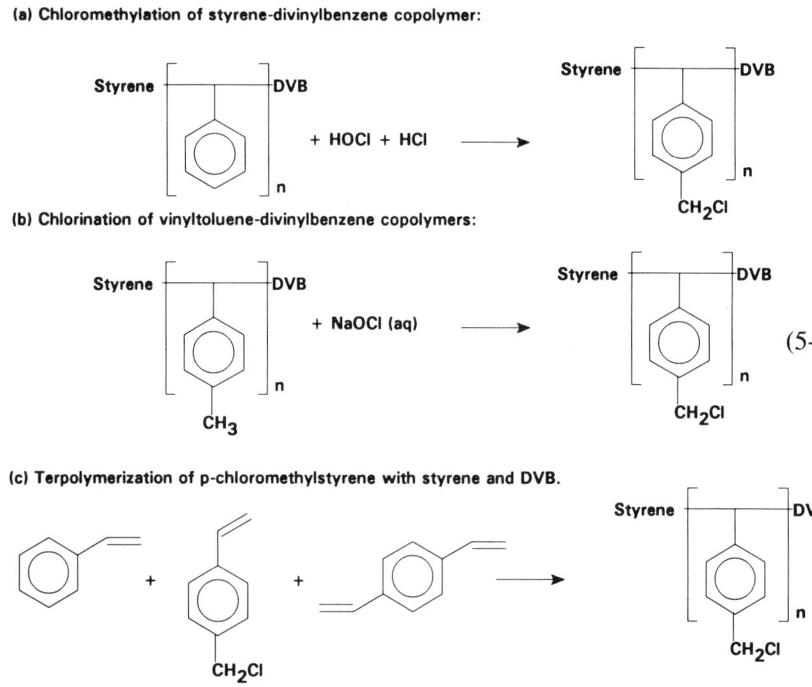

(a) Chloromethylation of styrene-divinylbenzene copolymer:

[Styrene—DVB—phenyl]$_n$ + HOCl + HCl ⟶ [Styrene—DVB—phenyl-CH$_2$Cl]$_n$

(b) Chlorination of vinyltoluene-divinylbenzene copolymers:

[Styrene—DVB—phenyl-CH$_3$]$_n$ + NaOCl (aq) ⟶ [Styrene—DVB—phenyl-CH$_2$Cl]$_n$ (5-4)

(c) Terpolymerization of p-chloromethylstyrene with styrene and DVB.

styrene + chloromethylstyrene + divinylbenzene ⟶ [Styrene—DVB—phenyl-CH$_2$Cl]$_n$

Resins can have various levels of crosslinking, various ratios of chloromethylated rings to nonsubstituted rings ["percent ring substitution" (RS)], and various degrees of macroporosity, all of which are usually important factors affecting catalyst efficiency.

A convenient preparation of the basic resin involves free-radical suspension polymerization of the styrene, chloromethylstyrene, and divinylbenzene without solvent to produce microporous resin, and with solvent (typically isooctyl alcohol) to produce macroporous resin. A procedure is given by Shan, Kang, and Li [52]:

Macroporous resin: A solution of 1.0 g of gelatin and 0.6 Ml of 0.1% methylene blue in 100 g water was added to a 250 Ml round-bottom flask fitted with a reflux condenser, a magnetic stirrer, and a water bath temperature controller. A solution of 13.4 g of styrene, 5.7 g chloromethylstyrene, 0.9 g of 43.2% divinylbenzene, 6.0 g of isooctyl alcohol, and 0.2 g of BPO initiator was added. The flask was purged with nitrogen for 30 min, and a nitrogen atmosphere was maintained throughout polymerization. Stirring was started at room temperature. The size of the organic droplet suspended in water can be determined by sampling the mixture and examining it under a microscope. If smaller droplets are desired, the stirring speed can be increased. After the droplet size was established, the mixture was heated with stirring to 72°C for 2 h and then to

Table 5-2. Pore size as a function of crosslinking in SDV.

Crosslink density (%)	Pore size (Å)
4	700–1500
6	500–1000
8	150–380
10	50–130

95°C for 4 h. Insoluble polymer was collected on sieves and washed thoroughly with water. Particles of 100–200 mesh were used. To make microporous resin the isooctyl alcohol is omitted.

The amount of each ingredient used in polymerization is important. Increasing chloromethylstyrene and decreasing styrene produces catalysts with a higher level of ring substitution (PT-group density). If percent RS is too high the resulting catalyst may tend to be too highly hydrophilic around the active site for some kinds of reactions in that it inhibits diffusional approach of hydrophobic organic reagents into the catalyst zone. This hydrophilic–organophilic balance also depends on the nature of the PT group, so both must be selected to obtain a final resin that will accommodate easy diffusion of both aqueous and organic reagents to the catalyst site.

The amount of crosslinking agent used is also important. If too little is used the resin will be soft and flexible, particularly at higher temperatures, and it will crumble, lose its shape, and mechanically deteriorate. If too much crosslinker is used the resin has smaller pores, is less easily swollen by liquids, retards diffusion into the resin, and leads to a less active catalyst. Highest catalyst activities are usually observed with catalysts containing about 2% crosslinking, but good mechanical stability of the resin occurs at about 8–10% crosslinking.

Shan et al. measured the size of pores in their resins using an electron microscope, with the results listed in Table 5-2.

Polymerization in the presence of isooctyl alcohol, in which the resin is insoluble, produces a resin containing small pockets of the alcohol. This alcohol ultimately gets washed out, leaving voids, or *macropores,* and this increases the internal porosity and surface area of the resin. Resins that do not have such macropores are referred to as *micropore* or *gel-type* resins.

Additional porosity increases the catalyst efficiency for some reactions but has little effect on others, depending on the diffusion patterns of the reagents. If the resin contains too many voids it will lose mechanical strength and be more easily ground into dust by liquids flowing past. When used to support PTC catalysts, resin swelling by various solvents is an important consideration. Increased swelling facilitates diffusion of low molecular weight reagents to catalyst sites in and on the resin. The effects of crosslinking and kind of solvent on swelling of the resins of Table 5-2 are shown in Figure 5-3.

Ruckenstein and Hong have explored the polymerization procedure for produc-

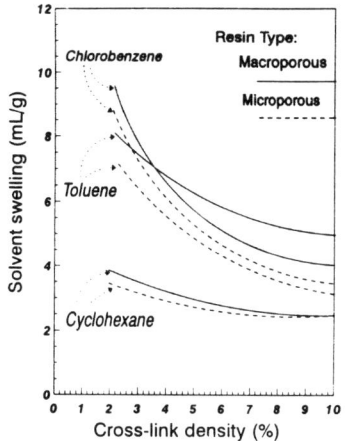

Figure 5-3. Effect of crosslink percentage and micro/macro porosity on the swelling of styrene–divinylbenzene resins by chlorobenzene, toluene, and cyclohexane.

tion of styrene–divinylbenzene–vinylbenzyl chloride as a method for producing resin-bound catalysts with different properties. In one report [53] they described preparation of resins by three different procedures:

1. Concentrated emulsions of vinylbenzyl chloride in water were polymerized to obtain particles in the micrometer range.
2. A small amount of vinylbenzyl chloride is added to a partially polymerized styrene–divinylbenzene emulsion in water. After polymerization is allowed to go to completion the copolymer obtained consists of particles having a nonuniform shell of polyvinylbenzyl chloride that covers a crosslinked poly(styrene–divinylbenzene) core.
3. A solution of vinylbenzyltrialkylammonium chloride is adsorbed on, then polymerized on, the surface of a poly(styrene–divinylbenzene) particles in an emulsion.

After conversion of the first two resins to quaternary ammonium salts, the three different types of triphase catalysts were tested for their activity for alkylation of isopropylidene malonate. The catalyst prepared by the second procedure was more effective than that prepared by the first, and the catalyst from the third procedure exhibited a low activity because of the small amount of supported -onium salt it contained.

In a second article [54], these workers described starting the polymerization of styrene–divinylbenzene lattices, then added vinylbenzyl chloride to the polymerization mixture at various stages of polymerization.

This procedure allowed preparation of resin-bound quaternary salts wherein the dual surface characteristics due to the (1) hydrophilic bound cations and (2) hydrophobic polystyrene could be varied by controlling the partial polymerization time and the amount vinylbenzyl chloride introduced. If vinylbenzyl chloride is added after short prepolymerization of styrene and divinylbenzene, the vinylbenzyl chloride is partly incorporated into the resin backbone and partly homopolymerized to the surface of the resin particles. When the partial polymerization is long, only vinylbenzyl chloride formed as homopolymers grafted to the resin surface. After conversion to quaternary ammonium salts, these products were tested as triphase catalysts for alkylation of isopropylidene malonate, where it was found that the second type of catalyst, that is, long prepolymerization before addition of vinylbenzyl chloride were the most effective, apparently because the quaternary ammonium groups was more accessible and required less diffusion of the reagents. However, the ratio of mono- to dialkylation products increased with decreasing accessibile hydrophilicity of the catalysts.

Several basic resins types other than styrene–divinylbenzene may also be used to immobilize phase-transfer functions, such as for example, crosslinked acrylate and methacrylate esters, glycidyl ether resins, and calixarene resins, as listed in Table 5-1. Organophilic dextran, crosslinked with epichlorohydrin [55], perfluoroethylene [56], carbon [57], and cellulose [58], have also been used as insoluble supports for PTC functional groups.

b. Spacer Chains

Spacer chains typically have 8–20 carbon atoms and serve to separate the PTC active function from the resin backbone, and to provide a reaction environment thought to more nearly duplicate conditions when soluble PTC catalysts are used. Some types of reaction, particularly those characterized as having a slow organic-phase reaction rate, usually are 1.5–4.0 times faster when the resin contains a spacer chain. Spacers move the reaction center away from the polymer chain and facilitate transport of molecules near the reaction center. Spacers also separate ionic reaction centers away from each other to better reduce the aggregation of PTC functional groups [59].

Several approaches to attachment of spacer units to resins have been successful. [45,60–65,131]. A particularly convenient and inexpensive synthetic procedure for attachment of spacer chains to styrene-containing resins involves alkylation of the resin with an ω-bromoalkene in the presence of a Freidel–Crafts catalyst [58,61,66,67]. For example:

$$\text{Polymer}-\text{C}_6\text{H}_5 + CH_2=CH(CH_2)_nCH_2Br \xrightarrow{CF_3SO_3H} \text{Polymer}-\text{C}_6\text{H}_4-\overset{CH_3}{\underset{}{CH}}(CH_2)_nCH_2Br$$

$$n = 9-21 \quad (5\text{-}5)$$

A one-pot synthesis of PTC resins for phase-transfer catalysts employs bromoalkylation of crosslinked styrene–divinylbenzene resins with ω-bromoalkenes catalyzed by CF_3SO_3H [68]. For example, preparation of a support from ω-bromo-

1-undecene, followed by functionalization with trialkylphosphine, gave immobilized phosphonium salts that had higher catalytic activity than that of the corresponding salts derived from chloromethylated polystyrene supports. The effects of crosslinking, percent, and particle size were also important.

Catalysts having spacer chains made of repeat $-CH_2CH_2-O-$ groups increase the activity of resin-bound crown ether catalysts, apparently induced by the cooperative coordination of the ether plus crown-unit complexing of alkali metal cations [69].

c. The Active PTC Functional Group

As seen in Table 5-1 all common PTC group types such as quaternary ammonium, phosphonium groups, polyethylene glycol (PEG) chains, crown ether groupings, cryptands, calixarenes, and dipolar-aprotic functions can be bound to the resin through appropriate chemical reactions to produce the final triphase catalyst. The kind of functional group chosen is usually of great importance for insoluble PTC catalysts, just as in choice of soluble PTC catalysts, and needs to match the requirements of the reaction. These are described in further detail in Section 4.

Catalyst stability is also particularly important with resin-bound PTC catalysts, since these are expensive to produce and they are required to repeatedly perform without significant loss of activity. Some common types of resin-bound catalysts used in laboratory work are represented by the following structures:

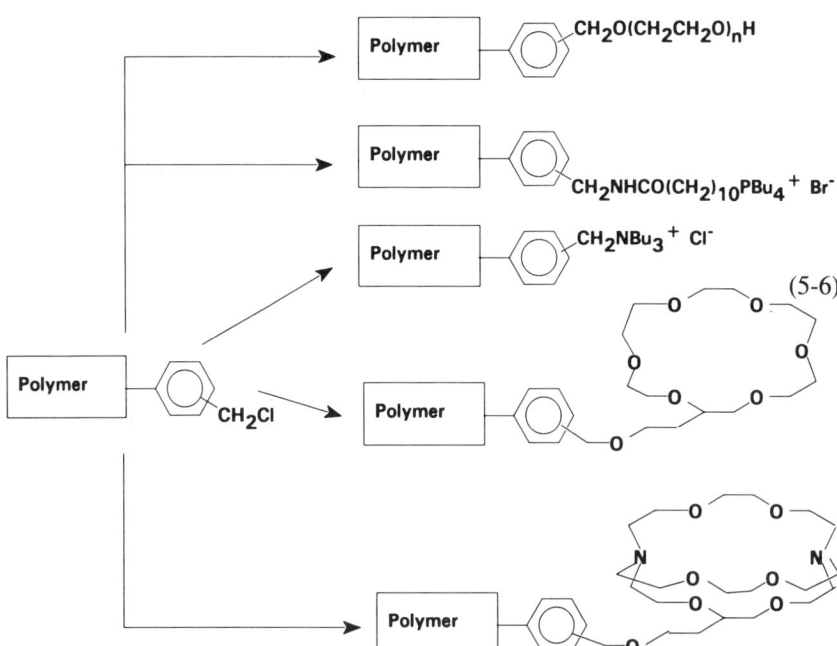

(5-6)

A general review of analytical methods for characterization of resin-bound phase-transfer catalysts has been assembled by Tomoi and Ford [3]. A review on the methods for preparation and uses of resins containing poly(ethylene oxide) groups has been published [70].

4. Effects of Reaction and Catalyst Parameters on Triphase Catalyst Effectiveness

This section deals with an analysis of the parameters and characteristics of triphase catalyst systems, and how these parameters affect rates and selectivity of PTC reactions. The effect of catalyst characteristics and of experimentally controllable reaction parameters has been extensively investigated and reviewed [4,63,71,72]. Ford and Tomoi, for example, studied the reaction of an aqueous NaCN phase with neat 1-bromooctane, using a quaternary phosphonium catalyst group, $-PBu_3Br$, bound to insoluble styrene–divinylbenzene resins. They sorted out the sequence of steps involved, as illustrated in Figure 5-1, and showed it be similar to ordinary PTC except that extra diffusion steps are involved.

The usual factors that affect reaction rates in simple PTC are also important in catalysis with resin-bound PTC catalysts. Thus, in bromide ion displacement on 1-octyl methanesulfonate, catalyzed by polymer-bound phosphonium ion at 60–90°, the intrinsic rate of reaction was proportional to the concentration of octyl methanesulfonate, and gradually increased as the concentration of KBr in the aqueous phase was increased [73]. In benzyl butyl ether synthesis using a polymer-supported quaternary salt catalyst, the reaction rate depended on the concentrations of benzyl alcohol and 1-bromobutane in the organic phase, and on concentration of KOH in the aqueous phase [74].

Aside from these usual features expected with PTC reactions, the introduction of a new solid-phase catalyst into a two-phase reaction mixture causes many parameters and characteristics of the system to be become much more significant in the overall success of the catalyst. At least thirteen features have been identified:

1. Agitation
2. Catalyst particle size
3. Active site density in/on the catalyst particle (with SDV resins, termed percent ring substitution, or percent RS)
4. Extent of crosslinking in the resin
5. Chemical composition of the resin
6. Resin porosity (macroporous vs. microporous or gel resins)
7. Presence and characteristics of a spacer chain
8. Kind of solvent, if one is used, and prereaction swelling of the catalyst particles

9. Amount of catalyst used
10. Chemistry of the active sites
11. Size of the organic substrates
12. Ratio of liquid phases
13. Catalyst degradation and recycle issues

Any one of these, if not sufficiently adequate or appropriate for the reaction system under study, can severely reduce the effectiveness of the catalysis process. Not all of the parameters and characteristics in the above list have been carefully and fully examined, but most have received enough attention that the important issues can be understood. Mostly the effects of the parameters and characteristics in the above list are interdependently related to their effect on diffusivity of reactants to active sites.

a. Agitation

Agitation or stirring is a critically important factor for simple two-phase PTC systems using soluble catalysts. Agitation increases interfacial area as dispersed droplet size decreases with increased agitation, and correspondingly this increased surface area allows increased transfer to occur between phases. Use of resin-bound catalysts introduces a new, but related, feature, that is, a thin stagnant layer of the continuous-phase liquid that exists around the catalyst particle, known as the "Nernst layer" in electrochemistry, or "laminar sublayer" in fluid mechanics for flowing systems. All reactants must diffuse through this layer to active sites on or in the catalyst matrix. According to Tomoi and Ford [4]:

> The thickness of this layer depends on the degree of agitation in the mixture. The more turbulent the mixing or the faster the liquid flow, the thinner the film at the solid surface. Molecular transport through the film occurs as diffusion through a quiet liquid. . . . In catalysis by spherical particles, the rate of mass transfer refers to the rate at which reactant molecules from bulk liquid [e.g., 1-bromooctane from its dispersed droplet] reach the surface of the catalyst. This rate depends on both the degree of agitation of the mixture, which affects the thickness of the quiet film, and the diffusion coefficient of the reaction in the quiet film around the catalyst. Whenever mass transfer limits the observed reaction rate, there is a gradient of decreasing concentration of reagent from bulk liquid across the quiet film to the catalyst surface.

These authors demonstrated the effect of agitation on reaction rates, as shown in Figures 5-4 and 5-5, with catalysts having different degrees of crosslinking and different particle sizes.

Ultrasonic mixing in triphase reaction systems produces two- to fourfold faster reaction rates compared with magnetic, mechanical, or turbine stirring [5]. The polymer-bound tributylbenzylammonium chloride catalyst used in a continuous

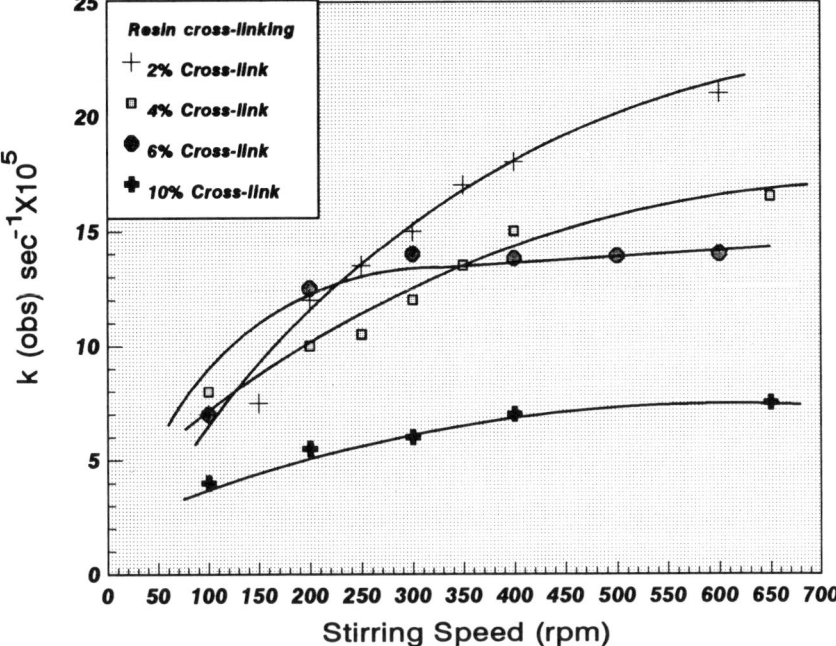

Figure 5-4. Stirring speed and percent crosslinking effect on reaction rate of 1-bromooctane displacement with NaCN, catalyzed by polymer–Pbu$_3^+$ catalyst, of -100 to -200 mesh.

fixed bed reactor with an ultrasonic mixer of suitable geometry can reach a reactivity close to that shown by the soluble analogue benzyltriethylammonium chloride in the monoalkylation of phenylacetonitrile. For this reaction ultrasonic mixing allows easy catalyst recycling with no appreciable loss of activity through thermal decomposition.

b. Catalyst Particle Size

For the reaction of 1-bromooctane with aqueous sodium cyanide, catalyzed by a polystyrene-bound benzyltributylphosphonium catalyst, particle size exhibits a significant effect on reaction rates, as shown in Figures 5-6 and 5-7.

Figure 5-6 shows that rate constants for the cyanide displacement reaction increase as particle size decreases but approach a plateau at $1/r$ values between 500 and 1000 cm^{-1}. If mass transfer alone were the limiting factor, then all reactions could take place at active sites on or near the surface, and observed rate constants would be proportional to surface area, and therefore to the reciprocal of the radius. Likewise, the extent to which reactants can diffuse to interior sites of a catalyst particle depends on particle surface area. Thus, a reaction whose

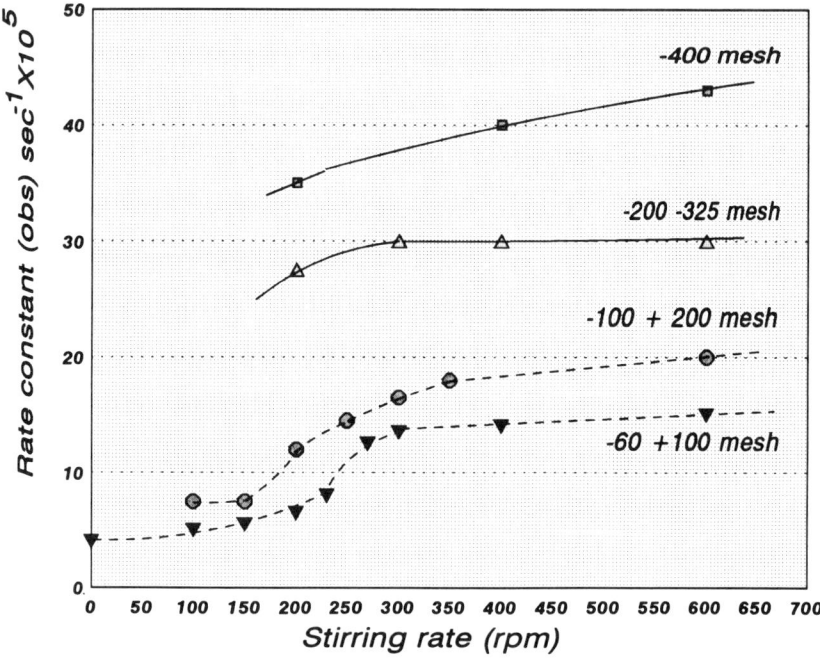

Figure 5-5. Stirring speed and particle size effect on rate of 1-bromooctane displacement reaction with NaCN, 2% crosslinking. Particle sizes are screen mesh sizes.

rate was limited only by diffusion into the catalyst particle would also show a linear dependence on the reciprocal of particle radius. Since the experiments represented in Figure 5-6 were carried out at agitation levels above which mass transfer rates are limiting, the curves of Figure 5-6 show the transition from a regimen (on the right side of the graph) with small particle size where only the intrinsic chemical reaction at the active site is rate controlling to a regimen of large particle size (left side of graph) where both intrinsic chemical reactivity and intraparticle diffusion of reactants to active sites determine the value of the observed rate.

Anelli et al. [75] found that for iodide–bromide exchange with 1-bromooctane using a polystyrene-bound benzyltributylphosphonium catalyst with 2% crosslinking and 36.1% ring substitution, the reaction rate is linearly dependent on the reciprocal of the particle diameter in the range between 125 and 53 μm as illustrated in Figure 5-7. The line passes through the origin, meaning that catalytic activity is essentially zero for a hypothetical particle of infinite diameter, as expected. However, in the same plot, the catalyst with only 1% crosslinking and 8.7% RS is relatively insensitive to particle size, indicating that here factors other than diffusion are limiting the reaction rate.

Use of emulsion polymerization to make styrene–divinylbenzene resin gives

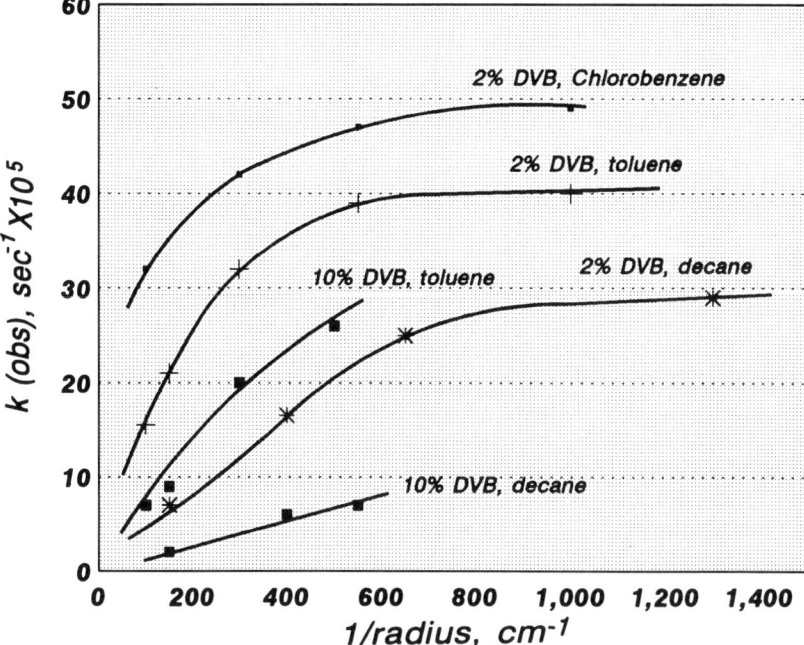

Figure 5-6. Effect of mean particle size plotted as the reciprocal of average particle radius, and solvent on rate of 1-bromooctane displacement reaction with NaCN, stirring at 600–650 rpm, with resins of 2 and 10% crosslinking.

particle sizes in the range of 0.12–0.27 μm diameter, compared to 10–30 μm at the smaller end of suspension polymerized catalysts [51]. For cyanide displacement on benzyl bromide the emulsion-based resin catalysts are about four times more active than suspension-based resin.

c. Active Site Density in/on the Resin (Percent RS)

In general, one would expect that increasing the density of active sites on an insoluble PTC catalyst would increase catalytic activity, and indeed this is true for reactions where transfer is the rate-limiting step. However, for reactions limited by the rate of intrinsic organic reaction the active site density of the catalysts depends more on the extent to which diffusion affects rates. When diffusion is rate-limiting the loading or percent RS must be balanced between a need for higher activity provided by more numerous active sites, or higher activity provided by higher diffusion rates for nonpolar reactants which occurs by reducing the density of active PTC sites.

For example, Figures 5-8 and 5-9 show rates of cyanide displacements on 1-chlorooctane and 1-bromooctane, respectively [76].

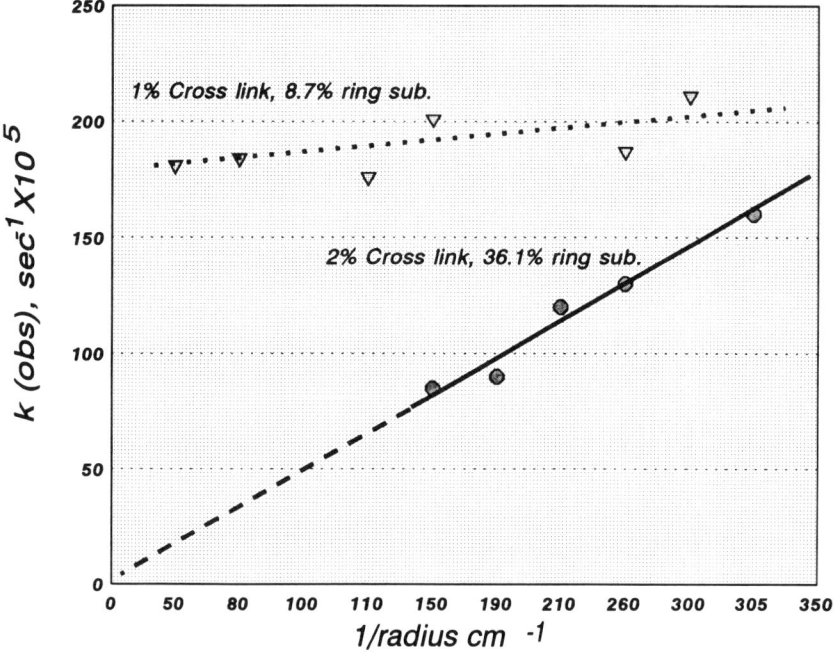

Figure 5-7. Dependence of rate constant on particle size and percent crosslinking for Br–I exchange for 1-bromooctane, in toluene–water, 90°C, 0.01 mol equiv of catalyst 15 h conditioning without stirring.

Rates for the two-phase 1-chlorooctane-sodium cyanide reaction decreased as percent RS increased, reflecting greater difficulty for the 1-chlorooctane to diffuse to the active site, apparently because the increased site density tended to make the catalyst more "aqueous." With more reactive 1-bromooctane catalyst activity does initially increase with site density, reaches a maximum at about 20 percent RS, then decreases at ring substitution greater than about 30%, reflecting a low diffusion of rate of cyanide ion when percent RS is low (matrix too lipophilic) and low bromooctane diffusion when percent RS is high (matrix too hydrophilic). In a similar study, in which bromide ion displacement on 1-octyl methanesulfonate is catalyzed by polymer-bound phosphonium ion at 60–90°, catalyst activity decreased with increased phosphonium ion content [69,71].

In contrast to the cyanide displacement reactions, the alkylation of phenylacetonitrile with 1-bromobutane and 50% aqueous NaOH, where diffusion and transfer of hydroxide anion onto the active catalyst site is the limiting step, catalysts with RS up to 50% resulted in increasing rate with increasing percent RS of the catalyst, as illustrated in Figure 5-10 [50].

With resin-bound crown ethers as catalysts, the effect of percent ring substitution depends on the nucleophile [21,77,78,89]. For example, with soft nucleo-

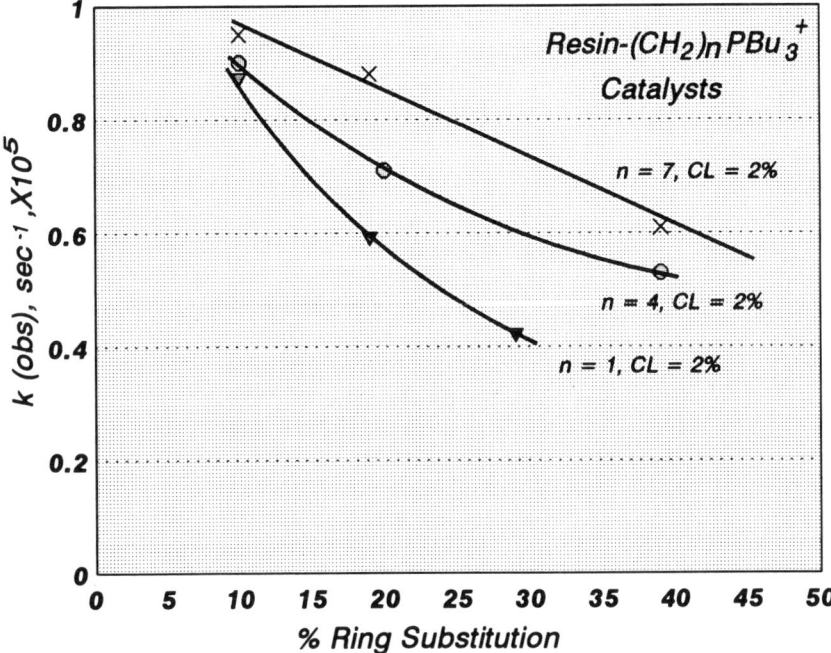

Figure 5-8. Rate 1-$C_8H_{17}Cl$ with NaCN(aq) reactions vs. percent RS; with 100/200 mesh S/DVB resin, with $-CH_2PBu_3^+$ PTC, with three spacer chain lengths.

philes such as iodide, log k_{obs} for displacement on 1-octyl methanesulfonate linearly decreases about five times on going from 5 to 60% RS. But, with harder anions, for example, bromide, log k_{obs} shows a maximum at 30% RS. This was explained by the different extent of complexation of crown ethers not only on complexed cation, but also on the anionic counterpart. Swelling and hydration measurements of polymer-supported crown ethers in toluene/aqueous KI or KBr showed that the content of water in the imbibed solvent increases with the loading. This leads to a progressive polarity increase within the polymers and to a better crown-ether complexing capability, more relevant for hard anions. In all cases catalysts having spacer arms were two to four times more reactive than directly bonded catalysts.

d. Crosslinking Effect

The effect of crosslinking is illustrated in Figure 5-4 in conjunction with stirring speed, showing that catalyst activity decreased with increased crosslinking. This effect is due to greater resistance to reactant diffusion to active sites in the catalyst particle because of increased tortuosity and rigidity, as reflected by decreased catalyst swelling with increased crosslinking (Table 5-3). These data also show

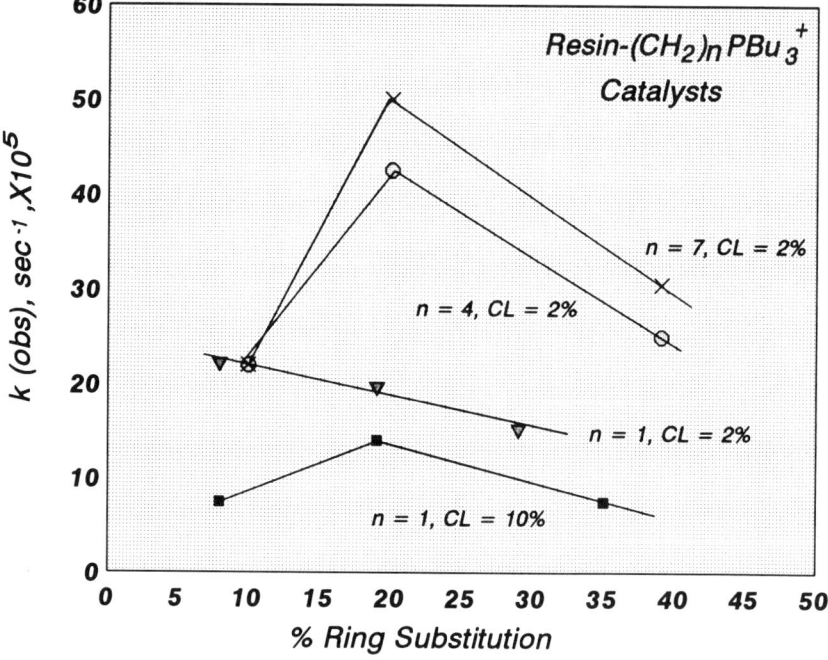

Figure 5-9. Same as Figure 5-8, except with 1-bromooctane instead of 1-chlorooctane.

catalyst swelling to be highly sensitive to the kind of solvent used, and imply that reactants that look more like decane than benzene would suffer from lower diffusion rates. In the bromide ion displacement reaction of 1-octyl methanesulfonate catalyzed by polymer-bound phosphonium ion at 60–90°, intraparticle diffusion of methanesulfonate was an important factor controlling total rate of reaction [69]. The effective diffusivity of methanesulfonate in polymer resin decreased remarkably with increases in the degree of crosslinking.

In nucleophilic displacement reactions catalyzed by glycidyl methacrylate–ethylene dimethacrylate copolymers functionalized with tributylammonium groups, both swelling and diffusion were restricted severely as the percentage of crosslinking was increased [79]. This restriction was of particular importance with macroporous resins. Comparison of kinetic data obtained with various polymers suggested that gel polymers with very small bead sizes and therefore high external surface areas provided much higher reactivities than their macroporous counterpart.

e. Effect of Resin Chemical Composition

It is difficult to demonstrate experimentally the effect of resin composition on catalyst activity because so many factors, such as porosity, elasticity, hydration,

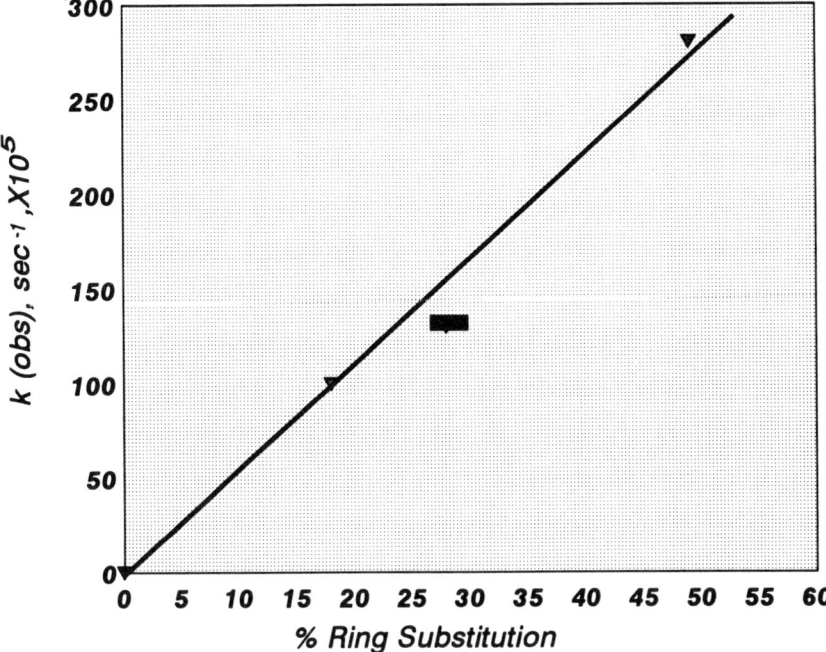

Figure 5-10. Effect of percent RS on rate of butylation of phenylacetonitrile with aqueous KOH using resin-bound trimethylammonium PTC catalyst.

etc., cannot be kept constant between resins made from various monomer types. Speculatively, however, chemical composition of the resin would be expected to importantly affect catalytic activity. Resins that are most like the organic reactant to be used will be most easily swollen by the organic reaction component and would therefore be expected to favorably affect diffusion rates or organic species. Likewise, inclusion of second and third comonomers that could place dipolar aprotic groups, such as $-CON(CH_3)_2$ or $-CH_2SOCH_3$, or other functional groups that can act as cocatalysts, near the active site would also be expected to speed the reaction. Changing the chemistry of the resin polymer so that it favors predominantly an aqueous environment or predominantly an organic environment can affect product selectivity. For example, resins containing $-NMe_3^+$ groups promote a predominantly aqueous active site environment and favor monobutylation of phenylacetonitrile and retard dibutylation.

f. Resin Porosity: Microporous vs. Microporous Resins

Macroporous resins have greater porosity and surface areas than nonporous material, and for catalyst particles of the same size, consequently allow faster diffusion of reagents to active sites. For example, Ford et al., [80] calculated 1

Table 5-3. Catalyst swelling at 25°C in different solvents.

Active site	Percent-Crosslink	Solvent swollen volume/dry volume			
		Water	Toluene	Chloro-benzene	Decane
$-P(Bu)_3^+$	2	1.8	2.2	3.0	1.0
$-P(Bu)_3^+$	4	1.2	2.0	2.7	1.0
$-P(Bu)_3^+$	6	1.0	1.6	1.9	1.0
$-P(Bu)_3^+$	10	1.0	1.4	1.7	—
$-N(Me)_3^+$	2	2.0	1.6	3.0	—
$-N(Me)_3^+$	10	1.0	1.4	2.0	1.0

ml of cubes with edges 100 μm long has a surface area of 0.06 m^2/g, which was approximately equal to the total surface area of the gel-type catalysts used in their study. Their macroporous resins had surface areas up to 588 m^2/g. For triphase-catalyzed cyanide displacement on 1-bromooctane their macroporous-supported catalysts gave slightly lower rates than catalysts based on nonmicroporous resins. They concluded that the pores in the macroporous resins were completely filled with organic-phase reactants, limiting the diffusion of ionic reactants to active sites, and consequently limiting cyanide displacement reaction rates.

In oxidation of benzyl alcohol to benzaldehyde by aqueous sodium hypochlorite, catalyzed by anion-exchange resins, the effective surface diffusivity for the macroporous resin was 1.7 times that of the gel-type resin [81]. Microporous resins were claimed to be more active than gel-type resins for aqueous sodium cyanide reaction with benzyl chloride in toluene to produce phenylacetonitrile in 94% yield at 95°C after 5 h [82].

g. Effect of a Spacer Chain

Spacer chains can increase some reactions by removing the active site away from the polymer chain, and from other active sites. When active sites, particularly quaternary onium ions, are located close to one another they join to form doublets, triplets, and higher aggregates that are less active catalyst centers, and that tend to present an "aqueous" face to the reactants. Thus the use of spacer chains increases the rates of some reactions, such as nucleophilic displacements, by two-to fourfold [58,63,64,71].

When the spacer chain also contains complexable ether oxygen atoms, using 15-crown-5 as the PTC functional group, catalyst activity is even greater, as observed in halide exchange of KI with 1-bromooctane [83].

h. Kind of Solvent Used

Tomoi and co-workers [84] suggest that solvents may affect rates of triphase-catalyzed reaction in three ways: intrinsic chemical reactivity; solvent effect on ion-exchange rate; and overall activity, including diffusion effects due to swelling, of polymer-supported phosphonium salts under three-phase conditions. They found with resin-bound phosphonium salts:

1. The intrinsic activity of the catalysts, as well as of soluble phosphonium salts, depended only slightly on organic solvents for cyanide displacement reactions.
2. The exchange rate of chloride ion in the catalysts vs. that of acetate depended on the solvents when the degree of ring substitution was < 16%. With 30% ring-substituted catalysts, the rate increased and hardly depended on the solvents.
3. The overall catalyst reactivity for the reaction of organic halides with NaCN depended on the substrate structure and organic solvents. For 1-bromooctane the catalysts were more reactive in good solvents (e.g., toluene or chlorobenzene) than in poor solvents, such as octane. For benzyl chloride the catalysts exhibited the opposite effect. In poor solvents, benzyl chloride, rather than solvent, absorbed preferentially into the catalysts.

In alkylation of phenylacetonitrile with 1-bromobutane and 50% aqueous sodium hydroxide solution, reaction rates were up to eight times faster when the phenylacetonitrile was added before the 1-bromobutane, than when the opposite order of addition was employed [50]. This was shown to be due to greater rate and extent of swelling of the catalyst by the nitrile.

Shan and co-workers [52] explored the use of resin-bound PEG as triphase catalysts, varied by macro- and microporosity, crosslink density, and size of PEG chain added. The effect of solvent on catalyst activity for reaction of benzyl bromide with potassium acetate is shown by the data in Table 5-4.

Wang and Wu [85] examined the effects of solvents and other resin-bound catalyst parameters on the rates of $NaOCH_2CF_3$ displacement on hexachlorocyclotriphosphazene, and found similar results.

The diffusion environment within triphase catalyst particles is changed by the relative volume of the different reagents within the reaction mixture, and such changes in diffusion can have a marked effect on reaction rates [86]. Drastic kinetic rate improvements can be achieved, by for example, using a polar solvent that can expand the resin matrix [87], or by careful adjustment of the volumes and concentrations of the reagents in both phases to keep the catalyst matrix expanded [88].

i. Amount of Catalyst used

Dependence of rates of reactions catalyzed by polymer-bound crown ethers has been reported by Anelli and Co-workers [89] to be a linear function of the

Table 5-4. *Effect of solvent on percent ester yield from benzyl bromide and aqueous KOAc under standardized conditions.*

Catalyst[b]	Solvent (ε)[a]	Percent ester yield
Macro (6%)–400	Cyclohexane (2.02)	13.2
	Toluene (2.38)	16.5
	Chlorbenzene (5.62)	23.3
	Benzy ethyl ketone (17.4)	94.5
	Nitrobenzene	100.0
Micro (6%)–400	Cyclohexane	9.2
	Toluene	13.5
	Chlorobenzene	18.2
	Benzyl ethyl ketone	86.0
	Nitrobenzene	100.0

[a]Dielectric constant of solvent.
[b]6% cross-linking; PEG 400 used for active sites on catalyst.

amount of catalyst used, as illustrated in Figure 5-11 for the reaction of *n*-octyl methanesulfonate with iodide. Likewise, reaction rates of phenol alkylation with 1-bromobutane, catalyzed by poly(dibenzo-18-crown-6), polymer-supported octaethylene glycols, and benzyltributylammonium chloride increase linearly with the mass of added catalyst [89].

However, Pugia and co-workers [91] found the reaction rate of 1-bromooctane with potassium cyanide to be linear with the amount of catalyst up to about 0.3 mol equivalent, but then reach a plateau, as illustrated in Figure 5-12.

j. Chemistry of the Active Site

Quaternary Salt Catalysts

As with soluble phase-transfer catalysts, intrinsic reactivity of active sites in triphase catalysts is expected to be a function of the kind of reaction being catalyzed, and its need for an "activated" anion. Additionally, organophilicity of the PTC functional groups attached to the active site, that is, whether -onium functionality was made from trimethylamine, tributylamine, trioctylamine, etc., can affect diffusion rates of ionic and organic species, and thereby affect reaction rates [2,72,88].

To determine the effect of alkyl group size attached to the quaternary atom, Tomoi and Ford [90] compared reaction rates for four alkyl halides with different catalysts, as listed in Table 5-5. (See page 234.)

These data show:

- Rates depend on catalyst particle size, and therefore diffusion is an important factor, except for the reactions of 1-bromooctane and benzyl chloride

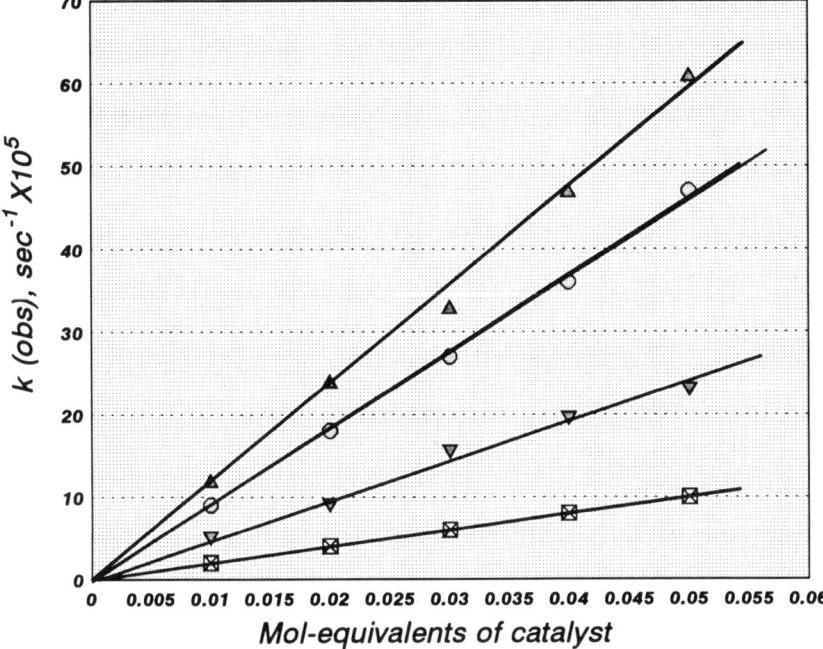

Figure 5-11. Dependence of rates on mole equivalent of various polymer-supported crown ether PTC catalysts for iodide displacement on 1-octyl methanesulfonate.

on the trimethylammonium catalyst. Here, it is believed that diffusion is fast relative to intrinsic reactivity.

- Rates of 1-bromohexadecane are one-third to one-half that of 1-bromooctane, a difference that is clearly attributable only to differences in diffusion rates, since in homogeneous solutions or in simple PTC reactions using soluble catalyst these two alkyl halides undergo SN2 reactions at about the same rate.
- The tributylphosphonium group is a substantially more active reaction site than trimethylammonium.

Catalysts prepared by reaction of 1% crosslinked chloromethylated polystyrene with R_3N (R = Et, Pr, Bu), $RNMe_2$ (R = Et, Bu, n-dodecyl, n-hexadecyl), or Bu_3P were tested for activity with several anions with benzyl bromide and n-octyl methanesulfonate at 70° and 90°C [91]. In every reaction tested the catalysts' activities increased with increasing size of the immobilized quaternary cation. Variation of the structure of the immobilized cation modified the catalytic activity not only by changing the anion–cation interaction energy but also by changing the reaction environment around the active site.

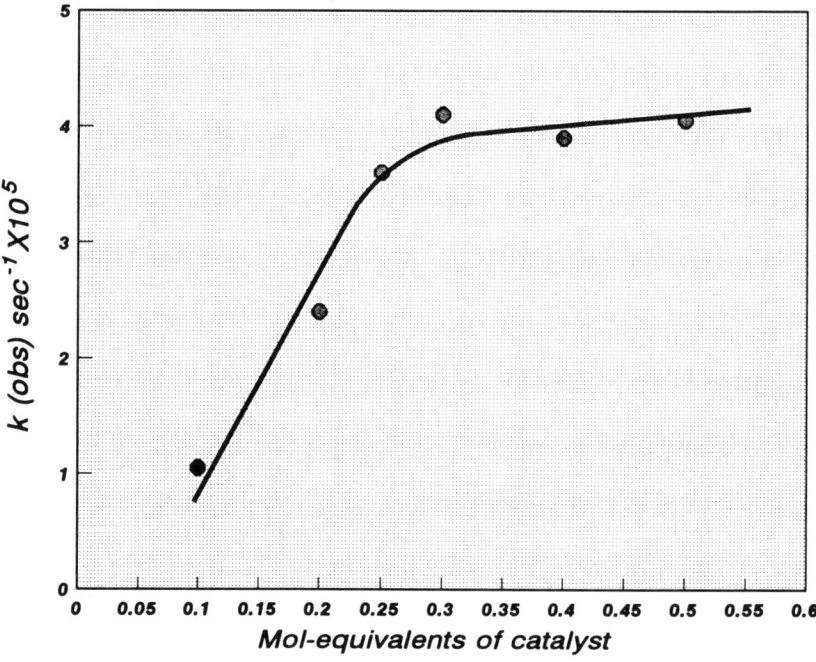

Figure 5-12. Rate constants vs. mole equivalent of polymer-bound crown ether for KCN displacement reaction with 1-bromooctane at 90°C.

Table 5-5. Comparison of cyanide displacements With Triphase Catalysts

Percent crosslinking	Mesh size	Active group	Rate Constant, $10^5 k_{obs}$ sec^{-1}			
			$1-C_8H_{17}Br$	$1-C_{16}H_{33}Br$	$C_6H_5CH_2Br$	$C_6H_5CH_2Cl$
2	−100+200	−PBu$_3$	21	—	125	—
2	−200+325	−PBu$_3$	31	13	—	—
2	−325+400	−PBu$_3$	38	16	400	—
2	−400	−PBu$_3$	40	17	580	—
10	−100+200	−PBu$_3$	9	—	47	—
10	−200+325	−PBu$_3$	26	—	170	—
2	−60+100	−NMe$_3$	1.4	0.21	44	5.9
2	−400	−NMe$_3$	1.4	0.52	112	6.1
10	−100+200	−NMe$_3$	1.0	—	26	6.3
10	−200+325	−NMe$_3$	1.8	—	99	7.8

Ragaini and co-workers [5] compared both soluble and solid-bound phase-transfer catalysts for monobutylation of phenylacetonitrile, as shown by the data in Table 5-6. They found the catalysts made from the bulkiest trialkylamines to be the best. However, in the alkylation of phenylacetonitrile with 1-bromobutane and 50% aqueous sodium hydroxide, commercial macroporous anion-exchange resins containing $-NMe_3^+$ active sites were as active as specially prepared triphase catalysts. Commercially available anion-exchange resins usually perform poorly as resin-bound PTC catalysts. However, Arrad and Sasson [92] demonstrated for formate displacement on benzyl chloride that some anion-exchange resins may be relatively good triphase catalysts when used with solid inorganic reagents, and when the amount of water present was severely restricted, as shown by the data in Table 5-7. (See page 237.) Some resins were found to have similar or even superior activity in comparison with the usual lab-produced triphase catalysts. In this situation, restriction of water acted to keep the active sites from becoming too hydrophilic to prevent diffusion of organic substrate to the active site. The ion-exchange resins were chemically stable under the reaction conditions, but their mechanical breakdown could be avoided only by use of ultrasonic agitation rather than mechanical agitation. The low-water technique was demonstrated also for other reactions including halogen exchange; nucleophilic displacements with potassium acetate, cyanide, nitrite, methoxide, phthalimide; permanganate oxidations; and for alkylation reactions. However, under the conditions used most reactions required about 1 day to obtain good yields, even with reactive substrates such as benzyl chloride.

Resin-Bound Crown Ethers, Cryptands

Many workers have reported on the use of resin-bound crown ethers as phase-transfer catalysts, often with very good results. The complexation of cations by crown ether groups bound to insoluble resins has been explored by Kahana et al. [17], who observed a spontaneous elution of cations when the temperature was raised to 60°C.

Montanari and co-workers [93] compared resin-bound crown ethers with their soluble equivalents.

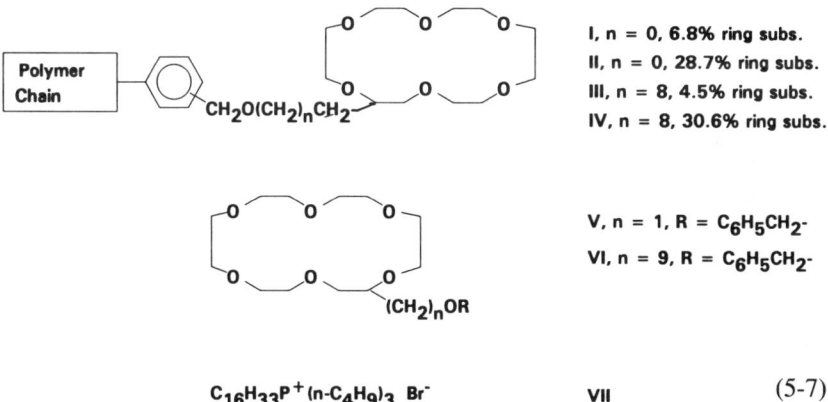

Using these catalysts they measured the rate constants for four different displacement reactions on octyl methanesulfonate with iodide, thiocyanate, bromide, and cyanide as nucleophiles. The results are given in Table 5-8. (See page 238.) Regarding these data Montanari and co-workers concluded:

1. Reactivity depends on the nature of the nucleophile.

2. Catalyst activity depends on percent RS, varying with the kind of nucleophile. With soft nucleophiles rates diminish as loading increases, following a linear semilogarithmic plot, whereas with harder nucleophiles the rates reach a maximum at 30% RS.

3. Catalysts having a spacer chain are on the average two to four times more reactive than directly bonded catalysts.

4. Complexation of potassium salts depends largely on the anion (high degree for soft nucleophiles such as iodide and thiocyanate; lower degree for smaller and less polarizable nucleophiles such as bromide and cyanide), and rates depend on the degree of complexation that occurs.

5. All reactions follow pseudo-first-order kinetics, and rates are linearly dependent on molar equivalents of polymer-supported crown ethers.

6. Hydrophilicity of catalysts and the extent of complexation increase with the extent of loading. Phenol is exclusively O-alkylated, even in the presence of most hydrophilic catalysts.

As a whole, they concluded that the reactions occur in the organic shell surrounding a complexed crown ether, following a mechanism analogous to that demonstrated for immobilized quaternary salts.

Table 5-6. *Comparison of catalysts for monobutylation of phenylacetonitrile [5].*

Catalyst	Temp. (°C)	Rate constant $\times$ 100 min^{-1} mol^{-2} L^{-2a}
Soluble catalysts		
Bu_4N^+	70	3043
$PhNEt_3^+$	70	68.9
	60	57.6
Lab-produced resin-bound catalysts		
Poly-NBu$_3^+$	70	14.4
Poly (12%)−NEt$_3^+$	70	4.21
	60	2.55
Poly (2%)−NEt$_3^+$	60	2.15
Poly−(CH$_2$CH$_2$O)$_4$Me	70	2.26
Commercial ion-exchange resins		
Duolite A101	60	1.28
Duolite A171	60	1.24
Duolite A161	60	0.54
IRA 904	60	0.10

[a] Rate constant based on molar equivalents of catalyst.

Table 5-7. Benzyl formate yield from benzyl chloride and potassium formate under standard conditions catalyzed by various polymer-bound quaternary salts [72,88].

Catalyst	Macro or Gel	Percent	Percent Cross-links	Functionality	Formate percent yield at 24 h
Commercial ion-exchange resins					
Amberlyst A27	M	40	—	$-CH_2NMe_3Cl$	97
Amberlite IRA 904	M	40	—	$-CH_2NMe_3Cl$	98
Dowex 1×2	G	60	2	$-CH_2NMe_3Cl$	4
Dowex 1×4	G	60	4	$-CH_2NMe_3Cl$	2
Dowex 1×8	G	60	8	$-CH_2NMe_3Cl$	9
Amberlite IRA 900	M	87	—	$-CH_2NMe_3Cl$	2
Amberlyst A26	M	87	—	$-CH_2NMe_3Cl$	20
Amberlite IRA 410	G	67	—	$-CH_2NMe_2(C_2H_4OH)Cl$	9
Amberlite IRA 910	M	83	—	$-CH_2NMe_2(C_2H_4OH)Cl$	8
Lab-synthesized resin-bound catalysts					
Polymer-Ph-CH$_2$NMe$_3$Cl	G	14	2	$-CH_2NMe_3Cl$	83
	G	20	2	$-CH_2NMe_3Cl$	36
Polymer-Ph-CH$_2$NEt$_3$Cl	G	12	2	$-CH_2NEt_3Cl$	93
	G	16	2	$-CH_2rNEt_3Cl$	91
Polymer-Ph-CH$_2$NPr$_3$Cl	G	10	2	$-CH_2NPr_3Cl$	8
Polymer-Ph-CH$_2$NBu$_3$Cl	G	8	2	$-CH_2NBu_3Cl$	19
	G	9	1	$-CH_2NBu_3Cl$	20
	G	14	2	$-CH_2NBu_3Cl$	4
Polymer-Ph-CH$_2$PBu$_3$Cl	G	10	1	$-CH_2PBu_3Cl$	75
Polymer-(CH$_2$)$_6$PBu$_3$Cl	G	12	2	$-(CH_2)_6PBu_3Cl$	97

Hodge, et al. [94] prepared a number of resin-bound benzo-crown-ethers (both 18-crown-6 and 15-crown-5 derivatives), joined by acetal groups to standard resins, and compared their catalytic activity for liquid–liquid PTC displacement reactions on 1-bromooctane with (1) potassium iodide, (2) sodium p-thiocresylate, and (3) potassium cyanide with the results is given in Table 5-9.

Bartsch and co-workers [95] studied several polymer-bound crown ethers and their soluble equivalents, as catalysts for cyanide displacement on 1-bromooctane at 90°C, 1300 rpm stirring, with 10 mol% equivalent of catalyst. In this system rates for the polymer bound catalysts were 5–30 times faster than for the soluble analogs, as shown in Table 5-10.

The resin support used in these studies was a chloromethylstyrene–styrene copolymer, crosslinked with divinylbenzene with 67% of the rings containing a chloromethyl group. After substitution of the crown ether groups for chloro-, the catalysts were ring-substituted to the extent of 58–65%.

Comparison of resin-bound crown ethers with the corresponding resin-bound quaternary phosphonium salts show for halide displacement reactions on 1-octyl methanesulfonate that the crown ether catalysts are more active by a factor of

Table 5-8. Rate constants for octyl methanesulfonate displacement reactions catalyzed by catalysts I–VII [89].

Catalyst	$k_{obs} \times 10^6$ sec^{-1} for nucleophiles Y			
	Y = I$^-$	Y = SCN$^-$	Y = Br$^-$	Y = CN$^-$
For solvent system = chlorobenzene–water				
Resin-bound PTC catalysts				
I	30.0	4.9	—	—
II	—	—	2.9	5.9
III	31.0	6.2	—	—
IV	—	—	4.7	8.3
Soluble PTC catalysts				
V	146.0	21.9	12.4	22.7
VI	90.0	17.0	8.3	14.9
VII	47.6	18.0	61.2	320.6
For solvent system = toluene–water				
Resin-bound PTC catalysts				
I	47.8	10.0	—	—
II	—	—	2.8	6.5
III	91.9	10.1	—	—
IV	—	—	8.0	10.4
Soluble PTC catalysts				
V	93.0	27.8	2.6	6.1
VI	194.0	30.9	4.4	6.7
VII	110.2	23.6	106.6	410.0

Table 5-9. Comparison of soluble and insoluble phase-transfer catalysts for three displacement reactions on 1-bromooctane [90].

Catalyst type	Percent yield after 24 h under standard conditions[a]		
	Rxn. 1 Y = I$^-$	Rxn. 2 Y = $-SC_6H_4CH_3-p$	Rxn. 3 Y = CN$^-$
Polymer-bound crown ethers			
(8 different catalysts)	20–75	26–64	0–4
Polymer–PPh$_3^+$Br$^-$	23–65	52–61	—
15-Crown-5	12–28	6–9	2
$n-C_{16}H_{33}P^+Bu_3Br^-$	97	56	99
$(C_8H_{17})_4N^+Br^-$	97	43	99

[a]Yield data taken after 5 h of reaction time paralleled the 24-h sample results. These reactions were conducted in slurry-type reactions with vigorous (1000 rpm) stirring.

Table 5.10. *Comparison of crown ether groups in soluble catalysts and in polymer-bound PTC catalysts on rates of cyanide displacement on 1-bromooctane.*

Crown-ether Structure	Rate Constant, sec^{-1}, ×10^6		
	Polymer-bound Crown	Soluble Crown	
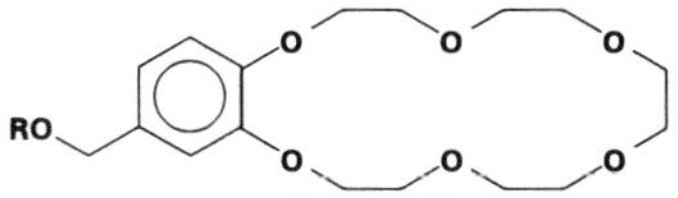 RO—[benzo-18-crown-6]	17.8	R = H R = C$_6$H$_5$CH$_2$-	0.6 1.8
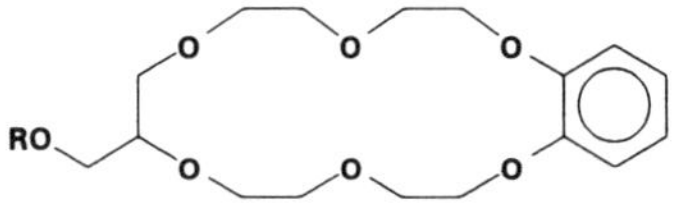 RO—[benzo-18-crown-6 isomer]	44.8	R = H R = C$_6$H$_5$CH$_2$-	2.1 1.5
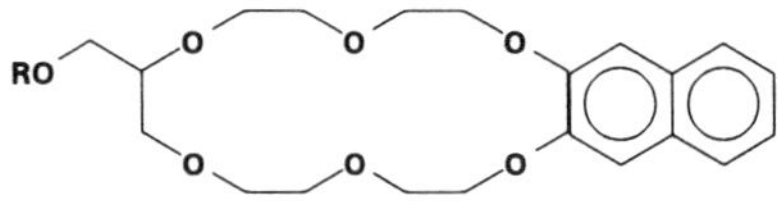 RO—[naphtho-crown]	53.0	—	—
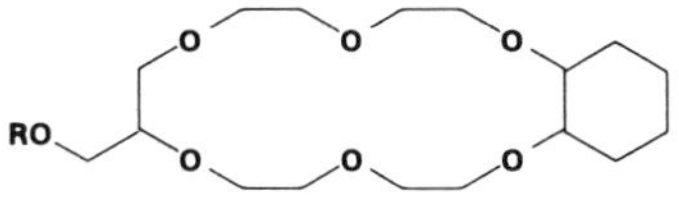 RO—[cyclohexyl-crown]	165.0	R = H	23.0
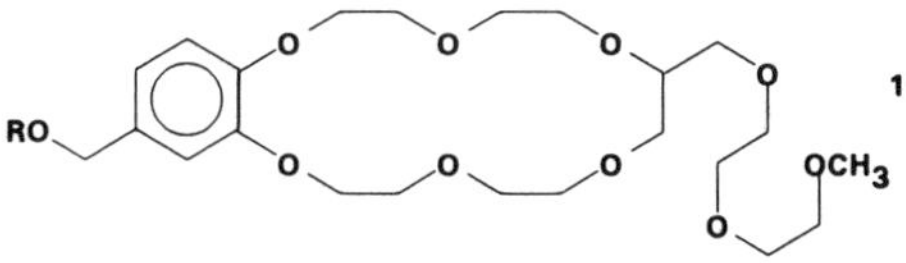 RO—[benzo-crown with OCH$_3$ sidechain]	15.0	R = C$_{12}$H$_{25}$-	3.3
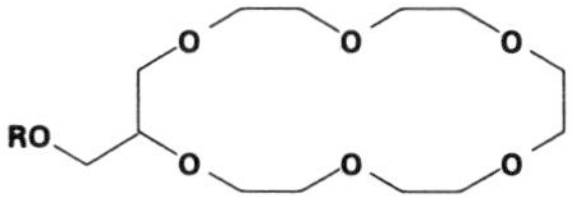 RO—[18-crown-6]	18.0	—	—
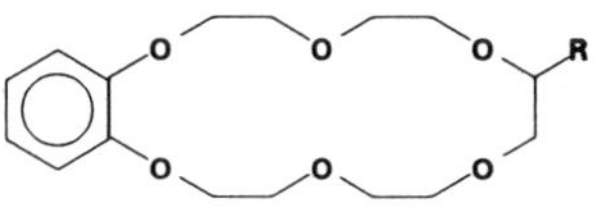 [benzo-crown]—R	—	R = H R = MeO(CH$_2$CH$_2$O)$_2$-	1.0 0.9
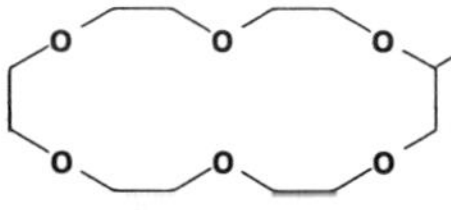 [crown]—R	—	R = -(CH$_2$)$_9$-O-CH$_2$C$_6$H$_5$	7.2

2.1–3.3 when the nucleophile is iodide, but only 0.07–0.74 as active when the nucleophile is bromide. This trend is in agreement with results found for soluble phase-transfer catalysts [44,89].

Resin-bound cryptands are generally better catalysts than the corresponding quaternary phosphonium salts, by a factor of 1.3–4 for iodide displacement, and 0.6–2.7 for bromide displacement on 1-octyl methanesulfonate [96,97]. Like other resin-bound catalysts, the cryptand analogs lose activity as percent RS increases.

PEG Sites

Resin-bound PEGs are effective catalysts for aqueous–organic two-phase reactions [2,3,26,27,46,52,98–103]. An especially interesting catalyst is made by radiation of poly(ethylene oxide) with high-energy radiation to produce crosslinked gels. These gels have been found to be highly active insoluble third-phase catalysts for reaction of 1-bromooctane with aqueous cyanide in the absence of organic solvent [191]. The reaction follows pseudo-first-order kinetics with rates linearly dependent on the amount of PEO gel. The reaction rate is not limited by intraparticle diffusion and mass transfer at stirring speeds above 200 rpm. Crosslinked PEO is easily produced in large quantities from commercially available polymer so it may become successful as a gel medium for phase transfer catalysis in various applications. Methacrylate esters of polyethylene glycol copolymerize with butyl acrylate to yield PEG groups supported by a polymer structure, suitable for use as phase transfer catalysts [192]. Polyethylene oxide moities also react with chlorinated poly(vinyl chloride) to give polymner-supported PEGs that are said to be useful as phase-transfer catalysts in the preparation of ethers and other commercial derivatives [193]. Shan et al. [52] explored the use of resin-bound PEG as triphase catalysts for reaction of benzyl bromide with potassium acetate. Figures 5-13 to 5-15 show the effects of crosslink density, RS, and PEG molecular weight, respectively. It was a concluded that:

- Some PEG is on the surface of the catalyst particle, but most is in the resin interior.

- Both surface and interior sites are active. However, the greater the crosslink density and the smaller the resin pores, the greater difficulty for reactants to diffuse to interior sites.

- The more swollen the resin (a function of crosslinking and solvent) the easier for reactants to diffuse to interior sites.

- Increasing the PEG segment molecular weight from 400 to 600 caused the ester yield to increase, but then to fall as the molecular weight was further increased to 800. Increasing the ethoxylate chain length is believed to increase the concentration of inorganic species at the PEG sites to such an extent that diffusion of organic reactants to the sites is reduced.

- Ester yield reaches a maximum when the catalyst contains about 30% RS, where presumably the hydrophilic–hydrophobic characteristics of the active sites are optimally balanced.

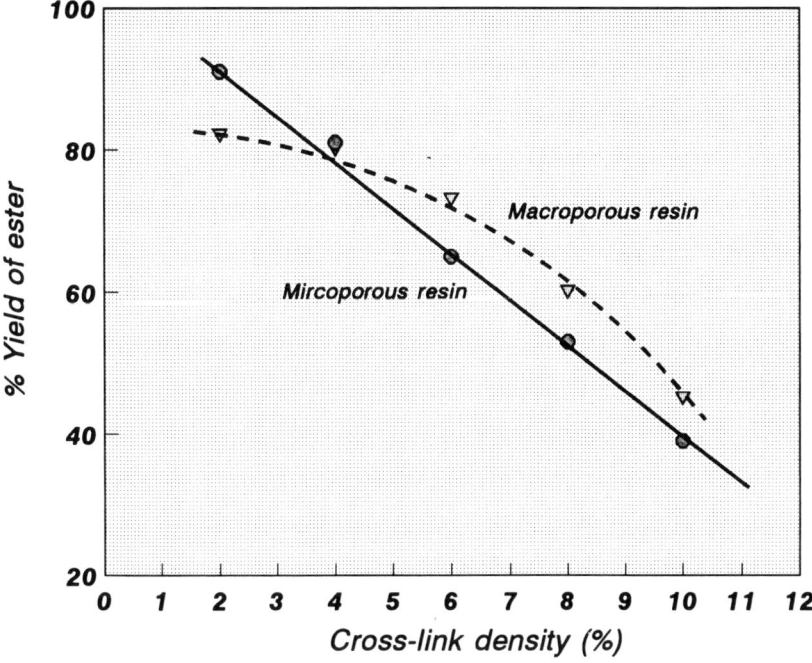

Figure 5-13. Crosslinking effect on percent ester from reaction of benzyl bromide with KOAc on polymer-bound PEG (PEG mol wt = 400), at 130°C, 10 h.

Polymers Containing Dipolar Aprotic Functional Groups and Other Special Catalyst Structures

With soluble phase-transfer catalysts it has been shown that special catalyst structures can sometimes significantly enhance catalytic activity, and this carries over also into use of insoluble PTC catalysts. For example, strategically placed hydroxyl groups in resin-bound catalysts can enhance reaction rates, just as in soluble catalysts. Ogawa and Tomoi [104] found that resin-bound aza-crown groups with a neighboring hydroxyl can easily form betaine-like structures, which exhibit high activity for dehydrobromination of phenethyl bromide.

(5-8)

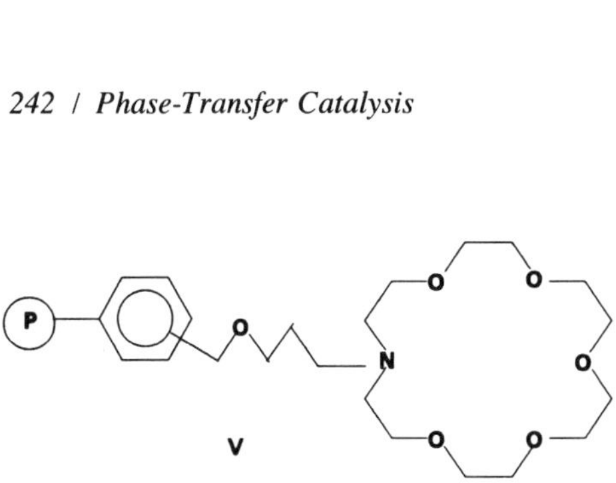

(5-9)

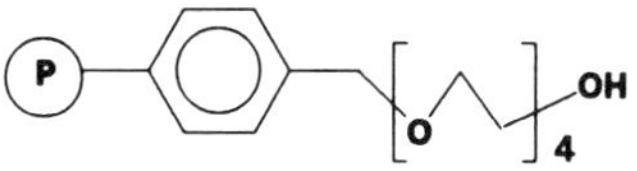

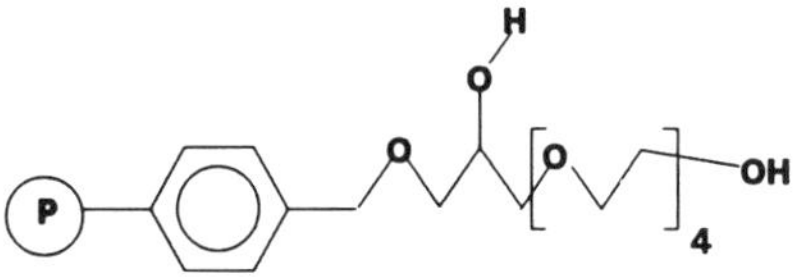

These workers compared the activity of these polymer-bound catalysts for dehydrohalogenation of phenethyl bromide with KOH, to achieve the results shown in Figure 5-16. (See page 245.)

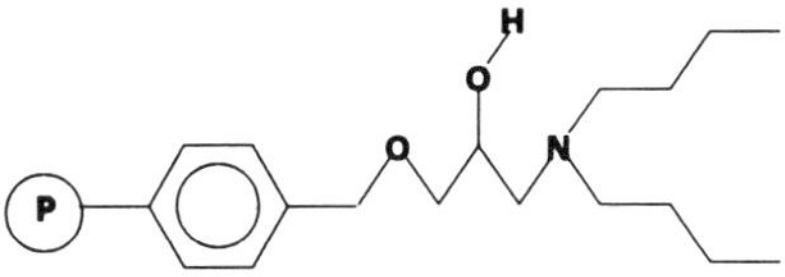

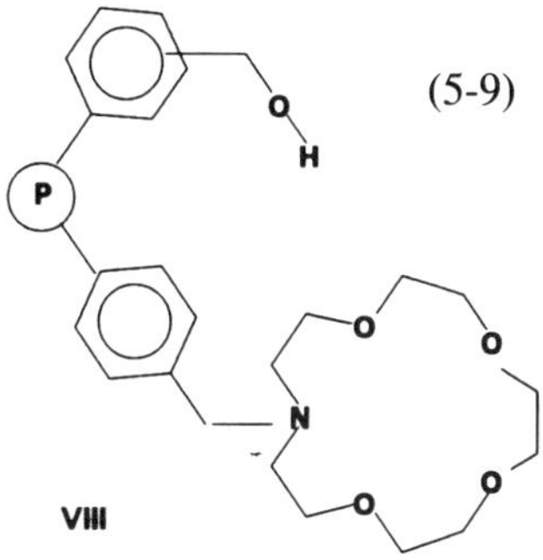

The PTC catalytic activity of crown ethers bound to insoluble resins for the reaction of KI or NaI with 1-bromooctane is enhanced by having spacer chains containing two or more oxygen atoms in the chain [105] or by having a hydroxyl group adjacent to the crown ether rings [106].

Polymeric analogs of triphenylphosphine oxide were excellent, stable, and reusable phase-transfer catalysts for the nucleophilic substitution reactions of 1-bromohexane with thiocyanate, iodide, or phenoxide anions, and for butyl bromide alkylation of phenylacetonitirile, and other reactions [107].

N-Vinylpyrrolidone, copolymerized with styrene and divinylbenzene, exhibits high activity as a phase-transfer catalyst [108].

Polymer-supported ureas obtained by polymerization of p-CH$_2$ = CHC6H4CH$_2$NRCONR'Me and divinylbenzene, show PTC catalytic activity thiocyanate substitution on 1-bromooctane with solid KSCN, although the corresponding monomeric ureas are not active catalysts. [194]. The catalytic activity is enhanced remarkably by replacing the amino hydrogen for a methyl group,

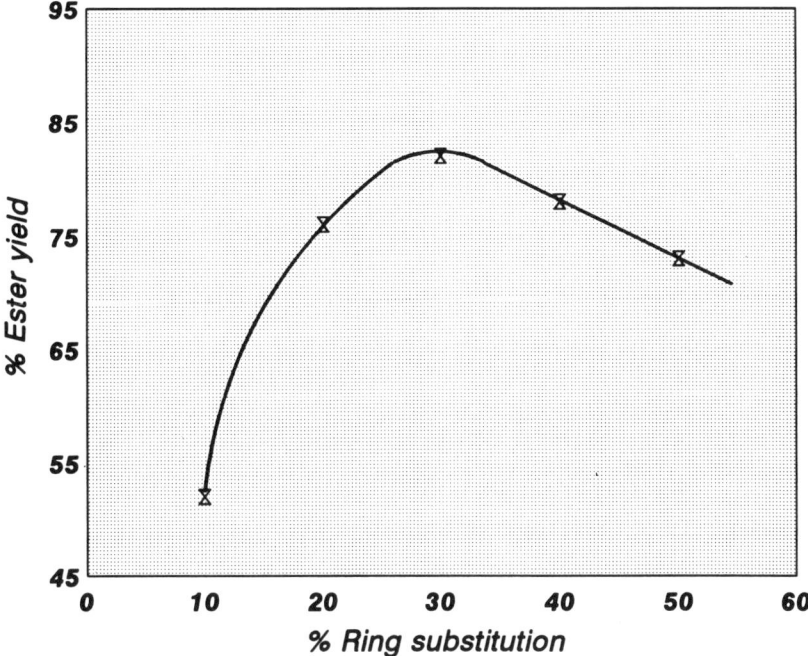

Figure 5-14. Effect of ring substitution on percent ester yield from reaction of benzyl bromide with KOAc on polymer-bound PEG (PEG mol wt = 400), at 130°C, 10 h.

and activity is affected by stirring, catalyst particle size of the catalyst, degree of crosslinking, and solvent.

Insoluble and Resin-Bound Chiral Triphase Catalysis

The best results for inducing chirality using insoluble catalysts has been with poly(amino acids) such as in the oxidation of chalcone in toluene with aqueous hydrogen peroxide and NaOH using poly(S-alanine) or poly(S-valine) as catalyst to give epoxide in 96–97% enantiomeric excess [109–111]:

(5-10A)

Chiral epoxidation:

Poly(S-alanine) catalyst: $n_{avg} = 10$

$H-[NHCH(CH_3)C(O)]_n-OR$

$\underset{H}{\overset{Ph}{>}}C=C\underset{COPh}{\overset{H}{<}}$ + H_2O_2 $\xrightarrow[\text{NaOH (aq)}]{\text{Toluene}}$ Ph,,,△COPh (O)

97% Enantiomeric excess

Polymer-bound poly(S-alanine):

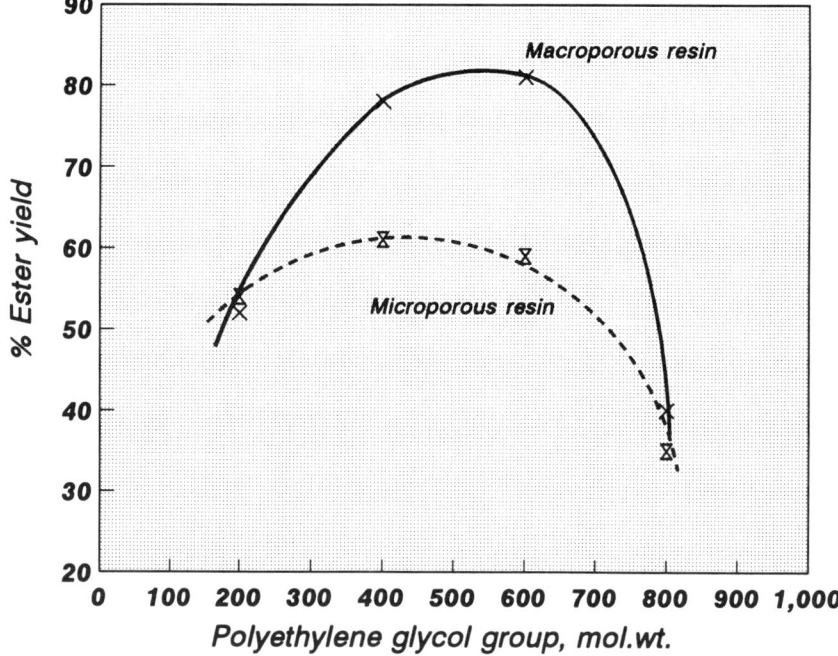

(5-10B)

Highest percent ee's are obtained when the average degree of polymerization, n, is greater than 10, and when the carboxylic acid end group is esterified with an n-butyl group rather than methyl. Copolymers of poly(p-hydroxymethylstyrene-styrene) esterified with poly(S-alanine) [$P_{avg} = 10$] gave epoxide in 84%ee, with lower optical yields with recycled catalyst, 75% ee after the first and 52% ee after the second recycle.

Only limited success (about 50–60% ee) has been achieved with chiral triphase catalysts to produce chiral products from nonchiral reactants [112–122]. For

Figure 5-15. Effect of PEG group molecular weight, bound to SDV resin, ester yield from KOAc and benzyl chloride at 130°C after 10 h.

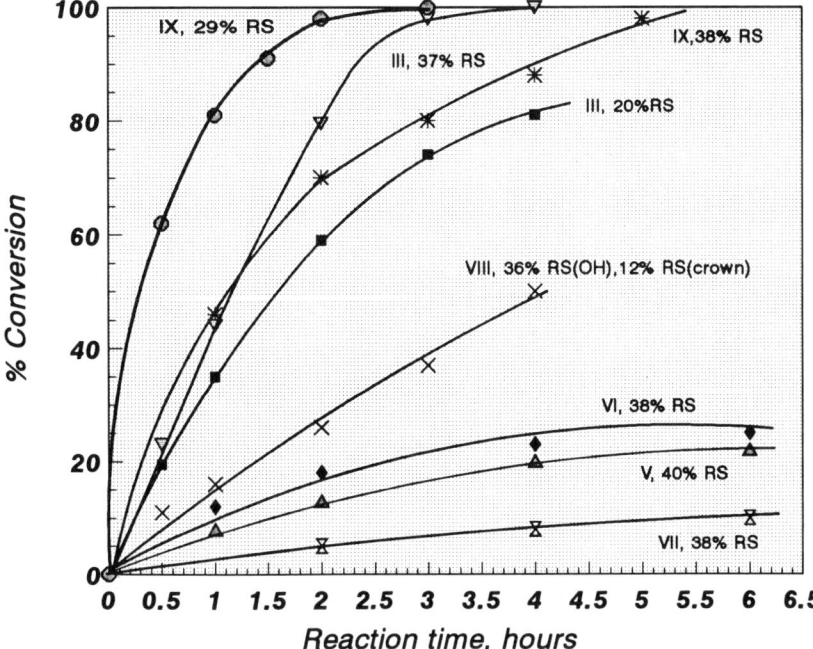

Figure 5-16. Comparison of polymer-bound catalysts containing a β-hydroxyl group for conversion of phenethyl bromide to styrene in a dioxane–aq. KOH two-phase system.

example, α-amino acids functionalized on styrene–divinylbenzene resins [123], and polymers of (−)−menthyl methacrylate or (−)−menthyl vinyl ether with vinyl compounds are effective phase-transfer catalysts in the reduction of ketones with sodium borohydride but induce little optical activity in the products [124]. Linear and crosslinked polymers containing chiral active centers were catalytically active for borohydride reduction of ketones, alkylation of phenylacetonitrile, and oxidation of racemic alcohols, but none of these were able to exert enantioselective discrimination on any of the prochiral or chiral substrates examined [117].

k. Size of the Organic Substrate

Unlike PTC reactions with soluble catalysts, the size of the organic reagents involved may have a significant effect on the rate of reactions catalyzed by polymer-bound PTC groups because of lower diffusion rates for larger molecules. For example, it was shown earlier (see Table 5-5) in data of Tomoi and Ford [86] that the rate of 1-bromohexadacane reaction with NaCN is only one-third to one-half that of the same reactions with 1-bromooctane using a polymer-bound tributylphosphonium triphase catalyst. These differences are clearly attributable to differences in diffusion rates since both alkyl bromides react at about the same rate when soluble PTC catalysts are employed. Similar diffusion-slowed results have also been obtained using polymer-bound tributylphosphonium catalysts for reactions of 1-bromooctane, benzyl chloride, and 3-phenyl-1-bromopropane with sodium cyanide [125].

l. Ratio of Volumes of Aqueous and Organic Phases

In the reaction between 1-bromooctane and potassium cyanide, catalyzed by polystyrene-bound Bu_3P^+ groups, the liquid-phase ratio has a significant effect on the kinetics [126]. The organic phase must be the dispersed phase to carry out the reaction effectively with microporous, low level of cross-linking and low-percent RS polystyrene chloromethylated resins. Drastic kinetic rate improvement was achieved with proper adjustment of phase ratios and the amount of catalyst for a given organic phase volume. Likewise, in the substitution reactions between hexachlorocyclotriphosphazene, $(NPCl_2)_3$, 2,2,2-trifluoroethanol, and sodium hydroxide, catalyzed by resin-bound tri-*n*-butylammonium, dramatic kinetic rate improvement was achieved by careful adjustment of the volume of the aqueous solution which was a low concentration of salt in the aqueous phase [127].

m. Catalyst Degradation and Catalyst Recycle

Degradation of resin-bound catalysts occurs from both chemical and physical causes. Quaternary ammonium of phosphonium salts undergo chemical decomposition reactions more or less rapidly depending on the reaction temperature and the presence of strong hydroxide solution. Resins with -onium groups may be used for extended periods, or repeated cycles, only if the reactions catalyzed occur under sufficiently mild conditions to avoid degradation. This condition is not as restrictive as it seems, however, since even in the presence of 50% aqueous NaOH used for the alkylation of phenylacetonitrile with 1-bromobutane, catalysts having a benzyltrimethylammonium structure bound to resin retain about 88% of their activity on recycle to subsequent batches [5]. Other authors have reported high catalytic activities and extensive recycling (at least five cycles) of triphase -onium catalysts [128,129]. Onium groups bound to porous ultrathin (5 μm) nylon capsule membranes were three times more efficient PTC catalysts for azide displacement on benzyl chloride than soluble quaternary ammonium salts [130, 131].

Resin-bound PEGs, crown ethers, and cryptands are more chemically stable than the corresponding −onium salts [2,132–136] and have been recycled repeatedly in laboratory reactions with only slight loss of activity [137,138].

Mechanical degradation of catalysts bound to insoluble resins occurs because of abrasive motions and swelling. Although this is not a problem in the laboratory, it is often a significant problem in industrial scale operations, where turbulent flows and flow fluctuations exert considerable grinding action on packed beds and slurries and even greater damage when using efficient mechanical agitation.

Table 5-11. Effect of catalyst recycle on yield of benzyl acetate from benzyl bromide and potassium acetate under standard reaction conditions [45,131].

Cycle	Percent of ester from catalyst		
	Ma (4%)−400[a]	Ma (6%)−400	Ma (10%)−400
1	100.0	82.3	55.6
2	97.2	81.1	55.0
3	93.4	80.5	54.2
4	88.4	78.3	53.6
5	80.4	73.0	53.5
6	67.5	65.2	53.0
Percent loss of initial activity after 6 cycles	32.5	20.7	4.6

[a]The notation Ma (4%)−400 represents a macrocyclic resin, 4% crosslinked, and functionalized with PEG of mol wt 400.

Ragaini and Verzella [139] have shown that for phase-transfer catalysts bonded to silica this mechanical degradation may produce a gel that is practically inseparable from the liquid phases.

Fixed bed reactors with recycling pump and ultrasonic mixing have been used to increase contact and reaction time with two liquid reagents (1-bromooctane and aq. KI) with resin- or silica-bound phase-transfer catalysts to model continuous industrial-type reactors [131]. These reactors were shown to avoid catalyst attrition, common to slurry reactors, and gave reactivity levels of about 70% that observed for slurry reactors with high mixing efficiency. In another study to determine the catalytic activity of various catalysts Ragaini and co-workers [5] also used the continuous fixed-bed reactor with ultrasonic agitation to minimize mechanical degradation. Even so, the activity of a resin-bound PEG catalyst for the reaction of benzyl bromide with potassium acetate declined steadily on sequential recycling, as shown by the data in Table 5-11 [52]. These tests to determine if the catalysts could be recycled showed a constant decline in activity that is especially noticeable with lower crosslinking. Under these reaction conditions chemical degradation of the ethoxylate group is unlikely and therefore loss of activity is probably due to mechanical degradation. This is supported by lower catalyst activity loss with highest crosslinking, due to increased mechanical strength of the highest crosslinked resins. Several studies, as shown in Table 5-1, indicate that resin-bound catalysts can be recycled many times without loss of activity, but these studies were normally not subjected to the kind of mechanical abrasions usually observed in industrial-scale reactors.

5. Kinetics of Reactions Catalyzed by Resin-Bound PTC Groups

As is evident from the previous discussion, reaction rates for triphase-catalyzed reactions depend on several factors that generally affect mass transfer, diffusion of reactants to active sites, and intrinsic reactivities for chemical reactions at the active sites. Marconi and Ford [140] have analyzed kinetic data for cyanide displacement reactions of 1-bromooctane and benzyl bromide, and derived the following kinetic expression:

$$-\frac{dC_A}{dt} = \frac{1}{\left[\frac{RV_{org}}{3V_{cat}k_L} + \frac{V_{org}}{k\lambda M_c \eta}\right]} C_A$$

where
C_A = concentration of species A, mol cm^{-3}
R = radius of swollen catalyst particles, cm
V_{org} = volume of organic phase, cm^3
V_{cat} = volume of swollen catalyst, cm^3
k = intrinsic reaction rate constant, cm^{3}mol^{-1}sec^{-1}
k_L = mass transfer coefficient based on unit exterior surface of particle, cm sec^{-1}
λ = partition coefficient, dimensionless
M_c = molar equivalents of catalyst, mol
η = intraparticle catalytic effectiveness, dimensionless

Although this approach requires an empirical estimate of some constants, the equation gives excellent correlation between catalyst parameters and observed rate constants.

Wang and Yang [141] have developed a more elaborate and detailed kinetic model of triphase catalytic reactions which considers the mass transfer of reactants in the bulk phases, the diffusion of reactants within the catalyst pores, and the intrinsic reactions involved. Approximately 30 parameters relating to the characteristics of the catalyst and to experimentally controlled parameters such as concentrations and volumes are used. This model is also highly consistent with experimental results.

C. Phase-Transfer Catalysts Bound to Inorganic Solid Supports

Excellent methods for the preparation of high-surface-area insoluble inorganic solids are known to catalyst chemists, and have been used extensively for making catalysts and supports for many kinds of reactions. It is natural that this wealth of experience and information would also be used to make suitable insoluble supports for phase-transfer catalysts. Indeed, two types of catalysts on inorganic supports have been explored as PTC catalysts: (1) those having a PTC catalyst simply adsorbed onto the inorganic support, and (2) those having the PTC function chemically bound to the inorganic support.

Table 5-12. *Comparison of rates for quaternary salt-hectorite catalysts with rates of unsupported and dissolved quaternary salts for the two-phase cyanide displacement reaction with 1-bromopentane.*

Quaternary cation	$k_{obs}hr^{-1}$	
	Quat-Hectorite	Dissolved quat
$n-C_{14}H_{29}NMe_3^+$	0.14	0.07
$n-C_{16}H_{33}NMe_3^+$	0.16	0.07
$n-C_{14}H_{29}NMe_2CH_2Ph^+$	0.16	0.38
$n-C_{16}H_{33}NMe_2CH_2Ph^+$	0.16	0.37
$n-C_{18}H_{37}NMe_2CH_2Ph^+$	0.16	0.42
$(n-C_{12}H_{25})_2NMe_2^+$	0.16	0.54
$(n-C_8H_{17})_3NMe^+$	0.18	1.10

^{a}All reactions at 90°C with 2 mmol RBr, 6 mL toluene, 20 mmol KCN in 3 mL water, 0.073 meq organo-clay or quaternary salt.

Inorganic compounds themselves have been shown to have some activity as insoluble phase-transfer catalysts. Alumina and alumina impregnated with the inorganic reagent to be used have been extensively tested as an insoluble phase-transfer catalyst [142–152]. This approach is useful, although alumina without prior impregnation is a sluggish PTC catalyst. Barium hydroxide is a successful solid phase-transfer catalyst for ylide reactions and for the Horner reaction [153, 154]. Cadmium iodide has likewise been used as a phase-transfer catalyst for reaction of benzyl iodide with potassium acetate, although here the cadmium salt formed soluble complexes [155].

1. PTC Catalysts Adsorbed on Inorganic Supports

The most generally successful and commercially useful of all adsorption-type (and perhaps most useful for all solid insoluble PTC catalysts) have been made by simple adsorption of quaternary salts on organophilic clays [158, 159]. Most interesting from the commercial viewpoint have been catalysts obtained by adsorption of long-chain quaternary ammonium cations on particular forms of smectite clay (hectorite), as described by Pinnavaia and co-workers [158]. These are efficient, inexpensive, stable, and recyclable catalysts. For example, alkyl bromides in toluene react in nucleophilic displacement reactions with aqueous NaCN, NaSCN, Na$_2$S, and alcohols-NaOH to give the expected nitriles, thiocyanates, sulfides, and ethers, respectively, in the presence of $[(n-C_8H_{17})_3NMe]^+$-hectorite as an insoluble catalyst [159]. These catalysts also facilitate permanganate oxidation of alcohols, C-alkylation of nitriles, and dehalogenation of *vic*-dibromides. Some results comparing reaction rates obtained adsorbed quaternary salts as catalysts vs. soluble quaternary salts for cyanide displacement on 1-bromopentane are given in Table 5-12.

In the hectorite-quaternary salt systems the quaternary cations cause the clay to form thin, membrane-like assemblies of platelets at the liquid–liquid interface of an oil-in-water systems. Pinnavia suggested that the surface of the clay becomes covered with quaternary salt with long alkyl chains extended outward in a cooperative oily coverage that is permeable to both aqueous and organic reactant phases. If the chains of the quaternary ammonium groups are too short to orient vertically on the hectorite surfaces, the clay layers do not assemble as part of the emulsion-like particles, and the triphasic rate constants are reduced by up to an order of magnitude. For example, hectorite exchanged with tetrabutylammonium is a relatively poor triphase catalyst, even though dissolved tetrabutylammonium cation is a good catalyst without the clay. The emulsion formed by use of quaternary salt-hectorite catalysts can easily be broken by gravity or by low-speed centrifugation, thus allowing for efficient recovery of catalyst and its removal from reaction products.

PTC catalysts adsorbed on other high-surface-area inorganic supports function as PTC catalysts with varying degrees of effectiveness. For example, use of SiO_2, Al_2O_3–KF, SiO_2–KF, Al_2O_3, C, or sand as solid supports for $PhCH_2NEt_3^+$ Cl^- as a phase-transfer catalyst for N-alkylation of 2-oxazolidone was reported [160]. Inorganic supports including metal oxides, microporous glass, and clay [161–163] have been used in place of polymers as supports for triphase catalysts, but these suffer from low reactivity or structural instability under the reaction conditions. Alumina itself has been shown to promote phase-transfer reactions for alkylation [164], condensation [165], hydrolysis [166], halide exchange [167], cyanide and acetate displacements [168], sulfide displacements [169], and carbonylation of benzyl bromide [169]. Intercalated transition metal complex salts in layered silicate clays have been applied as phase-transfer catalysts [170].

2. Catalysts with PTC Functions Chemically Bonded to Inorganic Supports

Solid insoluble PTC catalysts made by chemical attachment of PTC functions to solid inorganic supports appear to perform with about the same effectiveness as catalysts supported on polymeric resins, although the inorganic-based materials can have better mechanical strength and abrasion resistance. Chemically binding of PEG chains to silica and other refractory metal oxides [170–174], to cellulose [175], "Silochrome" or "Teflon" [176] gives materials having moderate catalytic activity. Sawicki [173] patented the preparation and use of catalysts made by reacting a porous refractory oxide, such as silica gel, containing surface hydroxyl groups with a polyoxyalkylene oxide or monoalkyl ether of a PEG, to give a surface with -O-$(CH_2CH_2O)_n$R groups. These catalysts matched the reactivity as resin-bound catalysts. Silica-gel-bound PTC catalysts were said to have been used repeatedly through seven cycles of reaction without loss of activity [177].

Tundo and co-workers [178–180] examined the kinetics of reactions of anions associated with cations bonded to high-surface silica and alumina. The reactions of these anions with n-octyl methanesulfonate was studied using a full molar

Table 5-13. Comparison of rate constants, $k_{obs} \times 10^3$ M^{-1} sec^{-1}, for reaction of anions with n-octylmethanesulfonate in wet toluene and anhydrous toluene.

Soluble cation or support	Wet toluene			Anhydrous toluene		
	Br$^-$	I$^-$	SCN$^-$	Br$^-$	I$^-$	SCN$^-$
Soluble systems						
[n-C$_{16}$H$_{33}$PBu$_3^+$]	7.7	3.6	—	10.7	25.7	—
C$_{14}$Cry[2.2.2]K^{+a}	—	—	—	54.4	—	—
Supported systems						
60-Å SiO$_2$-(CH$_2$)$_3$ NHCO(CH$_2$)$_{10}$PBu$_3^+$	—	6.4	—	0.94	13.6	1.7
60-Å SiO$_2$-K$^+$	0.1	2.8	—	—	—	—
60-Å SiO$_2$-(CH$_2$)$_3$NH$_3^+$	—	—	—	2.2	5.3	—
150-Å Al$_2$O$_3$-SiO$_2$-(CH$_2$)$_3$-NHCO-C$_{10}$H$_{20}$PBu$_3^+$	—	15.	0.69	118.	170.	—
150-Å Al$_2$O$_3$-K$^+$	2.4	5.2	—	—	—	—

^{a}C$_{14}$-Cry[2.2.2]K$^+$ is a tetradecyl-cryptate, potassium salt.

amount of supported catalyst plus anion, sometimes under anhydrous conditions. For example, supported catalyst structures such as

$$\text{—Si—O—Si—(CH}_2\text{CH}_2\text{CH}_2\text{)P}^+\text{Bu}_3\ \text{Br}^-$$

(with Si—O—Si and O—Si substituents on the central Si)

(5-11)

$$\text{Al}_2\text{O}_3\text{—Si—(CH}_2\text{)}_3\text{NHCO(CH}_2\text{)}_{10}\text{-PBu}^+\ \text{Br}^-$$

were found to undergo displacement reactions very well, as shown by the data in Table 5-13 comparing rates of supported reagent with rates of soluble reagents.

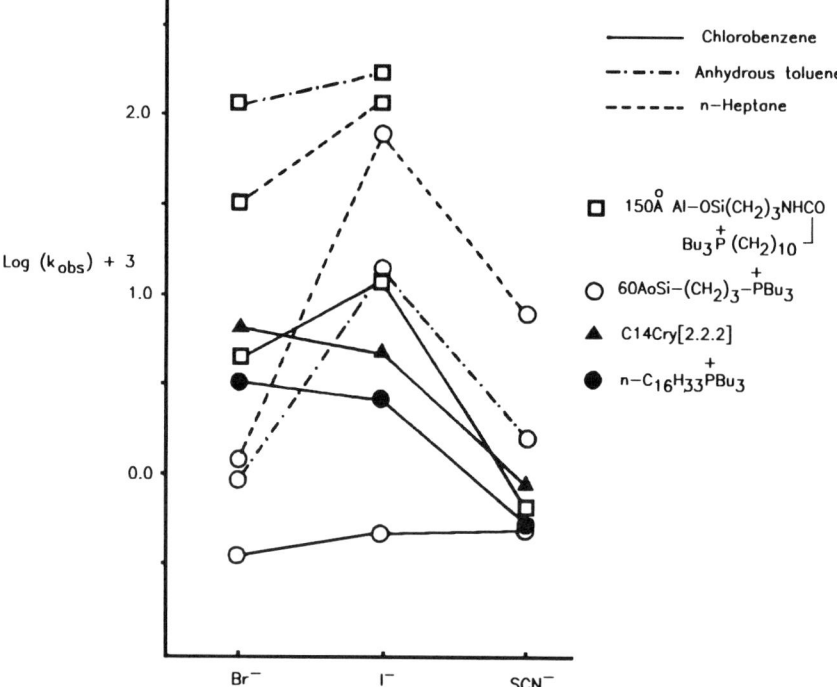

Figure 5-17. Comparison of rates for reactions with *n*-octylmethanesulfonate with three anions.

Further comparative rate data are shown in Figure 5-17. The important point to be recognized here is the high reactivity of the anions when bound to a high-surface-area alumina. This is believed to be due to the highly polar environment provided by the alumina. In previous work the organic cations bonded to high-surface-area silica and alumina showed a high affinity for hydrophilic anions such as bicarbonate, sulfate, phosphate, and fluoride, a behavior that is very rare for polymer-bound quaternary groups [181]. The polar nature of the support was believed to alter the microenvironment of the reaction site, providing a highly favorable situation for inducing substitution reactions.

The activity of silica-bound quaternary salts in displacement reactions and for reduction of ketones with sodium borohydride is equivalent or slightly higher than for many highly effective soluble quaternary salt catalysts.

D. PTC Catalysts Contained in a Separate Liquid Phase (Third-Liquid-Phase Catalyst)

Although use of solid insoluble catalysts is clearly an industrially useful chemical processes, insoluble liquid PTC catalyst phases can be even more useful. In

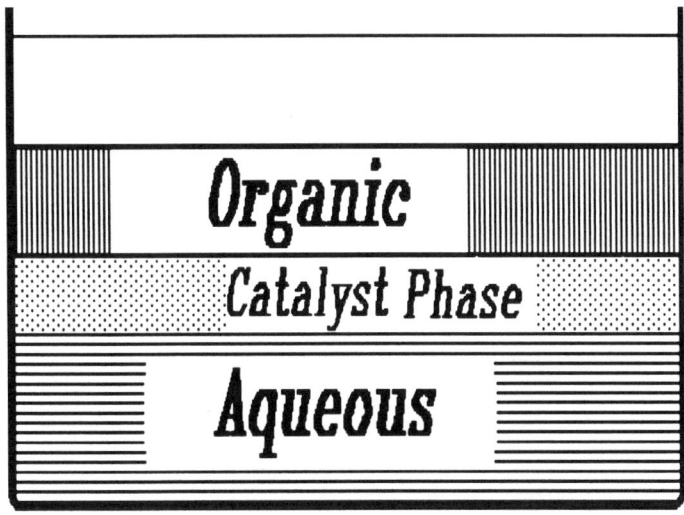

Figure 5-18. Representation of a three-liquid-phase PTC system, with catalyst in the middle phase.

third-liquid phase systems the phase-transfer catalyst is found concentrated in neither the organic or aqueous phase, but in a third liquid layer of its own that can be easily recovered and recycled. The few examples of this type of system indicate that even higher reaction rates can be achieved with third-phase catalysts than with insoluble solid catalysts and even soluble catalysts. The most formidable problem for third-liquid-phase catalysts is the much more difficult process-development activity required to obtain third-phase conditions that provide high reactivity yet do not cause significant loss of active catalyst by its being extracted into either of the other phases.

In principle, formation of a third catalyst could be accomplished either by (1) use of a phase-transfer catalyst that has limited solubility in both the organic and aqueous phases to be used, or (2) by use of a special solvent, perhaps a fluorocarbon, having great affinity for catalyst but only modest affinity for both of the reactants. Figure 5-18 shows a schematic representation of a three-phase system where the middle liquid phase contains the catalyst. Figure 5-19 is a more realistic representation showing for example, under dynamic reaction conditions, formation of droplets of the organic phase surrounded by a catalyst phase [182]. Under the conditions of third phase formation, the system of reactions occurring is somewhat different than in simple PTC, but the general principles are the same. Tetrabutylammonium salts frequently form third layers when used in conjunction with an organic phase that has little polarity (e.g., neat 1-chlorooctane, toluene, or cyclohexane as solvent, but not methylene chloride) and with a concentrated aqueous solution of inorganic salts.

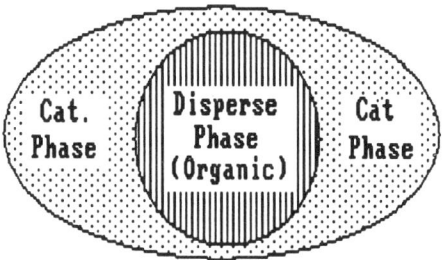

Figure 5-19. Expected positioning of third liquid catalyst phase during agitation.

Table 5-14. Composition of third phase in Bu_4NBr/toluene/water/NaBr system.

Component	Grams in whole system	Grams in third phase
Toluene	17	0.5
Bu_4NBr	5	3.8
Water	50	1.5
NaBr	20	0.04

An excellent example of third-phase PTC behavior has been provided by Wang and Weng [183] for the chloride–bromide exchange of benzyl chloride with sodium bromide.

$$PhCH_2Cl + NaBr\ (aq) \xrightarrow[75°]{Bu_4N^+\ Br^-} PhCH_2Br + NaCl\ (aq) \quad (5\text{-}12)$$

Using toluene as a solvent for the organic phase, these workers showed that the first small amounts of catalyst added to the reaction behaved as rather slow phase-transfer catalyst, probably because of limited solubility of the catalyst in the organic phase. However, addition of more than a critical amount of catalyst caused to rate to accelerate rapidly, as illustrated by the interesting curves in Figure 5-20. At each level of sodium bromide addition, there is a critical quaternary salt concentration, above which the rate increases sharply. Independently, they found that a third phase formed when either the NaBr concentration or tetrabutylammonium bromide concentration exceeded critical values. Such critical values for Bu_4NBr addition to cause third phase formation in the reaction system are marked in Figure 5-20 by small vertical dashed lines at the intersecting points. It is seen that quaternary salt concentrations at the point at which rates increase rapidly correspond almost identically with the critical point where third phase formation starts. The comparison of the third phase in this system is typified by the data in Table 5-14. The concept for simple PTC needs to be modified a small amount to accommodate the third-phase process. Wang and

Weng suggest that in the third-phase process, both organic and inorganic reagents are transferred to the third phase where most of the intrinsic reaction occurs:

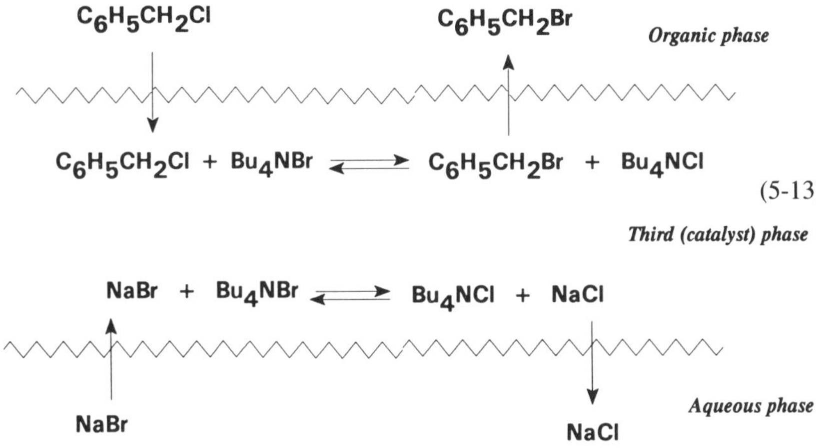

$$C_6H_5CH_2Cl \qquad\qquad C_6H_5CH_2Br \qquad \textit{Organic phase}$$

$$C_6H_5CH_2Cl + Bu_4NBr \rightleftharpoons C_6H_5CH_2Br + Bu_4NCl \qquad (5\text{-}13)$$

$$\textit{Third (catalyst) phase}$$

$$NaBr + Bu_4NBr \rightleftharpoons Bu_4NCl + NaCl$$

$$NaBr \qquad\qquad\qquad NaCl \qquad \textit{Aqueous phase}$$

The substantial increase in rates with third-phase formation supports this suggestion, although alternatives could also be suggested.

In commercial practice this system will allow reaction to proceed rapidly, with easy separation of the organic and aqueous phases, but leaving the catalyst phase for further batches. As shown in Figure 5-21 approximately 25% of the catalyst would be lost in each batch for this particular reaction system, mostly to the aqueous phase. However, tetrabutylammonium salts may be extracted from aqueous salt solutions by polar organic extractants such as methylene chloride or ethyl acetate.

Mason, Magdassi, and Sasson [182] showed that tetrabutylammonium bromide uniquely forms a third liquid phase in the toluene/$Bu_4N^+Br^-$/aq. NaOH system, whereas tetraethyl-, tetrapropyl-, tetrapentyl- and tetrahexylammonium salts formed only two-phase systems. They showed further in the PTC dehydrobromination of phenethyl bromide with aqueous NaOH to form styrene, that a clear discontinuity was evident in the behavior of the tetrabutylammonium salt compared to its neighbors, as in Figure 5-21. In forming the third phase, the catalytic behavior of the system is increased significantly, by about fourfold, similar to the behavior observed by Wang and Weng in the bromide–chloride exchange with benzyl chloride. Mason et al. also observed that when the concentration of NaOH in the aqueous phase was gradually increased from 42–44% to 46–49% there was between 46% and 49% NaOH concentration, an abrupt 10-fold decrease in the rate of phenethyl bromide elimination. On further examination, they were able to show this to be due to dehydration of the catalyst phase by the more concentrated NaOH and consequent precipitation of anhydrous solvent-free tetra-

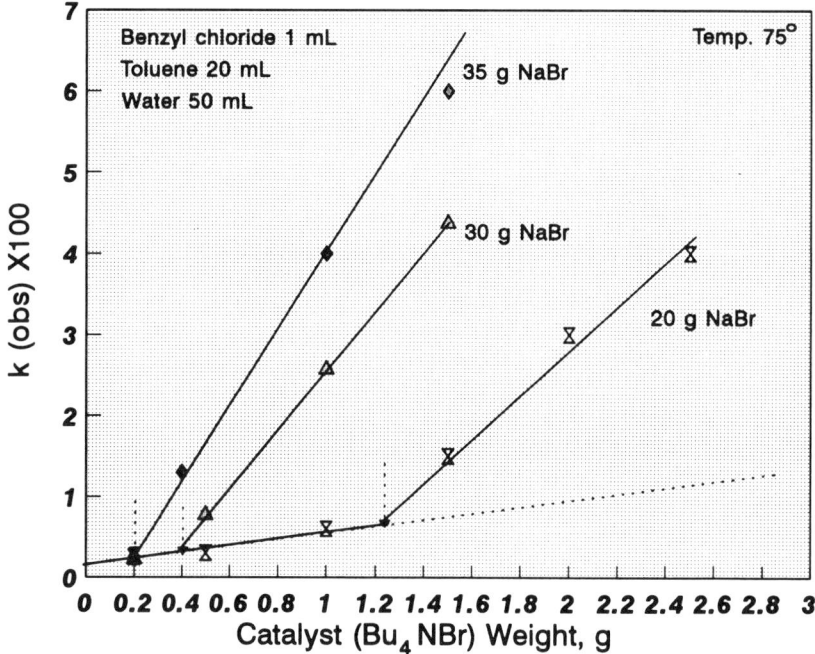

Figure 5-20. Effect of NaBr and catalyst on benzyl chloride–bromide exchange, showing rate acceleration due to third liquid phase formation.

butylammonium bromide with resulting loss of active catalyst. Analysis of the third-phase composition is given in Table 5-15.

The alkylation of pentaerythritol to produce ethers is believed to be a three-phase process, dependent on having the sodium hydroxide solution sufficiently concentrated that the catalyst is forced into formation of a third phase, rich in pentaerythritol [184].

Correia [185] found that aqueous sodium hypochlorite and tetrabutylammonium hydrogen sulfate, in the presence of sodium bromide, produce a third phase consisting largely of $Bu_4N^+Br_3^-$ but also containing water, hypohalite, halide ions, and possibly Br_2. Treatment of this with cyclohexene produces trans-1,2-dibromo- and 1-bromo-2-chlorocyclohexane, and treatment of 1-alkynes produces 1-bromo-1-alkynes.

Another interesting third-phase process has been described for epoxidation of perfluoroolefins with sodium hypochlorite using toluene and coco benzyl bis (β-hydroxypropyl)ammonium chloride to form the third phase [186]. A three-liquid PTC process has been used in the synthesis of fluoro- and chlorocyclobutenes [195].

PEG forms a third liquid phase between toluene and aqueous KOH solutions, especially when some methanol is added. Neumann and Sasson [187] examined

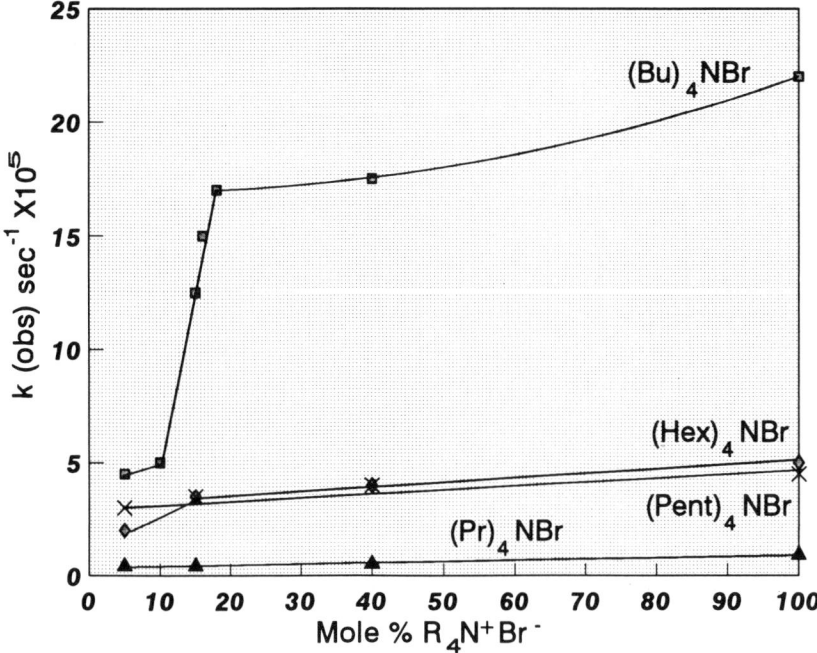

Figure 5-21. Rates of phenethyl bromide conversion to styrene by NaOH (aq) with $R_4N^+Br^-$ catalysts.

Table 5-15. Composition of the third phase formed in the presence of toluene, aqueous sodium hydroxide, and tetrabutylammonium bromide [164].

	Wt %	Molar ratio
Toluene	44.4	3.95
Water	2.2	1.00
$Bu_4N^+Br^-$	52.6	1.34
OH^-	50 ppm	—

the use of PEG as a phase-transfer catalyst for the isomerization of allylanisole to anethole.

$$\underset{OCH_3}{\underset{|}{C_6H_4}}-CH_2CH=CH_2 \xrightarrow[\text{KOH (aq)}]{\text{PEG}} \underset{OCH_3}{\underset{|}{C_6H_4}}-CH=CHCH_3 \quad (5\text{-}14)$$

Here, the reaction mixture consisted of three phases: an organic solvent phase, a PEG–potassium hydroxide complex phase, and a basic aqueous phase. The reaction steps included diffusion of the substrate from the solvent to the complex phase where reaction isomerization occurred, and then back-diffusion of the product. When the aqueous phase is not completely saturated with KOH, no reaction occurs, evidently because either no hydroxide anion is transferred into the catalyst phase, or if it is, then it is too highly hydrated to cause isomerization. Concentrations above saturation cause a steady rate increase. Short-chain catalysts were more effective per gram, but not per mole, of catalyst, and etherification of the terminal hydroxyl group reduced the catalyst activity. Use of anhydrous alkoxide species rather than hydroxide as base reversed the catalyst trends, long-chain catalysts being more effective.

A PEG three-phase system has been used for dehydrobromination of 2-bromooctane in toluene with aqueous KOH [189]. Here, some methanol must be added to reduce the amount of PEG that otherwise would dissolve in the toluene phase. Successive dehydrobrominations could be accomplished by removal and replacement of the organic and aqueous phases after completion of reaction. After four cycles no catalyst was lost. When too much methanol was added the aqueous phase disappeared [190].

Hydrogenative dehalogenation of polychlorinated aromatic halides can be accomplished by hypophosphite reduction using a quaternary ammonioum salt as a phase-transfer catalyst in conjunction with a palladium-on-carbon co-catalyst [196]. The quaternary salt, being insoluble in both reactant phases, coats the Pd/C catalyst forming a third phase. The strongly alkaline medium and the phase-transfer agent are synergic. Thus, at 50°, 1,2,4,5-tetrachlorobenzene gives 99% benzene after 2.5 hours. Enhancement of the reaction rate, was attributed (1) to phosphite transfer from the aqueous phase to the catalyst surface, (2) to partition of halogenated compounds between the hydrocarbon solution and the liq. phase of the phase-transfer agent and (3) to the rapid removal of HCl adsorbed on Pd/C by the alkaline medium.

References

1. D.C. Sherrington, Chem. Ind. (Lond.), 15 (1991).
2. S.L. Regen, Angew. Chem., Int. Ed., Engl., **18**, 421 (1979).
3. M. Tomoi and W.T. Ford, Polymeric Phase Transfer Catalysts, in Syntheses and Separations Using Functional Polymers, D.C. Sherrington and P. Hodge, eds., J. Wiley & Sons, 1988, p. 183.
4. M. Tomoi and W.T. Ford, J. Am. Chem. Soc., **103**, 3821 (1981).
5. V. Ragaini, G. Colombo, P. Barzaghi, E. Chiellini, and S. D'Antone, Ind. Eng. Chem. Res., **27**, 1382 (1988).
6. N. Zhong and Z. Liang, Gaofenzi Xuebao (5) 624-7 (1989) [CA: 114.063298].
7. S.M. Kumar, Synth. Commun., **21** (21), 2121-7 (1991).

8. S.M. Kumar, Synth. Commun., **17**, 1015 (1987).
9. S. Kobayashi, M. Suzuki, T. Sakaya, and T. Saegusa, Makromol. Chem., **188**, 457 (1987).
10. S. Kondo, T. Mori, H. Kunisada, and Y. Yuki, Macromol. Chem. Rapid Commun. **11**, 309 (1990).
11. S. Kondo, T. Murayama, Y. Taleda, and K. Tsuda, Makromol. Chem., Rapid Commun., **9**, 625 (1988).
12. Y. Gui and D. Lin, Youji Huaxue, 346 (1987) [CA 108.204602].
13. J.H. Shim, K.B. Chung, S.H. Chang, D.K. Song, and K.Y. Sung, Taehan Hwahakhoe Chi., **32**, 593 (1988) [CA 111.022847].
14. M. Tomoi, N. Yanai, S. Shiiki, and H. Kakiuchi, J. Polym. Sci., Polym. Chem. Ed., **22**, 911 (1984).
15. Z. Huang, J. Shu, and Z. Yu, Gaofenzi Tongxun, 220 (1984) [CA 102.167856].
16. C. Zhan, S. Dong, and Y. Xu, Gaodeng Xuexiao Huaxue Xuebao, **5**, 802 (1984) [CA 102.114057].
17. N. Kahana, A. Deshe, and A. Warshawsky, J. Polym. Sci., Polym. Chem. Ed., **23**, 231 (1985).
18. E. Blasius, K.P. Janzen, H. Klotz, and A. Toussaint, Makromol. Chem., **183**, 1401 (1982).
19. Y. Gui, B. Liu, J. Liu, Z. Zhong, and H. Lin, Youji Huaxue, 373 (1986) [CA 107.007181].
20. C. Zhan, Y. Xu, and S. Dong, Gaofenzi Tongxun, 445 (1983) [CA 101.112207].
21. F. Montanari and P. Tundo, J. Org. Chem., **46**, 2125 (1981).
22. Y. Xu, S. Dong, X. Bai, and H. Wu, Gaofenzi Tongxun, (4) 266 (1983) [CA 100.121716].
23. P.L. Anelli, F. Montanari, V. Pollak, and S. Quici, Gazz. Chim. Ital., **116**, 127 (1986) [CA 106.196418].
24. Y. Chen and F. Zeng, Gaofenzi Tongxun, 409 (1986) [CA 106.214501].
25. S.J. Harris, J.M. Rooney, and J.G. Woods, to Loctite Ltd., Eur. Pat. Appl. EP 196895 (1986) [CA 106.067887].
26. S. Itsuno, I. Moue, and K. Ito, Polym. Bull. (Berl.), **21**, 365 (1989) [CA 111.097840].
27. M. Tomoi, T. Suzuki, and H. Kakiuchi, React. Polym., **10**, 27 (1989).
28. Yu, Shanxin- Yang, Jianwen, Huaxue Shijie 1991, **32**, 308–10 [CA: 116.108762].
29. T. Wakui, W.Y. Xu, C.S. Chen, and J. Smid, Makromol. Chem., **187**, 533 (1986).
30. T. Wakui, W.Y. Xu, and J. Smid, Polym. Prepr. (Am. Chem. Soc., Div. Polym. Chem.), **25**, 126 (1984).
31. O.E. Filippova, Deposited Doc., VINITI 3676-83 (1982) [CA 101.230088].
32. E. Yaacoub and P. LePerchec, React. Polym., Ion Exchange Sorbents, **4**, 285 (1988) [CA 110.056805].

33. V. Janout, J. Kahovec, H. Hrudkova, F. Svec, and P. Cefelin, Polym. Bull. (Berl.), **11**, 215 (1984).
34. S. Kondo, M. Ozeki, N. Nakashima, K. Suzuki, and K. Tsuda, Angew. Makromol. Chem., 163, 139 (1988).
35. S. Kondo, M. Iwasaki, and K. Tsuda, J. Macromol. Sci., Chem., **A27**, 1053 (1990).
36. S. Kondo, T. Yamamoto, H. Kunisada, and Y. Yuki, J. Macromol. Sci., Chem., **A27**, 1515 (1990).
37. M. Hasaanein, A. Akelah, and F. Abdel-Galil, Eur. Polym. J., **21**, 475 (1985).
38. N. Ohtani, Kagaku Kogyo, **39**, 331 (1988) [CA 109.099365].
39. Yu. Sh. Gol'berg, and M.V. Shimanskaya, Zh. Vses. Khim. O-va. im. D.I. Mendeleeva, **31**, 149 (1986) [CA 105.171419].
40. F. Svec, J. Kahovec, and J. Hradil, Chem. Listy, **81**, 183 (1987) [CA 106.139343].
41. N. Ohtani, Yuki Gosei Kagaku Kyokaishi, **43**, 313 (1985) [CA 102.220197].
42. Yu. Sh. Gol'berg, Usp. Geterog. Katal. Khim. Geterotsikl. Soedin, 110 (1984) [CA 102.203318].
43. D.C. Sherrington and K.J. Kelly, Org. Coat. Appl. Polym. Sci., Proc., **46**, 278 (1981) [CA 100.051116].
44. F. Montanari, D. Landini, A. Maia, S. Quici, and P.L. Anelli, ACS Symp. Ser., **326**, 54 (1987).
45. H. Molinari, F. Montanari, S. Quici, and P. Tundo, J. Am. Chem. Soc., **101**, 3920 (1979).
46. W.T. Ford and M. Tomoi, Adv. Polym. Sci., **55**, 49 (1984).
47. K.W. Pepper, H.W. Paisley, and M.A. Young, J. Chem. Soc., 4093 (1953).
48. S. Mohanraj and W.T. Ford, Macromolecules, **19**, 2470 (1986).
49. R.C.Jones and Y. Matsubayashi, Polymer, **33**, 1069 (1992).
50. T. Balakrishnan and W.T. Ford, J. Appl. Polym. Sci., **27**, 133 (1982).
51. M. Bernard, W.T. Ford, Macromolecules, **177**, 1812 (1984).
52. Y. Shan, R. Kang, and W. Li, Ind. Eng. Chem. Res., **28**, 1289 (1989).
53. L. Hong and E. Ruckenstein, Polymer **33**, 1968 (1992).
54. E. Ruckenstein, L. Hong, J. Catal., **136**, 378 (1992).
55. H. Kise, K. Araki, and M. Seno, Tetrahedron Lett., **22**, 1017 (1981).
56. A.V. Bogatsky, N.G. Lukyanenanko, and V.N. Pastushok, Dokl. Akad. Nauk SSSR, **283**, 628 (1985).
57. V.F. Loktev, I. Yu. Morozova, A.K. Tashmukhamedova, and V.A. Likholobov, Izv. Akad. Nauk SSSR. Ser. Khim., 1257 (1990).
58. A. Akelah and D.C. Sherington, Eur. Polym. J., **18**, 301 (1982).
59. M. Tomoi, N. Kori, and H. Kakiuchi, Makromol. Chem., **187**, 2753 (1986).
60. J.M. Brown and J. Jenkins, J. Chem. Soc., Chem. Commun., 458 (1976).

61. J.G. Heffernan, D.C. Sherrington, and C.J. Suckling, Reactive Polymers, **1**, 357 (1986).
62. M. Tomoi, S. Shiki, and H. Kakiuchi, Makromol. Chem., **187**, 357 (1986).
63. M.S. Chiles, D.D. Jackson, and P.C. Reeves, J. Org. Chem., 45, 2915 (1980).
64. J.P. Idoux, R. Wysocki, S. Young, J. Turcot, C. Ohlaman, and R. Leonard, Synth. Commun. **13**, 139 (1983).
65. M. Tomoi, E. Ogawa, Y. Hosokawa, and H.J. Kakiuchi, J. Polym. Sci., Polym. Chem. Ed., **20**, 3015 (1982).
66. P. Tundo, Synthesis, 315 (1978).
67. H. Kakiuchi and M. Tomoi, Kenkyu Hokoku - Asahi Garasu Kogyo Gijutsu Shoreikai, **42**, 237 (1983) [CA 100.067589].
68. M. Tomoi, N. Kori, and H. Kakiuchi, React. Polym., Ion Exchange, Sorbents, **3**, 341 (1985) [CA 104.006287].
69. K.B. Chung and M. Tomoi, J. Polym. Sci. Part A- Polym. Chem., **30**, 1089 (1992).
70. H. Xie, P. Zhou, W. Sun, J. Xia, J. Liu, and D. Xie, Dong, Chin. J. Polym. Sci., **9**, 1 (1991) [CA. 116.152429].
71. F. Montanari, S. Quici, and P. Tundo, J. Org. Chem., **48**, 199 (1983).
72. J.G. Heffernan and D.C. Sherington, Tetrahedron Lett., **24**, 1661 (1983).
73. H. Takeuchi, Y. Miwa, S. Morita, and J. Okada, Chem. Pharm. Bull., **32**, 409 (1984) [CA 101.054175].
74. H. Takeuchi, Y. Miwa, S. Morita, and J. Okada, Chem. Pharm. Bull., **33**, 3101 (1985) [CA 104.147985].
75. P.L. Anellli, F. Montanari, and S. Quici, J. Chem. Soc., Perkin Trans. II, 1827 (1983).
76. M. Tomoi, E. Ogawa, T. Hosokama, and H. Kakiuchi; J. Polym. Sci., Polym. Chem., Ed., **20**, 3421 (1982).
77. M. Cinquini, S. Colonna, H. Molinari, F. Montanari, and P. Tundo, J. Chem. Soc., Chem. Commun., 394 (1976).
78. H. Molinari, F. Montanari, and P. Tundo, J. Chem. Soc., Chem. Commun., 112 (1977).
79. J. Hradil, F. Svec, and J.M.J. Frechet, Polymer, **28**, 1593 (1987).
80. W.T. Ford, J. Lee, and M. Tomoi, Macromolecules, **15**, 1246 (1982).
81. T. Ido, S. Hajime, K. Sakurai, and S. Goto, Kagaku Kogaku Ronbunshu, **10**, 287 (1984) [CA 101.090171].
82. G. Popov, L. Feistel, and G. Schwachula, to VEB Chemiekombinat Bitterfield, Ger. (East) DD 249,199 (1986) [CA 109.055872].
83. K.B. Chung and M. Tomoi, J. Polym. Sci., Part A: Polym. Chem., **30**, 1089 (1992).
84. M. Tomoi, E. Nakamura, Y. Hosokawa, and H. Kakuuichi, J. Polym. Sci., Polym. Chem. Ed., **23**, 49 (1985).

85. M.-L. Wang and H.-S. Wu, Ind. Eng. Chem. Res., **31**, 490 (1992).
86. M.L. Wang and H.S. Wu, Ind. Eng. Chem., Res., **31**, 2238 (1992).
87. M.L. Wang and H.S. Wu, Ind. Eng. Chem. Res., **31**, 490 (1992).
88. M.L. Wang and H.S. Wu, J. Polym. Sci., Part A., Polym. Chem., **30**, 1393 (1992).
89. A. Roska, M. Klavins, and A. Zicmanis, Latv. PSR Zinat. Akad. Vestis, Kim. Ser., 97 (1988) [CA 109.229838].
90. M. Tomoi and W.T. Ford, J. Am. Chem. Soc., **103**, 3828 (1981).
91. H. Takeuchi, Y. Miwa, S. Morita, and J. Okada, Chem. Pharm. Bull., **32**, 823 (1984) [CA 101.109951].
92. O. Arrad and Y. Sasson, J. Org. Chem., **54**, 4993 (1989).
93. P. Anelli, B. Czech, F. Montanari, and S. Quici, J. Am. Chem. Soc., **106**, 861 (1984).
94. P. Hodge, E. Khoshdel, and J. Waterhouse, J. Chem. Soc., Perkin Trans., **1**, 2451 (1984).
95. M.J. Pugia, A. Czech, B.P. Czech, and R.A. Bartsch, J. Org. Chem., **51**, 2945 (1986).
96. P.L. Anelli and S. Quici, Synthesis, 1070 (1985).
97. P.L. Anelli, F. Montanari, and S. Quici, J. Org. Chem., **51**, 4910 (1986).
98. K. Hiratani, P. Reuter, and G. Manecke, J. Mol. Catal., **5**, 241 (1979).
99. W.M. MacKenzie and D.C. Sherrington, Polymer, **21**, 791 (1980).
100. J.G. Heffernan, W.M. Mackenzie, and D.C. Sherrington, J. Chem. Soc., Perkin Trans II, 514 (1981).
101. Y. Kimura and S.L. Regen, J. Org. Chem., **48**, 195 (1983).
102. Y. Kimura and S.L. Regen, Synth. Commun., **13**, 443 (1983).
103. J. Hradil and F. Svec, Polym. Bull., **11**, 159 (1984).
104. N. Ogawa and M. Tomoi, React. Polym., **13**, 55–62 (1990).
105. J.H. Shim., K.B. Chung, and M. Tomoi, Pollimo, **16**, 291 (1992) [CA. 117.152331].
106. J.H. Shim., K.B. Chung, and M. Tomoi, Bull. Korean Chem. Soc., **13**, 274 (1992) [CA.117.152343].
107. S. Kondo, K. Furukawa, and K. Tsuda, J. Polym. Sci., Part A: Polym. Chem. **30**, 1503 (1992).
108. S. Kondo, M. Ozeki, N. Nakashima, K. Suzuki, and K. Tsuda, Angew. Makromol. Chem., **163**, 139 (1988) [CA 111.114700].
109. S. Julia, J. Guixer, J. Masana, J. Rocas, S. Colonna, R. Annuzita, and H. Molinari, J. Chem. Soc., Perkin Trans. I, 1317 (1982).
110. S. Banfi, S. Colonna, H. Molinari, S. Julia, and J. Guixer, Tetrahedron, **40**, 5207 (1984).
111. S. Colonna, H. Molinari, S. Banfi, S. Julia, J. Massana, and A. Alvarez, Tetrahedron, **39**, 1635 (1983).

112. T. Yamashita, H. Yasueda, Y. Miyauchi, and N. Nakamura, Bull. Chem. Soc., Japan, **50**, 1532 (1977).
113. T. Yamashita, H. Yasueda, N. Nakatani, and N. Nakamura, Bull. Chem. Soc., Japan, **51**, 1183 (1978).
114. K. Herman and H. Wynberg, Helv. Chim. Acta., **60**, 2208 (1977).
115. N. Kobayashi and K. Iwai, J. Am. Chem. Soc., **100**, 7071 (1978); J. Polym. Sci., Chem. Ed. 18, 223 (1980); Makromol. Chem. Rapid Commun. **2**, 105 (1981).
116. P. Hodge, E. Khoshdel, and J. Waterhouse, J. Chem. Soc., Perkin Trans. I, 2205 (1983).
117. P. Hodge, E. Khoshdel, J. Waterhouse, and J.M.J. Frechet, J. Chem. Soc., Perkin Trans. I, 2327 (1985).
118. N. Kobayashi, Brit. Polym. J., **16**, 205 (1984).
119. N. Kobayashi and K. Iwai, Macromolecules, **13**, 31 (1980).
120. J.M.J. Frechet, J. Kelley, and D.C. Sherrington, Polymer, **25**, 1491 (1984).
121. S. D'Antone, M. Penco, R. Solaro, and E. Chiellini, Reactive Polymers, **3**, 107 (1985).
122. J. Kelly and D.C. Sherrington, Polymer, **25**, 1499 (1984).
123. B. He, W. Wang, Q. Zhang, and X. Li, Gaofenzi Tongxun 219 (1982) [CA 97.217156].
124. S. D'Antone, R. Solaro, M. Penco, and E. Chielini, Conv. Ital. Sic., Macromol. [Atti] 6th., 2, 381 (1983) [CA 101.056147].
125. M. Tomoi, E. Nakamura, Y. Hosokawa, and H. Kakiuchi, Makromol. Chem., Rapid Commun., **5**, 281 (1984).
126. S. Telford, P. Schlunt, and P.C. Chau, Macromolecules, **19**, 2435 (1986).
127. M.L. Wang and H.S. Wu, J. Polym. Sci., Part A: Polym. Chem., **30**, 1393 (1992).
128. M. Tomoi and W. Ford, J. Am. Chem. Soc., **103**, 3828 (1981).
129. J.P. Idoux, R. Wysocki, S. Young, J. Turcot, C. Ohlman, and R. Leonard, Synth. Commun., **13**, 139 (1983).
130. Y. Okahata, K. Ariga, and T. Seki, J. Chem. Soc., Chem. Commun., 920 (1985).
131. Y. Okahata and K. Ariga, J. Org. Chem., **51**, 5064 (1986).
132. K. Funkunishi, B. Czech, and S.L. Regen, J. Org. Chem., **46**, 1218 (1981).
133. F. Montanari and P. Tundo, J. Org. Chem., **47**, 1298 (1982).
134. F. Montanari and P. Tundo, Tetrahedron Lett., 5055 (1979).
135. H.J.M. Dou, R. Gallo, P. Hassanaly, and J. Metzger, J. Org. Chem., **42**, 4275 (1977).
136. T. Kamentani, K. Kigasawa, M. Hiiragi, N. Wayatsuma, and K. Wakisaka, Tetrahedron Lett., 635 (1969).
137. S. Yu and J. Yang, Huaxue Shijie, **32**, 308 (1991) [CA. 116.108762].
138. C.S. Chen, W.C. Hwang, S.C. Chang, Huaxue **49**, 99 (1991) [CA. 116.150922].

139. V. Ragaini, G. Verzella, A. Ghignone, and G. Colombo, Ind. Eng. Chem., Process Des. Dev., **25**, 878 (1986).
140. P.F. Marconi and W.T. Ford, J. Catalysis, **83**, 160 (1983).
141. M.-L. Wang and H.-M. Yang, Ind. Eng. Chem., Res., **31**, 1868 (1992).
142. G.H. Posner, Angew. Chem., Intl. Ed. Engl., **17**, 487 (1978).
143. K.-T. Liu and Y.-C. Tong, Synthesis, 660 (1978).
144. S. Quici and S.L. Regen, J. Org. Chem., **44**, 3436 (1979).
145. B. Czech, S. Quici, and S.L. Regen, Synthesis, 110 (1980).
146. S.L. Regen, B. Czech, and S. Quici, Pol. J. Chem., **55**, 841 (1981).
147. T. Ando, T. Kawate, J. Ichihara, and T. Hanafusa, Chem. Lett., 725 (1984).
148. J.R. Dalton and S.L. Regen, J. Org. Chem., **44**, 4443 (1979).
149. R. Ballini and M. Petrini, Synthesis, 1024 (1986).
150. V.V. Veselovsky, A.S. Gybin, A.V. Lozanova, A. Moiseenkov, W.A. Smit, and R. Caple, Tetrahedron Lett., **29**, 175 (1988).
151. H. Alper, S. Ripley, and T.L. Prince, J. Org. Chem., **48**, 250 (1983).
152. T.C. Jempty, L.L. Miller,and Y. Mazur, J. Org. Chem., **45**, 749 (1980).
153. J.V. Sinisterra, M.E. Borredon, Z. Mouloungui, M. Delmas, and A. Gaset, React. Kinet. Catal. Lett., **29**, 41 (1985) [CA 105.208207].
154. J.V. Sinisterra, Z. Mouloungui, M. Delmas, and A. Gaset, Synthesis, (12) 1097 (1985).
155. S. Szakacs, M. Jaky, S. Gobolos, and F. Nagy, Magy. Kem. Foily., **89**, 402 (1983) [CA 100.051186].
156. A. Cornelis and P. Laszio, Synthesis, (2) 162 (1982).
157. P. Monsef-Mirzai and W.R. McWhinnie, Inorg. Chim. Acta, **52**, 211 (1981).
158. C.L. Lin and T.J. Pinnavaia, Chem. Mater, **3**, 213–5 (1991).
159. T.J. Pinnavaia and C.L. Lin, to Michigan State University, U.S. Patent US 5,099,054 (1992).
160. H.D. Nguyen, N.B. Hiep. T.N.H. Lee, and P.N.S. Chu, C.R. Acad. Sci., Ser. 2, **300**, 799 (1985).
161. A. Kadkhoduyan and T.J. Pinnavaia, J. Mol. Catal., **21**, 283 (1983).
162. P. Tundo, P. Venturello, and E. Angeleti, Israel J. Chem., **26**, 283 (1985).
163. A. Cornelius and P. Laszlo, Synthesis, 162 (1982).
164. G. Bram and T. Fillebeen-Khan, J. Chem. Soc., Chem. Comm., 522 (1979).
165. J. Muzart, Synthesis, 60 (1982).
166. S.L. Regen and A.K. Mehrotra, Synth. Commun., 413 (1981).
167. S. Quici and S.L. Regen, J. Org. Chem., **44**, 3436 (1979).
168. S.L. Regen, B. Czech, and S. Quici, Pol. J. Chem., **55**, 84 (1981).
169. N.C. Pradhan and M.M. Sharma, Ind. Eng. Chem. Res., **31**, 1610 (1992).

170. R.A. Sawicki, ACS Symp. Ser., **326**, 143 (1987).
171. A. Kadkhodayan, Diss. Abstr. Int. B. 1985, **46**, 516 [CA 104.121821].
172. R.A. Sawicki, to Texaco Inc., U.S. Patent 4,474,704 (1984).
173. R.A. Sawicki, to Texaco Inc., U.S. Patent 4,421,675 (1983).
174. R.A. Sawicki, Tetrahedron Lett., **23**, 2249 (1982).
175. A. Akelah and D.C. Sherrington, Eur. Polym. J., **18**, 301 (1982).
176. A.V. Bogatskii, N.G. Lukyanenko, V.N. Pastushok, and M.N. Parfenova, Dokl. Akad. Nauk SSSR, **283**, 628 (1985).
177. R. Kang, F. Fan, Y. He, and L. Zhang, Yingyong Huaxue, **6**, 78 (1989) [CA 112.76355].
178. P. Tundo, J. Chem. Soc., Chem. Commun., 641 (1977).
179. P. Tundo and P. Venturello, J. Am. Chem. Soc., **101**, 6066 (1979); **103**, 856 (1981); **104**, 655 (1982).
180. P. Tundo and M. Badiali, React. Polym., **10**, 55–65 (1989).
181. P. Tundo, P. Venturello, and E. Angeletti, J. Am. Chem. Soc., **104**, 6547 (1982).
182. D. Mason, S. Magdassi, and Y. Sasson, J. Org. Chem., **56**, 7229 (1991).
183. Der-Her Wang and Hung-Shan Weng, Chem. Eng. Sci., **43**, 2019 (1988).
184. R. Nouguier and M. Mchich, Tetrahedron, **44**, 2477 (1988).
185. J. Correia, J. Org. Chem., **57**, 4555 (1992).
186. J.R. Lawson, to Dupont, Eur. Pat. Appl., EP 414,569 (1991) [CA 115.008555].
187. R. Neumann and Y. Sasson, J. Org. Chem., **49**, 3448 (1984).
188. R. Nouguier and M. Mchich, Tetrahedron, **44**, 2477 (1988).
189. T. Ido, M. Saiki, and S. Goto, Kagaku, Kogaku Ronbunshu, **15**, 403 (1989) [CA 111.056994].
190. T. Ido, Y. Kitamura, and S. Goto, Kagaku Kogaku Ronbunshu, **16**, 388 (1990) [CA 113.008361].
191. T.S. Tsanov and T. Tsvetanov, Polymer, **34**, 616 (1993).
192. J. Guo, H. Xie, X. Liu, and Yingvong Huaxue, **9**, 1 (1992), [CA 116:196962].
193. S. Yu, Hunan Shifan Daxue Ziran Kexue Xuebuo, **14**, 32 (1991) [CA 116:213716].
194. S. Kondo, T. Okamura, M.T. Masakazu, H. Kunisada, Y. Hideo, Y. Yuki, and J. Makromol. Chem. **193**, 2265 (1992).
195. N.C. Craig, S.S. Borick, M.A. Fisher, T.R. Tucker, and Y. Xiao, J. Fluorine Chem., **59**, 215 (1992).
196. C.A. Marques, M. Selva, and P. Tundo, J. Chem. Soc., Perkin Trans. I, 529 (1993).

6

Variables in Reaction Design for Laboratory and Industrial Applications of Phase-Transfer Catalysis

Many factors, known and unknown, contribute to the effective selection of a catalyst and other conditions for a given phase-transfer catalysis (PTC) application. Therefore, no simple single guideline exists for a universal effective choice of reaction conditions. However, several useful, though noncomprehensive, approaches exist for separately considering the optimization of various components of PTC reactions under a variety of conditions. It is recommended that the reader be familiar with the various thought approaches (presented below) and choose reaction conditions based on a combination of the approaches. In all cases, the factors that are thought to enhance, or otherwise affect, reaction rates in PTC will be put into the perspective of the PTC Rate Matrix.

A. Choice of Catalyst

The criteria for selecting a phase-transfer catalyst always include structure–activity relationships, usually include stability and catalyst separation considerations, and, in industrial environments, also include other attributes such as cost, toxicity, availability, recycle, waste treatment, and others.

The following discussion focuses primarily on the mechanistic and other chemical factors to be considered when choosing a phase-transfer catalyst. However, it should be noted that the logistic needs of the application constitute a prime determining factor for effective selection of candidate catalysts (a single "optimal" catalyst may not be able to be determined even for a single given application). For example, effective catalyst selection will be quite different if the goal is (1) to prepare a compound in high yield and high purity in the 10th step of a 23-step total synthesis in the laboratory or (2) to manufacture a large volume commodity chemical by a process in which reactor turnaround, catalyst cost, separation, recycle, toxicity, and environmental fate are key factors. For the former application, a crown ether may be a suitable candidate, using column chromatography for separation after a lengthy reaction time. In the latter application a polyethylene glycol (PEG) or quaternary ammonium salt may be a better catalyst candidate in a continuous process.

In all applications, the catalyst must effectively promote the desired reaction. The next section describes structure–activity relationships of the most common family of phase-transfer catalysts, the quaternary ammonium salts.

1. Structure–Activity Relationships of Quaternary Ammonium Catalysts

   a. Quaternary Ammonium Cation–Anion–Solvent Polarity Relationships

The chemical species that are usually involved in the rate-determining step are (1) the catalyst, (2) the anion, (3) the solvent, and (4) water (see Chapter 3). In addition, the organic substrate and other species are involved in the rate-determining step if the reaction is intrinsic reaction rate limited. Electrostatic interactions between the chemical species govern most of the thermodynamic and kinetic factors that determine the progress and outcome of the PTC reactions. In addition, physical factors such as temperature and agitation play roles in determining reactivity. All of these factors are addressed below.

The course of most organic reactions is heavily dependent on electrostatic interactions. The polarity and polarizability of the chemical species play a major role in their affinity for one another. An oversimplified, but useful, first-pass approach in considering quaternary -onium cation–anion–solvent interactions is the Hard and Soft Acids and Bases, "HSAB", empiricism [1]. Ions or sites within molecules that possess high charge density are termed "hard" and ions or sites within molecules that possess low charge density are termed "soft". For example, iodide is considered a soft anion (low negative charge-to-volume ratio), fluoride is considered a hard anion (high negative charge-to-volume ratio), tetraoctylammonium is considered a soft cation (low positive charge-to-volume ratio), and Li^+ is considered a hard cation (high positive charge-to-volume ratio). It is thought that hard anions prefer to pair with hard cations and soft anions prefer to pair with soft cations. Thus, it may be considered that preferential pairing of quaternary -onium cations and anions may be illustrated by the diagonal (lower left to upper right corner) in Figure 6-1.

If only HSAB is taken into account (more factors are introduced later), fluoride would rather pair with tetramethylammonium relative to tetraoctylammonium and iodide would rather pair with tetraoctylammonium relative to tetramethylammonium. Hard anions would like to get as close to the nitrogen as possible for electrostatic stabilization, whereas softer anions can assume positions of greater distance from the nitrogen and can be stabilized by polarization. The concept of a *soft* quaternary cation is consistent with the concept of the *anion activating* quaternary cation, and the concept of a *hard* quaternary cation is consistent with the concept of the *accessible* quaternary cation. (See Chapter 4, Section B.)

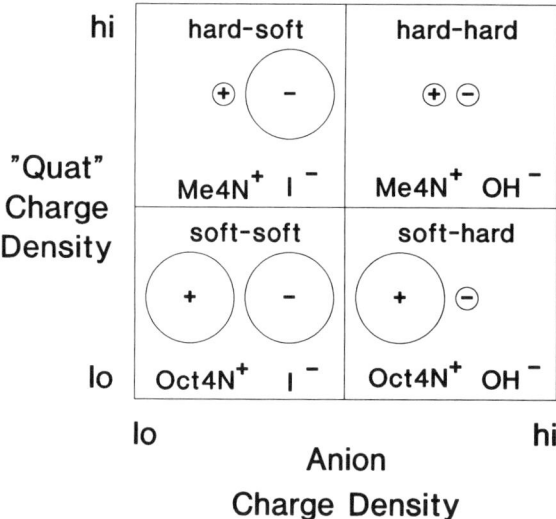

Figure 6-1. Hard and soft acids and bases theory applied to quaternary cation–anion pairs.

In addition to the mutual affinities of the quaternary -onium cation, anion, and water, the polarity of a solvent determines its affinity toward the ion pair. For example, in reactions using tetramethylammonium fluoride under PTC conditions, the highly polar ion pair would probably not be very soluble in typical nonpolar organic solvents. Thus, if the use of tetramethylammonium fluoride was feasible, a very polar solvent would enhance the solubility of this ion pair (solvent effects other than solubility are important in PTC reactions; see Section B below). Indeed, the reaction of *p*-chloronitrobenzene with fluoride in the presence of tetramethylammonium has been patented using dimethyl sulfoxide (DMSO) as the solvent [2]. The solubility of the hard–hard ion pair in the polar solvent is thought to be a key factor, which would otherwise limit reactivity, in this reaction.

The interaction between catalyst and solvent is exemplified by the shift of the optimal catalyst to larger quaternary ammonium cations as the polarity of the solvent decreases from methylene chloride to benzene in the ethylation of deoxybenzoin [3]. Examples are shown below in which "optimal" choices of catalyst and solvent may be opposite in nature for two reactions involving the same anion.

Figure 6-2 illustrates a first-pass guideline for matching quaternary -onium cation and solvent with anion based on polarity, ONLY in cases where a hard anion is used in a transfer rate limited reaction or a soft anion is used in an intrinsic reaction rate limited reaction. Reactions that involve hard anions in intrinsic reaction rate limited reactions or soft anions in transfer rate limited

for hard anions in transfer rate limited reactions and soft anions in intrinsic reaction rate limited reactions

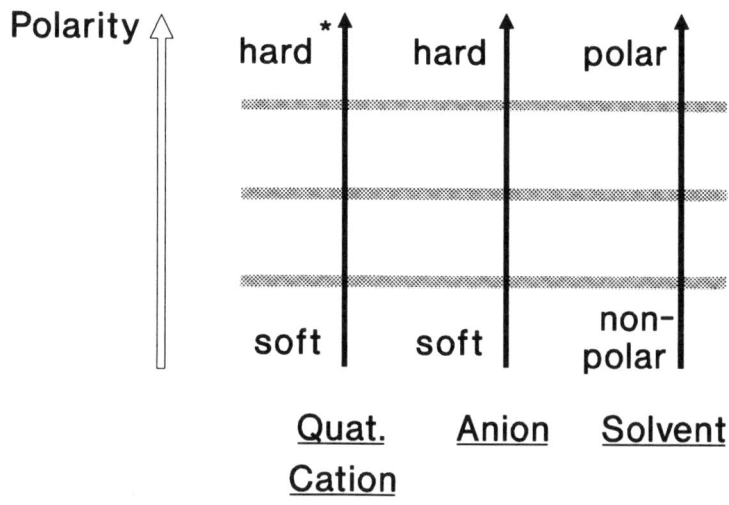

* quaternary cation "hardness" may be related to the accessibility of the charge on the nitrogen

Figure 6-2. Matching quaternary cation–Anion–solvent: first-pass guideline.

reactions require a different set of guidelines and are addressed in Section C of this chapter.

The two factors that are most often preset in typical PTC applications are the anion and the organic substrate, as they determine the desired reaction. Often, these two factors play the major role in determining whether the reaction is transfer rate limited or intrinsic reaction rate limited. An anion that is difficult to transfer (usually a hard anion) but may react rapidly with the substrate once transferred (e.g., hydroxide into benzene, deprotonating an activated methylene compound) will lead to a transfer rate limited reaction. An anion that is readily transferred (usually a soft anion) but needs significant energy of activation to react with the substrate (e.g., benzoate into methylene chloride reacting with methylene chloride) will lead to intrinsic reaction rate control.

Since the anion and substrate are usually given, the catalyst and solvent need to be chosen. The catalyst and solvent are chosen based on their polarity which should be matched to the charge density/distribution of the specific anion and the identity of the rate-determining step. It is important to note that when choosing catalyst and solvent it is necessary but insufficient to consider only one of the factors of anion and rate-determining step. One must consider *both* factors simultaneously. For example, although hard anions are more likely to be difficult to transfer to the organic phase than soft anions, a hard anion may be involved in an intrinsic reaction rate limiting reaction if the organic phase reactions are slower than the transfer step.

When the rate-determining step occurs in the organic phase, the reactivity of the anion will be affected by all species associated with the anion, regardless of whether the anion is soft or hard. A quaternary -onium cation with an accessible nitrogen (quaternary -onium cation accessibility is defined below) will electrostatically bind the anion (or the site on the anion with the highest charge density) and reduce the reactivity of the anion. Likewise, water of hydration will reduce anion activity (see Chapter 2). Reducing hydration may be achieved by saturating the aqueous phase with salt and/or a metal hydroxide, thereby providing competition between the ions in the aqueous and organic phases for hydration. In general, water prefers to remain in the aqueous phase, although some water of hydration almost always accompanies the anion into the organic phase [4–6].

b. Quantitative Parameters for Characterizing Quaternary Ammonium Cation Structure

Number of Carbons, C#

In a classic paper by Landini and Montanari [5], it was suggested that "organophilic" quaternary -onium cations served as more effective phase-transfer catalysts than quaternary -onium cations with small alkyl chains. For example, Figure 6-3 shows a qualitative trend of increasing observed rate of reaction with increasing number of carbons, C#, for the displacement reaction of bromide + octylmethanesulfonate [5]. This reaction was shown to be intrinsic reaction rate limited. As such, as quaternary -onium cation organophilicity increases, $[Q^+Br^-]_{org}$ increases (due to enhanced solubility in the chlorobenzene phase) and/or the chemical rate constant k_{chem} increases due to less electrostatic attraction between quaternary -onium cation and the reacting anion [7].

C# may be considered an effective quantitative parameter for characterizing quaternary -onium cation structure in certain cases as evidenced in Figures 6-4 and 6-5. The figures show that the extraction constants of Q^+Br^- into dichloroethane and $Q^+picrate^-$ into methylene chloride are related to C#. The Q^+Br^- extraction dependence on C# is consistent with the results obtained for the reaction of Br^- with octylmethanesulfonate [5].

Although most of the early hydroxide ion applications were reported using quaternary ammonium cations with small alkyl chains, some hydroxide ion

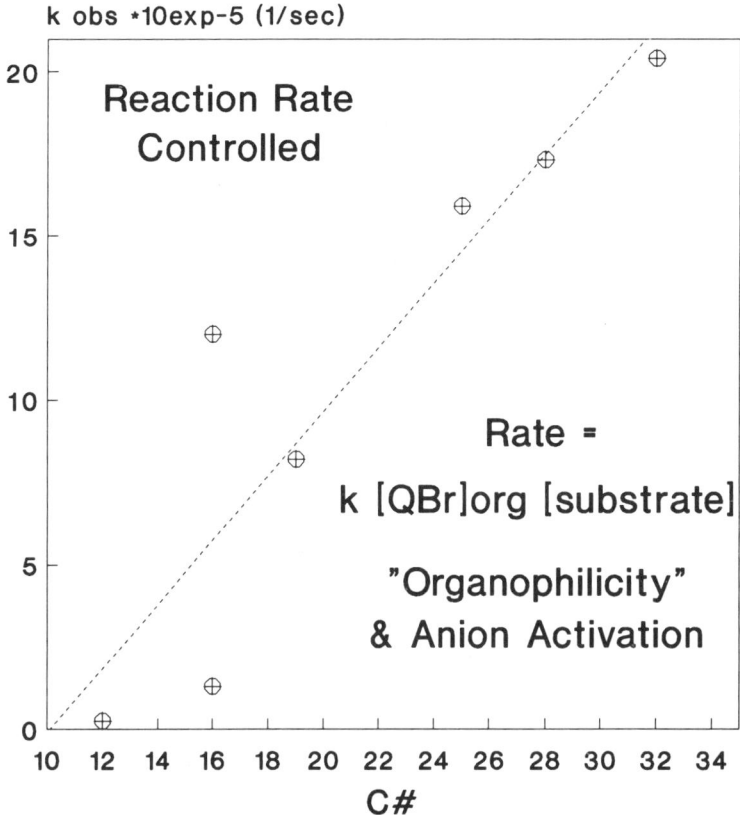

Landini, D., Maia, A., Montanari, F.
J. Amer. Chem. Soc., 1978, *100*, 2796
0.6M sub./0.04M cat./PhCl/4.2M KBr/60C

Figure 6-3. Structure–activity relationship: quaternary cation C# for intrinsic reaction rate limited reactions.

reactions exhibit higher reactivity with more organophilic quaternary ammonium cations. The first kinetic study of such a system involved the isomerization of allylbenzene [8]. Figure 6-6 describes the reaction, which involved the deprotonation of an organic substrate by hydroxide (pK_a difference between water and allylbenzene = 18). Figure 6-7 shows the dependence of the observed rate constants of the isomerization upon the length of the alkyl chains in the symmetrical tetraalkylammonium cations (proportional to C#) and for methyltrioctylammonium. Again, as the alkyl chains increase in length, reactivity increases due to the increased presence of the $[Q^+OH^-]$ in the organic phase and/or the reduced electrostatic attraction between the quaternary ammonium cation and the hydroxide. In the case of methyltrioctylammonium, the reactivity is smaller than in the

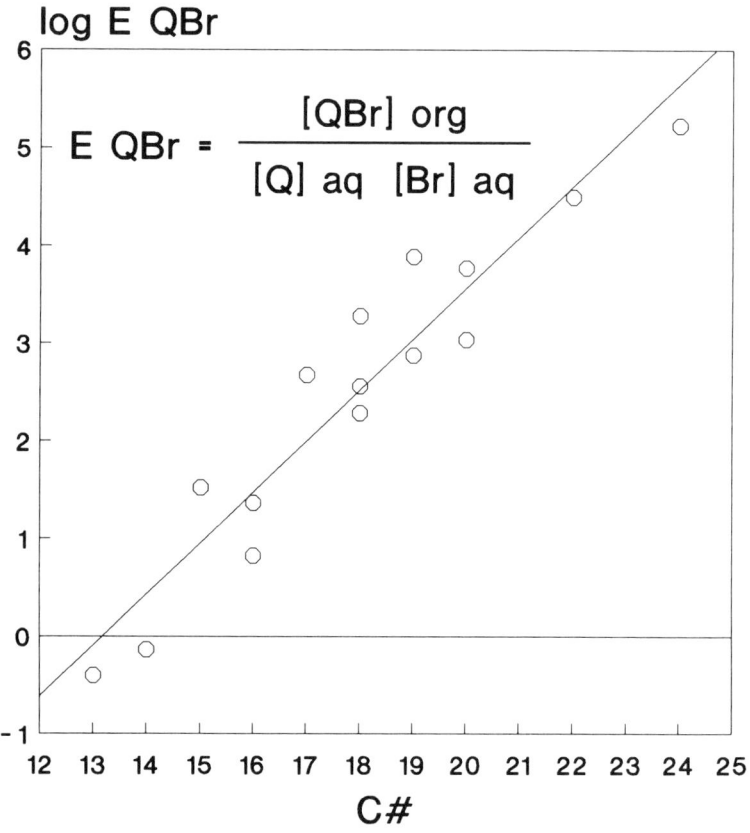

Czapkiewicz, J., quoted in
"Phase-Transfer Catalysis", Dehmlow E.,
Dehmlow, S., Verlag Chemie, 1983, p. 13

Figure 6-4. Extraction of quaternary ammonium bromide into 1,2- dichloroethane.

presence of quaternary ammonium cations with roughly the same C#, such as tetrahexylammonium. This is probably due to the tighter electrostatic attraction of the "accessible" positive charge of this quaternary ammonium cation with the hydroxide anion, thereby decreasing its reactivity. Similar behavior is exhibited in the displacement reaction of thiophenolate with octyl bromide (see Figure 6-8) [9]. In this case, the homologous series of alkyltriethylammonium induces higher reactivity as the alkyl chain increases, but the three short ethyl groups render the quaternary ammonium cation disproportionately "accessible" relative to tetrabutylammonium. In other words, tetrabutylammonium is a much better anion activator than alkyltriethylammonium (see Chapter 3).

In general, reactions that are intrinsic reaction rate limited show increased

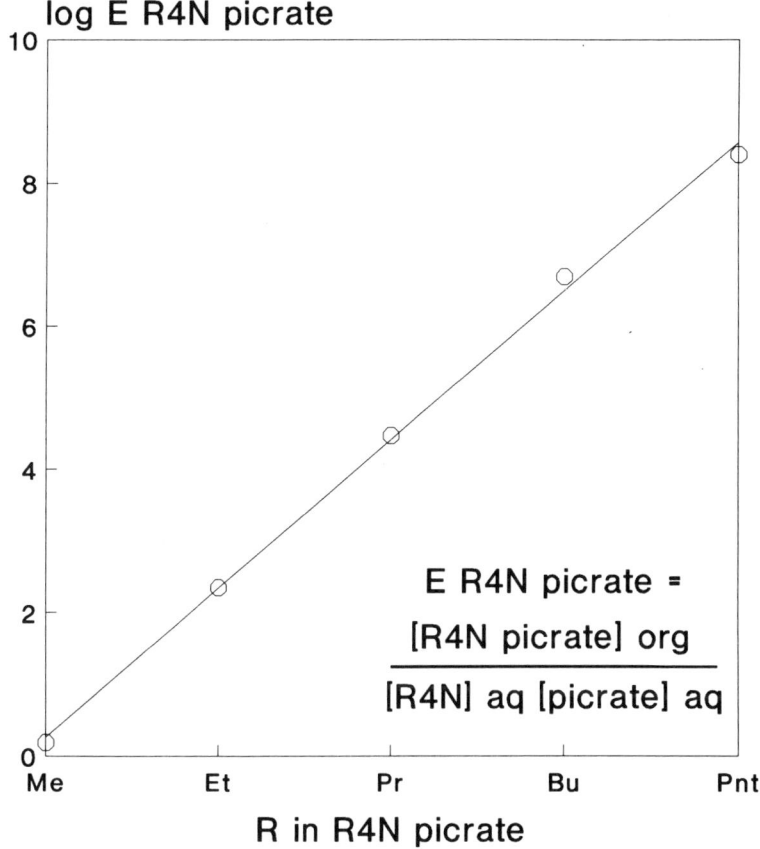

Gustavii, K.,
Acta Pharm. Suec., 1967, *4*, 233

Figure 6-5. Extraction of quaternary ammonium picrate into methylene chloride.

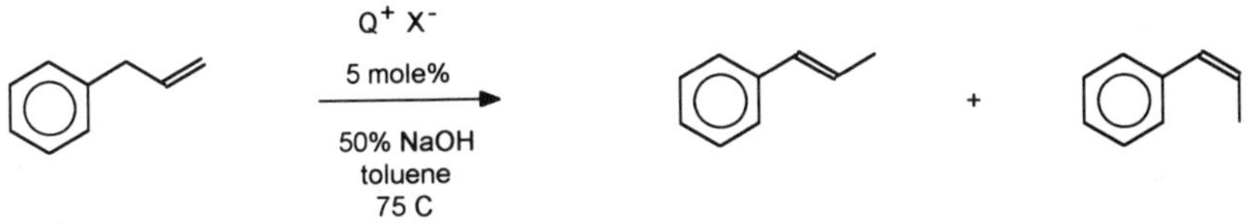

Figure 6-6. Isomerization of allylbenzene.

273

274 / *Phase-Transfer Catalysis*

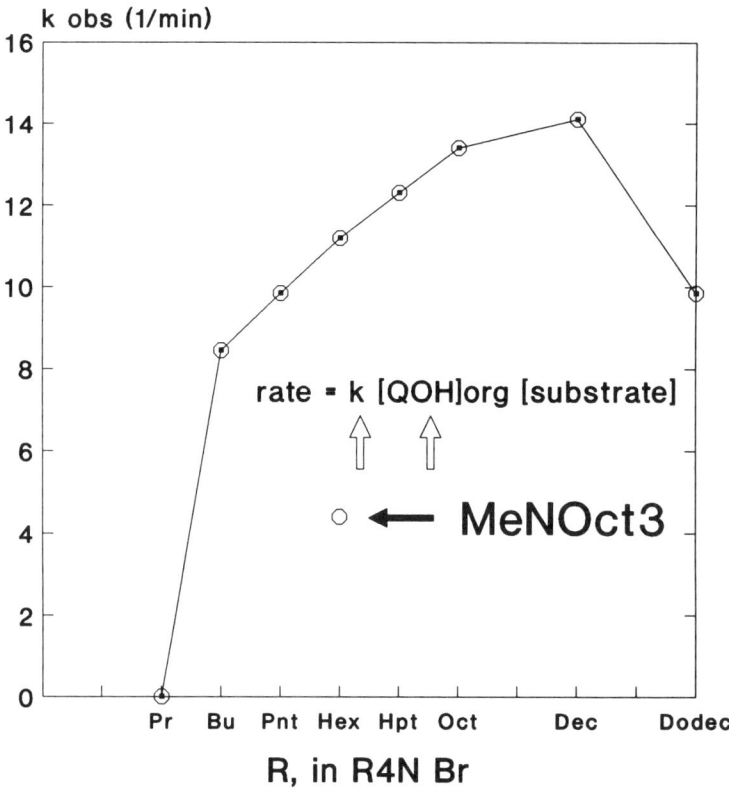

Halpern, M., Sasson, Y., Rabinovitz, M.,
J. Org. Chem., 1983, *48*, 1022
1.42M sub/0.07M cat/PhCH3/50% NaOH/75C

Figure 6-7. Structure–activity relationship: quaternary cation C# for a PTC/OH intrinsic reaction rate limited reaction.

reactivity (not linear) with increasing C# in homologous series (see Figure 6-9). The key factors in intrinsic reaction rate limited reactions are:

Key Factor	*Affects*
C#, Organophilicity	$[Q^+X^-]_{org}$
Low-accessible nitrogen	k_{chem} (anion activation)

when: Rate $= k_{chem} [Q^+X^-]_{org}$ [substrate]

Accessibility

Not all PTC reactions are optimally enhanced by organophilic quaternary ammonium cations. Makosza has published extensive reports using triethylbenzylammonium (TEBA) chloride as the phase-transfer catalyst [10]. In particular,

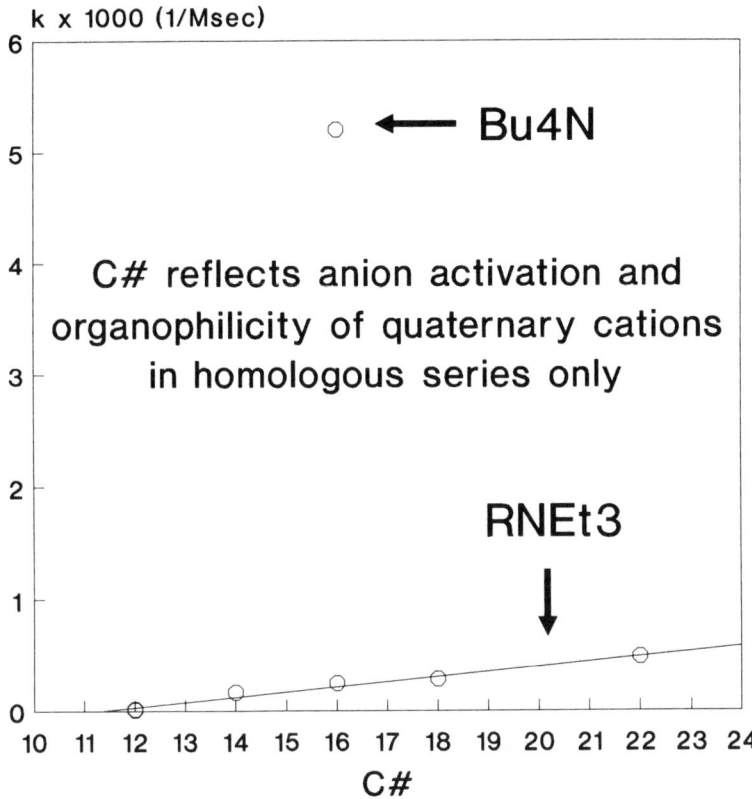

Herriott, A., Picker, D.
J. Amer. Chem. Soc., 1975, 97, 2345

Figure 6-8. Structure–activity relationship: quaternary ammonium cation C# may be representative of structure in homologous series (only) for an intrinsic reaction rate limited reaction.

hydroxide ion initiated reactions were thought to have maximum reactivity with alkyltriethylammonium cations [11]. Halpern suggested that the key structural factor in determining the reactivity in many hydroxide ion initiated reactions is "accessibility" [11]. The nitrogen of a quaternary ammonium cation is considered increasingly accessible as the chains of the quaternary ammonium cation are shorter with particular significance to the shortest alkyl chain. This section shows: (1) how the PTC Rate Matrix can aid in anticipating which reactions are likely to use accessible quaternary ammonium cations and (2) how accessibility can be quantitatively characterized.

In this chapter, when the term "transfer limited" is used, it will mean that the transfer rate plays a major role in the overall kinetics (see Table 3-1).

In the previous section it was shown that intrinsic reaction rate limited reactions

276 / Phase-Transfer Catalysis

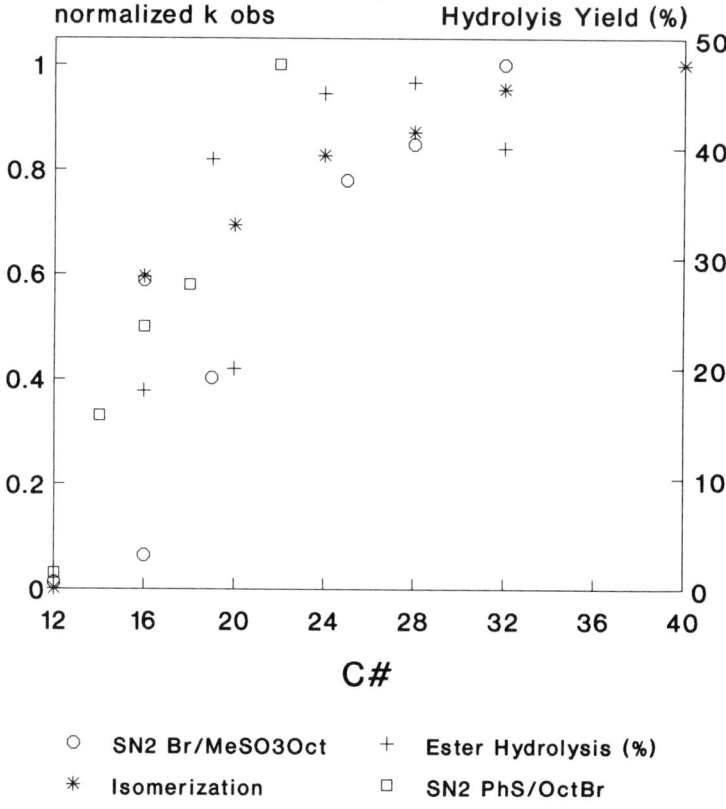

Figure 6-9. Structure–activity relationship: quaternary ammonium cation C# vs. normalized reactivity for four intrinsic reaction rate limited reactions.

have enhanced reactivity in the presence of organophilic quaternary -onium cations. Transfer rate limited reactions usually occur when (1) the anion that needs to be transferred greatly prefers to be in the aqueous phase (e.g., hard anions) or (2) the organic phase reaction is very fast (relative to the transfer rate). Anions such as hydroxide and fluoride have a high charge density and are difficult to transfer to a nonpolar organic phase. As described in Section A.1.a., hard anions (such as hydroxide and fluoride) would prefer to pair with hard cations. As the alkyl chains of the quaternary ammonium cation decrease in size the charge density on the quaternary cation is higher and the positive charge on the quaternary cation can electrostatically interact more closely with the hard anion. Thus, accessible quaternary ammonium cations are effective in pairing

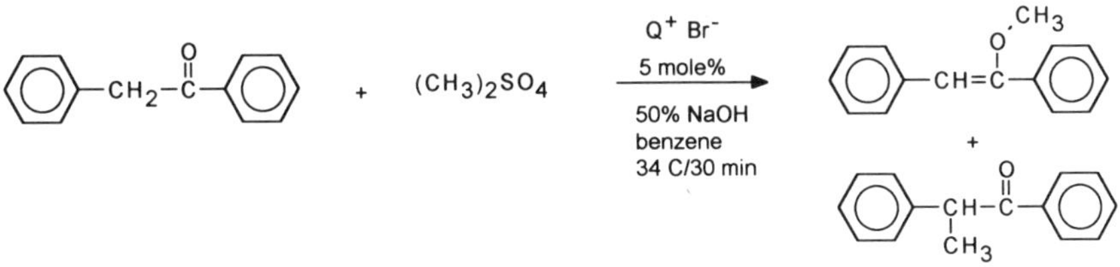

Figure 6-10. Methylation of deoxybenzoin.

with hydroxide and other hard anions. This effect of quaternary ammonium cation structure is important because hydroxide ion initiated reactions account for over half of the PTC applications.

An example of a reaction that is likely to be transfer rate limited is the methylation of deoxybenzoin shown in Figure 6-10. The conversion of this reaction at 30 min is shown in Figure 6-11 for symmetrical quaternary ammonium cations and in Figure 6-12 for nonsymmetrical quaternary ammonium cations with one of the four alkyl groups being different from the other three. With the exception of tetramethylammonium (see below), as the length of the alkyl chains decreases, the reactivity increases, for both the symmetrical and nonsymmetrical quaternary ammonium cation series. For example, Et_4N^+ enhances reactivity more than Oct_4N^+ and within the alkyltrioctylammonium series, reactivity increases as the smallest alkyl chain decreases $MeNOct_3^+ > EtNOct_3^+ > BuNOct_3^+ > NOct_4^+$.

Until recently, accessibility was a qualitative concept. A quantitative parameter for characterizing accessibility was suggested [12] based on the strong dependence of electrostatic interaction on the distance of closest approach between cation and anion (which is determined by steric factors). This parameter, termed q, is simply the sum of the reciprocals of the length of the linear alkyl chains attached to the central nitrogen of the quaternary ammonium cation (see Figure 6-13).

$$q = 1/C\#_1 + 1/C\#_2 + 1/C\#_3 + 1/C\#_4$$

where C# is the number of carbon atoms in each of the four alkyl chains in the quaternary cation.

For example, methyltrioctylammonium would have a q value of $1/1 + 1/8 + 1/8 + 1/8 = 1.38$. This quaternary ammonium cation has one accessible face due to the methyl group. Quaternary ammonium cations with a q value > 1 are considered accessible. Examples of accessible quaternary ammonium cations are methyltrioctylammonium (1.38), triethylbutylammonium (1.25), and methyltributylammonium (1.75).

Figure 6-14 illustrates the utility of the accessibility parameter, q, in correlating reactivity (conversion at 30 min) of the methylation of deoxybenzoin described above. Both symmetrical and nonsymmetrical quaternary ammonium cations are

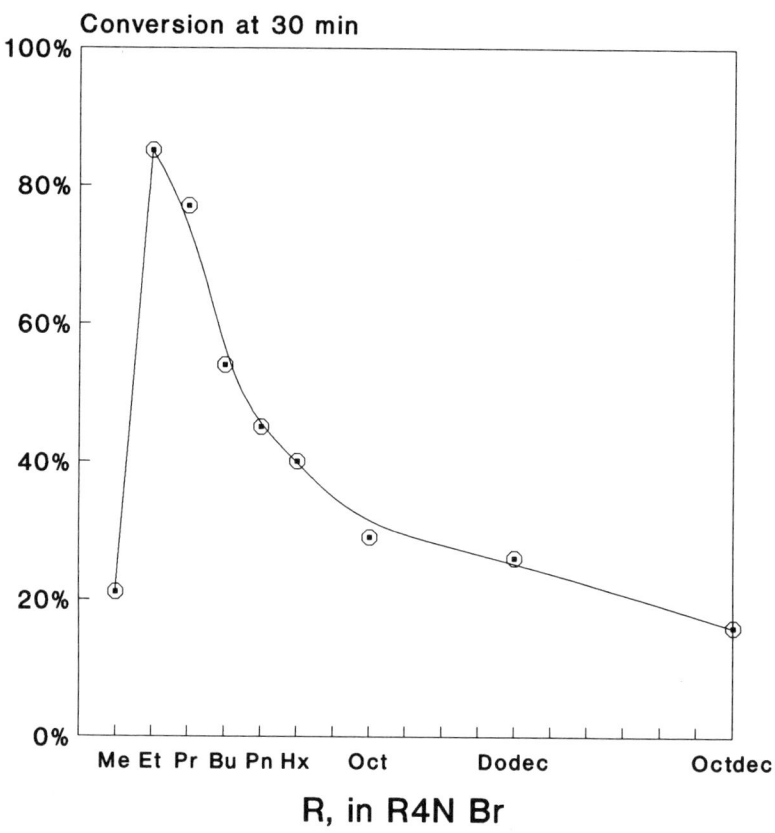

Halpern, M., Sasson, Y., Rabinovitz, M., Tetrahedron, 1983, *38*, 3183

Figure 6-11. Structure–activity relationship: quaternary ammonium cation accessibility for a reaction likely to be transfer rate limited—symmetrical tetralkylammonium cations.

included in the figure (tetramethylammonium has been left out). Figure 6-15 shows the normalized yield (i.e., the ratio of the yield at a fixed reaction time in the presence of a given catalyst vs. the highest yield reported for the best catalyst) of three hydroxide ion initiated PTC reactions as a function of q [13]. It is observed that a rough bell-shaped curve is obtained, with reactivity increasing up to $q = 1.5$–2 (i.e., usually alkyltriethylammonium), then decreasing to $q = 4$ (i.e., tetramethylammonium).

Figures 6-14 and 6-15 suggest that for several common alkylation and carbene reactions, *q may be considered a structural characteristic of the quaternary ammonium cation* adequate in describing much of the behavior of the systems. If accessibility was the only structural factor determining the outcome of transfer rate limited reactions, then tetramethylammonium, or even ammonium (NH_4^+) or

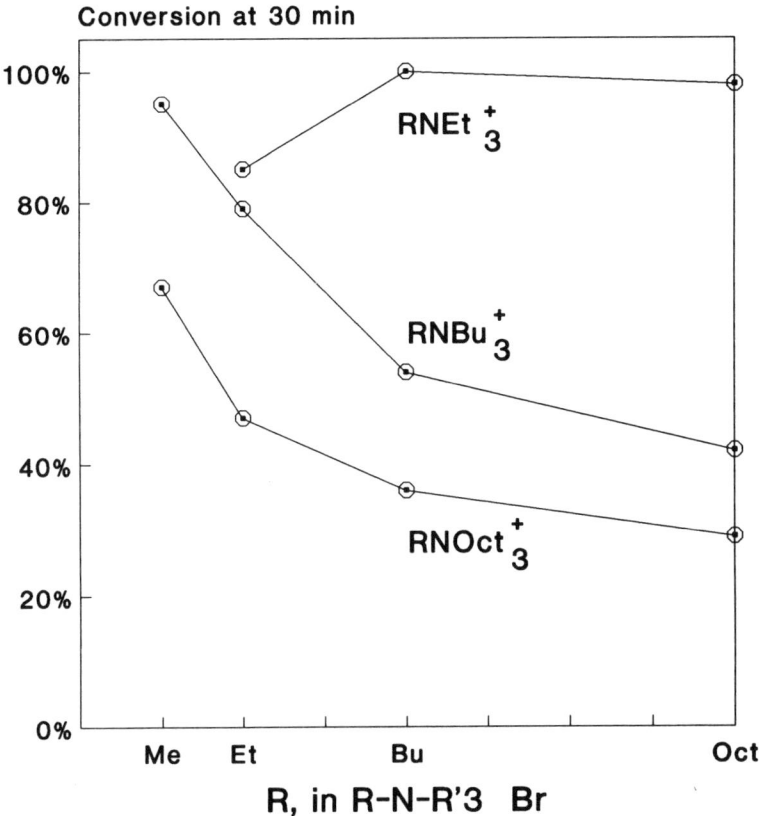

Halpern, M., Sasson, Y., Rabinovitz, M., Tetrahedron, 1983, *38*, 3183

Figure 6-12. Structure–activity relationship: quaternary ammonium cation accessibility for a reaction likely to be transfer rate limited—nonsymmetrical tetralkylammonium cations.

simply a proton would be expected to be the optimal catalyst. In reality, a quaternary ammonium cation needs to have some minimal organophilic nature associated with it, in order to form an ion pair with an anion that would be soluble to some extent in the organic phase. Thus, transfer rate limited reactions will react with enhanced reactivity as the accessibility increases until a point where the cumulative hydrophilicity of the ion pair is too great to allow mediation of the organic phase. The threshold at which the hydrophilicity of the quaternary ammonium cation renders it an ineffective catalyst will depend on the organophilic nature of the anion. For example, the organophilicity threshold will be different for the transfer of hydroxide in comparison to alkoxide.

Figure 6-16 provides further support for q being a reasonable measure of

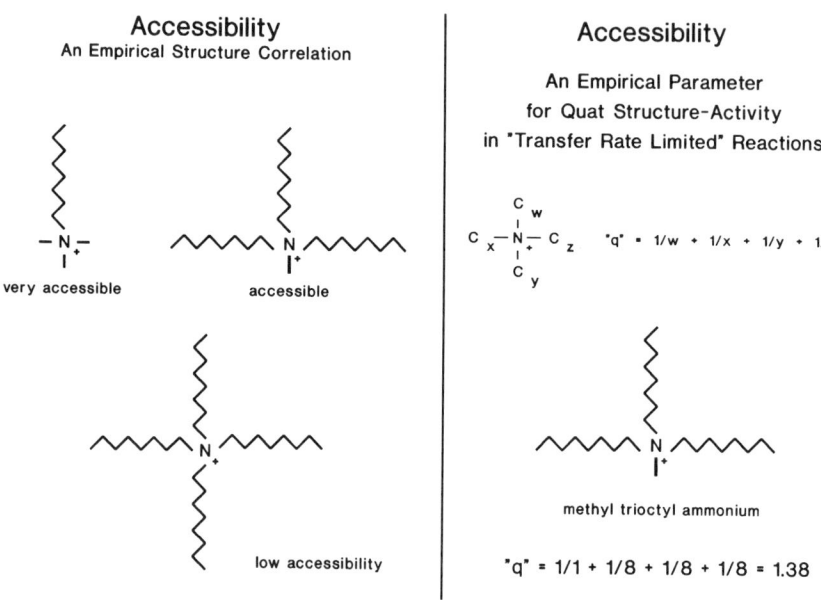

Figure 6-13. "Accessibility": an empirical parameter useful in structure–reactivity relationships for quaternary ammonium cations in transfer rate limited reactions.

quaternary ammonium cation accessibility. Figure 6-16 describes the ionization equilibria of quaternary ammonium cation picrates in ethylene dichloride [14]. The y-axis represents $\log K_i$ and the two x-axes represent q and C#. It is observed that q gives a reasonably good correlation with $\log K_i$ whereas C# does not. This may be an indication that, as the quaternary ammonium cation becomes more accessible, as defined by q, it binds the picrate anion more tightly and dissociates less.

In summary, the key quaternary ammonium cation structural factors for many transfer rate limited reactions are:

Key Factor	*Affects*
Accessibility of nitrogen positive charge	Enhanced ability to pair with "hard" anions and transfer into the organic phase
Organophilicity threshold	Enough to be somewhat soluble (as an ion pair) in a suitable organic solvent.

Many other considerations, other than quaternary -onium cation structure, must be taken into account when optimizing a PTC reaction and include, solvent, pK_a of the organic substrate, nature of the anions/leaving groups, agitation,

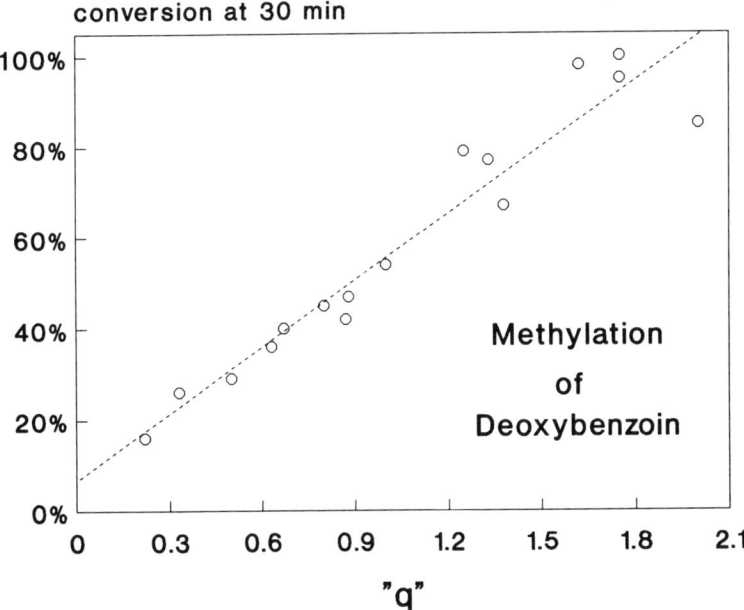

Figure 6-14. "Accessibility", q, and reactivity in the methylation of deoxybenzoin.

hydration, and salt effects (see Chapter 2). All of these factors are interactive with quaternary -onium cation structure and are addressed in later sections of this chapter.

One of the pitfalls in choosing quaternary -onium cations in PTC has been choosing quaternary -onium cation based on historical preference. Many have chosen alkyltriethylammonium for hydroxide ion initiated reactions because Makosza reported much success with such quaternary ammonium cations in hydroxide ion reactions [10]. Others have chosen organophilic quaternary onium cations based on the generic recommendations of Landini et al. [5] and Herriott et al. [9] in the classic papers in J. Am. Chem. Soc. of the 1970s. Two reactions have been described in this section that are both initiated by the hydroxide ion but that exhibit opposite behavior (see Figures 6-7 and 6-11). Lets examine these two reactions in light of the PTC Rate Matrix.

282 / Phase-Transfer Catalysis

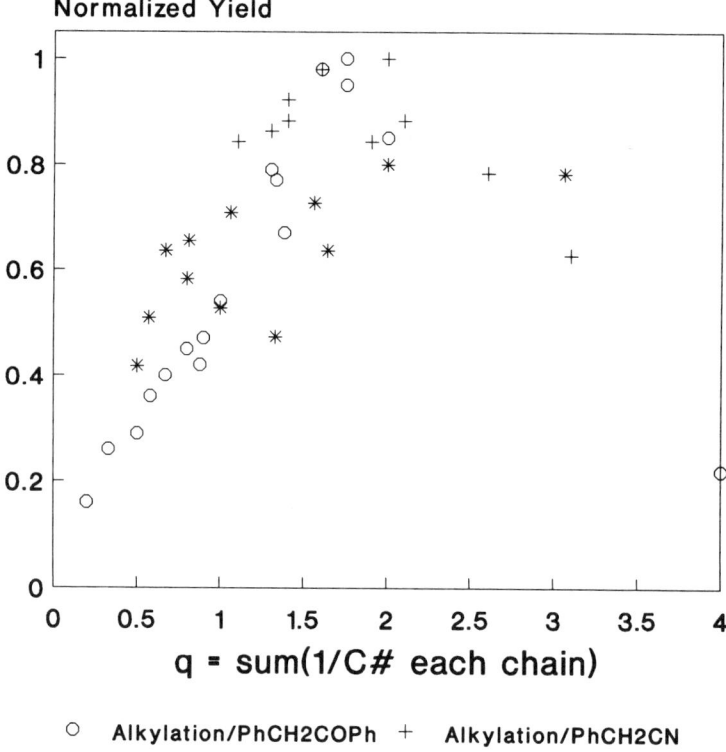

Figure 6-15. Structure–activity relationship: quaternary ammonium cation "accessibility, q" vs. normalized reactivity for three reactions likely to be transfer rate limited.

Two PTC/OH Reactions

Steps
1. Hydroxide transfer
2. Deprotonation
3. Subsequent reactions

Isomerization of allylbenzene		Methylation of deoxybenzoin
34	pK_a	16
toluene	Solvent	benzene
75°C	Temperature	34°C
1-2 h	Time	30 min

The first step of both reactions is the transfer of hydroxide to toluene/benzene followed by deprotonation. In the alkylation reaction, deoxybenzoin has a pK_a

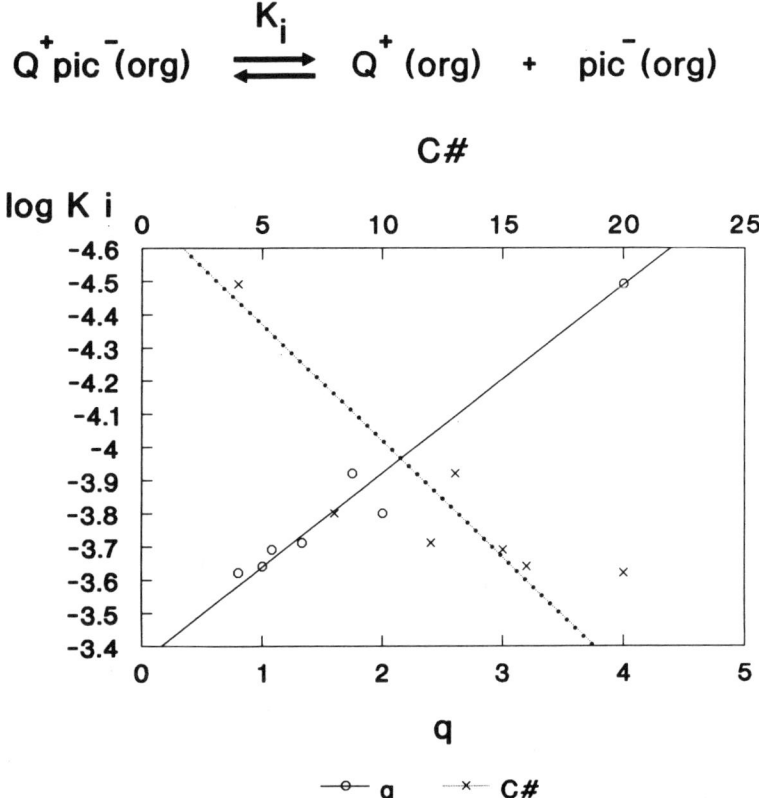

Figure 6-16. Good correlation of "accessibility, q" with quaternary ammonium picrate ionization equilibria.

of 16 and is very readily deprotonated by the hydroxide ion. The subsequent reaction with dimethyl sulfate is likely to be rapid. In the isomerization reaction, allylbenzene has a pK_a of 34 (nearly 18 orders of magnitude less acidic than water/[OH$^-$]) and requires a higher temperature to proceed. The subsequent reaction is conjugation of a double bond (movement of electrons and protonation). From the comparison of the reaction conditions needed to perform the two reactions, it may speculated that the deprotonation of deoxybenzoin is more likely to be faster than the hydroxide transfer rate into benzene. In contrast, the deprotonation of allylbenzene is more likely to be slower than the hydroxide transfer rate into toluene. Thus, the alkylation of deoxybenzoin is probably transfer rate limited and the isomerization of allylbenzene is probably intrinsic reaction rate limited. In the case of the isomerization of allylbenzene, the deprotonation reaction is the step that needs to be enhanced. Therefore, even though the "hard" hydroxide may preferably pair with the smaller quaternary ammonium cations, it is the ability of the larger quaternary ammonium cations to (1) activate the hydroxide by nontight pairing (k_{chem}) and (2) provide more organic phase solubility in toluene ($[Q^+OH^-]_{org}$) which enhances reactivity in this particular rate determining step.

When the alkylating agent in the alkylation of deoxybenzoin is changed from dimethyl sulfate (highly reactive) to ethyl bromide, the optimal catalyst changes as a function of solvent [3]. When methylene chloride is used as the solvent, tetraethylammonium affords the highest reactivity. When benzene is used as the solvent, tetrabutylammonium affords the highest reactivity. The alkylation of deoxybenzoin with dimethyl sulfate in benzene is speculated to be transfer limited (transfer slower than deprotonation and alkylation). Ethylbromide is significantly less reactive alkylating agent than dimethyl sulfate and may change the rate-limiting step (in benzene) to an intrinsic reaction rate limited reaction. This would result in the need for more anion activating (organophilic) quaternary ammonium cations to be more effective. Many other factors contribute to the behavior of these reactions. Magdasi et al. suggest that the surface pressure is related to this behavior, and that the reaction proceeds though an interfacial mechanism. The relative solubilities of the quaternary ammonium cation–anion pairs in the two solvents may play a role as well as anion solvation, aggregation states, and solvent separation of the ion pairs. It is interesting to note that the maximum reaction rates in benzene for tetramethylammonium through tetrabutylammonium were at least twice as large of those in methylene chloride.

c. Choosing Quaternary Ammonium Cation Structure

Figure 6-17 provides a first-pass guideline at choosing quaternary onium cation structure based on the PTC Rate Matrix and the structure–activity relationships shown above. Intrinsic reaction rate limited reactions usually exhibit high reactivity with quaternary -onium cations of moderately high to high anion activation ability: $C\# = 16$–32, such as tetraoctylammonium ($C\# = 32$) > methyltrioctylammonium ($C\# = 25$), tetrabutylammonium ($C\# = 16$). Transfer rate limited reactions usually exhibit high reactivity with moderately accessible quaternary onium cations: $q = 1$–2, such as butyltriethylammonium ($q = 1.75$), methyltributylammonium ($q = 1.75$), methyltrioctylammonium ($q = 1.38$) > tetrabutylammonium ($q = 1.0$). Appropriate solvents must be chosen to provide the proper solubility in the organic phase of the quaternary -onium cation–anion pairs.

Tetrabutylammonium (TBA) is the most cited quaternary ammonium cation in the patent as well as general PTC literature (>500 patents and >2000 references through mid-1991). From a reactivity standpoint, it is usually not the optimal

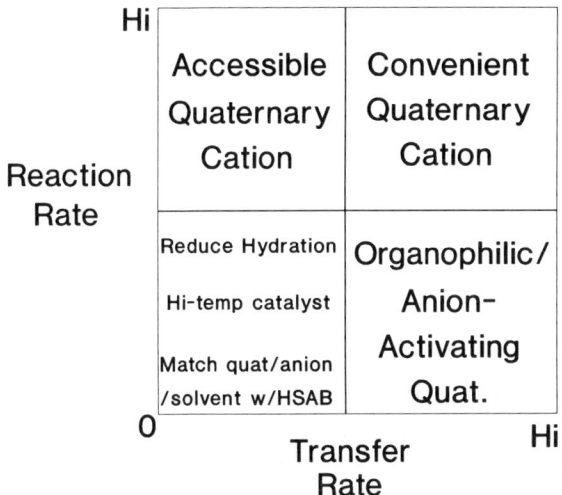

Figure 6-17. Choosing quaternary ammonium cation structure based on the PTC Rate Matrix.

quaternary-onium cation, but it almost always works when other quaternary ammonium cations work. In addition, tetrabutylammonium bromide is commercially available in bulk as a moderately priced fine chemical (see Table 6-2, in which tetrabutylammonium bromide is used as the cost reference) as well as in research quantities in a wide variety of anion forms. As shown in a later section in this chapter, tetrabutylammonium salts can be easily separated and recovered by extraction, then recycled. Therefore, tetrabutylammonium is the quaternary-onium cation of choice in many reactions and processes. Another popular quaternary ammonium cation is methyltrioctylammonium. It is also commercially available (see Table 6-2) and is anion activating/organophilic enough to catalyze most intrinsic rate limited reactions and accessible enough to catalyze most transfer rate limited reactions. Methyltributylammonium is used in commercial applications and will grow as a PTC quaternary ammonium cation due to its high reactivity in transfer rate limited reactions, its low price (see Table 6-2), its lower toxicity than most quaternary ammonium cations [24] (see Table 6-2), its low molecular weight, and its ease of separation (extraction into water).

The choice of quaternary-onium cation for fast PTC reactions (upper right corner of PTC Rate Matrix) should be based on convenience in the laboratory or economic and environmental considerations for commercial processes. For slow reactions it is important to match, as closely as possible, the "hardness/softness" of the quaternary onium cation with that of the anion to be transferred as well as choosing a solvent that will solubilize the quaternary onium cation–anion pair. Hydration of the anion should be minimized by using moist solid salts or saturated aqueous solutions of the anion to be transferred. It may also be necessary to use thermally stable quaternary -onium cations (see Section A.3).

Another catalyst family worthwhile to consider, although no quidelines currently exist for its choice, is the β-hydroxyalkyl quaternary ammonium cation family (see also Chapter 4, Section 2.e.). β-Hydroxyalkyl quaternary ammonium salts often provide selectivity and/or enhanced reactivity, for PTC/OH reactions in particular. The β-hydroxyalkyl group may impart special properties via hydrogen bonding to the substrates or anions. β-Hydroxyalkyl quaternary ammonium salts have been found to be useful in chiral alkylation (see Chapter 12), dehydrohalogenation (see Chapter 8, Section E), carbene addition (see Chapter 8, Section F), reduction (see Chapter 11), aldol and Darzens condensations (see Chapter 8, Section G), and other reactions.

Discussion of the effect of catalyst structure on selectivity of PTC reactions may be found in Chapter 4, Section B.1.f, and in Chapter 8, Section A.1.

The catalyst is the key component of PTC systems. Much of its optimization relies on reactivity. Other factors are important as well and are the focus of the next sections of this chapter. Without minimizing the importance of catalyst reactivity and the other catalyst factors, solvent, agitation, etc., it is important to reemphasize that logistic factors may override other considerations. To illustrate this point, let us examine Figure 6-18. In this figure are compiled normalized reactivity results as a function of C#. The S_N2 reaction represents normalized second-order rate constants for the reaction of bromide + octylmethanesulfonate [5]; the hydrolysis represents yields at a fixed time of the hydrolysis of dimethyladipate [15]; the isomerization represents normalized pseudo-first-order rate constants of the isomerization of allylbenzene [8]; the alkylation represents the normalized conversion at 30 min of the methylation of deoxybenzoin [11]; and the carbene addition represents the normalized yield of the addition of dichlorocarbene to cyclohexene, after a fixed time [16]. A wide variety of quaternary ammonium cations is represented in Figure 6-18 with gross reactivities for most quaternary ammonium cations spanning less than an order of magnitude. A key observation is that most quaternary ammonium cations induce some reactivity in a given reaction. In a commercial application, it may be prudent to choose a quaternary ammonium cation that induces half the reactivity of another quaternary -onium cation, and use twice as much, if it is more readily available, or has lower cost, or more easily separable from product, or less toxic, etc., or all of the above.

2. Structure–Activity Relationships—Other Catalysts

The properties of the PEG derivatives, crown ethers, and other phase-transfer catalysts are extensively covered in Chapter 4, Section D. Chapter 4 describes

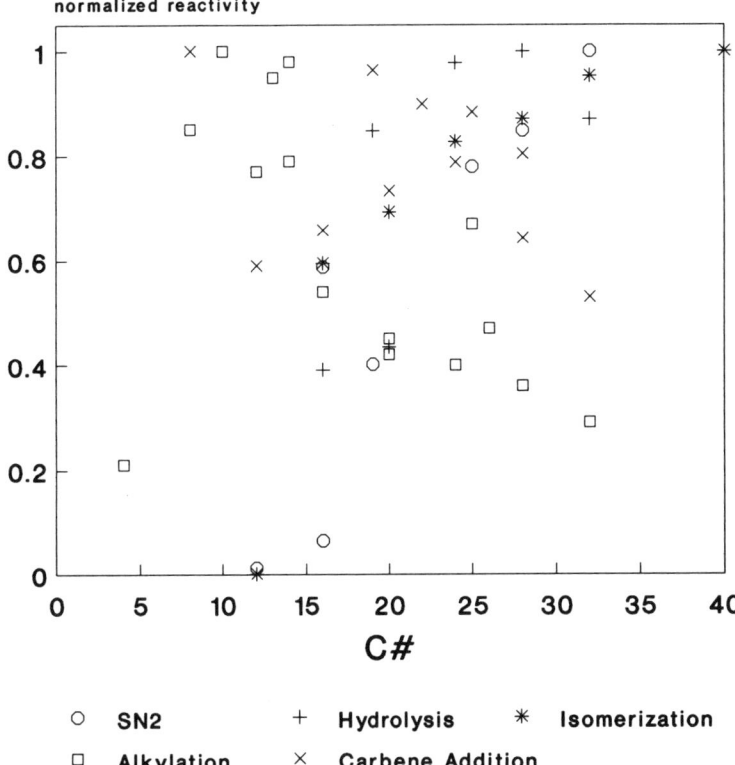

Figure 6-18. No simple single structure–activity relationship exists for quaternary ammonium catalysts for all PTC reactions.

the many variations and conditions to achieve selective complexation of cation/anions with PEG derivatives and crown ethers. Choice of a specific PEG ether or crown ether will usually be based upon factors other than structure–activity relationships. Catalyst separation or cost (see Sections A.4 and A.10 of this chapter) will usually be the determining factor. One should screen a range of these catalysts for a new application.

For industrial processes, PEG derivatives should be included in standard screening programs. In particular, PEGs (e.g., PEG-400) should always be considered for hydroxide ion initiated reactions. The hydroxy terminated PEGs have been shown to be particularly effective catalysts in dehydrohalogenations.

Tris(2,5-dioxaheptyl)amine (TDA-1) should be considered in cases where particularly high temperature, highly basic reactions need to be performed or when metals need to be solubilized in the organic phase (Na metal, Cu salts for the Ullmann reaction, etc.). Due to the hydrophilicity of TDA-1, it is usually used in solid–liquid PTC systems.

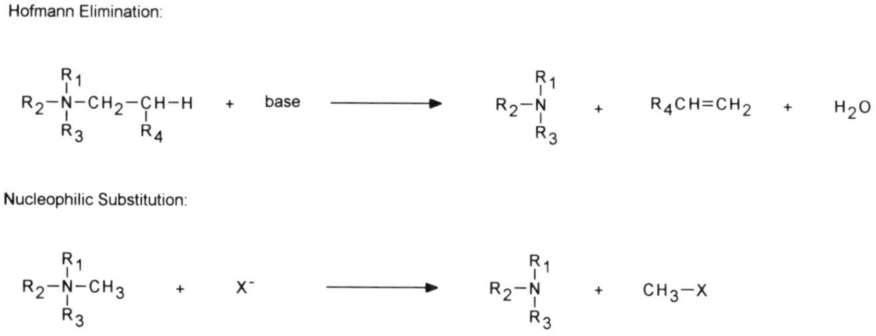

288 / Phase-Transfer Catalysis

Hofmann Elimination:

$$R_2-\underset{R_3}{\underset{|}{N}}-CH_2-\underset{R_4}{\underset{|}{CH}}-H \;+\; \text{base} \;\longrightarrow\; R_2-\underset{R_3}{\underset{|}{N}}^{R_1} \;+\; R_4CH=CH_2 \;+\; H_2O$$

Nucleophilic Substitution:

$$R_2-\underset{R_3}{\underset{|}{N}}-CH_3 \;+\; X^- \;\longrightarrow\; R_2-\underset{R_3}{\underset{|}{N}}^{R_1} \;+\; CH_3-X$$

Figure 6-19. Decomposition of quaternary ammonium cations.

3. Catalyst Stability

Quaternary ammonium salts decompose by two major mechanisms, the Hofmann elimination and nucleophilic displacement (see Figure 6-19 and Chapter 4-B.1.e). In the Hofmann elimination, a base, such as hydroxide, removes a β-proton (by a concerted E2 mechanism) and liberates a trialkylamine, an alkene, and water. This reaction is the key concern when performing base-catalyzed PTC reactions at high temperatures in strong base (e.g., 50% NaOH). Nucleophilic displacement occurs when the attacking anion performs an S_N2 displacement on the quaternary ammonium cation, liberating trialkyl amine as the leaving group while alkylating the nucleophile. For example, quaternary ammonium cations containing methyl groups (such as methyltrioctylammonium) behave as CH_3X, where X is the good leaving group trioctyl amine. In the presence of a good nucleophile, such as thiophenoxide, this path of quaternary ammonium cation decomposition can be highly significant [17].

Reliable stability data for quaternary-onium cations have been reported mainly for quaternary-onium cations in the presence of base, since this is a common reaction matrix, and is likely to be a candidate condition for decomposition. Figures 6-20 and 6-21, and Tables 6-1, 4-5, and 4-6 illustrate the stability of quaternary-onium cations in the presence of base. The most comprehensive study of quaternary-onium cation stability in the presence of base was reported by Landini et al. (Figure 6-20) [18]. In the presence of 50% NaOH longer alkyl chains tend to be more stable (e.g., tetrahexyl ammonium > tetrabutylammonium), benzyl groups tend to reduce stability (tetrahexyl ammonium > benzyltrihexyl ammonium), and ethyl groups are particularly suited for Hofmann elimination. When using quaternary ammonium cations with benzyl groups for C- or O-alkylations, the C-benzyl and benzyl ethers, respectively, are detected [19]. In such cases the benzyl quaternary ammonium cation acts as an alkylating agent (where X in $PhCH_2X$ is a trialkylamine).

The decomposition of tetrabutylammonium hydrogen sulfate in the presence

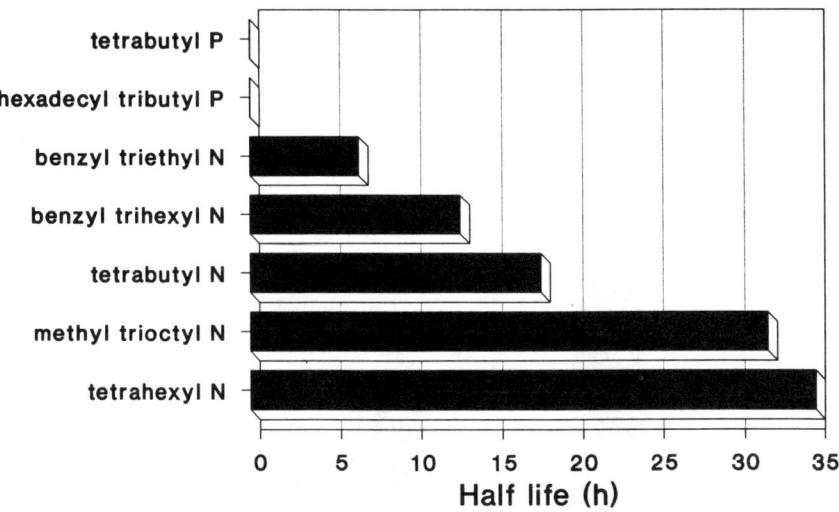

Figure 6-20. Stability of quaternary -onium chlorides in 50% NaOH at 25°C.

of 50% NaOH was found to have an energy of activation of 17.4 kcal/mol [20]. Since many PTC/OH reactions have energies of activation in the range of 10–15 kcal/mol, these reactions can be successfully performed without significant loss of catalyst, if performed at lower temperatures.

In Chapter 4, Section B.1.e we saw that the rate of quaternary cation decomposition in the presence of concentrated NaOH decreases in the presence of bromide relative to chloride [18]. The reason for this is that less hydroxide is associated with the quaternary cation. This means that the rate of the desired reaction would also be decreased, and addition of bromide to retard catalyst decomposition is not an effective technique.

Although the data for the quaternary phosphonium cations show them to be very unstable in the presence of strong base [18] (liberating the phosphine oxide as the byproduct), the quaternary phosphonium cations are more stable than the quaternary ammonium cations in other reaction matrices [21]. As a rule of thumb, quaternary ammonium cations, such as tetrabutylammonium or methyltrioctylammonium, may be used in reactions for up to an hour at 75°C in the presence of 50% NaOH. Under harsher conditions, high recovery of the catalyst (intact) should not be expected.

The particular lability of the ethyl group in the presence of base is illustrated in Figure 6-21 [22]. The high stability of tetramethylammonium hydroxide relative to the larger symmetrical quaternary ammonium cations is due to the ability of the higher quaternary ammonium cations to undergo Hofmann elimination, whereas the tetramethylammonium can decompose only by nucleophilic displace-

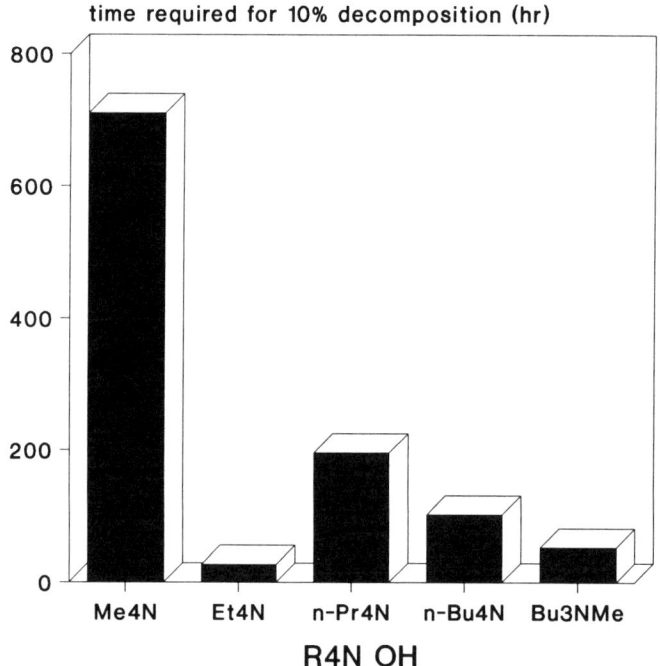

Courtesy of J. Dutcher (Nov 92)
SACHEM

Figure 6-21. Decomposition of quaternary ammonium hydroxides (1 M aq. solution at 120°C). (Courtsey of J. Dutcher, SACHEM.)

Table 6-1. Stability of 10 w/v% Aliquat 336 in xylene and 50% NaOH

	4 h		8 h	
	3° Amine	Quaternary salt	3° Amine	Quaternary salt
25°C	0.28	10.1	0.37	10.0
60°C	1.14	9.4	1.77	8.7
100°C	9.21	1.3	9.28	1.2
Control (0 hr)	0.22	10.2		

Data provided by Henkel Corporation.

ment. Methyltributylammonium is slightly less stable than tetrabutylammonium, presumably due to greater accessibility of the β-hydrogens for Hofmann elimination.

The displacement of a benzyl group from benzylpyridinium chloride by neutral nucleophiles was studied [23]. The activity of the neutral nucleophiles toward benzylpyridinium chloride (in the absence of solvent at 150°C/1.3 h) was surprising: $n\text{-Bu}_3\text{N} > n\text{-Bu}_2\text{NH} > n\text{-BuNH}_2 > \text{HOAc} > \text{RSH} \gg \text{NH}_3$.

Sometimes it is desired to work at higher temperatures than the conventional quaternary ammonium cations can tolerate. Several other phase-transfer catalysts are available that are effective at higher temperatures of 120–200°C (see Figure 6-22). They are the crown ethers and their open-chain analogues, PEGs (and

More Likely Than Classical Quats

to Function at 120 - 200°C

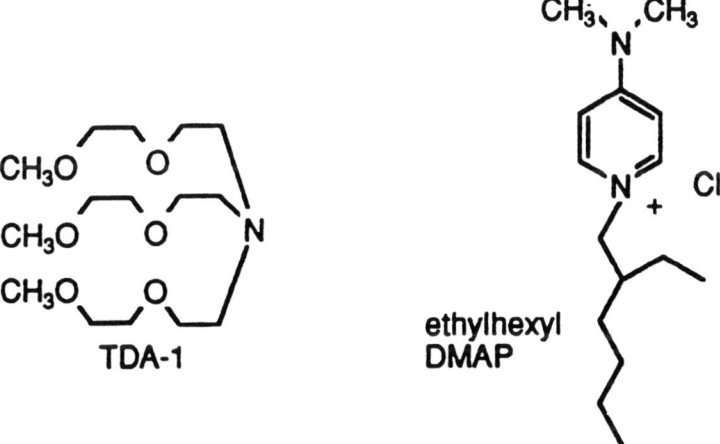

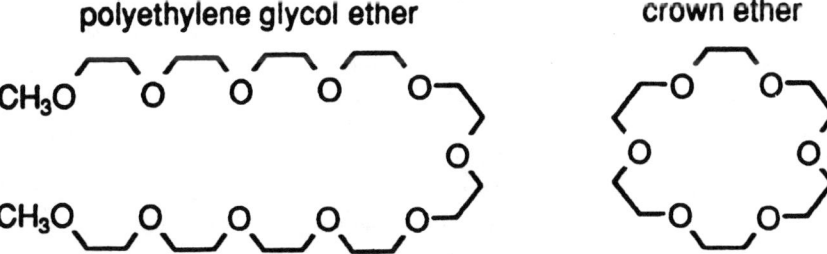

Figure 6-22. "High temperature" phase-transfer catalysts.

(see Table 6-2). With an LD_{50} [24] of 4500 mg/kg, and the potential for recovery and recycle, TDA-1 may be suitable for commercial applications [25]. TDA-1 is also stable in the presence of strong base and high temperature. The PEGs and their capped ethers (see Chapter 4 for an in-depth discussion) are more stable than the quaternary -onium salts, are the lowest priced catalysts (see Table 6-2), and are the least toxic (LD_{50} [24] > 30,000 mg/kg). The PEGs are attractive for processes in which the use of an excess of the PEGs is not a problem (e.g., may be separated by extraction or product distillation) [26]. 18-Crown-6 and its derivatives have recently become available in ton quantities (by two U.S. suppliers) in various grades of purity. Owing to the stability of crown ethers, especially in the presence of base at 150–250°C, their reduced price and their high reactivity at low mol%, the crown ethers should be seriously considered for commercial PTC applications. EtHexDMAP is a catalyst invented by Brunelle at General Electric for use in polymer and monomer synthesis [27,28]. This catalyst is effective for the attack of phenolates on activated haloaromatics (e.g., the reaction of bisphenol A with bischlorophenylsulfone) and for fluorination of halonitroaromatics. Unfortunately, EtHexDMAP is readily attacked by hydroxide and sulfur nucleophiles and decomposes. General Electric owns composition of matter patents on aminopyridine-based quaternary ammonium cations. Further discussion of aminopyridine catalysts may be found in Chapter 4, Section B.2.a.

4. Catalyst Separation and Recycle

There are four major methods for the separation of phase-transfer catalysts from the reaction matrix: extraction, distillation, adsorption, and binding to an insoluble support [29].

a. Extraction

Extraction is probably the most common form of catalyst separation used in commercial processes. The separation usually involves the extraction of the phase-transfer catalyst into water. An example of a typical separation by extraction is shown in Figure 6-23. Tetrabutylammonium bromide, for example, may form a 27% solution in 1% NaOH but is soluble only 0.07% in 15% NaOH [30]. Thus, tetrabutylammonium bromide (and other salts) may be effectively separated from a reaction mixture, then isolated as a wet oil (of relatively small volume for handling). The wet tetrabutylammonium oil can then be recycled or sent for incineration (if the wet oil contains contaminants such as small quantities of potential "catalyst poisons" originating from the leaving group or other byproducts). In principle, any phase-transfer catalyst that is soluble in water may be separated and recycled via extraction. Methyltributylammonium is an example of a catalyst that is often amenable to the above procedure (though it is more water-soluble than tetrabutylammonium). Catalysts such as benzyltriethylammonium and benzyltrimethylammonium can be easily separated by extraction, but the significant extent to which these catalysts sometimes decompose negatively affects the economics of recycle. Examples illustrating the recovery of catalyst by extraction have been patented for methyltributylammonium chloride [31] and triethylphenylammonium chloride [32]. Ethylhexyldimethylaminopyridine may also be recovered by extraction. Brunelle [33] has reported that ethylhexyldimethylaminopyridine can be extracted approximately 20:1 into water from toluene and 6:1 into methylene chloride from water. TDA-1 is usually used under nearly anhydrous conditions and may be extracted into water using an acid such as aqueous HCl and then recovered (by basification) for recycle. Crowns and PEGs can often be separated form the reaction matrix by extraction into water.

Another form of catalyst separation by extraction is used after polymerization (see Figure 6-24). In such cases, the solvent is washed away from the polymer. This may be achieved by using the reaction solvent, for cases in which the polymer precipitates during the reaction, or by a secondary solvent, for cases in which the polymer is soluble in the reaction solvent.

When using high salt concentrations and nonpolar organic solvents, the catalyst may form a third layer during the PTC reaction. This has been observed for PEGs, quaternary -onium salts, and crowns. In such cases, catalyst recovery involves a simple phase separation. In industrial kettles (with limited visibility in the reactor) or when a "rag" layer distorts an otherwise sharp phase boundary, a phase separation operation may not be simple. Nevertheless, choosing conditions in which the catalyst separates as a third layer is usually advantageous. As mentioned in Chapter 4, a dehydrohalogenation was performed four times consecutively, with no loss in catalytic activity by simply replacing the organic phase and replenishing the aqueous base phase after each use, leaving the third phase containing PEG in the reactor.

Waste Treatment

Material Safety Data Sheets for phase-transfer catalysts recommend incineration as the preferred method of disposal. In some cases, biooxidation ponds are used for waste treatment. Quaternary -onium salts sometimes affect the effectiveness of the biological organisms in their ability to consume other organics in aqueous waste streams. Quaternary ammonium cations containing benzyl groups will often affect the microorganisms more than alkyl quaternary ammonium cations. Therefore, some engineers prefer to avoid the use of benzyl quaternary ammonium cations or minimize the concentration of quaternary ammonium cations in the waste streams going to bioponds. In some cases, the large volume of a PTC process can justify using dedicated bioponds for treatment of the quaternary -onium cation. Other oxidative processes may be used to decompose the quaternary -onium cations, such as chlorination. Chlorination may not be complete and chlorination produces volatile chlorinated hydrocarbons, which themselves must be contained, treated, and monitored. This discussion of waste

Composition at the end of a typical alkylation

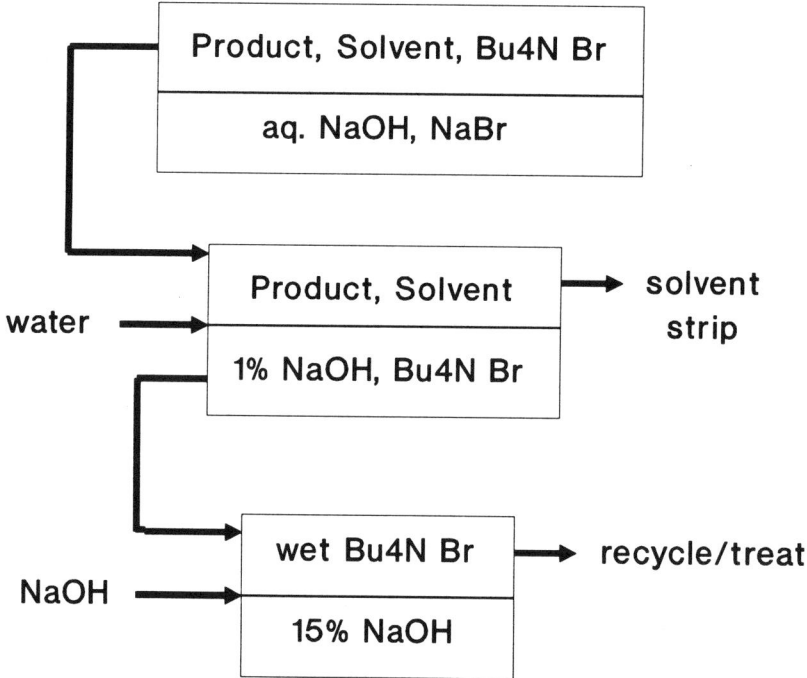

Figure 6-23. Catalyst separation by extraction.

treatment is intended to suggest methods for consideration but does not imply that these methods are suitable. All waste treatment methods should be fully and comprehensively evaluated before testing and implementation and must comply fully with local, state, federal, and other regulations.

b. Distillation

Some quaternary -onium salts and other phase-transfer catalysts are not water-soluble, such as methyltrioctylammonium salts. One method of separating these quaternary -onium salts from the product is by distillation of product overhead,

Composition at the end of a typical polymerization

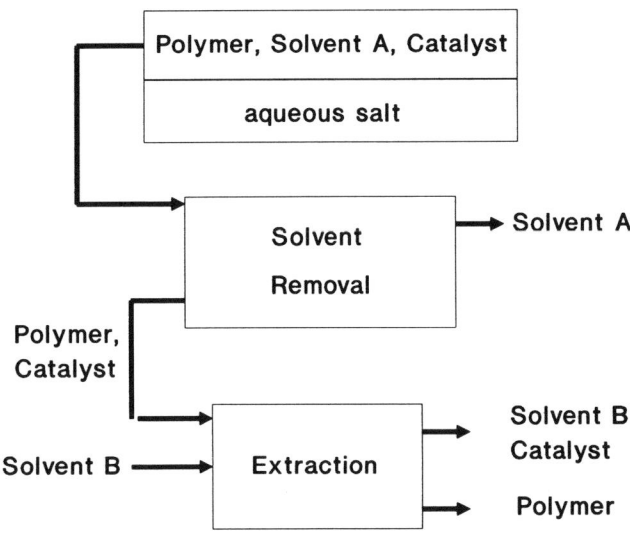

When to Use: To wash catalyst out of polymer

Figure 6-24. Extraction of catalyst from polymer.

with the catalyst remaining in the residue. For example (see Figure 6-25), at the end of a typical cyanation, the organic phase of the reaction product mixture may be distilled. As usual, a solvent must be chosen that will easily be condensed apart from the product. In the case of a quaternary ammonium cation such as methyltrioctylammonium, volatile quaternary ammonium cation decomposition products may be formed during the distillation (e.g., trialkylamines or alkenes). It should be noted that the structure of the quaternary ammonium cation may be modified (attenuated) such that the volatile quaternary ammonium cation decomposition products will not interfere with the clean recovery of product or solvent. Capped PEGs (PEG ethers) will often be separated by distillation. Obviously, catalyst separation by distillation is applicable only if the product is distillable.

c. Adsorption

When the catalyst is not water soluble and the product is not distillable, the catalyst may have to be separated by adsorption (see Figure 6-26). Quaternary -onium salts are polar compounds and salts of methyltrioctylammonium, for example, will adsorb onto silica or ion-exchange resins. The catalyst may be

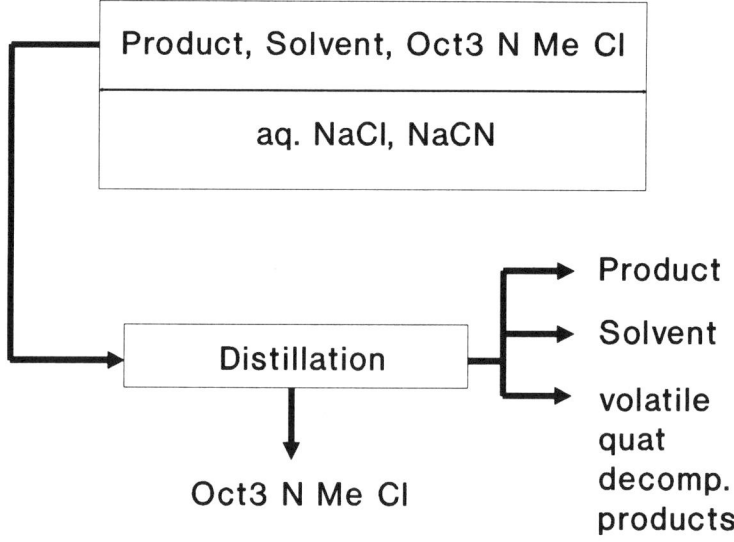

Figure 6-25. Catalyst separation by distillation of product.

recovered by elution using a well chosen polar solvent. Catalysts may also be amenable to adsoprtion on carbon. The authors are not aware of commercial use of adsorption for the separation of catalyst, but this option is described in principle [34].

d. Insoluble Catalysts (See Chapter 5 for review of insoluble catalysts.)

Principles, Advantages, and Disadvantages of Insoluble Catalysts

A schematic of the interactions required for successful phase-bound PTC is shown in Figure 6-27. It can be seen that several additional factors must be

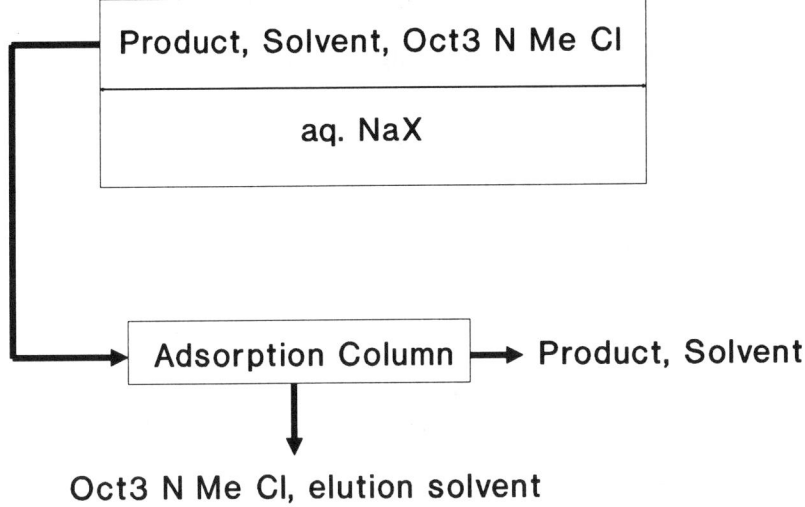

When to Use: Catalyst not water soluble and product is not distillable

Figure 6-26. Catalyst separation by adsorption.

balanced for successful PTC using insoluble catalysts relative to conventional soluble PTC. The key factor affecting reactivity in a phase-bound PTC system is the mutual affinity between the various components in the various phases. The affinities are affected primarily by polar and steric effects. Polar factors include the charge densities of the catalyst and anion as well as the charge distribution in the organic substrate. The polarity of the solvent and the dielectric constant of the aqueous phase are very important. The solvent and water mediate the anion exchange and determine the ability of the reactants to approach each other and interact with the supporting phase. The polarity of the aqueous phase is in turn affected by salt and other concentration effects. Structural and steric effects related to the catalyst support include degree and nature of crosslinking, particle size and porosity, the separation between the catalyst and the support (e.g., by alkyl chain spacers), swelling characteristics, and substrate structure. In addition, the method and efficiency of agitation also affect the reactivity and the mechanical stability of the system.

In light of the above discussion, it is easy to understand that achieving the proper balance for successful PTC using insoluble catalysts is not a trivial endeavor. But

FACTORS AFFECTING REACTIVITY:

Mutual affinity of components/polarity of three phases

multifactor charge densities
selective solvation
salt effects
concentration

Structure/steric effects

crosslinking
particle size, porosity
separation between catalyst and backbone
swelling
substrate structure

Agitation

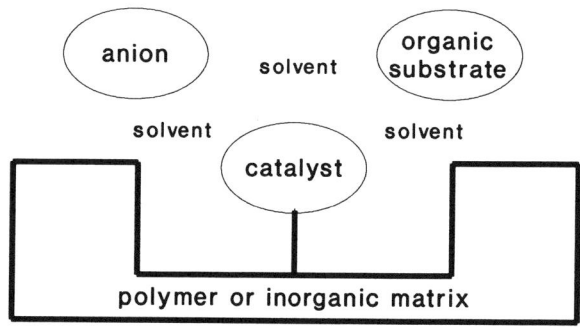

Figure 6-27. Factors affecting reactivity in insoluble PTC.

it is not impossible, and when successful, offers unique advantages. As seen in the previous section, catalyst separation can pose a serious challenge in an operating manufacturing process. Catalyst recycle is also a desired attribute, bearing positive economic impact on the process. PTC using insoluble catalysts offers the opportunity to easily separate the catalyst from the product and recycle the catalyst. Assuming that the leaching of catalyst from the support is not a problem, PTC using insoluble catalysts may be advantageous for preparing high purity products, such as for pharmaceuticals and other applications. Most, though not all, conventional PTC operates in batch mode. PTC using insoluble catalysts appears to offer particular suitability for continuous PTC.

In order to reap the benefits of PTC using insoluble catalysts, the disadvantages and limitations of the method must be overcome. The cost of insoluble phase-transfer catalysts are higher than that of the classical soluble phase-transfer catalysts. The insoluble catalysts are less active than the soluble analogues due to the steric and polar requirements discussed above. The capacities of the insoluble resin catalysts are usually low, and therefore much resin needs to be used. Finally, insoluble resin catalysts are susceptible to stability problems, mostly thermal degradation of the polymer backbone as well as the bound catalyst, and also mechanical attrition to some extent.

Guidelines for the Design and Choice of Insoluble Catalysts

Insoluble phase-transfer catalysts are described in Chapter 5. The following discussion will focus on the limited guidelines currently known for choosing a polymer-bound PT catalyst.

The most studied polymer-bound phase-transfer catalysts are covalently bound to chloromethylated polystyrene and to silica. These polymer-bound catalysts are most similar to commercially available materials which are chloromethylated polystyrene quaternized with trimethyl, triethyl, or tributylamine. These materials, meeting various specifications, are available from the suppliers of ion-exchange resins (e.g., Dow, Rohm and Haas, Sybron). Most of the research in this area has focused on systems in which the insoluble catalyst is stirred as a third-phase catalyst in the reaction vessel [35–47]. Ideally, a continuous reactor configuration would be highly desirable.

An example of the use of a catalyst covalently bound to a polymer in a fixed-bed reactor is illustrated in Figure 6-28 [48]. This study, described in Chapter 5, investigated a useful reaction, the alkylation of phenylacetonitrile. In this particular case, the use of a very highly crosslinked polymer (12%) gave low reactivity relative to soluble catalysts (for the quaternary ammonium salt), though reactivity was enhanced by the presence of a supplementary bound PEG-type catalyst. It is anticipated that the fixed-bed reactor will be the configuration of choice in many commercial applications in the future (especially those performed near room temperature). This prediction is contingent upon gaining a better understanding of balancing the interactions responsible for reactivity in polymer-bound systems and ongoing improvements in the stability of resin supports. The continuous nature of the fixed-bed system also partially compensates for the low capacity these resins usually possess. Even though a large bed must be used to achieve the desired conversion, the unit operations for separation of the resins are reduced to maintenance (occasional backwash skimming of the fine particles due to mechanical attrition and occasional topping off of the bed with fresh resin).

As noted above, most of the research published described batch reactions in which the catalyst was stirred with the two phases. Such systems have been very well characterized. Despite the extensive characterization, the guidelines for

Fixed-Bed Reactors

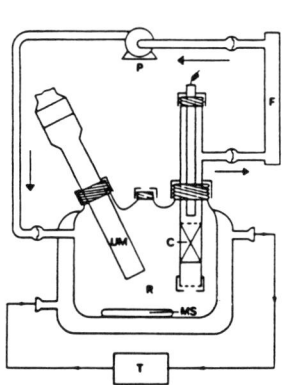
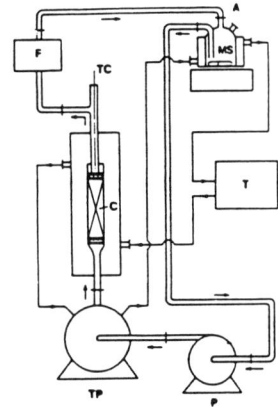

Figure 1. Reaction apparatus with fixed-bed reactor and ultrasonic mixer (FB-UM).

Figure 2. Reaction apparatus with fixed-bed reactor and in-line turbine stirrer acting partially as a pump (FB-TP).

Reproduced with the permission of the American Chemical Society

Ragaini et al, J. Org. Chem., 27, 1382 (1988)

Figure 6-28. Fixed bed reactor with polymer-bound phase-transfer catalyst.

choice of optimal polymer-bound catalyst are individual for each type of reaction and for each type of support and pendant catalyst studied. Nevertheless, some generalizations may be made for choosing certain insoluble catalyst parameters.

Since most of the reactions studied appear to be diffusion controlled (migration of the reacting species into the resin beads), low crosslinking density is recommended, typically 1–3% [45,46,49–52] (although the same reaction can be intrinsic reaction rate controlled if a different pendant quaternary ammonium cation is used [51] or if the chemical reaction is extremely slow [45]). This allows the resin bead to "breathe", i.e., expand and contract in response to osmotic pressures. In addition, as the crosslink density decreases the hydrophobicity decreases, thereby making it easier for inorganic materials to diffuse into the bead but more difficult for organic materials to diffuse into the bead. One drawback of low crosslinking (not emphasized in the literature) is the relatively low resistance to oxidation of the divinylbenzene crosslinks. Decomposition of these crosslinks is a concern in commercial ion-exchange resins used for purification of water with high chlorine levels. Decomposition of low-crosslink resins by oxidation results in the formation of "mush". Additional considerations: Low crosslink resins that are not subject to oxidation are quite resilient to osmotic shock. But since low crosslinked resins swell to several times their size during the reaction cycle, large-scale use of these resins in fixed beds could result in stress on the column if not sized properly.

The solvents used in most of the studies were either toluene or one of the reactants (e.g., bromooctane). Since diffusional factors are key in these reactions, choosing the ratio between the organic and aqueous phases is important to optimize reactivity [49]. If diffusion coefficients are known for the reaction, the optimum phase ratio may be calculated. If not, trial and error or experimental design may be used to optimize the phase ratio.

Most commercial ion-exchange resins strive for maximum loading of quaternary sites on the polymer backbone. In contrast, most polymer-bound PTC applications exhibit maximum reactivity at much lower ring substitution levels, typically <50% of the theoretical complete substitution [35,43,49] (optimum loading is dependent on anion softness with crown ethers) [53]. The reactions reported in the four previous references were simple nucleophilic substitutions. Ford [45] investigated the triphase alkylation of phenylacetonitrile and found that the higher ring substitutions (up to 75%) afforded higher reactivities. A major difference between this alkylation system and the other substitutions described is the need to transfer the hydroxide ion. As shown in Section A.1.b, hydroxide ion reactions often exhibit anomalous behavior due to its high hydrophilicity. Therefore, when considering the use of polymer-bound PTC for hydroxide reactions, one should also screen high loading catalysts. One should also be aware of the increased potential for quaternary ammonium cation decomposition in the presence of hydroxide (especially with ethyl or benzyl containing quaternary ammonium cations).

Since diffusion is a key factor, agitation is often an important variable. Effective agitation will enhance reactivity primarily if it depends on diffusion at the outer surface of the bead (where much of the reaction may take place according to the "alternating shell model" [49]). Special agitation techniques, such as ultrasound and vibromixing, do not seem to be advantageous in the alkylation of phenylacetonitrile [54].

To further overcome diffusion limitations and to balance polar effects, spacers should sometimes be considered for optimizing the catalyst [53,55] (see Chapter 5).

Smaller particle size may also be considered to enhance the physical availability of the exchange/reaction sites [45]. A drawback to smaller particle size is the difficulty in handling and filtration, especially on larger scales.

Choice of the pendant catalyst is specific to each reaction. Phosphonium groups offered the highest reactivity for the simple nucleophilic displacements using inorganic nucleophiles (Cl-, CN-, acetate) with alkyl bromide or methanesulfonate [35,56]. In these reactions the phosphonium groups performed better than the ammonium or PEG groups. The reusability of the phosphonium catalysts appeared to be quite high in these cases. As stated in Section A.3, the phosphonium groups should not be used in the presence of hydroxide.

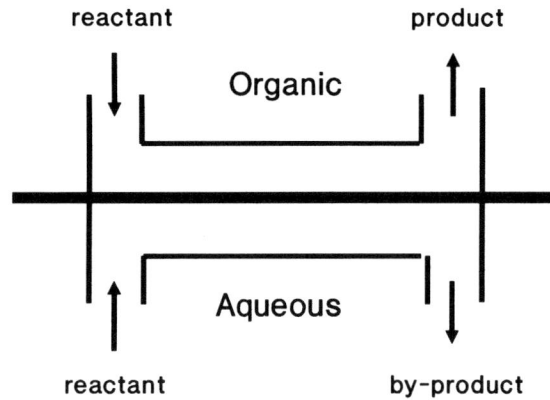

Figure 6-29. PTC membrane reactor.

Again, a comprehensive discussion of insoluble catalysts, used under a wide variety of conditions and reactions, may be found in Chapter 5.

e. The Use of Membranes in PTC

Stanley and Quinn reported the use of a membrane reactor for performing PTC reactions [57]. In the membrane reactor described, the organic and aqueous phases were passed, cocurrently (same direction), over opposite sides of the membrane (see Figure 6-29). Theoretical models and calculations were performed to predict the kinetic behavior of the system. Experimental results confirmed the model developed for the reaction of bromooctane in chlorobenzene with aqueous iodide in the presence of tetrabutylammonium bromide (introduced into the organic phase). This membrane configuration does not actually aid in the separation of the catalyst from the reaction matrix. The membrane system offers the advantage of helping to keep the phases apart, which minimizes the potential for forming emulsions. Emulsions are occasionally a problem in PTC systems. Matson at Sepracor investigated the commercial feasibility of such membrane systems [58].

A novel membrane system was described by Okahata and Ariga [59].This system was prepared by starting with a large porous nylon 2, 12 capsule membrane of 2.5 mm diameter and 5 μm membrane thickness (prepared by interfacial polycondensation). A PEG dimethacrylate was then grafted onto the capsule membrane to introduce vinyl groups which could then react with acrylamides. The pendant acrylamides were then quaternized with trimethylamine or tributylamine. The result was a hollow capsule with pendant quaternary ammonium groups and PEG groups (functioning also as spacers) on the outside surface.

The reaction performed with the grafted membrane catalyst was between sodium azide and benzyl bromide. The organic phase diffused into the hydrophobic interior of the capsule and the aqueous phase remained on the outside. The reaction followed pseudo-first-order kinetics and the rate-limiting step was reported to be the nucleophilic substitution to produce benzyl azide. The reactivity was largely dependent on the hydrophobic property and the spacer chain length of the -onium salt, the concentration of NaN_3, and the graft amount of the polymer on the capsule. The advantage of this system was the easy separation of the catalyst by filtration.

5. Commercial Catalyst Reference

Table 6-2 summarizes information relating to selected commercial catalysts. These data were obtained from catalyst suppliers and MSDS sheets. These data can be used as very general guidelines only. No representation is made regarding the safety, health, environmental, reactivity, stability, or cost aspects of these catalysts. In particular, regarding the toxicity data, before using any phase-transfer catalyst, an individual must obtain the most recent Material Data Safety Sheet and be sure that he/she knows the toxicity of the materials being worked with. The authors are not responsible for the accuracy of the toxicity data reported. All evaluations of catalysts for actual use should be performed by using good manufacturing and research practices as well as complying with all applicable local, state, and federal laws and guidelines.

The cost per weight factor was calculated by comparing the cost of the catalyst to the approximate cost of tetrabutylammonium bromide (cost during the early 1990s). During this period, tetrabutylammonium bromide was a moderately priced fine chemical.

The authors strongly suggest that individuals considering using phase-transfer catalysts contact the catalyst suppliers for up-to-date information regarding the catalysts.

B. Choice of Solvent

During the 1970s and early 1980s, the key driving force for considering PTC for the preparation of organic chemicals, both in the lab and in industry, was to

Table 6-2. Phase-transfer catalyst summary sheet

Catalyst	organophilic accessible	Cost/wt factor	Ease of separation	Toxicity orl-rat	Stability	Suppliers (alphabetical)
Bu$_4$NBr	intermediate	1.0	hi	~500 mg/kg	med	SACHEM, Schweizerhall, Zeeland Chemicals
Oct$_3$NMeCl	org/acc	0.5	med	223 mg/kg	med	Henkel, Witco
Bu$_3$NMeCl	acc	0.5	hi	933 mg/kg	med-lo	SACHEM
Et$_3$NCH$_2$PhCl	acc	1	hi	700 mg/kg	lo	Zeeland Chemicals
Me$_3$NCH$_2$PhCl	acc	0.2	hi	200 mg/kg cont. poison	lo	Sybron Chemicals, Witco
Me$_3$NC$_{16}$H$_{33}$Br	acc	1.3	med	410 mg/kg	med-lo	Schweizerhall, Zeeland Chemicals
Bu$_4$PCl	org/hyd	2	med	~1000 mg/kg	med (no OH)	Cytec
Polymer bound (Ion Exchange Resins)			hi		lo	Rohm & Haas, Dow, Sybron
18-Crown-6		50 (95%)	hi	710 mg/kg	hi	Parish, PCR
PEG	hyd	0.15	hi	33,750 mg/kg	hi	Dow, Union Carbide
TDA-1	hyd	5	hi	4500 mg/kg	hi	Rhone-Poulenc
EtHexDMAP	intermediate		hi		hi (no OH, RS)	

Cost, toxicity, and supplier information was provided by suppliers or MSDSs (at time of publication). Ease of separation, stability, and org/acc/hyd are rough estimates made by the authors to be used only as general guidelines; any evaluation of catalyst should include comprehensive study of safety, health, and environmental and economic factors specific for each case.

Table 6-3. Criteria for selecting a PTC solvent.

Nature of the chemical reaction
Environmental considerations
Recoverability/recyclability
Toxicity
Flammability
Boiling point
Cost
Polarity
Special handling (e.g., anhydrous)
Choosing NO solvent

perform the synthesis in high yield under mild and simple conditions. Since the late 1980s, environmental considerations became increasingly dominant in decisions made in the chemical industries because they often affect the economics of a decision as much as nameplate productivity. Some of the major issues facing the decision makers are air emissions, wastewater treatment, total organic content in waste streams, exposure of workers to hazardous materials, and other industrial hygiene and product stewardship concerns. Choice of solvent is involved in all of these issues as well as productivity and, therefore, an important factor in PTC reactions.

One of the major advantages of PTC is the flexibility provided in choosing a solvent. The criteria for choosing a solvent for a PTC application are shown in Table 6-3.

1. Choice of Solvent and the Nature of the Chemical Reaction

In principle, there are several ways in which solvent can affect the rate of a PTC reaction. As a basis for discussion, solvent can affect k_{chem} (equivalent to k_2 in Chapter 3) and $[Q^+X^-]_{org}$ in the simplified rate expression, rate $= k_{chem} [Q^+X^-]_{org} [R-Y]$, for organic-phase limited reactions, or $k_{transfer}$ for transfer rate limited reactions (see Table 6-4). These effects include: (1) stabilization of the transition state (affects k_{chem}), (2) solvation of the anion, reducing anion reactivity (affects k_{chem}), (3) solvent separation of the ion pair, enhancing anion reactivity (affects k_{chem}), (4) solubility of the catalyst–anion pair/complex in the organic phase (affects $[Q^+X^-]_{org}$), (5) rate of transfer of the catalyst–anion pair/complex from the aqueous or solid phase to the organic phase (determines rate in transfer rate limited reactions), (6) determining the extent of aggregation of the catalyst-anion pair/complex (affects k_{chem} and effective $[Q^+X^-]_{org}$), (7) decreasing the rate of competitive side reactions, and (8) many other unknown factors such as determining the nature of the interface. Since any combination of these factors may be important in a given reaction, choice of solvent will probably remain an art. Nevertheless, several effects of solvent can be rationalized and are discussed below. Please refer to Chapter 3 for the discussion of solvent

Table 6-4. Solvent effects in PTC—intrinsic reaction or transfer rate control.

Solvent factors affecting $k_{(chem)}$
- Stabilization of the transition state
- Solvent separation of the ion pair
- Solvation of the anion
- Aggregation

Solvent factors affecting $[QX]_{(org)}$
- Solubility/extraction coefficient of QX

Solvent factors affecting transfer rate
- QX transfer rate
- Unknown interfacial factors

Conclusion: polar/nonpolar not universally good

---

separation of the ion pair (factor 3 above) and a rigorous mathematical treatment of aggregation (factor 6 above). Data characterizing the effect of solvent on aggregation can be found in Chapter 2. An example and guidelines for the effect of solvent on selectivity of competitive side reactions can be found in Chapter 8, Section 1.a. Following is a discussion of the remaining factors 1, 2, 4, and 5.

2. Stabilization of the Transition State and Solvation of the Anion

The effect of solvent on the rate of an S$_N$2 reaction has been conceptually described by Ingold [60]. This treatment is based on the change of the charge distribution of the transition state relative to the starting materials. To illustrate, in most PTC S$_N$2 reactions an anion reacts with a neutral molecule. The negative charge which was concentrated on the anion in the starting material $[X^- + R-Y]$, would be dispersed in the transition state $[X\text{----}R\text{----}Y]$. As a result, the transition state is slightly less polar than the starting materials. An increase in the polarity of the solvent would stabilize the starting materials slightly more than the transition state. Therefore, if the reaction is intrinsic reaction rate limited, an increase in solvent polarity would result in a small decrease in the rate of reaction. This would affect the term k_{chem} in the simplified rate expression:

$$\text{rate} = k_{chem} [Q^+X^-]_{org} [R-Y]$$

Strong solvation of the anion (including hydration) reduces the reactivity of the anion. Solvent separation and ionization of the ion pair increases the reactivity of the anion. Referring to Chapter 3, Table 3-2, it is evident that as the solvent polarity increases, ionization increases for tetrabutylammonium nitrate and acetate and for tetramethylammonium picrate. Both of these effects may be considered to be subsets of the Ingold treatment described in the last paragraph.

3. Solubility of the Catalyst–Anion Pair/Complex in the Organic Phase

The most oversimplified, though somewhat useful, guideline for predicting solubility is "like dissolves like." This guideline is similar to incorporating solvent polarity into the HSAB thought process (described in Section A.1.a). Accordingly, it may be expected that tetramethylammonium fluoride would be more soluble in DMSO than in cyclohexane. Likewise, it may be expected that tetraoctylammonium iodide (consisting of two large polarizable ions) would be more soluble in nonpolar solvents than tetramethylammonium fluoride.

In the simplified rate expression of an intrinsic reaction rate: rate = k_{chem} [Q$^+$X$^-$]$_{org}$[R–Y], the concentration of [Q$^+$X$^-$]$_{org}$ is an important factor in determining the rate of reaction. Therefore, the solubilizing ability of the solvent, relative to water, toward the specific ion pair Q$^+$X$^-$ will affect the rate of reaction. Many parameters have been defined to characterize solvent polarity, such as dielectric constant, Z-values, E_T, acceptor number, donor number, and others. None of these parameters provides a universal predictor of solubilizing ability for the range of catalyst–anion pairs/complexes typically used in PTC. For example, Figure 6-30 shows that dielectric constant may not be used as a predictor of extraction of tetrabutylammonium bromide from water into organic solvents. On the other hand, Figure 6-31 shows that for the lipophilic quaternary ammonium picrates, common solvent parameters (dielectric constant, Z-values, and E_T) show that as the polarity of the solvent increases, extraction of that ion pair increases. Figure 6-32 shows that E_T does not significantly affect anion selectivity constants (cation = methyltrioctylammonium), though the nature of the anion is observed to be important. Figure 6-33 [61] shows the solubility of three quaternary ammonium cation–anion pairs in a variety of solvents (100 g/ 100 g solvent refers to >100 g salt soluble per 100 g solvent). If these data are correct, they would suggest that tetrabutylammonium bromide is more soluble than tetraethylammonium bromide or tetrabutylammonium chloride in less polar solvents. This would be consistent with the trend of more lipophilic/polarizable anions and quaternary cations components rendering the ion pair more soluble in less polar solvents.

Since catalyst–anion pairs/complexes contain at least some ionic character, these species are at least somewhat polar. As the polarity of the solvent decreases, aggregation of the ion pairs/complexes increases. Aggregation reduces reactivity by decreasing the "effective concentration" of the reacting anion in the organic phase and increasing the energy barrier for making all of the anions in the organic phase available for reaction (see Chapter 3 for mathematical treatment).

As noted in Chapter 4, PEG derivatives function best in the presence of a polar solvent. PEG derivatives are not very soluble in nonpolar media.

4. Rate of Transfer

Not a lot of data are available describing the rates of transfer of catalyst–anion pairs/complexes. Transfer-limited reactions are mostly postulated. Transfer

308 / Phase-Transfer Catalysis

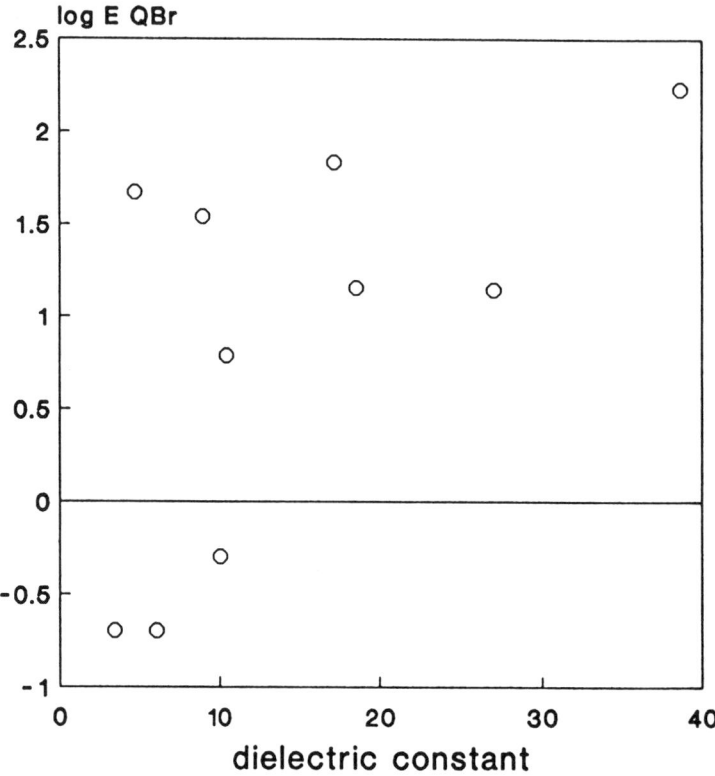

A. Brandstrom, Adv. Phys. Org. Chem.
15, 276 (1977)

Figure 6-30. Effect of solvent on extraction of tetrabutylammonium bromide; dielectric constant is not a good predictor [Data taken from Brandstrom, Adv. Phys. Org. Chem., **15**, 276 (1977).]

limitation is an obvious possibility that is derived from the nature of the extraction mechanism. As seen in Chapter 3, the behavior of simulated transfer-limited systems agrees well with the behavior exhibited by actual reactions.

In the absence of experimental data, one can only speculate about the effect of solvent on the rate of transfer of a catalyst–anion pair/complex. The relative affinity of a catalyst–anion pair/complex toward a solvent versus water will determine the extraction equilibrium, which is a thermodynamic property. The rate of transfer is a kinetic property. The question is: could the factors governing thermodynamic solubility also affect rate of transfer?

The Hammond postulate relates thermodynamic factors to kinetic factors under certain constrained conditions. This postulate suggests that as the transition state of a chemical reaction is more similar in structure to the starting materials relative to the products, the energies of the transition state and starting materials will be

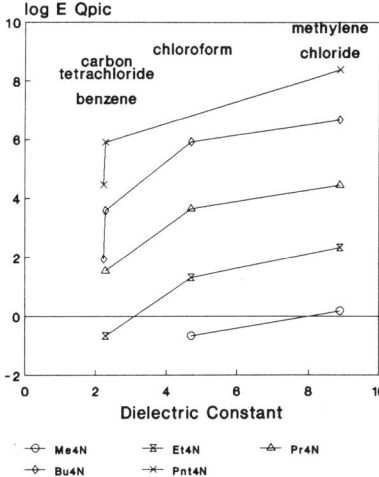

data taken from:
K. Gustavii, Acta Pharm. Seuc.,
4, 233 (1967) and 3, 241 (1966)

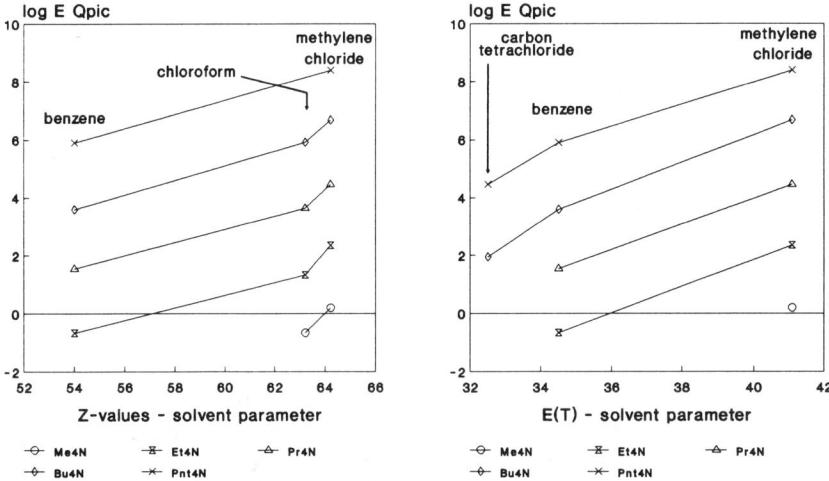

Figure 6-31. Effect of solvent and quaternary ammonium cation structure on quaternary ammonium picrate extraction.

more similar. During the transfer of a catalyst–anion pair/complex from an aqueous to an organic phase the transferred species undergoes a change in its state of solvation (qualitatively and quantitatively). We have already seen that, thermodynamically, soft ion pairs will be more soluble in nonpolar solvents than hard ion pairs. It may be speculated that, kinetically, soft ion pairs will be transferred more rapidly to nonpolar solvents than hard ion pairs.

It may be helpful to visualize a comparison between the benzene–water inter-

310 / Phase-Transfer Catalysis

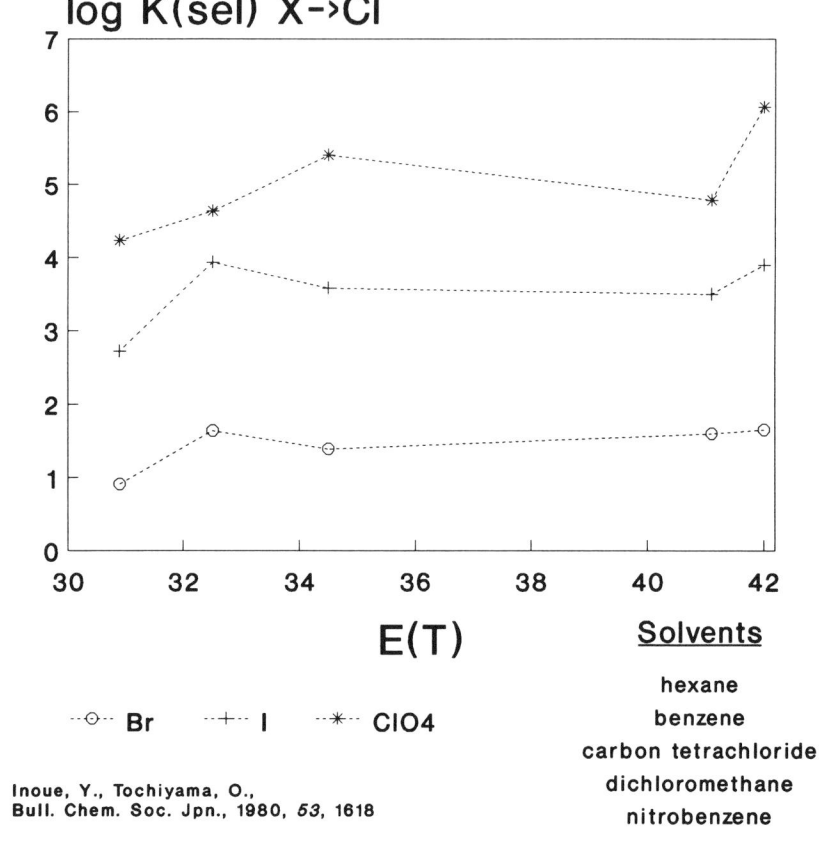

Figure 6-32. K_{sel} is more anion dependent than solvent dependent.

face as opposed to a methylisobutylketone–water interface. In both cases, the transfer of the "hard" tetramethylammonium fluoride, for example, would need to take place through the organic–water interface. It would seem that more energy would be needed to transfer the hard ion pair through the benzene–water interface than the more polar methylisobutylketone–water interface. Therefore, current thinking would suggest (though no evidence at all is available) that "like dissolves like" can be applied to the kinetics of transfer as well as the thermodynamics of solubility, since the natures of the molecular interactions are similar.

5. Solvent and the Nature of the Two Phases

When choosing solvent for a reaction, the nature of the chemical reaction (i.e., intrinsic reaction rate limited or transfer rate limited) is the first key criterion to consider. The nature of the two phases (e.g., liquid–liquid, solid–liquid, etc.)

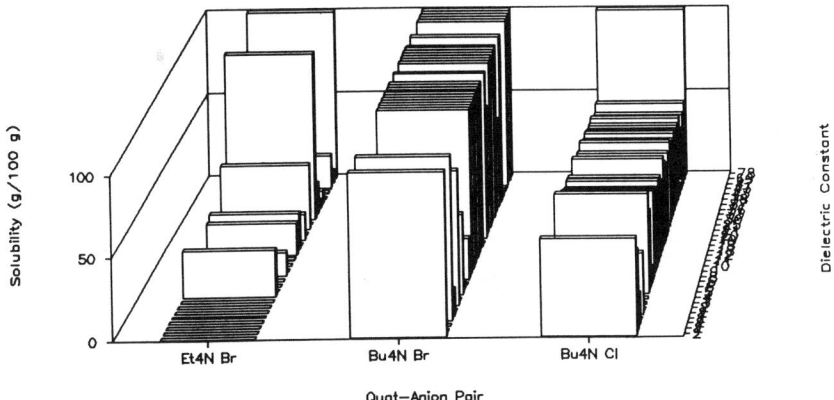

Figure 6-33. Effect of solvent on the solubility of three quaternary ammonium halide ion pairs.

should also be considered when choosing solvent. If the reaction is a conventional liquid–liquid PTC system then the solvent chosen should be immiscible with the aqueous phase. Water-immiscible solvents such as methylene chloride and toluene were chosen very often out of convenience. Indeed, historically, the choice of solvent was an art or not even considered a critical factor. As shown in the following sections (C and F), the choice of solid–liquid PTC conditions is sometimes advantageous. In such cases it is not necessary that the solvent be immiscible with water, and a solvent such as acetonitrile could be considered. In many cases, regardless of the use of liquid, solid, or gas phases, one of the reactants may advantageously be chosen as solvent (more about this attractive option later). In principle, the only constraint for choosing a solvent for a PTC system, which is dependent on the nature of the phases, is that at least two phases are formed. Therefore, in principle, a very wide range of solvents may be considered for any given PTC reaction.

It should be noted that the nature of the phases may change as a result of the concentration ratios chosen for the various components. The formation of a third phase and/or an omega phase was discussed in Chapter 3. A case has been described in which the concentration of tetrabutylammonium bromide was increased in a toluene–water system (S$_N$2 reaction of NaBr with benzyl chloride) [62]. On increasing the tetrabutylammonium bromide beyond a specific level, a third phase separated out that consisted of approximately two-thirds tetrabutylammonium bromide, one-fourth water, and the rest toluene and salt. The reaction was thought to take place primarily in the third phase. The appearance and reactivity within the third phase were temperature and concentration dependent.

6. Examples of Effect of Solvent

Figure 6-34 shows that less polar solvents are more effective for the nucleophilic displacement of methanesulfonate by bromide [5], which is an intrinsic reaction

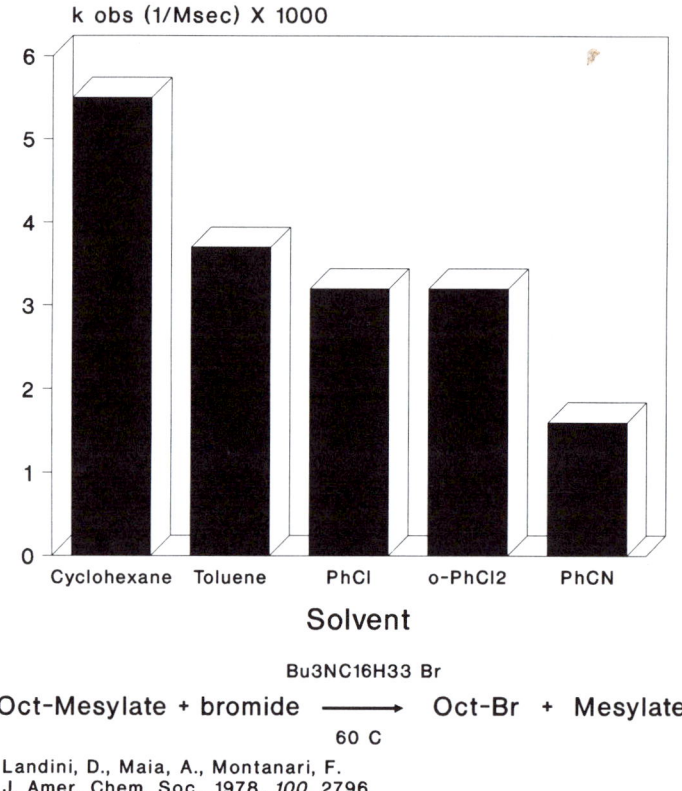

Figure 6-34. Effect of solvent on the displacement of methanesulfonate by bromide.

rate limited reaction. From a mechanistic standpoint, the rate is limited by the organic phase reaction. Thus, factors that promote this step will enhance the reaction rate. The solvent can affect the rate-determining step by (1) providing for little solvation of the anion, (2) reducing the amount of hydration allowed to accompany the anion into the organic phase by hydrophobic effect and (3) increasing the concentration of Q^+X^- in the organic phase characterized by distribution constants. It is obvious that the first two factors are directly enhanced by nonpolar solvents. As for the third factor, it may be speculated that, according to the rule-of-thumb "like dissolves like," the ion pairs present with the larger organophilic quaternary -onium cations (characteristic of intrinsic reaction rate limited reactions) are more likely to be in more organophilic nonpolar solvents.

In contrast, Figure 6-35 shows that polar solvents promote the ethylation of deoxybenzoin by ethyl bromide more than nonpolar solvents [63]. Indeed, a very polar solvent, such as methylisobutyl ketone (MIBK), which has a dielectric

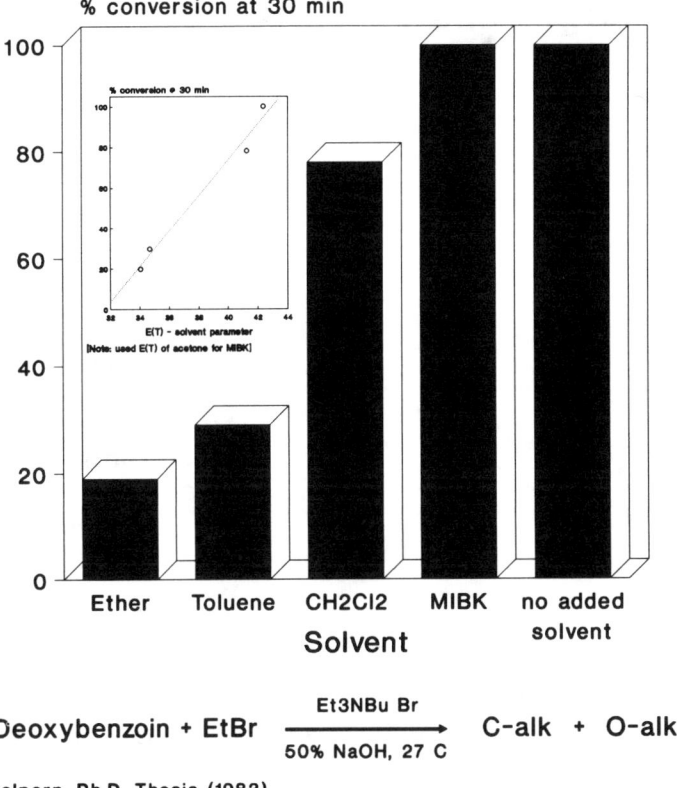

Figure 6-35. Effect of solvent on the ethylation of deoxybenzoin.

constant of 30.1, is highly effective. Transfer rate limited reactions usually apply to reactions involving the transfer of anions of high charge density, such as hydroxide. As mentioned above, it has been theorized that a "hard" quaternary ammonium cation is needed to effectively associate with the "hard" hydroxide. If so, a polar solvent would be needed to stabilize this ion pair, and thus facilitate the transfer of the polar ion pair into the organic phase. Indeed, the solubility of an ion pair composed of a hard anion and hard cation usually is very limited in nonpolar solvents. In the case of the ethylation of deoxybenzoin, using excess ethyl bromide as the solvent gave the highest conversion in 30 min. This may indicate that the reaction is not solely transfer limited, but may be governed by more than one PTC process, such as that described in Chapter 3, Case 4.

In the nucleophilic substitution of trifluoroethoxide on chlorophosphazene [64] more polar solvents promote the overall reaction (see Figure 6-36). This reaction was reported to be controlled by both mass transfer and the organic phase reaction

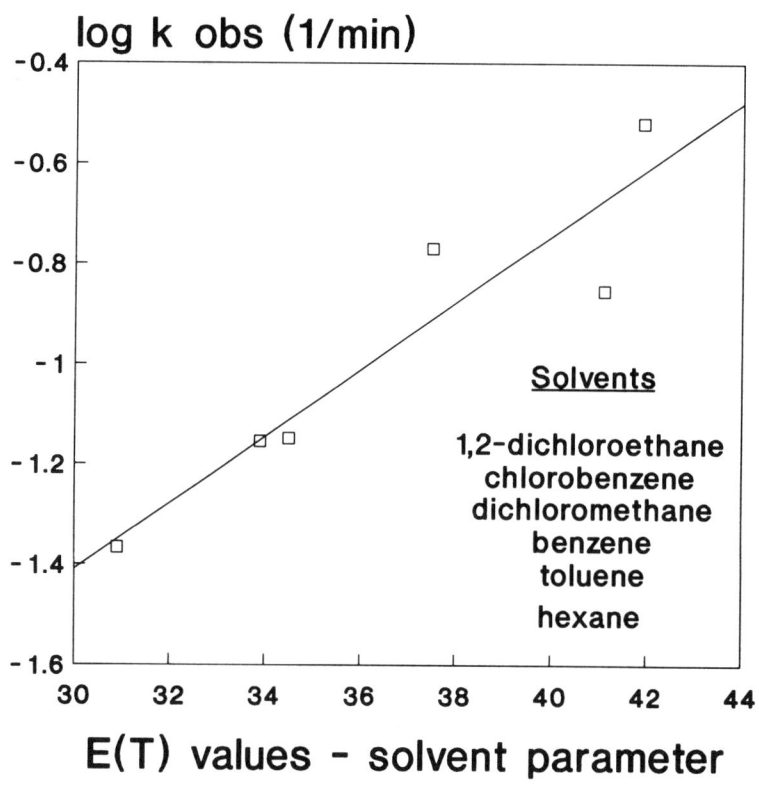

Wang, M., Wu, H.,
J. Org. Chem., 1990, *55*, 2344

Figure 6-36. Effect of solvent on the displacement of chloride by trifluoroethoxide.

(Case 4, Chapter 3, Table 3-1). In this case, many of the solvent effects described above are operational and it would be difficult to predict, a priori, which effects would predominate.

7. "Solvent-Free" PTC

One of the attractive options in choosing solvent is to use one of the reactants (or product, if liquid) as the reaction solvent. A review article citing over 50 references has been published describing the wide use of reactants or products as the solvent [65]. The reactions reported include applications from almost every reaction family within the scope of PTC. Several advantages are associated with using one of the reactants as the solvent. Increased reaction rates are achieved when an excess of the reactant is used and the kinetic expression for the rate-limiting step includes the reactant term. Although recovery of the excess of reactant is still necessary, not introducing a new chemical entity into the reaction matrix eliminates the possibility of a new impurity. In a commercial process, not introducing a separate solvent eliminates the need for extra hardware (holding tanks, extra fractionation during distillation) and extra logistics (purchasing, shipping, etc.).

The criteria for considering "solvent-free" (i.e., not adding an *additional* solvent) are as follows (choose [1a or 1b] AND [2a or 2b or 2c]). (1a) The crude reaction products and/or starting materials are liquids at the reaction temperature OR (1b) one reactant, which is a liquid, may be used in excess as solvent and recycled; AND (2a) the anionic reactant may be used as an aqueous solution OR (2b) solids handling is feasible OR (2c) use some of the reaction mixture for the next batch.

Bram et al. have championed the use of "solvent-free" PTC [65–76]. Table 6-5 shows representative examples of Solvent-Free PTC. It is highly recommended that "solvent-free" PTC be considered as an option for commercial PTC applications.

8. Choice of Solvent and Process Aspects

The process aspects of choice of solvent are environmental concerns, industrial hygiene factors, and processability. First and foremost are the environmental implications of choosing a solvent. Reduction of air emissions is already one of the reasons that chemists and engineers use to initiate evaluation of PTC alternatives in new process development or retrofitting existing processes. For example, it is interesting to note that DMSO is used as the solvent for many commercial displacement reactions. Even though it has a high boiling point, is difficult to distill, must be kept anhydrous, and is expensive, its processability has been tolerated as an industrial solvent until sulfur emissions became a problem. Replacing DMSO with toluene, for example, and a phase-transfer catalyst offers the advantages of eliminating sulfur emissions as well as easy recoverability, low cost, eliminating possible impurities due to reaction of DMSO (toluene is inert), and eliminating the need for strict drying of the reactor.

Prior to PTC, nucleophilic displacements with inorganic salts as the source of the attacking anion required the use of DMSO, DMF, NMP, and the like or a cosolvent such as ethanol. The polar aprotic solvents mentioned have high boiling points (189°C, 153°C, and 210°C, respectively) and usually require anhydrous or other special handling. The wide range of potential solvents to choose from under PTC conditions allows choice of boiling point that would not interfere with the distillation of the product (or require many theoretical plates in a column to separate the polar aprotic solvent from the product boiling at ca. 190°C). The vapor pressure of the solvent could be attenuated by choice to reduce air emissions but also allow for easy recovery and recycle. The use of protic cosolvents often suffers from low solubility of the inorganic salt in the mixed solvent system.

Table 6-5. Examples of "solvent-free" PTC ("solvent-free" means no additional solvent other than the reactants in any molar ratio).

Reaction/reactants	Catalyst (mol%)	Conditions	Yield	References
Esterification				
K acetate + *n*-BuBr	Bu$_4$NBr (1)	2 h/60°C	100%	[66]
K acetate + *n*-RBr	PEG-400	30 min/110°C	90–92%	[77]
K acetate + *n*-RX	Aliquat 336 (5–10)	1 min/microwave	95–98%	[67]
Cephalosporin/penicillin salts	18-crown-6	1–2 d/r.t.	80–100%	[78]
Transesterification				
PhCo$_2$Me + 2-ethylhexanol (and others)	Aliquat 336 (5)	8 h/r.t./K$_2$CO$_3$	100%	[72]
Etherification				
MANY aliphatic and aromatic ethers prepared in this ref. such as: *n*-OctOH + *n*-OctBr	Aliquat 336 (2)	2 h/20°C/KOH	98%	[68]
PhOK + *n*-OctBr	PEG-400	30 min/110°C	94%	[77]
Phenol + Br(CH$_2$)$_5$Br	Bu$_4$NBr (3)	1 d/80°C/KOH	98%	[75]
trichloropyridinate Na + ethylchloroacetate	Bu$_4$NBr (3)	11 h/70°C	93%	[79]
C-Alkylation				
malononitrile + benzyl bromide	Bu$_4$NBr (4)	5 h/K$_2$CO$_3$	99%	
N-Alkylation				
Indole + Et$_2$SO$_4$	Bu$_4$NBr(1)	10 min/r.t./KOH	98%	[70]
p-NH$_2$PhOH + *n*-RX	Aliquat 336 (5)	0.5–24 h/60°C/NaOH or KOH	31–98%	[75]
Azoles + CH$_2$Cl$_2$	Bu$_4$NBr	1 h/KOH	93%	[80]

Nucleophilic aromatic substitution
Phenol or phenylacetonitrile or benzyl alcohol or alkoxide were deprotonated with base and reacted with halobenzene, nitrohalobenzene or dinitrohalobenzene in the presence of 10 mol% Aliquat 336 or TDA-1 (10). Notable examples include KOMe + PhF (TDA-1/24 h/16°C) -> 91%; p-ClPhNO$_2$ gave 99% yield w/PhOH (TDA-1/4 h/120°C) and 80% yield w/PhCH$_2$CN (Aliquat 336/1 h/80°C) [73]
2-bromopyrimidine + PhOH, PhCH$_2$OH, PhCH$_2$CN [74]

Other nucleophilic aliphatic substitution

KCN + Br(CH$_2$)$_3$Br	Aliquat 336 (2)	8 h/20°C	100%	
KCN + n-RBr	PEG-400	3 h/110°C	90-94%	[77]
KI + n-BuBr	PEG-400	0.5 h/110°C	88%	[77]
KF + n-OctCl	Bu$_4$NBr	6 h/120°C	93%	
ArSO$_2$Na + RX	Aliquat 336	2-24 h/r.t.	35-95%	[71]
X = Cl, Br; R = nitrile, ester, amide, aryl, halo				

Chiral Michael addition
N-Acetamidomalonate + α-enone/ephedrine-based catalysts [76]

Oxidation

K$_2$Cr$_2$O$_7$ + PhCH$_2$Br → PhCHO	PEG-400	2 h/110°C	85%	[77]

Reduction

NaBH$_4$ + CH$_3$COC$_6$H$_{13}$	PEG-400	2 h/25°C	90%	[77]

For initial screening of a new PTC reaction, a wide range of solvents should be considered. If the chemist is pressed to quickly screen the reaction without regard to solvent, toluene (or other suitable aromatic or aliphatic hydrocarbon) could be chosen for a reaction in which the intrinsic reaction contributes significantly to the overall rate. Likewise, a polar solvent, such as methylisobutylketone, should be initially chosen for a reaction that involves an anion that is difficult to transfer (such as hydroxide or fluoride). If one just wants to demonstrate that a PTC reaction works, without regard to optimization, a chlorinated hydrocarbon (such as methylene chloride or chlorobenzene) may be chosen. Ultimately, optimization of solvent should be performed by screening a variety of acceptable polar and nonpolar solvents.

One of the most common solvents in the PTC literature is methylene chloride. There is a tendency in the industry to eliminate methylene chloride and other volatile chlorinated hydrocarbons due to stricter emissions standards. In addition, if a PTC alkylation is being performed, methylene chloride often/occasionally acts as an alkylating agent (yielding a few percent of methylene bridged disubstrate). For these synthetic and/or regulatory reasons, it is worthwhile to consider replacing methylene chloride by methyl isobutyl ketone, or better yet, one of the reactants. If the product is a liquid, then stoichiometric quantities of reactants can be used with no additional solvent.

In all cases, the use of no additional solvent should be considered as an option [65]. In the case of solvent choice, it is feasible to screen a wide range of solvents for chemical reactivity using the environmental, industrial hygiene, and process convenience criteria as preliminary screens for the study.

C. Presence of Water

As shown in Chapter 3, hydration of the reacting anion has a profound effect on the reactivity of the PTC system. The optimal water concentration for a given PTC system needs to be evaluated on a case-by-case basis, but from the standpoint of reactivity, optimal water concentration is usually low. For example, Sasson showed that for a specific fluoride exchange reaction, the optimal rate was observed when 0.33 molecules of water were available for hydration of one KF ion pair [81].

Water effects are particularly great when transferring and/or reacting hard anions. "Observed kinetic orders" of ca. 5 have been found for hydroxide ion reactions (see Chapter 2) [82]. Such behavior may be attributed to factors such as the great dependence of the basicity of the hydroxide ion upon hydration. Solaro et al. were the first to speculate that salting out effects may also increase the observed reactivity of the hydroxide ion. Landini [83] showed that "Surprisingly, by increasing the base concentration in the aqueous phase (up to 50% aqueous NaOH) the extractability of OH^- was found to diminish (2.4–5 times) for all the anions." Landini attributed the dramatic increase in hydroxide reactivity (despite lower extractability) to lower hydration. Landini [84] further showed that an increase in NaOH concentration from 5 M to 20 M reduces the hydroxide hydration sphere of the tetrahexylammonium hydroxide dissolved in chlorobenzene from 11 to 3.5 molecules of water.

The remarkable hydration behavior of hydroxide was further illustrated by Landini et al. [85]. The inorganic salt concentration in the aqueous phase of chloride, azide, bromide, and iodide did not affect the hydration of these anions in the organic phase, when using up to saturated aqueous solutions of these anions. However, in the presence of 50% NaOH or 60% KOH, the anions transferred were essentially nonhydrated and followed homogeneous anion reactivity trends (rate enhancement of 13 for chloride, 4 for azide, 2.6 for bromide, and 1.4 for iodide). Therefore, a common "trick" for enhancing nucleophilicity or basicity of transferred anions in PTC systems is to add salt to the aqueous phase to reduce hydration of the transferred anion in the organic phase.

From a processing standpoint, these kinds of systems would be treated as solid–liquid systems. Solids handling is not usually a desired process element, but the tradeoff needs to be evaluated: solids handling versus enhanced reactivity. Sometimes solid–liquid systems also offer the added advantage of reduced impurities due to reduced hydrolysis.

D. Agitation

Agitation efficiency is expected to have a significant effect on the reactivity of a PTC system. Conceptually, in a PTC system, an ion pair must physically move from at least one bulk phase or interface into another bulk phase. Any of these phases or interfaces may be liquid, solid, or gas, though most PTC systems involve two liquid phases or a solid and a liquid phase. If the two (or more) phases are completely stationary and nonagitated, then the interfacial area is minimal and transfer rate is minimized. When stationary, this rate will be very slow and will definitely limit the rate of reaction. As agitation is commenced and its efficiency increased (by stirring speed, baffles, impeller shape, and positioning, etc.) the ion pair transfer rate will increase. As the ion pair transfer rate increases, it can approach, meet, or surpass the maximum possible rate of the organic-phase reaction. Since the transfer rate (k_1 from Chapter 3) is solvent dependent, different stirring speeds/agitation efficiencies may be required to maintain the same transfer rate for a given reaction performed in different solvents. Overall, depending on the agitation efficiency, EVERY PTC reaction has the potential to be dominated by either the transfer rate limited or the organic phase reaction, or both.

In most literature reports, experiments and kinetic studies have been conveniently performed at 500–1000 rpm in round-bottomed flasks (no baffles) with

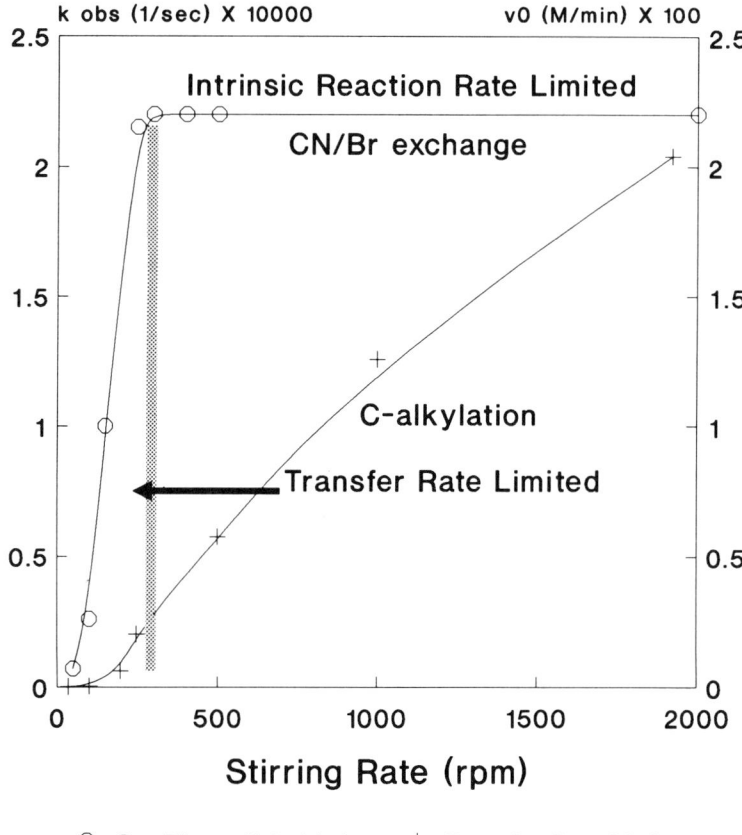

Figure 6-37. Effect of agitation. [From Starks et al., J. Am. Chem. Soc., **95**, 3613 (1973) and Solaro et al., JOC, **45**, 4179 (1980).]

half-moon Teflon blades. Under these typical conditions, most of the common anions, which do not have particularly high charge densities, are thought to be transferred into the organic phase at rates faster than the rates at which they react with the substrate. Classic and well researched reactions such as the cyanation of alkyl bromide ("o"'s in Figure 6-37) [4] and the nucleophilic displacement of methanesulfonate by bromide [5], have been shown to have rates independent of stirring speed above 300–500 rpm. In contrast, the ethylation of phenylacetonitrile is highly dependent on stirring speed and method, up to 1950 rpm ("+"'s in Figure 6-37) [86]. In this reaction, hydroxide, a high charge density anion, is transferred into the organic phase. In such a case, the physical movement of an ion pair, Q^+X^-, is the rate-determining step. Hydroxide is difficult to transfer into a hydrophobic phase and the subsequent reactions are relatively fast. A third behavior is observed in the isomerization of allylbenzene [8]. In this reaction, allylbenzene, which has a pK_a of 34, is deprotonated by the hydroxide ion. Water has a pK_a of about 16, which is approximately 18 orders of magnitude less basic than allylbenzene. It might then be expected that the deprotonation of allylbenzene would be show and that the transfer of anions might be faster than the deprotonation. It is observed that this reaction is independent of stirring speed above 500 rpm.

Transfer rate limited as well as intrinsic reaction rate limited reaction behaviors have been classically observed and determined under convenient agitation conditions. Since the inherent maximum potential rate of an organic chemical reaction is constant (at a given temperature, hydration, etc)., then the agitation efficiency will determine whether the overall system will be transfer rate limited, i.e., whether the transfer rate will be greater, lower, or equal to the chemical contribution to the reaction rate.

The obvious ramifications of the effect of agitation on reaction rate are that one should choose the best agitation efficiency possible (to include baffles and many impeller blade/s variables) within economic constraints for transfer limited reactions, and one need not invest in elaborate agitation schemes for organic phase rate limiting reactions.

In some cases, two parallel reactions may take place, one PTC and one not. For example, in the reaction of sodium phenolate with benzoyl chloride [87], the esterification is PTC and the hydrolysis of benzoyl chloride may occur by both a PTC as well as by a noncatalyzed interfacial process. Although no data are available, it may be speculated that it would be advantageous to perform the reaction at an agitation efficiency at which the chemical reaction and ion pair transfer are nearly the same. Beyond this point, "adding a little more agitation efficiency for good measure" would probably enhance the undesired yield loss due to the noncatalyzed hydrolysis without increasing the rate of the desired reaction.

Another case of interest is the E2 dehydrochlorination of α-phenethyl chloride. The dehydrochlorination proceeds faster with tetraoctylammonium bromide as the catalyst compared to benzyltriethylammonium bromide [88]. In this reaction, some polymerization was observed. This could have been caused by side reactions resulting from deprotonation. It may then be speculated that the elimination is limited by the rate of reaction in the organic phase. A possible side reaction may occur from the deprotonation of the α-phenethyl chloride to form a short-lived anion that could initiate an anionic polymerization of the resulting styrene. If the PTC deprotonation/polymerization initiation is a transfer rate limited reaction and the elimination is an organic phase limited reaction, then increased agitation efficiency would adversely affect the selectivity of the reaction.

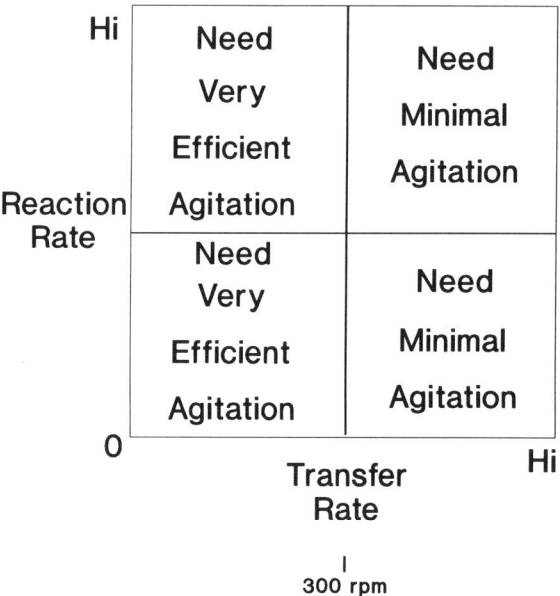

Figure 6-38. Choosing agitation and the PTC Rate Matrix.

Agitation may also be used to regulate the rate of heat liberation of an exothermic reaction, by regulating the rate of reaction in the transfer rate limited regime. The choice of agitation in relation to the PTC Rate Matrix is shown in Figure 6-38. The 300 rpm level shown at the "midpoint" of the transfer rate scale is typical for small laboratory round-bottomed flasks with no baffles.

E. Choice of Anion, Leaving Group, and Counteranion

The mechanistic discussions in the previous sections focused upon the effects of several reaction variables on two steps: the transfer of the reacting anion and the subsequent chemical reaction. The complete PTC catalytic cycle includes other steps, such as the exchange of anions that associate with the catalyst. The nature of the anions involved in a PTC system has a significant effect on the outcome of the reaction, since it plays a role in all of the steps of the catalytic cycle. The source of the anions in question may be the desired nucleophile (or other reacting anion), the leaving group, or the counteranion introduced with the catalyst. The nature of the anion may affect the chemical reaction and/or the affinity for association with the catalyst. The affinity of the anion for association with the catalyst may, in turn, affect the rate, extent, or even blockage of transfer of the desired anion to the organic phase. This section discusses the relationship between the nature of the anion and the steps of the catalytic cycle.

When the rate-determining step is the organic phase reaction, the nature of the anions will affect the outcome as reactant or product. Typical anionic reactants in PTC include nucleophiles, bases, oxidizing agents, and nonreactants such as anionic transition metal complexes acting as cocatalysts. Anionic products generally are leaving groups. No attempt will be made here to discuss nucleophilicity, nucleofugicity, basicity, or oxidizing capability or other reactivity of anions. These characteristics of the anions do play a role in the choice of PTC reaction conditions, but are not the only important factors. For example, we have seen in Chapter 2 that a classical leaving group such as iodide may shut down the ability of the catalyst to function. The characteristics of various anions in PTC displacement reactions are summarized in Chapter 7, Table 7-1.

Other than the classic effects of anions on chemical reactivity, the key factor in determining the effect of anions on the outcome of a PTC system is the relative affinity of the anion for association with the catalyst. The implications of HSAB on the choice of leaving group anion may be illustrated by considering the PTC halide exchange reaction. A soft organophilic quaternary-onium cation, such as tetraoctylammonium, would readily associate with the soft and highly polarizable iodide which would be highly soluble in an organic solvent. The iodide would react with the alkyl chloride liberating Cl^-, which would form an ion pair with the quaternary-onium cation. The ion pair composition Q^+Cl^- would not be as preferable as Q^+I^-, and the quaternary-onium cation would readily exchange anions to generate the iodide form, which would be ready for another catalytic cycle. The rate-determining step of such a cycle would be the attack of the iodide on the alkyl chloride in the organic phase. Mathematical treatment of this situation can be found in Case 3 in Chapter 3.

For contrast, the reverse reaction will be considered, that is, an attempt to perform the displacement reaction using chloride as the nucleophile and alkyl iodide as the substrate. From the chemical standpoint, a low-hydrated chloride, as is typical for a PTC system, is more reactive than the iodide from the first example. In addition, the iodide in this second example is a better leaving group than the chloride from the first example. On the other hand, the ion pair containing the desired reacting anion, Q^+Cl^-, is made from a large quaternary ammonium cation and a relatively hard anion. After the reaction, the ion pair, Q^+I^-, is formed and is the preferred form relative to Q^+Cl^-. Since the quaternary ammonium cation prefers to be associated with the iodide relative to the chloride, anion exchange will be hindered, in turn hindering the catalytic cycle. This phenomenon, of reduced reactivity (or complete halting of the reaction) due to strong and preferential association of the catalyst with one species, is known as catalyst poisoning. Mathematical treatment of catalyst poisoning can be found in Case 2 in Chapter 3.

This principle can be used to enhance displacement reactions such as esterification or alkylation. For example, an esterification was performed by the reaction of a substituted benzoic acid with benzyl chloride in the presence of 2 mol%

iodide (used in the preparation of a polyvinyl chloride plasticzer in 99.9% yield) [89]. The iodide was preferentially extracted into the organic phase and reacted with the benzyl chloride to form benzyl iodide. The benzyl iodide then served as a more reactive alkylating agent in the attack by the carboxylate.

For quaternary ammonium ions, Dehmlow has compiled an order for lipophilicity of anions, in which anions at the left will have a greater affinity for association with the quaternary ammonium cation than those at the right [90] (see also Tables 2-2 and 2-3 in Chapter 2). Anions at the extreme left may act as catalyst poisons.

picrate $>> MnO_4^- > ClO_4^- > SCN^- > I^- > (ClO_3^-$, toluenesulfonate)
$> NO_3^- > Br^- > (CN^-, BrO_3^-, PhCOO^-) > (NO_2^-, Cl^-) > HSO_4^- > HCO_3^- > CH_3COO^- > (F^-, OH^-) > SO_4^{-2} > CO_3^{-2} > PO_4^{-3}$

When hydrophilic hard anions need to be used, such as hydroxide, HSAB would predict that smaller and harder quaternary ammonium cations would be preferable. Tetrabutylammonium appears to be a quaternary ammonium cation with sufficient lipophilicity and sufficient hydrophilicity to be able to perform reasonably well with the widest range of anions. Methyltrioctylammonium also performs well with a wide range of anions, since its organophilic nature allows it to associate well with polarizable anions and its accessible face, containing the methyl group, allows it to associate well with anions of a higher charge density.

The effect of catalyst counteranion on suppression of hydroxide ion extraction is illustrated by the isomerization of allylbenzene in the presence of 40% NaOH, 5 mol% tetrabutylammonium salt, and toluene at 75°C [8]. In this case a 100-fold excess of hydroxide is present relative to the quaternary ammonium cation counteranion. Nevertheless, a 45-fold increase in rate can be obtained by changing the quaternary ammonium cation counteranion from bromide to hydrogen sulfate.

Sometimes catalyst poisoning is so severe that stoichiometric quantities of quaternary ammonium cation need to be used, and the quaternary ammonium cation must be introduced with a hydrophilic counteranion (or a counteranion capable of conversion to a hydrophilic anion). For example, the dehydrohalogenation of 1,2-dibromo-1,2-diphenylethane in the presence of 50% NaOH requires one equivalent of tetrabutylammonium hydrogensulfate for every HBr eliminated [88]. Upon addition of the catalyst, the hydrogen sulfate anion is deprotonated and the more hydrophilic sulfate is formed, facilitating hydroxide ion association with the quaternary ammonium cation and subsequent transfer. In this case, the bromide leaving group accumulates and is sufficiently lipophilic to poison the tetrabutylammonium cation.

It is interesting to observe that the dehydrobromination of 1,2-dibromo-1,2-diphenylethane in the presence of solid KOH proceeds in the presence of catalytic quantities of tetraoctylammonium [91]. Other dehydrobrominations are known to occur catalytically. These results show that choice of anion, and other reaction variables, is still a mixture of art and science. At this point in the discussion, it

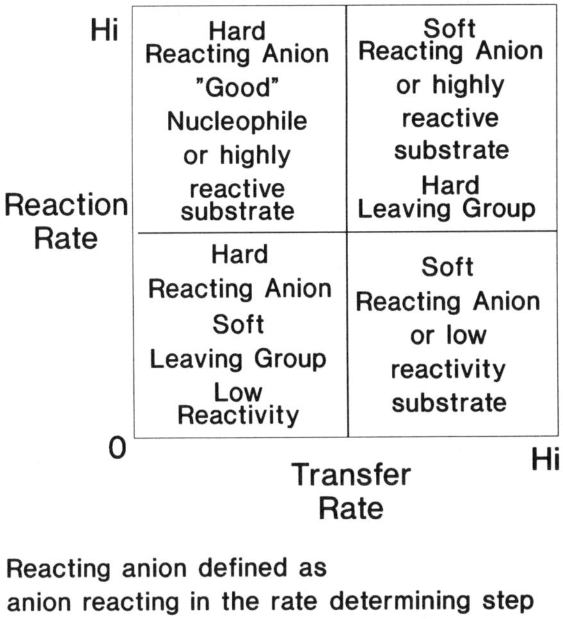

Hi	Hard Reacting Anion "Good" Nucleophile or highly reactive substrate	Soft Reacting Anion or highly reactive substrate Hard Leaving Group
Reaction Rate	Hard Reacting Anion Soft Leaving Group Low Reactivity	Soft Reacting Anion or low reactivity substrate
0	**Transfer Rate**	**Hi**

Reacting anion defined as
anion reacting in the rate determining step

Figure 6-39. Effect of anion in the PTC Rate Matrix.

should be apparent that factors such as the nature of the anion, the nature of the quaternary-onium cation, hydration, and solvent effects are closely interrelated and play a combined significant role in determining the outcome of a PTC reaction.

Many *C*- and *N*-alkylations use accessible quaternary ammonium cations, such as alkyltriethylammonium, and involve the hydroxide ion. The alkylating agents used are almost exclusively alkyl bromides (except for benzyl or allyl chloride). If the rate-determining step is the transfer of hydroxide across the interface, it may be that the hard and accessible quaternary ammonium cation in the ion pair, Q^+OH^-, may preferably associate with OH^- relative to the softer Br^-. Empirically, it appears that alkylations (C-, N-, O-, and S-) may be performed successfully on acidic substrates ($pK_a < 23$) using alkyl bromides as the alkylating agent.

The likely scenarios for the effect of anion on PTC reactions are shown in Figure 6-39, within the framework of the PTC Rate Matrix.

F. Choice of Base

No firm guidelines exist for the choice of base for base-promoted PTC reactions. The major choices are aqueous and solid NaOH, aqueous and solid KOH, and solid K_2CO_3. The ramifications of choice of base are expressed in terms of reactivity, catalyst stability, and side reactions (such as hydrolysis). In many cases there is a tradeoff between the three factors.

As noted in Chapter 2, Landini showed that as the aqueous hydroxide concentration increases, the quantity of hydroxide extracted decreases. The overall activity of the hydroxide actually increases due to the extremely strong dependence of hydroxide basicity on hydration. Increased basicity means that the base-promoted reaction will be accelerated but so will the Hofmann decomposition of the quaternary ammonium cation. In order to evaluate the effect of base strength on the outcome of the reaction, the relative rates of the desired reaction and quaternary ammonium cation decomposition must be measured.

Not enough is yet known about the effect of K^+ vs. Na^+ as a parameter for choice of countercation. From an economic standpoint NaOH is much more inexpensive than KOH. At room temperature, higher concentrations of KOH are achievable relative to NaOH. Therefore, KOH may offer a slight process advantage in a specific case that requires the use of a high concentration base pumpable at room temperature.

The use of solid K_2CO_3 may be advantageous when trying to avoid hydrolysis of a reactant or product while maintaining strong basic conditions (for examples, see Chapter 8, Section A.3). From a process standpoint, the use of solid K_2CO_3 may require a filtration step which may pose a disadvantage to be weighed against the benefits gained. In addition, K_2CO_3 may not be a strong enough base to deprotonate organic substrates with very low acidity. Indeed, concentrated or solid NaOH or KOH may be needed to even initiate base-promoted reactions involving organic substrates with very high pK_a (see Section G of this chapter). If the use of solid NaOH or KOH in a process is feasible, the excess base could then be removed by a water wash to avoid a filtration step.

Another "trick" for performing PTC/OH reactions requiring base is to add a little alcohol. Small amounts of methanol or ethanol will form the alkoxide which is much more easily transferred to the organic phase than hydroxide and is at least as basic as hydroxide. In some cases, these alkoxides can cause side reactions due to nucleophilic attack by the alkoxide. In these cases, one can use non-nucleophilic substrates which can deprotonate to form an adequate base, while not having the ability to easily perform nucleophilic substitution (such as trityl alocohol, pinacol, or a hindered indole) [92]. Further discussion of the addition of alcohols for the enhancement of PTC/OH reactions may be found in Chapter 4, Section F.2, and in Chapter 8 Section E.

G. Guidelines for Exploring New PTC Applications

At this stage of the discussion, it is apparent that a large array of variables interact in obtaining "optimal" reaction conditions for a given PTC application.

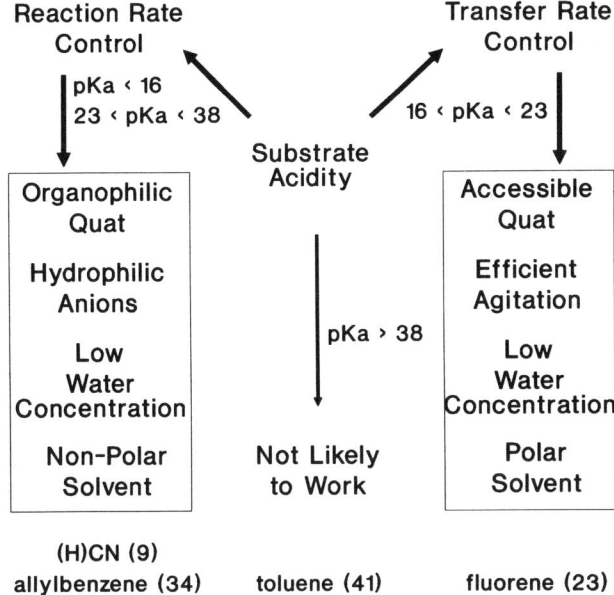

Figure 6-40. Guideline for exploring New PTC reactions.

Furthermore, the kinetic treatment of Chapter 3 showed that any PTC reaction can be transfer rate limited, organic phase reaction limited, or a mathematically complicated combination of both. It is, therefore, conceivable that no simple single guideline could exist for the evaluation and optimization of PTC reaction conditions.

Nevertheless, when attempting to explore the potential of a reaction that may be a candidate for PTC application, the chemist is confronted with an overwhelming choice of reaction conditions, of which only one combination will be tried first. Fortunately, a strictly empirical relationship has been found that suggests a first-pass approach to choosing reaction conditions for PTC candidate reactions [93]. This approach, illustrated in Figure 6-40, is based on "substrate acidity" and, as a very rough approximation, has been related to the PTC Rate Matrix. The basis and application of these guidelines are explained below.

1. Definitions

The majority of the PTC literature deals with the transfer and reaction of anions. This discussion deals only with the transfer and reaction of anions, mostly in nucleophilic substitution reactions. This discussion also relates to the nucleophilic anions in terms of their conjugate acids. For the purpose of this discussion, an acid is termed "very acidic" if it has a pK_a of up to approximately 16, "moderately

Table 6-6. Anions derived from "very acidic" acids (pK_a <16) commonly used in PTC nucleophilic displacements.

PTC nucleophilic anion	Approximate pK_a of conjugate acid [94]
Iodide	−10
Bromide	−9
Azide	
Nitrate	−1
Nitrite	3
Carboxylate (unactivated)	5
Hydrosulfide	7
Hypochlorite	8
Thiophenoxide	8
Cyanide	9
Phthalimide	
Phenoxide	10
Thiolate	12
Deprotonated methylene activated by:	
Ketone and ketone (acetylacetone)	9
Ketone and ester	11
Nitrile and nitrile (malononitrile)	12
Ester and ester (malonic ester)	13
Sulfone and sulfone	14
N- and C-anions derived from:	
Imidazole	
Pyrrole	
Cyclopentadiene	15 (not in this category since the anion is not water soluble)
Alkoxide	16–19 (in this category if water soluble)

Note: Hydroxide (pK_a of water = 16) and fluoride (pK_a of HF = 3) are excluded from this category since they are very "hard" anions (see Section A of this chapter) and are difficult to transfer.

acidic" for pK_a values between 16 and 23, "weakly acidic" for pK_a values between 23 and 38, and "very weakly acidic" for pK_a values greater than 38 [94].

It should be emphasized that the following discussion does not constitute a rigorous analytical approach for defining PTC kinetics. Gross assumptions and simplifications will be used to relate the mathematical kinetic analysis offered in Chapter 3 with commonly encountered nucleophilic substitution reactions. In addition, it is expected that there are no "sharp borderlines" between the pK_a ranges which empirically characterize gross reaction behaviors.

2. Reactions Involving Nucleophilic Anions Derived from "Very Acidic" Conjugate Acids with pK_a<16

Anions in this category readily dissociate, at least to some extent, in water or in the presence of weak base (see Table 6-6). In the overall rate expressions for nucleophilic displacements involving these anions, the rate of deprotonation (i.e., dissociation) is usually not even considered. Therefore, the predominant rates to take into account are the rates of transfer and of the organic phase nucleophilic substitution. Most of the anions in this category are soft inorganic anions (e.g., iodide) or are organic anions (e.g., carboxylate, phenoxide, thiolate, acetylacetonate). Therefore, transfer to the organic phase by quaternary -onium cations is not likely to be limiting. The rate of the organic phase nucleophilic substitution is likely to be involved in the rate expression.

If indeed the organic phase reaction is expected to be predominate, a simplified rate expression may be assumed:

$$\text{Rate} = k_{\text{chem}} [Q^+X^-]_{\text{org}} [\text{substrate}]$$

In light of the discussion in Sections A.1.b, C, and E of this chapter and the kinetic discussions of Chapters 2 and 3 (in which k_{chem} is termed k_2), such reactions are promoted by: the use of an organophilic/anion activating quaternary -onium cation (enhancing both k_{chem} and $[Q^+X^-]_{\text{org}}$); hydrophilic anions as leaving groups and catalyst counteranions (to avoid catalyst poisoning); some minimal agitation (no more than needed for mass transfer); and low water concentration to increase nucleophilicity by reduced organic phase hydration of the anion (affecting k_{chem}) and possibly increasing ion pair extraction due to salting out. It is difficult to predict the effect of solvent, though if very organophilic and anion activating quaternary -onium salts are used, nonpolar solvents may enhance reactivity more than polar solvents.

In order for an anion to be in this category, it should be water soluble. If the anion is not water soluble, then it would require either interfacial deprotonation before nucleophilic attack or transfer of hydroxide to the organic phase. In such cases, the choice of reaction conditions would be complicated by the rate of hydroxide transfer or the rate of interfacial deprotonation.

3. Reactions Involving Nucleophilic Anions Derived from "Moderately Acidic" Conjugate Acids with: $16<pK_a<23$

Anions in this category are generally organic compounds with somewhat activated methylene groups and are not water soluble and do not readily dissociate in the absence of strong base (see Table 6-7). In order to generate such anions, the transfer of the hydroxide ion is usually required. Since the hydroxide ion is a "hard" anion, and may be difficult to transfer, the transfer rate is likely to play a major role in the overall rate. In the overall rate expressions for nucleophilic displacements involving these anions, the rate of deprotonation should be considered, and is treated as k_2 in Chapter 3. Since the substrates that generate these anions are "moderately acidic," it is speculated (based upon the observed empirical reaction behaviors) that the transfer rate of hydroxide may play a greater role in the overall rate than the rate of deprotonation.

Table 6-7. Anions derived from "moderately acidic" acids ($16 < pK_a < 23$) commonly used in PTC nucleophilic displacements.

C- and N-anions derived from non-aqueous soluble methylene compounds activated by [94]:
Phenyl and nitrile (e.g., phenylacetonitrile); $pK_a \sim 16$
Phenyl and ketone (e.g., phenylacetone, deoxybenzoin); $pK_a = 16–18$
Imine and ketone (e.g., O'Donnell's alkylations; see Chapter 8, Section A.4)
Ketone or aldehyde; $pK_a = 17–20$
Three halogens (e.g., chloroform)
Aromaticity (e.g., indene, $pK_a = 20$; fluorene, $pK_a = 23$)
Other highly conjugated hydrocarbon (e.g., phenylacetylene, $pK_a = 20$)
Phenyl and amide (e.g., acetanilide, benzanilide)

Usually, an additional reaction, such as alkylation, takes place, with a rate constant treated as k_2 in Chapter 3. Therefore, it may be speculated that the predominant rates to take into account are the rates of transfer and of the organic-phase nucleophilic substitution. Usually these reactions, which are usually alkylations, are performed at temperatures that promote the alkylation at "acceptable" rates. For the purpose of these oversimplified guidelines, mass transfer of $[Q^+OH^-]$ to the organic phase is assumed to play the major role in the overall rate for this category. In light of the discussion in Sections A.1.b, C, and E of this chapter and the kinetic discussions of Chapters II and III such reactions are promoted by: the use of an accessible quaternary-onium cation to transfer the hard hydroxide ion; hydrophilic anions as leaving groups and catalyst counteranions (to avoid catalyst poisoning); much agitation (to promote mass transfer); low water concentration (i.e., high NaOH concentration) to increase the basicity of the hydroxide ion and the nucleophilicity of the organic ion by reduced hydration so that neither deprotonation nor alkylation is rate limiting (remember that Landini showed that less hydroxide is extracted at high [NaOH], therefore salting out does not apply here); and a polar solvent to enhance Q^+OH^- extraction into the organic phase, since the Q^+OH^- ion pair is likely not very organophilic with the accessible quaternary ammonium cation and a more polar solvent is likely to reduce the interfacial tension, thereby increasing interfacial area and hence the rate of transfer.

The rate equation from Chapter 3 relating to PTC/OH reactions can be written as follows, using the term K_a for the ratio (k_2/k_{-2}):

$$\text{Rate} = \frac{k_1 K_a k_3 [QX]_{org}[OH^-]_{aq}[RH]_{org}[R'X]_{org}}{k_{-1}[X^-]_{aq} + (k_{-1}k_3/k_{-2})[R'X]_{org}[X^-]_{aq} + k_3 K_a[R'X]_{org}[RH]_{org}}$$

It can be shown (from the PTC/OH kinetic equation shown in Chapter 3) that under certain circumstances, which would be expected to be associated with slow hydroxide transfer and rapid deprotonation, the rate expression would be reduced to:

Table 6-8. Anions derived from "weakly acidic" acids (23 < pK_a < 38) used in PTC/OH reactions.

Substrate	Approximate pK_a of conjugate acid [94]
Xanthene	27
Dibenzocycloheptadiene	31
Triphenylmethane	32
Tetrafluorobenzene	32
Diphenylmethane	34
Allylbenzene	34
Halo and nitrotoluenes	
Benzofuran	36
Benzothiophene	37
Thiophene	38

Acetylene (pK_a = 25), acetonitrile (pK_a = 25), diphenylamine, unactivated amides (pK_a = 25), and unactivated anilines (pK_a = 25) may be considered at the lower end of this pK_a range and nonaqueous soluble unactivated sulfones (pK_a = 33) may be considered in this category.

$$\text{Rate} = k_1[QX]_{org}[OH^-]_{aq}$$

This would be consistent with increasing the overall rate by increasing the aqueous hydroxide concentration.

Many of the alkylations of activated ketones [95], nitriles [96,97], and hydrocarbons [98], first reported by Makosza, using triethylbenzylammonium chloride, fall into this category.

4. Reactions Involving Anions Derived from "Weakly Acidic" Conjugate Acids with: 23<pK_a<38

Anions in this category are very difficult to deprotonate, except in the presence of very strong base (see Table 6-8). In the overall rate expressions for reactions involving these anions, the rate of deprotonation is likely to predominate. Based on empirical observations of the behavior of reactions in this category, it may be speculated that the rate of deprotonation is slower than hydroxide transfer. Additional reactions may occur, with deuteration, oxidation, or isomerization being the most commonly reported. The rate constant associated with deuteration, oxidation, or isomerization is termed k_3 in the kinetic analysis of Chapter 3. The rates of deuteration, isomerization, and oxidation are typically high, and may be assumed to be faster than deprotonation of these weakly acidic substrates.

If indeed the deprotonation in the organic phase is expected to be predominate, a simplified rate expression may be assumed:

$$\text{Rate} = k_{obs}[Q^+OH^-]_{org}[\text{substrate}]$$

In light of the discussion in Sections A.1.b, C, and E of this chapter and the kinetic discussions of Chapters 2 and 3, such reactions are promoted by: the use of an organophilic/anion activating quaternary -onium cation (enhancing both k_{obs} and $[Q^+OH^-]_{org}$; hydrophilic anions as leaving groups and catalyst counteranions (to avoid catalyst poisoning); some minimal agitation (no more than needed for mass transfer); and low water concentration to increase the basicity of the hydroxide ion by reduced organic phase hydration of the hydroxide ion by reduced organic phase hydration of the hydroxide ion (affecting k_{obs}). It is difficult to predict the effect of solvent, due to the likely mixed hard–soft nature of the ion pair consisting of the hard hydroxide and an organophilic/anion activating quaternary ammonium cation.

It can be shown (from the PTC/OH kinetic equation shown in Chapter 3) that under certain circumstances, which would be expected to be associated with hydroxide transfer and slow deprotonation (very low substrate acidity; high pK_a), the rate expression for isomerization would be reduced to:

$$\text{Rate} = \frac{k_1 k_2 [QX]_{org}[OH^-]_{aq}[RH]}{k_{-1}[X^-]_{aq} + k_2[RH]_{org}}$$

and the rate expression for alkylation would be reduced to:

$$\text{Rate} = \frac{k_1 k_2 k_3 [QX]_{org}[OH^-]_{aq}[RH]_{org}[R'X]_{org}}{k_{-1} k_{-2}[X^-]_{aq}}$$

The expression for isomerization is indeed consistent with the observed kinetic behaviors observed for the isomerization of allylbenzene [8].

5. "Very Weakly Acidic" Substrates with $pK_a > 38$

Organic substrates with $pK_a > 38$ have not yet been reported to have been deprotonated under PTC/OH conditions. For example, numerous attempts at deuterating unsubstituted toluene have failed [99]. It is likely that some creative organic chemist will figure out how to use PTC to expand the pK_a range of successfully deprotonated organic substrates. Until then, the current pK_a range definition will include reactions that are termed "not likely to work."

6. A "Quick and Dirty" Method to Screen New Potential PTC Applications

Based on the above guidelines, it is possible to design a "quick and dirty" screening experimental program for new PTC applications. The screening design is illustrated in Figure 6-41. Tetrabutylammonium hydrogen sulfate is a catalyst that will usually work for almost all PTC reactions that will eventually proceed well (not all but almost all). If one assumes that there will be much catalyst

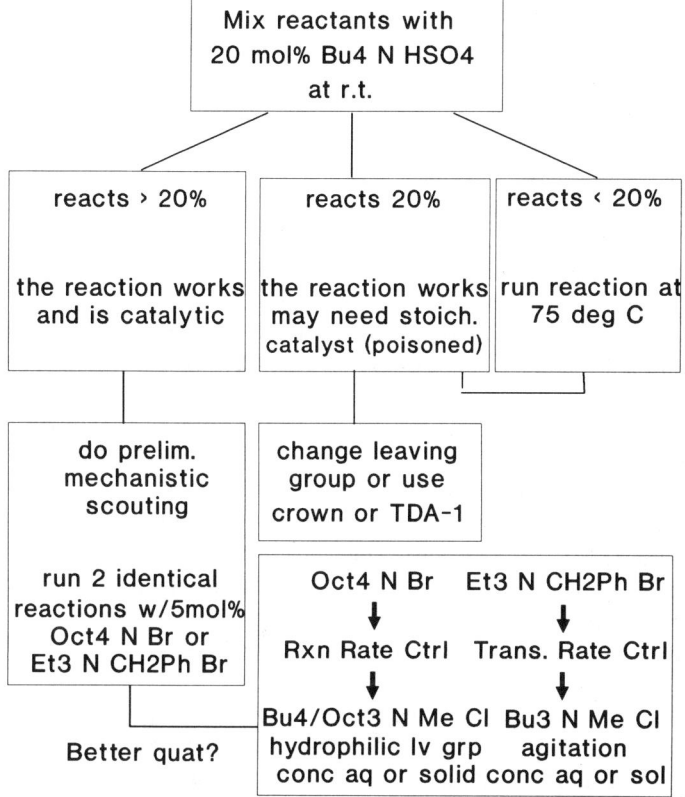

Figure 6-41. An experimental approach to rapid evaluation of new PTC reactions.

decomposition and/or catalyst poisoning, the screening can be performed with a large, though catalytic, quantity of catalyst, such as 20–50 mol%. It is recommended that screening begin with 20 mol% tetrabutylammonium hydrogen sulfate. If the reaction works (proceeds to high conversion well over 20%) then it can be said that the reaction is catalytic and is a good PTC candidate and preliminary screening can begin.

Once the reaction has been determined to be catalytic, the reaction should be performed in two identical side-by-side vessels. One reaction should contain 5 mol% tetraoctylammonium bromide and the other should contain 5 mol% benzyltriethylammonium bromide. The rationale here is to use one organophilic/anion activating catalyst and one accessible catalyst. If the reaction with the tetraoctylammonium proceeds faster than the reaction with the benzyltriethylammonium, then the reaction is likely to be dominated by the intrinsic reaction rate. In such a case, further optimization should be continued by choosing a desirable organophilic/anion activating quaternary -onium salt such as tetrabutylammonium or methyltrioctylammonium, choosing hydrophilic leaving groups, and using concentrated aqueous or solid salt containing the reacting anion. A good example in which this method was actually used to rapidly evaluate and partially "optimize" a reaction was for the preparation of dihalobenzopheneones by oxidation of dihalodiphenylmethanes ($pK_a \sim 32$) [100]. Within fewer than 10 experiments it was found that maximum conversion was obtained for the catalysts/anions Aliquat 336 (Cl) > tetraoctylammonium bromide > tetrabutylammonium hydrogensulfate > tetrabutylammonium bromide > triethylbenzylammonium chloride >> no catalyst. The criteria for determining the desirability of the catalyst are discussed in detail in Section A of this chapter.

If the reaction with the benzyltriethylammonium proceeds faster than the reaction with the tetraoctylammonium, then the reaction is likely to be dominated by the transfer rate. In such a case, further optimization should be continued by choosing a desirable accessible quaternary -onium salt such as methyltributylammonium, assuring as efficient agitation as possible, using concentrated aqueous or solid base, and choosing a polar solvent.

If the reaction, in the presence of 20 mol% tetrabutylammonium hydrogen sulfate, proceeds to < 20% conversion, then the reaction, obviously, does not work well and may be due to the difficulty of the reaction. In such a case it is worthwhile to perform the reaction again at higher temperature. If the reaction works at the higher temperature, perform the preliminary mechanistic scouting described above.

If at the higher temperature the reaction proceeds sluggishly or the reaction proceeds to only 20% conversion (in the presence of 20 mol% tetrabutylammonium hydrogen sulfate), then the reaction is likely to need stoichiometric quaternary -onium salt because of catalyst poisoning or possibly the catalyst decomposed. In such a case it is worthwhile to consider changing the leaving group (to an anion that is less likely to poison the catalyst) or use a PEG, crown, or TDA-1, all of which are more thermally stable and are less prone to poisoning.

References

1. T. Ho, Hard and Soft Acids and Bases Principle in Organic Chemistry, Academic Press, New York (1977).
2. White, C., Mallinckrodt, U.S. Patent 4,642,399 (1987).
3. D. Mason, S. Magdasi, and Y. Sasson, J. Org. Chem., **55**, 2714 (1990).
4. C. Starks and R. Owens, J. Am. Chem. Soc., **95**, 3613 (1973).
5. D. Landini, A. Maria, and F. Montanari, J. Am. Chem. Soc., **100**, 2796 (1978).
6. M. Halpern, Y. Sasson, I. Wilner, and M. Rabinovitz, Tetrahedron Lett., 1719 (1981).
7. A discussion of ion pairing based on ESCA studies of PT catalysts is found in: R.

Moberg, F. Bokman, O. Bohman, and H. Siegbahn, J. Am. Chem. Soc., **113**, 3663 (1991).
8. M. Halpern, Y. Sasson, and M. Rabinovitz, J. Org. Chem., **48**, 1022 (1983).
9. A Herriott and D. Picker, J. Am. Chem. Soc., **97**, 2345 (1975).
10. M. Makosza, Pure Appl. Chem., **43**, 439 (1975).
11. M. Halpern, Y. Sasson, and M. Rabinovitz, Tetrahedron, **38**, 3183 (1983).
12. M. Halpern, unpublished (1988).
13. Data taken from: M. Halpern, Y. Sasson, and M. Rabinovitz, Tetrahedron, **38**, 3183 (1982); E. Dehmlow and S. Dehmlow, Phase Transfer Catalysis, 2nd edit., Verlag Chemie, Weinheim, pp. 48–49 (1983); M. Makosza and B. Serafinowa, Rocz. Chem., **39**, 1223 (1965).
14. Data taken from: M. McDowell and C. Draus, J. Am. Chem. Soc., **73**, 3293 (1951).
15. E. Dehmlow and S. Naranjo, J. Chem. Res. (S), 238 (1979).
16. E. Dehmlow and S. Dehmlow, Phase-Transfer Catalysis, 2nd edit., Verlag Chemie, Weinheim, p. 48 (1983).
17. H. Dou, R. Gallo, P. Hassanaly, and J. Metzger, J. Org. Chem., **42**, 4275 (1977).
18. D. Landini, A. Maria, and A. Rampoldi, J. Org. Chem., **51**, 3187 (1986).
19. Additional work reported: E. Dehmlow, M. Slopianka, and J. Heider, Tetrahedron Lett., 2361 (1977).
20. J. de la Zerda, R. Neumann, and Y. Sasson, J. Chem. Soc. Perkin Trans. II, 823 (1986).
21. C. Starks, J. Am. Chem. Soc., **93**, 195 (1971).
22. J. Dutcher, SACHEM, unpublished results.
23. T. Lane and J. Speier, J. Org. Chem., **41**, 2714 (1976).
24. LD_{50}'s were provided by suppliers. Before using any phase-transfer catalyst, you must obtain the most recent Material Data Safety Sheet and be sure that you know the toxicity of the materials you work with. The authors are not responsible for toxicity data reported.
25. Rhone-Poulenc, commercial literature.
26. Review article: G. Totten and N. Clinton, JMS-Rev. Macromol. Chem. Phys., **C28**, 293 (1988).
27. D. Brunelle, Phase-Transfer Catalysis, ACS Symposium Series 326, C. Starks, ed., p. 38, American Chemical Society, Washington (1987).
28. D. Brunelle and D. Singleton, Tetrahedron Lett., 3383 (1984).
29. PT catalyst separation techniques are discussed as part of a general economic evaluation of the use of phase-transfer catalysts in industrial applications in: B. Zaidman, Y. Sasson and R. Neumann, Ind. Eng. Chem. Prod. Res. Dev., **24**, 390 (1985).
30. K. Sjoberg, Aldrichimica Acta, **13**, 55 (1980).
31. B. Berris, Ethyl, U.S. Patent 5,030,757 (1991).
32. E. Taracon, Ercros, Spanish Patent ES 2,023,606 (1992); CA 117:48107b.

33. Brunelle, D., private communication.
34. PT catalyst separation techniques are discussed as part of a general economic evaluation of the use of phase-transfer catalysts in industrial applications in: B. Zaidman, Y. Sasson and R. Neumann, Ind. Eng. Chem. Prod. Res. Dev., **24**, 390 (1985).
35. S. Regen, J. Heh, and J. McLick, J. Org. Chem., **44** 1961 (1979).
36. S. Telford, P. Schlunt and P. Chau, Macromolecules, **19**, 2435 (1986).
37. S. Regen and J. Besse, J. Am. Chem. Soc., **101**, 4059 (1979).
38. A. Au and H. Freedman, U.S. Patent 4,173,693 (Dow Chemical) Nov 6 (1979).
39. M. Tomoi and W. Ford, J. Am. Chem. Soc., **102**, 7140 (1980).
40. S. Regen, J. Am. Chem. Soc., **97**, 5956 (1975).
41. P. Anelli, B. Czech, F. Montanari, and S. Quici, J. Am. Chem. Soc., **106**, 861 (1984).
42. T. Nishikubo, T. Iizawa, M. Shimojo, T. Kato, and A. Shiina, J. Org. Chem., 55, 2536 (1990).
43. S. Regen, J. Am. Chem. Soc., **98**, 6270 (1976).
44. P. Tundo and P. Venturello, J. Am. Chem. Soc., **103**, 856 (1981).
45. T. Balakrishnan and W. Ford, J. Org. Chem., **48**, 1029 (1983).
46. M. Tomoi and W. Ford, J. Am. Chem. Soc., **103**, 3828 (1981).
47. P. Tundo, P. Venturello, and E. Angeletti, J. Chem. Soc. Perkin Trans. I, 2157 (1987).
48. V. Ragaini, J. Org. Chem., **27**, 1382 (1988).
49. S. Telford, P. Schlunt, and P. Chau, Macromolecules, **19**, 2435 (1986).
50. S. Regen and J. Besse, J. Am. Chem. Soc., **101**, 4059 (1979).
51. M. Tomoi and W. Ford, J. Am. Chem. Soc., **102**, 7140 (1980).
52. S. Regen, J. Am. Chem. Soc., **102**, 7140 (1980).
53. P. Anelli, B. Czech, F. Montanari, and S. Quici, J. Am. Chem. Soc., **106**, 861 (1984).
54. T. Balakrishnan and W. Ford, J. Org. Chem., **48**, 1029 (1983).
55. P. Tundo and P. Venturello, J. Am. Chem. Soc., **103**, 856 (1981).
56. S. Regen and J. Besse, J. Am. Chem. Soc., **101**, 4059 (1979).
57. T. Stanley and J. Quinn, Chem. Eng. Sci., **42**, 2313 (1987).
58. S. Matson, Advances in Catalytic Technologies Seminars, Catalytica, Mountain View, CA (1988).
59. Y. Okahata and K. Origa, J. Org. Chem., **51**, 5064 (1986).
60. Ingold, Structure and Mechanism in Organic Chemistry, Cornell University Press, Ithaca, pp. 345–350 (1953).
61. Figure compiled from data provided by Fluka AG.
62. D. Wang and H. Weng, Chem. Eng. Sci., **43**, 2019 (1988).

63. M. Halpern, Ph.D. Thesis, Hebrew University of Jerusalem (1983).
64. M. Wang and H. Wu, J. Org. Chem., **55**, 2344 (1990).
65. G. Bram, A. Loupy, and J. Sansoulet, Israel. J. Chem., **26**, 291 (1985).
66. J. Barry, G. Bram, G. Decodts, A. Loupy, P. Pigeon, and J. Sansoulet, Tetrahedron, **39**, 2673 (1983).
67. G. Bram, A. Loupy, and M. Majdoub, Synth. Commun., **20**, 125 (1990).
68. J. Barry, G. Bram, G. Decodts, A. Loupy, P. Pigeon, and J. Sansoulet, Tetrahedron, **40**, 2945 (1984).
69. A. Loupy, J. Sansoulet, E. Diez-Berra, and J. Carrillo, Synth. Commun., **21**, 1465 (1991).
70. J. Barry, G. Bram, G. Decodts, A. Loupy, P. Pigeon, and J. Sansoulet, Tetrahedron Lett., 5407 (1982).
71. G. Bram, A. Loupy, M. Roux-Schmitt, J. Sansoulet, T. Strzalko, and J. Seyden-Penne, Synthesis, 56 (1987).
72. J. Barry, G. Bram, and A. Petit, Tetrahedron Lett., 4567 (1988).
73. A. Loupy, N. Philippon, P. Piogeon, and J. Sansoulet, Synth. Commun., **20**, 2855 (1990).
74. A. Loupy, N. Philippon, P. Pigeon, and H. Galons, Heterocycles, **32**, 1947 (1991).
75. A. Loupy, J. Sansoulet, E. Diez-Barra, and J. Carrillo, Synth. Commun., **21**, 1465 (1991).
76. E. Delee, I. Jullien, L. Le Garrec, A. Loupy, J. Sansoulet, and A. Zaparucha, J. Chromatogr., **450**, 183 (1988).
77. E. Santaniello, A. Manzocchi, and P. Sozzani, Tetrahedron Lett., 4581 (1979).
78. I. Ganboa and C. Palomo, Synthesis, 52 (1986).
79. T. Adaway, Dow Chemical U.S. Patent 4,701,531 (1987).
80. E. Diez-Barra, A. De la Hoz, A. Sanchez-Migallon, and J. Tejeda, Heterocycles, **34**, 1365 (1992).
81. S. Dermeik and Y. Sasson, J. Org. Chem., **50**, 879 (1985).
82. R. Solaro, S. D'Antone and E. Chiellini, J. Org. Chem., **45**, 4179 (1980) and M. Halpern, Y. Sasson, and M. Rabinovitz, J. Org. Chem., **48**, 1022 (1983).
83. D. Landini, A. Maia, and A. Rampoldi, J. Org. Chem., **51**, 5465 (1986).
84. D. Landini and A. Maia, J. Chem. Soc. Chem. Commun., 1041 (1984).
85. D. Landini, A. Maia, and G. Podda, J. Org. Chem., **47**, 2264 (1982).
86. R. Solaro, S. D'Antone, and E. Chiellini, J. Org. Chem., **45**, 4179 (1980).
87. V. Krishnakumar and M. Sharma, Ind. Eng. Chem. Proc. Des. Dev., **23**, 410 (1984).
88. M. Halpern, Ph.D. Thesis (1983).
89. M. Yeh, H. Guo, L. Chen, and Y. Shih, Ind. Eng. Chem. Res., **27**, 1582 (1988).
90. E. Dehmlow and S. Dehmlow, Phase-Transfer Catalysis, Verlag Chemie, Weinheim, pp. 14–15 (1983).

91. E. Dehmlow and M. Lissel, Tetrahedron, **37**, 1653 (1981).
92. M. Makosza and W. Lasek, Tetrahedron, **47**, 2843 (1991).
93. M. Halpern, Ph.D. Thesis, Hebrew University of Jerusalem (1983).
94. pK_a values may be found in the following references:
    A. Streitwieser, E. Ciuffarin, and J. Hammons, J. Am. Chem. Soc., **89**, 63 (1967).
    A. Streitwieser and P. Scannon, J. Am. Chem. Soc., **95**, 6273 (1973).
    R. Pearson and R. Dillon, J. Am. Chem. Soc., **75**, 2439 (1953).
    The Chemists Companion Handbook of Practical Data, Techniques and References, A. Gordon and R. Ford, eds., Wiley, New York, pp. 61–64 (1972).
95. A. Jonczyk, B. Serafin, and M. Makosza, Tetrahedron Lett., 1351 (1971).
96. M. Makosza and B. Serafinowa, Rocz. Chem., **39**, 1223 (1969).
97. M. Makosza and B. Serafinowa, Rocz. Chem., **39**, 1401 (1965).
98. M. Makosza, Tetrahedron Lett., 4621 (1966).
99. D. Feldman, M. Halpern and M. Rabinovitz, unpublished results.
100. M. Halpern and Z. Lysenko, J. Org. Chem., **54**, 1201 (1989).

7

Phase-Transfer Catalysis Displacement Reactions with Simple Anions

A. General Considerations

Phase-transfer catalysis (PTC) is an excellent technique both in liquid–liquid and liquid–solid (and sometimes in liquid–vapor) modes for conducting displacement reactions with anions.

INSERT EQUATION (7-1)

A variety of anions, phase-transfer catalysts, reaction conditions, and organic substrates can be used. Most work has been done on displacements with alkyl chlorides, bromides, and methanesulfonates, although others, for example, aryl halides containing strongly electron-withdrawing o- or p-substituents and dialkyl carbonates can also be used [1,2].

PTC operations, particularly on an industrial scale, are generally superior to conventional methods using dipolar aprotic solvents such as N,N-dimethylformamide, dimethyl sulfoxide, and hexamethylphosphoramide because of:

- Convenience of conducting two-phase reactions
- Excellent methods for control of PTC reactions
- Absence of side reactions related to solvolysis products and impurities that accumulate in solvents
- Ability to avoid environmental problems and toxicity problems, and solvent purification costs encountered in use of solvents
- Ability to conduct sequential PTC reactions by simple removal of the aqueous or solid phase and replacing it with another reagent to conduct the second reaction.

340 / Phase-Transfer Catalysis

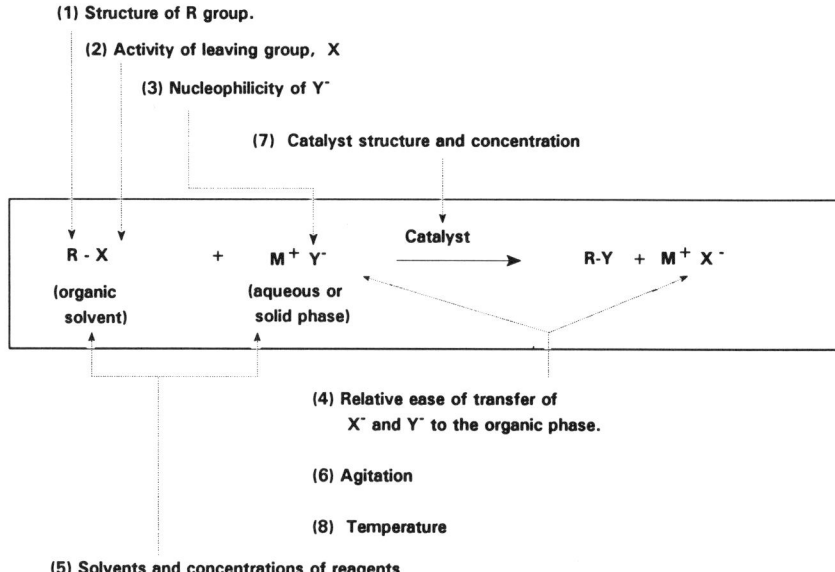

Figure 7.1. Important factors in PTC displacement reactions.

If a solvent for the organic phase must be used the PTC method usually allows the chemist a much wider choice of possible solvents.

1. Important Factors in PTC Displacement Reactions

Many aspects of simple PTC displacement reactions affect the outcome and the rate of these processes. Most of these are noted in Figure 7.1. These factors fall into several subgroups, which sometimes overlap.

1. Factors affecting the organic-phase reaction:
   a. Structure of the R group: For example, rates of displacement generally follow the order:

   $PhCH_2 >$ allyl $>$ primary alkyl $>$ secondary alkyl

   b. Leaving group activity: For example,

   $MeSO_3^- > Br^- > Cl^-$

   c. Nucleophilicity of displacing group. For example,

   $RS^- > SCN^-, I^- > CN^- > Cl^- > F^-, OH^-$

These factors depend mostly on the kind of reaction system being used and are not generally affected by whether the reaction system is in homogeneous solution or in a PTC system. These orders of reactivity may be reversed, depending on the amount of water available for anion hydration.

2. Transfer efficiency factors:
   d. Relative ease of transfer of nucleophile, Y^-, from aqueous or solid phase to the organic phase, compared to ease of transfer of displaced anion, X^- from the organic to the aqueous or solid phase. Once reaction has started, displaced anion X^- from the organic-phase reaction competes with nucleophile Y^- for association with the catalyst. Best catalytic activity is realized with fast transfer of Y^- and slow transfer of X^-. For example, displacement reactions on alkyl chlorides, bromides, and methanesulfonates occur readily and go to completion. The displaced Cl^-, Br^-, and $CH_3SO_3^-$ anions do compete with most anions for association with the catalyst, but not so strongly as to inhibit or poison the reaction. However, the same displacement reactions with alkyl iodides often fail to go to completion. This occurs because the iodide anion produced associates so strongly with most phase-transfer catalysts that negligible catalyst is available in the active Q^+Y^- form. (Trace amounts of iodide in some PTC reactions accelerate the rate by intermediate formation of more active alkyl iodides. Even so, iodide anion concentration must be kept to low levels.)

   e. Solvents and concentration of reagents: Use of polar solvents, for example, methylene chloride, has a levelling effect on the ease of transfer of anions from aqueous environments to organic solutions, whereas nonpolar solvents tend to magnify the differences in extractability between anions. Highly nonpolar solvents may reduce the solubility of the phase-transfer catalyst in the organic phase to such an extent that it slows or inhibits reaction. High concentrations of salts in the aqueous phase (of liquid–liquid PTC reactions) reduce the degree of hydration of anions and also level out the differences in ease of transfer of different anions. Too little water in the inorganic phase may reduce transfer rates by limiting access of the catalyst to inorganic reagents.

   f. Agitation: Agitation is an important factor in all phase-transfer systems. With no agitation, reaction rates are limited by interfacial diffusion, and are slow (on the order of days or weeks). For reaction systems where the intrinsic organic-phase reaction rate is slow compared to nucleophile transfer to the organic phase, some agitation is essential but as transfer rate increases with agitation, the overall conversion rate will reach a point where it becomes independent of agitation rate. (See Chapter 3.) This point occurs when the rate of anion transfer into the organic phase is greater than the rate of organic-phase reaction. For PTC systems where the intrinsic organic-phase reaction is much faster than the net rate of transfer of anion to the organic phase, overall conversion rates will continuously increase with agitation for all practical agitation rates, as described in Chapter 3.

3. Factors that affect both transfer rates and intrinsic reaction rates:

   g. PTC catalyst structure and concentration: The kind and concentration of the phase-transfer catalyst is frequently the most important factor in the use of the PTC technique. For most displacement reactions the catalyst must function both to transfer the nucleophile and to activate the transferred nucleophile for displacement reaction. Catalyst requirements and options are extensively reviewed in Chapters 2–6.

   h. Temperature: Increasing the reaction temperature usually increases the rate of organic-phase displacement reactions, and this is a useful approach to try with reactions limited by the rate of the organic phase reaction.

   The effect of temperature on transfer rates of anions from aqueous or solid phases to organic phases is more complex. For quaternary salts, increasing reaction temperatures does not significantly affect catalyst efficiency, except when quaternary salt decomposition becomes significant (see Chapter 5), or where temperature can significantly affect the degree of hydration for the anions involved. For polyethers and crown ethers as phase-transfer catalysts, where catalyst efficiency depends on formation of thermally reversible complexes, increasing temperature can reduce complexation and therefore catalyst efficiency, although this is not usually observed.

2. Characteristics of Various Anions for Simple PTC Displacement Reactions

As noted in Figure 7-1, anions have three important characteristics for PTC displacement reactions: ease of anion transfer to the organic phase relative to the displaced anion, nucleophilicity, and relative activity as a leaving group. These three characteristics are qualitatively ranked in Table 7-1 for several common anions under generally used reaction conditions.

3. PTC Catalysts for Simple Displacement Reactions

In general, bulky quaternary salts such as tetrabutyl- or higher tetraalkylammonium or phosphonium salts perform very well for displacement reactions. Smaller quaternary salts, tetramethyl to tetrapropyl, have less solubility in organic phases and less ability to activate anions, leading to less efficient performance as catalysts, although this also depends on the polarity of the organic phase. Crown ethers function well as catalysts in most solid–liquid displacement reactions, as do cryptands. Polyethylene glycols (PEGs) also function reasonably well, particularly with potassium salts with minimal water concentration in liquid–solid PTC. Some generalizations regarding utility of various phase-transfer catalysts for displacement reactions are given in Table 7-2.

Table 7-1. Characteristics of various anions for simple PTC displacemments.

Anion	Relative ease of transfer	Relative nucleophilicity	Relative activity as leaving group
Cl^-	++	++	+
Br^-	++	+++	++
I^-	++++	++++	+++
F^-	+	+	—
CN^-	++	++	nd[a]
SCN^-	++++	++++	nd
N_3^-	++++	+++	nd
NCO_2^-	+	+	nd
$CH_3CO_2^-$	++	++	nd
RCO_2^-	++	++	nd
OH^-	—	+	nd
RO^-	++	+++	nd
ArO^-	++	+++	nd
S^{2-} or HS^-	++	+++	nd
RS^-	++++	++++	nd
ArS^-	++++	++++	nd
$ArSO_2^-$	++	+++	nd
$CH_3SO_3^-$	+	—	+++
$p\text{-}CH_3C_6H_4SO_3^-$	++++	—	+++
SO_4^{-2}	—		++
PO_4^{-3}	—		

[a]ND, not practically displaceable.

B. Behavior of Various Anions in PTC Displacement Reactions

1. Cyanide Displacements

Many types of PTC catalysts have been used for cyanide displacements on alkyl halides to yield nitriles, as represented by the equation:

$$\text{R-X} + \text{M}^+\text{CN}^- \xrightarrow{\text{PTC Catalyst}} \text{R-CN} + \text{M}^+\text{X}^- \quad (7\text{-}2)$$
(org)   (aq or s)

A summary of some cyanide displacement reactions is listed in Table 7-4 (see p. 348) with soluble PTC catalysts, and in Table 7-5 (see p. 349) for insoluble PTC catalysts.

Using aqueous NaCN (liquid–liquid PTC), simple tetraalkylammonium or phosphonium salts (akyl groups are, e.g. butyl to decyl) or triakylmethylammonium salts perform excellently to give high yields of products at convenient reaction times at moderate temperatures (75–105°C). Most quaternary ammonium salts having reasonable solubility in the organic phase, and that are bulky enough to provide some activation for the cyanide anion, are good PTC catalysts for cyanide displacement. However, tetramethylammonium, tetraethylammonium, tetrapropylammonium, and benzyltriethylammonium salts are usually not good PTC

Table 7-2. General catalysts for PTC displacement reactions.

Catalyst types		Performance as PTC Catalyst
Quaternary salts		
R_4N^+ or R_4P^+	[R = C_5 to C_{10} alkyl groups]	Excellent catalysts
R_3NMe^+	[R > C_6 alkyl groups]	Usually excellent catalysts
Bu_4N^+, Bu_3NMe^+		May form third phase
R_4N^+	[R = Me, Et, Pr]	May be satisfactory for easily transferred anions with fast intrinsic reactions; not generally good unless polar solvent used.
R_4N^+	[R > C_{11} alkyl groups]	Activity declines with greater chain length because of less efficient anion transfer
Special catalysts (See Chapter 5.)		Higher temperature stability, chiral, stable to hydroxide, etc.

PEGs and tris (2,5-dioxaheptyl)amine (TDA-1):
  These are often satisfactory catalysts for displacement reactions when used with nearly anhydrous salts, and are particularly good when used with potassium salts. PEGs are especially effective with strong hot hydroxide or alkaline reaction systems. PEGs may form a third phase with highly concentrated aqueous solutions and nonpolar organic phases. PEG can often be recovered from organic products by extraction into water; TDA-1 can be recovered by extraction into dilute aqueous acids.

Soluble polymers containing dipolar groups:
  These phase-transfer agents are good catalysts for most displacement reactions.

Crown ethers and cryptands:
  These materials are excellent catalysts for PTC displacement reactions.

Insoluble catalysts:
  Resin-bound catalysts, catalysts on inorganic, solids and third-phase liquid catalysts are often excellent catalysts for displacement reactions, but usually require substantial experimental work to develop the best catalyst. See Chapter 5.

catalysts for liquid–liquid cyanide displacement reactions, unless the organic phase is highly polar, and/or the alkyl halide is of exceptionally high reactivity, such as with benzyl or allyl chloride or bromide.

Tetrabutylammonium salts are often good catalysts for cyanide displacements since they can be reasonably well separated from most organic products by washing with water. A patent [3] for preparation of an ibuprofen intermediate by a nitrile displacement reaction provides an example of this kind of separation.

Another good example of a commercial process for cyanide displacement is described in three Japanese patents for conversion of 1,6-dichlorohexane to suberonitrile, an intermediate in the manufacture of 1,8-diaminooctane and of suberic acid, for production of Nylon-8.

$$\text{Cl-(CH}_2)_6\text{-Cl} + \text{NaCN} \xrightarrow{Bu_4N^+ \; X^-} \text{NC-(CH}_2)_6\text{-CN} + \text{NaCl} \quad (7\text{-}3)$$

This process [4–6] also uses tetrabutylammonium salts as catalyst so that much of the catalyst can be removed and recovered from the organic phase by extraction of catalyst into water. The crude suberonitrile product from the reaction, after water washing, is then initially flash distilled at high temperature to take crude dinitrile overhead, along with fragments resulting from decomposition of any remaining catalyst. This distillate, which no longer contains thermally decomposable materials, can then be carefully fractionally distilled to produce high-purity suberonitrile. This flash-then-fractionate procedure avoids complications from a continuous slow decomposition of Bu_4N^+ salts that occurs when crude organic product is fractionated directly without prior decomposition of residual catalyst.

With highly reactive alkyl halides, such as benzyl halides, cyanide displacement occurs rapidly to produce easily alkylated phenylacetonitriles, but these can lead to alkylated by-products: Slow addition of NaCN to keep the aqueous phase pH in the range of 8–9.5 prevents these secondary alkylations [7]. This problem is not observed in liquid–solid crown ether-catalyzed cyanide displacement reactions.

$$C_6H_5CH_2Cl + CN^-(aq) \xrightarrow{R_4N^+} C_6H_5CH_2\text{-}CN \quad (7\text{-}4)$$

$$\xrightarrow[pH > 9.5]{PhCH_2Cl}$$

$$C_6H_5\underset{CH_2C_6H_5}{-}CH\text{-}CN \xrightarrow[pH > 9.5]{PhCH_2Cl} C_6H_5\underset{CH_2C_6H_5}{\overset{CH_2C_6H_5}{-}}C\text{-}CN$$

Cyanide displacements also proceed well with solid NaCN or KCN (i.e., solid–liquid PTC) when catalyzed by quaternary salts, provided a small amount of water [8,9] or alcohol [10] is present. When catalyzed by crown ethers, it is not necessary to add any water to the reaction mixture [11], although it has been demonstrated [12] that maximum rates are obtained when minute amounts of water are present. Under these conditions, most of the catalyst is no longer present in the organic phase, but is located on the surface of the solid salt particles. The initial water added to the system coats the surface of the solid salt phase, known as the *omega* phase, which extracts the 18-crown-6 from the organic phase onto the surface of the salt. Cyanide displacements may be run under strictly anhydrous conditions by the ion-pair extraction technique, where a full molar amount of anhydrous $Q^+ CN^-$ is prepared in an organic solvent [13–16].

Use of poly(ethylene glycols) as phase-transfer catalysts for cyanide displacement on 1-chlorooctane has been demonstrated by Gokel and co-workers [17], where kinetics were shown to be independent of the molecular weight of the PEG.

Use of insoluble phase-transfer catalysts, particularly polymer-bound quaternary salts and crown ethers have been extensively investigated for cyanide displacement reactions, as summarized in Table 7-5. Work by Ford and Tomoi, as summarized in Chapter 5, has provided a basic understanding of the physical principles involved in using insoluble catalysts for cyanide displacements. By careful adjustment of the reaction conditions, the insoluble-resin support, and the PTC groups, rather good catalytic efficiencies can be achieved. For commercial applications, however, the most cost-effective insoluble catalysts appear to be those supported on organophilic clays, as described by Lin and Pinnavaia [18] and by Choudary, Rao, and Prasad [19], who demonstrate that simple treatment of the clays with low-cost commercially available quaternary salts produce insoluble catalysts that have good activity, are easily separated by centrifugation, and reusable, as described in Chapter 5.

Reactions of acid chlorides with NaCN in liquid–liquid PTC reactions yield the corresponding acyl-nitrile derivatives [20],

$$C_6H_5COCl + NaCN \xrightarrow[\text{(aq)}]{Bu_4N^+ X^-} C_6H_5COCN + NaCl \quad (7\text{-}5)$$

Yields of simple acylnitrile products are reduced because of dimer formation,

$$PhCOCN + CN^- \xrightarrow{PTC} Ph\underset{CN}{\overset{O^-}{\underset{|}{-C-}}}CN \xrightarrow{PhCOCl} Ph\underset{CN}{\overset{OCOPh}{\underset{|}{-C-}}}CN \quad (7\text{-}6)$$

as listed in Table 7-3. The dimer-forming reaction, starting with 3,4-dichlorobenzoyl chloride, has been patented as a commercial route to pesticidal compounds [21].

Carbonyl cyanide, $CO(CN)_2$ can be prepared from phosgene and KCN in a liquid–solid phase transfer system [22], and cyanoformic acids and cyanomonothioformic acids, ROCOCN and RSCOCN, can be prepared from the corresponding acid chlorides, ROCOCl or RSCOCl [23].

Vinyl halides do not undergo simple cyanide displacement reactions, but 18-crown-6 catalyzed PTC reactions in the presence of $Pd(PPh_3)_4$ as a cocatalyst yield substituted acrylonitriles in excellent yields and in a stereospecific manner [24]:

$$\underset{R}{\overset{R}{>}}C=C\underset{Br}{\overset{R}{<}} + KCN \xrightarrow[Pd(PPh_3)_4]{18\text{-Crown-}6} \underset{R}{\overset{R}{>}}C=C\underset{CN}{\overset{R}{<}} \quad (7\text{-}7)$$

Cyanide reaction with epoxides in the presence of potassium hydroxides gives 3-hydroxynitriles in a regioselective manner [266]. Thus, (R)-styrene oxide in

Table 7-3. Product distribution from NaCN reaction with substituted benzoyl chlorides.

Benzoyl chloride substrate	Percent yield of PhCOCN	Percent yield of Dimer
H	60	35
p-CH_3-	72	—
p-CH_3O-	60	—
p-Cl-	22	46

the presence of KOH and KCN, catalyzed by tetrabutylammonium salts, give (S)-3-phenyl-3-hydroxypropane nitrile.

2. Halide Displacement and Exchange Reactions

a. Equilibrium Features of Halide Exchange and Displacement Reactions

Halide displacement reactions, with a few exceptions, work well under PTC conditions. These reactions may be represented by the equilibrium:

At equilibrium, the concentration of the product R−Y is determined by two subequilibria, one involving anion transfer, and the other involving the relative rates of the forward and reverse organic-phase reactions. For anion transfer where the equilibrium is governed by:

$$\text{R-X}_{(org)} + \text{NaY}_{(aq\ or\ s)} \underset{}{\overset{PTC}{\rightleftarrows}} \text{R-Y} + \text{NaX} \quad (7\text{-}8)$$

$$(\text{Catalyst})^+ \text{X}^-_{(org)} + \text{Y}^-_{(aq)} \overset{K_p}{\rightleftarrows} (\text{Catalyst})^+ \text{Y}^-_{(org)} + \text{X}^-_{(aq)} \quad (7\text{-}9)$$

$$K_p = \frac{[(Catalyst)^+\ Y^-]_{org}[X^-]_{aq}}{[(Catalyst)^+\ X^-]_{org}[Y^-]_{aq}}$$

where K_p represents the selectivity of the catalyst cation to be associated with anion Y^- relative to anion X^-.

The organic phase equilibrium,

$$\text{R-X} + (\text{Catalyst})^+\ Y^- \overset{K_o}{\rightleftarrows} \text{R-Y} + (\text{Catalyst})^+\ X^- \quad (7\text{-}10)$$

is represented by

Table 7-4. General survey of PTC cyanide displacements with soluble catalysts.

Substrate type	Product	Catalyst type	References
$CH_3(CH_2)_n-X$ $X = -Br, -Cl,$ $-O_3SMe$	$CH_3(CH_2)_n-CN$ (ca. 100%)	R_4N^+, R_3NMe^+, R_4P^+ ($R = C_4-C_{12}$ alkyl groups)	[25, 26]
		$PhCH_2NMe_2(C_{12}H_{25})^+$	[27]
		$C_{16}H_{33}NMe_3^+$	[28]
		Crown ethers	[29–32]
		Cryptands	[30]
		PEGs	[17]
		Polypod ethers	[33]
		Ethoxylated bis-quats	[34]
		Soluble poly(N,N-dimethyl-acrylamides)	[35]
$CH_3(CH_2)_nCH_2CHR-X$	$CH_3(CH_2)_nCH_2CHR-CN$ (70–85% yields) $CH_3(CH_2)_nCH = CHR$ (15–30%)	$R_4N^+, R_4P^+,$ Crown ethers	[25] [29]
$(CH_3)_3C-Br$	$(CH_3)_2C = CH_2$ (100%)	R_4P^+, crown ethers	[36]
Bromocyclohexane	cyclohexene (100%)	R_4P^+, crown ethers	[36]
$RCH_2C(Me) = CHCH_2X$	$RCH_2C(Me) = CHCH_2CN^a$	Quats	[37]
$R-C_6H_4CH_2-Cl$	$R-C_6H_4CH_2CN$ (90–98%) Alkylated byproducts (2–10%)	R_4N^+ Amines[b]	[25, 38–41] [42–44]
$(C_6H_5)_2CH-Br$	$(C_6H_5)_2CH-CN$	R_4N^+	[45]
$4-R-C_6H_4CHR'Cl$	$4-R-C_6H_4CHR'CN^c$	Bu_4N^{+d} $C_6H_5NEt_3^+$	[3] [46]
$X-(CH_2)_n-X$	$NC-(CH_2)_n-CN$	R_3NMe^+ $PhCH_2NMe_3^+$ Crown ethers	[4, 6, 25] [47–49] [29]
	$X-(CH_2)_n-CN$	$C_{16}H_{33}NMe^+$	[50–52]
$X-(CH_2)_n-Y$ $Y = -CN, -CO_2Et,$ $-CONH_2$	$NC-(CH_2)_n-Y$	$C_6H_5CH_2NMe_3^+$	[47]
$S(CH_2CH_2-Cl)_2$	$S(CH_2CH_2CN)_2$	$PhCH_2NBu_3^+$	[53]

[a]Products used in formulation of perfumes.

[b]Quaternary salts generated in situ.

[c]Products used for manufacture of ibuprofen and for precursors for nonsteroidal antiflamatory α-arylaliphatic acids.

[d]Method for recovery of tetrabutylammonium catalysts by extraction is described.

$$K_o = \frac{[R-X]_{org}\,[(Catalyst)^+\,Y^-]_{org}}{[R-Y]_{org}\,[(Catalyst)^+\,X^-]_{org}}$$

From the two equilibrium equations the concentration of product, R–Y, in the equilibrated mixture will be given by

Table 7-5. *General survey of PTC cyanide displacements on n-alkyl halides and methanesulfonates with insoluble PTC catalysts.*

Catalyst type	References
Quaternary groups, crown ethers, and cryptands bound to insoluble polymer resins	[54]
Poly(styrene-divinyl benzene)bound $-PBu_3^+$ groups	[45, 55–59]
with spacer chains[a]	[60]
Poly(styrene-divinyl benzene) bound crown ethers[b]	[61–63]
Poly(N,N-dimethylacrylamido-co-styrene-co-DVB)	[64]
Poly (p-phenyl)trimethylenephosphonium salts]	[65]
Quaternary salts on organophilic clays	[18, 66]
Fiber-bound phosphonium salts[c]	[67]

[a]Addition of spacer chains produces much higher catalyst activity.

[b]Polymer-bound crown ethers were found to be more efficient catalysts than the corresponding soluble crown ethers.

[c]Fiber-bound phosphonium salts were claimed to be easier to remove by filtration than bead-bound catalysts.

$$[R-Y] = \frac{K_p [R-X]_{org} [Y^-]_{aq}}{K_o [X^-]_{aq}}$$

On a practical basis it is found that when $X = {}^-O_3SCH_3$ as a leaving group, such that although K_0 is large, the value of K_p is still sufficiently larger that displacements yield essentially 100% conversion to product $R-Y$. At the other extreme, when X = iodide as a leaving group, the value of K_p (but not K_0) is so small that Cl^- and Br^- displacements proceed only to a few percent (ca. 10–20%) conversion before iodide concentration builds up to a sufficient concentration as to almost completely stop the reaction. However, if iodide is the nucleophile, that is, $Y^- = I^-$, and chloride or bromide is the leaving group, $X = Br$ or Cl, reaction to produce $R-I$ goes almost to 100% conversion. Between these two extremes conversions of $R-X$ to $R-Y$ are determined by equilibrium conditions. Conversion can be driven to near 100% by adjustment of the Y^-/X^- ratios in the aqueous (or solid phase), or by periodic replacement of equilibrated aqueous or solid phase with fresh NaY solution or solid.

b. Bromide–Chloride Exchanges

As can be seen from the equations above the concentration of RY in the product will depend directly on the relative concentrations of the inorganic anions in the aqueous or solid phase. For example, for the simple bromide–chloride exchange, represented by Eqs. (7-9) and (7-10), the value of $K_p = 16$ and $K_0 = 13$. However, beyond the simple concentration dependencies on anions X^- and Y^-, the fraction of 1-bromooctane produced may also depend on (1) the nature of the catalyst, (2) on the nature of the cation M^+, and (3) on the amount of water available in the aqueous or solid phase to hydrate anions and cations. Catalyst dependency results from unequal abilities of different catalyst structures to activate anions and unequal abilities to transfer anions

$$1\text{-}C_8H_{17}\text{-}Cl + M^+ Br^- \underset{}{\overset{C_{16}H_{33}PBu_3^+ \ Br^-}{\rightleftarrows}} 1\text{-}C_8H_{17}\text{-}Br + M^+ Cl^- \quad (7\text{-}11)$$

(neat org)   (aq)

into the organic phase. Cation dependence results from unequal (or nonproportional) solubilities of M^+Cl^- and M^+Br^- for different kinds of M^+ in the aqueous phase (assuming saturated solutions are used in the aqueous phase.) The degree of anion hydration affects both the transferability of an anion from aqueous or solid to organic phase, and reactivity once it has been transferred. For example [68], excess water available for hydration significantly reduces the reaction rates for chloride and bromide displacement on 1-octyl methanesulfonate, but hydration had little effect on the corresponding iodide displacement on 1-octyl methanesulfonate.

In another study the equilibrium position of the PTC reaction of a primary alkyl chloride with M^+Br^- (M = Li, Na, K, Ca) to give the alkyl bromide was found to depend strongly on the nature of cation, M^+, and on the amount of water present [69]. Primary alkyl chlorides were converted to the corresponding bromides in 80–94% yield by treatment with a slight excess of LiBr containing 1% water and 5 mol% Aliquat 336 at 85–110°C. Alkyl bromides were converted to the corresponding chlorides by treatment with 25% aq. solutions of the metal chlorides containing Aliquat 336 at 95°C. Equilibrium constant data for these reactions have been published [70]. These constants depend on both cation and inorganic phase composition.

Bulky quaternary salts are preferred catalysts [71–73] although even cetyltrimethylammonium bromide in water at 100°C for 48 h will work [28]. Crown ethers, cryptands [30,68,74], and bis-crown ethers [75] are all excellent catalysts for bromide–chloride exchange reactions [76].

An interesting commercial application of the PTC halide exchange processes is manufacture of 1-chloroalkanes from 1-alkenes, involving first the anti-Markinikov addition of HBr to the olefin in the presence of a free-radical initiator to yield largely ($\sim$95%) 1-bromoalkane, which then undergoes exchange with gas-phase HCl to produce 1-chloroalkane and regenerate HBr.

$$CH_3(CH_2)_nCH=CH_2(\text{liq}) + HBr(g) \xrightarrow{X^\cdot} CH_3(CH_2)_nCH_2CH_2\text{-}Br$$

$$CH_3(CH_2)_nCH_2CH_2\text{-}Br + HCl(g) \xrightarrow{R_4N^+ X^-} CH_3(CH_2)_nCH_2CH_2\text{-}Cl + HBr(g) \quad (7\text{-}12)$$

The HCl–HBr mixture recovered the second step of this sequence may be recycled into the first step, since the HCl present does not interfere or participate in the free-radical addition of HBr to 1-alkene [77]. Halide exchange reactions can also be conducted in the vapor phase between alkyl halides and hydrogen chloride or hydrogen bromide by use of PTC catalysts such as PEG 6000 adsorbed on potassium carbonate, or phosphonium salts coated onto zeolites [78].

PTC displacement reactions provide an excellent method for analysis of trace amounts of bromide, chloride, iodide, fluoride, and many other anions in aqueous solutions, since the resulting alkyl bromide, chloride, or iodide produced can then be analyzed by simple gas chromatographic methods. Pentafluorobenzyl methanesulfonate [79] in the presence of quaternary ammonium salts constitutes a convenient reagent for reaction with aqueous anion solutions for this analysis. Also poly(pentafluorobenzyl p-styrenesulfonate) is an extremely convenient pentafluorobenzylating reagent for inorganic anions [80]. These methods allow anions to be quantitatively analyzed down to the ng/mL levels.

Aizenberg et al. [81] showed that a variety of inorganic bromide salts could be prepared from the respective chloride salts by reaction with 1,2-dibromoethane under convenient PTC conditions. Conversion and yield are made quantitative by driving the reaction equilibrium by in situ distillation of the organic product, 1-bromo-2-chloroethane.

Chen et al. [82] have developed an extensive mathematical model for the kinetics of liquid–liquid PTC batch reaction of benzyl chloride with aqueous sodium bromide, catalyzed by tetrabutylammonium bromide catalyst. Results from these calculations emphasize a number of quantitative features of the PTC process, including extractive equilibria for the quaternary salt, and an explicit determination of the respective contributions of the reaction and the mass transfer to the overall rate. At high concentrations of inorganic salts in the aqueous phase, the tetrabutylammonium salt catalyst forms a third layer, which alters the distribution coefficients of the phase-transfer catalysts and which can be explained by the model. Other kinetic and mechanistic studies [83,84] featuring bromide–chloride exchange reactions, with various metal halides, show the catalyst to have a twofold role: as activator of the nucleophile ion pair and as acceptor of departing halide anion.

Kinetic studies dealing with some insoluble catalysts, such as quaternary ammonium salts bound to insoluble resins [85], show the efficiency of the bound catalyst to be higher than that of an analogous soluble catalyst. Catalysts made by immobilization of bulky -onium salts on inorganic matrices (silica and alumina) allow high nucleophilic activity in bromide displacement on octyl methanesulfonate [86]. The reactions follow second-order kinetics with observed rate constants dependent on the matrix (alumina > silica gel), on the nature of the alkyl chains bonded to the quaternary atoms and to the support, and on the solvent used in the reaction (n-heptane > toluene > chlorobenzene). Some reaction rates are higher by more than an order of magnitude than the same reactions carried out in a homogeneous phase in the presence of strong anion activators [lipophilic -onium salts and lipophilic [2.2.2]cryptands]. The dramatic rate increase is believed to result from both anion activation and adsorption of the substrate by the inorganic matrix.

Kinetic studies with resin-bound crown ethers on various displacement reactions on 1-octyl methanesulfonate by bromide [63] show PTC catalytic efficiency to depend on a combination of three factors: (1) nature of the nucleophile, (2) percent ring substitution of the resin by PTC functional groups, and (3) the presence of a spacer chain. (See Chapter 5.) Cryptand groups, attached to lightly loaded chloromethyl polystyrenes crosslinked with divinylbenzene, and ω-aminodecyl[2.2.2]cryptand, attached to carboxylated polystyrenes, catalyze nucleophilic substitution of octyl bromide by iodide and of octyl methanesulfonate by iodide, bromide, and chloride. Catalytic activities decrease with increasing percent of ring substitution in the polymer [87].

Substitution of iodide by bromide is normally a difficult PTC process because of the tendency of the generated iodide anion to "poison" the catalyst. With use of a molar amount of quaternary bromides, conversion of fluorinated alkyl iodides to bromides can be accomplished [88]. An oxidative PTC method for replacement of bromo-groups on aryl halides by chloride using aqueous hypochlorite salts has been demonstrated to occur at pH 8.5. Aromatic compounds having electron-withdrawing substituents are less reactive and 4-nitrobromebenzene is almost inert [89].

c. Iodide Displacements

Iodide displacement reactions on various alkyl chlorides, bromides, and methanesulfonates have been frequently reported in the literature, and are often employed to study PTC kinetics and mechanisms. As noted above, the exceeding ease of iodide transfer into organic phases and its strong association with catalyst cations, plus its strong nucleophilicity and negligible hydration, make such systems particularly simple to use as model reactions.

Catalysts used for iodide displacements include crown ethers [90], macrocyclic tetra-esters similar to crown ethers [91], phosphazenic polypodants for both liquid–liquid and liquid–solid PTC systems, PEG, [92], and quaternary ammonium and phosphonium salts [25,28]. Methylene chloride was converted to methylene iodide by treatment with NaI using $PhCH_2NEt_3^+Cl^-$ catalyst at 100–110°C [93].

Polymer-bound crown ethers [62,94–96] polymer-bound quaternary salts [63,97], crown ethers immobilized on silica [98], bulky -onium salts immobilized on inorganic matrixes (silica gel and alumina) [86] polymer-supported macrocyclic polyethers with pendant quaternary ammonium salts [99], polymer-supported cryptands [87], and polymer-bound phosphazene polypodants [100] have all been shown to have excellent catalytic activity for iodide displacement on alkyl chlorides, bromides, or methanesulfonates. Frequently, these insoluble catalysts have higher activities than those observed with the corresponding monomeric catalysts. Kinetics and reaction sensitivities are similar to other PTC reactions catalyzed by solid-bound PTC catalysts [101]. Catalysts prepared by adsorption of simple quaternary ammonium salts on organophilic clays are good catalysts for iodide displacement reactions [102].

A practical application of halide exchange reactions involves radioactive hadide and astatine exchange labeling of 6β-iodomethyl-19-norcholest-5(10)-en-3β-ol with ^{82}Br, ^{131}I, and ^{123}I using benzo-12-crown-4 or 18-crown-6 as a catalyst. The labelled products are of use as adrenal therapeutic drugs [103]. Also, a method for iodine labeling of DOPA involving halide exchange reactions using an ethoxylated phosphonium salt, $Cl(CH_2CH_2O)_2P^+$ $(NMe_2)_2$ PF_6^-, has been reported [104].

d. Fluoride Displacement Reactions

PTC fluoride displacement reactions can be conducted rather well, as indicated by the data in Table 7-6, although transfer of fluoride into organic solutions is difficult, and fluoride when hydrated is not a particularly powerful nucleophile.

$$RCH_2\text{-}Cl + M^+ F^- \xrightarrow{R_4N^+} RCH_2\text{-}F + M^+ Cl^- \quad (7\text{-}13)$$

Alkyl chlorides produce higher equilibrium yields of alkyl fluoride products than do alkyl bromides because of greater difficulty in transferring fluoride against bromide as a produced anion, compared to the transfer of fluoride against chloride. Even when formamide is used instead of water, fluoride exchange with octyl chloride gave a 72% yield of octyl fluoride, compared to < 50% yield when octyl bromide is used as starting material [105].

Water concentration is a critical factor in fluoride displacements [106,107]. With quaternary ammonium salts as catalysts, maximum conversions are obtained when the system contains about 0.33 mol of water per mole of KF. Better transfer of fluoride occurs when it is drier, but decreasing hydration increases the rate of catalyst decomposition. The selectivity to alkyl fluorides depends on the composition of the inorganic salts and the temperature.

Much work has been devoted to exploration of fluoride substitution of aromatic halides activated by one or more strongly electron-withdrawing substituents. For example, fluoride substitution of 2,4-dinitrochlorobenzene yields the corresponding fluoro-compound in high yield [110]:

$$KF + \underset{NO_2}{\underset{|}{\overset{Cl}{\underset{|}{C_6H_3}}}}\text{-}NO_2 \xrightarrow{18\text{-}Crown\text{-}6} \underset{NO_2}{\underset{|}{\overset{F}{\underset{|}{C_6H_3}}}}\text{-}NO_2 + KCl \quad (7\text{-}14)$$

Table 7-6. Fluoride exchange, aliphatic substrates.

Substrate	Product	Catalyst	References
$1-C_8H_{17}-X$	$1-C_8H_{17}-F$	$C_{16}H_{33}PBu_3^+$	[108]
(X = Cl, Br, O$_3$SMe)	80–92%	Amberlyst A26	[104, 106, 109, 110]
		18-crown-6	[111]
$2-C_8H_{17}-Br$	$2-C_8H_{17}-F$, 32%	18-crown-6	[110]
	Octenes (68%)		
$2-C_8H_{17}-X$	$2-C_8H_{17}-F$ (50–70%)	$C_{16}H_{33}PBu_3^+$	[112, 107]
Cyclohexyl Cl or Br	Cyclohexene (100%)	$C_{16}H_{33}PBu_3^+$	[47, 107]
		18-crown-6	[110]
$(CH_3)_3CCH_2Br$	No reaction	18-crown-6	[110]
$PhCH_2-X$	$PhCH_2F$ (90–100%)	$C_{16}H_{33}PBu_3^+$	[107]
		Bu_4N^+	[104, 109]
		Crown ethers	[113, 110]
		PEG-200	[114]
$PhCH_2CH_2Cl$	$PhCH_2CH_2F$	R_4N^+	[104]
$PhCHCl_2$	PhCHO, 100%	$C_{16}H_{33}PBu_3^+$	[107]
$n-C_6H_{13}CHCl_2$	$n-C_5H_{11}CH = CHCl$	$C_{16}H_{33}PBu_3^+$	[107]
$EtO_{02}CCH_2Br$	EtO_2CCH_2F (65%)	Amberlyst A26	[108]
$PhCOCH_2CH_2Br$	$PhCOCH_2F$	Amberlyst A26	[108]
$CHF_2OCH(CF)_3Cl$	$CHF_2OCH(CF)_3F^a$	Me_4N^+	[115]
CH_3COCl	CH_3COF	18-crown-6	[110]

^{a}Product used as an anesthetic.

Similar substitutions have been done with other nitro-substituted halobenzenes [116–121], chlorocyanobenzenes [122,123], polychloro- and polyfluorohalobenzenes [124,127], chlorobenzaldehydes [128], and cyanuric trichloride [129] and its derivatives [130]. Although these reactions proceed at moderate rates with catalysts such as tetrabutylphosphonium salts [131], more convenient reaction times are achieved at high temperatures, requiring high-temperature catalysts such as tetraphenylphosphonium salts [115,125], 4-(N,N-dialkylamino)pyridinium salts [116,117], tetramethylammonium chloride in aprotic solvents [132], or crown ethers [129].

Fluoride displacements on sulfonyl chlorides to produce sulfonyl fluorides can also be accomplished using solid–liquid PTC, for example [133]:

$$\underset{CH_3}{C_6H_4}-SO_2Cl + M^+F^- \xrightarrow{PEG} \underset{CH_3}{C_6H_4}-SO_2F + M^+Cl^- \quad (7\text{-}15)$$

Several metal fluorides (Li, Na, K, Cs) were tested in this process, although the most convenient reaction was with KHF_2, using PEG-1000 as phase-transfer catalyst.

3. Displacement with Carboxylate Anions

Displacement of carboxylate anions on alkyl halides or methanesulfonates under PTC conditions is usually straightforward both in liquid–liquid and liquid–solid systems [134]. Transfer of formate and acetate is somewhat difficult but higher carboxylates having more oleophilic structures are easily transferred.

a. Formate Ion Displacements

$$R-X + HCO_2^- \text{(aq)} \xrightarrow{\text{PTC}} R-O_2CH + X^- \quad (7\text{-}16)$$

Sasson and co-workers [135] examined formate displacement on 1,6-dichlorobutane, catalyzed by quaternary ammonium salts. Kinetic data for the reaction under solid–liquid phase-transfer conditions show the reaction to follow two consecutive first-order steps. Activity of the quaternary ammonium salts varied with catalyst counterion: $Cl^- > Br^- > I^- > HSO_4^-$. The reaction rate was linearly dependent on catalyst concentration up to 12 mol% of catalyst relative to the substrate, above which the rate no longer depended on catalyst concentration. In related work, "catalyst poisoning" in esterification of alkyl chlorides by formate (i.e., chloride anion product preventing transfer of formate to the organic phase) was minimized by using highly concentrated aqueous formate solutions, and use of potassium formate rather than sodium formate. Quantitative conversion and first-order kinetics were observed using 98% wt/wt sodium formate solution and tetrabutylammonium bisulfate as the PTC catalyst [136]. Cetyltrimethylammonium bromide has also been reported as a catalyst for formate displacement reactions on alkyl chlorides [28].

b. Acetate Ion Displacements

$$RX + CH_3CO_2^- \xrightarrow{\text{PTC}} ROAc + X^- \quad (7\text{-}17)$$

The poor nucleophilicity of acetate ion toward various substrates in condensed systems has been attributed to a combination of polarizability, basicity, and solvation factors. However, Liotta and co-workers [137] reported that acetate solubilized as the potassium salt in acetonitrile or benzene containing 18-crown-6 became sufficiently nucleophilic to react smoothly and quantitatively, even at room temperature, with a wide variety of organic substrates under liquid–sol.-PTC conditions.

An extensive investigation into a commercial-type process for conversion of benzyl chloride to benzyl acetate [138] includes studies of the kinetics, effect of agitation, structure and amount of catalyst, temperature, relative amount of sodium acetate, and phase volume ratios. Benzyltributylammonium chloride, which can be produced in situ by addition of tributyl amine, was the recommended catalyst for the process. Ultrasound accelerates the reaction of potassium acetate with alkyl halides using tetrabutylammonium or tributylmethylammonium bromide as catalysts [139].

Selective displacement of the bromo-group from α-bromo-ω-chloroalkanes by acetate to yield α-acetoxy-ω-chloroalkanes can be accomplished using tetraalkylammonium salts, such as Bu_4N^+. For example, 1-acetoxy-3-chloropropane was obtained in 94% selectivity at 98% conversion [140]. Selective displacement of the labile chlorine atoms in poly(vinyl chloride) by PTC using an 18-crown-6 catalyzed reaction with acetate has been employed as a technique to identify the chemistry and structure of sites on PVC bearing a labile chlorine [141].

Alkyl halides can only rarely be hydrolyzed by hydroxide anion directly to alcohols because of fast subsequent reaction of the alcohol product with alkyl halide to yield dialkyl ethers. For conversion of alkyl halides to alcohols, a two-step procedure involving first acetate displacement, then hydrolysis of the acetate ester has been suggested as the best procedure for this conversion. Both steps can be accomplished using PTC reactions, and the acetate-containing aqueous phase from the second step can be recycled to the first for more complete utilization of acetate [142].

PEGs are good catalysts for liquid–solid PTC reactions of acetate with alkyl halides, as for example, in esterification of 1-bromobutane with potassium acetate to give butyl acetate in yields up to 92% [143]. Catalytic activity of the PEGs increased with increasing molecular weight. Higher polarity solvents also increase ester yields. Alkyl iodides could not be used because of "catalyst poisoning" [144]. In the PTC reaction of benzyl chloride and potassium acetate, PEGs were more effective catalysts than were crown ethers; in 1-butanol the effectiveness of polyethylene oxides and crown ethers as phase-transfer catalysts was the same [145].

Other soluble catalysts used for acetate displacement reactions include cetyltrimethylammonium bromide [28], bis-quaternary ammonium salts [146], and dibenzo-18-crown-6 derivatives [89]. Catalytic activity of diastereomers of dicyclohexano-18-crown-6 was tested in acetonitrile for the reaction of benzyl chloride with solid potassium acetate [147]. Under the conditions of the reaction the activities of the three isomers were listed in Table 7-7.

Various insoluble PTC catalysts have also been examined for use with acetate displacement reactions. Polymethacrylate resins with bound PEG chains are effective catalysts for substitution of benzyl bromide with solid alkali acetates at 100°C [148]. Reaction rates with the metal acetates increase in the order Na < K < Cs [149]. In another study with similar catalysts, the conversions were all pseudo-first-order reactions and the observed rate constants were proportional to the amount of catalyst. Temperature and solvents also affect rates. A

Table 7-7. *Catalytic activities of the stereoisomers of dibenzo-18-crown-6 for liquid–solid PTC acetate displacement on benzyl chloride under standard conditions* [145].

Stereoisomer of dibenzo-18-crown-6	Yield of benzyl acetate (%)
cis, syn, cis	85.4
cis	63.3
trans, syn, trans	35.3

slight loss in catalytic activity after reuse of this resin-bound catalyst may be due to the cleavage of ether linkage on the polymer-based resin. In further work with copolymers, reaction rates decreased with increasing crosslink density of the catalyst, but catalysts with longer oligooxyethylene chains had considerably higher activity [150]. Incorporation of > 50 mol% of a non-PTC active comonomer decreased catalytic activity but the structure of the comonomer had no significant effect. The catalysts were easily synthesized in large quantities from commercially available monomers.

Detailed kinetic studies have also been made using resin-bound quaternary ammonium and phosphonium salts for acetate displacements. The catalyst particle size, percent ring substitution, polymer morphology, and use of spacer chains all affect reaction rates significantly [59]. For 1-bromooctane and benzyl chloride, reaction rates with macroporous, 7–25% ring-substituted catalysts increased with increasing ring substitution. Rates with macroporous catalysts increased as a $(CH_2)_7$ spacer chain was introduced between the active site and the polymer backbone. Rates of ion-exchange with macroporous catalysts were enhanced by increasing ring substitution or by introduction of a spacer chain. The high activity of the catalyst with long spacer chains was explained by the increased reactivity at the active site, induced by the increased organophilicity of the catalyst and by facilitated ion exchange [60,151].

Crown ethers immobilized on a modified Silochrome C-120 solid or incorporated into Teflon particles catalyze the substitution reaction of benzyl and potassium acetate, but not as effectively as the soluble crown ether analogs [152]. Use of phase-transfer catalysts adsorbed on high-surface-area inorganic solids [153], or use of inorganic compounds themselves as insoluble phase-transfer catalysts [154] has been reported for acetate displacement reactions.

Ester interchange of methyl or ethyl carboxylate esters with primary or secondary alcohols occurs when solid–liquid PTC is used in the absence of added organic solvent [155]. Vinyl and isopropenyl acetates undergo ester interchange, even in the absence of PTC catalyst, with primary alcohols and phenols.

c. Displacements with Other Carboxylate Anions

Displacement reactions with some other carboxylate salts are indicated by the general survey of ester-forming reactions in Table 7-8 (see p. 360).

Raths and Dehmlow [185] found that PTC trichlorooacetate anion displacements yield an unusual coupling product, as illustrated for the reaction with allyl chloride:

$$CH_2=CHCH_2Cl \text{ (org)} + CCl_3CO_2Na \text{ (aq)} \xrightarrow{R_4N^+} CH_2=CHCH_2\text{-}CCl_2CO_2CH_2CH=CH_2 \quad \text{61-82\% Yield} \quad (7\text{-}18)$$

Phase-transfer catalysts efficiently promote transesterification reactions, as for example in the reaction of trilaurin and methyl palmitate at 100°C in the presence of catalytic amounts of tetraalkylammonium bromides and sodium methoxide [186]. The amount of dilaurin produced increased with increasing chain length of the phase-transfer catalyst up to tetrahexyl, and the amounts of lauroylpalmitin and dipalmitin increased up to tetraheptyl. Enrichment of palmitic acid in the 2-position of the triglycerides formed was by a factor of 1.51. A technique for rapid conversion of the fatty acids in oils and fats to methyl esters is based on a two-step process: PTC saponification of the fat or oil, then PTC esterification with methyl tolsylate [187], or with dimethyl sulfate [188]. The methyl esters could then be easily analyzed by chromatographic methods.

The effect of iodide ion on the kinetics of esterification of sodium carboxylates with benzyl chloride was studied using quaternary ammonium salts as phase-transfer catalysts [189]. Iodide at a suitable concentration accelerates the reaction, whereas iodide in excess poisons the catalyst. The optimum concentration varies with the reaction system. (See Chapter 4 on cocatalysts.)

Surprisingly, PTC allows displacement from the vinyl carbon of 2-propenyl bromide with carboxylate salts such as potassium cis-9-octadeconate in good yields [190]. 2-Propenyl chloride reacts sluggishly and gives only low yields of ester.

1,4-Dichlorobutene (I) reacts quantitatively with salts of carboxylic acids under PTC conditions to form 1,4-diesters of butene [191], 3,4-Dichloro-butene (II) does not react with carboxylate nucleophile, but instead is isomerized to I by quaternary ammonium halide catalysts. By proper adjustment of the reaction conditions, isomer mixtures of dichlorobutene can be directly converted to pure 1,4-diesters of 2-butene.

Yazawa and Ishikame [192] have demonstrated that pivalate anion PTC displacement with tetrabutylammonium bromide catalyst on chlorobromoomethane can be partially stopped at the monoester stage, producing chloromethyl pivalate in 34% yield.

4. Azide Displacements

Azide, a powerful nucleophile, undergoes PTC displacements very well, even with alkyl iodides [193], and produces azide derivatives even with cyclohexyl bromide or cyclohexyl iodide in about 80% yield. These alkyl halides normally yield 100% cyclohexene in PTC displacement reactions. Azide displacement reactions with *n*-octyl choride and bromide using tetrabutylammonium halides and Aliquat 336 as catalysts give excellent yields of alkyl azides [194].

$$R\text{-}X + NaN_3 \text{ (aq)} \xrightarrow[100°C]{R_4N^+} R\text{-}N_3 + NaX \quad \text{90-95\% Yield} \tag{7-19}$$

$X = Cl, Br, I, -O_3SCH_3$

Poly(vinyl chloride) normally undergoes displacement reactions very poorly with most nucleophiles, but reacts well with azide in the presence of dipolar aprotic solvents and quaternary-onium salts to produce substitution products with negligible side reactions [195].

(7-20)

PVC in THF solution

[Structure: PVC chain with Cl, Cl substituents] + NaN$_3$ (suspended solid) $\xrightarrow{R_4N^+ X^-}$ [Structure: chain with N$_3$, N$_3$ substituents]

Acid chlorides and sulfonyl chlorides react with azide under PTC conditions, providing acyl azides or sulfonyl azides with soluble PTC catalysts [196–198], and with polymer-supported PTC catalysts [199]. Acyl azides, when isolated into an inert and dry solvent, such as benzene, undergo rearrangement on heating under reflux to give isocyanates in 50–90% yields [200].

$$RCO\text{-}Cl + NaN_3 \xrightarrow{PTC} RCO\text{-}N_3 + NaCl$$

(7-21)

$$RCO\text{-}N_3 \longrightarrow R\text{-}N=C=O + N_2$$

C_{8-24}-Alkylbenzenesulfonyl azides, useful as diazo group transfer reagents, are conveniently prepared in a one-pot procedure by first conversion of the sulfonic acids with $SOCl_2$, $POCl_3$, or $COCl_2$ to the alkylbenzenesulfonyl chloride; then the resultant sulfonyl chlorides react with aqueous sodium azide in the same solvent and in the presence of a phase-transfer catalyst, such as Aliquat 336 to give sulfonyl azides in near-quantitative yields [201].

A simple one-pot method for preparation of diazo carbonyl compounds using polymer-bound phase-transfer catalysts involves first reaction of methanesulfonyl chloride with sodium azide to give $MeSO_2N_3$, which without isolation can be used to treat active methylene compounds, for example, ethyl acetoacetate, to give the corresponding diazo-compound, $CH_3COC(N_2)CO_2Et$, in 69–94% yields [202].

Table 7-8. General survey of carboxylate displacements.

Carboxylate salt	Alkyl halide	Product	Catalyst	References
RCO_2Na^+	$PhCOCH_2Br$	Phenacyl esters	R_4N^+	[156]
	Me_2SO_4	Methyl esters (solv.-free)	R_4N^+	[157]
	$PhCH_2-X$	esters (kinetics)	R_4N^+	[158]
	Allyl chloride	Allyl esters	R_4N^+	[159]
	Poly[(chloromethyl)styrene]	Esters	R_4N^+	[160]
Na acrylate	$C_6Br_5CH_2Br$	$C_6Br_5CH_2O_2CCH = CH_2^a$	Aliquat 336	[161]
	Haloalkylsilanes	$CH_2 = CRCO_2(CH_2)_nSiR_m(OR)_{3-m}$	R_4N^+	[162]
Na methacrylate	Epichlorohydrin	Glycidyl methacrylate^b	R_4N^+	[163]
Na stearate	Epichlorohydrin	Glycerol monostearate^b	R_4N^+	[164, 165]
Fatty acid salts	9-Bromomethylacridine	acridinyl esters^c	R_4N^+	[166]
$(CH_3)_3CCO_2Na$	CH_2ClBr	$(CH_3)_3CCO_2CH_2Cl$ (34%)	Bu_4N^+	[167]
$HOCH_2CO_2Na$	RBr	$HOCH_2CO_2R$ (89%)^d	PEGs	[168, 169]
Na 1-adamantane carboxylate	Epichlorohydrin	Glycidyl esters^e	18-C-6	[11]
Na salt of carboxymethylcelulose	α-Halo-α-ethoxyacetic acid ester	carboxymethylcellulose esters	R_4N^+	[170]
Na carboxylates (in waste water)	Et_2SO_4	Ethyl esters (org. sol)^f	R_4N^+	[171]
$Na^+C_6H_5CO_2^-$	Benzyl bromide	Benzyl benzoate	Polymer-bound-NR_3^+	[172]
	$PhCH_2Cl$ (kinetics)	Benzyl benzoate	Bu_4N^+	[173]
$Na^+o-X-C_6H_4CO_2^-$	$PhCOCH_2Br$	$o-XC_6H_4CO_2CH_2COPh$	$C_{16}H_{33}NMe_3^+$	[174]
$Na^+RC_6H_4CO_2^-$	CH_2Cl_2	$RC_6H_4CO_2(CH_2)_nCl$	R_4N^+	[175]
	$ClCH_2CH_2Cl$	$RC_6H_4CO_2(CH_2)_nO_2CC_6H_4R^g$		
Na nitrobenzoates	$PhCH_2Br$	$O_2NC_6H_4CO_2CH_2Ph$	Bu_4N^+	[176]
	$Br(CH_2)_3Cl$	$[4-O_2NC_6H_4CO_2]_2(CH_2)_3$	Bu_4N^+	[177]
Salts of hindered acids (e.g., mesitoic acid)	$R-X$, or Et_2SO_4	hindered esters	R_4N^+	[178]

			PTC	
4-MeOC$_6$H$_4$CO$_2$Na$^+$	PhCH$_2$-X	Esters	R$_4$N$^+$	[179]
p-HOC$_6$H$_4$O--CHMeCO$_2$Na$^+$	EtCl	Et ester (98%)	R$_4$N$^+$	[180]
Na monobutyl phthalate	PhCH$_2$Cl	Butyl benzyl phthalate[g]	R$_4$N$^+$	[181]
Na phthalate	Allyl bromide	Diallyl phthalate	R$_j$$_4N^+$, crowns	[182]
Hydroxybenzoic acid salts	R-X	Esters of hydroxybenzoic acid (82–95% yield)	R$_4$N$^+$	[183]
K$^+$ 3,4,5-trimethoxybenzoate	Br(CH$_2$)$_3$Cl	(MeO)$_3$C$_6$H$_4$CO$_2$(CH$_2$)$_3$Cl[h]	PhCH$_2$Me$_3^+$	[184]

[a]These products used for fire resistant polymers.
[b]Products useful for preparation of surfactants with low skin irritancy.
[c]Esters can be automatically analyzed by liquid chromatography down to very low concentrations by use of the acridine group for fluorometric detection.
[d]Product is used as cerebral vascdilator and nootropic.
[e]Product is used as an insecticide and acaricide.
[f]This method allows 99% removal of the acids from waste water.
[g]When R = nitro, product reduced to diamine, used to make polyurethanes.
[g]Kinetics, and iodine anion cocatalyst effect studied.
[h]Product esters used as intermediates in the preparation of 1,4-bis[3-(3,4,5-trimethoxybenzoyloxy) propyl]-perhydro-1,4-diazepine, a coronary vasodilator.

5. Sulfide and Disulfide Displacements

Sulfide, disulfide, or bisulfide anions appear to be relatively easy to transfer from aqueous or solid to organic phases, and because of their exceedingly high nucleophilicity, they undergo fast displacement reactions.

a. Preparation of Thiols

Mercaptans can be prepared in near quantitative yields by reaction of alkyl halides with ammonium or alkali metal hydrogen sulfides in the presence of a quaternary ammonium halide phase-transfer catalyst [28,203].

$$\text{R-X} + \text{NaHS(aq)} \xrightarrow{\text{PTC}} \text{R-SH} + \text{NaX} \quad (7\text{-}22)$$

Substituted benzyl mercaptans, useful as intermediates for agrochemicals, are prepared by this reaction using tetrabutylammonium halides as phase-transfer catalysts [204]. Sulfur-containing fatty acids and their esters, useful in extreme-pressure lubricants, are prepared by treatment of water-insoluble halogenated fatty acid esters with aqueous sulfide solutions in the presence of phase-transfer catalysts, for example, tetrabutylphosphonium bromide [205]. For example, 2-ethylhexyl thioglycolate was prepared from 2-ethylhexyl chloroacetate in 98% yield.

b. Preparation of Dialkyl Sulfides

Use of alkali metal sulfides gives thio-ethers in high yields [206]:

$$2\,\text{R-X} + \text{Na}_2\text{S (aq)} \xrightarrow{\text{PTC}} \text{R-S-R} + 2\text{NaCl} \quad (7\text{-}23)$$

Since these reactions proceed so readily, most all types of phase-transfer catalysts are suitable. Thiacrown ether polymers, an interesting kind of PTC catalyst, have high complexing capacities for sodium and potassium salts and proved to have good phase-transfer catalytic activity in the reaction of 1-chlorobutane with sodium sulfide to produce dialkyl sulfides [207]. Use of several types of phase-transfer catalysts (quaternary -onium salts, crown ethers, cryptands, polymer-supported catalysts) for displacement by sulfide and other anions on octyl methanesulfonate has been reviewed [75]. Cyclic and acyclic sulfides were potent cocatalysts with quaternary ammonium salts for reactions of α-ω-dihaloalkanes and sodium sulfide to produce polymers with unusually high molecular weights [208].

Dimethallyl sulfide [209], or bis(2-vinyloxyethyl) sulfide from 2-chloroethylvinyl ether [210], and heat-resistant poly(vinylene sulfides) from 1,2-dichloroethylene [211], are commercial products, producible by use of phase-transfer catalysts. α, ω-Dibromopolythioether oligomers, themselves useful as phase-transfer catalysts, result from polycondensation of 2,6-dibromopyridine with Na_2S under

Table 7-9. *Effect of catalyst on rates of benzyl chloride and p-chlorobenzyl chloride with sodium sulfide [213].*

Catalyst(s)	k for $PhCH_2Cl$ (mol cm^{-3} sec^{-})	k for $p\text{-}ClC_6H_4Cl$ (mol cm^{-3} sec^{-1})
None	0.096×10^{-8}	0.081×10^{-8}
5 wt% Al_2O_3	1.40×10^{-8}	0.408×10^{-8}
0.4 wt/vol $Bu_4N^+Br^-$	27.0×10^{-8}	12.2×10^{-8}
Al_2O_3 (5%) + Bu_4NBr	$189. \times 10^{-8}$	39.7×10^{-8}

solid–liquid PTC conditions. Under certain conditions these oligomers cyclize to yield macrocylcic thioethers [212].

Insoluble PTC catalysts, that is, ion-exchange resin, basic aluminum oxide, and tetrabutylammonium bromide on alumina, catalyze solid–liquid displacement reactions of Na_2S with benzyl chloride and p-chlorobenzyl chloride to produce the corresponding dialkyl sulfides, useful as additives for high-pressure lubricants, antiwear additives for motor oils, stabilizers for photographic emulsions, recovery of precious metals, and in anticorrosive formulations. Data in Table 7-9 show the effect of catalyst compositions on the rate of the sulfide displacement reactions.

Clearly, alumina can function by itself as an insoluble PTC catalyst, better when used as a cocatalyst with quaternary salt. The anion-exchange resin Amberlyst A27 enhanced the reaction rate to about the same extent as alumina. The cocatalytic effect of alumina was believed to be due to adherence of Na_2S on the alumina surface, making it more readily available for displacement.

PTC aromatic displacement reactions have been employed to produce diaryl sulfides useful in the production of certain high-value polymers. For example, bis(phthalimide) sulfides result from PTC reaction of 4-chloro-*N*-methylphthalimide with sodium sulfide using a phosphonium salt as catalyst [214].

$$\text{Cl-phthalimide-N-CH}_3 + Na_2S \xrightarrow{R_4N^+ X^-} CH_3\text{-N-phthalimide-S-phthalimide-N-CH}_3$$

84% Yield (7-24)

The reaction of *o*- and *p*-chloronitrobenzenes with sodium sulfide under solid–liquid conditions, catalyzed by either tetrabutylammonium bromide or PEG-400, gave bis(*o*- or *p*-chlorophenyl) sulfide in yields up to 98%, and/or *o*- or *p*-chloroaniline in yields up to 100% depending on the amount of solid Na_2S present, the amount of water present, and the intensity of agitation [215]. In liquid–liquid reactions with aqueous sodium sulfide only the reduction products were obtained.

$$\text{(7-24A)}$$

PTC melt synthesis of diaryl sulfides starting from chlorobenzene and chlorophthalic acid, using tetraalkylammonium and phosphonium salts, has been reported [216], and a method for removing halogenated aromatics, such as PCBs from transformer oils by PTC reaction with sodium sulfide, has been patented [217].

PEGs have been used as phase-transfer catalysts in the solid–liquid synthesis of diacyl sulfides in 91–98% yields from acid chlorides and sodium sulfide [218].

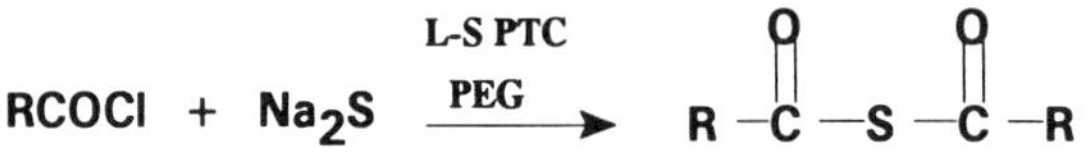

$$\text{RCOCl} + \text{Na}_2\text{S} \xrightarrow[\text{PEG}]{\text{L-S PTC}} \text{R}-\overset{\overset{\text{O}}{\|}}{\text{C}}-\text{S}-\overset{\overset{\text{O}}{\|}}{\text{C}}-\text{R} \quad \text{(7-24B)}$$

c. Preparation of Dialkyl Disulfides

Sodium disulfide, prepared from reaction of elemental sulfur with sodium sulfide, reacts under PTC conditions with alkyl halides to produce dialkyl disulfides.

$$\text{Na}_2\text{S (aq)} + \text{S} \longrightarrow \text{Na}_2\text{S}_2 \text{ (aq)}$$

$$2\text{ R-X} + \text{Na}_2\text{S}_2 \xrightarrow{\text{R}_4\text{N}^+ \text{ X}^-} \text{R-S-S-R} + 2\text{ NaX}$$

(7-25)

Dibenzyl disulfide [219] and diallyl disulfide [220] are obtained in high yields by PTC reaction of benzyl chloride or allyl chloride with Na_2S_2.

Synthesis of diacyl disulfides from acid chlorides and Na_2S_2, optionally employing phase-transfer catalysts, has been described [221].

6. Thiocyanate Displacement

Thiocyanate ion is easily transferred into organic phases, and it is a highly reactive nucleophile. PTC displacement reactions with this anion are therefore readily accomplished although the products obtained may depend on reaction conditions. For example, solid–liquid PTC conversions of KSCN and benzyl chloride at 180°C give both benzyl thiocyanate and benzyl isothiocyanate in ratios that depend on catalyst structure [222]. Liquid–liquid phase-transfer calalysis at 100°C yields benzyl thiocyanate exclusively.

$$C_6H_5CH_2\text{-}Cl + KSCN(aq) \xrightarrow[100°C]{R_4N^+ X^-} C_6H_5CH_2\text{-}SCN + KCl$$
100% Yield   (7-26)

$$C_6H_5CH_2\text{-}Cl + KSCN(s) \xrightarrow[100°C]{R_4N^+ X^-} C_6H_5CH_2\text{-}SCN \text{ and } C_6H_5CH_2\text{-}NCS + KCl$$

Kinetics of the benzyl chloride displacement reaction with aqueous thiocyanate and various PTC catalysts follow pseudo-first-order dependence on $PhCH_2Cl$ concentration, with rates linearly related to the concentration of catalyst, temperature and solvent [223].

Examples of thiocyanate displacement include benzyl chloride with tetrabutylammonium salts as catalysts [224], alkyl chlorides with cetyltrimethylammonium catalyst [28], displacement on butyl iodide using crown ether catalysts [145], and reaction of 2-(fluoroalkyl)ethyl iodides catalyzed by Aliquat 336 [225]. Reaction of 9-bromoacridine gave 63–90% acridinylisothiocyanates [226]; acetylated bromosugars with alkali metal thiocyanates in the presence of crown ethers gave products with glycosidically bound isothiocyanate derivatives [227]; and $S(CH_2CH_2Cl)_2$ was converted quantitatively into $S(CH_2CH_2SCN)_2$ by KSCN using benzyltributylammonium chloride as PTC catalyst [228]. Even the bromogroup of bromo-alkyl acrylates undergoes simple thiocyanate displacement when a small amount of potassium iodide is used as a cocatalyst [229]. Cyclophosphazenic polypodans are useful and recoverable catalysts for thiocyanate displacement reactions [91].

Thiocyanate displacement reactions have been extensively used to test and compare phase-transfer catalysts. These include resin-bound crown ethers [63], quaternary salts adsorbed on clays [18,101], polystyrene-supported PEG [230], and -onium salts immobilized on inorganic matrixes (silica gel and alumina) [86]. Resin-bound macrocyclic polyethers containing pendant quaternary ammonium groups show high activity for thiocyanate displacement reaction on hexyl bromide [98]. Insoluble polystyrenes containing formamide and acetamide groups are good phase-transfer catalysts for reaction of n-octyl bromide with potassium thiocyanate in toluene–water. Nylon-66 is also a PTC catalyst when some of the amide hydrogens are replaced with N-alkyl groups [231]. Resin-supported tetraphenylphosphonium bromides are more effective phase-transfer catalysts than the soluble analogs, Ph_4PBr and benzyltriethylammonium chloride, for thiocyanate displacement on 1-bromooctane [232].

tert-butyl chloride undergoes PTC displacement to yield three products with varying selectivity depending on the reaction conditions [233]. Formation of tert-butyl thiocyanate and isothiocyanate and hydrolysis of tert-butyl chloride are the main parallel reactions occurring in these systems. The main factors affecting product selectivity are concentration of reagents in one of the reaction phases, the ratio of volumes of both liquid phases, and the presence of phase-transfer catalysts. The PTC catalyst Bu$_4$NBr decreases the rate of *tert*-butyl chloride solvolysis but only slightly influences the selectivity between thio- and isothiocyanate formation.

Thiocyanate displacement on bromoalkyl methacrylate esters can be accomplished in 71–95% yield in the presence of a phase-transfer catalyst and a catalytic amount of KI [234]. Acyl chlorides also undergo dispacement with thiocyanate to give the corresponding acyl isothiocyanates [235].

Photogenerated aryl vinyl cations in a two-phase system of methylene chloride and water, using tetrabutylammonium salt as a phase-transfer catalyst with thiocyanate, gave products derived from S-attack, vinyl thiocyanates, and also products derived from N-attack, vinyl isothiocyanates or isoquinolinethiones [236]. The corresponding reaction with cyanate ion gave only isoquinolone derivatives.

7. Sulfite Displacement

Sulfite anion participates readily in PTC reactions with alkyl halides or methanesulfonates to yield sodium alkyl sulfonates using quaternary ammonium salts such as tetraethyl- to tetrabutylammonium salts or dodecyltrimethylammonium salts as catalysts [237,238]. Product yields are usually high (70–98%).

$$\text{R-X} + \text{Na}_2\text{SO}_3 \text{ (aq)} \xrightarrow{\text{R}_4\text{N}^+ \text{ X}^-} \text{RSO}_3\text{Na} + \text{NaCl} \qquad (7\text{-}27)$$

The products, when starting from longer-chain (C_7–C_{11}) alkyl halides, are excellent specialty surfactants. Even 1,2-dichloroethane reacts smoothly with aqueous sodium sulfite to yield sodium 1,2-ethanedisulfonate, when catalyzed by cetylpyridinium chloride [239]. Perfluoroalkyl iodides give the corresponding perfluoroalkanesulfonates [240]. Aromatic displacement by sulfite on 1,4-diamino-2,3-dichloroanthraquinone using benzyltriethylammonium chloride as catalyst gave 1,4-diaminoanthraquinone-2,3-disulfonic acid, useful in dye chemistry [241].

8. Nitrite Displacement

Reaction of alkyl halides with nitrite anion under phase-transfer conditions gives both nitro-compounds and nitrite esters, the latter of which may decompose under reaction conditions [77]. Treatment of 1-bromopropane with aqueous sodium nitrite and sodium carbonate in the presence of tetrabutylammonium sulfate gave 61% 1-nitropropane [242]. Similarly, displacement reactions of various *n*-alkyl chlorides and bromides with potassium nitrite in the presence of 18-crown-6 gave the corresponding nitro-compounds in yields of 50–70% [237]. PEGs are good phase-transfer catalysts for nitrite displacement reactions [243].

The variable nature of nitrite PTC reactions is apparently due to the amount of water present in these reactions, as has clearly been demonstrated for reactions of 2,4-dinitro-chlorobenzene with nitrite [244]. Dry (or at least a low water content) PTC conditions produces 1,2,4-trinitrobenzene as product, whereas wet solid–liquid PTC conditions produce 1,2,4-trinitrobenzene and 2,4-dinitrophenol as product.

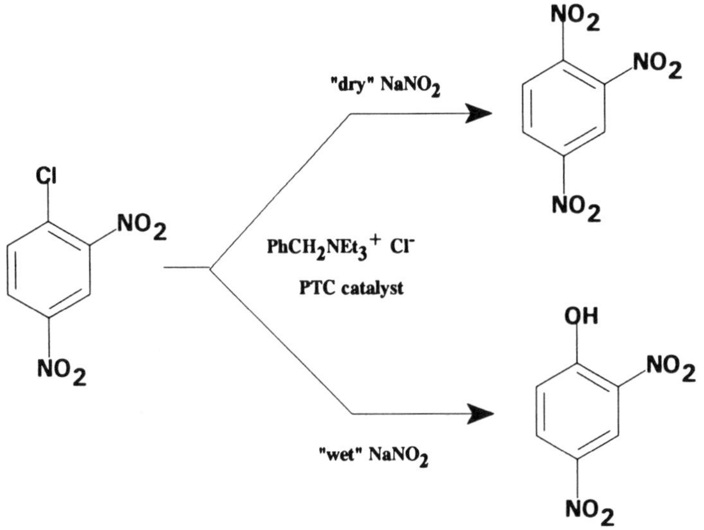

$$
\text{2,4-dinitrochlorobenzene} \xrightarrow[\text{PTC catalyst}]{\text{PhCH}_2\text{NEt}_3^+ \text{Cl}^-}
\begin{cases}
\text{"dry" NaNO}_2 \rightarrow \text{1,2,4-trinitrobenzene} \\
\text{"wet" NaNO}_2 \rightarrow \text{2,4-dinitrophenol}
\end{cases}
\quad (7\text{-}28)
$$

The threshold value of water, in terms of moles water per mole of nitrite salt, for formation of trinitrobenzene was 0.24 for KNO_2 and 0.22 for $NaNO_2$.

Displacement of nitrite on pentafluorobenzyl methanesulfonate and related substrates provides a method for quantitative analysis of ng/mL levels of nitrite in water [79,80].

Sodium nitrite in reaction with poly(chloromethylstyrene) beads, catalyzed by tetrabutylammonium bromide, resulted in oxidation of the chloromethyl groups to carboxylic acid groups [245]. This result probably was due to initial formation of nitrite ester groups, which decomposed under the reaction conditions with the formation of oxygenated derivatives. The product contained hydroxyl, ester, and carbonyl groups, in addition to carboxylic acid groups.

9. Hydroxide Anion Displacements

Since hydroxide anion is difficult to transfer from aqueous to organic phases, simple PTC displacements with hydroxide to yield alcohol are usually difficult to accomplish. Use of large, bulky quaternary salts does allow transfer of hydroxide so that alcohols are produced, but most of the ROH formed is rapidly consumed to generate ethers, which are the main products. Some dehydrohalogenation of the alkyl halide to olefin also occurs. For example, in the reaction of 1-bromooctane with aqueous sodium hydroxide [246]:

$$1\text{-}C_8H_{17}\text{-}Br + OH^- (aq) \xrightarrow{PTC} \begin{cases} 1\text{-}C_8H_{17}\text{-}OH \;(20\%) \longrightarrow (1\text{-}C_8H_{17})_2O \;(75\%) \\ C_6H_{13}CH=CH_2 \;(5\%) \end{cases}$$

(7-29)

Dioctyl ether occurs as the principle product because the $C_8H_{17}O^-$ anion is much more readily transferred into the organic phase than the highly hydrophilic hydroxide anion. With 2-bromooctane as substrate, dehydrohalogenation becomes the major reaction path, giving 86% yield of 1- and 2-octenes. Hydroxide anions, transferred from dilute aqueous NaOH solution into chlorobenzene using tetrahexylammonium hydroxide, bring along 11 molecules of water of hydration. By increasing the sodium hydroxide concentration to 50% the hydration number of OH^- decreased to a value of 3.5, leading to enhanced reactivity of OH^- [75,105].

The kinetics and basic physical chemistry of the reactions involved in hydroxide anion displacement on benzyl chloride to produce dibenzyl ether (ca. 100% yield) have been reported in some detail [247]. Reaction of (chloromethyl)styrene with aqueous potassium hydroxide in the presence of tetrabutylammonium bromide as PTC catalyst gives bis(vinylbenzyl) ether in 70% yields [248].

Some exceptions to the usual ether formation are sometimes observed. For example, 1-adamantol was obtained in 92% yield by PTC hydrolysis of 1-bromoadamantane [249]. 2-Hydroxyphenylacetic acid can be made by direct hydrolysis of 2-chloro-2-phenylacetic acid [250]. Polyfluorobenzenes hydrolyze with aqueous sodium hydroxide in the presence of tetrabutylammonium salts to give tri-, tetra-, and pentafluorophenol [251]. Hydrolysis of 2,6-dichloropyridines with benzyltriethylammonium chloride PTC catalyst yields 2-hydroxy-6-chloropyridines in high (97%) yields [252]. Hydrolysis of haloaromatic ketones to the corresponding phenols for use as monomers has been patented [253].

10. Carbonate and Bicarbonate Anion Displacement

The preparation of dialkyl carbonates via the PTC alkylation of alkali metal carbonate and bicarbonate salts has been reported [254,255]. Best yields are obtained when a phase-transfer catalyst is used in conjunction with a dipolar aprotic solvent, as illustrated by the data in Table 7-10.

Use of activated aryl halides in these reactions led to the formation of diaryl

Table 7-10. PTC alkylation with alkyl halides of potassium bicarbonate to produce dialkyl carbonates [250].

Alkyl halide	PTC-Catalyst	Solvent[a]	Percent Yield
PhCH$_2$Cl	None	toluene	0
	Bu$_4$N$^+$Br$^-$	toluene	33
	None	DMAC	29
	Hex$_4$N$^+$Cl$^-$	DMAC	45
	Aliquat 336	DMAC	52
	Me$_4$N$^+$Br$^-$	NMP	24
	Bu$_4$N$^+$Br$^-$	NMP	79
	Bu$_4$P$^+$Br$^-$	NMP	80
	Hex$_4$N$^+$Br$^-$	NMP	79
	18-C-6	NMP	53
CH$_3$Cl	Bu$_4$N$^+$Br$^-$	DMAC	63
	Bu$_4$P$^+$Br$^-$	DMAC	53
	Hex$_4$N$^+$Cl$^-$	DMAC	86
		Sulfolane	46
		NMP	69
	Ph$_3$P(C$_{12}$H$_{25}$)	DMAC	56
	Cyclohex$_3$P$^+$(C$_{12}$H$_{25}$)	DMAC	45
1,2-C$_2$H$_4$Cl$_2$	Hex$_4$N$^+$Br$^-$	DMAC	15[b]
1-C$_8$H$_{17}$Br	Bu$_4$P$^+$Br$^-$	NMP	47
Allyl Br	Bu$_4$P$^+$Br$^-$	NMP	50

[a]DMAC = N,N-dimethylacetamide; NMP = N-methyl-2-pyrrolidone.

[b]Product was ethylene carbonate. Reaction conditions: twofold excess of KHCO$_3$ over alkyl halide, at 150°C at reflux, with 5 mol% PTC catalyst using 1M concentration of alkyl halide in solvent.

ethers, α,α'-Dibromoxylene is reported to react with dry potassium carbonate, using crown ether as a PTC catalyst, to yield polycarbonates [256].

11. Displacement with Peroxide and Superoxide Anions

Solid–liquid PTC reaction of alkyl hydroperoxides with primary alkyl bromides, using solid potassium hydroxide as base and benzyltriethylammonium salts as phase-transfer catalyst, provides a simple and quick method for synthesis of dialkyl peroxides in fair yields [257].

$$\text{ROOH} + \text{R'Br} + \text{KOH} \xrightarrow{\text{PTC}} \text{R-OO-R'} + \text{KBr} \quad (7\text{-}30)$$

Similarly, magnesium peroxide, in the presence of tetrabutylammonium chloride as catalyst, reacts under liquid–solid conditions with benzoyl chlorides to give magnesium peroxybenzoates in high yields, for use as bleaching components in detergent formulations. This procedure obviates the need to isolate the intermediate peroxy acids, which are prone to explosion [258].

Folgia and Silbert have found that direct displacement reactions of potassium superoxide with alkyl bromides in the presence of tetraethylammonium bromide as a phase transfer catalyst, and using N,N-dimethylformamide as a solvent yields dialkyl peroxides in yields generally above 80–90% [267]. Both symmetrical and unsymmetrical could be prepared.

$$R-Br + K_2O_2 \xrightarrow[DMF]{Et_4N^+Br^-} R-O-O-R + KBr$$

Dry potassium superoxide, in the presence of crown ethers in solvent (benzene, tetrahydrofuran, or N,N-dimethylformamide), undergoes reaction with primary and secondary alkyl bromides to form dialkyl peroxides [259].

$$\text{R-Br} + \text{KO}_2 \text{ (s)} \xrightarrow{\text{18-Crown-6}} \text{R-O-O-R} + \text{KBr} \quad (7\text{-}31)$$

Small yields of alcohols and alkenes are reported as side products; however, alcohols can be obtained in yields of 25–75% using DMSO, DMF, DME, and diethyl ether as solvents [260,261].

12. Phosphide and Phosphinite Anion Displacements

Phosphide and phosphinite ions, generated directly from elemental phosphorus under superbasic conditions, react with organic halides under PTC conditions to afford triorganophosphines and phosphine oxides in good yields [262].

13. Cyanate Anion Displacements

Cyanate anion is reasonably well transferred by most phase–transfer catalysts and undergoes displacement reactions at moderate rates to produce isocyanates as initial products. At temperatures greater than about 100°C, however, the isocyanates in the absence of water tend to undergo dimerization and trimerization reactions to cyanuric acid derivatives [263,264], or in the presence of water to be hydrolyzed to urea derivatives [77]:

$$\text{R-X} + \text{Na}^+ \text{NCO}^- \quad (7\text{-}32)$$

$$\downarrow \text{PTC}$$

$$\text{R-NHCONHR} \xleftarrow{H_2O} \text{R-N=C=O} \xrightarrow{\text{Dry}} \text{(cyanurate trimer)}$$

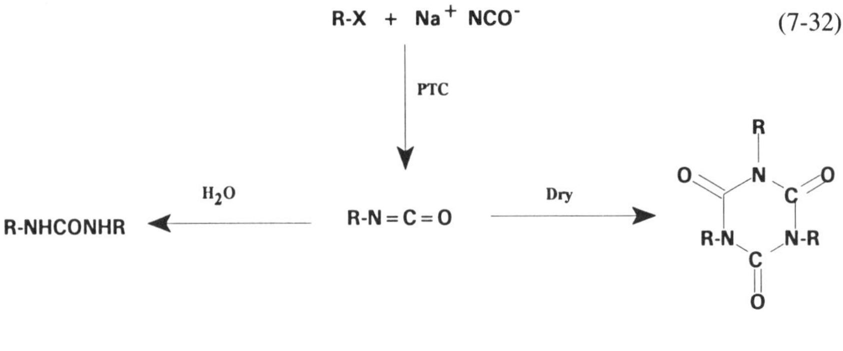

In the presence of ethanol and absence of water, cyanate anion reaction with allyl halides using triphenylethylphosphonium salt catalyst gave N-alkylurethanes [77]. Liquid–solid displacement by cyanate in the presence of $Et_4N^+I^-$ at 170°C on benzyl chloride produces benzyl isocyanate [265].

References

1. P. Tundo, F. Trotta, and G. Moraglio, React. Polym. **10**, 185–8 (1989) [CA: 112.098147].
2. P. Tundo, F. Trotta, G. Moraglio, and F. Ligorati, Ind. Eng. Chem. Res., **27**, 1565–71 (1988).
3. F. Hampl, et al. Czech. Patent CS 265,360 (1990) [CA:114.206808].
4. H. Nanba, N. Takahashi, K. Abe, and M. Saito, to Mitsubishi Gas Chemical Co., Inc., Jpn. Kokai Tokkyo Koho JP 63/196549 (1988) [CA: 110.023360].
5. H. Nanba, N. Takahashi, K. Abe, and M. Saito, to Mitsubishi Gas Chemical Co., Inc., Jpn. Kokai Tokkyo Koho JP 63/196547 (1988) (CA: 110.023358].
6. H. Nanba, K. Abe, M. Saito, to Mitsubishi Gas Chemical Co., Inc., Jpn. Kokai Tokkyo Koho JP 631196550 (1988) [CA: 110.023361].
7. H. Coates, R. L. Barker, R. Guest, and A. Kent, to Albright and Wilson, Ltd. British Patent 1,336,883 (1973).
8. C. M. Starks and C. L. Liotta, *Phase Transfer Catalysis, Principles and Techniques*, Academic Press, New York (1978).
9. G. Bram, A. Loupy, and M. Pedoussaut, Bull. Soc. Chim. Fr., (1), 124–8 (1986).
10. S. S. Shavanov, G. A. Tolstikov, and G. A. Viktoro, Zh. Obshch,. Khim, **59**, 1615–19 (1989).
11. C. L. Liotta, F. L. Cook, and C. W. Bowers, J. Org. Chem., **39**, 3416 (1974).
12. C. L. Liotta, E. M. Burgess, C. C. Ray, E. D. Black, B. E. Fair, ACS Symp. Ser., 326 (Phase Transfer Catal.; New Chem., Catal. Appl.) 15–23 (1987) [CA: 107.096254].
13. A. Brandstrom, *Preparative Ion Pair Extraction,* Apotekarsocieteten/Hassle, Lakemedel, Sweden, p. 145 (1974).
14. G. Simchen and H. Kobler, Synthesis, 605 (1975).
15. J. Solodar, Synth. Inorg. Metal-Org. Chem., **1**, 141 (1971).
16. D. A. White and M. M. Baizer, J. Chem. Soc., Perkin Trans. I, 2230 (1973).
17. G. W. Gokel, K. Arnold, T. Cleary, R. Friese, V. Gatto, D. Goli, C. Hanlon, M. Kim, S. Mill, et al., Prepr.-Am. Chem. Soc., Div. Pet. Chem., **30**, 374–7 (1985).
18. C. L. Lin and T. J. Pinnavaia, Chem. Mater. **3**, 213–15 (1991) [CA: 114.142309].
19. B. M. Choudary, Y. V. S. Rao, and B. P. Prasad, Clays Clay Miner, **39**, 329–32 (1991) [CA: 115.231444].
20. K. E. Koenig and W. P. Weber, Tetrahedron Lett., 2275 (1974).

21. R. Lantzsch and H. Krall, to Bayer A.-G., Ger. Offen. DE 3626411 (1988) (CA: 109.054497].
22. B. I. Rosen, to UOP Inc., U.S. Patent 4,401,639 (1983).
23. B. Sturm and G. Gattow, Z. Anorg. Allg. Chem., **510**, 136–42 (1984) [CA:101.038119].
24. K. Yamamura and S.-I. Murahashi, Tetrahedron Lett. 4429 (1977).
25. Reference [8] pp. 96–99.
26. M. Cinouni, S. Colonna, H. Molinari, and F. Montanari, J. Chem. Soc., Chem. Comm. 112 (1977).
27. F. Devinsky, I. Lacko, L. Krasnec, and R. Kurillova, Czech. Patent CS 214154 (1985) [CA: 103.022179].
28. B. Jursic and J. Branko, Chem. Res., Synop. (4), 104–5 (1989).
29. Reference [8], p. 101.
30. M. Cinquini, F. Montanari, and P. Tundo, Chem. Commun., 393 (1975).
31. G. Jannes, P. Vanderwegen, L. Elsen, and F. Carette, J. Mol. Catal., **62** 255–63 (1990).
32. E. Dehmlow and H. C. Raths, J. Chem. Res., Synop. (12) 384 (1988).
33. R. Fornasier and F. Montaneri, Tetrahedron Lett. 1381 (1976).
34. F. Devingky and I. Lacko, Acta Fac. Pharm. Univ. Comeniane **44**, 119–26 (1990) [CA: 116.040884].
35. S. Kondo, N. Nakashima, H. Hado, and K. Tsuoa, J. Polym. Sci., Part A: Polym. Chem. **28** 2229–32 (1991).
36. 8, p. 101.
37. G. Lauterbach, R. Becker, and K. Janse, to BASF, Eur. Patent Appl., EP 395,982 (1990) [CA: 114.206588].
38. H. Coates, R. L. Barker, R. Guest, and A. Kent, to Albright and Wilson, Ltd., British Patent 1,336,883 (1973).
39. P. A. Verbrugge and E. W. Urbanus, to Shell Oil Co., U.S. Patent 4,056,509 (1978).
40. K. Fukunaga, Yuki Gosei Kagaku Kyokai Shi., **33**, 774 (1975) [CA: 85.177013]; K. Fukunaga, S. Ide, M. Mori and M. Kimura, Nippon Kagaku Kaishi 1379 (1977) [CA: 88.005818].
41. H. Lehmkuhl, F. Rabet, and K. Hauschild, Synthesis 184 (1977).
42. N. Sugimoto, T. Fujita, N. Shigematsu, and A. Ayada, Chem. Pharm. Bull, **10**, 427 (1962); Japanese Patent 19961/63.
43. H. Leuchs and K. H. Schmidt, to Farbenfabriken Bayer, British Patent 1,200,970 (1970) [CA: 73.087662].
44. V. L. Afanas'eva, M. R. Bagreeva, A. V. Lyubeshkin, V. M. Pechenina, N. A. Epshtein, and R. G. Glushkov, Khim.-Farm. Zh., **21**, 1114–19 (1987) [CA: 108.149703].
45. P. Schlunt and P. C. Chau, J. Catal., **102**, 348–56 (1986).

46. N. P. Singh and A. Kumar, Indian J. Chem., Sect. B, **24B,** 1195–6 (1985) [CA: 106.032497].
47. N. Sugimoto, T. Fujita, N. Shigematsu, and A. Ayada, Chem. Pharm. Bull., **10,** 427 (1962).
48. E. G. Woods, L. H. Shepherd, Jr., and E. P. Breidenbach, to Ethyl Corp., U.S. Patent 4,022,813 (1977).
49. R. P. Johnson, to Michigan Chemical Co., U.S. Patent 3,941,827 (1976).
50. N. Sugimoto, T. Fujita, N. Shigematsu, and A. Ayada, Chem. Pharm. Bull., **10,** 427 (1962).
51. E. G. Woods, L. H. Shepherd, Jr., and E. P. Breidenbach, to Ethyl Corp., U.S. Patent 4,022,813 (1977).
52. R. P. Johnson, to Michigan Chemical Co., U.S. Patent 3,941,827 (1976).
53. D. Martinetz and K. Lohs, Leipzig DDR-7050/15, Ger. Dem. Rep. (1985) [CA: 106.083984].
54. M. Cinouni, S. Colonna, H. Molinari, and F. Montanari, J. Chem. Soc., Chem. Commun. **394** (1976).
55. J.P. Idoux and J.T. Gupton, Prepr.-Am. Chem. Soc., Div. Pet. Chem., **30,** 429–35 (1985) [CA: 104.005289].
56. M. Tomoi, E. Nakamura, Y. Hosokawa, and H. Kakiuchi, J. Polym. Sci., Polym. Chem. Ed., **23,** 49–61 (1985).
57. M. Bernard, W. T. Ford, and T. W. Taylor, Polym. Prepr. (Am. Chem. Soc., Div. Polym. Chem.), **25,** 170–1 (1984) [CA: 101.071910].
58. W. T. Ford, J. Lee, and M. Tomoi, Polym. Prepr. (Am. Chem. Soc., Div. Polym. Chem.), **23,** 183–4 (1982).
59. M. Tomoi, Y. Hosokawa, and H. Kakiuchi, J. Polym. Sci., Polym. Chem. Ed., **22,** 1243–50 (1984).
60. M. Tomoi, N. Kori, and H. Kakiuchi, Makromol. Chem., **187,** 2753–61 (1986).
61. M.J. Pugia, A. Czech, B. Caech, and R.A. Bartsch, J. Org. Chem. **51,** 2945 (1986).
62. P. Hodge, E. Khoshdel, and J.J. Waterhouse, J. Chem. Soc., Perkin Trans. **1,** (11), 2451–5 (1984).
63. P.L. Anelli, B. Czech, F. Montanari, and S. Quici, J. Am Chem Soc., **106,** 861–9 (1984) [CA: 100.084976].
64. S. Kondo, N. Nakashima, H. Hado, and K. Tsuoa, J. Polym. Sci., Part A: Polym. Chem. **28,** 2229–32 (1991).
65. S. Kobayashi, M. Suzuki, T. Sakaya, and T. Saegusa, Makromol. Chem., **188,** 457–62 (1987).
66. B.M. Choudary, Y.V.S. Rao, and B.P. Prasad, Clays Clay Miner. **39,** 329–32 (1991) [CA: 115.231444].
67. M. Tomoi and H. Kakiuchi, Makromol. Chem., Rapid Commun., **5,** 685–8 (1984).
68. D. Landini, F. Montanari, and F.M. Prisi, Chem. Commun. 879 (1974).

69. Y. Sasson, M. Weiss, A. Loupy, G. Bram, and C. Pardo, Israel J. Chem. Soc., Chem. Commun. (16), 1250–2 (1986).
70. M. Yonovich-Weiss and Y. Sasson, Israel J. Chem., **26,** 243–7 (1985).
71. D. Forster, J. Chem. Soc., Chem. Commun. 918 (1975).
72. C.M. Starks and R. D. Gordon, to Continental Oil Co., U.S. Patent 3,899,542 (1975).
73. R.D. Gordon, to Continental Oil Co., U.S. Patent 3,696,084 (1972).
74. D.J. Sam and H.C. Simmons, J. Am. Chem. Soc., **96,** 2252 (1974).
75. F. Montanari, D. Landini, A. Maia, S. Ouici, and P.L. Anelli, Prepr. -Am. Chem. Soc., Div. Pet. Chem., **30,** 387–93 (1985) [CA: 104.108925].
76. E.V. Dehmlow and J. Stuetten, Liebigs Ann. Chem., (2), 187–9 (1989).
77. C.M. Starks, unpublished data.
78. E. Angeletti, P. Tundo, P. Venturello, and F. Trotta, Br. Polym. J., **16,** 219–21 (1984) [CA: 103.070921].
79. M. Tanaka, H. Takigawa, Y. Yasaka, T. Shono, K. Funazo, and H.L. Wu, J. Chromatogr., **404,** 175–82 (1987).
80. M. Tanaka, Y. Yasaka, M. Kamino, T. Shono, K. Funazo, and J. Wu, J. Chromatogr., **438,** 253–61 (1988).
81. R. Aizenberg, O. Arrad. and Y. Sasson, Ind. Eng Chem. Res., Res. **31,** 431–4 (1992).
82. C.T. Chen, C. Hwang, and M.Y. Yeh, J. Chem. Eng. Japan, **24,** 284–90 (1991) [CA:115.113820].
83. S.S. Yufit, I.A. Esikova, and O.I. Danilova, Dokl. Akad. Nauk SSSR, **295,** 621–4 (1987).
84. S.S. Yufit and I.A. Esikova, J. Phys. Org. Chem., **4,** 336–40 (1991).
85. H.S. Weng, W. Huang, and C. Wei, J. Chin. Inst. Chem. Eng., **18,** 109–15 (1987) [CA: 107.136283].
86. P. Tundo and M. Badiali, React. Polym., **10,** 55–65 (1989).
87. P.L. Anelli, F. Montanari, and S. Quici, J. Org. Chem., **51,** 4910–14 (1986).
88. A. Winterfeldt, G. Bartels, and R. Knieps, Ger. Patent Offen- DE 4,025,227 (1992) [CA: 116,173550].
89. J.T. Arnold, T.O. Bayraktaroglu, R.G. Brown, C.R. Heiermann, W.W. Magnus, A.B. Ohman, and R.G. Landolt, J. Org. Chem. **57,** 391–3 (1992).
90. A.V. Bogatskii, N.G. Luk'yanenko, and V.N. Pastushok, Dokl. Akad. Nauk SSSR, **271,** 1392–4 [CA: 100.0855665].
91. G. Podda, C. Anchisi, L. Corda, and A. Maccioni, Gazz. Chim. Ital., **116,** 515–17 (1986) [CA: 107.059008].
92. D. Landini, A. Maia, L. Corda, A. Maccioni, and G. Podda, Tetrahedron, **47,** 7477–88 (1991).
93. X. Zhang, Huaxue Shiji, 5(5), 308, 304 (1983) [CA: 100.102691].

94. E.E. Ergozhin, M. Kurmanaliev, and N.A. Asanov, Zh. Prikl. Khim. (Leningrad), **62**, 1679–81 (1989) [CA: 112.177981].
95. J.H. Shim, K.B. Chung, S.H. Chang, D.K. Song, and Y.K. Sung, Taehan Hwahakhoe Chi, **32**, 593–602 (1988) [CA: 111.022847].
96. M. Tomoi, N. Yanai, S. Shiiki, and H. Kakiuchi, J. Polym. Sci., Polym. Chem. Ed., **22**, 911–26 (1984) [CA: 100.210601].
97. N. Zhong and Z. Liang; Gaofenzi Xuebao (5) 624-7 (1989) [CA: 114.063298].
98. Y. Chen, J. Luo, and X. Lu, Yingyong Huaxue, **6**, 16-21 (1989) [CA: 113.023995].
99. Y. Chen, X. Lu, M. Cai, and X. Song, Gongneng Gaofenzi Xuebao, **3**, 122–6 (1990) [CA: 115.030053].
100. L. Bonsignore, L. Corda, E. Maccioni, G. Podda, and M. Gleria, Gazz. Chim. Ital., **121**, 341–4 (1991) [CA: 115.280649].
101. H. Takeuchi, Y. Miwa, S. Morita, and J. Okada, Chem. Pharm. Bull., **32**, 409–17 (1984) [CA: 101.054175].
102. B.M. Choudary, Y.V.S. Rao, and B.P Prasad, Clays Clay Miner. **39**, 329–32 (1991) [CA: 115.231444].
103. B.L. Liu, Y.T. Jin, Z.H. Liu, C. Luo, M. Kojima, and M. Maeda, Int. J. Appl. Radiat. Isot., **36**, 561–3 (1985) [CA: 104.149252].
104. B. Escoula, I. Rico, A. Lattes, J. Simon, and R. Guiraud, New J. Chem., **15**, 75–8 (1991).
105. B. Escoula, I. Rico, and A. Lattes, Bull. Soc. Chim. Fr., (2), 256–9 (1989).
106. D. Landini and A. Maia, Tetrahedron, **47**, 1285 (1991).
107. S. Dermeik and Y. Sasson, J. Org. Chem., **50**, 879–82 (1985).
108. D. Landini, F. Montanari, and F. Rolla, Synthesis 428 (1974).
109. G. Cainelli and F. Manescalchi, Synthesis 472 (1976).
110. G. Bram, A. Loupy, and P. Pigeon, Synth. Commun., **18**, 1661–7 (1988).
111. Reference [8] p. 116.
112. D. Forster, J. Chem. Soc., Chem. Commun., 918 (1975).
113. M. Gingras and D.N. Harpp, Tetrahedron Lett., **29**, 4669–72 (1988).
114. T.J. Mason, J.P Lorimer, A.T. Turner, and A.R. Harris, J. Chem. Res. Synop., (8) 300 (1986) [CA: 106.032430].
115. T. Kawai, to Central Glass Co., Ltd., Brit. UK Pat. Appl., GB 2,219,292 (1989) [CA: 112.234804].
116. J.H. Clark and N. Boechat, Chem. Ind. (London), (12) 436 (1991).
117. G.L. Cantrell, to Mallincrokodt, Inc., U.S. Patent, 4,973,771 (1990).
118. G. Cantrell, to Mallinckrodt, Inc., U.S. Pat. 4,973,772 (1990).
119. E. Kysela, R. Braden, to Bayer, A.-G., Ger. Offen. DE 3,827,436 [CA: 113.040127].
120. C.R. White, to Mallinckrodt, Inc., U.S. Patent 4,642,399 (1987).

121. C.R. White, to Mallinckrodt, Inc., U.S. Patent 4,418,229 (1983).
122. J. Venci, V. Chmatal, Z. Zuzanek, and Z. Vidner, Chech. Patent, CS 269,080 (1990) [CA: 115.028855].
123. O. Kaieda, K. Hirota and T. Nakamura, to Nippon Shokubai Kagaku Kogyo Co., Ltd. Jpn. Kokai Tokkyo Koho JP 61/43126 A2 [86/43126], 1986 [CA: 105, 152681].
124. J. Hackenbruch and T. Papenfuhs, to Hoechst A.G. PCT Int. Patent Appl. WO 91 00,266 (1991) [CA: 114.206809].
125. S. Yonemori, Y. Hayashi, S. Humai, and A. Wada, Nippon Kagaku Kaishi, (8) 1146–8 (1991) [CA: 115.159279].
126. S. Kumai, M. Sasabe, and H. Matsuo, to Asahi Glass Co., Jpn. Kokai Tokkyo Koho JP 60/228436 (1985) [CA: 104.148468].
127. Shokubai Kagaku Kogyo Co., Ltd. Jpn Kokai Tokkyo Koko JP 60/149566 A2 [85/149566], 7 Aug 1985, Appl. 84/5629, 18 Jan 1984.
128. O. Kaieda, K. Hirota, and T. Nakamura, to Nippon Shokubai Kagaku Kogyo Co., Ltd. Jpn Kokai Tokkyo Koho JP 61/43126 A2 [86/43126], 1986.
129. J. Vencl, Z. Vidner, V. Chmatal, Czech. Patent CS 247,969 (1987) [CA: 108.188926].
130. S.V. Nikolaeva, A.M. Kolbin, Yu E. Sapozhnikov, R.B. Valitov, V.F. Ivanov, Khim Geterotsikl. Soedin. (10) 1370–2 (1990) [CA: 114.143365].
131. S. Yonemori, Y. Hayashi, S. Humai, and A. Wada, Nippon Kagaku Kaishi, (8) 1146–8 (1991) [CA: 115.159279].
132. J. Wild and N. Goetz, to BASF A.-G., Ger. Patent Offen. DE 3,820,979 (1989) [CA: 113.040150].
133. E.N. Krylov and T.A. Buslaeva, Zh. Obsch. Khim, **61,** 974 (1991) [CA: 115.255718].
134. Ref. No. [8], p. 141 et seq.
135. H.A. Zahalka, and Y. Sasson, Can. J. Chem., **67,** 245–9 (1989).
136. Y. Sasson and H.A. Zahalka, J. Chem. Soc., Chem. Commun., (22), 1347–9 (1983).
137. C.L. Liotta, H.P. Harris, M. McDermott, T. Gonzalez, and K. Smith, Tetrahedron Lett. 2417 (1974).
138. T.T. Wang and T.C. Huang, Chem. Eng. Commun., **100,** 135–47 (1991) [CA: 115.052224].
139. X. Sun, L. Zhang, H. Liu, J. Guo, and X. Jia, Hebei Shifan Daxue Xuebao, Ziran Kexueban, (3), 21–6, 15 (1988) [CA: 111.038668].
140. S. Matsura and O. Miyano, to Tosoh Yuki Kagaku KKY, Jpn. Patent Kokai Tokkyo Koho JP 03,188,044 (1991) [CA: 116.006165].
141. J. Lewis, M.K. Naqvi, and G.S. Park, Polym. Prepr. (Am. Chem. Soc., Div. Polym. Chem.), **23,** 140–1 (1982).
142. H. Ali Zahalka and Y. Sasson, J. Mol. Catal., **18,** 57–60 (1983) [CA: 98.125092].

143. N. Yang, S. Wang, and L. Pan, Jilin Daxue Ziran Kexue Xuebao, (3), 113–17 (1986) [CA: 106.158256].
144. O.E. Filippova, I.N. Topchieva, and V.P. Zubov, Vestn. Mosk. Univ., Ser. 2: Khim., **24,** 590–4 (1983) [CA: 100.102832].
145. O.E. Filippova, I.N. Topchieva, V.V. Lutsenko, and V.P. Zubov, Vysokomol. Soedin., Ser. A, **26,** 402–8 (1984) [CA: 101.022686].
146. H. Schiefer, J. Beger, and U. Lwenz, J. Prakt. Chem., 327(3), 383–98 (1985) [CA: 104.068420].
147. S.Z. Kusov, E.G. Lubenets, V.S. Kobrin, and O.P. Sheremet, Khmel'nitskii, A.G.; Akad. Nauk SSSR, Ser. Khim., (3), 709–10 (1986) [CA: 106.084100].
148. T. Wakui, W.Y. Xu, and J. Smid, Polym. Prepr. (Am. Chem. Soc., Div. Polym. Chem.), **25,** 126–7 (1984).
149. C.S. Chen, W.C. Hwang, and S.C. Chang, Huaxue 1991, **49,** 99–106 [CA: 116.150922].
150. T. Wakui, W.Y. Xu, C.S. Chen, and J. Smid, Makromol. Chem., **187,** 533 (1986).
151. N. Zhong and Z. Liang, Huaxue Shiji, **12,** 300–1 (1990) [CA: 115,008177].
152. A.V. Bogatskii, N.G. Lukyanenko, V.N. Pastushok, and M.N. Parfenova, Dokl. Akad. Nauk SSSR, **283,** 628–9 [CA: 104.68501].
153. R.A. Sawicki, Prepr. Am. Chem. Soc., Div. Pet. Chem., **30,** 415–20 (1985).
154. S. Szakacs, Sandor, M. Jaky, S. Gobolos, and F. Nagy, Magy. Kem. Foly. 89(9) 402–6 (1983) [CA: 100.051186].
155. J. Barry, G. Bram, and A. Petit, Tetrahedron Lett., **29,** 4567–8 (1988).
156. X. Huang and L. Xie, Hangzhou Daxue Xuebao, Ziran Kexueban, **13,** 332–4 (1986) [CA: 106.156002].
157. C.A.R. Skow and M.K.L. Bicking, Chromatographia, **21,** 157–60 (1986) [CA: 106.084094].
158. Y.P. Shih and M.Y. Yeh, K'o Hsueh Fa Chan Yueh K'an, **12,** 88–98 (1984) [CA: 101.129903].
159. L. Cerveny, A. Marhoul, and L. Kovarova, Parfuem. Kosmet. **72,** 502–4, 507 (1991) [CA: 116.020664].
160. T. Iizawa, T. Nishikubo, Y. Masuda, and M. Okawara, Macromolecules, **17,** 992–8 (1984).
161. T.M. Fisher, M. Peled, and L.M. Shorr, to Bromine Compounds Ltd., Eur. Pat. Appl. EP 390,042 (1990) [CA: 114.123257].
162. G. Bernhardt, J. Amort, M. Haas, H. Hansich, and H. Kragl, to Huels A.G., U.S. Patent 4,946,977 (1990).
163. X. Sun, D. Kong, G. Lu, and Y. Fu, Shiyou Huagong, **15,** 766–9 (1986) [CA: 107.078276].
164. A. Aserin, N. Garti, and Y. Sasson, Ind. Eng. Chem. Prod. Res. Dev. **23,** 452–4 (1984).

165. Y. Lu and M. Tao-Riyong Huaxue Gongye, (2), 49–52 (1988) [CA: 110.010047].
166. F.A.L. Van der Horst, M.H. Post, J.J.M. Holthuis, and U.A.T. Brinkman, J. Chromatogr., **500**, 443–52 (1990).
167. N. Yazawa and K. Ishikame (Ihara Chemical Industry Co Ltd.) Eur. Pat. Appl. EP 453,993 (1991) [CA: 116.058764].
168. Q. Zheng and L. Zeng, Huaxue Shiji, **13**, 252–3 (1991) [CA: 115,255599].
169. Q. Zheng and L. Zeng, Zhongguo Yiyao Gongye Zazi, **21**, 76–7 (1990) [CA: 114.101088].
170. D. Klemm and G. Geschwend, Ger. Patent (East) DD 222,879 (1985) [CA: 105.008204].
171. T. Nishikimi and H. Shintaku, to Nissan Chemical Industries, Ltd., Jpn. Kokai Tokkyo Koho, JP 02009836 (1990) [CA: 113.005318].
172. N. Zhong and Z. Liang, Gaofenzi Xuebao (5) 624-7 (1989) [CA: 114.063298].
173. J.R. Chang, M.Y. Yeh, and Y.P. Shih, J. Chin. Chem. Soc. (Taipei), **31**, 185–90 (1984) [CA: 101.110010].
174. S.S. Mishra, G.B. Behera, and S. Nayak, Indian J. Chem., Sect. A., **24A**, 771–2 (1985) [CA: 105.005905].
175. N.N. Sukhanov and V.N. Chetverikov, Arm. Khim. Zh., **40**, 323–6 (1987) [CA: 108.221370].
176. J. Barry, G. Bram, G. Decodts, A. Loupy, C. Orange, A. Petit, and J. Sansoulet, Synthesis, 40–5 (1985).
177. S.M. Seekings and C.E. Whitten, to Polaroid Corp. Eur. Pat. Appl. EP 171762 (1986) [CA: 105.0060424].
178. A. Loupy, M. Pedoussaut, and J.J. Sansoulet, J. Org. Chem., **51**, 740–2 (1986).
179. A. Furaengen, Acta Pharm. Suec., **22**, 197–208 (1985) [CA: 104.129232].
180. D. Nobatov, A.M. Kobin, E.A. Amanov, R.B. Valitov, and N.F. Popova, Izv. Akad. Nauk Turkm. SSR, Ser. Fiz.-Tekh., Khim. Geol. Nauk, (3), 99–102 (1988) [CA: 111.057190].
181. M.Y. Yeh, H.H. Guo, L.C. Chen, and Y.P. Shih, Ind. Eng. Chem. Res., **27**, 1582–7 (1988).
182. A.A. Akopyan, S.A. Grigor, G.O. Torosyan, and A. Babayan, Zh. Prikl. Khim. (Leningrad), **62**, 1904–6 (1989) [CA: 112.158991].
183. S. Pavlov, M. Bogavac, L. Arsenijevic, and V. Arsenijevic, Arh. Farm., **36**, 161–5 (1986) [CA: 107.154023].
184. T. Balogh, et al., Hung. Teljes Patent HU 52,032 (1990) [CA: 114.206790].
185. H.C. Raths and E.V. Dehmlow, Chem. Ber., **120**, 647–8 (1987).
186. J. Cast, J. Hamilton, J. Podmore, and L. Porecha, Chem. Ind. (Lond) (20), 763 (1991).
187. P.I. Dem'yanov, N. Malo, and I.V. Petrosyan, Vestn. Mosk. Univ., Ser. 2: Khim., 28~5). 484–8 (1987) [CA: 108.114614].

188. P.I. Dem'yanov, N. Malo, and V.S. Petrosyan, Vestn. Mosk. Univ., Ser. 2: Khim., 26(2), 223 (1985) [CA: 103.038938].
189. M.Y. Yeh, D.H. Hwu, C. Hwang, T.K. Hwang, and Y.P. Shih, J. Chin. Chem. Soc. (Taipei), **38** 221–30 (1991) [CA: 115.135239].
190. P. Klan and P. Benovsky, Monatsh. Chem. **123**, 469 (1992).
191. H.A. Zahalka, Z. Neiroukh, O. Arrad, and Y. Sasson, Ind. Eng. Chem. Res., **31**, 2062 (1992).
192. N. Yazawa, K. Ishikame (Ihara Chemical Industry Co Ltd.) Eur. Pat. Appl. EP 453,993 (1991) [CA: 116.058764].
193. W.P. Reeves and M.L. Bahr, Synthesis, 823 (1976).
194. S.C. Choi, J.K. Kim, and C.B. Kim, Taehan Hwahakhoe Chi, **30**, 556–8 (1986) [CA: 107.096103].
195. M. Takeishi, R. Kawashima, and M. Okawara, Makromol. Chem., **167**, 26 (1973).
196. A. Brandstrom, B. Lamm, and I. Palmertz, Acta Chem. Scand., Ser. B, **28**, 699 (1974).
197. V.A. Zlobin, Izv. Vyssh. Uchebn. Zaved., Khim. Khim. Tekhnol., **31**, 104–5 (1988) [CA: 109.189579].
198. A. Draganov and V. Bozhinov, Dokl. Bolg. Akad. Nauk, **40**, 61 (1987) [CA: 108.186280].
199. S.M. Kumar, Synth. Commun., **17**, 1015 (1987).
200. V.A. Zlobin, V.T. Kosolapov, and A.K. Tarasov, Izv. Vyssh. Uchebn. Zaved., Khim. Khim. Tekhnol., **27**, 169 (1984) [CA: 100.209083].
201. E.F. Roberts, S.H. Pines, W.K. Russ, Jr., to Merck and Co., Inc., Eur. Pat. Appl. EP 160218 (1985) [CA: 104.148502].
202. S.M. Kumar, Synth. Commun. **21**, 2121 (1991).
203. H.W. Mark, to Phillips Petroleum Co., Eur. Pat. Appl EP 268261 A1, 25 May 1988 [CA: 109.230275].
204. H. Suzuki, M. Kawashima, Y. Kawamura, and T. Ogura, Jpn. Patent Kokai Tokkyo Koho JP 62/294652 (1986) [CA: 109.092460].
205. K. Imamura, H. Miura, Y. Uno, and E. Watanabe, to Nippon Chemical Industrial Co., Ltd., Jpn. Patent Kokai Tokkyo Koho Patent JP 02,304,061 (1990) [CA: 114.206583].
206. Y. Chen, J. Ni, and J. Yso, Huaxue Shji, **12**, 382–3 (1990) [CA: 114.184752].
207. Y. Chen and F. Zeng, Gaofenzi Tongxun, 409–14 (1986) [CA: 106.214501].
208. T.D. Shaffer and M.C. Kramer, Polym. Prepr. (Am. Chem. Soc., Div. Polym. Chem.), **30**, 171 (1989).
209. J.M. Castex, J.F. Roussel, G. Parc.J.L. Mieloszynski, and G. Kirsch, Sulfur Lett. **2**, 77 (1984) [CA: 101.229951].
210. V.I. Lavrov, L.N. Parshina, and L.A. Lapkanova, Zh. Prikl. Khim. (Leningrad), **62**, 1190–1 (1989).

211. Y. Ikeda, M. Ozaki, and T. Arakawa, Jpn. Patent Kokai Tokkyo Koho JP 61/148235 (1986) [CA: 106.085271].

212. H. Ben Romdhane, M. Bartholin, B. Sillion, and S. Boileau, C.R. Acad. Sci., Ser. 2, **311,** 525 (1990).

213. N.C. Pradhan and M.M. Sharma, Ind. Eng. Chem., Res., **31,** 1610 (1992).

214. General Electric Co., Jpn. Patent Kokai Tokkyo Koho JP 60/188368 (1985) [CA: 104.130428].

215. N.C. Pradhan and M.M. Harma, Ind. Eng. Chem., Res. **31,** 1606 (1992).

216. T.L. Evans, Synth. Commun., **14,** 435 (1984) [CA: 101.191235].

217. D.J. Brunelle, to General Electric Co., U.S. Patent 4,410,422 (1983).

218. J. Wang and L. Geng, Xibei Shffan Xueyuan Xuebao, Ziran Kexueban, (3), 36–40, 57 (1988) [CA: 111.057014].

219. J. Simandl, J. Lorenc, B. Kutil, and J. Cuda, Czech. Patent CS 215,492 (1985) [CA. 104.068578].

220. V.V. Nosyreva and S.V. Amosova, Zh. Org. Khim., **27,** 1412–15 (1990).

221. J. Wang, W. Cui, Y. Hu, and K. Zhao, Chin. Chem. Lett., **1,** 193–6 (1990).

222. E.V. Dehmlow and G.O. Torossian, Z. Naturforsch. B: Chem. Sci., **45,** 1091–2 (1990) [CA: 114.005984].

223. S.H. Chang, M.H. Yoon, C.S. Kim, K.B. Chung, and B.J.H. Kwang, Taehan Hwahakhoe Chi, **33,** 651 (1989) [CA: 112.197284].

224. Y. Dai and J. Chen, Xibei Shifan Xueyuan Xuebao, Ziran Kexueban, (2), 32–5 (1983) [CA: 101.054658].

225. F. Szonyi and A. Cambon, J. Fluorine Chem., 42(1), 59–68 (1989) [CA: 111.023060].

226. M. Vlassa, M. Kezdi, and M. Bogdan, J. Prakt. Chem., **327,** 1007 (1985).

227. T. Hassel and H.P. Mueller, to Bayer A.-G., Ger. Patent Offen. DE 3,341,018 (1985) [CA: 104.051081].

228. D. Martinetz and K. Lohs, Leipzig DDR-7050/15, Ger. Dem. Rep., (1985) [CA: 106.083984].

229. M.C. Berthe and Y.-C. Fort, Synth. Commun., **22,** 617 (1992).

230. S. Yu and J. Yang, Hunan Shifan Daxue Ziran Kexue Xuexao **4,** 61 (1991) [CA: 116:150833].

231. S. Kondo, Y. Inagaki, H. Yasui, M. Iwasaki, and K. Tsudar. J. Polym. Sci. Part A: Polym. Chem., **29,** 243–9 (1991) [CA: 114.206539].

232. S. Kondo, T. Mori, H. Kunisada, and Y. Yuki, Macromol. Chem. Rapid Commun. **11,** 309–13 (1990).

233. J. Silhanek and J. Bartl. Sb. Vys. Sk. Chem.-Technol. Praze, C.: Org. Chem. Technol., **C31,** 81 (1991) [CA: 117.130674].

234. M.C. Berthe, Y. Fort, and P. Caubere, Synth. Commun. **22,** 617 (1992).

235. W.P. Reeves, A. Simmons, Jr., J.A. Rudis and T.C. Bothwell, Synth. Commun. **11,** 781–5 (1981).
236. T. Kitamura, S. Kobayashi, and H. Taniguchi, J. Org. Chem., **55,** 1801 (1990).
237. R. Lantzsch, A. Markhold, and K.-F. Lehment, to Beyer A.G., Ger. Patent 2,545,644 (1977).
238. R. Miethchen, K. Kohlheim, H. Stock, and U. Roth, Wiss. Z. Ernst-Moritz-Arndt-Univ. Greifsw., Math.-Naturwiss. Riehe, **34,** 40–2 (1985) [CA: 105.208983].
239. N.V. Korotchenkova, A.D. Bulat, N.R. Reznikova, and B.V. Passet, USSR Patent SU 1,616,906 (1990) [CA: 114,249731].
240. I. Tabuse and N. Morioka, Nippon Oil Fats Co., Ltd., Jpn. Kokai Tokkyo Koho JP 60/228453 (1985) [CA: 105.081150].
241. M. Hattori, A. Taguma, T. Morimitsu, and A. Takeshita, to Sumitomo Chemical Co., Eur. Pat. Appl. EP 105762 (1984) CA: 101.173037].
242. P.V. Sane and M.M. Sharma, Org. Prep. Proced. Int. **20,** 598–9 (1988) [CA: 110.192211].
243. K.J. Matsunaga and T. Yamashita, Kogyo Kayaku, **41,** 3 (1980) [CA: 94.31679].
244. N. Bhati, A. Kabra, C.K. Narang, and N.K. Mathur, J. Org. Chem., **56,** 4967–9 (1991).
245. K.T. Howang, M. Takahashi, K. Iwamoto, and M. Seno, Makromol. Chem., **188,** 1383 (1987).
246. A.W. Herriott and D. Picker, Tetrahedron Lett., 4521 (1972).
247. T.T. Wang, T.C. Huang, C. Ting, and M.Y. Yeh, J. Mol. Catal. **57,** 271–89 (1990) [CA: 112.197299].
248. S. Tanaka and K. Miura, to Tokuyama Soda Co., Ltd., Jpn. Patent Kokai Tokkyo Koho JP 61/215340 (1986) [CA: 106.101872].
249. Ya. M. Slobodin; Zh. Org. Khim., **24,** 2621 (1988) [CA: 110.231162].
250. A.G. Williams, to Imperial Chemical Industriesl PLC, Ger. Patent Offen. DE 4,118,444 (1991) [CA: 116.058970].
251. D. Feldman, D. Segal-Lew, and M. Rabinovitz, J. Org. Chem., **56,** 7350 (1991) [CA: 115.279515].
252. Seitetsu Kagaku Co., Ltd., Jpn. Patent Kokai Tokkyo Koho JP 58/154561 (1983) [CA: 100.120891].
253. J. Hackenbruch and T. Papenfuhs, to Hoechst A.-G., Ger. Offen. DE 3,921,449 (1991) [CA: 115.093143].
254. M. Lissel and E.V. Dehmlow, Chem. Ber., **114,** 1210 (1981).
255. J.A. Cella and S. Bacon, J. Org. Chem. **49,** 1122 (1984).
256. K. Soga, S. Hosoda, and S. Ikeda, J. Polym. Sci., Polym. Lett. Ed., **15,** 611 (1977); J. Polym. Sci., Polym. Chem. Ed., **17,** 517 (1979).
257. J. Moulines, M.J. Bougeois, M. Campagnole, A.M. Lamidey, B. Maillard, and E. Montaudon, Synth. Commun., **20,** 349 (1990).

258. J.C. Dyer, to Procter and Gamble Co., Eur. Pat. Appl. EP 206740 A1, 30 Dec. 1986, US Appl. 747469 [CA: 106.213416].
259. R.A. Johnson and E.S. Nidy, J. Org. Chem., **40** 1680 (1975).
260. E.J. Corey, K.C. Nicolaou, M. Shibaski, Y. Machida, and C.S. Shiner, Tetrahedron Lett., **37,** 3183 (1975).
261. J. San Filippo, Jr., C. Chern, and J.S. Valentine, J. Org. Chem., **40,** 1678 (1975).
262. B.A. Trofimov, N.K. Gusarova, et al., Phosphorus, Sulfur, Silicon Relat. Elem., **55,** 271–4 (1991) [CA: 115.071753].
263. K.F. Zenner and H. Appel, Ger. Pat. 2,126,296 (1972) [CA: 78.058473].
264. Y. Nadachi and M. Kokura, Japan Kokai 7595289 (1976) [CA: 84.164852].
265. B. Graham, U.S. Patent, 2,866,802 (1958).
266. D. Mitchell (to Eli Lilly Co.), U.S. Patent 5,136,079 (1992).
267 T.A. Foglia and L.S. Silbert, Synthesis 545 (1992).

8

Phase-Transfer Catalysis Reaction with Strong Bases

Among the vast scope of phase-transfer catalysis (PTC) applications, the largest sector involves the reaction of anions that are reacted and/or generated under PTC conditions with species containing electrophilic centers. Among the most widespread and useful anions reacted under PTC conditions are those generated by deprotonation of "acidic" or "moderately acidic" organic compounds. PTC advantageously uses convenient, inexpensive, and relatively safe bases such as NaOH and K_2CO_3, instead of the classical organic alkoxides, amides and hydrides, to generate these anions. Approximately 40% of PTC patents involve the hydroxide ion and it has been estimated [1] that approximately 60% of commercial PTC applications involve the hydroxide ion.

Such common useful organic anions include enolates, alkoxides, anions derived from activated methylene compounds (benzylic and allylic ketones, imines, sulfones, and nitriles), anions derived from activated $N-H$ compounds, and even activated hydrocarbons. Typically, these activated organic acids have acidities in the pK_a range of 15–23. The alkylation reaction is among the most useful synthetic transformations in organic chemistry and is advantageously performed under PTC conditions using these anions.

Depending on semantic definition, one may include in the "deprotonated organic substrate" category much more acidic organic acids such as phenols ($pK_a \sim 9$), carboxylic acids (pK_a 4–5), mercaptans ($pK_a \sim 11$), and the like. Phenols and mercaptans are discussed in Section C, and carboxylates are discussed in Chapter 7.

Other advantageous uses for NaOH for performing common organic chemistry include its use as a base for elimination reactions such as dehydrohalogenation. Another application of PTC/OH base catalyzed reactions is for initiating other types of reactions by deprotonation, such as the conversion of chloroform to the trichloromethide anion which spontaneously decomposes into dichlorocarbene, which may react further in the synthesis of a wide variety of compounds. Carbon chains can be further built by using PTC/OH conditions to generate organic anions for an even wider variety of organic reactions such as the Michael addition, the Wittig reaction, the aldol condensation, and others.

The focus of this chapter is the application of base-catalyzed PTC reactions which involves the removal of a proton from a carbon, nitrogen, oxygen, or sulfur, of pK_a generally greater than 9, during any stage of the reaction. In addition, discussion will sometimes include speculation relating to the reaction conditions chosen in the publications reported. In limited cases, speculation relating to the process aspects of the reactions and choice of catalyst is also offered. In addition, most (not all) PTC/OH reactions involve nucleophilic displacements as well as deprotonation. A discussion of the factors that affect the nucleophilic displacement aspects of these reactions is discussed in Chapter 4 Section A.1.

A. C-Alkylation

C-Alkylation is one of the most fundamental organic reactions for building the carbon skeleton of organic compounds. The carbon-hydrogen bond of typical hydrocarbons does not easily lend itself to dissociation into a proton and a carbanion. Either an extremely strong base is needed to deprotonate the C—H bond of an unactivated hydrocarbon or the C—H bond must be activated by strong electron-withdrawing groups to stabilize the carbanion to be formed. Examples of activating groups for deprotonation of the C—H bond by hydroxide under PTC conditions are carbonyl, nitrile, imine, or sulfone located in the α position to the C—H bond. The carbanion may be further stabilized if the C—H bond, activated by the aforementioned functional groups, is also located in a benzylic position. Generally, the location of a C—H bond in a benzylic position by itself (e.g., as in toluene) does not render it sufficiently stabilized to be deprotonated under PTC/OH conditions and remain so long enough to be alkylated. As a rule of thumb, "carbon acids" with a pK_a of up to approximately 23 may be successfully deprotonated AND alkylated (e.g., with ethyl bromide, dimethyl sulfate, etc.) under typical PTC/OH conditions.

C-alkylation of activated methylene compounds under catalytic conditions was pioneered by Makosza and co-workers. Three classic PTC C-alkylations were described in Makosza's series of articles entitled "Reactions of Organic Anions" (I and II—alkylation of activated benzylic nitriles [2,3]; XI—alkylation of activated hydrocarbons [4]; XXVI—C-alkylation of activated ketones [5]). These articles are a good starting point for the following discussion of PTC/OH C-alkylation.

1. Ketones

The general case of PTC C-alkylation of activated ketones is exemplified by the reaction of phenylacetone, *1*, with *n*-BuBr (90%) in the presence of triethylbenzylammonium chloride and no solvent [5]. The pK_a of phenylacetone is likely to be approximately 15–18 [6], and the reaction is likely to be hydroxide transfer rate limited. Thus the choice of an accessible quaternary ammonium cation, such as triethylbenzylammonium, seems appropriate according to the guidelines of Chapter 6, Section G. Phenylacetone has also been C-alkylated by methyl iodide by "extractive alkylation." This technique was developed by Brandstrom [7] and involves the formation of the quaternary ammonium salt of the organic anion by taking a stoichiometric quantity of, for example, tetrabutylammonium HSO_4^- and phenylacetone in the presence of 200 mol% NaOH and extracting the resulting quaternary ammonium cation–organic anion salt into a solvent such as methylene chloride. Addition of methyl iodide to the isolated organic phase yielded 92% of the monomethylated phenylacetone (in the benzylic position) within 10 min at room temperature. Under identical conditions, the β-diketone diethyl malonate, 2 (pK_a = 13.3) [8] was monoethylated with ethyl iodide in 100% yield and acetylacetone (pK_a = 9) [8], 3, was monomethylated with methyl iodide in 98.5% yield [9].

$$\text{C}_6\text{H}_5-\text{CH}_2-\overset{\overset{\text{O}}{\|}}{\text{C}}-\text{CH}_3 \qquad 1$$

$$\text{C}_2\text{H}_5\text{O}-\overset{\overset{\text{O}}{\|}}{\text{C}}-\text{CH}_2-\overset{\overset{\text{O}}{\|}}{\text{C}}-\text{OC}_2\text{H}_5 \qquad 2$$

$$\text{CH}_3-\overset{\overset{\text{O}}{\|}}{\text{C}}-\text{CH}_2-\overset{\overset{\text{O}}{\|}}{\text{C}}-\text{CH}_3 \qquad 3$$

Phenylacetone was alkylated with n-BuBr in 95% yield using a polymer-bound phosphonium catalyst (using a long spacer for enhanced reactivity) [10]. Using a polyethylene glycol (PEG) bound to a polymer the same reaction was performed in 75% yield [11]. Further discussion of the use of polymer-bound PTC for C-alkylation can be found in Section A.5 of this chapter.

Deoxybenzoin has been the subject of alkylation mechanistic studies due to its ease of alkylation and the ability to monitor C-/O-alkylation selectivity (see Figure 8-1) [12–16]. The concept of "accessibility" of quaternary onium cations was first proposed based on the reactivity and selectivity of the alkylation of deoxybenzoin; a discussion of the effect of quaternary -onium cation structure on reactivity of the alkylation of deoxybenzoin is found in Chapter 6, Section A.1.b. Regarding selectivity, deoxybenzoin undergoes more O-alkylation than phenylacetone and other activated ketones, since a driving force exists to conjugate the two phenyl groups through the enolate. The C-/O-alkylation selectivity of deoxybenzoin is affected by quaternary ammonium cation accessibility (see Figure 8-2), hydration, alkylating agent, NaOH concentration (see Figure 8-3), solvent (see Figure 8-4), and other factors.

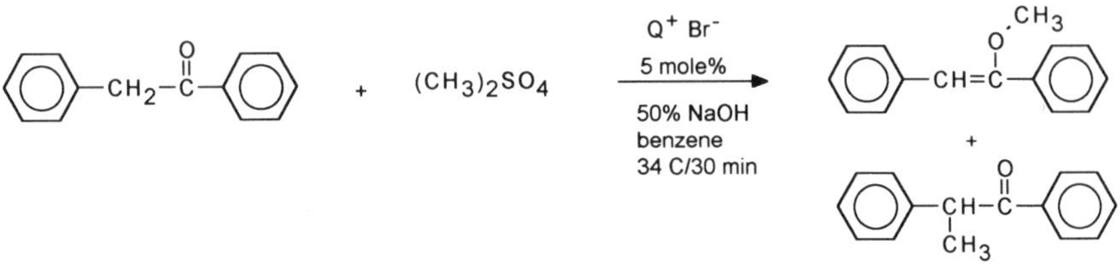

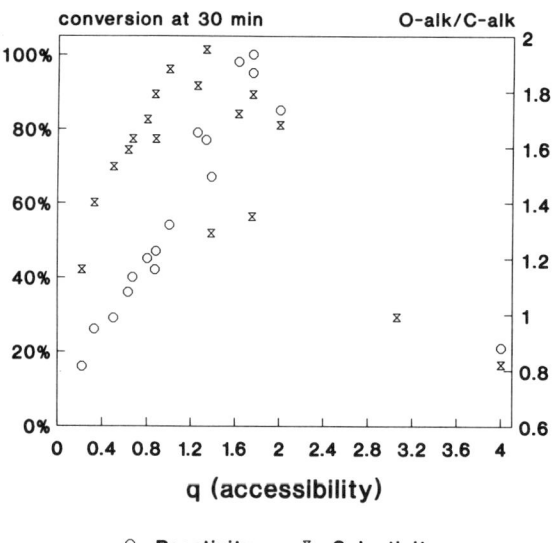

Figure 8-1. Methylation of deoxybenzoin.

a. Effect of Water on Selectivity in Catalytic Phase-Transfer C-/O-alkylation

As the system is starved for water, that is, increasing the aqueous NaOH concentration from 20% to solid NaOH, the amount of O-alkylation increases. This may be explained by the selective hydration of the oxygen of the enolate [17] (higher electron density) which increases as the base concentration decreases, allowing for more hydration of the enolate (less competition with Na^+ and OH^- for association with water molecules). As the oxygen of the enolate is more hydrated, the C-alkylation alternative increases.

"q" is defined in Chapter VI-A.1.b.2.

Halpern, unpublished

Figure 8-2. Effect of quaternary ammonium cation structure on selectivity of alkylation of ambident anion. q is defined in Chapter 6, Section A.1.b. (Halpern, unpublished.)

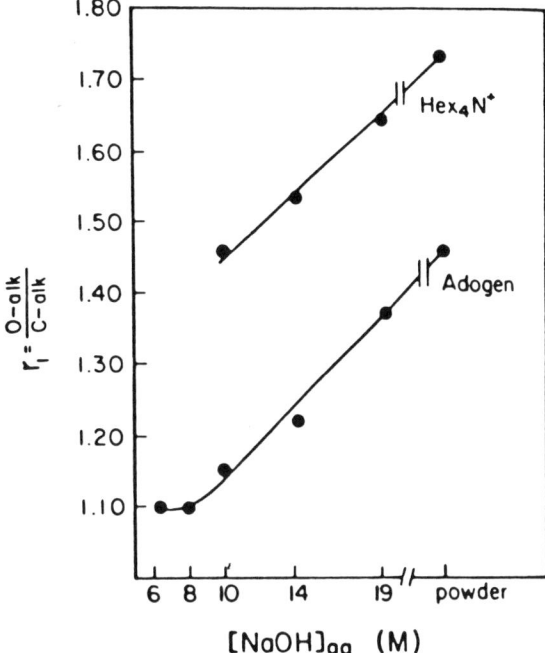

Figure 8-3. Effect of hydration on selectivity of alkylation of ambident anion. The more electronegative site attracts more accessible quat and water (selective solvation). (M. Halpern, Ph.D. Thesis, 1983.)

b. Effect of the Structure of the Quaternary -Onium Cation on Selectivity in Catalytic Phase-Transfer C-/O-Alkylation

Similarly, for very accessible quaternary -onium cations, the extent of O-alkylation decreases with increasing accessibility. Tetramethylammonium binds tighter to the oxygen of the enolate than tetraethylammonium and less O-alkylation is observed. In addition, methyltrioctylammonium gave more C-alkylation than ethyltrioctylammonium, and methyltributylammonium gave more C-alkylation than ethyltributylammonium. Thus, quaternary -onium cations with very accessible methyl groups can bind to the oxygen tighter than those with no methyl groups and result in more C-alkylation.

c. Effect of Solvent and Alkylating Agent on Selectivity in Catalytic Phase-Transfer C-/O-Alkylation

When using dimethyl sulfate as the alkylating agent, in benzene, methylene chloride, or methyl isobutylketone, alkylation is rapid and the C-/O-alkylation ratio is in the range of 0.3–1.2 (more polar solvents giving more O-alkylation).

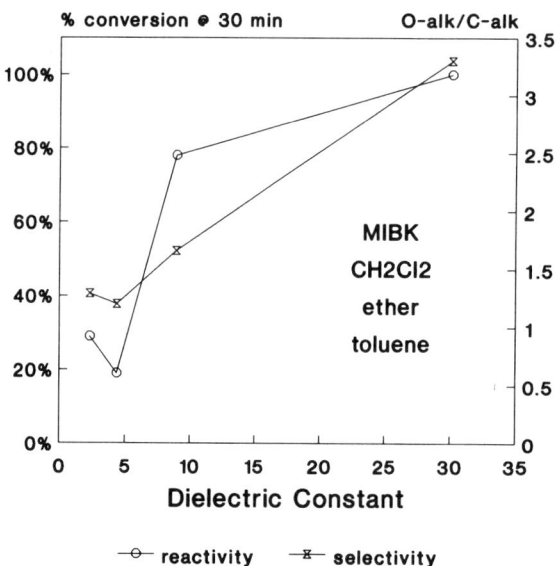

Figure 8-4. Effect of solvent on selectivity of alkylation of ambident anion. (M. Halpern, Ph.D. Thesis, 1983.)

Methyl iodide gives exclusively C-alkylation. In this case, no effect of the counteranion was observed on selectivity, using tetrabutylammonium bromide or methyltrioctylammonium bromide (pure, not Aliquat 336) as catalysts (although the tetrabutylammonium gave more O-alkylation). When using ethyl bromide as the alkylating agent, in benzene, xylene, or methylene chloride, the C-/O-alkylation ratio is in the range of 4–6. It is interesting to note that only a 25 mol% excess of dimethyl sulfate is required for complete conversion of deoxybenzoin, since the highly reactive dimethyl sulfate reacts more rapidly with the organic substrate than it hydrolyzes.

In summary, if one wants to obtain alkylation at the more electronegative site of an ambident anion, one should strive to minimize association at that site (see Figure 8-5).

The C-/O-alkylation selectivity trends found for the PTC alkylation of deoxybenzoin were similar to those using the "extractive alkylation" method of Brandstrom, with β-diketones as substrates [18–21]. As the steric hindrance of the alkyl iodide increases (MeI < EtI < n-BuI < iso-PropI), the percent C-alkylation of tetrabutylammonium$^+$ [α-benzoyldimethylmalonate]$^-$, *4*, decreases from 100% to 14%. As the solvent polarity increases (toluene to dimethyl sulfoxide) in the homogeneous isopropylation of tetrabutylammonium$^+$ [acetylacetonate]$^-$,

If you want more alkylation at the electronegative atom:

Quat structure: not accessible nitrogen

Bu4N, Pr4N > Et4N > MeNBu3 > Me4N

Solvent: more polar solvent

MIBK > CH2Cl2 > PhH

Hydration: less hydration

solid base > aqueous base

Alkylating agent: harder leaving group

CH3SO4 > Cl > Br

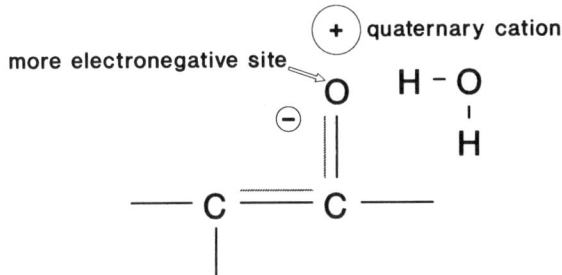

Figure 8-5. Guidelines for selectivity of alkylation of ambident anion.

5, the percent C-/O-alkylation ratio decreases from 13.8 to 0.72. The C-ethylation of tetrabutylammonium$^+$ [acetylacetic acid ethyl ester]$^-$, 6, was also studied. C-Alkylation decreased as the leaving group changed from iodide to bromide to tosylate. Using ethyl bromide as the alkylating agent, the C-/O-alkylation ratio also decreased as the size of the cation increased (Li$^+$, 70; Na$^+$, 60; K$^+$, 41; Cs$^+$, 10.3; tetrabutylammonium, 2.9). These results are again consistent with selective solvation/association between polar species (cation, water molecules, other solvent specific interactions) and the site with the higher electron density (i.e., oxygen). This selective solvation diverts the alkylating agents to the site on the ambident anion with the lower electron density. This principle will be applied later in the chapter when discussing selectivity of N-/O- and N-/C-alkylation. In summary, *when attempting to optimize selectivity of alkylation of ambident nucleophiles, one should use a combination of the following factors: quaternary ammonium cation accessibility, hydration, other solvation effects, and concentration/kinetic order effects, according to their anticipated polarities and specific interactions.*

$$\text{Bu}_4\text{N}^+$$
$$\text{CH}_3\text{O}-\overset{\overset{\text{O}}{\|}}{\text{C}}-\overset{\ominus}{\underset{\underset{\underset{\displaystyle \bigcirc}{|}}{\text{C}=\text{O}}}{\text{C}}}-\overset{\overset{\text{O}}{\|}}{\text{C}}-\text{OCH}_3 \quad 4$$

$$\text{Bu}_4\text{N}^+$$
$$\text{CH}_3-\overset{\overset{\text{O}}{\|}}{\text{C}}-\overset{\ominus}{\underset{\text{H}}{\text{C}}}-\overset{\overset{\text{O}}{\|}}{\text{C}}-\text{CH}_3 \quad 5$$

$$\text{Bu}_4\text{N}^+$$
$$\text{CH}_3-\overset{\overset{\text{O}}{\|}}{\text{C}}-\overset{\ominus}{\underset{\text{H}}{\text{C}}}-\overset{\overset{\text{O}}{\|}}{\text{C}}-\text{OC}_2\text{H}_5 \quad 6$$

Gas–liquid PTC was used to alkylate ethyl malonate, ethyl acetoacetate, and acetylacetone [22]. The alkylations were performed at 170°C and 20 torr so that the reactants and products are gases. The reactants were passed over molten catalyst. The catalysts used were (in decreasing order of reactivity) $\text{Bu}_4\text{P}^+\text{Br}^-$, crown > PEG and were prepared by making a suspension of K_2CO_3 or NaHCO_3 in a solution of the catalyst in methylene chloride, then evaporating the solvent. Pure C- or O-alkylation was not obtained and the C-/O-alkylation ratios depended on the substrate. An intriguing and, as yet unexplained, phenomenon was observed in the C-butylation of diethylmalonate under solid–liquid PTC/K_2CO_3 conditions [23]. Higher yields were obtained by combining Aliquat 336 with either crown or PEG in cocatalyst systems, relative to performing the reaction with quaternary ammonium cation, crown, or PEG separately. Possible explanations may involve enhancement of transfer or omega-phase formation by one catalyst and anion activation (by charge separation during the actual displacement reaction) by the other.

Acetophenone, 7, which is a slightly activated ketone ($pK_a = 19$) [24], has been alkylated with allyl chloride for use in augmenting or enhancing the aroma of perfume compositions [25]. An attempt to perform the Favorski reaction under PTC/OH conditions resulted in a C-alkylation [26]. α-Chlorodibenzylketone was reacted with aq. NaOH (10%, 20%, 50%) in the presence of triethylbenzylammonium chloride and tetrabutylammonium HSO_4^- at varying concentrations in methylene chloride and benzene in an attempt to produce 2,3-diphenylpropanoic acid by the Favorski rearrangement. The only product that could be identified was

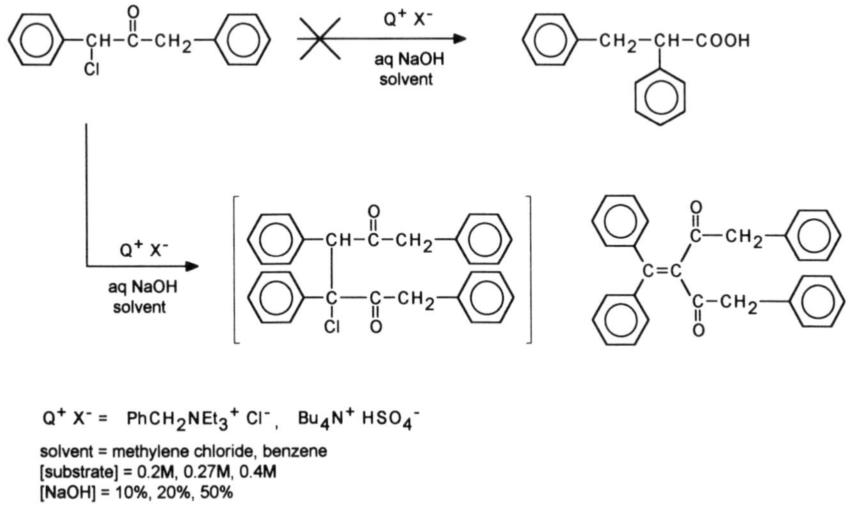

$Q^+X^- = PhCH_2NEt_3^+ Cl^-, Bu_4N^+ HSO_4^-$
solvent = methylene chloride, benzene
[substrate] = 0.2M, 0.27M, 0.4M
[NaOH] = 10%, 20%, 50%

Figure 8-6. Attempt to perform Favorski rearrangement results in C-alkylation.

the product of the displacement of the chloride by the enolate of the chloroketone, followed by dehydrohalogenation (see Figure 8-6).

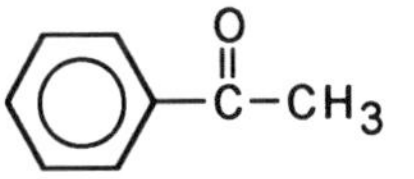

$$\text{Ph-C(=O)-CH}_3 \qquad 7$$

Among the most elegant PTC work ever performed is the chiral C-alkylation of indanone derivatives performed at the Merck, Sharpe and Dohme laboratories [27,28]. This work is reviewed in Chapter 12. A recurring theme in this chapter is the use of quaternary ammonium salts containing the 2-hydroxyalkyl group, such as the cinchoninium salts, in selective PTC/OH alkylations and additions (see below). The successful use of 2-hydroxyethyl quaternary ammonium salts for enantioselective C-alkylation of a ketone was reported as early as 1979 [29].

2. Aldehydes

The C−H group α to a carbonyl has a pK_a of approximately 20 [30]. Therefore, in principle, ketones and aldehydes, without special activation, should be readily deprotonated under PTC/OH conditions. Indeed, the hindered aldehyde, isobutyraldehyde, can be C-alkylated by benzyl chloride, in the presence of 14 mol% tetrabutylammonium iodide and 30% NaOH in toluene at 70°C to yield 96% of the monobenzylated isobutyraldehyde without aldol products [31] (see Figure 8-7). The choice of the iodide counteranion may be surprising at first glance, but probably is a cocatalyst, converting the benzyl chloride into benzyl iodide, in situ. Propanal and butanal were C-alkylated by 2-chloromethylfuran [32].

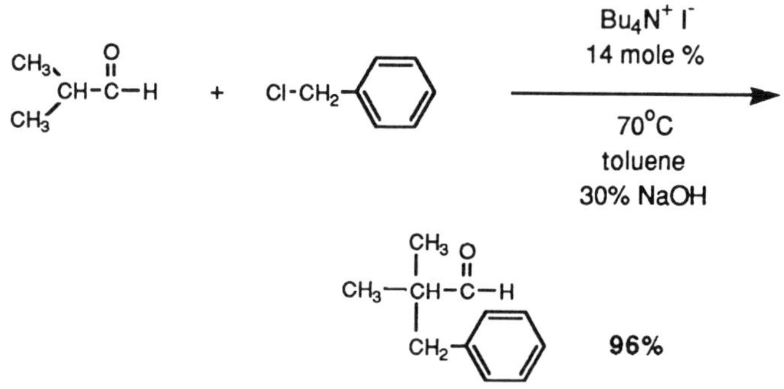

$$\underset{CH_3}{\overset{CH_3}{\diagdown}}CH-\overset{\overset{O}{\|}}{C}-H \;+\; Cl-CH_2-\!\!\!\!\bigcirc \quad \xrightarrow[\substack{70°C \\ \text{toluene} \\ 30\% \text{ NaOH}}]{\substack{Bu_4N^+\ I^- \\ 14\ \text{mole }\%}} $$

$$CH_3-\underset{\underset{CH_2-\!\!\!\bigcirc}{|}}{\overset{\overset{CH_3}{|}}{CH}}-\overset{\overset{O}{\|}}{C}-H \qquad 96\%$$

Figure 8-7. C-Alkylation of a nonactivated hindered aldehyde.

3. Esters and Carboxylic Acids

C—H bonds that are α to the ester and carboxylic acid functional group (pK_a = 24–25) [24] can be alkylated. In these cases, complications may arise if the esters hydrolyze faster than the α carbon can alkylate. In addition, carboxylic acids have pK_a values of ca. 4–5 (roughly 10^{15} times more acidic than unactivated C—H α to carbonyl), and will usually alkylate first.

An enlightening study reported the selective C-alkylation of malonic esters, using *catalytic* quantities of quaternary ammonium salts, while minimizing hydrolysis [33]. The key to the success of this reaction (without resorting to "extractive alkylation"; see Section A.1) was to use just enough water (5–10 mol%) to promote the reaction, but not enough to promote hydrolysis. Another key factor was the choice of K_2CO_3 as the base. In this manner dialkyl cyclopropane-1,1-dicarboxylates were obtained in 73–91% yield.

A very interesting triple PTC reaction was described for the C-alkylation of phenylacetic acid and derivatives [34]. Phenylacetic acid was reacted with 3.3 equivalents of dimethyl sulfate to yield 98% α-monomethylphenylacetic acid, in the presence of 31 mol% triethylbenzylammonium chloride, 10 equivalents of $KOH_{(s)}$, and methylene chloride as solvent (see Figure 8-8). In this reaction, the carboxylic acid (most acidic proton) is first deprotonated and esterified to the methyl ester, followed by deprotonation and C-alkylation in the α-position and finally hydrolyzed by the excess KOH. All three reactions are phase-transfer catalyzed using hydroxide. Naproxen and ibuprofen are obtained in 84% and 74% yields, respectively, from 6-methoxynaphthaleneacetic acid and *p*-isobutylphenylacetic acid. The C-alkylation of the diactivated methylene group is likely to be hydroxide transfer rate limited; therefore, triethylbenzylammonium was an appropriate choice of catalyst (see Chapter 6, Section G).

4. Imines

O'Donnell has done excellent work on chiral alkylations of imines for the synthesis of amino acids since the late 1980s (see Chapter 12) [35–37]. O'Donnell's

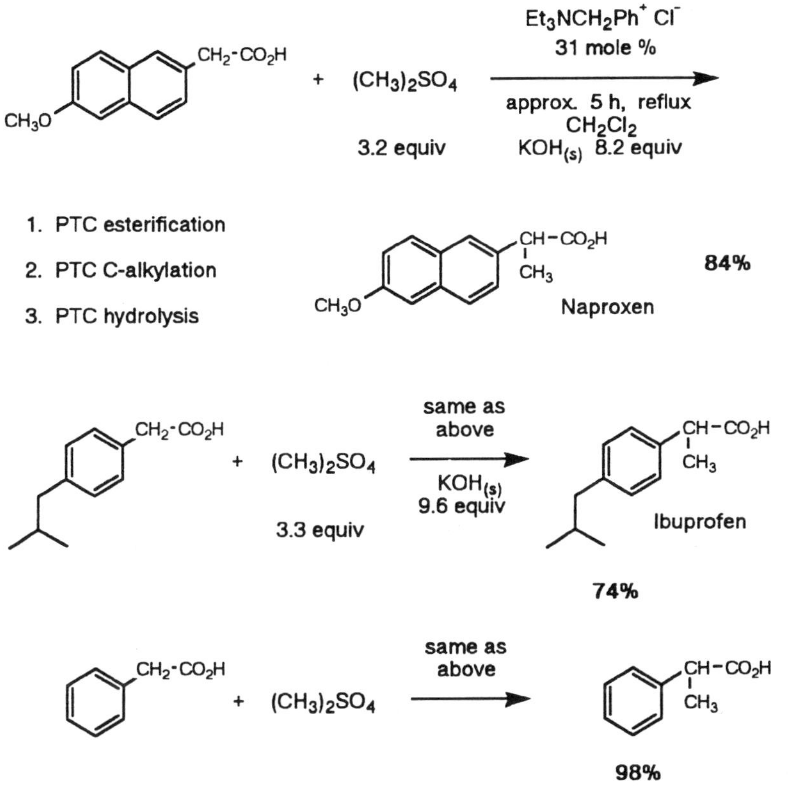

Figure 8-8. C-Alkylation of esters.

original work described a variety of alkylations of methylene groups activated by an imine and ester (see Figure 8-9) [38]. In most cases, only monoalkylation was obtained (59–89% yield depending on the alkyl halide used) followed by hydrolysis of the alkylated imine to yield the corresponding amino acid. The PTC C-alkylation was compared to anhydrous alkylation and was found to be comparable in yield though much more simple and convenient to perform, and in addition, would allow recycle of the catalyst (tetrabutylammonium). Yields improved to 76–95% when the methylene group was activated by the imine and a nitrile group [39]. The catalysts used in these reactions were tetrabutylammonium HSO_4^- and triethylbenzylammonium chloride. Considering the pK_a range of the diactivated imines (14.5–24.3, mostly 17–21, for a wide variety of ketimines and aldimines activated by benzylic, ester, and/or nitrile groups [36, 40]), the reactions may be hydroxide transfer rate limited and, therefore, accessible quaternary ammonium cations (and tetrabutylammonium as always) were good choices for these reactions (see Chapter 6, Section G).

This reaction has been expanded to include the synthesis of α-amino acids

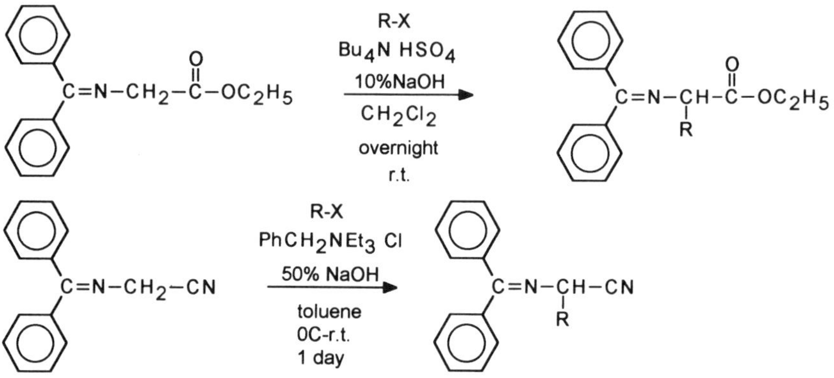

Figure 8-9. C-Alkylation of a methylene group activated by an imine and an ester and by an imine and a nitrile.

with two different alkyl groups, by using a monoalkylated aldimine as the starting material (see Figure 8-10) [40]. Since the monoalkylated aldimine is less acidic and more sterically hindered than the ketimines O'Donnell used, a stronger base and less hydration are needed to enhance the reaction. The author used solid–liquid PTC conditions by using solid KOH, K_2CO_3, or Na_2CO_3. The reaction was scaled up to a 2 mol scale. Fifteen α-amino acids were prepared in 41–97% yield by this method. The deprotonated aldimines were also successfully reacted with Michael acceptors under similar solid–liquid PTC conditions (triethylbenzylammonium as catalyst) in high yield. Reaction of the deprotonated aldimines with aldehydes (addition to carbonyl) resulted in the formation of a mixture of several of the possible products. A mechanistic study was performed and showed an induction period followed by rapid benzylation. As expected, decreasing particle size of the solid base increases the reaction rate. An interfacial mechanism (solid–liquid interface) was proposed by the authors. The results reported would also be consistent with a transfer rate limited reaction.

Some of the most striking advantages of PTC were exemplified by work with highly expensive radioisotopes with very short half-lives. In such a case, one must choose reaction conditions that have "state of the art" speed, work-up, and high yield. Figure 8-11 shows the methylation of an imine, additionally activated by a carbonyl, using $^{11}CH_3I$ (half-life of 20 min) [41]. The methylation reaction

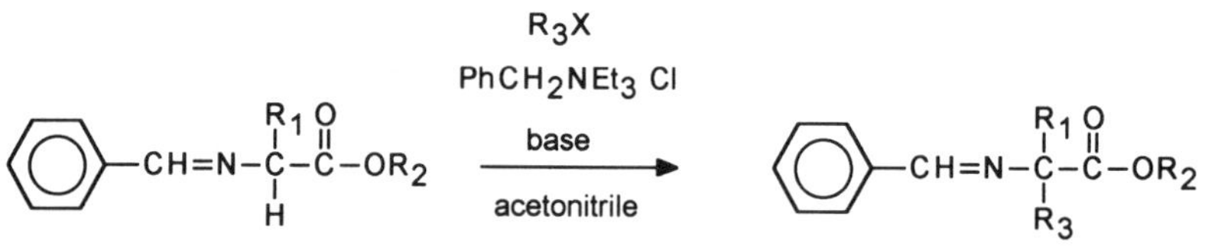

Figure 8-10. Synthesis of an α-amino acid with two different alkyl groups.

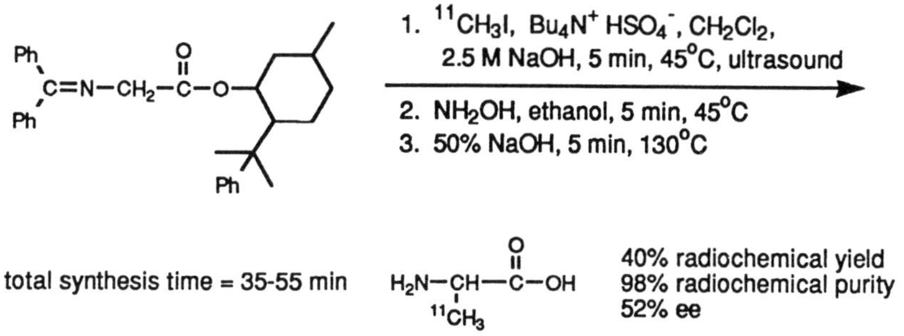

Figure 8-11. Rapid and easy C-alkylation.

Reaction scheme: Ph₂C=N-CH₂-C(=O)-O-menthyl reacted with:
1. $^{11}CH_3I$, $Bu_4N^+ HSO_4^-$, CH_2Cl_2, 2.5 M NaOH, 5 min, 45°C, ultrasound
2. NH_2OH, ethanol, 5 min, 45°C
3. 50% NaOH, 5 min, 130°C

total synthesis time = 35–55 min

Product: $H_2N-CH(^{11}CH_3)-C(=O)-OH$

40% radiochemical yield
98% radiochemical purity
52% ee

time was 5 min, followed by reaction with hydroxylamine, then, hydrolysis, with a total synthesis time (including work-up) of 35–55 min, to produce a simple ^{11}C-radiolabelled amino acid with high radiochemical yield. This is a good example of the effectiveness of PTC as a synthetic procedure as well as an advantageous unit operation.

Imines can also be activated by other functional groups. α-Amino phosphonic acids (the phosphonic analogues of α-amino acids) have been prepared under solid–liquid PTC conditions using 5-mol% Aliquat 336 and solid KOH (see Figure 8-12) [42]. The alkylations were obtained in 55–98% yield, and the α-amino phosphonic acids were obtained in 60–85% yields from the phosphonic ketimines by hydrolysis. Aldimines, activated by the 2-pyridyl group (similar to benzylic activation), were also C-alkylated under solid–liquid (KOH) PTC conditions [43]. Hydrolysis of the imine group afforded α-substituted 2-pyridylmethyl amines. O'Donnell's imines also underwent chiral alkylation by bromomethylpyridine derivatives [44].

5. Nitriles

The C-alkylation of phenylacetonitrile, "PAN," published by Makosza in 1965, may be considered a foundation reaction for all PTC/OH C-alkylations [2,3]. The ethylation of PAN by ethyl chloride (see Figure 8-13) was systematically and thoroughly investigated under a wide range of conditions. Makosza synthe-

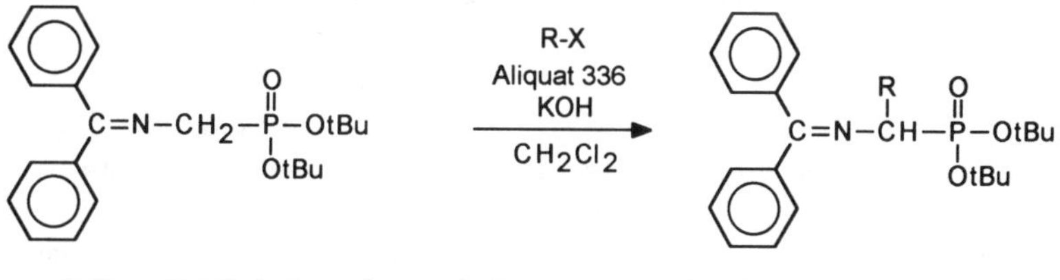

Figure 8-12. C-Alkylation of a methylene group activated by an imine and a phosphonate.

Reaction scheme: Ph₂C=N-CH₂-P(=O)(OtBu)₂ with R-X, Aliquat 336, KOH, CH_2Cl_2 → Ph₂C=N-CHR-P(=O)(OtBu)₂

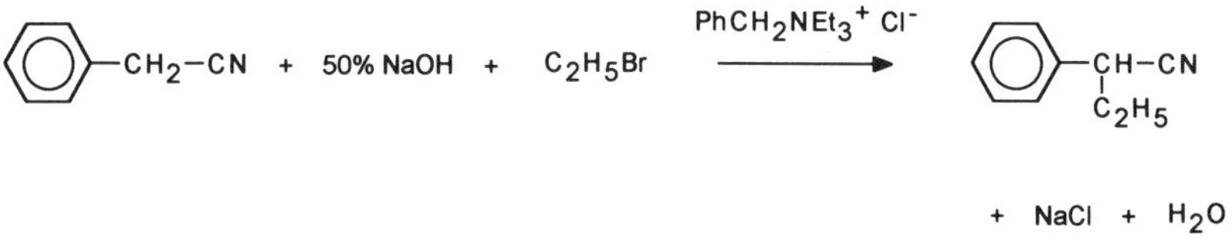

PhCH₂-CN + 50% NaOH + C₂H₅Br →[PhCH₂NEt₃⁺ Cl⁻] PhCH(C₂H₅)-CN + NaCl + H₂O

Figure 8-13. Ethylation of phenylacetonitrile.

sized 18 accessible quaternary ammonium salts in high yield, mostly containing benzyl and triethyl functionalities. The triethylbenzylammonium cation was subsequently used in thousands of reactions. Makosza was the first to find that the presence of iodide (and perchlorate) slowed down the reaction (later to be termed "catalyst poisoning"). He also showed that ethyl bromide required a higher NaOH/substrate ratio than ethyl chloride to achieve higher conversion. At a molar ratio of approximately 4:1 NaOH/PAN, maximum reactivity was achieved for both alkylating agents. Makosza also found that at low NaOH concentrations (e.g., 30%), conversions were very low, though the conversions were higher for ethyl bromide than for ethyl chloride (probably due to the higher reactivity of ethyl bromide). At higher NaOH concentrations (crossover at 40%) higher conversions were obtained with ethyl chloride relative to ethyl bromide. This may be due to the higher nucleophilicity of the PAN anion (less hydration) and the poisoning effect of the bromide relative to chloride. The reaction was performed without solvent, since the starting materials and products are liquids. This is a useful process feature. The addition of solvents decreased the reaction rate, probably due to slower kinetics due to dilution.

The ethylation of PAN, using ethyl bromide, was the subject of a kinetic profile study [45]. The alkylation, in the presence of tetrabutylammonium bromide, was found to be highly dependent on agitation, was 0.6 order in catalyst, and the alkylation rate seemed to be linearly dependent upon PAN and the alkylating agent. The alkylation rate was also found to be very heavily dependent upon hydroxide concentration ("5.3" order). The energy of activation was approximately 20 kcal/mol. The mechanistic aspects of this reaction were discussed in Chapter 3.

The alkylation of PAN was used in the preparation of a variety of pharmaceuticals. Alkylation of PAN by ethyl chloride was used in the preparation of oxeladine; 2-bromobutane for pentapiperide; bis(chloroethyl)aminophenylsulfone for phenoperidine; and 1,5-dibromopentane for dicyclonine [46]. A series of coronary vessel dilation agents, *8*, were prepared by the patented *C*-alkylation of PAN derivatives in 85–95% yields by γ'-halo, β'-aryl tertiary amines using small catalytic quantities of tetrabutylammonium iodide [47]. The synthesis and use of Reissert compounds, *9* (for the synthesis of alkaloids and other biologically active compounds), and the alkylation of the PAN-like site, in particular, has been reviewed and expanded upon [48]. Reaction rates, and to some extent yields, appeared to be enhanced by using ultrasound mixing.

$$\underset{R}{\overset{CN}{\underset{|}{Ar-C}}}-CH_2-CH_2-CH_2-\underset{}{\overset{CH_3}{\underset{|}{N}}}-CH_2-CH_2-Ar' \qquad 8$$

9

Brandstrom ethylated phenylacetonitrile in 100% yield by extractive alkylation (described above) [49].

The buytlation of PAN using polymer-bound PTC was first studied comprehensively by Ford (see Chapter 5) [50]. Factors that were found to affect the rate of reaction included order of addition of reactants, stirring speed, particle size, crosslinking, and degree of ring substitution. The 2% crosslinked resin was found to be the best commercial anion-exchange resin for the monoalkylation of PAN. Another extensive study of the butylation of PAN using polymer-bound PTC was performed by Ragaini [51]. Batch slurry reactors and fixed bed reactors were used under continuous flow conditions and various stirring modes (see Chapter 6, Section A.4.d). A polymer-bound triethylbenzylammonium analogue at 12% crosslink exhibited approximately 10% of the reactivity of the soluble triethylbenzylammonium reaction (probably due to difficult diffusion through the highly crosslinked material). Polymer-bound tributylbenzylammonium chloride catalyst that was used in a fixed bed apparatus with an ultrasonic mixer reached a reactivity close to that shown by the soluble analogue of triethylbenzylammonium in the monoalkylation of PAN.

When using very active alkylating agents, such as benzyl chloride, it is difficult to stop the reaction at the mono-C-alkylation stage. The benzylation of malononitrile was studied [52] and as expected, monoalkylation was obtained only at low conversion. Nevertheless, high yield with a 4:1 mono-/di-benzylation ratio was obtained under solid–liquid PTC conditions using K_2CO_3 as base and no added solvent. Malononitrile and phenylacetonitrile were also benzylated in the presence of a modified silica [53]. PAN derivatives were reacted with gem-dichlorocyclopropanes to yield di-PAN derivatives of cyclopropane [54].

The pK_a of acetonitrile is approximately 25 [24]. Therefore, under the proper PTC/OH conditions, it should be possible to deprotonate an aliphatic C–H group α to a nitrile. A patent described the cyclization of X-$(CH_2)_4$–CN to cyclopropanecarbonitrile (X = Cl, Br) [55]. A yield of 99.3% was obtained under solid–liquid conditions, using NaOH as the base, Aliquat 336 as the catalyst, and methylene chloride as the solvent. Intramolecular C-alkylation was performed to convert 3-(chloropropylthio)acetonitrile into 2-cyanotetrahydrothiphene [56]. In this case, the α-CH bond was activated only by the nitrile and slightly by an adjacent sulfur. The 2-cyanotetrahydrothiphene was oxidized to the sulfone (not using PTC), then the highly activated C—H bond (α to both the cynao and sulfone) was further alkylated using PTC/K_2CO_3 conditions.

6. Sulfones

Sulfones are important compounds as products and as intermediates [57]. Sulfones are acidic enough to be deprotonated under PTC/OH conditions (pK_a of dimethylsulfone = 23) [8] and the resulting carbanions are nucleophilic enough to react with electrophiles leading to the formation of new C—C bonds. After alkylation, Michael addition, etc., additional desired compounds can be obtained by removing the SO_2 group by reduction, elimination, or substitution.

Activated phenylsulfones ($ArSO_2CHX_2$; activation by two halogens) were first C-alkylated by Makosza, using triethylbenzylammonium chloride as the catalyst [58]. A series of α-iodophenylsulfones were prepared by the C-alkylation of $PhSO_2CH_2I$ in the presence of triethylbenzylammonium chloride and benzene [59]. Similarly, α-bromo- and chlorophenylsulfones were alkylated in the presence of triethylbenzylammonium chloride and 50% NaOH [60]. Benzylsulfones and benzylsulfonamides were C-alkylated in 56–90% yield in the presence of tetrabutylammonium bromide, NaOH, and HMPT [61]. Allylphenylsulfones ($PhSO_2CH_2CH=C\ (CH_3)_2$) were C-alkylated also in the presence of tetrabutylammonium bromide, NaOH, and HMPT [62]. A relatively nonactivated sulfone, $PhCH_2CH_2SO_2CH_2Br$, was cyclized by intramolecular C-alkylation to give a four-membered ring containing the sulfone [63]. The reaction was performed in the presence of tetrabutylammonium bromide, NaOH, and methylene chloride in 83% yield. α-Methylene groups of phenylsulfones have also been activated for C-alkylation by nitro [64], phosphonate [65], and isonitrile groups [66] using tetrabutylammonium bromide or triethylbenzylammonium chloride as catalyst and benzene or methylene chloride as solvent. Solid–liquid PTC/K_2CO_3 conditions were employed to alkylate the diactivated tolylsulfonyl acetate ethyl ester in 61–81% yield [67]. This reaction was reported again employing a commercial microwave oven for rapid reaction [68]. A patent describes the C-alkylation of an activated benzylsulfone with methyl chloride in the presence of 5 mol% triethylbenzylammonium chloride, methylene chloride, and two equivalents KOH in 94% yield [69] (see Figure 8-14). The benzylsulfones and activated phenylsulfones described above probably have pK_a values in the range of 16–23, and may be speculated to be hydroxide transfer rate limited. Thus the choice of an accessible quaternary ammonium cation, such as triethylbenzylammonium, was probably a good choice (see Chapter 6, Section G). As always, tetrabutylammonium is usually also a good choice for PTC/OH reactions. The C-alkylation of a highly activated cyanosulfone is described in Section A.5 of this chapter [56].

7. Hydrocarbons

Activated hydrocarbons, such as cyclopentadiene (pK_a = 15) [70], indene (pK_a = 21) [71], fluorene (pK_a = 23) [71], and diphenylmethane (pK_a = 33)

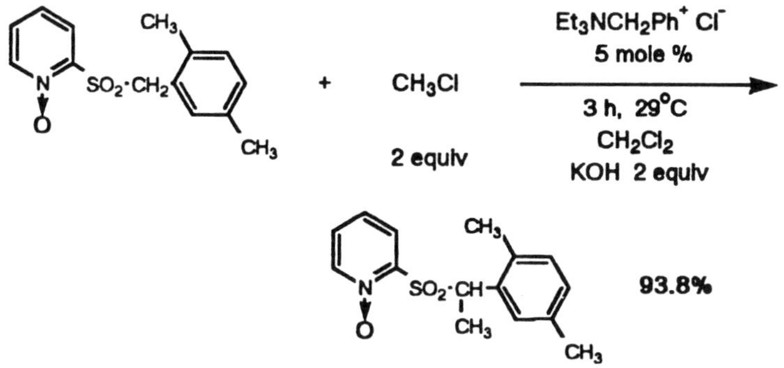

Figure 8-14. C-Alkylation of an activated sulfone.

[71], can be deprotonated under PTC/OH conditions. The carbanions formed from these hydrocarbons may undergo reactions such as oxidation, deuteration, and C-alkylation with varying degrees of difficulty. In Section H of this chapter, we shall see that all of these substrates are acidic enough to undergo deuteration, but not all of their derived carbanions are "nucleophilic" enough to undergo C-alkylation.

Cyclopentadiene, *10* [72, 73], and indene, *11* [74], were easily C-alkylated in the presence of aqueous NaOH and triethylbenzylammonium chloride. Using dihaloalkanes as alkylating agents, high yields of a very wide variety of spiro C-alkylations of cyclopentadiene were obtained when using nonpolar solvents. Elimination predominated when polar solvents were used [75]. Makosza was first able to alkylate fluorene, *12*, in the presence of 50% NaOH and triethylbenzylammonium chloride (dibenzylate in 80% yield) but needed to add 5% dimethyl sulfoxide and perform the reaction at 70–90°C in order for the reaction to proceed [76]. Since fluorene may be easily deprotonated at room temperature (deuteration is 80% complete in 6 min in the presence of 16.1 M NaOD/D_2O and 5 mol% triethylbenzylammonium chloride [77]), the need to add dimethyl sulfoxide and use high temperature in order for the reaction to proceed must, therefore, be due to some aspect of reactivity of the fluorenyl anion. The reactivity of the fluorenyl anion may be altered by manipulating its state of hydration. Fluorene was C-alkylated in the presence of powdered NaOH, 5 mol% triethylbenzylammonium chloride, 2.5 equivalents of dimethyl sulfate, nitrogen (to prevent oxidation to fluorenone) in benzene to yield 85% 9-methylfluorene, and 15% 9,9-dimethylfluorene (no unreacted fluorene) after 18 h at room temperature [78]. Steric hinderance may also play a role since under these conditions, at room temperature, pure 9-methylfluorene did not react readily with dimethyl sulfate. It may be speculated that Makosza needed to add dimethyl sulfoxide to the first alkylation above in order to modify the hydration of the fluorenyl anion and use a higher temperature in order to obtain dialkylation.

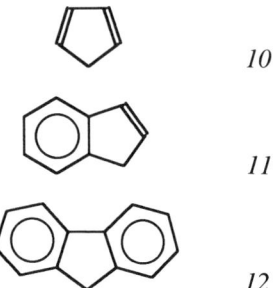

10

11

12

Under a wide variety of PTC/OH conditions, including using powdered NaOH, triphenyl methane (pK_a = 31.5) [71] and diphenyl methane (pK_a = 33) could not be alkylated with dimethyl sulfate [79]. Diphenyl methane was easily oxidized and deuterated (see Section H of this chapter) under PTC/OH conditions.

Cyclopentadiene surprisingly underwent "alkylation" with *tert*-butyl bromide in the presence of 50% KOH and Aliquat 336 to yield 44% of a mixture of dialkylated cyclopentadiene [80]. It was previously thought that a compound such as *tert*-butyl bromide would undergo dehydrohalogenation to an extent that would preclude alkylation. The authors invoked a single-electron transfer mechanism.

B. *N*-Alkylation

Many nitrogen-containing compounds, especially heterocycles, have protons acidic enough to be deprotonated under PTC/OH conditions, and the resulting N-anions are often nucleophilic enough to effectively participate in displacements. Solid–liquid PTC conditions are being increasingly used for *N*-alkylations of all types of N−H-containing compounds. The advantages of performing *N*-alkylation under base-catalyzed PTC conditions, coupled with the importance of *N*-alkylation in pharmaceuticals, agricultural chemicals, and other products, have led *N*-alkylation to be among the most widely used applications of PTC.

1. Nitrogen-Containing Heterocycles

The alkylation of indole, *13*, constitutes a foundation reaction for the PTC *N*-alkylation reaction. *N*-Methylindole is obtained in 98% yield from indole and dimethyl sulfate in the presence of 5 mol% tetrabutylammonium HSO_4^-, 50% NaOH, and benzene at 33°C in 6 h [81]. Tetrabutylammonium, as well as accessible quaternary ammonium cations, can be used for this reaction. When the nitrogen heterocycle is acidic enough, preformed salts, such as potassium phthalimide, can easily be alkylated under PTC conditions (95% yield with hexadecylbromide) [82]. Based on the pK_a of phthalimide and the nature of the reaction, one would expect this alkylation to be intrinsic reaction rate limited, and organophilic quaternary -onium salts would be preferred for this reaction (see Chapter 6, Section G). The actual catalyst used was hexadecyltributylphosphonium bromide, which indeed is quite organophilic.

The use of solid–liquid PTC conditions has become particularly prevalent in both lab and industrial N-alkylations. N-Alkyl pyrazoles may be obtained in 91–98% isolated yield by reacting equimolar amounts of n-alkyl (C_1–C_{16}) bromides or iodides with pyrazole, *14*, and two-equivalents of finely ground KOH in the presence of 5 mol% tetrabutylammonium bromide and no solvent at 0°C to room temperature [83]. Under these solid–liquid PTC conditions, catalyst poisoning did not take place, even in the presence of iodide. When propargyl bromide is the alkylating agent, the resulting N-propargylpyrazole isomerizes under these PTC/OH conditions to the allene. Less isomerization was obtained when K_2CO_3 was used as the base. Solid–liquid PTC/OH conditions advantageously allowed the use of stoichiometric quantities of the highly toxic 2-chloroethylamine in the N-alkylation of pyrrole, *15*; carbazole, *16*; imidazole, *17*; and benztriazole, *18* in the presence of 4 mol% tetrabutylammonium HSO_4^- and a solvent [84]. Tetrabutylammonium salts were appropriately chosen for all of the above reactions. Solid–liquid PTC conditions were also used in the N-alkylation of caprolactam, *19*, in the presence of PEG as the catalyst and KOH/K_2CO_3 as the base [85] and in the N-alkylation of a β-lactam in the presence of powdered KOH and 2–10 mol% of benzyltriethylammonium chloride (K_2CO_3 was not an effective base) [86]. 2-Acetylpyrrole, *20*, was sequentially and selectively N-alkylated and then C-alkylated (with different alkyl groups on each site) using finely powdered KOH and n-alkyl iodides in the presence of 18-crown-6 and a solvent [87]. The side chain C-alkylation required higher temperature and longer reaction time.

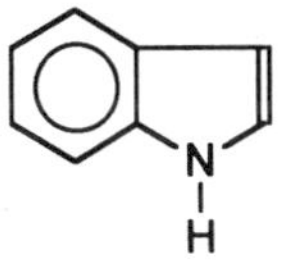

13

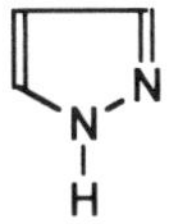

14

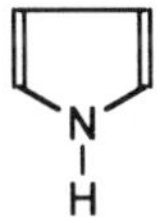

15

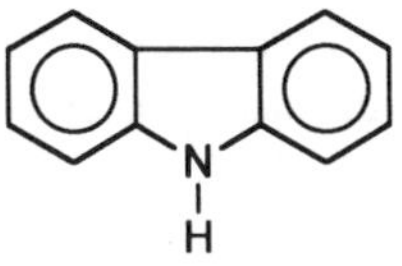

16

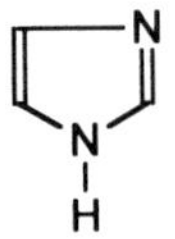

imidazole 17

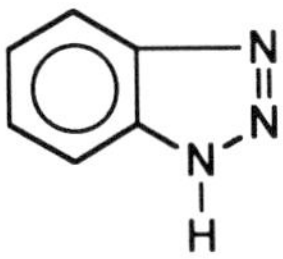

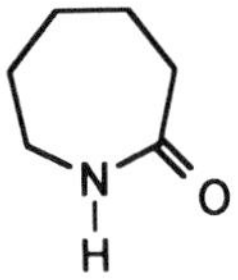

benzotriazole 18

caprolactam 19

2-acetylpyrrole 20

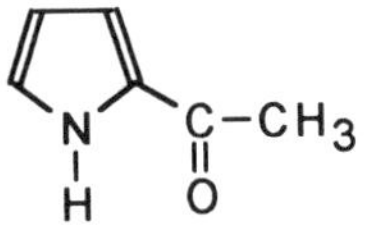

Early work by Makosza, on the *N*-alkylation of carbazoles, used dimethyl sulfoxide in addition to triethylbenzylammonium chloride [88]. Later work showed that dimethyl sulfoxide was not required [89]. The use of a capped PEG/KOH system gave the highest yields and probably the most inexpensive conditions for the *N*-alkylation of carbazole [90]. *N*-Vinylcarbazole (specialty monomer for photoconductors) was prepared by PTC *N*-alkylation of carbazole with dichloroethane, followed by PTC dehydrochlorination, in one step [91]. Tetrabutylammonium bromide effectively catalyzed both reactions and the excess dichloroethane served as the solvent. *N*-(Hydroxyethyl)carbazole, a material used to initiate the polymerization of *N*-vinyl carbazole, was prepared in 92% yield by the reaction of carbazole with 1.1 equivalents of ethylene oxide, under solid–liquid PTC/OH conditions [92]. Pentaglyme (17 mol%) was used as the catalyst and 1.5 equivalents of solid NaOH was used as the base, in toluene as the solvent.

Solvent-free PTC conditions (using the alkylating agent as the solvent, sometimes followed by recovery) have also been gaining popularity for PTC/OH *N*-alkylations. For example, indole was reacted with diethyl sulfate and KOH in the presence of 1 mol% tetrabutylammonium bromide for 10 min at room temperature to yield 98% *N*-ethylindole [93]. Imidazole and pyrazole likewise have been alkylated without the addition of solvent other than the alkylating agent [94,95]. Antiulcer agents were prepared by the *N*-alkylation of pyrazole by 4,6-dichlorpyrimidine [96]. Solid–liquid conditions were used with tetrabutylammonium bromide as the catalyst and KOH as the base.

Other pharmaceutical heterocycles that have been *N*-alkylated include phenothiazine, dibenzazepine, purines, and piperazines. Phenothiazine, *21*, and derivatives were *N*-alkylated in high yield using tetrabutylammonium HSO_4^- as the catalyst and methylene chloride or methylisobutyl ketone (sometimes preferably) as the solvent [97]. Dibenzazepine, *22*, and dihydrodibenzazepines, *23*, which have strained seven-membered ring systems in the N-anion form, were *N*-alkylated in high yields under similar conditions [98]. Substituted piperazines were alkylated to yield antitumor agents [99]. Tetrabutylammonium bromide was used as the catalyst and K_2CO_3 was the base.

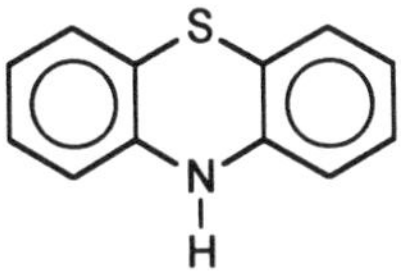

21

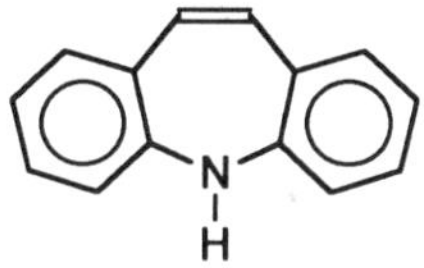

22

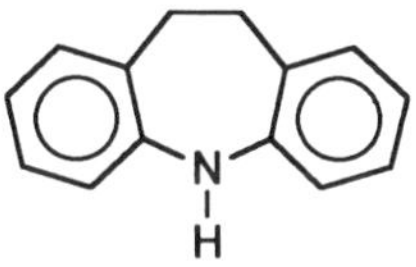

23

Purines are commonly *N*-alkylated under PTC conditions. 6-Chloro-9*H*-purine was reacted with benzyl alcohol in a nucleophilic aromatic substitution, under solid–liquid PTC conditions to yield 6-benzyloxy-9*H*-purine (see Figure 8-15) [100]. The product was then *N*-alkylated with a variety of alkylating agents in the presence of 10 mol% tetrabutylammonium bromide, 50% NaOH, and benzene. A mixture of approximately 1:1 of the N7 and N9 alkylated products were obtained that could be separated. An interesting feature of this reaction, which may be applicable occasionally to other specific liquid–liquid PTC systems, is that the termination of the reaction could be monitored visually by the disappearance of the solid starting material suspended in the system. When only one nitrogen of an activated purine has an available hydrogen, such as in 3,7-dimethylxanthine (i.e., theobromine), *24*, *N*-alkylation can be easily performed in high yield under

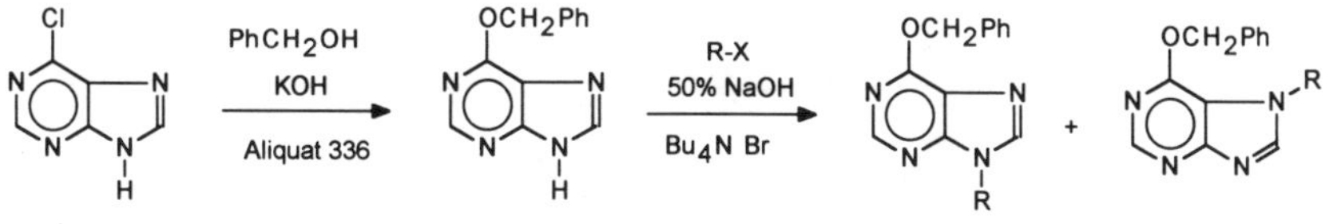

Figure 8-15. *N*-Alkylation of a purine.

solid–liquid PTC/OH conditions (see Figure 8-16) [101]. Tetrabutylammonium HSO$_4^-$ was used as the catalyst, no solvent was used, and a wide range of compounds were prepared, such as radiolabelled caffeine, *25*, and immunosuppression drugs. Other compounds containing the purine nucleus have been *N*-alkylated (on the ring nitrogen) such as adenine [102,103], *26* (to form a nucleoside) [104], xanthine, *27*, theobromine, *24*, theophylline [103,105], *28*, and fluorouracil, *29* [106].

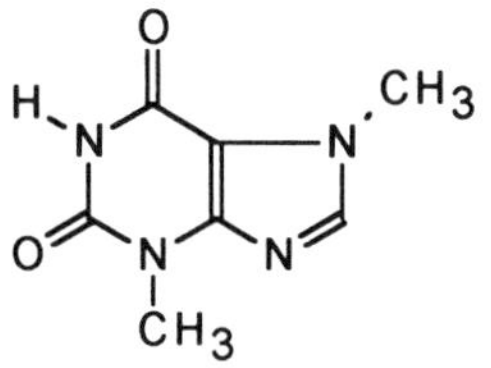

24

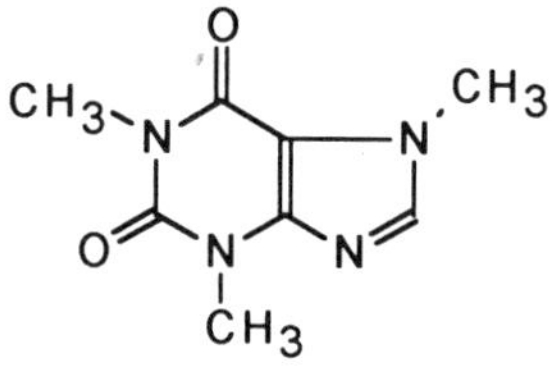

25

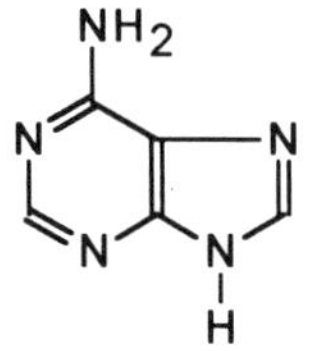

26

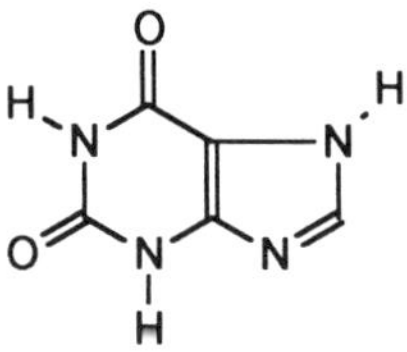

27

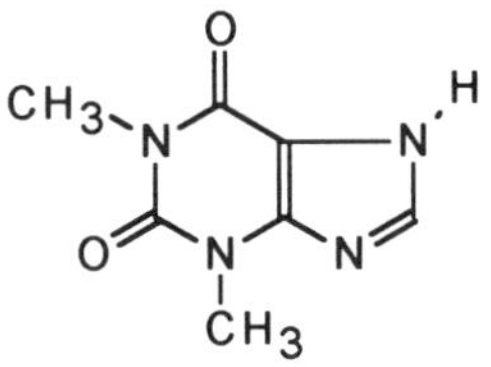

28

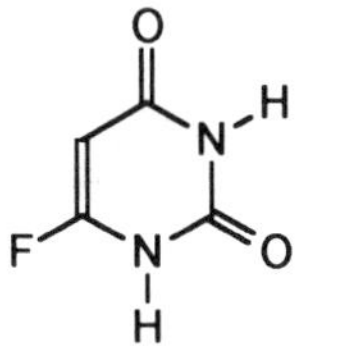

                                    29

Some of the activated purines mentioned above could theoretically undergo both *N*- and *O*-alkylation. In the case of the purines mentioned above, only *N*-alkylation is obtained when the N−H group is activated by two carbonyls. *O*-Alkylation may be significant primarily in systems in which the N−H is α to one carbonyl. One of the early interesting *N*-/*O*-alkylation studies described the alkylation of 2-pyridone in the presence of tetrabutylammonium bromide, 50% NaOH, and a solvent (see Figure 8-17) [107]. In all cases, *N*-alkylation predominated (65–85%) but more *O*-alkylation could be obtained by using hexane as solvent or using hindered alkylating agents (such as isopropylbromide). The addition of three chlorines to the pyridone ring (as in the 3,5,6-trichloropyrid-2-one intermediate for a commercial herbicide; see Section C.2 below) results in the predominance of *O*-alkylation [169, 170]. The minimization of either *N*-alkylation or *O*-alkylation of pyridones (depending upon the product desired) has required significant effort in the development of commercial pharmaceutical and agricultural chemical processes based on pyridones.

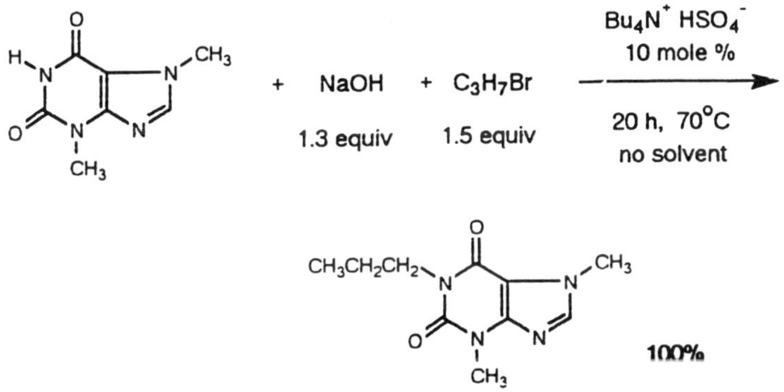

Figure 8-16.   *N*-Alkylation of a purine.

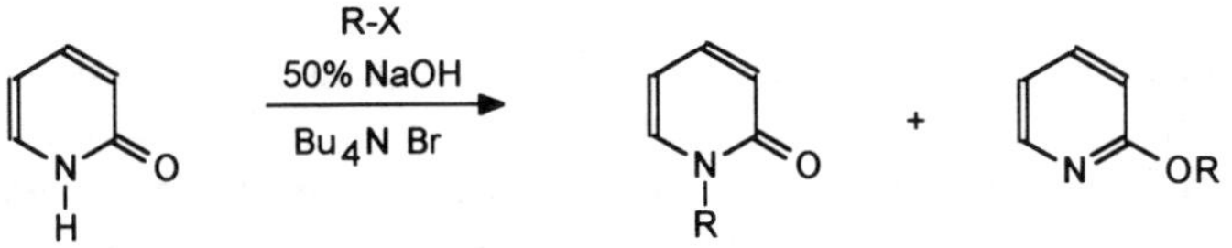

Figure 8-17.   Alkylation of an *N*-/*O*-ambident anion.

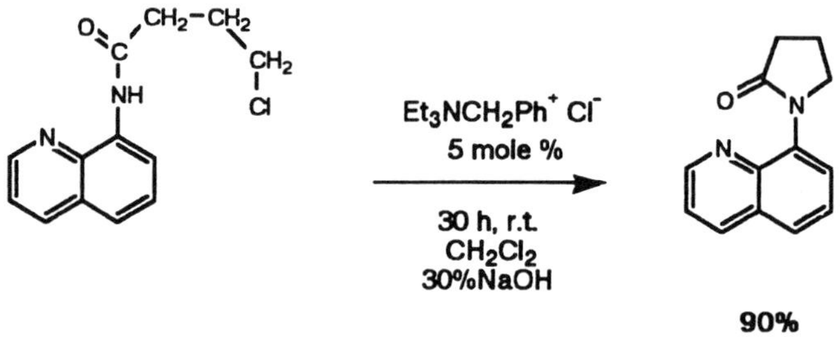

Figure 8-18. Intramolecular *N*-alkylation.

2. Amides

Deprotonation of the N—H group of amides is possible under PTC/OH conditions, with pK_a values in the range of 20–24 [108]. Acetanilide, *30*, has an N—H group that is activated by being α to a carbonyl and being benzylic (acetanilide is the nitrogen analogue of phenylacetone discussed above). Acetanilide was alkylated in high yield (82–95%) by a variety of alkylating agents with various leaving groups (including iodide), in the presence of 1–10 mol% triethylbenzylammonium chloride, 50% NaOH, and benzene [109]. A derivative of acetanilide underwent intramolecular *N*-alkylation to cyclize to a substituted pyrrolidone (see Figure 8-18) [110]. Again, triethylbenzylammonium chloride (5 mol%) was chosen as the catalyst to give 90% yield in the presence of 30% NaOH. In light of the pK_a of the N—H group in these cases, an accessible quaternary ammonium cation was an appropriate choice (see Chapter 6, Section G).

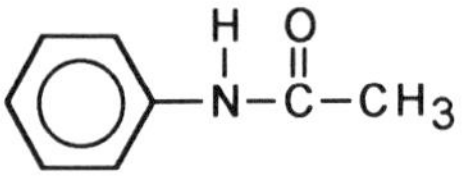

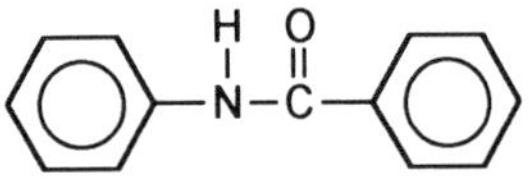

The *N*-alkylation of benzanilide, *31*, has been performed under solid–liquid PTC conditions, using a mixture of powdered NaOH/K$_2$CO$_3$, in the presence of tetrabutylammonium HSO$_4^-$ and benzene [111]. Yields of 88–98% were achieved to produce 16 *N*-alkylated benzanilides. In the presence of 50% NaOH, only 5% yield was obtained. It may be speculated that the nucleophilicity of the highly delocalized anion is significantly affected by hydration, and thus requires relatively dry conditions to proceed. Benzanilide itself is the nitrogen analogue of deoxybenzoin (discussed in Chapter 6, Section A.1.b, and Section A.1 of this chapter). A key difference is that the charge in benzanilide is distributed between two sites of more similar electronegativity (N and O) relative to deoxybenzoin (C and O). Benzanilide did not yield *O*-alkylated products as did deoxybenzoin (the products would have been imines).

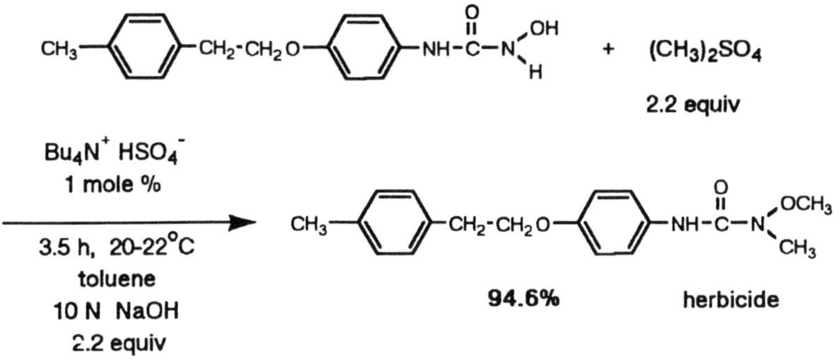

Figure 8-19. N-Alkylation and O-alkylation of aryl N-hydroxyurea.

A substituted phenyl N-hydroxy urea was both N-alkylated and O-alkylated with dimethyl sulfate in the presence of tetrabutylammonium HSO_4^- and 10 N NaOH to prepare a herbicide in 95% yield (see Figure 8-19) [112]. It is interesting to observe that the acidic N-H group, which is benzylic and α to the carbonyl, is not the one alkylated.

Benzamide ($pK_a \sim 25$) [113] is less acidic than benzanilide or acetanilide. Benzamide was dialkylated also under solid–liquid PTC conditions (see Figure 8-20) [114]. Dibutylbenzamide was obtained in 98% yield in the presence of $NaOH/K_2CO_3$, tetrabutylammonium HSO_4^-, and toluene. Benzsulfonamide ($pK_a = 10.1$) [115] was also dialkylated in > 90% under the same conditions (see Figure 8-20). 4-Methylbenzsulfonamide was N-dialkylated with dibromoalkanes to produce N-containing cyclic compounds [116]. Selectivity between intramolecular dialkylation (to yield monosubstituted heterocycles) and intermolecular dialkylation (to yield disubstituted heterocycles) was affected by the leaving group as well as the length of the dibromoalkane.

A useful and very interesting alkylation of sulfonamide diuretics was reported using "extractive alkylation" (see Section A.1 of this chapter) [117]. Sixteen sulfonamides could be alkylated and rapidly analyzed after the PTC derivatization, using gas chromatography/mass spectrometry. Organophilic quaternary ammonium salts were found to be most effective, using methyl iodide as the alkylating agent.

a. Miscellaneous Amides

Trihaloacetamide was N-alkylated by alkyl 2-bromocarboxylic esters [118, 119]. Mono-N-alkylation worked better for trichloro- vs. the trifluoroacetamide and could obtained in up to 95% yield. The competing dehydrohalogenation occurred only when the alkylating ester contained a (2-bromoethyl)benzene group (possibility of forming a styrenic conjugated system). Cyanamide was dialkylated by activated alkyl chloride or nonactivated alkyl bromides in 32–96% yield in

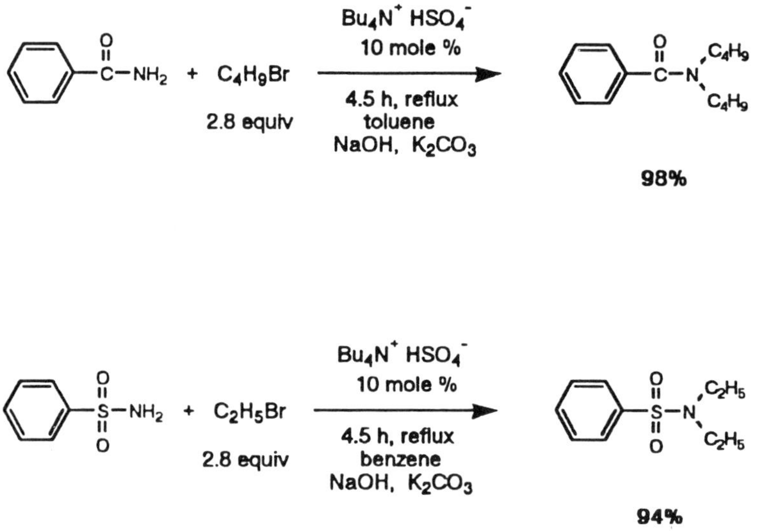

Figure 8-20. N-Alkylation of amide and sulfonamide.

the presence of 50% NaOH, no solvent, and Aliquat 336 [120]. Dialkyl cyanamide could be hydrolyzed to produce pure dialkyl amines. Potassium phthalimide was N-alkylated with asymmetric induction using benzylcinchoninium chloride as catalyst [121]. An interesting two-step PTC reaction was patented in which bis(hydroxymethyl) tetrahydrofuran was reacted with potassium cyanate followed by N-alkylation with octadecylbromide to yield the monocarbamate [122]. Butylurea was mono-N-alkylated by ethylbromide under solid–liquid PTC/K_2CO_3 conditions using tetrabutylammonium chloride as the catalyst [123]. $N-N'$-Diarylureas (nitrogen analogue of diibenzylketone discussed above) were N-alkylated in high yield [124]. Butyltriethylammonium chloride, a very accessible quaternary ammonium cation (q = 1.75; see Chapter 6, Section A.1.b), was used with dimethyl sulfate as the alkylating agent and a highly activated/acidic N–H as the deprotonated group. According to the pK_a guidelines (see Chapter 6, Section G), it may be speculated that the reaction is transfer rate limited and that the choice of the accessible quaternary ammonium cation was appropriate.

3. Amines

Aniline, *32*, is the model compound for aromatic amines and has a pK_a of approximately 27 [111]. 2-Nitroaniline would be expected to have a lower pK_a and be a good PTC/OH substrate if the side NO_2/OH reactions could be controlled. 2-Nitroaniline was N-alkylated in the presence of triethylbenzylammonium chloride, 50% NaOH, and activated substituted benzyl halides [125]. The mono-N-alkylation of 1-aminoanthraquinone (dye), *33*, required the use of solid–liquid

PTC conditions (powdered NaOH/K$_2$CO$_3$) and a very reactive alkylating agent (dimethyl sulfate) [126]. Tetrabutylammonium bromide gave a higher yield than tetrapropylammonium bromide or butyltriethylammonium bromide. This is consistent with the pK_a-quaternary ammonium cation structure guidelines from Chapter 6, Section G. When the pK_a of the aniline is lowered by the proximity of electron-withdrawing groups (on the ring and in the β position), triethylbenzylammonium chloride/NaOH was able to catalyze the formation of aziridines by intramolecular S$_N$2 using N-(2-chloroethyl)-2-cyanoanilines, *34*, as the substrates [127].

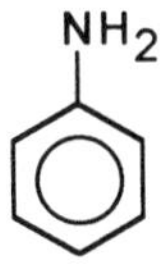

32

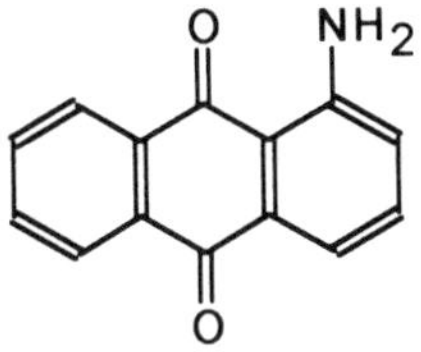

33

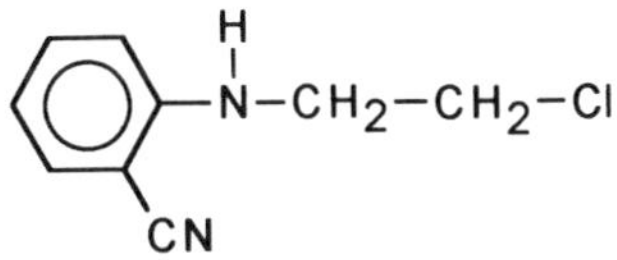

34

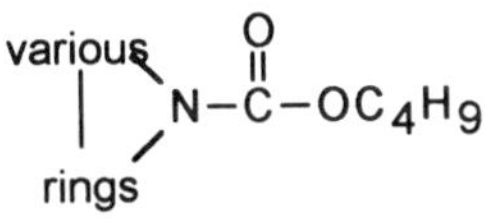

35

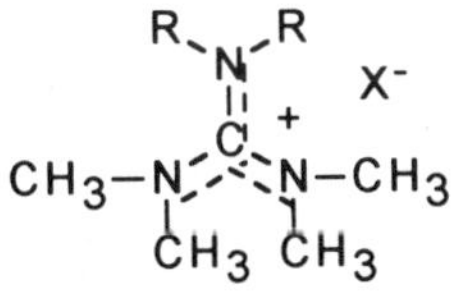

36

An attempt was made to prepare tertiary amines from secondary dialkyl amines (unactivated) and alkyl halides using solid–liquid PTC conditions in the presence of K$_2$CO$_3$ as the base and tetrabutylammonium HSO$_4^-$ as the catalyst [128]. PTC was not able to catalyze the formation of the 3° amine (pK_a of the unactivated 2° amine probably too high, resulting in an N-anion that is probably too short lived to alkylate; pK_a of NH$_3$ = 35 [113]), but surprisingly, the carbamates, *35*, were formed. Further investigation revealed the presence of dibutylcarbonate when tetrabutylammonium HSO$_4^-$ was mixed with butyl chloride and K$_2$CO$_3$ in heptane in the absence of an amine. When using stoichiometric quantities of tetrabutylammonium HSO_4^- and the 2° amine, 55–82% of the carbamates were obtained, presumably through the tetrabutylammonium butylcarbonate intermediate. The 3° amine was formed in 85–94% yield in the absence of the catalyst, using butyl bromide and acetonitrile as the solvent. Thus, selectivity was manipulated by solvent, leaving group, and quaternary ammonium salt.

Brunelle used solid–liquid PTC conditions to prepare hexaalkylguanidinium salts, *36*, from tetraalkylguanidine [129]. It was suggested that the tetrabutylammonium bromide and K_2CO_3 facilitates the neutralization of the intermediate pentaalkylguanidinium salt, since in the absence of the catalyst, the reaction stopped at an earlier stage, giving a mixture of the penta- and hexaalkylguanidinium salts. Acetonitrile was the preferred solvent; *n*-alkyl bromides and iodides were the alkylating agents; solid K_2CO_3 was the preferred base; and 2 mol% tetrabutylammonium bromide was sufficient to afford 82–98% product. These salts were found to be useful for the PTC nucleophilic aromatic substitution of bisphenol A with nitrophthalimide.

Hexamethylenetetraamine was *N*-alkylated with chloromethylketones in the presence of tetrabutylammonium bromide and *no* base to yield an intermediate for the preparation of antidotes for herbicides [130].

C. *O*-Alkylation—Etherification

1. *Etherification of Alkoxides*

The Williamson ether synthesis is one of the most basic organic reactions that has been greatly simplified by PTC, and that has become the method of choice in many cases. Merz was the first to show that methylation of a wide variety of primary alcohols and activated primary, secondary, and tertiary alcohols could be performed in high yield (85–97%) under PTC conditions (see Figure 8-21) [131]. Dimethyl sulfate was used as the alkylating agent in the presence of catalytic tetrabutylammonium iodide, 50% NaOH, and petroleum ether. Addi-

$$R-O-H + (CH_3)_2SO_4 \xrightarrow[\text{petroleum ether}]{\substack{Bu_4N\ I \\ 50\%\ NaOH}} R-O-CH_3$$

$$R-O-H + R'\text{-Cl} \xrightarrow[\substack{\text{excess R'Cl} \\ \text{as solvent}}]{\substack{Bu_4N\ HSO_4 \\ 50\%\ NaOH}} R-O-R'$$

Figure 8-21. Etherification.

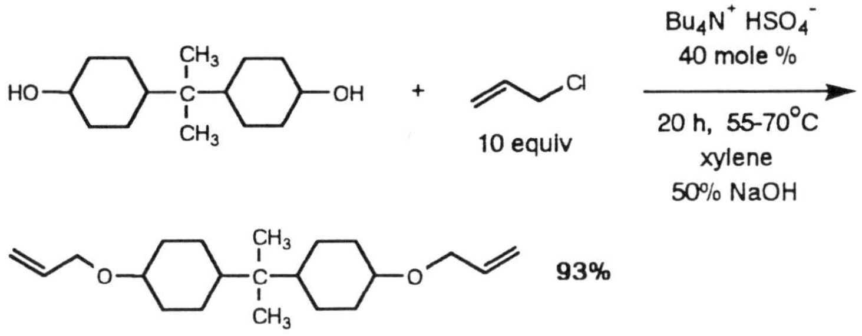

Figure 8-22. Etherification of polyols.

polyvinyl alcohol (10 mole % TBAH/75-85 C/24 h) -> 78% of theoretical double bonds
mannitol (10 mole% TBAH/75-80 C/14 h) -> 83% mannitol hexaallyl ether

tionally, some nonactivated secondary and tertiary alcohols could be methylated in moderate yields under these conditions. PTC *O*-alkylation was expanded by Freedman to prepare mixed ethers from generic primary and secondary alcohols and primary alkyl chlorides (see Figure 8-21) [132]. Optimum conditions consisted of 50% NaOH, the alkylating agent as the solvent, and 3–5 mol% tetrabutylammonium HSO_4^- at 25–70°C. Freedman provided supporting data consistent with the extraction mechanism of this PTC/OH reaction and showed that the reaction was sensitive to catalyst poisoning by soft anions.

Capped PEGs are an important class of phase-transfer catalysts (see Chapter 4). The capped PEGs are themselves prepared preferentially under PTC conditions, optionally using tetrabutylammonium bromide as catalyst [133]. Monoalkylated PEG can be obtained by the reaction of any primary alkyl chloride or bromide with a slight excess of 50% NaOH and 5–10 equivalents of the glycol at 100°C/24 h, in 79–90% yield. The glycol itself serves as catalyst. Tetrabutylammonium bromide catalyzes the reaction but also forms byproducts that need to be separated. Tetrabutylammonium HSO_4^- (9 mol%) has also been used as the catalyst for capping PEGs [134]. *O*-Alkylation of alkoxides has been applied to the capping of other polyols (tetrabutylammonium HSO_4^- as catalyst) such as polyvinyl alcohol and the saturated analogue of bisphenol A (see Figure 8-22) [135].

Carbohydrates usually readily undergo PTC *O*-alkylation at the least hindered hydroxyl available [136]. Carbohydrates are sometimes difficult to alkylate under PTC conditions when they (and their derived alkoxides) are water soluble and difficult to transfer to the organic phase. This may be solved by modifying the carbohydrate chain with THP (tetrahydropyranyl) groups which lend organophilicity to the molecule [137]. The THP-protected carbohydrate is easily etherified, hydrolyzed, then reetherified under PTC conditions, all in overall >90% yield. Ethyl and methylcellulose have been prepared by PTC *O*-alkylation of cellulose [138].

Much work has been done on the preparation of glycidyl ethers from alcohols

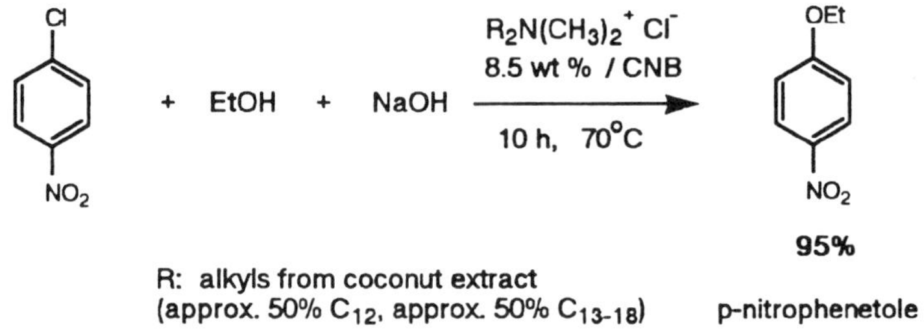

412 / Phase-Transfer Catalysis

$$\text{Cl-C}_6\text{H}_4\text{-NO}_2 + \text{EtOH} + \text{NaOH} \xrightarrow[\text{10 h, 70°C}]{\substack{R_2N(CH_3)_2^+ \, Cl^- \\ 8.5 \text{ wt \%} / \text{CNB}}} \text{EtO-C}_6\text{H}_4\text{-NO}_2$$

R: alkyls from coconut extract
(approx. 50% C_{12}, approx. 50% C_{13-18})   95%
                                                                                        p-nitrophenetole

Figure 8-23. Nucleophilic aromatic substitution with alkoxide.

and phenols (see next section) under PTC conditions. Twenty-four primary and secondary alcohols were etherified with epichlorohydrin in 67–99% and 30–91% yield, respectively, in the presence of tetrabutylammonium HSO_4^- at room temperature [139]. A patent described the reaction of hydroxyalkyl-modified lignin with epichlorohydrin in the presence of solid NaOH or KOH and a variety of phase-transfer catalysts, to produce useful prepolymers [140]. Lignin is the second most abundant chemical in wood, and its modification may lead to highly significant commercial opportunities. It may be anticipated that PTC will be increasingly applied in the modification of abundantly available natural materials.

A series of 14 α-trichloromethyl alkyl and benzyl alcohols was etherified with allyl bromide derivatives in 71–92% yield in the presence of 5 mol% tetrabutylammonium iodide, 6 N NaOH, and methylene chloride [141]. Dehydrohalogenation afforded allyl dichlorovinylethers.

The viability of a PTC *O*-alkylation, using *tert*-butyl bromoacetate, was the key factor in the successful synthesis of the pharmaceutical calcitriol [142]. *p*-Nitrophenetole was prepared by the nucleophilic aromatic substitution reaction of ethoxide with *p*-chloronitrobenzene (see Figure 8-23) [143]. In this reaction, a dimethyldialkylammonium was used, in which the alkyl groups were long alkyl chains (C_{12-18}) derived from natural extract. More recently, the PTC reaction of ethoxide with *p*-nitrochlorobenzene has been shown to be catalyzed (rate enhancement of two orders of magnitude) by microwave irradiation [144]. PTC is also useful in preparing *o*-nitrophenetole without making azoxy byproducts [145]. The phenetoles are useful intermediates for dyes and pharmaceuticals. A hindered substituted neo-pentyl alcohol was alkylated in 86% yield in the presence of tetrabutylammonium hydrogen sulfate and KOH to yield an insecticide [146]. An oxirane was prepared by the intramolecular attack of the alkoxide derived from $PhCH(X)CH(X)CH_2OH$, on the α-C−X group [147]. Attack at the β-C−X group would have given the four-membered ring, but the oxirane predominated. Ethoxide and phenoxide were reacted with gem-dichlorocyclopropanes to yield diether derivatives of cyclopropane [54]. A monomer, useful in the preparation of low surface energy polystyrene, was synthesized by the etherification of a highly fluorinated alcohol with *p*-chloromethylstyrene [148].

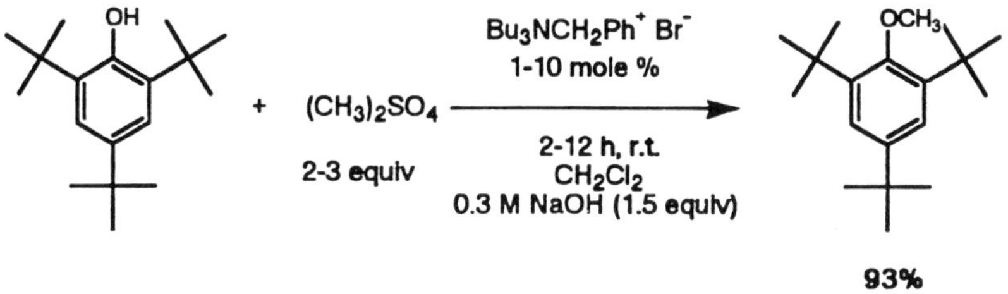

Figure 8-24. Methylation of a hindered phenol.

2. Etherification of Phenoxides

a. Using Quaternary -Onium Salts

McKillop published the classic first paper which demonstrated the power of PTC in etherification of phenoxides [149]. Many simple phenol derivatives, *37*, were O-alkylated in high yield (typically 70–95%), using dimethyl sulfate, MeI, benzyl chloride, allyl bromide, epichlorohydrin, n-BuBr, and even cyclohexyl bromide as the alkylating agents; benzyltributylammonium bromide as the catalyst; and methylene chloride as the solvent. The reaction rates were high using dimethyl sulfate at room temperature (10 mol% catalyst gave a 3-min half-life and 1 mol% gave a 40-min half-life). Even highly hindered phenols, such as 2,4,6-tri-*tert*-butylphenol, were etherified in high yield (see Figure 8-24). When the alkylating agent is not used, the methylene chloride solvent acts as the alkylating agent, and diaryloxymethane derivatives are obtained [150].

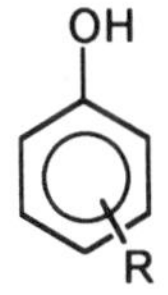

The most extensive study of the etherification of phenoxide (2,4,6-tribromophenol + allyl bromide; allyltribromophenol useful in flame-retardant polymers) in the presence of quaternary ammonium salts showed the reaction to be intrinsic chemical reaction rate controlled [151]. The key features consistent with intrinsic chemical reaction rate control and optimal conditions found for this reaction were: tetrabutylammonium bromide and benzyltributylammonium bromide served as better catalysts than triethylbenzylammonium chloride; the reaction was first order in catalyst and tribromophenol; no effect of enhanced agitation was observed above 600 rpm; KOH worked slightly better than NaOH (no explanation); more hydration decreased the reaction rate; and the order of reactivity in solvents was methylene chloride > chlorobenzene > toluene. This article is a good example of how to go about characterizing and optimizing a new application of a classic well established PTC reaction.

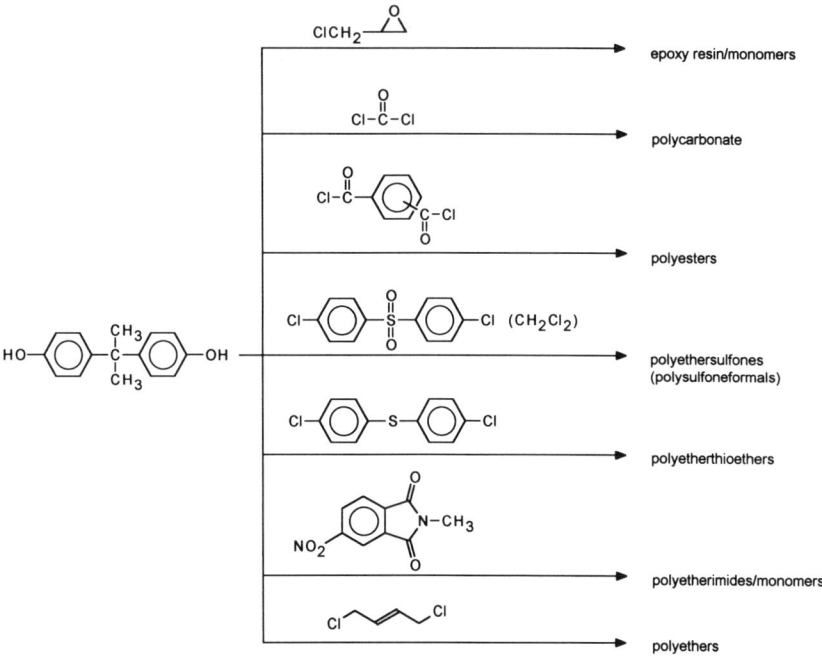

Figure 8-25. Nucleophilic reactions of bisphenol A.

The most widely reacted phenoxide reported in the general and patent PTC literature is bisphenol A (see Figure 8-25). Bisphenol A has been reacted (with mono- and difunctional electrophiles, in the preparation of monomers and polymers, respectively. As described in Chapter 9, polymers have been made using bisphenol A (and derivatives) as the starting material under PTC conditions to prepare polyethers [152], polyesters [153,154], polyetherimides [155], polyethersulfones [156], polyethersulfoneformals [157], polythioethers [158], polycarbonates [159,160], and epoxy resins [161]. These reactions include simple *O*-alkylation as well as nucleophilic aromatic substitution and esterification.

Monomers have been prepared by the *O*-alkylation of bisphenol A with 2-chloroethyl vinyl ether [162] and epichlorohydrin [161] (see Figure 8-26). The reaction with 2-chloroethyl vinyl ether illustrates several practical ramifications of PTC systems. The tetrabutylammonium bromide catalyst was recovered as tetrabutylammonium chloride, in 94.5% recovery, by adding water to the solid NaOH phase followed by extraction of the aqueous phase with methylene chloride. Toluene was used as the solvent to facilitate the separation of the organic and aqueous phases during work-up. Aliquat 336 was almost as effective a catalyst as the tetrabutylammonium bromide, whereas dibenzo-18-crown-6, PEG, and tetrabutylphosphonium bromide were not as effective. The authors noted that, since the product could not be distilled and Aliquat 336, dibenzo-18-crown-

6, and PEG did not have appreciable solubility in the aqueous phase used for the recovery of tetrabutylammonium chloride, the separation from product would have been difficult with other than tetrabutylammonium salts. The second O-alkylation was shown to proceed faster than the first, presumably due to the greater organophilicity of the O-anion of the mono-ether intermediate relative to the bisphenol A dianion. Potential side reactions included the possibility of elimination or hydrolysis of the 2-chloroethyl vinyl ether. Indeed, very small amounts of divinyl ether and acetaldehyde were detected but the conditions were chosen under which the desired diether formation predominated. The diether was obtained in 91% yield after crystallization.

A highly hindered tetraphenolic calixarene, 38 (used for conformational studies), was converted to the dimethyl ether with dimethyl sulfate in the presence of tetrabutylammonium bromide [163]. Phenol was alkylated with α,α'-dibromo-p-xylene in the presence of K_2CO_3, 18-crown-6, and tetrahydrofuran to yield diphenoxyxylene to prepare a model compound for study of polyethers [164]. The same article described the preparation of oligomeric ethers from diphenols (containing long-chain aliphatic or xylenyl spacers) and dibromododecane or dibromoxylene, using tetrabutylammonium bromide as catalyst, 50% NaOH, and dichlorobenzene as solvent. A polyether was prepared by the nucleophilic aromatic substitution of a difunctional halophenol [165]. Thus, 4-bromo-2,6-dimethyl phenol was reacted with itself in the presence of tetrabutylammonium HSO_4^-, 6 M NaOH, toluene, and a small amount of a secondary amine (to regulate molecular weight), to achieve polymers of varying intrinsic viscosity.

38

Selectivity of etherification was easily achieved in which a phenolic OH and an allylic OH were present on the same molecule [166]. Thus, 3-ethylmorphine was prepared in 95% yield by the reaction of morphine with diethyl sulfate in the presence of tetrabutylammonium bromide, K_2CO_3, and toluene.

Phenoxide and naphthoxide may act as ambident C-/O-nucleophiles. Montanari investigated the benzylation of phenoxide under PTC/OH conditions using polymer-bound quaternary phosphonium and ammonium cations [167]. When long organophilic spacers were used between the polymer backbone and the quaternary -onium cation, much O-alkylation was obtained, consistent with reduced

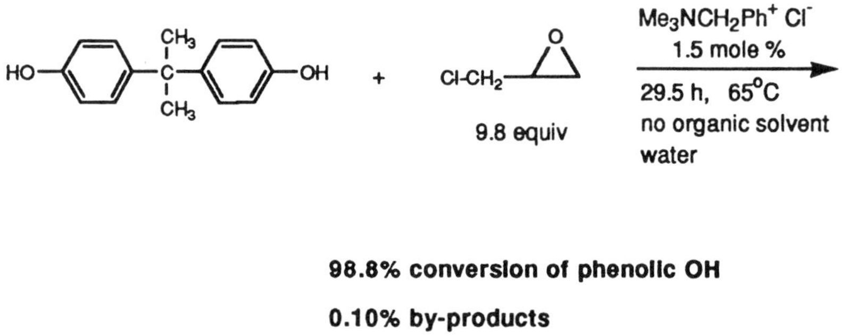

Figure 8-26. Etherification of bisphenol A.

98.8% conversion of phenolic OH
0.10% by-products

hydration (of the O-anion) within the organic solvation shell firmly surrounding the catalytic site. In a more polar and hydrophilic environment, that is, polymer-bound benzyltrimethylammonium, more hydration of the O-anion is present, and therefore more C-alkylation is obtained (see above). The same predominance of O-alkylation of phenol by benzyl chloride was found using a hydrophilic commercial ion-exchange resin [168]. This exemplifies a recurring theme in this book, that hydration, charge accessibility, and the polarity of the reaction environment determine much of the behavior of PTC systems.

Selective O-Alkylation of "Pyridones"

Several patents describe the PTC O-alkylation and O-phosphorylation of the ambident pyridinates, using quaternary ammonium salts and polyglycols, which are used as commercial herbicides and insecticides (see Figure 8-27) [169–173]. In an early patent, the pyridinate sodium salt was reacted with methyl chloroacetate (excess or stoichiometric in toluene) in the presence of triethylbenzylammonium chloride to yield 90–95% O-alkylated product [169]. In the later environmentally improved patent [170], no solvent was used (only 18 mol% excess of ethyl chloroacetate) and Na_2CO_3/$NaHCO_3$/NaCl was added, and 3 mol% tetrabutylammonium bromide was used. The solvent-free phosphorothiolation of chlorinated pyridinates was also patented [173]. In principle, these pyridinates may be viewed as pyridonates (negative charge on the nitrogen α to carbonyl) and would be expected to undergo some N-alkylation. The high yield of O-alkylated product indicates that under these conditions, N-alkylation has been overcome. Less hydration, polar "solvent," and nonaccessible quaternary -onium cations would favor O-alkylation over N-alkylation (see Section A.1 of this chapter for guidelines on selectivity of alkylation of ambident anions). O-Alkylation of 2-amino-3-hydroxypyridine was performed using a wide variety of benzyl chloride derivatives in the presence of 40% NaOH, 1 mol% Adogen 464 (trademark of Henkel), and methylene chloride [174]. When using methylene chloride in PTC nucleophilic substitutions/alkylations, one should always be aware of the

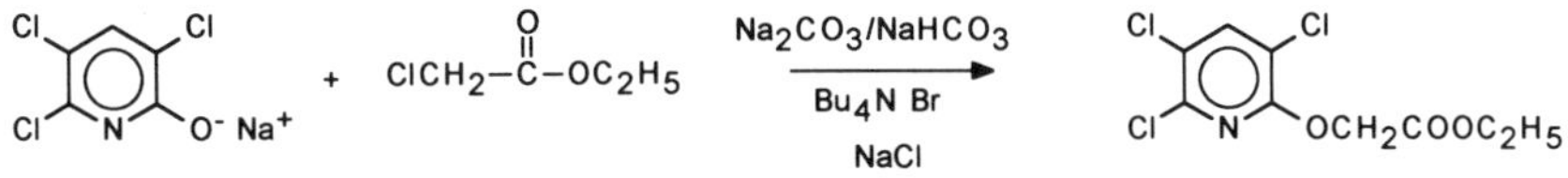

Figure 8-27. Solvent-free selective etherification.

possibility of this solvent unintentionally serving as a difunctional alkylating agent (which indeed was detected in this case).

b. Using PEGs

The methylation of phenol using methyl iodide is very rapid in the presence of PEG [175]. In this first report of using PEG for etherification of phenol, it was found that PEG with 12 oxyethylene units catalyzed the reaction faster than PEG with 6 oxyethylene units. Using solid KOH as the base and either methylene chloride or chloroform as the solvent, the reaction was complete within 30 min. 2,4,6,-Trimethylphenol gave only 60% ether after 30 min.

Svec performed extensive studies to characterize the O-butylation of phenol in the presence of PEG [176] and polymer-bound PEG [177]. A comparative study showed that PEG catalyzed the reaction less than other catalysts (in parentheses are the relative rate constants related to the noncatalyzed reaction): tetrabutylammonium bromide (58.6), Dowex 11 (1.3), PEG-1550 (3.9), polymer-bound PEG-600 (2.5), dibenzo-18-crown-6 (6.2). Interestingly, in the presence of tetrabutylammonium bromide or polymer-bound PEG, when KOH was used as base, etherification rates three to four times greater were observed relative to using NaOH as base. Polar solvents enhanced the reaction more than nonpolar solvents in the presence of PEG. The optimal polymer-bound PEG was PEG-2000 and was found to induce slightly more reactivity than optimal soluble PEG-1550. The activities of these PEGs are likely related to their selectivity toward sorption capacity toward alkali metals. The enhanced activity of KOH relative to NaOH in the octylation of phenoxide catalyzed by polymer-bound PEG was exemplified in a patent (twice the yield with KOH vs. NaOH under otherwise identical conditions) [178]. This patent also described the preparation of the polymer-bound PEG by reaction of chloromethylated polystyrene with a monocapped PEG in the presence of tetrabutylammonium bromide.

A more comprehensive study was reported for the allylation of phenol in the presence of PEG-1500. A significant finding that may have practical ramifications for planning PEG-based reactions is that the reaction was promoted in *both the aqueous and organic phases* due to the presence of PEG. Not only did the PEG increase the concentration of phenoxide in the organic phase (enhancing reactivity), but the PEG, which is somewhat water soluble, actually solubilized the allyl chloride in the aqueous phase (allyl chloride water solubility data shown). The allyl chloride in the aqueous phase could then react with the water-soluble phenoxide. The reaction rate was found to be insensitive to $[NaCl]_{aq}$ or to agitation

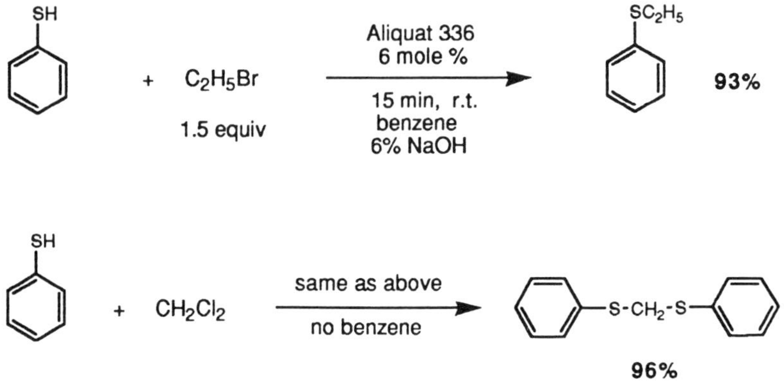

Figure 8-28. S-Alkylation.

above 150 rpm. At high ratios of alkali metal hydroxide to phenol, the reaction rate increased much more dramatically for KOH relative to NaOH. The authors noted the formation of a third liquid phase at these high base concentrations, which may be responsible for this behavior. The reaction rate was found to increase somewhat with larger PEGs, but reactivity was constant with PEG-2000 and higher. Although the authors claim that PEG is preferable for use at high phenol concentrations, tetrabutylammonium bromide catalyzed the reaction more than PEG at all phenol concentrations.

The above studies describing the etherification of phenoxides represent typical thought processes and aspects encountered when trying to evaluate and optimize new potential applications using quaternary -onium salts, PEGs, and polymer-bound catalysts. It is recommended that academic and industrial chemists alike refer to this section when grappling with choosing reaction conditions for PTC reactions.

D. S-Alkylation—Thioetherification

The application of PTC conditions for the preparation of thioethers is particularly advantageous for two reasons: (1) the S-anions are soft and polarizable and, therefore, have a high affinity toward the soft quaternary -onium cations and (2) the transferred thiolates are very nucleophilic. Catalyst poisoning poses less of a threat when the reacting anion is an alkyl or aryl thiolate.

The extreme ease with which thioethers can be prepared was exemplified by the reaction of thiophenol with ethyl bromide (see Figure 8-28) [179]. The reaction was complete within 15 min at room temperature in the presence of 6 mol% Aliquat 336 in benzene. When methylene chloride was used as the solvent/alkylating agent, the thioacetal was obtained in 96% yield. Even the hindered *iso*-butylthiol could be converted to the thioacetal or thioether in 95% and 89%

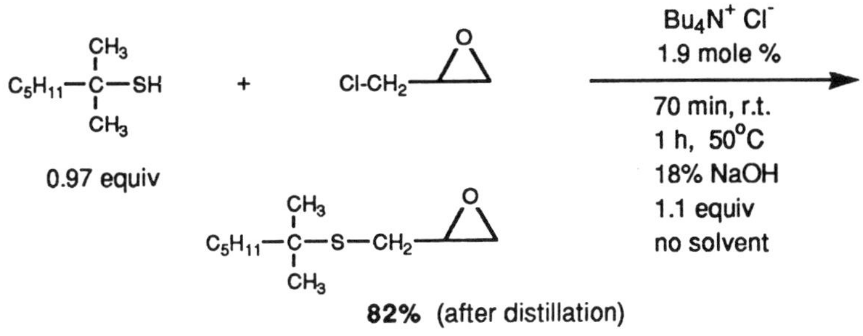

Figure 8-29. S-Alkylation of a hindered thiol.

Reactants: $C_5H_{11}-C(CH_3)_2-SH$ (0.97 equiv) + $Cl-CH_2-$(epoxide)

Conditions: $Bu_4N^+ Cl^-$ 1.9 mole %, 70 min, r.t., 1 h, 50°C, 18% NaOH, 1.1 equiv, no solvent

Product: $C_5H_{11}-C(CH_3)_2-S-CH_2-$(epoxide), 82% (after distillation)

yield, using methylene chloride or ethyl bromide as the reactant, respectively. Inorganic sodium sulfide was also used to form a variety of dialkyl sulfides in 90–100% yield from the n-alkyl chlorides and hindered alkyl bromides [180]. A highly hindered tertiary thiol was S-alkylated under very mild conditions (see Figure 8-29) [181].

Thiolates were reacted with polyhaloaromatics to afford products of nucleophilic aromatic substitution. Yields of > 80% were obtained from the reaction of o-dichlorobenzene with 1°, 2°, or 3° alkylthiols in the presence of crown ether [182]. Monosubstitution was more prevalent with the hindered alkylthiols. Lower yields were obtained for the meta- and para-isomers since nucleophilic aromatic substitution is highly dependent on the positive charge density of the carbon being attacked (needing more electron-withdrawing groups nearby). Tetrachlorobenzene and PCBs could be decomposed by nucleophilic aromatic substitution using alkylthiols in the presence of phosphonium salts [183] or PEGs [184]. This reaction is a good example of PTC for environmental applications. Thiophenoxide was reacted with gem-dichlorocyclopropanes to yield dithioether derivatives of cyclopropane [54].

In another nucleophilic aromatic substitution, 4-chlorobenzonitrile was quantitatively reacted with methyl mercaptan in the presence of solid KOH and 0.1 mol% 18-crown-6 in chlorobenzene to yield p-methylthiobenzonitrile [185]. Under the same reaction conditions, 20 mol% Aliquat 336 was needed to obtain high yield due to catalyst decomposition. This reaction was part of a three-step one-pot sequence (90% overall yield) in which chlorobenzene was the common solvent and the organic substrates remained in the vessel throughout by adjusting the concentrations of the aqueous phase work-ups and decanting.

Thioesters and carbamates have been prepared under a variety of PTC conditions. Dithiopyr was prepared in 100% yield by the reaction of methyl mercaptan with a substituted pyridine acyl chloride [186]. A number of water-soluble quaternary ammonium catalysts allowed for recycle of the aqueous layer containing the catalyst and excess reactants (thiolate and base) leading to waste minimization. Thiocarboxylates were reacted with methylene chloride in the presence of PEG and base to

Figure 8-30. Regioselective *S*-alkylation.

yield methylene bisxanthates in 90–95% yield [187]. These compounds are useful in the isolation of sulfide-containing minerals from ores. Solid–liquid PTC was employed for the selective *S*-alkylation (vs. *O*-alkylation) in the preparation of thiocarbamates [188] and *S*-alkyl dialkyl/diarylthiophosphinates [189]. An interesting case of regioselective *S*-alkylation was reported in which there was the potential for competition between *S*- and *O*-alkylation (see Figure 8-30) [190].

E. Dehydrohalogenation

Of all the base-catalyzed reactions discussed in this chapter, only dehydrohalogenation does not (usually) proceed via an organic anionic intermediate. Since a C-, N-, O- or S-anion is not formed during dehydrohalogenation, it is difficult to predict optimal reaction conditions based on the pK_a of the organic substrate. It appears that, as a rule of thumb, almost all compounds will dehydrohalogenate under PTC/OH (or other base) conditions if they have two adjacent carbon atoms, one bearing a halogen and one bearing a hydrogen. It may be generalized that such dehydrohalogenation will continue until the catalyst decomposes or is "poisoned" or if the basicity of the transferred hydroxide (or other base) is insufficient (e.g., too much hydration). Other factors that affect reactivity of PTC dehydrohalogenations are the relative thermodynamic stabilities of the unsaturated product vs. the starting material (e.g., conjugation of the resulting unsaturated site) and the ability of the system to remove the HX byproduct from the reaction site.

(2-Bromoethyl)benzene, *39*, was first reported to be dehydrobrominated to styrene using accessible alkyltriethylammonium cations [191]. This compound and its derivatives were later the focus of mechanistic studies [192–195]. These studies, which showed a wide variety of accessible, organophilic, and intermediate quaternary -onium cations to be effective, also showed wide variation in kinetic orders, energies of activation, quaternary -onium cation decomposition interference, and other behavior. Whereas much evidence supports a mechanism involving the formation of a complex such as Q^+X^-:HBr, the mechanism of anion/HX transfer is not fully resolved.

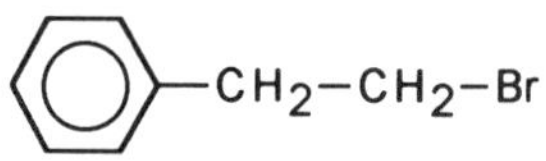

One of these studies [195] demonstrated the sometimes useful concept of using a third phase, containing the catalyst. The composition of the aqueous and organic phases can be designed such that a third phase is formed that contains approximately 50% catalyst (tetrabutylammonium bromide in this case). Phase diagrams and temperature dependence are shown for the toluene/40–43% NaOH/ tetrabutylammonium bromide system. Often, it is possible to attain high reactivity and easy catalyst separation and recycle when achieving catalyst–third-phase systems. (2-Bromoethyl)benzene was also used as the model compound to evaluate PEG and polymer-bound PEG, in the presence of 60% KOH, for catalytic dehydrohalogenation activity [196]. The reaction was found to be catalyzed only by PEG with free (noncapped) hydroxyl groups which served as base. The reaction was first order in catalyst. PEG with five oxyethylene groups had a catalytic activity 100 times greater than triethylbenzylammonium chloride and 40 times greater than 18-crown-6. Polymer-bound PEG had a catalytic activity up to 10 times greater than triethylbenzylammonium chloride. A number of dibromides were also converted into alkynes, by didehydrobromination, in >90% yield.

Herriott and Picker showed that hydroxide acts primarily as a base relative to being a nucleophile, in the PTC/OH reaction with 2-bromooctane (tetrabutylammonium as catalyst) [197]. Under homogeneous conditions the extent of elimination increased to 97% from 85% under the two-phase conditions. This may be due to the difference in hydration between the two systems. 1-Bromooctane did undergo nucleophilic substitution followed by etherification to yield dioctyl ether (75%), 1-octanol (20%), and a minor amount of 1-octene (5%).

Several of the early PTC/OH dehydrohalogenations reported required stoichiometric quantities of tetrabutylammonium HSO_4^- to perform the single and double dehydrohalogenations of a series of dibromoalkyl and dichloroalkenyl compounds [198, 199]. Mechanistic work by Halpern [200] showed that the single dehydrobromination of 1,2-dibromo-1,2-diphenylethane to α-bromostilbene (using 100 mol% tetrabutylammonium HSO_4^-) proceeds via an E2 mechanism with anti elimination (see Figure 8-31). The meso-starting material yielded only the *cis*-diphenyl product and the *d,l*-starting material yielded only the *trans*-diphenyl product. Further support for a nonionic intermediate during elimination is the relative insensitivity of the dehydrobromination of a series of β-arylethylbromides to the aromatic substituent [193].

Other catalysts used for dehydrohalogenation include a polymer-bound quaternary ammonium cation [201], cetyltributylphosphonium bromide/NaI using $Na_2S_2O_3$ instead of NaOH [202], and using nitrite as well as triethylbenzylammonium chloride as a cocatalyst [203]. Makosza showed that dehydrohalogenation can be promoted by the use of the addition of lipophilic bases, to serve as cocatalysts in the presence of tetrabutylammonium bromide and aqueous NaOH [204]. Lipophilic bases suitable for this purpose should be capable of deprotonation under PTC/OH conditions, such as nonnucleophilic alcohols, for example, trityl alcohol, and activated hindered heterocycles, for example, 2-methylindole. Cyclic ketene acetals have been synthesized (sensitive compounds; at 0°C) in

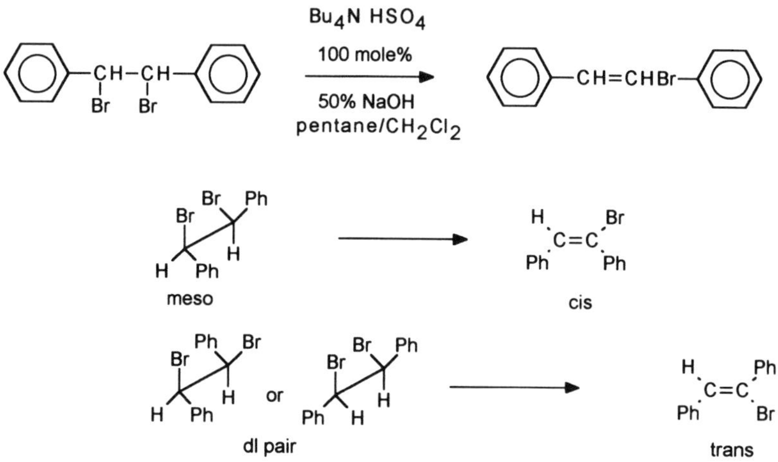

Figure 8-31. Stoichiometric dehydrobromination (E2).

high yields and short reaction times using 2 mol% Aliquat 336 and potassium *tert*-butoxide as base [205].

One of the largest scale PTC dehydrohalogenations reported is the DuPont chloroprene process (see Figure 8-32) [206]. A cascade reactor arrangement was used that incorporated recycle of NaOH and introduced fresh 20.5% NaOH into the middle of the train. Chloroprene was obtained from 3,4-dichlorbut-1-ene in 99.2% yield using 1115 ppm of a long-chain alkyl, benzyl, bis(2-hydroxypropyl) ammonium catalyst. An exhaustive review of similar reactions performed in the former Soviet Union may be found in Ref. 207. They found that when using $Me_3NR^+Cl^-$, the selectivity of this reaction toward chloroprene (vs. 1-chlorobuta-1,3-diene) increased rapidly as R increased from Me to a mixture of C_{12}–C_{16} [207].

When catalytic amounts of Katamin AB [(C_{10}–C_{18} alkyl)benzyldimethylammonium chloride] were used for the monodehydrohalogenation of 1,2-dihalo-1-phenylethane, 85% 1-halo-1-phenylethene and 5% 1-halo-2-phenylethene were

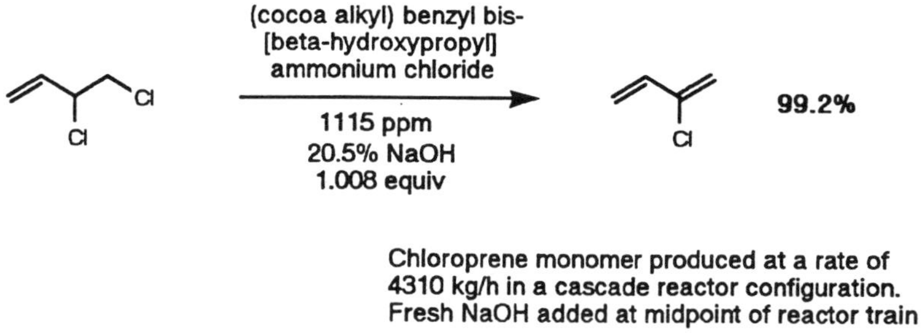

Figure 8-32. Large-scale dehydrochlorination (patent reports 4310 kg/h).

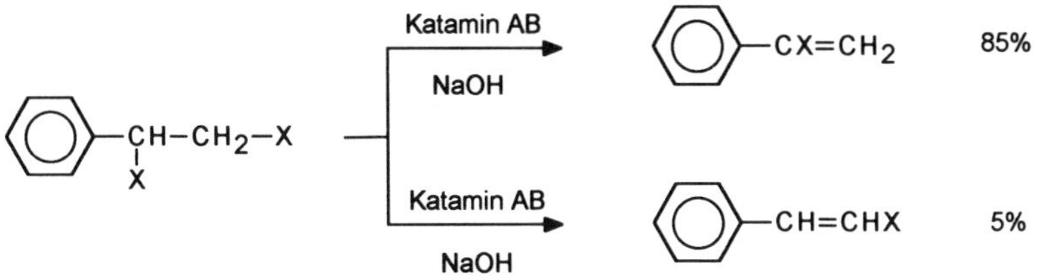

Figure 8-33. Selectivity of dehydrohalogenation.

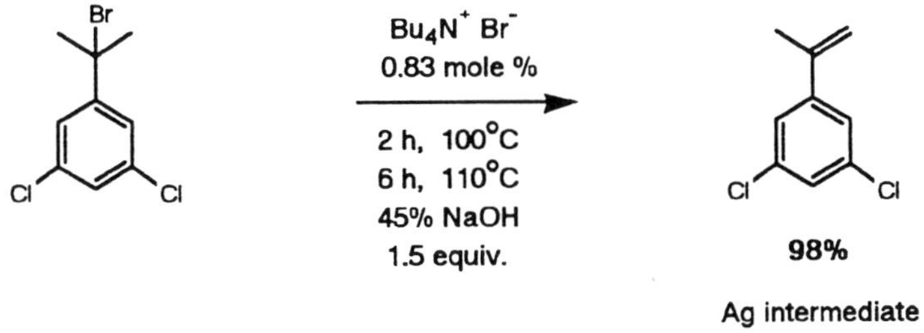

Figure 8-34. Dehydrobromination.

obtained, indicating that the more acidic benzylic hydrogen was preferably eliminated (see Figure 8-33) [207].

A key intermediate in the commercial manufacture of agricultural chemicals was prepared by the dehydrobromination of α-bromo-3,5-dichlorocumene (see Figure 8-34) [208]. 3,5-Dichloro-α-methylstyrene was obtained in 98% yield within 8 h at 110°C in the presence of 0.83 mol% tetrabutylammonium bromide and 1.5 equivalents of 45% NaOH.

An example of dehydroiodination was patented as being useful in the preparation of fluorine-containing monomers [209]. A quaternary bis(2-hydroxypropyl) ammonium salt was used in this application.

Alkynes can be prepared by double dehydrohalogenation of dibromides and dichlorides. Dehmlow used 1 mol% tetraoctylammonium bromide and solid KOH to catalytically didehydrobrominate, 1,2-dibromo-1-phenylethane, *40*, to phenylacetylene [210]. A yield of 98% was obtained within 1 h at 80°C. Chloroacetylene has been prepared by the dehydrochlorination of 1,1-dichloroethene [211].

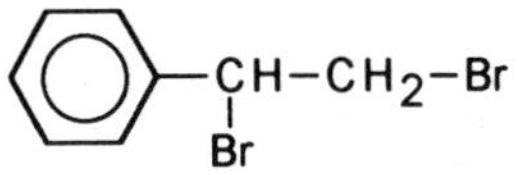

Long-chain alcohols and amines with internal triple bonds were prepared by (1) bromination of the internal double bonds of compounds such as oleyl alcohol followed by (2) dehydrobromination using PEG-300 [212]. Yields were approximately 80% after distillation.

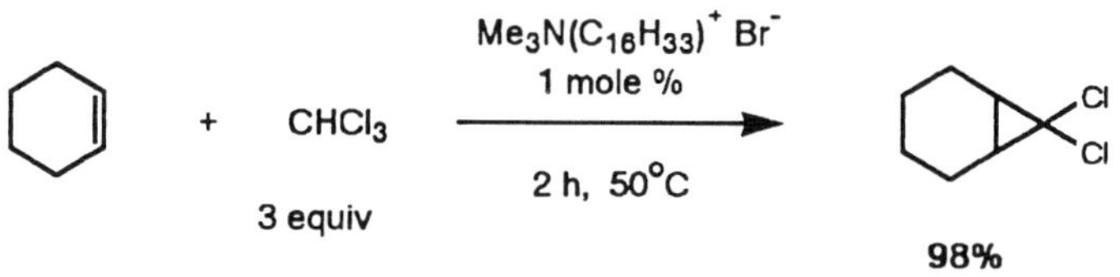

Figure 8-35. PTC/OH carbene addition.

F. Carbene Reactions

Alkylation and nucleophilic attack by C-, N-, O-, and S-anions described above are among the most common reactions used in organic synthesis. Carbene reactions are less widely used in general organic synthesis, but have received much attention in the PTC literature because of the ease and advantages of generating carbenes under PTC conditions. The general case simply involves mixing a haloform (usually readily available such as $CHCl_3$) in the presence of an inexpensive base such as 50% NaOH, a phase-transfer catalyst, and the substrate. Almost all typical phase-transfer catalysts have been shown to be effective for carbene reactions including a wide range of accessible or organophilic quaternary ammonium or phosphonium cations [213], crowns [213], PEGs [214], polymer-bound quaternary -onium cations [215] and sulfonium salts [216].

1. Dichlorocarbene Addition

The most common reaction of carbenes is addition of dihalocarbene to a double bond. The classic reaction reported by the pioneering PTC authors was the addition of dichlorocarbene to cyclohexene (see Figure 8-35). Starks used an organophilic catalyst in the presence of 25% NaOH to give 60% yield of carbene addition product vs. 5% without catalyst [217]. Makosza performed the reaction in 72% yield in the presence of triethylbenzylammonium chloride and 50% NaOH [218]. Soon thereafter, commercially available cetyltrimethylammonium salts were found to give 98% yield of this reaction [219]. A systematic study of the reactivity of this reaction by Dehmlow and Lissel showed that Et_4N bromide, cetylNMe_3 bromide, and cetylPBu_3 bromide gave similar reactivities and 18-crown-6 and $Ph_4P^+Cl^-$ gave even higher reactivity [213]. $Ph_3S^+Br^-$ induced twice the yield of that tetrabutylammonium bromide induced, and the use of $Ph_3S^+Cl^-$ gave 94% conversion in 2 h [220]. The use of PEG in the presence of powdered KOH for 3 h at 0°C to room temperature gave 100% conversion (93% isolated yield) in this reaction [214].

PEG molecular weights of 400–4000 gave similar reactivity; the ratio of ethylene oxide units to potassium ions was found to be optimal at 0.3, with lower reactivity at higher or lower ratios. Solid–liquid conditions were also used in the presence of triethylbenzylammonium chloride, giving 77% yield after 15 min, but the reaction gave no further yield at 30 min [221].

Most of the publications cited above reported the application of PTC/OH carbene addition to many other aliphatic, styrenic, acrylic, and terpenoid olefins. A series of 2,3-diaryl-1,1-dihalocyclopropanes, *41*, were prepared under PTC/OH conditions [222]. For example, dichlorocarbene was reacted with stilbene in the presence of triethylbenzylammonium chloride and 50% NaOH to yield 1,1-dichloro-*trans*-2,3-diphenylcyclopropane in 96% yield. PTC/OH conditions were used to synthesize gem-dichlorocyclopropane derivatives of long-chain fatty esters [223]. (The importance of cyclopropyl fatty acids has been postulated to be in the preservation of the configuration of ethylenic bonds of the fatty acid molecule and to prevent autooxidation of the phospholipids in microorganisms.)

The use of 2-hydroxyethyltrialkylammonium salts as the catalyst was found to improve the selectivity of mono-carbene addition to polyenes [224]. For example, 1,5,9-cyclododecatriene, *42*, gave only mono-addition in the presence of the 2-hydroxyethyl quaternary ammonium cation, whereas cetylMe$_3$N$^+$Br$^-$ gave tri-addition. Similarly limonene, 4-vinylcyclohexene, and a branched dienol gave selective mono-addition at the more substituted double bond. Though not fully explained, it is clear from preliminary results of asymmetric dichlorocarbene addition to styrene using the 2-hydroxyethyl catalyst that some specific interaction of the hydroxy group of the quaternary ammonium cation is occurring. Additional evidence for oxygen-directing interaction between the dichlorocarbene and hydroxyl groups comes from the highly selective syn-addition (13:1 syn:anti) of PTC generated carbene to 2-cyclohexenol [225]. Other examples of selective PTC/OH reactions of carbene or trihalomethide ions may be found in Chapter 4 Section B.1.f.

Dichlorocarbenes have been used for ring expansion via addition to olefins and rearrangement. The reaction of five-membered heterocycles with dichlorocarbene yields six-membered heterocycles (and minor amounts of addition products; selectivity is discussed below). For example, pure 3-chloro-4-methylquinoline was prepared from 2-methylindole in 68% yield in the presence of 1.5 mol% triethylbenzylammonium chloride, chloroform, 33% NaOH, and benzene (see

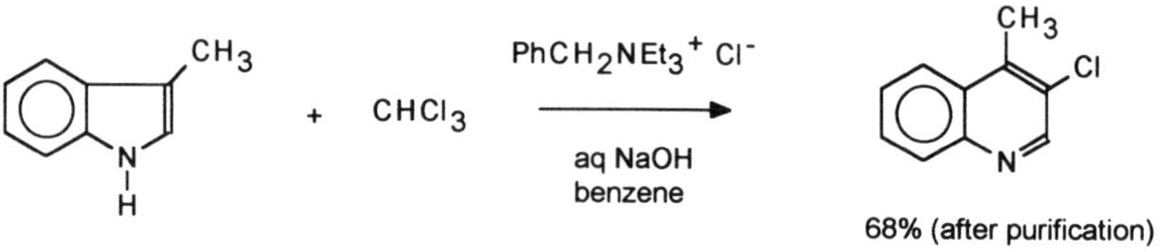

Figure 8-36. Ring expansion using carbene.

Figure 8-36) [226]. Polymethyl substituted pyrroles and imidazoles were converted to polymethylchloro pyridines and pyrimidines, respectively, under similar conditions ($C_{17-37}Me_2PhCH_2N^+Cl^-$ was also used as catalyst). Ring expansion of Reissert compounds was achieved after reacting it with PTC-generated carbene [227]. Mechanistic studies have been performed for the addition of dichloro- and dibromocarbene to norbornadiene [228, 229], *43*, and norbornene, *44* [230]. Two key differences were found between dichloro- and dibromocarbene addition to norbornadiene. The dibromocarbene gave a higher exo/endo addition ratio than dichlorocarbene and dibromocarbene gave no homo-addition.

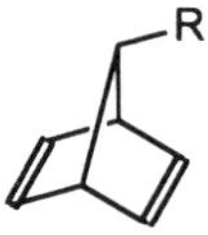

The reaction pathway of carbene reactions proceeds via deprotonation of chloroform to produce the trichloromethide anion, which subsequently releases a chloride anion to yield the dichlorocarbene. In principle, the trichloromethide anion, though not very nucleophilic, can add to a double bond or undergo other types of reactions. Indeed, enol esters, *45*, react with chloroform in the presence of triethylbenzylammonium chloride and aq. NaOH to yield two addition products: (1) a dichlorocyclopropyl derivative and (2) trichloromethide adds to the vinyl carbon adjacent to the ester and the deprotonated hydrogen adds to the other vinyl carbon [231]. Selectivity is determined by the substituents on the vinyl group and the ester.

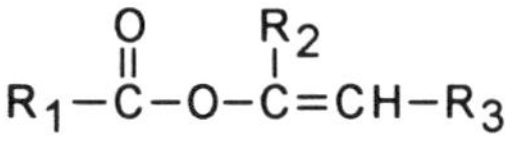

2. Dibromocarbene Addition

Makosza found that the addition of a little ethanol enhanced the yields of dibromocyclopropanes from bromoform [232]. Yields of 50–89% were obtained from the addition of dibromocarbene to derivatives of styrene, acrylates, and aliphatic alkenes. Dehmlow also employed the presence of ethanol in studies that showed that hydrolysis of the dibromocarbene is competing with addition of the dibromocarbene to the olefin [233]. $Me_4N^+Cl^-$ was not as effective as triethylbenzylammonium chloride as the catalyst, and $Oct_4N^+Cl^-$ was even slightly better. The use of ethanol did not appear to enhance the reactivity of the PTC/OH dichlorocarbene reactions. Dehmlow later found that the presence of pinacol can cocatalyze dichloro- and dibromocarbene reactions since it serves as an organophilic alkoxide under these conditions and does not participate in side reactions [234].

3. Mixed Dihalocarbene Addition

Carbenes have been generated and reacted from various combinations of halogens of haloform. Fluorochloro-, fluorobromo-, and fluoroiodocyclopropanes were prepared in 40–60% yield from fluorodichloromethane, fluorodibromomethane [235], and fluorodiiodomethane [236], respectively, in the presence of triethylbenzylammonium chloride and 50% NaOH and olefins (isobutene, 2,3-dimethylbutene, and styrene). A different approach to obtaining pure gem-bromochlorocyclopropanes involved the selective formation of bromochlorocarbene by the PTC/OH reaction of dibromomethane (deprotonation) with phenyltrichloromethane [237]. The bromochlorocarbene was reacted with a variety of aliphatic and styrenic alkenes to prepare the bromochlorocyclopropanes in 47–71% yield in the presence of tetrabutylammonium HSO_4^- and 60% KOH.

4. Other Reactions of Chloroform or Alternate Methods of Generating Carbenes Under PTC/OH Conditions

Carbenes may react in other, sometimes surprising, transformations. An intriguing reaction was reported involving the S_Ni mechanism of addition of dichlorocarbene to alcohols followed by rearrangement [238]. Thus, the bridgehead alcohol of 1-adamantyl alcohol, 46, was converted to the bridgehead chlorine, to yield 1-adamantyl chloride in 94%. This reaction was successfully applied to the conversion of 3-β-hydroxysteroids to 3-β-chlorosteroids with retention of configuration in 50–80% yield (see Figure 8-37) [239]. Replacement of OH by chloride using PTC/OH dichlorocarbene was useful in the synthesis of dyes derived from desyl esters (PhC(O)CH(OOCR)Ph) [240]. In this procedure, desyl alcohol was converted to the chloride in the presence of chloroform, triethylbenzylammonium chloride, and 50% NaOH. The chloride was then displaced by RCOONa, using PEG-400 as the phase-transfer catalyst. Subsequent non-PTC reactions produced the 2-substituted 4,5-diphenyloxazoles, of which some were found to be excellent fluorescent brightening agents for polyester fibers of various colors.

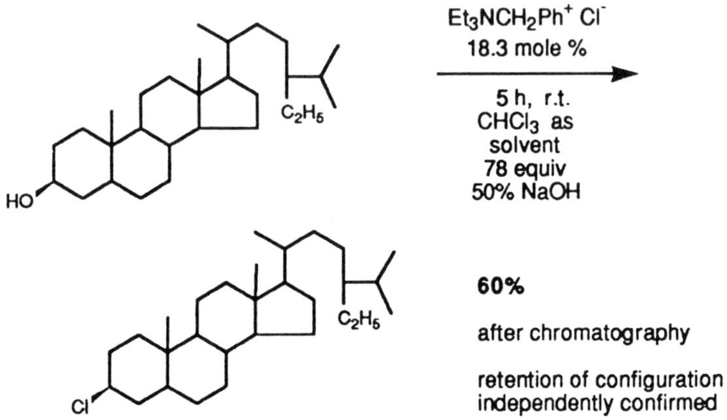

Figure 8-37. S_Ni substitution with retention of configuration.

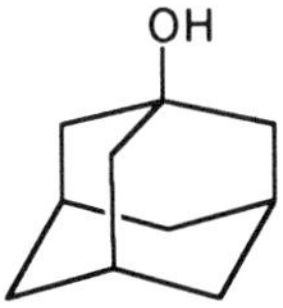

46

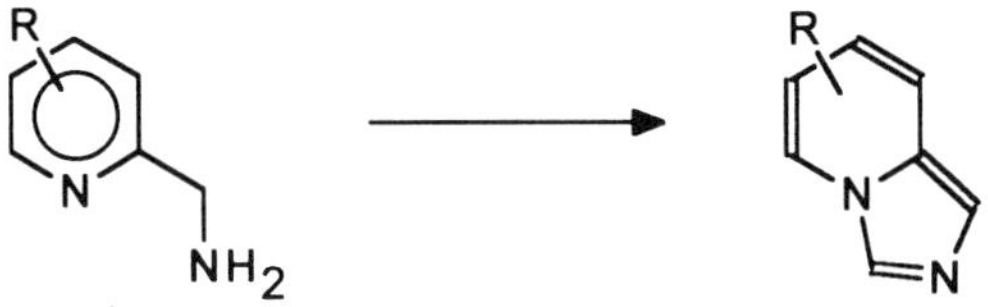

47

A novel patent described the preparation of fluorobenzene from cyclopentadiene via reaction of dichlorofluoromethane (1.2 mol% tetrabutylammonium bromide was the catalyst/base) followed by ring expansion and dehydrochlorination (see Figure 8-38) [241]. The presence of ethylene oxide was necessary for the reaction to proceed. A series of five imidazopyridines, 47, were prepared by the reaction of the nucleophilic α-(aminomethyl)pyridines with the electrophilic dichlorocarbene, followed by cyclization [242]. Dimethoxyethane was used as the solvent and tetrapropylammonium bromide was the catalyst. A mechanism was proposed based on the presence of minor byproducts/intermediates (such as isocyanide and formamide moieties). Dichlorocarbene has even been used as an oxidizing agent. In this reaction, a trithiolane (a five-membered ring with three adjacent sulfur atoms) was reacted with dichlorocarbene [243]. The carbon atom of the carbene was incorporated into the ring, via a series of intramolecular rearrangements. The products were dithiocarbonates and trithiocarbonates. Phenylcyclopropane rings were prepared by the addition of the carbene derived from $PhSCH_2Cl$ to a propenone [244].

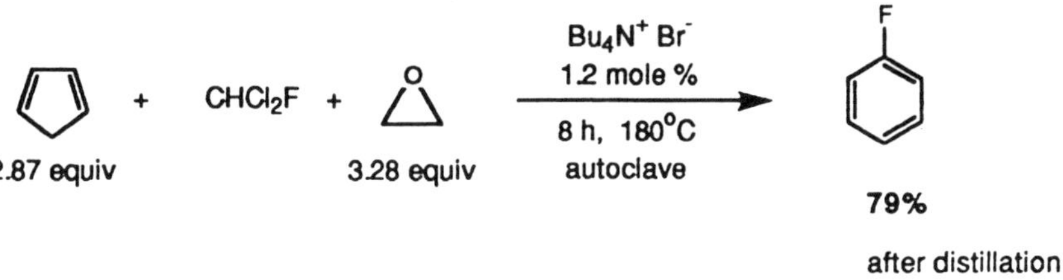

```
   ⌬       +   CHCl₂F   +   △(O)        Bu₄N⁺ Br⁻           F-⌬
                                        1.2 mole %
                                        ─────────→
                                        8 h, 180°C
2.87 equiv                 3.28 equiv   autoclave
                                                             79%
                                                          after distillation
```

Figure 8-38. Ring expansion using carbene.

Dichlorocarbene was generated using 5 mol% polymer-bound PEG and powdered NaOH at room temperature [245]. Addition to cyclohexene was 100% complete in 3.4 h; dehydration of benzamide to benzonitrile was 100% complete in 24 h; the SNi displacement of 1-octanol to give 1-chlorooctane was 65% complete in 5 h.

Dichlorocarbene has been generated under PTC/OH conditions using sodium trichloroacetate as the carbene source [246]. Tetrabutylammonium HSO_4^-, triethylbenzylammonium chloride, Aliquat 336, and tetraheptylammonium bromide were used as the catalysts. Chloromethylthiocarbene was generated from α,α-dichlorodimethyl sulfide, using 50% NaOH in the presence of triethylbenzylammonium chloride [247]. For the addition of the thiocarbene to tetramethylethylene, the ratio of cyclopropane formation to thiocarbene dimerization was approximately 2:1. In the presence of potassium *tert*-butoxide, the dimerization of the thiocarbene predominated.

Other references describing PTC carbene reactions may be found in a review article [248].

Some PTC/OH reactions performed in the presence of chloroform do not proceed via carbene intermediates. A PTC/OH reaction has been reported in which chloroform was deprotonated, but did not form or react as a carbene [249]. In this reaction the trichloromethide anion (formed in the presence of triethylbenzylamonium chloride and saturated aq. KOH) reacted with aromatic aldehydes in the presence of LiCl, and subsequently with ammonia, to yield α-amino acids after hydrolysis (see Figure 8-39). Trichloromethide was reacted with a gem-dichlorocyclopropane to yield a ditrichloromethide derivative of cyclopropane [54].

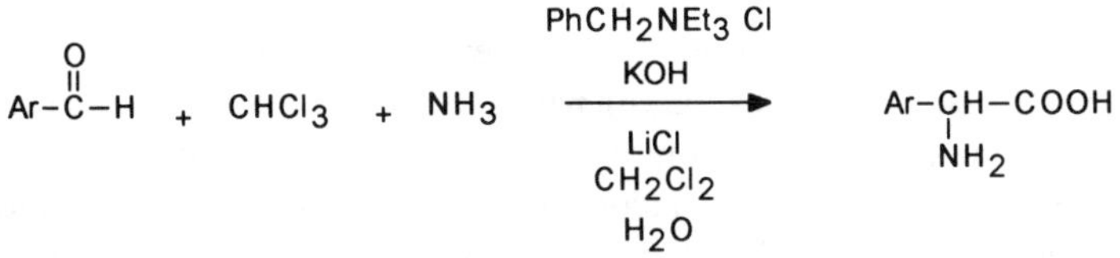

```
                                  PhCH₂NEt₃ Cl
     O                               KOH
     ‖                            ─────────→
Ar–C–H  +  CHCl₃  +  NH₃            LiCl            Ar–CH–COOH
                                    CH₂Cl₂               |
                                    H₂O                  NH₂
```

Figure 8-39. Synthesis of amino acids using chloroform.

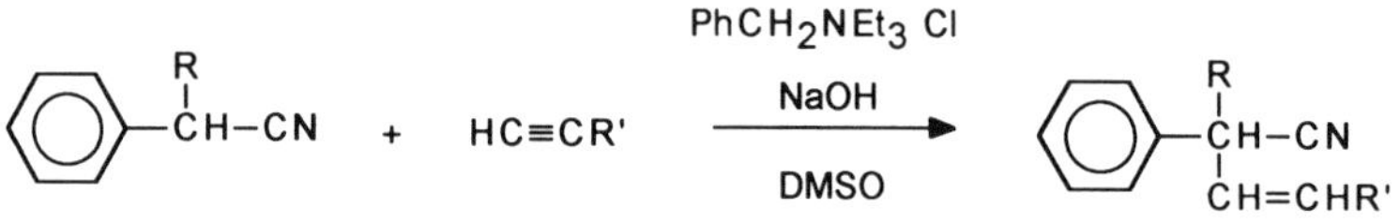

Figure 8-40. Addition to alkyne.

G. Condensation Reactions

1. Michael Addition

Many of the substrates containing activated methylene groups, which undergo PTC/OH C-alkylation (ketones, nitriles, imines, sulfones, etc.), are also good candidates for reaction with Michael acceptors. The addition of a nucleophile to an unsaturated carbon–carbon bond was pioneered by Makosza [250]. A series of α-monoalkylated phenylacetonitriles were deprotonated in the presence of triethylbenzylammonium chloride, NaOH, and some dimethyl sulfoxide and reacted with acetylene and phenylacetylene to give eight olefinic nitriles in 79–98% yield (see Figure 8-40). Later, Makosza reported the addition of α-alkylphenylacetonitriles (and Reissert compounds) to vinyl acetate, in up to 70% yield, in the presence of triethylbenzylammonium chloride and aq. NaOH, without the need for dimethyl sulfoxide [231].

The most notable PTC/OH Michael additions reported were chiral Michael additions for the preparation of amino acids from imines [35,36], Robinson annelations [251], and addition of indanone derivatives to methylethyl ketone [252]. These reactions used cinchona alkaloids as catalysts. Michael additions using ephedrine-based quaternary ammonium cations were also used for chiral Michael addition [253,254]. The use of "Solvent-Free" PTC (see Chapter 6, Section B.7), using ephedrine-based catalysts, was also shown to enhance enantiomeric excess of the chiral addition to α-enones [255]. Again, these selective catalysts contain the 2-hydroxyalkyl group on the quat. These reactions are described in Chapter 12.

Diaryl imines (ArCH=NAr') are highly reactive Michael acceptors, and react with activated methylene compounds ($Ph_2C=NCH_2CN$) within 10 min at room temperature to give 20–80% yield of the Michael adducts, in the presence of 5 mol% triethylbenzylammonium chloride (good choice based on pK_a; see Chapter 6, Section G), 33% NaOH, and benzene [256]. Selectivity decreased upon changing to more polar solvents or lengthening the reaction time to 60 min. When $PhCH=NCH_2CH_3$ was the Michael acceptor, a second reaction followed which was elimination of ethylamine to provide an olefinic nitrile imine (similar to a nitrogen analogue of the aldol condensation). The Michael addition of diethylmalonate to esters of 2-(1-hydroxyalkyl)propenoates, *48*, was performed with high stereoselectivity toward the syn-diastereomer (4:1 to 20:1) [257]. 18-

Crown-6 was the catalyst (5 mol%) and catalytic KF was used as the base in dimethyl sulfoxide or acetonitrile.

$$CH_3O_2C\diagup\!\!\diagdown CO_2CH_3 \;+\; \underset{48}{\overset{CO_2R}{\diagup\!\!\diagdown}\!\!\underset{OH}{\diagdown}\!\!R'} \quad\xrightarrow[\substack{\text{DMSO or}\\\text{acetonitrile}}]{\substack{\text{18-crown-6}\\(0.05\text{ equiv})\\KF\;(0.2\text{ equiv})}}\quad CH_3O_2C\!\!\diagdown\!\!\underset{CH_3O_2C}{}\!\!\diagup\!\!\diagdown\!\!\underset{OH}{\overset{CO_2R}{\diagup}}\!\!R'$$

(diastereomers)

The asymmetric Michael addition of *p-tert*-butylthiophenol on cyclohexene was thoroughly investigated, using quinine as the catalyst [258]. A tertiary quininium cation was formed. Adding tetrabutylammonium chloride decreased the enantiomeric excess dramatically, since the soft thiophenoxide associates preferably with the soft tetrabutylammonium cation, whereas the harder chloride preferably associates with the harder quininium center. The achiral tetrabutylammonium phenoxide could not induce asymmetric addition.

Allylphenylsulfones were added to activated alkenes in the presence of 50% NaOH and tetrabutylammonium bromide [62]. Yields of 72–76% were obtained when acrylonitrile or *tert*-butylacrylate were the Michael acceptors. Interestingly, when allyl bromide was used, 85% C-alkylation and 15% addition products were obtained.

Various thiolates were added to 6-cyano-1,3-dimethyluracil under solid–liquid PTC/K_2CO_3 conditions in the presence of tetrabutylammonium HSO_4^- [259]. The cyano group was not present in the final product, as a result of an additional S_N2 of thiolate at the cyano-bearing carbon followed by elimination of RSH. Direct elimination of HCN was shown not to be the mechanism, based on H/D exchange experiments performed together with the addition reaction.

2. Aldol Condensation

In general, carbon acids that are acidic enough to be deprotonated and nucleophilic enough to alkylate under PTC conditions are good candidates for addition to aldehydes in the aldol condensation. The aldol condensation was performed between phenylmethylsulfone and 4-aromatic aldehydes in 85–98% yield in the presence of 50% NaOH (see Figure 8-41) [260]. Triethylbenzylammonium chloride was an appropriate choice of catalyst based on the pK_a of the phenylsulfone (see Chapter 6, Section G). An insecticide intermediate (see Figure 8-42) was prepared by the double aldol condensation of both sides of acetone with an electron-deficient benzaldehyde [261]. A bis(2-hydroxyethyl) quaternary ammonium cation was used as catalyst, as is often the case for PTC/OH reactions.

Miller extended O'Donnell's work, on the chiral alkylation of imines, to the aldol condensation [262]. *N*-(Diphenylmethylene) glycinate *tert*-butyl ester was deprotonated in the presence of *N*-benzylcinchonidinium chloride, 5% NaOH, and methylene chloride, and added to *n*-heptanal in 86% yield and 43% diastereomeric

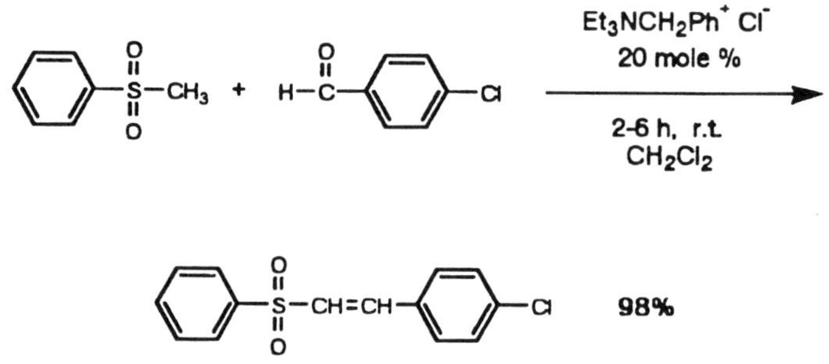

Figure 8-41. Aldol condensation with sulfone.

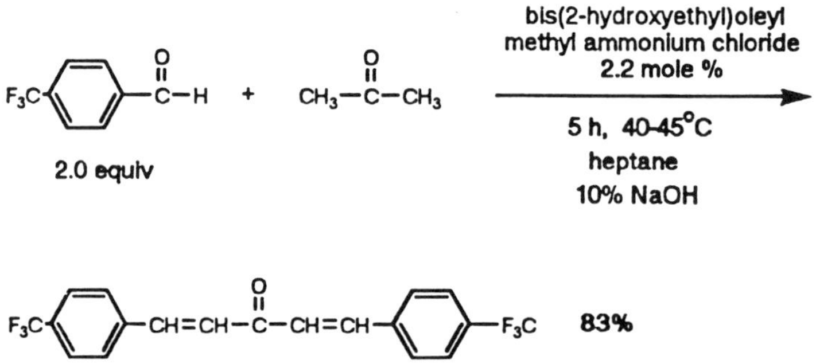

Figure 8-42. Double aldol condensation.

excess (see Figure 8-43). Enantioselectivity was low, due to the presumed weak interaction of the catalyst, enolate, and aldehyde.

A key intermediate in the synthesis of a fragrance from furfural was reported (see Figure 8-44) [32]. The first step of this one-pot two-step reaction was the C-alkylation of propanal or butanal by 2-chloromethylfuran. The resulting aldehyde was then deprotonated and condensed with another molecule of the starting aldehyde and subsequently underwent dehydration.

A patent was issued describing the preparation of the commercial antibiotic, chloramphenicol, in which glycine was added to 4-nitrobenzaldehyde (see Figure 8-45) [263]. The reaction also involved condensation of the amine with the aldehyde to form an imine which was subsequently hydrolyzed back to the

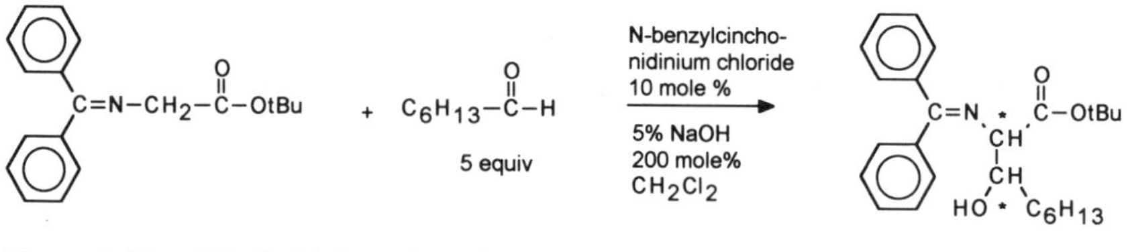

Figure 8-43. Chiral aldol condensation.

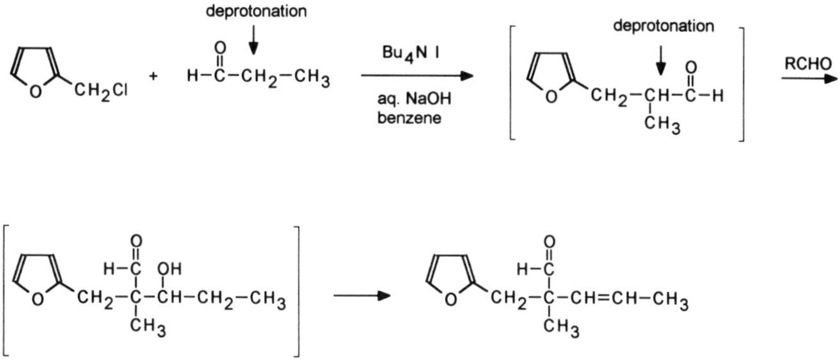

Figure 8-44. C-Alkylation followed by aldol.

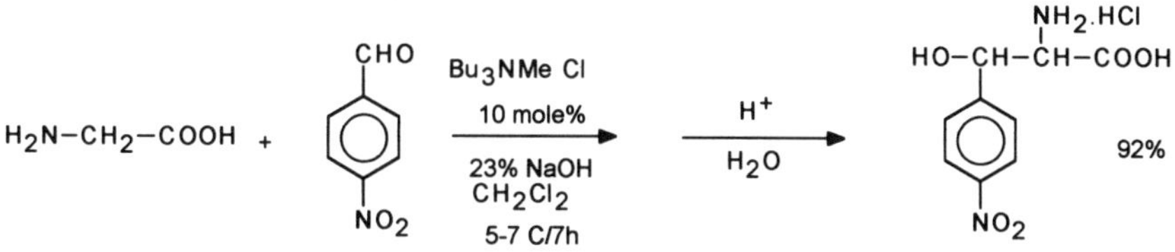

Figure 8-45. Aldol condensation without dehydration.

amine. Methyltributylammonium chloride was used as the catalyst and methylene chloride as solvent. An aldol condensation was patented that used a methide group activated by a sulfone and two fluorines to yield an aldol (no dehydration; no β-hydrogens; see Figure 8-46) [264]. Another sulfone–aldehyde aldol condensation was reported, which gave α,β-unsaturated sulfones in high yield, providing the aldehyde and sulfone were activated and not sterically hindered (see Figure 8-47) [260]. Triethylbenzylammonium chloride was used as the catalyst, and hexadecyltributylphosphonium bromide did not improve the yield (consistent with pK_a guidelines for choice of quaternary ammonium cation described in Chapter 6, Section G). Aliphatic aldehydes gave self-condensation and low yield of the unsaturated sulfone.

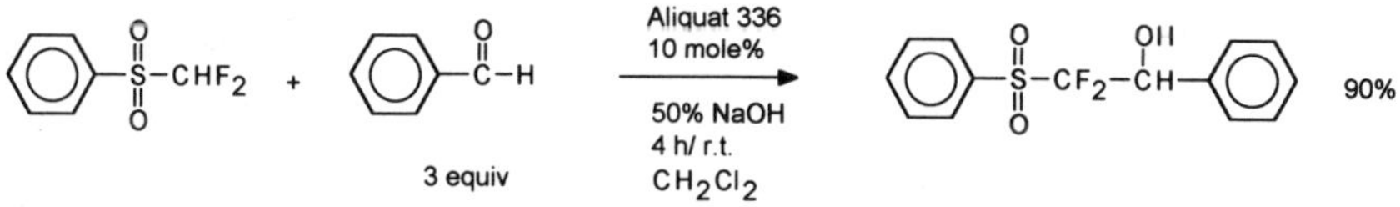

Figure 8-46. Aldol condensation with an activated sulfone.

3. Wittig

The early work by Starks showed that phosphonium salts were effective phase-transfer catalysts for reactions between inorganic anions and organic substrates.

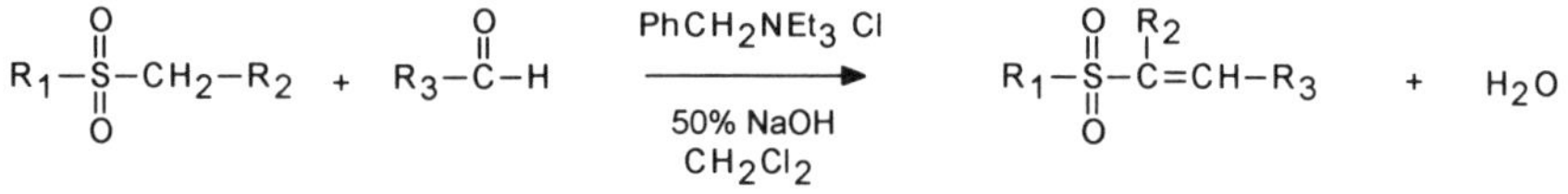

$$R_1-\overset{O}{\underset{O}{\overset{\|}{S}}}-CH_2-R_2 + R_3-\overset{O}{\overset{\|}{C}}-H \xrightarrow[\substack{50\% \text{ NaOH} \\ CH_2Cl_2}]{PhCH_2NEt_3\ Cl} R_1-\overset{O}{\underset{O}{\overset{\|}{S}}}-\overset{R_2}{\overset{|}{C}}=CH-R_3 + H_2O$$

Figure 8-47. Preparation of an unsaturated sulfone.

It was also known that in the presence of hydroxide, the phosphonium salts decompose to phosphine oxides (see Chapter 4). This combination of properties led to the possibility of adding aldehydes to a mixture of phosphonium salts and aqueous base to perform the Wittig reaction providing that the condensation could compete with phosphonium salt decomposition.

The foundation reaction for the PTC Wittig reaction was published in 1974 (see Figure 8-48) [265]. The model reaction consisted of reacting a reactive aldehyde, such as benzaldehyde, with methyltriphenylphosphonium iodide in the presence of 5 N NaOH and benzene to give styrene in 99% yield. As the reactivity of the aldehyde decreases (p-ClPhCHO > n-C$_7$H$_{15}$ > p-CH$_3$PhCHO > p-CH$_3$OPhCHO), the yield of the olefin decreases (95%, 73%, 55%, 38%, respectively), and the competing phosphonium salt decomposition predominates more. Ketones are unreactive (no olefin from acetophenone). In addition, the yield of olefin passed through a maximum as a function of base concentration and counteranion. At low base concentrations the reaction does not proceed well and the phosphonium salt can be recovered, but at concentrations above 5 N NaOH, the phosphonium salt decomposes more rapidly than condensation. As would be expected from $K^{sel}_{X \to OH}$, the phosphonium iodides tolerated a higher hydroxide concentration than the chlorides. This publication should be read by the PTC chemist before attempting the first Wittig candidate reaction. Dehmlow also showed that additional phase-transfer catalyst is not needed to perform the PTC Wittig reaction [266]. A Hammett plot was constructed for the Wittig reaction of substituted benzaldehydes [267]. A low sensitivity was found to the substituent ($\rho = 0.30$), indicating that the PTC Wittig reaction takes place via an intermediate of low polarity. Considering that the starting ylide is much more polar, it may

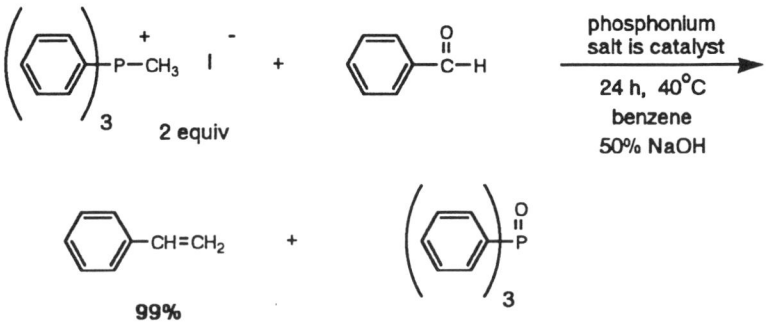

Figure 8-48. PTC Wittig foundation reaction.

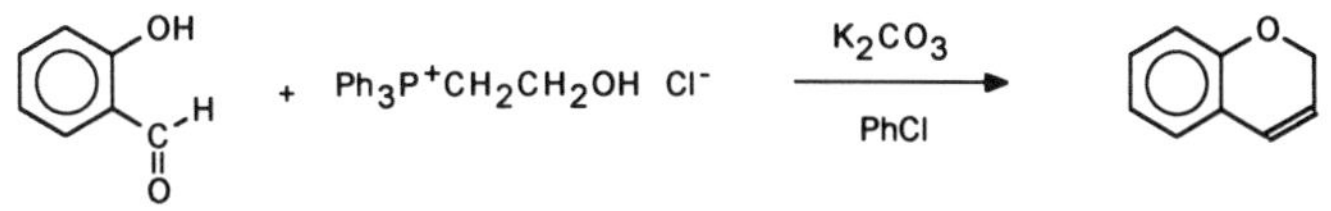

Figure 8-49. Intramolecular Wittig.

be speculated that nonpolar solvents may enhance the reaction rate, if the rate-determining step is the reaction of the ylide with the aldehyde (this may not be true if the reaction is transfer rate limited or if the deprotonation is the rate-determining step).

Phenanthroquinones and naphthoquinones were reacted with several bis-phosphonium salts to produce a series of fused aromatic hydrocarbons for study of multicharged aromatic compounds [268]. These bis-Wittig reactions were obtained in 16–32% yield using LiOH as the base and methylene chloride as the solvent. Phenanthroquinone can also be reacted under PTC conditions to give the mono-Wittig product [269].

Fused oxo-heterocycles were prepared by an innovative solid–liquid PTC/ K_2CO_3 Wittig reaction [270]. The 2-hydroxyethyl group of 2-hydroxyethyltriphenylphosphonium bromide was incorporated and used to make a pyran ring (see Figure 8-49). This phosphonium salt was reacted with salicylaldehyde to give the 2-hydroxyvinyl group. The 2-hydroxyvinyl group further reacted with the phenolic OH in the ortho position, by dehydration to form 2H-1-benzopyran in 37% overall yield.

The PTC Wittig reaction was performed without starting from the preformed phosphonium salt (see Figure 8-50) [271]. As a model reaction, chloroform and triphenylphosphine were reacted in situ to form the phosphonium salt which reacted in the same pot with benzaldehyde and 50% NaOH to give product and, likely, some aldol condensation. The authors then stepwise extended the reaction by replacing chloroform with dichlorodimethyl sulfide and then replaced benzaldehyde with a cyclopropyl aldehyde to produce a useful intermediate for the preparation of insecticidal pyrethroids.

Potassium carbonate has been used as the base in the PTC Wittig reaction [272]. Insect pheromones have been synthesized by PTC Wittig using a terminal hydroxyalkyl phosphonium salt [273]. Polymer-bound PEG was also used as the catalyst for the Wittig reaction [245]. A polymer-supported phosphonium salt could also be used in the PTC Wittig reaction [274].

4. Darzens

The Darzens condensation is a two-step reaction that begins with the deprotonation of an activated C–H group bearing a chlorine atom (activation by carbonyl, sulfone, benzyl position, nitrile, etc.). Deprotonation is followed by addition of the carbanion to a carbonyl, then intramolecular attack of the tetrahedral interme-

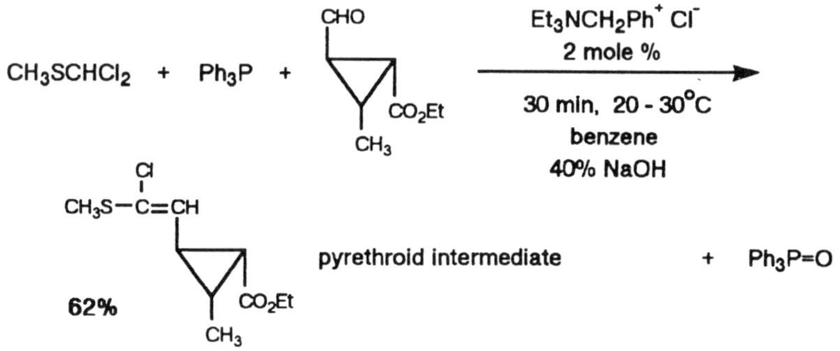

Figure 8-50. In situ Wittig.

diate O-anion to displace the chloride and form an oxirane (see Figure 8-51). Since PTC/OH conditions are suited for deprotonation of activated C−H bonds, the Darzens condensation is a natural candidate for PTC application.

Makosza reported the first example of PTC/OH Darzens condensation [275]. Chloroacetonitrile was reacted with cyclohexanone in the presence of 50% NaOH and benzyltriethylammonium chloride for 30 min at 15–20°C to yield 79% of the glycidic nitrile. Sulfonyloxiranes, 49, were prepared in the presence of triethylbenzylammonium and 50% NaOH by the deprotonation of α-chloro arylsulfones followed by addition of the carbanion to ketones (e.g., acetone and benzophenone) and aldehydes, and finally intramolecular displacement of the chlorine [276]. The ketones gave 90% yield and the aldehydes gave 60–65% yield (possibly due to competing aldol condensation). Similar yields were obtained by Colonna, with the addition of asymmetric induction using chiral phase-transfer catalysts containing the 2-hydroxyethyl group [277]. It was shown that when the chiral catalyst was bound to a polymer, it retained its catalytic efficiency even after three recycles. α-Chlorophenylacetonitrile was deprotonated and reacted with benzaldehyde in the presence of benzene, 50% NaOH, and triethylbenzylammonium chloride (see Figure 8-52) [278]. Stereochemical control was obtained, as the predominant product was *trans*-2,3-diphenylglycidonitrile (*trans/cis* 9:1).

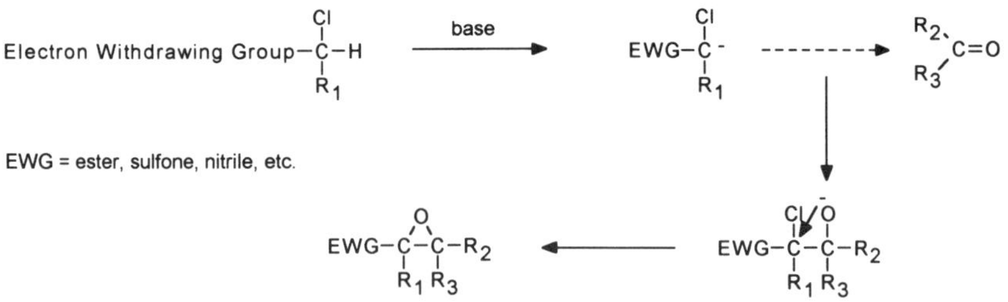

Figure 8-51. The Darzens condensation.

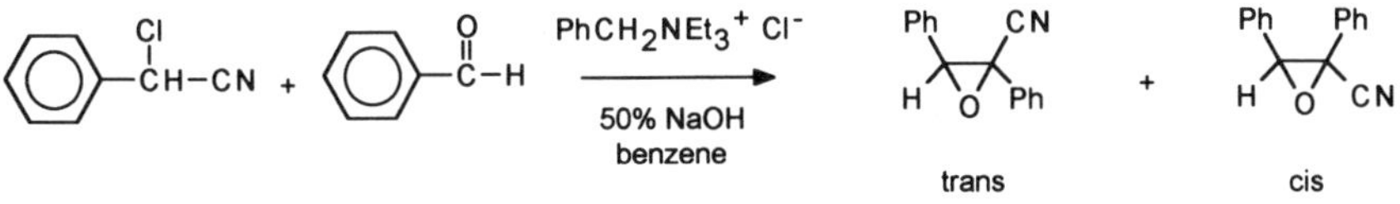

Figure 8-52. PTC reverses selectivity in the Darzens condensation.

In the absence of the quaternary ammonium catalyst, the *cis* isomer predominated (*cis/trans* 2:1).

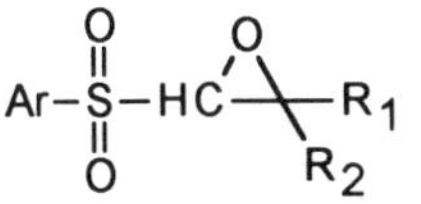

$$Ar-\overset{O}{\underset{O}{\overset{\|}{S}}}-HC\overset{O}{\diagdown}\overset{R_1}{\underset{R_2}{\diagup}} \quad 49$$

Other activated C—H/halo bearing carbon compounds that were successfully reacted with ketones/aldehydes in the Darzens condensation under PTC/OH conditions include: 9-chlorofluorene [279], ethyl chloroacetate [280], chloroacetonitrile [281, 282] (using polymer-bound PEG) [245], phenacyl halides [283, 284], and α-bromomalonic acid diethyl ester [285]. Polymer-bound PEG was also used in the Darzens condensation of benzophenone with chloroacetonitrile (61% yield; 5 mol% catalyst; 24 h; room temperature) [245].

5. Other Condensations

A multistep aldol-like condensation was performed [48] starting with the deprotonation of PAN-like Reissert compound (appropriately in the presence of triethylbenzylammonium chloride; see Chapter 6, Section G) and addition to aldehydes. The resulting tetrahedral intermediate then reacted, intramolecularly, with an adjacent amide to produce an ester. The negative charge remaining on the amide-source nitrogen was stabilized by the irreversible loss of cyanide (see Figure 8-53). A variety of aldehydes were reacted in 86–100% yield.

Another interesting reaction involved the addition of a carbanion (derived from pyrazolone) to carbon disulfide, followed by S-alkylation of the pendant dithiocarboxylate (see Figure 8-54) [286]. A similar reaction involving the reaction of alcohols with carbon disulfide and S-alkyltion in one pot was reported [287]. Triethylbenzylammonium chloride was chosen appropriately for the deprotonation of the activated ketone (see Chapter 6, Section G) and also worked

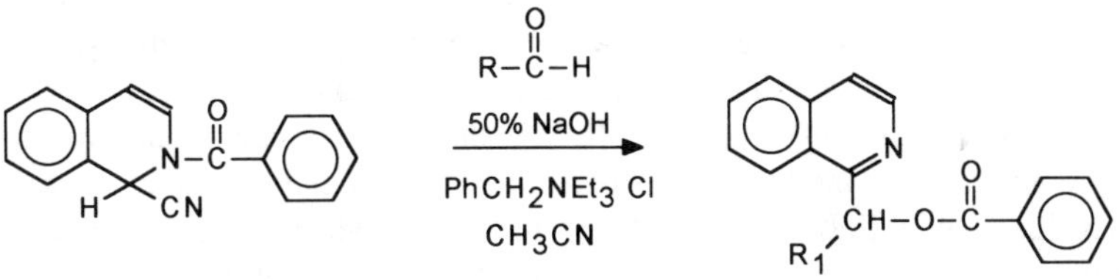

Figure 8-53. C-Condensation and intramolecular rearrangement.

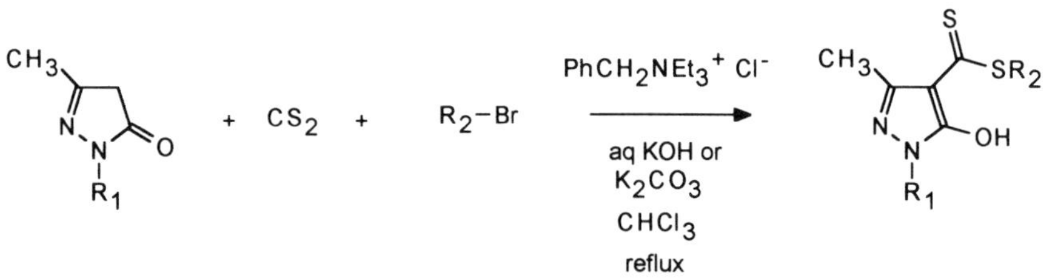

438 / Phase-Transfer Catalysis

Figure 8-54. Condensation with CS_2 and thioesterification.

satisfactorily for the thioesterification to prepare nine dithioesters in 52–85% yield. Solid–liquid as well as liquid–liquid PTC conditions were used. An open–chain precursor to chlorinated pyridines was prepared by condensation via the PTC/OH deprotonation of 1,2-dichloroacetone followed by addition to the double bond of acrylonitrile [288]. Cyanocarbonates were prepared by the PTC addition of cyanide to aldehydes in the presence of chlorofomates [289]. The O-anion of the tetrahedral complex, formed by the attack of the cyanide on the aldehyde, attacks the chlorofomate to produce the final product.

H. Deuterium Exchange, Isomerization, and Oxidation

The simplest reaction to perform on a deprotonated organic substrate is deuteration. Starks was the first to report hydrogen/deuterium exchange under PTC/OD conditions [290]. Using 5% $NaOD/D_2O$, complete equilibration of all of the α-protons of 2-octanone was achieved within 30 min at 30°C. After two H/D exchanges >99% deuteration was obtained. Another ketone deuterated was nonan-5-one [291]. Fluorene ($pK_a = 23$) [71] and indene ($pK_a = 21$) [71] were di- and trideuterated in 95% and 99%, respectively, in the presence of triethylbenzylammonium chloride, $NaOD/D_2O$, and methylene chloride [292]. 1-Methylindene was isomerized to 3-methylindene (more highly substituted double bond), under the same conditions (using NaOH, not NaOD). Later, the deuteration of fluorene was shown to be 80% complete within 8 min in the presence of triethylbenzylammonium chloride/bromide and benzene at 24°C [77].

Fluorene, and several other Ar-CH_2-Ar compounds (xanthene, acridane, dihydroanthracene, and others), have been deprotonated and oxidized to ketones under PTC/OH conditions and reaction with oxygen to yield benzophenones [293]. Fluorene was oxidized to fluorenone in 100% yield using dihexadecyldiethylammonium chloride in the presence of NaOH and benzene.

Allylbenzene, *50* ($pK_a = 34$) [294], was deprotonated and isomerized to the more conjugated α-methylstyrene (described in detail in Chapter 6, Section A.1.b) [295,296]. Subsequently, it was found that when the isomerization was performed in the presence of 18.3 M $NaOD/D_2O$ and tetrabutylammonium HSO_4^-, only one of the three terminal hydrogen atoms was exchanged, and none of the

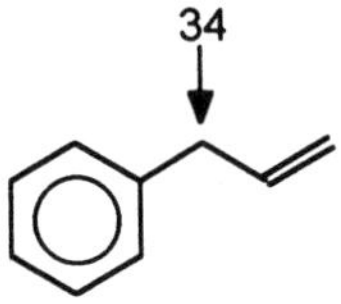

Structure	pKa	#
PhCH₂–CH=CH₂ (benzylic H)	34	50
thiophene (2-H)	38.4	51
benzofuran (2-H)	36.8	52
benzothiophene (2-H)	37	53
dibenzosuberane (CH₂)	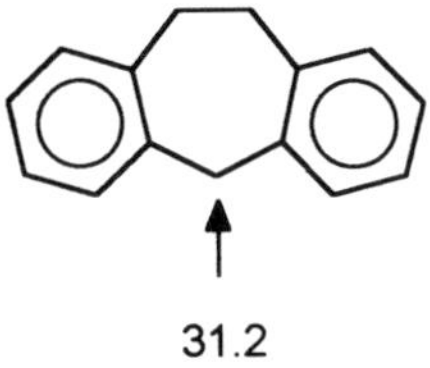 31.2	54
xanthene (9-H)	27	55
Ph₃C–H	31.5	56
Ph–CH₂–Ph	33.4	57

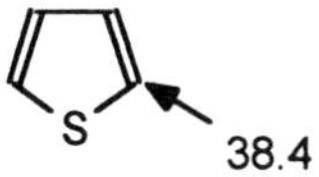

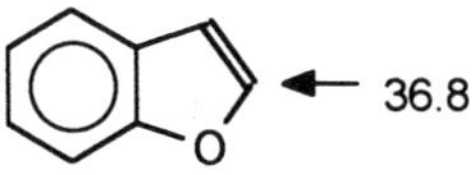

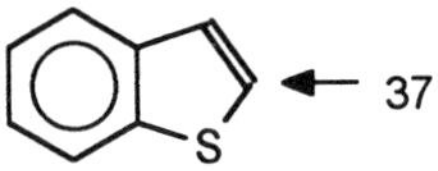

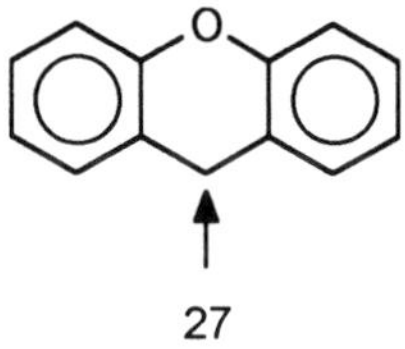

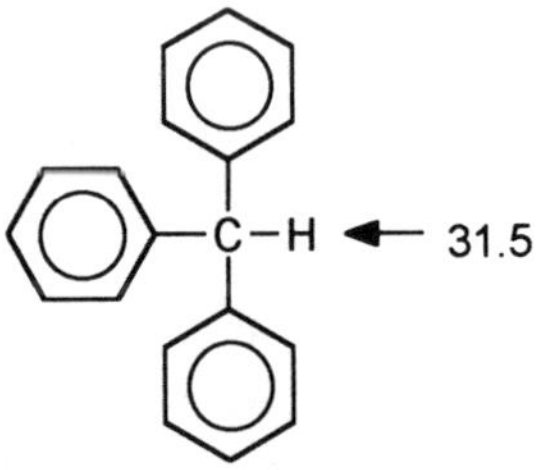

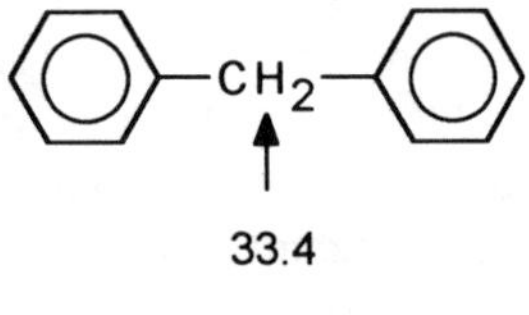

benzylic hydrogen atoms was exchanged [297]. It was concluded that isomerization was more rapid than deuteration. In addition, it was concluded that, whereas allylbenzene was acidic enough to be deprotonated by 18.3 M NaOD, α-methylstyrene was not, thereby defining the limit of acidity for deprotonation by 18.3 M NaOD. 4-Methoxyallylbenzene was isomerized in the presence of solid KOH and several PEG derivatives [298]. Since no nucleophilic reactions occur in this isomerization, the noncapped PEG (free hydroxyls at both ends) were the most effective due to the presence of alkoxidic end groups (in the presence of hydroxide the end groups are deprotonated and serve as the organic soluble base). Catalysts were most effective when seven or eight oxyethylene units per potassium ion were present (similar to crown). The thorough mechanistic study showed that the reaction was intrinsic chemical reaction rate controlled and the reaction was speculated to occur in a third phase (PEG/KOH).

During the alkylation of pyrazole with propargyl bromide, the resulting N-propargylpyrazole isomerized to the allene (using two equivalents of finely ground KOH in the presence of 5 mol% tetrabutylammonium bromide and no solvent at 0°C) [299]. Less isomerization was obtained when K_2CO_3 was used as the base.

Among the most extensive deuteration studies performed was the H/D exchange of 24 substituted thiazoles, pyridines, thiophenes, and imidazoles [300]. The effects of 11 catalysts, 17 solvents, kinetic orders, and temperature were also studied. This very comprehensive and systematic report was highly descriptive in nature, though firm conclusions, trends, or mechanisms were difficult to identify or substantiate. It was concluded that PTC/OD conditions are very effective for deuteration of a wide scope of compounds with activated hydrogen atoms. The least acidic compound deuterated was thiazole itself (pK_a = 29.5) [301].

To date, the least acidic compound deuterated was thiophene, 51, (91% H/D exchange on H-1; pK_a = 38.4) [302]. [Note: pK_a values reported below are cited with references in the respective articles described below. The arrows point to the C-H bonds in the structures, with those pK_a's.] Benzofuran, 52 (pK_a = 36.8), benzothiophene, 53 (pK_a = 37), and dihalobenzenes were among 22 very low acidity compounds reported to have been deuterated or oxidized. Organophilic quaternary ammonium cations and hydrophilic counteranions were crucial in obtaining high yield.

Other weakly acidic hydrocarbons deuterated in the presence of 18.3 M NaOD/D_2O and tetrabutylammonium HSO_4^- were 10,11-dibenzo[a,d]cycloheptadiene, 54 (pK_a = 31), xanthene, 55 (pK_a = 27), tetrafluorobenzene (pK_a = 31.5), and triphenylmethane, 56 (pK_a = 31.5) and many halo- and nitrotoluenes [297, 303]. Under these conditions, toluene (pK_a = 41) could not be deuterated. Diphenylmethane, 57 (pK_a = 33.4), was also completely deuterated in the presence of 18.3 M NaOD/D_2O and tetrabutylammonium HSO_4^- and oxidized in the presence of 50% NaOH, air, and tetrabutylammonium HSO_4^- [304]. Dihalodiphenylmethanes were oxidized to dihalobenzophenones (useful as a monomer for PEEK (polyetheretherketone) engineering thermoplastics) under PTC/OH conditions [305]. As expected (based on high pK_a of substituted diphenylmethanes; see Chapter 6, Section G), organophilic quaternary ammonium salts (e.g., tetraoctylammonium bromide) were more effective than accessible quaternary ammonium salts (tetrabutylammonium bromide). In addition, hydrogen sulfate was a better counteranion than bromide; solid NaOH was a more effective base than 50% NaOH; pseudo-first-order kinetics were observed; and chlorobenzene or toluene were suitable solvents for the oxidation. Diphenylmethane and picolines (pK_a = 29–33) were oxidized in the presence of PEG derivatives (capped PEG was best) and potassium *tert*-butoxide under solid–liquid PTC conditions [306]. The use of quaternary ammonium salts in the presence of butoxide gave low yield due to Hofmann elimination.

The deuteration of methylene chloride in 98.8% yield under PTC/OD conditions using Aliquat 336 was patented [307].

References

1. C. Starks, unpublished.
2. M. Makosza and B. Serafinowa, Rocz. Chem., **39**, 1223 (1965).
3. M. Makosza and B. Serafinowa, Rocz. Chem., **39**, 1401 (1965).
4. M. Makosza, Tetrahedron Lett., 4621 (1966).
5. A. Jonczyk, B. Serafin, and M. Makosza, Tetrahedron Lett., 1351 (1971).
6. The pKa of acetone is 20:R. Pearson and R. Dillon, J. Am. Chem. Soc., **75**, 2439 (1953); and the pK_a of acetophenone is 19: The Chemist's Companion Handbook of Practical Data, Techniques and References, A. Gordon and R. Ford, eds., Wiley, New York, 1972, p. 63; assuming the benzylic position lowers the pK_a by 2–3 units from acetone, the pK_a of phenylacetone is likely approximately 17.
7. A. Brandstrom and U. Junggren, Tetrahedron Lett., 473 (1972).
8. R. Pearson and R. Dillon, J. Am. Chem. Soc., **75**, 2439 (1953).
9. A. Brandstrom and U. Junggren, Acta Chem. Scand., **23**, 3585 (1969).
10. H. Molinari, F. Montanari, S. Quici, and P. Tundo, J. Am. Chem. Soc., **101**, 3920 (1979).
11. Y. Kimura, P. Kirszensztejn, and S. Regen, J. Org. Chem., **48**, 385 (1983).
12. M. Halpern, Y. Sasson, and M. Rabinovitz, Tetrahedron, **38**, 3183 (1982).
13. M. Halpern, Y. Sasson, I. Willner, and M. Rabinovitz, Tetrahedron Lett., 1719 (1981).
14. D. Mason, S. Magdasi, and Y. Sasson, J. Org. Chem., **55**, 2714 (1990).
15. M. Halpern, Ph.D. Thesis, Hebrew University of Jerusalem, Israel (1983) pp. 28–54.

16. A. Merz and R. Tomahogh, J. Chem. Res. S, **273**, M3070 (1977).
17. N. Kornblum, P. Berrigan, and W. LeNoble, J. Am. Chem. Soc., **85**, 1141 (1963).
18. A. Brandstrom and U. Junggren, Acta Chem. Scand., **25**, 1469 (1971).
19. A. Brandstrom and U. Junggren, Acta Chem. Scand., **23**, 3585 (1969).
20. A. Brandstrom and U. Junggren, Acta Chem. Scand., **23**, 2536 (1969).
21. F. Guibe, P. Sarthou, and G. Bram, Tetrahedron, **30**, 3139 (1974).
22. P. Tundo, P. Venturello, and E. Angeletti, J. Chem. Soc. Perkin Trans. I, 2159 (1987).
23. G. Szabo, K. Aranyosi, and C. Toke, Synthesis, 565 (1987).
24. The Chemist's Companion Handbook of Practical Data, Techniques and References, A. Gordon and R. Ford, eds., Wiley, New York, 1972, p. 63.
25. M. Hanna and M. Sprecker (International Flavors and Fragrances), U.S. Patent 4,582,634 (1986).
26. M. Halpern, Ph.D. Thesis, Hebrew University of Jerusalem, Israel, (1983) p. 100–101.
27. U. Dolling, P. Davis, and E. Grabowski, J. Am. Chem. Soc., **106**, 446 (1984).
28. D. Hughes, U. Dolling, K. Ryan, E. Schoenewaldt, and E. Grabowski, J. Org. Chem., **52**, 4745 (1987).
29. K. Saigo, H. Koda, and H. Nohira, Bull. Chem. Soc. Jpn., **52**, 3119 (1979).
30. R. Pearson and R. Dillon, J. Am. Chem. Soc., **75**, 2439 (1953).
31. T. Huang and S. Lin, J. Chin. I. Ch. E., **19**, 193 (1988).
32. J. Nowicki and J. Gora, Pol. J. Chem., **65**, 2267 (1991); CA 116:255415v.
33. J. Heiszman, I. Bitter, K. Harsanyi, and L. Toke, Synthesis, 738 (1987).
34. J. Canicio, A. Ginebreda, and R. Canellia, Anal. Quim. Ser. C., **81**, 181 (1985).
35. M. O'Donnell, W. Bennet, and S. Wu, J. Am. Chem. Soc., **111**, 2353 (1989).
36. M. O'Donnell, W. Bennett, W. Bruder, W. Jacobson, K. Knuth, B. LeClef, R. Polt, F. Bordwell, S. Mroczak, and T. Cripe, J. Am. Chem. Soc., **110**, 8520 (1988).
37. K. Lipkowitz, M. Cavanaugh, B. Baker, and M. O'Donnell, J. Org. Chem., **56**, 5181 (1991).
38. M. O'Donnell, J. Boniece, and S. Earp, Tetrahedron Lett., 2641 (1978).
39. M. O'Donnell and T. Eckrich, Tetrahedron Lett., 4625 (1978).
40. J. Yaozhong, Z. Changyou, W. Shengde, C. Daimo, M. Youan, and L. Guilan, Tetrahedron, **44**, 5343 (1988).
41. K. Fasth, G. Antoni, and B. Langstrom, J. Chem. Soc. Perkin Trans. I, 3081 (1988).
42. J. Genet, J. Uziel, M. Port, A. Touzin, S. Roland, S. Thorimbert, and S. Tanier, Tetrahedron Lett., 77 (1992).
43. Y. Wang, A. Mi, and Y. Jiang, Synth. Commun., **22**, 265 (1992).
44. B. Imperiali and S. Fisher, J. Org. Chem., **57**, 757 (1992).
45. R. Solaro, S. D'Antone, and E. Chiellini, J. Org. Chem., **45**, 4179 (1980).

46. L. Lindblom and M. Elander, Pharmaceutical Technology, **4**, 59 (1980).
47. W. Seitz, K. Scheib, and A. Michel (BASF), U.S. Patent 4,418,017 (1983).
48. J. Ezquerra and J. Alvarez-Builla, J. Heterocyclic Chem., **25**, 917 (1988).
49. A. Brandstrom and U. Junggren, Tetrahedron Lett., 473 (1972).
50. T. Balakrishnan and W. Ford, J. Org. Chem., **48**, 1029 (1983).
51. V. Ragaini, G. Colombo, P. Barzaghi, E. Chiellini, and S. D'Antone, Ind. Eng. Chem. Res., **27**, 1382 (1988).
52. E. Diaz-Barra, A. de la Hoz, A. Moreno, and P. Sanchez-Verdu, Synthesis, 391 (1989).
53. R. Jin and Y. Kurusu, J. Mol. Catal., **73**, 215 (1992).
54. M. Fedorynski, A. Dybowska, and A. Jonczyk, Synthesis, 549 (1988).
55. (Ciba-Geigy), GB Patent 1,570,319 (1980); CA 94: 65191y.
56. J. Wrobel and E. Hejchman, Synthesis, 452 (1987).
57. P. Magnus, Tetrahedron, **33**, 2019 (1977).
58. A. Jonczyk, K. Banko, and M. Makosza, J. Org. Chem., **40**, 266 (1975).
59. A. Jonczyk and I. Pytlewski, Synthesis, 883 (1978).
60. A. Jonczyk, K. Banko, and M. Makosza, J. Org. Chem., **40**, 266 (1975).
61. J. Golinski, A. Jonczyk, and M. Makosza, Synthesis, 461 (1979).
62. A. Jonczyk and T. Radwan-Pytlewski, J. Org. Chem. **48**, 910 (1983).
63. D. Scholz, Liebigs Ann. Chem., 98 (1983).
64. A. El-Khagawa, M. Ismail, A. Abdel-Wahab, Gazz. Chim. Ital., **112**, 235 (1982).
65. P. Ellingsen and K. Unheim, Acta Chem. Scand. B, **33**, 528 (1979).
66. K. Kurosawa, M. Suenaga, T. Inazu, and T. Yoshino, Tetrahedron Lett., 5335 (1982).
67. Z. Zhang, G. Liu, Y. Wang, and Y. Wang, Synth. Comm., **19**, 1167 (1989).
68. Y. Wang and Y. Jiang, Synth. Comm., **22**, 2287 (1992).
69. S. Ma, J. Minatelli, (Uniroyal), U.S. Patent 4,503,230 (1985).
70. R. Dessey, Y. Okozumi, and A. Chen., J. Am. Chem. Soc., **84**, 2899 (1962).
71. A. Streitwieser, E. Ciuffarin, and J. Hammons, J. Am. Chem. Soc., **89**, 63 (1967).
72. B. Berris (Ethyl Corporation), U.S. Patent 5,030,757 (1991).
73. O. Nefedov, L. Menchikov, A. D'yachenko, and S. Agre, Mendeleev Chemistry Journal (English Translation), **31**, 102 (English) (1986).
74. M. Makosza, Tetrahedron Lett., 4621 (1966).
75. O. Nefedov, L. Menchikov, A. D'yachenko, and S. Agre, Mendeleev Chemistry Journal (English Translation), **31**, 102 (English) (1986).
76. M. Makosza, Bull. Acad. Pol. Sci. Ser. Sci. Chem., **15**, 165 (1967).
77. M. Halpern, Y. Cohen, Y. Sasson, and M. Rabinovitz, New. J. Chem., **8**, 443 (1984).

78. M. Halpern, Ph.D. Thesis, Hebrew University of Jerusalem, Israel (1983) p. 110, 134.
79. M. Halpern, Ph.D. Thesis, Hebrew University of Jerusalem, Israel, (1983).
80. E. Dehmlow and C. Bollmann, Tetrahedron Lett., 5773 (1991).
81. A. Barco, S. Benetti, and G. Pollini, Synthesis, 124 (1976).
82. D. Landini and F. Rolla, Synthesis, 389 (1976).
83. E. Diez-Barra, A. de la Hoz., A. Sanchez-Migallon, and J. Tejeda, Synth. Comm., **20**, 2849 (1990).
84. A. Cuadro, M. Matia, J. Garcia, J. Vaquero, and J. Alvarez-Builla, Synth. Comm., **21**, 535 (1991).
85. R. Zhou, H. Zhang, and H. Ren, Xiamen Daxue Xuebao, Ziran Kexueban, **29**, 596 (1990); CA:116:151542t.
86. P. Mattingly and M. Miller, J. Org. Chem., **46**, 1557 (1981).
87. Y. Goldberg, E. Abele, and M. Shymanska, Synth. Comm., **23**, 557 (1991).
88. A. Jonczyk and M. Makosza, Rocz. Chem., **49**, 1203 (1975).
89. H. Nishi, H. Khono, and T. Kano, Bull. Chem. Soc. Jpn., **54**, 1897 (1981).
90. K. Sukata, Bull. Chem. Soc. Jpn., **56**, 280 (1983).
91. M. Fleming (Syntex), U.S. Patent 4,332,723 (1982).
92. H. Tappe (Cassella), U.S. Patent 4,342,688 (1982).
93. J. Barry, G. Bram, G. Decodts, A. Loupy, P. Pigeon, and J. Sansoulet, Tetrahedron Lett., 5407 (1982).
94. E. Diez-Barra, A. De la Hoz, A. Sanchez-Migallon, and J. Tejeda, Heterocycles, **34**, 1365 (1992).
95. S. Julia, P. Sala, J. Del Mazo, M. Sancho, C. Ochoa, J. Eleguero, J. Fayet, and M. Vertut, J. HeterocyclicChem., **19**, 1141 (1982).
96. T. Yamada and M. Onishi (Nisshin Shokuhin), Jpn Kokai Tokkyo Koho JP 04 18,090 (1992); CA 116:255631n.
97. I. Gozlan, D. Ladkani, M. Halpern, M. Rabinovitz, and D. Avnir, J. Heterocyclic Chem., **21**, 613 (1984).
98. I. Gozlan, M. Halpern, M. Rabinovitz, and D. Avnir, J. Heterocyclic Chem., **19**, 1569 (1982).
99. K. Suzuki, K. Yoshida, and H. Otaka (Kanebo), Jpn Kokai Tokkyo Koho JP 04,164,178 (1992); CA 117:212525j.
100. N. Ramzaeva, Y. Goldberg, E. Alksnis, and M. Lidaks, Synth. Comm., **19**, 1669 (1989).
101. G. Phillippossian (Nestle), U.S. Patent 4,450,163 (1984).
102. M. Hedayatullah, J. Heterocyclic Chem., **19**, 249 (1982).
103. M. Hedayatullah and A. Roger, J. Heterocyclic Chem., **26**, 1093 (1989).
104. F. Seela and H. Winkler, Angew. Chem., **93**, 105 (1981).

105. M. Hedayatullah, J. Heterocyclic Chem., **19**, 249 (1982).
106. T. Hoshiko, Y. Wanatabe, and S. Ozaki, Heterocycles, **20**, 2429 (1983).
107. H. Dou, P. Hassanaly, and J. Metzger, J. Heterocyclic Chem., **14**, 321 (1977).
108. N. Ayyanagar, A. Choudhary, U. Kalkote, and A. Natu, Synth. Comm., **18**, 2011 (1988).
109. R. Brehme, Synthesis, 113 (1976).
110. G. Hodgson (Burroughs Wellcome), U.S. Patent 4,366,158 (1982).
111. N. Ayyanagar, A. Choudhary, U. Kalkote, and A. Natu, Synth. Comm., **18**, 2011 (1988).
112. F. Fujita, N. Itaya, H. Kishida, and I Takenoto (Sumitomo), U.S. Patent 4,328,166 (1982).
113. The Chemist's Companion Handbook of Practical Data, Techniques and References, A. Gordon and R. Ford, eds., Wiley, New York, 1972, p. 62.
114. T. Gajda and A. Zwierzak, Synthesis, 1005 (1981).
115. The Chemist's Companion Handbook of Practical Data, Techniques and References, A. Gordon and R. Ford, eds., Wiley, New York, 1972, p. 61.
116. G. Isele, J. Martinez, and G. Schill, Synthesis, 455 (1981).
117. A. Lisi, G. Trout, and R. Kazlauskas, J. Chromatogr., **563**, 257 (1991).
118. D. Albanese, D. Landini, and M. Penso, J. Org. Chem., **57**, 1603 (1992).
119. D. Landini and M. Penso, J. Org. Chem., **56**, 420 (1991).
120. A. Jonczyk, Z. Ochal, and M. Makosza, Synthesis, 882 (1978).
121. S. Julia, A. Ginebreda, and J. Guixer, J. Chem. Soc. Chem. Comm., 742 (1978).
122. J. Tomesch, M. Prashad, and W. Houlihan (Sandoz), U.S. Patent 4,868,319 (1989).
123. K. Hackl and H. Falk (Chemie Linz), Ger. Offen. DE 4,028,040 (1992); CA 116:255625p.
124. U. Kalkote, A. Choudhary, and N. Ayyangar, Org. Prep. Proceed. Int., **24**, 83 (1992); CA 116:214104z.
125. Z. Vejdelek, V. Kmonicek, and J. Krelka, Czech Patent CS 272,922 (1991); CA 117:170958v.
126. U. Kalkote, A. Choudhary, A. Natu, and N. Ayyangar, Synth. Comm., **21**, 1129 (1991).
127. S. Appa Rao, A. Kumar, H. Ila, and H. Junjappa, Synthesis, 623 (1981).
128. V. Gomez-Parra, F. Sanchez, and T. Torres, Synthesis, 282 (1985).
129. D. Brunelle (General Electric), U.S. Patent 5,082,968 (1992).
130. L. Smith (Monsanto), U.S. Patent 4,962,212 (1990).
131. A. Merz, Angew. Chem. Int. Ed. Eng., **12**, 846 (1973).
132. H. Freedman and R. DuBois, Tetrahedron Lett., 3251 (1975).
133. T. Gibson, J. Org. Chem., **45**, 1095 (1980).
134. R. Bartsch, C. Cason, and B. Czech, J. Org. Chem., **54**, 857 (1989).

135. F. Lohse and C. Monnifer (Ciba-Geigy), U.S. Patent 4,433,179 (1984).
136. S. Rana, C. Piskorz, J. Barlow, and K. Matta, Carbohydr. Res., **83**, 170 (1980).
137. R. Nouguier, Tetrahedron Lett., 3505 (1982).
138. R. Hiraoka, H. Mokohyama, U. Saito, and S. Uhijima (Kohjin), U.S. Patent 4,547,571 (1985).
139. G. Mouzin, H. Cousse, J. Rieu, and A. Duflos, Synthesis, 117 (1983).
140. W. Glasser, W. Olivera, S. Kelley, and L. Nieh (Center for Innovative Technology), U.S. Patent 4,918,167 (1990).
141. T. Morimotot and M. Sekiya, Synthesis, 308 (1981).
142. G. Neef and A. Steinmeyer, Tetrahedron Lett., 5073 (1991).
143. H. Schubert and K. Baessler (Hoechst), U.S. Patent 4,454,355 (1984).
144. Y. Yuan, D. Gao, and Y. Jiang, Synth. Commun., **22**, 2117 (1992).
145. G. Folz, T. Papenfus, and H. Schubert (Hoechst), Eur. Patent Appl. EP 496,370 (1992).
146. J. Mindl, Czech Patent CS 270,368 (1991); CA 116: 151312t.
147. T. Hide, M. Ohshima, H. Sasaki, and T. Toda, Heterocycles, **33**, 507 (1992).
148. J. Hopken and M. Moller, Macromolecules, **25**, 1461 (1992).
149. A. McKillop, J. Fiaud, and R. Hug, Tetrahedron, **30**, 1379 (1974).
150. E. Dehmlow and J. Schmidt, Tetrahedron Lett., 95 (1976).
151. M. Wang and H. Yang, Ind. Eng. Chem. Res., **29**, 522 (1990).
152. O. Mahmat, M. Majdoub, F. Mechin, H. Sleiman, and H. Boileau, in Recent Advances in Mechanistic and Synthetic Aspects of Polymerization, M. Fontanille and A. Guyot, eds, Reidel, 1977, p. 191.
153. M. Asada, M. Matsuura, and K. Yonezawa (Kanegafuchi, U.S. Patent 4,426,511 (1984).
154. W. Reider (Isvolta), U.S. Patent 4,430,493 (1984).
155. D. Brunelle (General Electric), U.S. Patent 4,595,760 (1986).
156. D. Brunelle and D. Singleton, Tetrahedron Lett., 3383 (1984).
157. J. Carnahan (General Electric), U.S. Patent 4,310,554 (1982).
158. K. Yonezawa, M. Asada, and M. Matsuura (Kanegafuchi), U.S. Patent 4, 395,537 (1983).
159. K. Idel, Y. Serini, D. Freitag, and G. Fengler (Bayer), U.S. Patent 4,316,980 (1982).
160. D. Brunelle (General Electric), U.S. Patent 4,363,905 (1982).
161. G. Monnerat, G. Ham, J. Hairston, and M. Hatch (Dow Chemical), U.S. Patent 4,276,406 (1981).
162. R. Gallucci and R. Going, J. Org. Chem., **48**, 342 (1983).
163. E. Dahan and S. Biali, J. Org. Chem., **56**, 7269 (1991).
164. H. Jonsson, P. Werner, U. Gedde, and A. Hult, Macromolecules, **22**, 1683 (1989).

165. T. Shaffer, J. Bennet, and M. Denniston (General Electric), U.S. Patent 5,084,551 (1992).
166. N. Ayyangar, A. Choudhary, R. Uttam, and V. Sharma (Council of Scientific and Industrial Research India), Indian Patent IN 166,827 (1990); CA 116: 6798b.
167. F. Montanari, S. Quici, and P. Tundo, J. Org. Chem., **48**, 199 (1983).
168. O. Arrad and Y. Sasson, J. Org. Chem., **54**, 4993 (1989).
169. H. Freedman (Dow Chemical), U.S. Patent 3,969,360 (1976).
170. T. Adaway (Dow Chemical), U.S. Patent 4,701,531 (1987).
171. H. Freedman, S. McGregor, M. Yoshimine, and L. Kroposki (Dow Chemical), U.S. Patent 4,007,197 (1977).
172. Z. Cutie and M. Halpern (DowElanco), U.S. Patent 5,120,846 (1992).
173. S. Gatling (Dow Chemical), U.S. Patent 4,814,451 (1989).
174. J. Bristol, I. Gross, and R. Lovey, Synthesis, 971 (1981).
175. S. Slaoui, R. LeGoaller, J. Pierre, and J. Luche, Tetrahedron Lett., 1681 (1982).
176. J. Hradil and F. Svec, Polymer Bull., **10**, 14 (1983).
177. J. Hradil and F. Svec, Polymer Bull., **11**, 159 (1984).
178. A. Au and H. Freedman (Dow Chemical), U.S. Patent 4,173,693 (1979).
179. A. Herriott and D. Picker, Synthesis, 447 (1975).
180. D. Landini and F. Rolla, Synthesis, 565 (1974).
181. H. Wirth and H. Friedrich (Ciba-Geigy), Eur. Patent Appl. EP 166,695 (1986).
182. D. Landini, F. Montanari, and F. Rolla, J. Org. Chem., **48**, 604 (1983).
183. D. Brunelle, J. Org. Chem., **49**, 1309 (1984).
184. D. Brunelle (General Electric), U.S. Patent 4,410,422 (1983).
185. M. Halpern, unpublished results.
186. H. Janoski and M. Pulwer (Monsanto), U.S. Patent 5,071,992 (1991).
187. E. Saphores, A. Gleisner, J. Vega, and W. Mardones (Establecimientos Industriales Qumicos Oxiquim S.A.), U.S. Patent 5,011,965 (1991).
188. W. Hsien, Synthesis, 622 (1981).
189. C. Yuan, H. Feng, and Q. Lin, Synthesis, 48 (1989).
190. K. Majumdar, A. Khan, and S. Saha, Synth. Commun., **22**, 901 (1992).
191. J. Dockx, Synthesis, 441 (1973).
192. M. Halpern, Y. Sasson, and M. Rabinovitz, J. Org. Chem., **49**, 2011 (1984).
193. M. Halpern, H. Zahalka, Y. Sasson, and M. Rabinovitz, J. Org. Chem., **50**, 5088 (1985).
194. E. Dehmlow and M. Lissel, Tetrahedron, **37**, 1653 (1981).
195. D. Mason, S. Magdasi, and Y. Sasson, J. Org. Chem., **56**, 7229 (1991).
196. Y. Kimura and S., Regen, J. Org. Chem., **48**, 195 (1983).
197. A. Herriott and D. Picker, Tetrahedron Lett., 4521 (1976).

198. A. Gorgues and A. Le Coq, Tetrahedron Lett., 4723 (1976).
199. K. Mizuno, Y, Kimura, and Y. Otsuji, Synthesis, 688 (1979).
200. M. Halpern, Ph.D. Thesis, Hebrew University of Jerusalem, Israel (1983) p. 84–85, 131–132.
201. S. Regen, J. Org. Chem., **42**, 875 (1977).
202. D. Landini, S. Quici, and F. Rolla, Synthesis, 397 (1975).
203. S. Daren, D. Vofsi, and M. Asscher (Makhteshim Chemical Works), U.S. Patent 4,292,453 (1981).
204. M. Makosza and W. Lasek, Tetrahedron, **47**, 2843 (1991).
205. W. Bailey and L. Zhou, Tetrahedron Lett., 1539 (1991).
206. L. Maurin (E.I. DuPont de Nemours), U.S. Patent 4,418,232 (1983).
207. K. Kurginyan, Mendeleev Chemistry Journal (English Translation), **31**, 44 (English Translation) (1986).
208. K. Henneke, H. Diehl, and K. Wedemeyer (Bayer), U.S. Patent 4,594,467 (1986).
209. M. Hung (DuPont de Nemours), U.S. Patent 5,015,790 (1991).
210. E. Dehmlow and M. Lissel, Tetrahedron, **37**, 1653 (1981).
211. T. Kulinski and A. Jonczyk, Synthesis, 757 (1992).
212. J. Renga, A. Olivera, and M. Bosse (Henkel Research Corporation), U.S. Patent 5,093,536 (1992).
213. E. Dehmlow and M. Lissel, Tetrahedron Lett., 1783 (1976).
214. R. Le Goaller, S. Slaoui, J. Pierre, and J. Luche, Synch. Comm., **12**, 1163 (1982).
215. T. Saegusa, S. Kobayashi, A. Yamada, and S. Kashimura, Polymer J., **11**, 1 (1979).
216. S. Kondo, Y. Takeda, and K. Tsuda, Synthesis, 862 (1989).
217. C. Starks, J. Am. Chem. Soc., **93**, 195 (1971).
218. M. Makosza and W. Wawrzyniewicz, Tetrahedron Lett., 4659 (1969).
219. G. Joshi, N. Singh, and L. Pande, Tetrahedron Lett., 1461 (1972).
220. S. Kondo, Y. Takeda, and K. Tsuda, Synthesis, 862 (1989).
221. S. Julia and A. Ginebreda, Synthesis, 682 (1977).
222. E. Dehmlow and J. Schonefeld, Justus Leibigs Ann. Chem., **744**, 42 (1971).
223. M. Jie and C. Wong, Lipids, **27**, 59 (1992).
224. T. Hiyama, H. Sawada, M. Tsukanaka, and H. Nozaki, Tetrahedron Lett., 3013 (1975).
225. R. Ellison, J. Org. Chem., **45**, 2509 (1980).
226. F. De Angelis, A. Gambacorta, and R. Nicoletti, Synthesis, 798 (1976).
227. C. Perchonock, I. Lanyos, J. Finkelstein, and K. Holden, J. Org. Chem., **45**, 1950 (1980).
228. C. Jefford, V. de los Heros, and U. Burger, Tetrahedron Lett., 703 (1976).

229. P. Kwantes and G. Klumpp, Tetrahedron Lett., 707 (1976).
230. W. Kraus, G. Klein, H. Sadlo, and W. Rothenwohrer, Synthesis, 485 (1972).
231. M. Fedorynski, I. Gorzkowska, and M. Makosza, Synthesis, 120 (1977).
232. M. Makosza and M. Fedorynski, Synth. Comm., **3**, 305 (1973).
233. E. Dehmlow and J. Wilkenloh, J. Chem. Res. (S), 396 (1984).
234. E. Dehmlow, H. Raths, and J. Soufi, J. Chem. Res. (S), 334 (1988).
235. P. Weyerstahl, G. Blume, and C. Muller, Tetrahedron Lett., 3872 (1971).
236. P. Weyerstahl, R. Mathias, and G. Blume, Tetrahedron Lett., 611 (1973).
237. P. Balcerzak and A. Jonczyk, Synthesis, 857 (1991).
238. I. Tabushhi, Z. Yoshida, and N. Takahashi, J. Am. Chem. Soc., **93**, 1820 (1971).
239. R. Ikan, A. Markus, and Z. Goldschmidt, Israel J. Chem., **11**, 591 (1973).
240. G. Shenoy and D. Rangnekar, Dyes and Pigments, **10**, 165 (1989).
241. I. Tabushi, K. Shimokawa, D. Naito, S. Misaki, and T. Yoshida (Daikin Kogyo), Eur. Pat. Appl. EP-54274 (1982).
242. K. Langrey, J. Org. Chem., **56**, 2400 (1991).
243. T. Ghosh, J. Org. Chem., **55**, 1146 (1990).
244. D. Reddy, B. Seenaiah, V. Padmavathi, and T. Seshamma, Phosphorous, Sulfur Silicon Ralet. Elem., **69**, 31 (1992); CA 117:191644e.
245. S. Yanagida, K. Takahashi, and M. Okahara, J. Org. Chem., **44**, 1099 (1979).
246. E. Dehmlow, Tetrahedron Lett., 91 (1976).
247. L. Scott and J. Naples, Synthesis, 209 (1973).
248. R. Kostikov, Mendeleev Chemistry Journal (English Translation), **31**, 95 (English Translation) (1986).
249. D. Landini, F. Montanari, and F. Rolla, Synthesis, 26 (1979).
250. M. Makosza, Tetrahedron Lett., 5489 (1966).
251. A. Bhattacharya, U. Dolling, E. Grabowski, S. Karady, K. Ryan, and L. Weinstock, Angew. Chem. Int. Ed. Eng., **25**, 476 (1986).
252. R. Conn, A. Lovell, S. Karady, and Weinstock, J. Org. Chem., **51**, 4710 (1986).
253. R. Annunziata, M. Cinquini, and S. Colonna, Chem. Ind., 238 (1980).
254. S. Colonna, A. Re, and H. Wynberg, J. Chem. Soc. Perkin Trans. 1, 547 (1981).
255. E. Delee, I. Jullien, L. Le Garrec, A. Loupy, J. Sansoulet, and A. Zaparucha, J. Chromatogr., **450**, 183 (1988).
256. V. Dryanska and D. Tasheva, Synth. Comm., **22**, 63 (1992).
257. R. Lawrence and P. Perlmutter, Chem. Lett., 305 (1992).
258. H. Hiemstra and H. Wynberg, J. Am. Chem. Soc., **103**, 417 (1981).
259. S. Kumar and S. Chimni, J. Chem. Soc. Perkin Trans. I, 449 (1992).
260. G. Cardillo, D. Savoia, and A. Umani-Ronchi, Synthesis, 453 (1975).
261. N. Cortese and W. Gastrock (American Cyanamid) U.S. Patent 4,521,629 (1985).

262. C. Gasparski and M. Miller, Tetrahedron, **47**, 5367 (1991).
263. M. Koch and A. Magni (Gruppo Lepetit), U.S. Patent 4,501,919 (1985).
264. G. Stahly (Ethyl), U.S. Patent 4,999,429 (1991).
265. W. Tagaki, I. Inoue, Y. Yano, and T. Okonogi, Tetrahedron Lett., 2587 (1974).
266. E. Dehmlow and S. Barahona-Naranjo, J. Chem. Res. Synop, 142 (1981).
267. W. Huang, M. Ding, W. Xiao, and T. Wu, Chin Chem. Lett., 3, 411 (1992); CA 117:211921y.
268. A. Minsky and M. Rabinovitz, Synthesis, 497 (1983).
269. D. Nicolaides, S. Adampoulus, E. Hatzigrigoriou, and K. Litinas, J. Chem. Soc. Perkin Trans. I, 3159 (1991).
270. D. Billeret, D. Blondeau, and H. Sliwa, Tetrahedron Lett., 627 (1991).
271. R. Galli, L. Scaglioni, O. Palla, and F. Gozzo, Tetrahedron, **40**, 1523 (1984).
272. CA 112: 56083c; W. Huang, Y. Deng, and M. Fan (1988).
273. CA 114: 23571y; Y. Deng, H. Li, and H. Xu, Chin. Sci. Bull., **34**, 203 (1989).
274. S. Clarke, C. Harrison, and P. Hodge, Tetrahedron Lett., 1375 (1980).
275. A. Jonczyk, M. Fedorynski, and M. Makosza, Tetrahedron Lett., 2395 (1972).
276. A. Jonczyk, K. Banko, and M. Makosza, J. Org. Chem., **40**, 266 (1975).
277. S. Colonna, R. Fornasier, and U. Pfeiffer, J. Chem. Soc. Perkin Trans. I, 8 (1978).
278. A. Jonczyk, A. Kwast, and M. Makosza, J. Chem. Soc. Chem. Comm., 902 (1977).
279. CA 115 (11): 114276t; W. Tao, Y. Guo, and L. Xing, Chin. Chem. Lett., **2**, 359 (1991).
280. CA 112 (1): 7280f; R. Bansal and V. Sharma, Chim. Acta Turc., **16**, 227 (1988).
281. J. Roser and W. Eberbach, Synth. Comm., **16**, 983 (1986).
282. S. Akabori, M. Ohtomi, and S. Tatabe, Bull. Chem. Soc. Jpn., **53**, 1463 (1980).
283. R. Annunziata, Synth. Comm., **9**, 171 (1979).
284. J. Hummelen and H. Wynberg, Tetrahedron Lett., 1089 (1978).
285. J. McIntosh and H. Khalil, Can J. Chem., **56**, 2134 (1978).
286. A. Oliva, I. Castro, C. Castillo, and G. Leon, Synthesis, 481 (1991).
287. A. Lee, W. Chan, H. Wong, and C. Wong, Synth. Commun., **19**, 547 (1989).
288. M. Halpern, J. Orvik, T. Dietsche, and B. Barron (Dow Chemical), U.S. Patent 5,084,576 (1992).
289. A. Au, Synth. Commun., **14**, 743 (1984).
290. C. Starks, J. Am. Chem., Soc., **93**, 195 (1971).
291. E. Dehmlow and S. Baharona-Naranjo, J. Chem. Res. (S), 186 (1982).
292. I. Willner, M. Halpern, and M. Rabinovitz, J. Chem. Soc. Chem. Comm., 155 (1978).
293. E. Alneri, G. Bottaccio, and V. Carletti, Tetrahedron Lett., 2117 (1977).

294. K. Bowden and R. Cook, J. Chem. Soc. Perkin Trans. II, 1407 (1972).
295. M. Halpern, M. Yonowich-Weiss, Y. Sasson, and M. Rabinovitz, Tetrahedron Lett., 703 (1981).
296. M. Halpern, Y. Sasson, and M. Rabinovitz, J. Org. Chem., **48**, 1022 (1983).
297. M. Halpern, D. Feldman, Y. Sasson, and M. Rabinovitz, Angew. Chem. Int. Ed. Eng., **23**, 54 (1984).
298. R. Neumann and Y. Sasson, J. Org. Chem., **49**, 3448 (1984).
299. E. Diez-Barra, A. de la Hoz., A. Sanchez-Migallon, and J. Tejeda, Synth. Comm., **20**, 2849 (1990).
300. W. Spillane, P. Kavanaugh, F. Young, H. Dou, and J. Metzger, J. Chem. Soc. Perkin Trans. I. 1763 (1981).
301. A. Streitwieser and P. Scannon, J. Am. Chem. Soc., **95**, 6273 (1973).
302. D. Feldman and M. Rabinovitz, J. Org. Chem., **53**, 3779 (1988).
303. D. Feldman, M. Halpern, and M. Rabinovitz, J. Org. Chem., **50**, 1746 (1985).
304. M. Halpern, Ph.D. Thesis, Hebrew University of Jerusalem, Israel (1983) pp. 106–109, 132–133.
305. M. Halpern and Z. Lysenko, J. Org. Chem., **54**, 1201 (1989).
306. R. Neumann and Y. Sasson, J. Org. Chem., **49**, 1282 (1984).
307. C. Myers, R. Shan-Yu-King, Eur. Pat. Appl. EP 246,805 (1987).

9

Phase-Transfer Catalysis: Polymerization and Polymer Modification

A. Introduction

Early in the development of "small molecule" synthesis by phase-transfer catalytic techniques, the principles of PTC were applied to various aspects of polymer research. Indeed, during the last 15 years an enormous amount of work has been reported concerning the application of PTC techniques to polymer synthesis and to the chemical modification of polymers and polymer precursors. It appears that every new issue of the journals contains some novel and interesting application of these techniques to the broad areas of polymer research. As a consequence, no attempt is made to carry out an exhaustive survey of the published literature. The purpose of this chapter is to give the reader a broad view of the kinds of applications for which PTC has been used with enough details so as not to miss any of the subtleties. The first section deals primarily with *polymer synthesis*, including condensation polymerization, anionic polymerization, and free radical initiated polymerization. The second section addresses the *chemical modification of polymers*, including reactions at the terminal positions of a polymer chain, reactions on the backbone of the polymer molecule, and reactions on pendant groups attached to the polymer backbone.

B. Polymer Synthesis

1. Condensation Polymerization

There are several important characteristics of condensation polymerizations carried out under PTC conditions [1]. In nucleophilic displacement step-growth polymerization (1) reaction is rapid, reaching 100% conversion and high molecular weights; (2) the molecular weight of the polymer does not depend on the

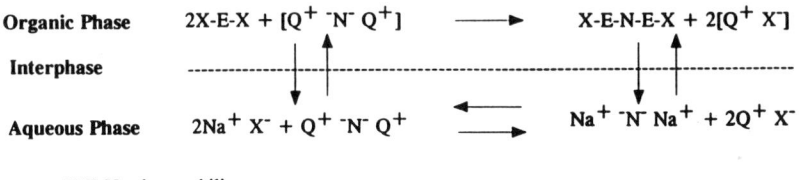

Figure 9-1

ratio of nucleophilic and electrophilic reactants as is the case in conventional polymerizations; and (3) the polymer almost always contains electrophilic functionalities as chain ends, independent of reaction yield and reactant ratio.

The general mechanism for PTC step-growth polymerization is presented in Figure 9-1, where N^{-2} is the bis-nucleophilic anion, X-E-X is the bis-electrophile, and Q^+ is the phase-transfer cation. According to the above mechanism, the bis-nucleophilic anion located in the aqueous phase ion pairs with two PTC cations and is transferred into the organic phase, where it reacts with the bis-electrophile. As in the case with "simple" PTC reactions, the bis-nucleophilic anion must be in a reactive form in the organic phase or the subsequent reactions will not take place. It must be emphasized that the concentration of the transferred bis-nucleophilic anion is dependent on the concentration of the phase-transfer catalyst. Since this is usually very small, the concentration of bis-electrophile is very high compared to that of the bis-nucleophile. These reaction conditions, as well as the high rates of the organic-phase reaction, provide the mechanistic origins of the observation that the chain ends always contain electrophilic functionalities. This is clearly illustrated in the above mechanistic scheme. There are, however, certain exceptions to the generalizations listed above. When the concentration of phase transfer catalyst is greater than approximately 10%, the concentration of the bis-nucleophilic anion is high in the organic phase and the resulting polymer may have both electrophilic and nucleophilic terminal functionalities. When the polymer that is formed becomes insoluble in the organic phase, the electrophilic terminal functionalities can hydrolyze or react with the bis-nucleophile to again produce polymers with both electrophilic and nucleophilic end groups.

As an aside, it should be noted that the mechanism shown above for PTC condensation polymerization shows a striking resemblance to the mechanism describing *interfacial polymerization* processes. Indeed, aromatic polyesters and polysulfonates prepared by interfacial polycondensation in the presence of -onium

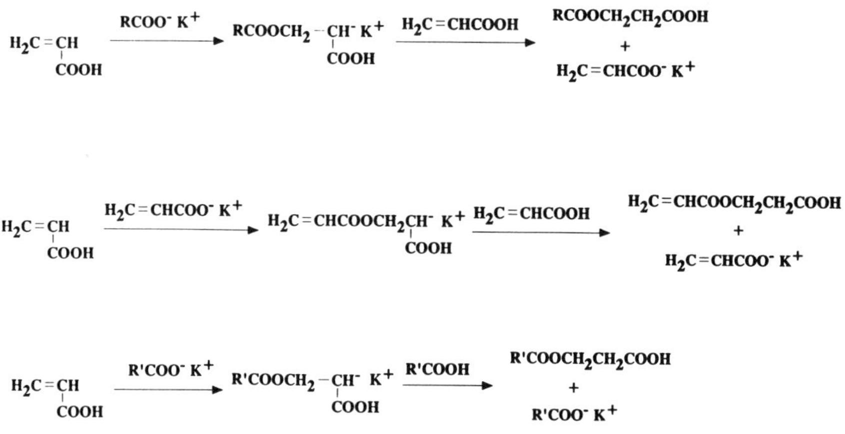

$$H_2C=CH-COOH \xrightarrow{RCOO^-K^+} RCOOCH_2-\underset{COOH}{CH^-}K^+ \xrightarrow{H_2C=CHCOOH} \begin{array}{c} RCOOCH_2CH_2COOH \\ + \\ H_2C=CHCOO^-K^+ \end{array}$$

$$H_2C=CH-COOH \xrightarrow{H_2C=CHCOO^-K^+} H_2C=CHCOOCH_2-\underset{COOH}{CH^-}K^+ \xrightarrow{H_2C=CHCOOH} \begin{array}{c} H_2C=CHCOOCH_2CH_2COOH \\ + \\ H_2C=CHCOO^-K^+ \end{array}$$

$$H_2C=CH-COOH \xrightarrow{R'COO^-K^+} R'COOCH_2-\underset{COOH}{CH^-}K^+ \xrightarrow{R'COOH} \begin{array}{c} R'COOCH_2CH_2COOH \\ + \\ R'COO^-K^+ \end{array}$$

Figure 9-2

salt accelerators [2] may be found in the literature years before the formal mechanisms for PTC were described.

a. Polyesters, Polysulfonates, Polyphosphonates, Polysulfones, Polythioesters, and Polyamides

The kinetics associated with the interfacial polycondensation of 2,2-bis(4-hydroxyphenyl)propane with isophthaloyl chloride and terephthaloyl chloride in dichloromethane with triethylbenzylammonium chloride as catalyst has been reported [3]. The reaction system employed was not an emulsion but a well defined two-phase system. As a consequence, the concentration of the bis-phenolate in the aqueous phase could be followed spectrophotometrically. The conversion was found to increase rapidly but the concurrent molecular weight increase was slow. For instance, at 95% conversion the polymer was soluble in methylene chloride, indicating that the molecular weight was low. At the final stages of the reaction, however, the molecular weight increased rapidly. It was observed that the mole ratio of 2,2-bis(4-hydroxyphenyl)propane to diacid chloride was critical to the molecular weight of the polymer. With mole ratios less than one, the molecular weight remained low. With mole ratios greater than one, very high molecular weights could be obtained.

Acrylic acid has been polymerized using sodium and potassium carboxylates in the presence and absence of 18-crown-6 to produce polyester [4]. The suggested mechanism is illustrated in Figure 9-2. In order to prevent competitive radical polymerization, *p*-methoxyphenol was added to the reaction mixture. In the presence of the crown ether the yield of polyester and the degree of polymerization were greater than in the absence of the crown over the same time period. In agreement with the above reaction mechanism, the resulting polymer appeared

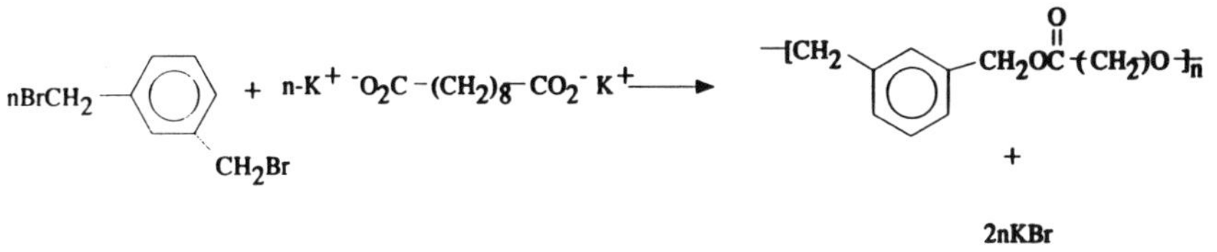

Figure 9-3

to have the following polyester structure: $R'-(CH_2CH_2COO)_n-H$ in which $R' = CH_2=CHCOO-$ or CH_3COO-. The polymerization, induced by the potassium acetate, was believed to be a stepwise reaction, and crown ether served to increase the yield and the number of monomer units in the polymer. In this study, sodium acetate and potassium and sodium benzoate were found to be as active as potassium acetate. The effect of the crown ether was found to be most important for potassium acetate. Attempts at polymerization of methacrylic acid under the same conditions failed [4].

Xylylene dihalides have been employed in PTC polymerization processes to form polyesters. In particular, the dipotassium salt of sebacic acid has been reacted with m-xylylene dibromide at 80°C for periods of approximately 4 days in the presence of hexadecyltriphenylphosphonium bromide, Aliquat 336, cryptand[2.2.2], and 18-crown-6 to produce the corresponding polyester (Figure 9-3) [5]. It was reported that the solvent system employed was critical to the success of the process. Acetonitrile and acetonitrile–benzene (1:1 by vol) were the only solvents that yielded polymers in these solid–liquid reaction systems. At temperatures of 60°C only low molecular weight materials were formed. Reaction of m-xylylene dibromide with potassium carbonate to form the corresponding polycarbonate was also reported. The reaction was conducted in an acetonitrile–benzene solvent system using 18-crown-6 as the catalyst. m-Xylylene dibromide was found to be more reactive than the para-isomer.

Aromatic polysulfonates with high molecular weights have been synthesized from aromatic disulfonyl chlorides and bisphenols by interfacial polycondensation using quaternary ammonium salts as accelerators [2,6–8]. The polycondensation reaction between 4,4'-oxydiphenylsulfonyl chloride with 2,2-bis(4-hydroxyphenyl)propane leading to the polysulfonate was carried out in an organic solvent–aqueous alkaline two-phase reaction system at room temperature in the presence of a variety of quaternary ammonium salts and crown ethers (Figure 9-4) [9,10] Tetra-n-butylammonium chloride, benzyltriethylammonium chloride, cetyltrimethylammonium chloride, 15-crown-5, 18-crown-6, dibenzo-18-crown-6, dicyclohexyl-18-crown-6, dibenzo-24-crown-8, dicylclohenyl-24-crown-8, and a variety of polyethylene glycols (PEGs) were among the catalysts employed. In the absence of catalyst, only low molecular weight polymer was produced even after long periods of time (24 h). In the presence of *all* the catalysts, high

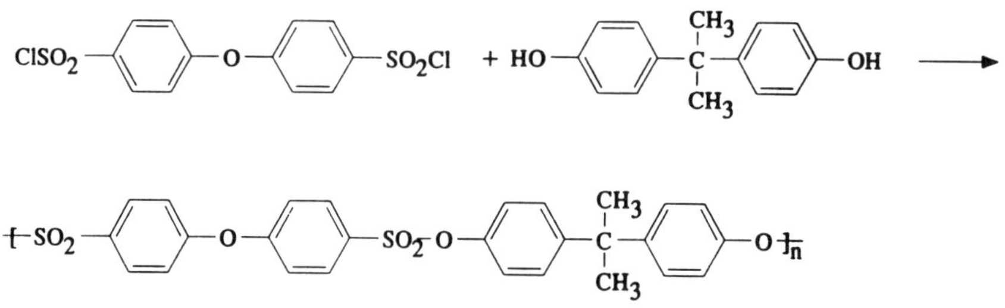

Figure 9-4

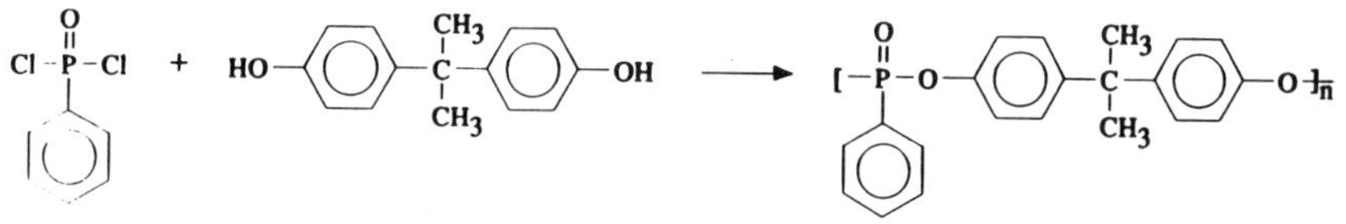

Figure 9-5

molecular weights of polymer were achieved after relatively short periods of time (3 h). Among these catalysts, however, tetra-*n*-butylammonium chloride and dicyclohexo-18-crown-6 were found to be highly efficient, leading to polysulfonates with high inherent viscosities. In the crown ether catalyzed reaction systems, potassium hydroxide was generally more effective than sodium hydroxide for preparing polymers of high molecular weight. Similar studies were reported in the preparation of polyphosphonates (Figure 9-5) [11].

The disodium salt of 4,4′-oxydibenzenesulfinate has been reacted with bis(chloromethyl) ketone in an organic solvent–aqueous base reaction system in the presence of a variety of quaternary ammonium salts and 18-crown-6 (Figure 9-6) [12]. Characterization of the resulting polymer revealed the presence of both sulfone and sulfinate linkages. The polymer with the highest molecular weight was obtained at 80°C after 48 h in nitrobenzene solvent in the presence of tetra-*n*-butylammonium bromide.

Aromatic polyamide-esters of moderately high molecular weight have been prepared from aminophenols [*m*- and *p*-aminophenol and 4-(4′-aminophenoxy)-phenol] and isophthaloyl and terephthaloyl chlorides in methylene chloride–water systems in the presence of tetra-*n*-butylammonium chloride [13].

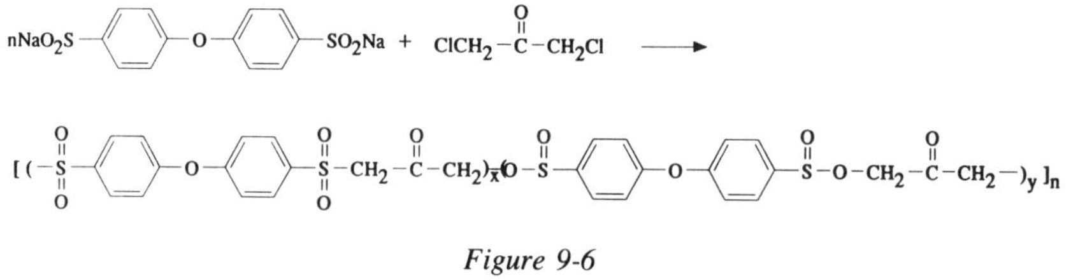

Figure 9-6

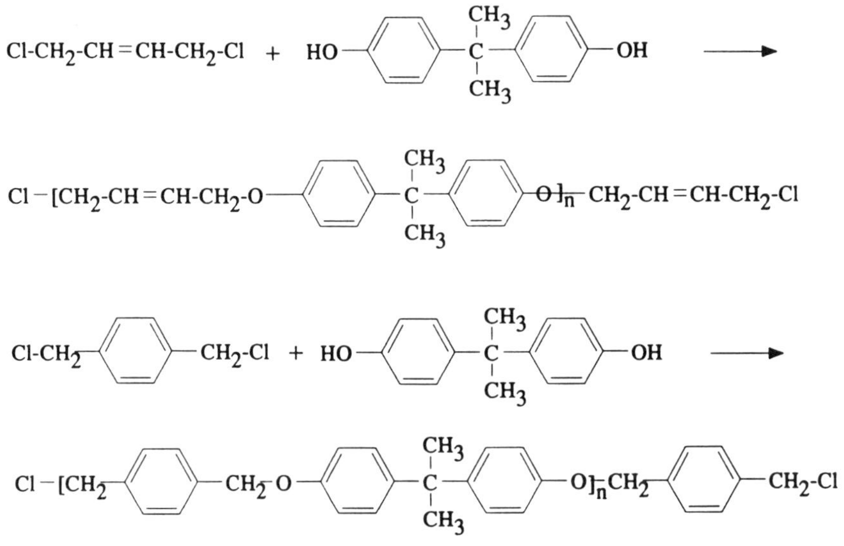

Cl-CH$_2$-CH=CH-CH$_2$-Cl + HO—⟨C$_6$H$_4$⟩—C(CH$_3$)$_2$—⟨C$_6$H$_4$⟩—OH ⟶

Cl—[CH$_2$-CH=CH-CH$_2$-O—⟨C$_6$H$_4$⟩—C(CH$_3$)$_2$—⟨C$_6$H$_4$⟩—O]$_n$—CH$_2$-CH=CH-CH$_2$-Cl

Cl-CH$_2$—⟨C$_6$H$_4$⟩—CH$_2$-Cl + HO—⟨C$_6$H$_4$⟩—C(CH$_3$)$_2$—⟨C$_6$H$_4$⟩—OH ⟶

Cl—[CH$_2$—⟨C$_6$H$_4$⟩—CH$_2$-O—⟨C$_6$H$_4$⟩—C(CH$_3$)$_2$—⟨C$_6$H$_4$⟩—O]$_n$CH$_2$—⟨C$_6$H$_4$⟩—CH$_2$-Cl

Figure 9-7

A series of polythioesters were prepared from bis(4-mercaptophenyl)ether and oxalyl, succinyl, adipoyl, suberoyl, and sebacoyl chlorides in an organic solvent–water two-phase reaction system. It was reported that the presence of benzyltriethylammonium chloride had unfavorable effects on the yield and reduced viscosity of the resulting polythioesters [14].

b. Polyethers and Polythioethers

There are many examples of polyethers prepared from the reaction of dialkaline metal salts of 2,2-bis(4-hydroxyphenyl)propane or a derivative of this bisphenol with a dihalide under homogeneous conditions in polar solvents at high temperatures. The use of PTC techniques to carry out such processes has also been reported. 2,2-Bis(4-hydroxyphenyl)propane was reacted with 1,4-dichloro-2-butene or 1,4-bis(chloromethyl)benzene in a toluene–aqueous sodium hydroxide two-phase system in the presence of tetra-n-butylammonium hydrogen sulfate, [2.2.2]-cryptand, and dicyclohexyl-18-crown-6 (Figure 9-7) [15]. It was observed that phenol end groups could not be detected in the polymer. In addition, stoichiometric quantities of 2,2-bis(4-hydroxyphenyl)propane and dihalide were not necessary in order to achieve high molecular weights—a distinct advantage of the PTC polycondensation technique.

The PTC polymerization between 2,2-bis(4-hydroxyphenyl)propane and epichlorohydrin in aqueous alkali–dioxane solvent systems at reflux temperatures of approximately 86°C in the presence of quaternary ammonium and phosphonium

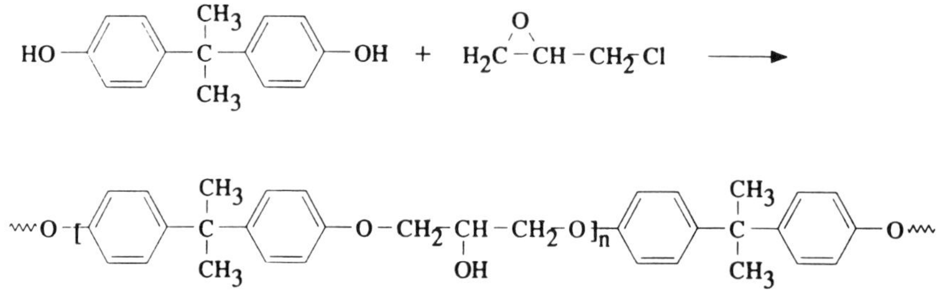

$$HO-\underset{}{\bigcirc}-\underset{CH_3}{\overset{CH_3}{\underset{|}{C}}}-\underset{}{\bigcirc}-OH \; + \; H_2\overset{O}{\overset{\diagup \diagdown}{C}}-CH-CH_2Cl \longrightarrow$$

$$\sim\sim O-[\underset{}{\bigcirc}-\underset{CH_3}{\overset{CH_3}{\underset{|}{C}}}-\underset{}{\bigcirc}-O-CH_2-\underset{OH}{\overset{}{\underset{|}{CH}}}-CH_2-O]_n-\underset{}{\bigcirc}-\underset{CH_3}{\overset{CH_3}{\underset{|}{C}}}-\underset{}{\bigcirc}-O\sim\sim$$

Figure 9-8

salts, dicyclohexyl-18-crown-6, and PEGs (mol wt 300–14,000) has been reported (Figure 9-8) [16]. Moderate molecular weight linear polyhydroxy ethers were produced. Benzyltriethylammonium chloride was found to be the most active of the quaternary salts. It is interesting to note that tetraethylammonium bromide showed moderate activity. In contrast, the catalytic activity of dicyclohexyl-18-crown-6 was only modest. In general, potassium hydroxide was a more effective base than sodium hydroxide. Increasing the concentration of tetraethylammonium chloride by a factor of 10 did not affect the intrinsic viscosity of the resulting polymer. Stoichiometric reactant balance seemed to be necessary but the effect was not quite as critical as is usually observed for other systems. As long as the polymer remained soluble or at least well dispersed, solvent composition (water–dioxane) was not observed to affect the intrinsic viscosity of the resulting polymer. PEGs were found to be effective phase-transfer catalysts in this heterogeneous reaction system.

Xylylene dihalides have been employed in PTC polymerization processes to form polyethers [17]. The PTC polymerization of α,α'-dichloro-*p*-xylene with 2,2-bis(4-hydroxyphenyl)propane was conducted in the presence of benzyltriethylammonium chloride in a two-phase system consisting of aqueous sodium or potassium hydroxide and an organic phase such as benzene, tetrahydrofuran, or *N*-methyl-2-pyrrolidone [18]. It was reported that addition of dimethyl sulfoxide to the system enhanced the reaction and decreased the side reactions (hydrolysis of the benzylic chlorides). The alkaline concentration greatly affected both yield and inherent viscosity of the resulting polymer. Concentration of sodium or potassium hydroxide below 30% (by wt) produced only low molecular weight polymer. Concentrations of aqueous base greater than 40% (by wt) were necessary for achieving high molecular weights in quantitative yield. Reaction temperature directly affected the inherent viscosity of the polymer. The effect of dimethyl sulfoxide is particularly interesting. Depending on the concentration of aqueous base, the dimethyl sulfoxide is either primarily located in the organic phase, the aqueous phase, or as a separate third phase. This third phase usually appeared at an intermediate concentration of base.

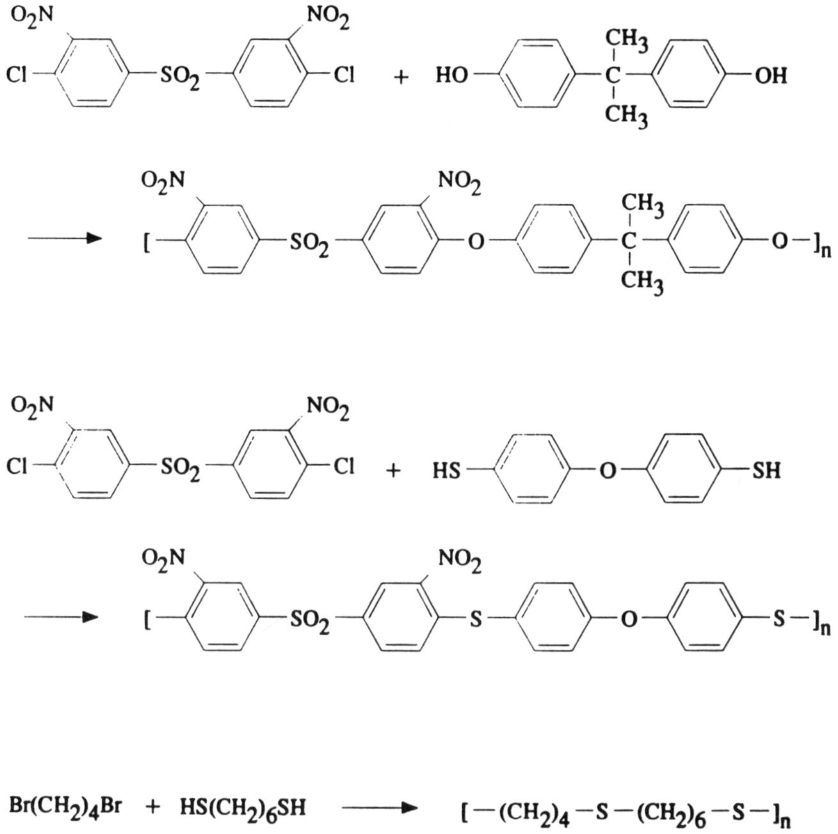

Figure 9-9

Aromatic polyethers and polythioethers and aliphatic polythioethers have been prepared in an organic solvent–aqueous base reaction system in the presence of a variety of quaternary ammonium salts and crown ethers (Figure 9-9) [11]. The aromatic polyethers and polythioethers were synthesized by an aromatic nucleophilic substitution pathway.

PTC Williamson ether synthesis has been applied to the synthesis of alternating block copolymers and regular copolymers. It has already been mentioned that the polycondensation of 2,2-bis(4-hydroxyphenyl)propane with *cis*- and *trans*-1,4-dichloro-2-butene in a toluene–aqueous base two-phase system in the presence of tetra-*n*-butylammonium hydrogen sulfate has been reported to produce unsaturated polyethers containing allylic chloride terminal functionalities. Condensation of this polymer with polyethersulfones containing terminal phenoxide groups in the presence of tetra-*n*-butylammonium hydrogen sulfate gave alternating block copolymers (Figure 9-10) [1].

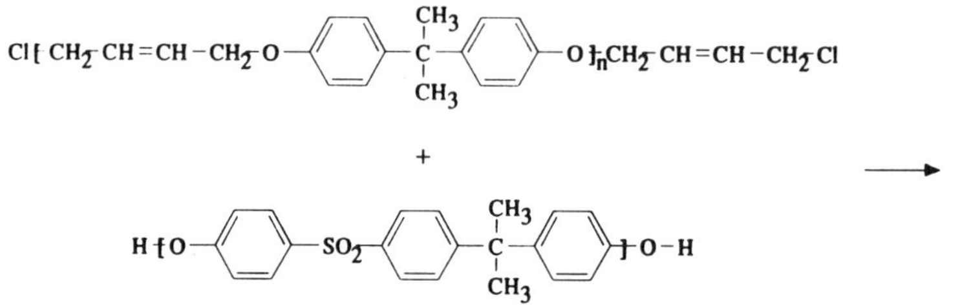

460 / Phase-Transfer Catalysis

$$Cl\text{-}[CH_2\text{-}CH=CH\text{-}CH_2\text{-}O\text{-}\underset{}{\bigcirc}\text{-}\underset{CH_3}{\overset{CH_3}{C}}\text{-}\underset{}{\bigcirc}\text{-}O\text{-}]_n CH_2\text{-}CH=CH\text{-}CH_2\text{-}Cl$$

$$+$$

$$H\text{-}[O\text{-}\underset{}{\bigcirc}\text{-}SO_2\text{-}\underset{}{\bigcirc}\text{-}\underset{CH_3}{\overset{CH_3}{C}}\text{-}\underset{}{\bigcirc}\text{-}]\text{-}O\text{-}H$$

$\longrightarrow$ Alternating Block Copolymer

Figure 9-10

Poly[oxy(2,6-dimethyl-1,4-phenylene)] with one 4-hydroxy-2,5-dimethylphenyl chain end was quantitatively etherified with aromatic polyethersulfones each containing two 1-chloroallyl or two bromobenzyl end groups to provide an ABA triblock copolymer. A chlorobenzene–50% aqueous sodium hydroxide reaction system was used in the presence of stoichiometric quantities of tetra-n-butylammonium hydrogen sulfate [19].

Polyisobutylene possessing terminal phenol functionalities has been reacted with α, ω-di(chloroallyl)- or α,ω-di(bromobenzyl) polyethersulfone in a chlorobenzene–3 N aqueous sodium hydroxide two-phase reaction system at 70°C in the presence of tetra-n-butylammonium hydrogen sulfate to form -$(A\text{-}B)_n$- block copolymers [20]. In addition, polyisobutylene and polyethersulfone, each possessing terminal phenol functionalities, were reacted with methylene chloride to form a -$[(A)_x\text{-}(B)_y]_n$- block copolymer [20].

The synthesis and characterization of a new class of thermotropic main-chain liquid-crystalline polyethers has been reported. 4,4′-Dihydroxy-α-methyl stilbene has been copolymerized with a series of dibromoalkanes, Br-$(CH_2)_n$-Br, where $n = 1$–11. The polymerization was carried out in an o-dichlorobenzene–50% aqueous sodium hydroxide two-phase system in the presence of catalytic quantities of tetra-n-butylammonium hydrogen sulfate [21–25]. No polymer was formed when 1,2-dibromoethane (n = 2) was used as a comonomer. Polyethers and copolyethers exhibiting thermotropic liquid crystalline properties have been prepared from 4,4′-dihydroxy-α-methylstilbene and α,ω-dichlorooligooxyethylenes having between 2 and 8.7 oxyethylene units [26]. Copolymers were prepared from an equimolar ratio of two dissimilar spacers.

Thermotropic liquid crystalline copolymers derived from 1,9-dibromononane and 4,4′-dihydroxydiphenyl by PTC methods have also been reported [27]. The related polythioethers and copolythioethers have been synthesized from 4,4′-dithiolbiphenyl and dibromoalkanes, Br-$(CH_2)_n$-Br, where $n = 2$–12 [28]. It was postulated that neighboring group participation by the sulfide linkage during the course of polymerization resulted in sulfonium ion intermediates that hydrolyzed and subsequently led to aliphatic ether structural units within the resulting polymer. Copolymers and alternating block copolymers containing a liquid crystalline polyether segment, based on 4,4′-diphenyldiol and 1,5-dibromopentane, and a

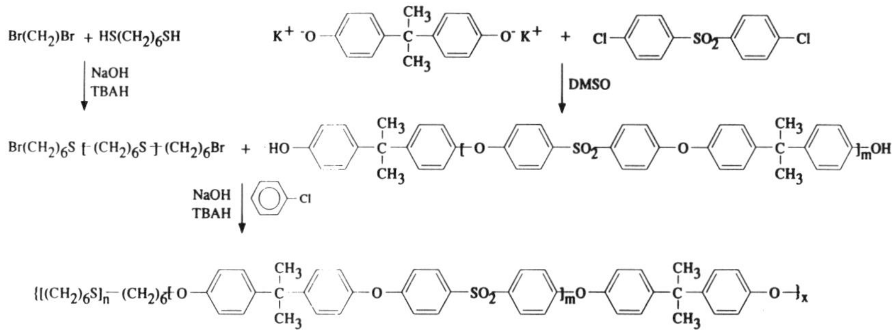

Figure 9-11

thermoplastic segment derived from an aromatic polyethersulfone, have been reported [29].

The PTC polyetherification of α,ω-diphenol aromatic polyether sulfones with *cis*-dichloro-2-butene and α,α'-dibromo-*p*-xylene in chlorobenzene–50% aqueous sodium hydroxide gave polymers in which the terminal positions contained allylic chlorine and benzylic bromine, respectively [30].

It has been reported that a successful PTC polyetherification of polyethersulfones terminated by phenol functionalities has been carried out using methylene chloride both as electrophilic monomer as well as solvent. It is interesting to note some of the details regarding this polyetherification. Precipitation resulted upon addition of the phenol-terminated polyethersulfone to 50% aqueous sodium hydroxide. With the subsequent addition of stoichiometric amounts of tetra-*n*-butylammonium hydrogen sulfate, the salt dissolved and polymerization ensued [31]. It was noted that the displacement of the second chlorine atom on methylene chloride was much faster than displacement of the first. An early example of an aromatic–aliphatic polyformal was that prepared from 2,2-bis(4-hydroxyphenyl)-propane and 1,4-bis(chloromethoxy)butane [32].

α,ω-Dibromopolyhexamethylene sulfides, prepared by PTC condensation, were polyetherified with α,ω-diphenol aromatic polyethersulfones in the presence of stoichiometric quantities of tetra-*n*-butylammonium hydrogen sulfate (Figure 9-11) [33].

Poly(aliphatic sulfide)s of high molecular weight have been prepared by the polycondensation of sodium sulfide with dibromoalkanes in the presence of quaternary ammonium and phosphonium salts [34]. The percent yields and inherent viscosities are summarized in Tables 9-1 and 9-2. Cetyltrimethylammonium chloride or bromide and cetyltri-*n*-butylphosphonium bromide were found to be particularly effective catalysts.

Reaction of bis(4-chloroacetylphenyl) ether with 4,4'-oxybis(benzenethiol) in an aromatic or chlorinated organic solvent–aqueous potassium hydroxide reaction system at 30°C in the presence of quaternary ammonium salts has been reported. It was found that polymerization took place even in the absence of catalyst.

Table 9-1. Polycondensation of sodium sulfide with 1,8-dibromooctane in the presence of various catalysts.

Catalyst	Percent yield	n
None	94	0.08
TBAC	99	0.09
CTMAC	91	0.71
CTBPB	99	0.42
CTMAB	86	0.70
BTEAC	99	0.19
BTPPC	96	0.08

TBAC = tetra-n-butylammonium chloride; CTMAC = cetyltrimethylammonium chloride; BTEAC = benzyltriethylammonium chloride; BTPPC = benzyltriphenylphosphonium chloride.

Table 9-2. Polycondensation of sodium sulfide with various dibromoalkanes in the presence of catalysts.

Alkane	Catalyst	Percent yield	n
1,6-dibromohexane	CTBPB	99	0.68
1,6-dibromohexane	CTMAC	88	0.64
1,8-dibromooctane	CTBPB	99	0.71
1,8-dibromooctane	CTMAC	99	0.70
1,10-dibromodecane	CTBPB	99	0.60
1,10-dibromodecane	CTMAC	99	0.83

CTBPB = cetyltri-n-butylphosphonium bromide; CTMAC = cetyltriemethylammonium chloride.

However, benzyltriethylammonium chloride and cetyltrimethylammonium chloride produced the highest molecular weight polymers as reflected by inherent viscosity measurements [35].

Poly(1,4-oxylyenethioethers) were synthesized from 1,4-bis(chloromethyl)-benzene, 1,4-bis(chloromethyl)-2,5-dimethoxybenzene, and 1,4-bis(chloromethyl)-2,5-dimethylbenzene and sodium sulfide in the presence of quaternary ammonium and phosphonium salts (Figure 9-12) [36]. The effects of catalyst, solvent, and temperature were studied. Polymerizations were carried out at 20°, 50°, and 70°C in toluene, chloroform, and nitrobenzene as the organic solvents.

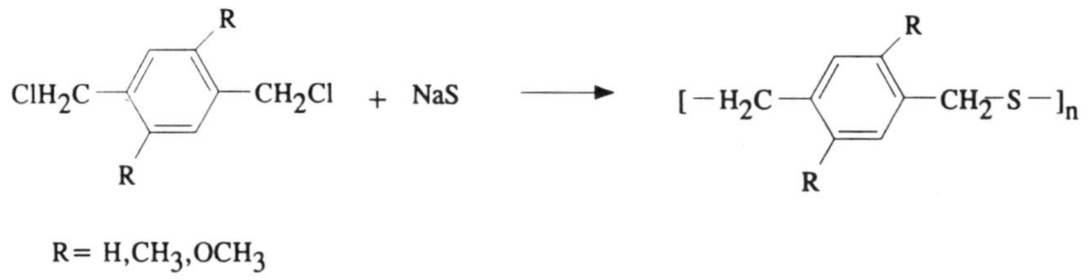

R = H, CH$_3$, OCH$_3$

Figure 9-12

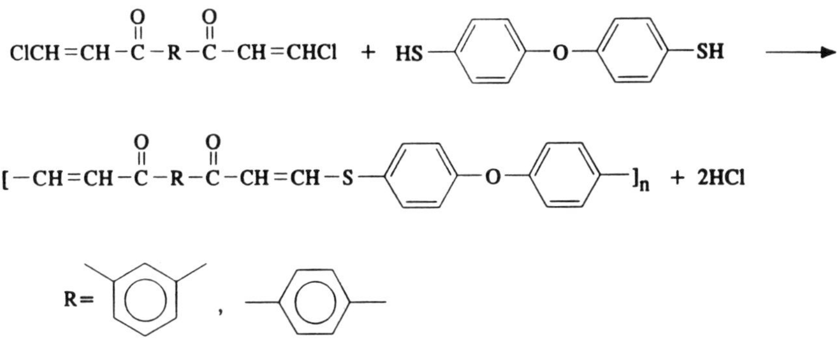

$$ClCH=CH-\overset{\overset{O}{\|}}{C}-R-\overset{\overset{O}{\|}}{C}-CH=CHCl + HS-\underset{}{\bigcirc}-O-\underset{}{\bigcirc}-SH \longrightarrow$$

$$[-CH=CH-\overset{\overset{O}{\|}}{C}-R-\overset{\overset{O}{\|}}{C}-CH=CH-S-\underset{}{\bigcirc}-O-\underset{}{\bigcirc}-]_n + 2HCl$$

$$R= \underset{}{\bigcirc}, -\underset{}{\bigcirc}-$$

Figure 9-13

Reactions conducted without catalyst did not produce polymer. In all solvents, the polythioethers precipitated in the course of reaction. This, of course, represented a serious limitation in obtaining high molecular weights. When toluene was employed as the organic solvent, the catalysts that produced the best yields were hexadecyltributylphosphonium bromide, hexadecyltrimethylammonium bromide, and tetra-n-butylammonium bromide. In contrast, when chloroform was used as the organic solvent, tetra-n-butylammonium bromide was relatively ineffective. In nitrobenzene, the yields of each of the polymers obtained at 70°C were essentially constant with all the catalysts employed. Increasing the temperature resulted in an increase in the yields of polymer. This was attributed primarily to an increase in the solubility of the polythioethers.

Poly(oxovinyl sulfides) have been prepared by vinylogous nucleophilic substitution polymerization of bis(3-chloroacryloyl) derivative of benzene with 4,4'-oxydibenzenethiol (Figure 9-13) [37]. The PTC polycondensations were carried out in a chlorinated aliphatic hydrocarbon–aqueous potassium hydroxide two-phase system under nitrogen at room temperature. Tetra-n-butylammonium chloride and dicyclohexyl-18-crown-6 were used as the phase-transfer catalysts.

A series of studies dealing with solid–liquid PTC aromatic substitution condensation polymerizations has been reported [38–40]. Polymerizations were carried out with a variety of bisphenols and bisthiophenols with hexafluorobenzene and with perfluorobiphenyl, perfluorodiphenyl ketone, and perfluorodiphenyl sulfide. Several of the synthesized polymers are illustrated in Figure 9-14. Equimolar quantities of co-monomers were reacted in the presence of an excess of anhydrous potassium carbonate and 27.7 mol% 18-crown-6 in a variety of organic solvents (dimethylacetamide, acetone, dimethyl sulfoxide, chlorobenzene, toluene, dioxane, acetonitrile, and tetrahydrofuran) at temperatures varying from 55° to 90°C. Dimethylacetamide, acetone, and acetonitrile were generally found to be the best solvents for achieving high molecular weights as reflected by inherent viscosity measurements. Rapid gelation was observed in some experiments and was attributed to branching/crosslinking reactions. The magnitude of the catalytic effect

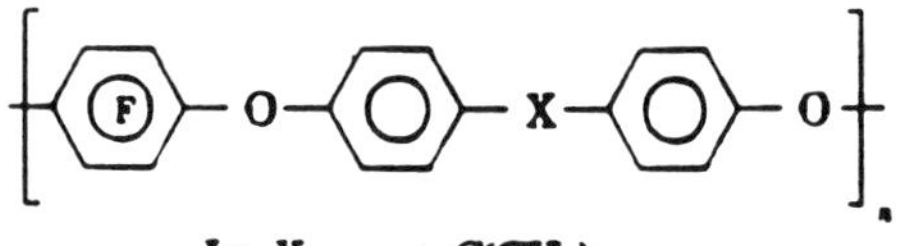

$$\left[\underset{F}{\bigcirc} - O - \bigcirc - X - \bigcirc - O \right]_n$$

Ia, X = $-C(CH_3)_2-$
b, X = $-S-$
c, X = $-SO_2-$

Figure 9-14

of 18-crown-6 appeared to depend on the nucleophile employed in the polymerization. The stronger the nucleophile, the smaller the catalytic effect of the crown. The presence of "trace" amounts of water in the organic phase was found to have profound effects on the inherent viscosity and percent yield of the resulting polymer.

c. Polycarbonates and Polythiocarbonates

The syntheses of polycarbonates under solid–liquid two-phase polycondensation conditions using crown ethers as catalysts are among the early examples of the use of phase-transfer catalysis. For instance, xylylene dihalides have been employed in PTC polymerization processes to form polycarbonates [41]. In particular, the PTC synthesis of poly(p-xylylene carbonate) from p-xylylene dibromide and potassium carbonate catalyzed by a variety of crown ethers and PEG derivatives has been reported (Figure 9-15) [42]. From infrared and nuclear magnetic resonance analyses, the product was identified as poly(p-xylene carbonate) having the structure shown in Figure 9-15. Crown ethers having 18-membered rings (18-crown-6, dicyclohexo-18-crown-6, and dibenzo-18-crown-6) were found to be the best catalysts. Using 18-crown-6 as the catalyst, the polymerization was studied in a variety of solvents, including benzene, toluene, tetrahydrofuran, dioxane, and diglyme, at temperatures ranging from 50 to 160°C. It was observed that the yield and the molecular weight of the polymer increased with increasing concentration of potassium carbonate. As the concentration of 18-crown-6 was increased, both the polymer yield and molecular weight first increased and then approached constant values. Lithium carbonate and sodium carbonate also produced polymer. Silver, zinc, barium, strontium, and calcium carbonates were unreactive under the same reaction conditions. Reaction temperature and solvent were also critical parameters in the yield and molecular weight of the polymer. It is interesting to note that it was postulated that the initial step

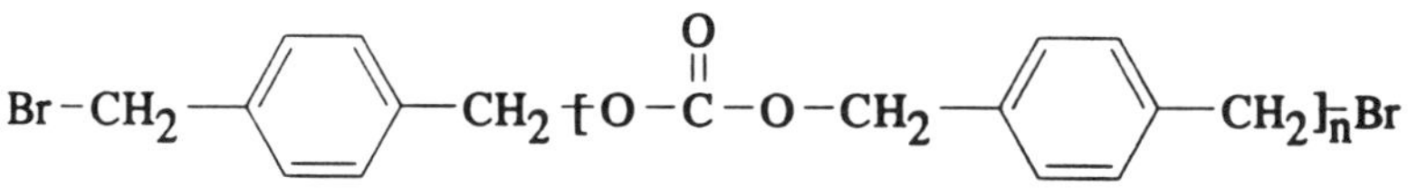

$$Br-CH_2-\bigcirc-CH_2 \left[O-\overset{O}{\underset{\|}{C}}-O-CH_2-\bigcirc-CH_2 \right]_n Br$$

Figure 9-15

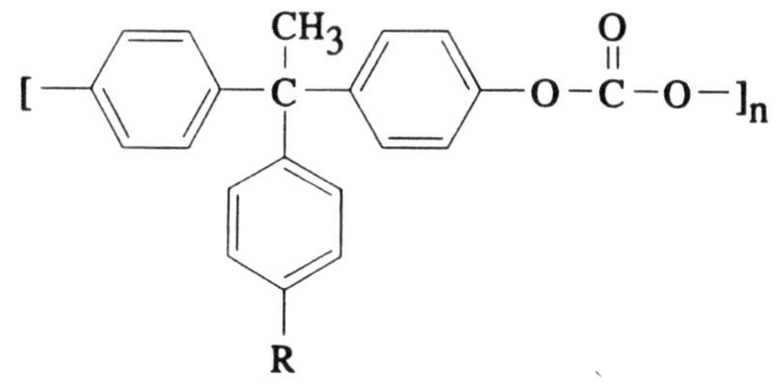

[structure: bisphenol-like polycarbonate unit with CH₃ and R-substituted phenyl group]

R = (I) -H, (II) -CH₃, (III) -Br

Figure 9-16

in the mechanism is the adsorption of the crown onto the surface of the potassium carbonate followed by the dissolution of the crown complex into the organic phase. The reactions of potassium carbonate with a series of aliphatic dibromides, Br-$(CH_2)_n$-Br, in the presence of 18-crown-6, were also investigated. Polycarbonates were obtained from dibromides where $n = 4, 5, 6$, and 10, and cyclic carbonates were obtained where $n = 2$ and 3.

The synthesis of polycarbonates from phosgene and substituted 2,2-bis-(4-hydroxyphenyl)-2-phenylethanes using several quaternary ammonium and phosphonium salts in a methylene chloride–aqueous sodium hydroxide two-phase system has been reported (Figure 9-16) [43]. In this investigation, the yield and inherent viscosity of product were studied as a function of the structure of the phase-transfer catalyst and time. It was pointed out that interfacial polycondensation between phosgene dissolved in the methylene chloride phase and the sodium salt of the diphenol in the aqueous phase could not be disregarded. Indeed, in the absence of catalyst and for reaction times of 150 min, the yields of polymer ranged from 43 to 71% with modest inherent viscosities. In the presence of catalyst, the percent yields and the inherent viscosities were measured as a function of time. The results are summarized in Tables 9-3, 9-4, and 9-5. In all cases reported in this study, the yields and the inherent viscosities initially increased and then subsequently decreased. The origins of these observations were attributed to the polymerization process being accompanied by competing hydrolysis. In all cases reported, it appears that benzyltriethylammonium chloride was the most effective of the catalysts employed; rapid polymerization with high inherent viscosities was achieved within short periods of time. This particular catalyst was found to be the poorest in promoting the hydrolysis of the polymer.

A series of patents have been issued describing the synthesis of polythiocarbonates-co-carbonates and polythiocarbonate-co-sulfonates in two-phase reaction

Table 9-3. Yields and inherent viscosities obtained for polycarbonate I.

Catalyst	10 min %	n	20 min %	n	30 min %	n	60 min %	n	90 min %	n	120 min %	n	150 min %	n
BTEAC	89	0.27	92	0.44	90	0.37	74	0.32	81	0.42	76	0.38	95	0.28
TBAB	84	0.17	—		86	0.12	89	0.20	90	0.48	77	0.22	66	0.12
HDTMAB	93	0.33	85	0.28	87	0.16	63	0.20	60	0.19	61	0.20	82	0.16
TBHDPB	85	0.14	—		89	0.13	62	0.15	56	0.10	59	0.11	94	0.33
Aliquat 336	94	0.22	91	0.24	87	0.21	91	0.10	52	0.11	79	0.19	88	0.16

BTEAC = benzyltriethylammonium chloride; TBAB = tetra-n-butylammonium bromide; HDTMAB = hexadecyltrimethylammonium bromide; TBHDPB = tri-n-butylhexadecylphosphonium bromide; Aliquat 336 = methyltri-n-octylammonium chloride.

Table 9-4. Yields and inherent viscosities obtained for polycarbonate II.

Catalyst	10 min %	n	20 min %	n	30 min %	n	60 min %	n	90 min %	n	120 min %	n	150 min %	n
BTEAC	82	0.67	82	0.89	91	0.44	76	0.42	79	0.39	73	0.18	87	0.23
TBAB	82	0.19	76	0.11	82	0.11	65	0.20	73	0.16	—		78	0.17
HDTMAB	80	0.19	77	0.32	66	0.26	47	0.21	67	0.22	54	0.23	70	0.21
TBHDPB	85	0.17	79	0.09	79	0.14	67	0.18	64	0.18	67	0.16	72	0.15
Aliquat 336	85	0.70	78	0.68	84	0.17	74	0.17	72	0.19	65	0.16	69	0.14

BTEAC = benzyltriethylammonium chloride; TBAB = tetra-n-butylammonium bromide; HDTMAB = hexadecyltrimethylammonium bromide; TBHDPB = tri-n-butylhexadecylphosphonium bromide; Aliquat 336 = methyltri-n-octylammonium chloride.

Table 9-5. Yields and inherent viscosities obtained for polycarbonate III.

Catalyst	10 min %	n	20 min %	n	30 min %	n	60 min %	n	90 min %	n	150 min %	n
BTEAC	93	0.75	96	0.27	43	0.15	91	0.15	82	0.20	94	0.15
TBAB	91	0.18	93	0.30	97	0.47	90	0.25	94	0.17	88	0.15
HDTMAB	79	0.12	86	0.13	62	0.19	75	0.16	66	0.18	80	0.12
TBHDPB	61	0.07	68	0.08	70	0.08	72	0.09	68	0.06	70	0.06
Aliquat 336	75	0.06	88	0.07	74	0.07	90	0.11	84	0.06	44	0.06

BTEAC = Benzyltriethylammonium chloride; TBAB = tetra-n-butylammonium bromide; HDTMAB = hexadecyltrimethylammonium bromide; TBHDPB = tri-n-butylhexadecylphosphonium bromide.

systems employing quaternary ammonium salts as catalysts in order to increase the molecular weight [44–50]. The polymers showed reasonably good properties and were claimed to have potential use as thermoplastics, adhesives, fibers, and films. It has been reported that polythiocarbonate derived from 2,2-bis(4-hydroxyphenyl)propane has been synthesized (1) in a two-phase system from an

Table 9-6. Reaction of 2,2-bis(4-hydroxyphenyl)propane and thiophosgene: effect of catalysts in the polycondensation reaction.

Catalyst	Yield (%)	n
None	50	0.05
TBAB	91	1.80
HDTMAB	89	0.59
TBHDPB	87	0.53
Aliquat 336	85	0.39
BTEAC	47	0.08
BTPPC	41	0.08
DCH-18-C-6	76	0.88
18-C-6	51	0.06
DB-18-C-6	30	0.06

TBAB = tetra-n-butylammonium bromide; HDTMAB = hexadecyltrimethylammonium bromide; TBHDPB = tri-n-butylhexadecylphosphonium bromide; Aliquat 336 = methyltri-n-octylammonium chloride; BTEAC = benzyltriethylammonium chloride; BTPPC = benzyltriphenylphosphonium chloride; DCH-18-C-6 = dicyclohexyl-18-crown-6; 18-C-6 = 18-crown-6; DB-18-C-6 = dibenzo-18-crown-6.

alkaline solution of the diphenolate and thiophosgene using benzyltriethylammonium chloride as the catalyst [51], and (2) under anhydrous conditions in the solvent dichloromethane from the diphenol, thiophosgene, and triethylamine catalyst at low temperatures [52]. In a subsequent series of reports poly(oxythiocarbonyloxy-1,4-phenylenepropylidene-1,4-phenylene) was synthesized from 2,2-bis(4-hydroxyphenyl)propane and thiophosgene in a methylene chloride–aqueous sodium hydroxide two-phase system at 20°C in the presence of a variety of quaternary ammonium salts, quaternary phosphonium salts, and crown ethers as phase-transfer catalysts [53,54]. The effects of catalyst structure and changes in organic solvent on the yield and inherent viscosity of the resulting polymer are summarized in Tables 9-6 and 9-7. It was observed that addition of the thiophosgene all at once produced polymer in highest yields and inherent viscosity. In addition, tetra-n-butylammonium bromide was found to be the most effective catalyst studied. Hexadecyltrimethylammonium bromide, tributylhexadecylphosphonium bromide, Aliquat 336 (methyltrioctylammonium chloride), and dicyclohexo-18-crown-6 produced polymers in good yields but lower inherent viscosities. Benzyltriethylammonium chloride, benzyltriphenylphosphonium chloride, 18-crown-6, and dibenzo-18-crown-6 produced materials in extremely low inherent viscosities. In fact, there appeared to be no difference between these latter "catalysts" and having no catalyst present. Toluene and 1,1,2-trichloroethane produced polymers in lower yields and lower inherent viscosities.

A series of polythiocarbonates were synthesized from the diphenols shown in Figure 9-17 and thiophosgene in methylene chloride–aqueous sodium or potassium hydroxide in the presence of quaternary ammonium and phosphonium salts and crown ethers as phase-transfer catalysts (Figure 9-17) [55]. In the absence

Table 9-7. Reaction of 2,2-bis(4-hydroxyphenyl)propane and thiophosgene: solvent effect in the polycondensation reaction.

Catalyst	CH_2Cl_2		$C_6H_5-CHCl_2$		$CHCl_2-CHCl_2$	
	%	n	%	n	%	n
TBAB	91	1.80	34	0.13	56	0.06
HDTMAB	89	0.59	81	0.34	51	0.09
TBHDPB	87	0.53	99	0.27	59	0.06
Aliquat 336	85	0.39	94	0.75	26	0.06
BTEAC	47	0.08	27	0.09	51	0.08
BTPPC	41	0.08	28	0.13	38	0.08
DCH-18-C-6	76	0.88	33	0.12	62	0.07

TBAB = tetra-*n*-butylammonium bromide; HDTMAB = hexadecyltrimethylammonium bromide; TBHDPB = tri-*n*-butylhexadecylphosphonium bromide; Aliquat 336 = methyltri-*n*-octylammonium chloride; BTEAC = benzyltriethylammonium chloride; BTPPC = benzyltriphenylphosphonium chloride; DCH-18-C-6 = dicyclohexyl-18-crown-6.

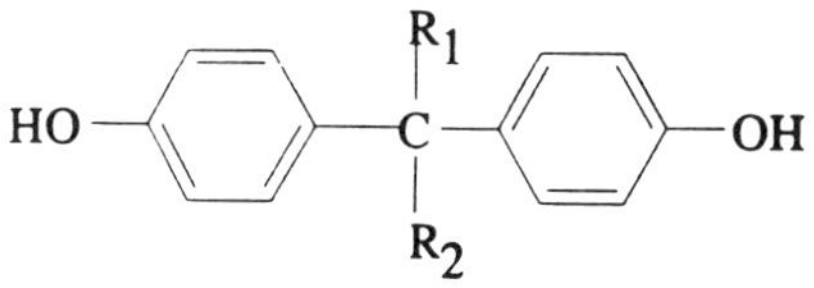

	R_1	R_2
1.	H	CH_3
2.	H	C_2H_5
3.	H	C_3H_7
4.	H	*iso*-C_3H_7
5.	H	$CH(C_2H_5)_2$
6.	H	$C_{10}H_7$
7.	CH_3	C_2H_5
8.	CH_3	$CH_2CH(CH_3)_2$
9.	$(CH_2)_5$	
10.	$(CH_2)_2CH(CH_3)(CH_2)_2$	
11.	CH_3	C_6H_5
12.	C_2H_5	C_6H_5

Figure 9-17

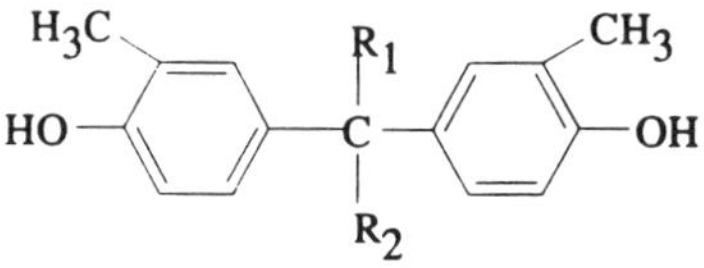

$$\text{HO} - \text{C}_6\text{H}_3(\text{CH}_3) - \overset{R_1}{\underset{R_2}{\text{C}}} - \text{C}_6\text{H}_3(\text{CH}_3) - \text{OH}$$

	R$_1$	R$_2$
I	-CH$_3$	-CH$_3$
II	-CH$_3$	-CH$_2$-CH$_3$
III	-CH$_3$	-C$_6$H$_5$
IV	-(-CH$_2$-)-$_5$	
V	-(-CH$_2$-)-$_2$-CH(CH$_3$)-(-CH$_2$-)-$_2$-	

Figure 9-18

of a catalyst, the yields spanned a range of 50–75% with low inherent viscosities. Tri-n-butylhexadecylphosphonium bromide, hexadecyltrimethylammonium bromide, and dicyclohexo-18-crown-6 (used with potassium hydroxide) were ineffective as catalysts. Aliquat 336 was found to be the best catalyst. It should be pointed out, however, that it is the only catalyst in this report that has a chloride as the counteranion. Tetra-n-butylammonium bromide showed intermediate catalytic activity.

Polythiocarbonates were synthesized from the five diphenols shown in Figure 9-13 and thiophosgene in the presence of quaternary ammonium and phosphonium salts (Figure 9-18) [56]. The yields and inherent viscosities as a function of time are summarized in Tables 9-8–9-12 for each of the bisphenol monomers. Benzyltriethylammonium chloride, hexadecyltri-n-butylphosphonium bromide, Aliquat 336, and tetra-n-butylammonium bromide were not particularly effective catalysts. Hexadecyltrimethylammonium bromide gave the best results.

Phenolphthalein polythiocarbonate was synthesized from phenolphthalein and thiophosgene in methylene chloride–aqueous sodium hydroxide in the presence of quaternary ammonium and phosphonium salts [57]. The yields and inherent viscosities of the resulting polymers are summarized in Table 9-13. Tetra-n-butylammonium bromide, hexadecyltri-n-butylammonium bromide, and Aliquat 336 were essentially ineffective as catalysts, whereas benzyltriethylammonium chloride and hexadecyltrimethylammonium bromide were substantially better. These latter catalysts represent more accesible quaternary cations and it is this property that may be critical in the reactivity between these reaction partners.

It was observed that during the course of polymerization of diphenols and

Table 9-8. Yields and inherent viscosities obtained for polythiocarbonate I.

Catalyst	10 min %	n	20 min %	n	30 min %	n	60 min %	n	90 min %	n	120 min %	n	150 min %	n
TBAB	76	0.05	91	0.05	75	0.11	92	0.12	85	0.07	77	0.39	79	0.10
MTAC	76	0.02	88	0.07	73	0.03	77	0.05	89	0.08	81	0.13	76	0.41
HDTMAB	76	0.15	87	0.18	89	0.34	79	0.28	91	0.40	77	0.53	84	0.64
TBHDPB	87	0.13	67	0.13	67	0.10	63	0.13	70	0.13	81	0.15	89	0.13
BTEAC	4	0.05	7	0.08	18	0.05	29	0.08	45	0.11	58	0.05	56	0.06

TBAB = tetra-n-butylammonium bromide; MTAC = Aliquat 336 = methyltri-n-octylammonium choride; HDTMAB = hexadecyltrimethylammonium bromide; TBHDPB = tri-n-butylhexadecylphosphonium bromide; BTEAC = benzyltriethylammonium chloride.

Table 9-9. Yields and inherent viscosities obtained for polythiocarbonate II.

Catalyst	10 min %	n	20 min %	n	30 min %	n	60 min %	n	90 min %	n	120 min %	n	150 min %	n
TBAB	66	0.07	69	0.09	69	0.08	59	0.07	70	0.09	55	0.06	81	0.09
MTAC	64	0.05	61	0.05	69	0.07	66	0.05	70	0.07	77	0.09	83	0.09
HDTMAB	88	0.19	88	0.23	88	0.17	84	0.09	84	0.61	94	0.47	91	0.49
TBHDPB	71	0.17	70	0.07	67	0.06	66	0.07	74	0.06	75	0.08	83	0.09
BTEAC	7	0.07	21	0.04	56	0.06	52	0.08	65	0.07	59	0.08	65	0.08

TBAB = tetra-n-butylammonium bromide; MTAC = Aliquat 336 = methyltri-n-octylammonium chloride; HDTMAB = hexadecyltrimethylammonium bromide; TBHDPB = tri-n-butylhexadecylphosphonium bromide; BTEAC = benzyltriethylammonium chloride.

Table 9-10. Yields and inherent viscosities obtained for polythiocarbonate III.

Catalyst	10 min %	n	20 min %	n	30 min %	n	60 min %	n	90 min %	n	120 min %	n	150 min %	n
TBAB	91	0.09	94	0.13	78	0.05	89	0.10	81	0.08	76	0.11	79	0.07
MTAC	84	0.05	86	0.07	87	0.06	94	0.10	87	0.09	89	0.10	89	0.09
HDTMAB	94	0.56	92	0.51	90	0.76	93	0.72	89	0.33	90	0.33	94	0.32
TBHDPB	94	0.13	95	0.16	96	0.11	94	0.10	92	0.10	88	0.13	89	0.26
BTEAC	79	0.07	85	0.08	85	0.05	83	0.08	80	0.09	80	0.07	76	0.07

TBAB = tetra-n-butylammonium bromide; MTAC = Aliquat 336 = methyltri-n-octylammonium chloride; HDTMAB = hexadecyltrimethylammonium bromide; TBHDPB = tri-n-butylhexadecylphosphonium bromide; BTEAC = benzyltriethylammonium chloride.

thiophosgene, the inherent viscosity of the polymer first increased and then decreased with time. These observations are consistent with hydrolysis processes competing with polymerization. In order to better understand the hydrolysis process, the effects of various catalysts on the rates of hydrolysis of the polythiocarbonate derived from 2,2-bis(4-hydroxyphenyl)propane as reflected by the

Table 9-11. Yields and inherent viscosities obtained for polythiocarbonate IV.

Catalyst	10 min %	n	20 min %	n	30 min %	n	60 min %	n	90 min %	n	120 min %	n	150 min %	n
TBAB	61	0.06	61	0.07	68	0.08	66	0.09	71	0.08	69	0.09	71	0.09
MTAC	90	0.11	70	0.09	68	0.08	74	0.09	84	0.11	77	0.09	75	0.09
HDTMAB	56	0.10	92	0.70	77	0.13	72	0.15	85	0.16	78	0.16	89	0.19
TBHDPB	78	0.04	91	0.05	76	0.08	79	0.08	88	0.05	87	0.07	86	0.08
BTEAC	63	0.08	75	0.11	77	0.06	74	0.08	77	0.07	80	0.06	74	0.08

TBAB = tetra-n-butylammonium bromide; MTAC = Aliquat 336 = methyltri-n-octylammonium bromide; HDTMAB = hexadecyltrimethylammonium bromide; TBHDPB = tri-n-butylhexadecylphosphonium bromide; BTEAC = benzyltriethylammonium chloride.

Table 9-12. Yields and inherent viscosities obtained for polythiocarbonate V.

Catalyst	10 min %	n	20 min %	n	30 min %	n	60 min %	n	90 min %	n	120 min %	n	150 min %	n
TBAB	68	0.05	63	0.05	73	0.07	76	0.07	71	0.07	81	0.08	80	0.07
MTAC	61	0.07	68	0.07	72	0.07	79	0.07	77	0.05	86	0.07	90	0.09
HDTMAB	60	0.09	61	0.06	76	0.06	80	0.08	81	0.07	96	0.10	87	0.09
TBHDPB	69	0.09	57	0.10	69	0.09	72	0.08	67	0.07	84	0.08	75	0.10
BTEAC	61	0.09	67	0.08	59	0.08	52	0.06	68	0.08	68	0.09	71	0.10

TBAB = tetra-n-butylammonium bromide; MTAC = Aliquat 336 = methyltri-n-octylammonium bromide; HDTMAB = hexadecyltrimethylammonium bromide; TBHDPB = tri-n-butylhexadecylphosphonium bromide; BTEAC = benzyltriethylammonium chloride.

Table 9-13. Yields and inherent viscosities obtained for phenolphthalein polythiocarbonate.

Catalyst	10 min %	n	30 min %	n	150 min %	n
TBAB	80	0.13	80	0.11	85	0.13
BTEAC	77	0.20	79	0.22	82	0.16
HDTMAB	85	0.22	96	0.19	89	0.18
HDTBPB	71	0.12	87	0.11	85	0.11
Aliquat 336	87	0.16	88	0.14	86	0.11

TBAB = tetra-n-butylammonium bromide; BTEAC = benzyltriethylammonium chloride; HDTMAB = hexadecyltrimethylammonium bromide; HDTBPB = hexadecyltri-n-butylphosphonium bromide; Aliquat 336 = methyltri-n-octylammonium chloride.

Table 9-14. Effect of the catalyst on the hydrolysis of 2,2-bis(4-hydroxyphenyl)propane polythiocarbonate.

Catalyst	Yield (%)	n
None	95	0.54
TBAB	69	0.18
BTEAC	74	0.42
HDTMAB	46	0.20
TBHDPB	62	0.18
Aliquat 336	71	0.15

TBAB = tetra-n-butylammonium bromide; BTEAC = benzyltriethylammonium chloride; HDTMAB = hexadecyltrimethylammonium bromide; TBHDPB = tri-n-butylhexadecylphosphonium bromide; Aliquat 336 = methyltri-n-octylammonium chloride.

change in inherent viscosity of the polymer were investigated [58]. Table 9-14 shows the results obtained in the absence and presence of catalyst. The initial polymer solution had an inherent viscosity of 0.67 dL g^{-1}. In the absence of catalyst, the viscosity decreased by approximately 15% at 20°C after a period of 2.5 h. Using tetra-n-butylammonium bromide, hexadecyltrimethylamonium bromide, hexadecyltri-n-butylphosphonium bromide, and Aliquat 336, the inherent viscosity decreased by approximately 80%. Of all the quaternary salts employed in this investigation benzyltriethylammonium chloride resulted in the least reduction in inherent viscosity (approximately 30%). For polycarbonates derived from diphenols in Figure 9-16, benzyltriethylammonium chloride also showed the slowest rate of hydrolysis compared with the other quaternary salts [43].

Copolycarbonate-thiocarbonates were synthesized from 2,2-bis(4-hydroxyphenyl)propane and mixtures of phosgene and thiophosgene in methylene chloride–aqueous sodium hydroxide two-phase systems in the presence of tetra-n-butylammonium bromide, hexadecyltrimethylammonium bromide, and Aliquat 336 (Figure 9-19) [59]. The reactions were studied at 0°, 10°, and 20°C as a function of time. Polymers of different ratios of carbonate to thiocarbonate were synthesized by simply varying the feed ratios of phosgene and thiophosgene, respectively. Specific ratios could be achieved because the reactivities of the two electrophiles

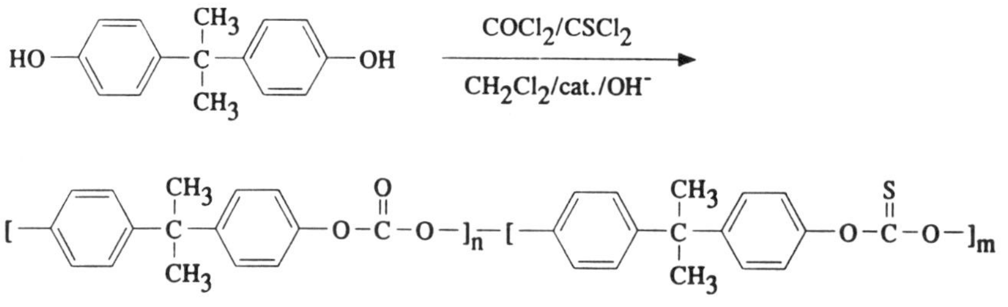

Figure 9-19

Table 9-15. Yields and inherent viscosities obtained for CO-76.

Catalyst temps. (°C)		10 min %	10 min n	20 min %	20 min n	30 min %	30 min n	60 min %	60 min n	120 min %	120 min n
TBAB	0	47	0.21	59	0.13	66	0.14	55	0.15	47	0.21
	10	39	0.13	54	0.17	53	0.19	44	0.27	66	0.23
	20	72	0.57	69	0.51	55	0.20	60	0.23	67	0.20
HDTMAB	0	53	0.37	66	1.05	65	1.15	74	0.63	71	0.30
	10	64	0.99	66	0.70	62	0.80	57	0.63	82	0.37
	20	76	0.37	69	0.37	73	0.42	65	0.34	69	0.23
Aliquat 336	0	78	0.24	74	0.58	80	0.90	81	0.42	85	0.25
	10	78	0.96	78	0.32	86	0.72	76	0.28	73	0.27
	20	82	0.32	80	0.30	59	0.25	74	0.22	42	0.12

TBAB = tetra-n-butylammonium bromide; HDTMAB = hexadecyltrimethylammonium bromide; Aliquat 336 = methyltri-n-octylammonium chloride.

are essentially identical with respect to the diphenoxide. It was observed that the temperature and the catalyst had important effects on the molecular weight of the polymer. Tables 9-15, 9-16, and 9-17 show the yield and inherent viscosity of polymer with respect to time. For polymers containing 76% carbonate (CO-76), tetra-n-butylammonium bromide was effective only at 20°C, whereas hexadecyltrimethylammonium bromide showed the best results at 0°C for short periods of time.

Polycarbonates and polythiocarbonates have been synthesized from phosgene and thiophosgene, respectively, and diphenols with chlorinated aromatic side groups in methylene chloride–aqueous sodium hydroxide in the presence of quaternary ammonium and phosphonium salts (Figure 9-20) [60].

Table 9-16. Yields and inherent viscosities obtained for CO-47.

Catalyst temps. (°C)		10 min %	10 min n	20 min %	20 min n	30 min %	30 min n	60 min %	60 min n	120 min %	120 min n
TBAB	0	19	0.12	32	0.12	42	0.12	30	0.17	50	0.20
	10	48	0.13	53	0.15	58	0.18	69	0.25	72	0.41
	20	54	0.40	60	0.27	60	0.34	59	0.16	37	0.12
HDTMAB	0	86	0.93	82	0.66	84	0.40	82	0.58	74	0.64
	10	79	0.78	77	0.42	82	0.44	75	0.41	76	0.69
	20	75	0.19	78	0.49	69	0.38	72	0.31	76	0.18
Aliquat 336	0	75	0.21	71	0.33	65	0.34	72	0.26	60	0.24
	10	73	0.27	74	0.22	73	0.29	60	0.25	64	0.20
	20	59	0.24	55	0.34	57	0.18	51	0.15	59	0.21

TBAB = tetra-n-butylammonium bromide; HDTMAB = hexadecyltrimethylammonium bromide; Aliquat 336 = methyltri-n-octylammonium chloride.

474 / Phase-Transfer Catalysis

Table 9-17. Yields and inherent viscosities obtained for CO-26.

Catalyst temps. (°C)		10 min		20 min		30 min		60 min		120 min	
		%	n	%	n	%	n	%	n	%	n
TBAB	0	21	0.08	31	0.09	38	0.10	57	0.18	61	0.14
	10	48	0.12	65	0.14	67	0.17	70	0.31	73	0.32
	20	60	0.22	70	0.44	68	0.27	68	0.24	57	0.21
HDTMAB	0	75	0.29	74	0.39	75	0.32	73	0.33	74	0.33
	10	77	0.49	82	0.36	77	0.36	80	0.34	77	0.36
	20	78	0.34	71	0.33	74	0.31	75	0.35	70	0.26
Aliquat 336	0	75	0.33	68	0.33	78	0.16	71	0.33	62	0.20
	10	72	0.24	72	0.20	70	0.26	63	0.22	68	0.24
	20	36	0.14	70	0.26	63	0.16	59	0.21	59	0.20

TBAB = tetra-n-butylammonium bromide; HDTMAB = hexadecyltrimethylammonium bromide; Aliquat 336 = methyltri-n-octylammonium chloride.

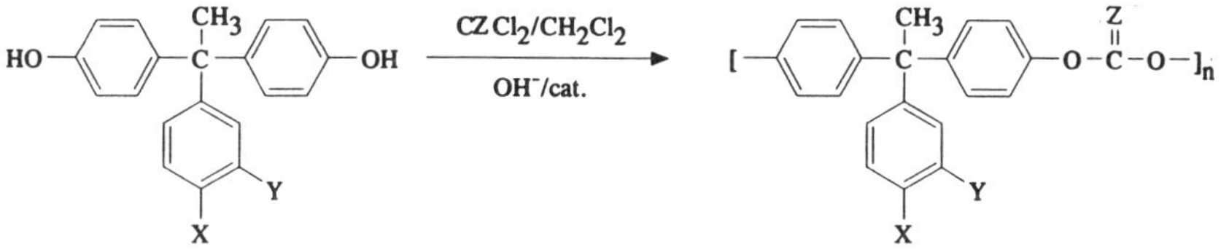

R = -H, -CH$_3$, -Br

Figure 9-20

Tables 9-18, 9-19, and 9-20 summarize the percent yield and inherent viscosity of polymer as a function of time.

d. Synthesis of Carbon–Carbon Chain Polymers

A successful condensation polymerization between α,α'-dichloro-p-xylene and t-butylcyanoacetate has been reported in benzene–50% aqueous sodium hydroxide at 50°C for 2 h in the presence of quaternary ammonium chlorides and crown ethers (Figure 9-21) [11,61,62]. The organic solvent anisole was found to be particularly effective; polymer was produced in reasonable yields with relatively high inherent viscosities. The polycondensation was strongly catalyzed by tetraethyl-, tetra-n-butyl-, benzyltriethyl-, and cetryltrimethylammonium chlorides. The crown ethers (including 15-, 18-, and 24-membered rings) were found to be less effective. High concentration of sodium hydroxide, and at least 50 mol% of the phase-transfer catalyst (base on monomers) were necessary to prepare polymers of high molecular weight. No hydrolysis of the ester functionality in the polymer was observed under the conditions of the polymerization.

Table 9-18. Yields and inherent viscosities obtained for polycarbonate 1a and polythiocarbonate 1b.

Catalyst	1a				1b			
	30 min		60 min		30 min		60 min	
	%	n	%	n	%	n	%	n
None	66	0.09	88	0.14	87	0.14	72	0.14
TBAB	70	0.08	78	0.12	87	0.12	81	0.14
HDTMAB	93	0.31	93	0.32	82	0.13	75	0.10
BTEAC	78	0.22	93	0.26	78	0.15	86	0.25
HDTBPB	88	0.23	83	0.22	94	0.18	89	0.15
Aliquat 336	89	0.29	88	0.18	71	0.10	60	0.07

TBAB = tetra-n-butylammonium bromide; HDTMAB = hexadecyltrimethylammonium bromide; BTEAC = benzyltriethylammonium chloride; HDTBPB = hexadecyltri-n-butylphosphonium bromide; Aliquat 336 = methyltri-n-octylammonium chloride.

Table 9-19. Yields and inherent viscosities obtained for polycarbonate IIa and polythiocarbonate IIb.

Catalyst	IIa				IIb			
	30 min		60 min		30 min		60 min	
	%	n	%	n	%	n	%	n
None	—	—	51	0.14	—	—	—	—
HDTMAB	52	0.24	66	0.46	89	0.24	69	0.11
BTEAC	41	0.56	43	0.47	80	0.18	77	0.19

HDTMAB = hexadecyltrimethylammonium bromide; BTEAC = benzyltriethylammonium chloride.

Table 9-20. Yields and inherent viscosities obtained for polycarbonate IIIa and polythiocarbonate IIIb.

Catalyst	IIIa				IIIb			
	30 min		60 min		30 min		60 min	
	%	n	%	n	%	n	%	n
None	63	0.12	80	0.12	66	0.09	68	0.14
TBAB	80	0.13	84	0.14	81	0.23	84	0.16
HDTMAB	92	0.31	87	0.29	82	0.23	70	0.14
BTEAC	90	0.16	90	0.25	78	0.36	78	0.34
HDTBPB	79	0.17	91	0.24	86	0.23	84	0.17
Aliquat 336	92	0.30	92	0.22	82	0.47	83	0.19

TBAB = tetra-n-butylammonium bromide; HDTMAB = hexadecyltrimethylammonium bromide; BTEAC = benzyltriethylammonium chloride; HDTBPB = hexadecyltri-n-butylphosphonium bromide; Aliquat 336 = methyltri-n-octylammonium chloride.

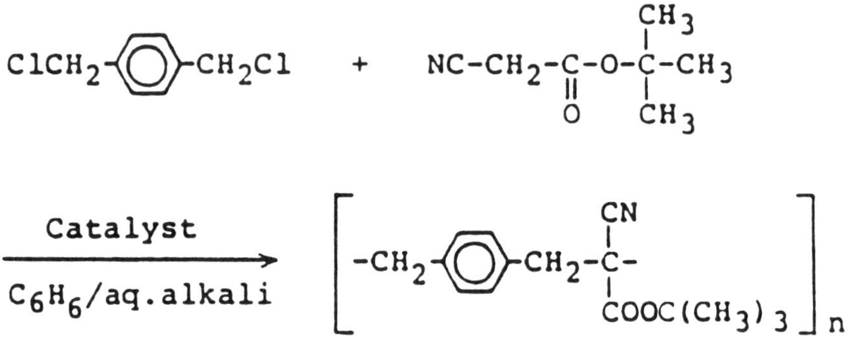

$$ClCH_2-\langle\bigcirc\rangle-CH_2Cl \quad + \quad NC-CH_2-\underset{\underset{O}{\|}}{C}-O-\underset{\underset{CH_3}{|}}{\overset{\overset{CH_3}{|}}{C}}-CH_3$$

$$\xrightarrow[C_6H_6/aq.alkali]{Catalyst} \left[-CH_2-\langle\bigcirc\rangle-CH_2-\underset{\underset{COOC(CH_3)_3}{|}}{\overset{\overset{CN}{|}}{C}}- \right]_n$$

Figure 9-21

e. Oxidative Coupling of Phenols

The first example of a PTC single-electron transfer polycondensation involved the polymerization of 4-bromo-2,6-dimethylphenol in a toluene–aqueous sodium hydroxide two-phase system in the presence of tetra-*n*-butylammonium hydrogen sulfate utilizing only oxygen from air as the oxidizing agent [63]. Figure 9-22 outlines two suggested propagation mechanisms. Initially, hydroxide ion deprotonates the phenol to form the corresponding phenoxide ion which ion pairs with the quaternary ammonium cation. An electron is then transferred from the phenoxide to molecular oxygen to produce the electron-deficient phenoxy radical. Attack of another phenoxide ion at the 4-position of the phenoxy radical produces a radical anion intermediate which eliminates a bromide ion to form a phenoxy radical dimer. The propagation mechanism involving attack of the phenoxy radical at the 4-position of the unionized phenol is not believed to be important. It was suggested that in this polymerization the role of the phase-transfer catalyst is not just the transfer of the highly reactive quaternary ammonium phenolate into the organic phase. At high conversions, the concentration of the phenolate is low and the phenoxy macroradicals are consumed by termination with oxygen of the air and water giving rise to phenol chain ends. At this point, the phase-transfer catalyst is believed to transfer the less organophilic hydroxide ion into the organic phase, where deprotonation of the terminal phenol takes place, thus allowing the oxidation and subsequent polymerization to continue. This condensation polymerization, which occurs by a radical reaction mechanism, has been termed a "reactive intermediate polycondensation."

The synthesis of poly(2,6-dimethyl-1,4-phenylene oxide) with one 2,6-dimethylphenol chain end and with well-defined molecular weight by PTC polymerization of 4-bromo-2,6-dimethylphenol in the presence of either 2,4,6-trimethylphenol or 4-*tert*-butyl-2,6-dimethylphenol as chain initiators has been reported [40,64,65]. The polymerization of 4-bromo-2,6-dimethylphenol in the presence

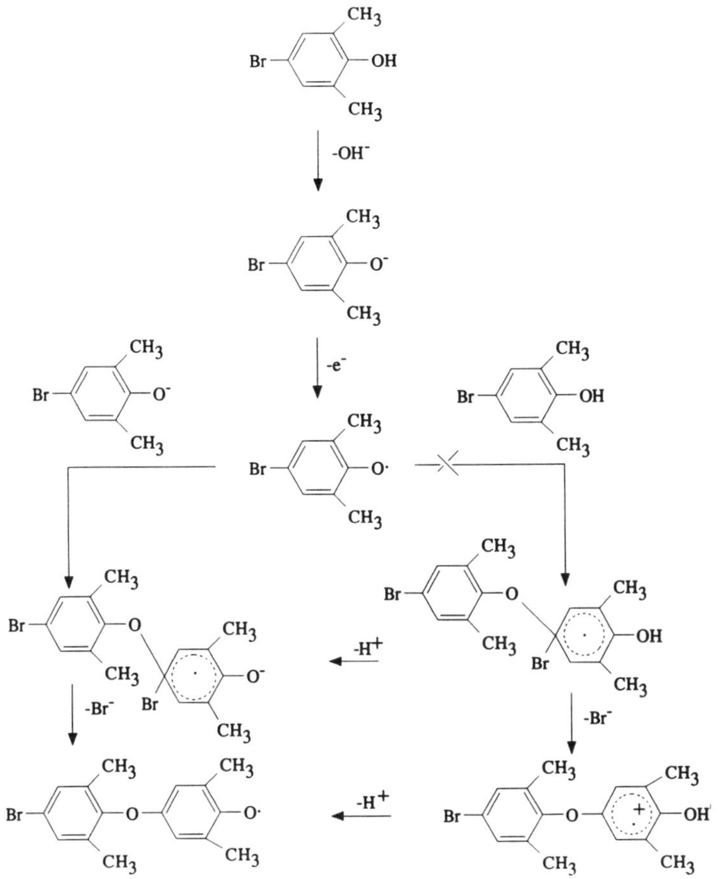

Figure 9-22

of 2,4,6-tri-*tert*-butylphenol or 4-bromo-2,6-di-*tert*-butylphenol has also been described [66].

The PTC depolymerization of poly(2,6-dimethyl-1,4-phenylene oxide) carried out in the presence of 2,4,6-trimethylphenol or 4-*tert*-butyl-2,6-dimethylphenol in a benzene–aqueous sodium hydroxide two-phase system in the presence of tetra-*n*-butylammonium hydrogen sulfate has been reported (Figure 9-23) [67].

Figure 9-24 outlines the suggested anion radical mechanism for the depolymerization process [68]. The polymeric phenoxide is primarily produced by deprotonation of the polymeric phenol by the quaternary ammonium salt of the monomeric phenol. According to the mechanism the monomeric phenoxide is oxidized by oxygen of the air to the corresponding phenoxy radical which attacks the para-position of the terminal phenolate unit of the polymer, resulting in a radical anion intermediate. The radical anion then partitions to starting monomeric radical

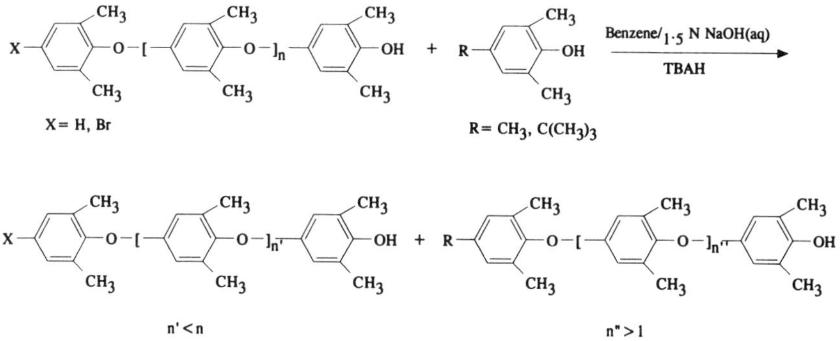

Figure 9-23

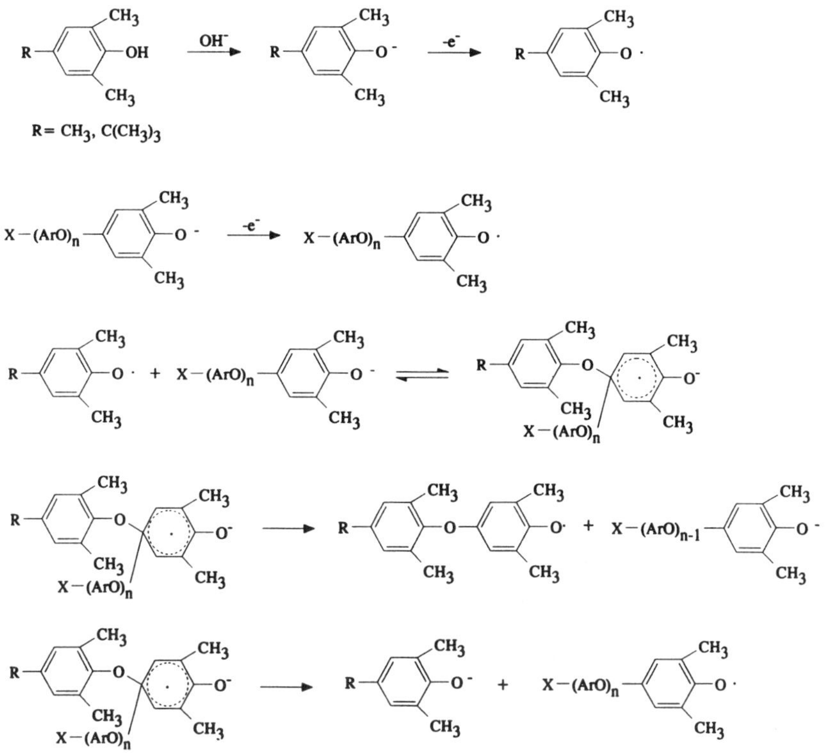

Figure 9-24

and polymeric phenolate or to dimeric radical and the polymeric phenolate diminished by one phenol unit. Other reactions contributing to the depolymerization process were considered to be relatively unimportant compared to those discussed above. The mechanism suggests that polymeric species derived from the depolymerization process should have either 2,4,6-trimethylphenoxy or 4-*tert*-butyl-2,6-dimethylphenoxy as chain ends. In addition, two polymer distributions will be generated. Both the molecular weights and the amounts of these two polymer distributions were found to be determined by the initial ratios of starting polymer to 2,4,6-trimethylphenol or 4-*tert*-butyl-2,6-dimethylphenol. It was experimentally observed that polymers obtained by depolymerization displayed a bimodal molecular weight distribution whereas those obtained by the polymerization of 4-bromo-2,6-dimethylphenol in the presence of 2,4,6-trimethylphenol and 4-*tert*-butyl-2,6-dimethylphenol displayed a monomodal molecular weight distribution.

The homopolymerization of 4-hydroxy-3,5-dimethylbenzyl alcohol and the copolymerization of 4-bromo-2,6-dimethylphenol in the presence of 4-hydroxy-3,5-dimethylbenzyl alcohol in a benzene–aqueous sodium hydroxide two-phase system in the presence of tetra-*n*-butylammonium hydrogen sulfate have been described [69]. The homopolymerization produced only a small yield of oligomer, the major product being the dimer. In contrast, the copolymerization produced poly(2,6-dimethyl-1,4-phenylene oxide) polymers in 71–89% yield. ^{1}H-NMR analysis indicated that a mixture of bifunctional and monofunctional polymers are formed. A radical–anion mechanism was proposed to rationalize the experimental observations.

2. Anionic Polymerizations

It has been reported that the polymerization of *n*-alkyl esters of acrylic acid initiated by tetra-*n*-butylammonium alkyl- and aryl-thiolates produced quantitative yields of the corresponding polyacrylate at room temperature (Figure 9-25) [70]. Isotactic polymers having a narrow molecular weight distribution were reported. The polymerization depended critically upon the solvent, the best choice often being acetonitrile, nitrobenzene, or a mixture of nitrobenzene and tetrahydrofuran. In the polymerization of *n*-butyl acrylate with tetra-*n*-butylammonium *n*-butylmercaptide, additional acrylate was added 2 h after completion of the polymerization at room temperature. It was observed that the polymerization continued with a quantitative formation of polymer having a slightly broader molecular weight distribution.

The use of the chiral macrocyclic multidentate ligands (Figure 9-26) in the anionic polymerization of ethyl, *tert*-butyl, or benzyl esters of methacrylic acid initiated by potassium *tert*-butoxide has been reported [71]. All three catalysts gave highly isotactic polymers with high optical rotations. The signs of the

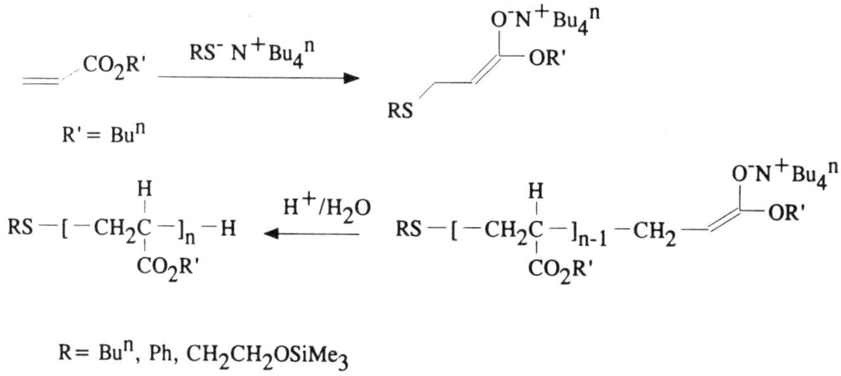

Figure 9-25

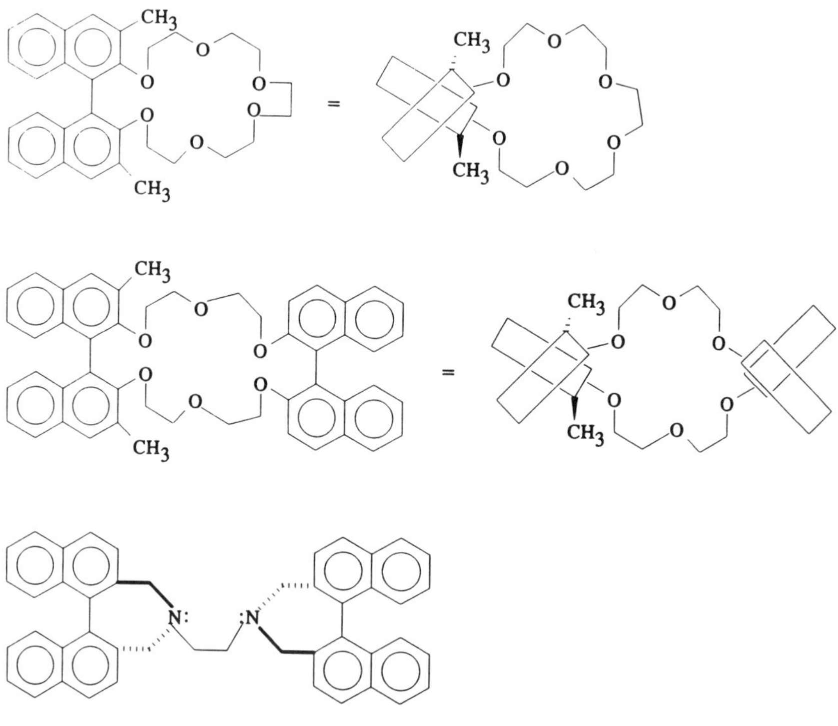

Figure 9-26

Table 9-21. Polymerization of n-butyl acrylate: effect of the phase-transfer catalyst on conversion.

Phase-transfer agent	Reaction time (h)	Percent conversion
n-Bu$_4$N$^+$HSO$_4^-$	21	97
Aliquat 336	21	97
CH3(CH2)$_{15}$C$_5$H$_5$N$^+$Br$^-$	24	92
CH3(CH2)$_{15}$ NMe$_3^+$Br$^-$	24	91
(CH3)$_4$N$^+$Cl$^-$	24	43
18-Crown-6	24	93
None	21	10

Aliquat 336 = methyltri-n-octylammonium chloride.

rotations were dependent on the configuration of the catalyst but independent of the nature of the alkyl group of the ester monomer.

3. Radical-Initiated Polymerizations

Potassium peroxydisulfate is a well known free radical initiator for the polymerization of vinyl monomers in aqueous media. The use of this inexpensive initiator is limited because of its modest solubility in aqueous media and its virtual insolubility in common organic solvents. It was reported that the 1:2 complex of potassium peroxydisulfate and 18-crown-6 was characterized. The complex was found to be soluble in methanol, dimethyl sulfoxide, and dimethylformamide [72]. Other reports indicated that peroxydisulfate may be phase transferred into a variety of solvents, including hydrocarbon solvents [73–75]. In addition, it was found that the transferred dianion could be utilized for the rapid polymerization of acrylic and methacrylic monomers, even at temperatures approaching ambient. For example, the PTC polymerization of n-butyl acrylate mediated by various crown ethers and quaternary ammonium salts was reported. The percent conversion of monomer to polymer for a specific reaction time (approximately 24 h) is summarized in Table 9-21.

Because it is well known that the activation energy for the decomposition of azobisisobutyronitrile is similar to that of potassium peroxydisulfate in 0.1 N aqueous base (31 kcal/mol and 33.5 kcal/mol, respectively), a comparison was made to determine the relative efficiencies of each of these initiators in the polymerization of n-butyl acrylate. Under conditions identical with those used with the crown/peroxydisulfate reactions, < 1% conversion to polymer was observed when azobisisobutyronitrile was used. The results indicate that the activation energy for peroxydisulfate decomposition in organic media must be markedly different from that in aqueous solution. It was conjectured that the apparent enhanced tendency toward fragmentation to sulfate radical anions may be due to Coulombic repusion between the anionic termini of the peroxydisulfate dianion.

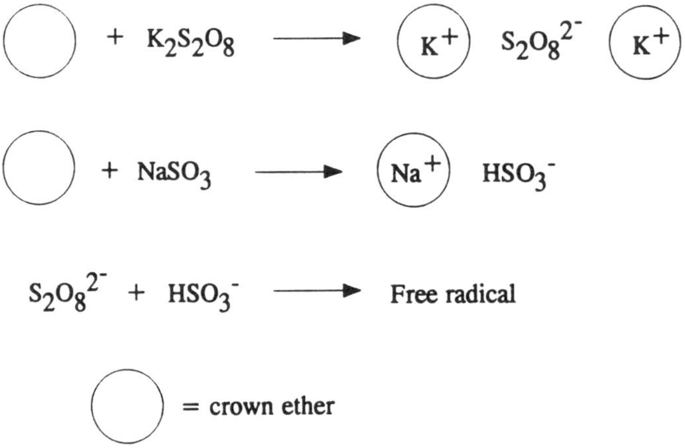

○ + $K_2S_2O_8$ ⟶ K^+ $S_2O_8^{2-}$ K^+

○ + $NaSO_3$ ⟶ Na^+ HSO_3^-

$S_2O_8^{2-}$ + HSO_3^- ⟶ Free radical

○ = crown ether

Figure 9-27

The effect of crown on the redox bulk polymerization of methyl methacrylate initiated by solid potassium persulfate together with solid sodium or potassium hydrogen sulfite has been reported [76]. The polymerization proceeded at 45°C only when the oxidizing agent, the reducing agent, and the crown ether were all present. In the absence of the crown, no reaction took place at this temperature. At temperatures > 55°C, the polymer was formed in the absence of the hydrogen sulfite ion. At temperatures < 50° C, polymerization took place very slowly in the absence of hydrogen sulfite; only trace quantities of polymer were detected. These results suggested that the crown solubilized both the persulfate and the hydrogen sulfite in the monomer phase, thus allowing the two inorganic reagents to interact and initiate the polymerization. The reaction sequence shown in Figure 9-27 illustrates this example of solid–liquid PTC. The polymerization of acrylonitrile was reported to proceed explosively.

The methylated β-cyclodextrins, heptakis(2,6-O-dimethyl)-β-cyclodextrin, and heptakis(2,3,6-O-trimethyl)-β-cyclodextrin have been found to promote the two-phase water/organic free radical polymerization of water-soluble vinyl monomers, such as acrylamide, sodium p-styrenesulfonate, acrylic acid, etc., initiated by water-insoluble free radical initiators such as azo or peroxide compounds [77]. Since the methylated β-cyclodextrins have high organophilic character and since it is well known that they form stable inclusion compounds with organic molecules, it was conjectured that the free radical initiator or the radical derived from the initiator was entrapped by the organophilic β-cyclodextrin in the organic phase and subsequently transported to the aqueous phase to initiate the polymerization of the water-soluble vinyl monomer.

In contrast to the above, it has been found that the two-phase polymerization of water-insoluble vinyl monomers using water-soluble free radical initiators is also accelerated by the addition of the organophilic β-cyclodextrins [78]. In particular, water/chloroform two-phase polymerization of benzyl methacrylate and phenyl methacrylate initiated by 2,2'-azoisobutyroamidine dihydrochloride, potassium peroxydisulfate, or ammonium peroxydisulfate was accelerated by the addition of small amounts of the organophilic methylated β-cyclodextrins. It was again conjectured that the organophilic methylated β-cyclodextrin traps a vinyl monomer in its hydrophobic cavity in the organic phase and transports it to the aqueous phase to react with the water-soluble initiator and then takes a monomer radical generated in the cavity again to the organic phase to initiate the polymerization. Experiments dealing with triphenylverdazyl appear to support the suggested mechanism.

The photochemically induced polymerization of methyl methacrylate employing an initiator system composed of methyl viologen–sodium dithionite–carbon tetrachloride was reported in an aqueous–organic two-phase system [79]. The initial rates of photopolymerization were found to be proportional to the square root of each of the initial concentrations of the methyl viologen, the sodium dithionite, and the carbon tetrachloride when each of these reagents was at relatively low concentration. At higher concentrations of the methyl viologen and the carbon tetrachloride, the initial rates reached a maximum and then began to decrease. This behavior was attributed to the generation of relatively high concentrations of radicals which increase the rate of polymerization termination. A PTC mechanism, involving the disproportionation of the methyl viologen cation radical, has been proposed. The details of the mechanism are outlined in Figure 9-28. According to the above mechanism, the methyl viologen dication was reduced by sodium dithionite in the aqueous phase to the cation radical. This species was then transferred from the aqueous to the organic phase where it disproportionated to give 1,1'-dimethyl-4,4'-bipyridylidene and methyl viologen. The former species reacted photochemically with carbon tetrachloride to produce the trichloromethyl radical initiator while the latter species transferred to the aqueous phase for further reaction with the dithionite.

The quaternary ammonium salts N-benzyl-3-(carboxyamide)pyridinium chloride and N-benzylpyridinium bromide and chloride photoinitiate the polymerization of methyl methacrylate. The primary photochemical step is believed to be an electron transfer from the halide ion to the pyridinium ion to generate a halogen atom that subsequently initiates the polymerization process [80–82].

The aqueous–organic two-phase polymerization of methyl methacrylate, employing the initiator system tetra-n-butylammonium chloride–sodium dithionite–carbon tetrachloride, has been reported [83]. It was observed that the initial rates of polymerization were proportional to the first power of the initial concentrations of tetra-n-butylammonium chloride and to the one-half power of the initial concentrations of sodium dithionite and carbon tetrachloride. In order to account for the experimental observation the mechanism outlined in Figure 9-29 was suggested.

The first step involved the transport of the dithionite dianion from the aqueous to the organic phase where, in the second step, it reacted with the carbon

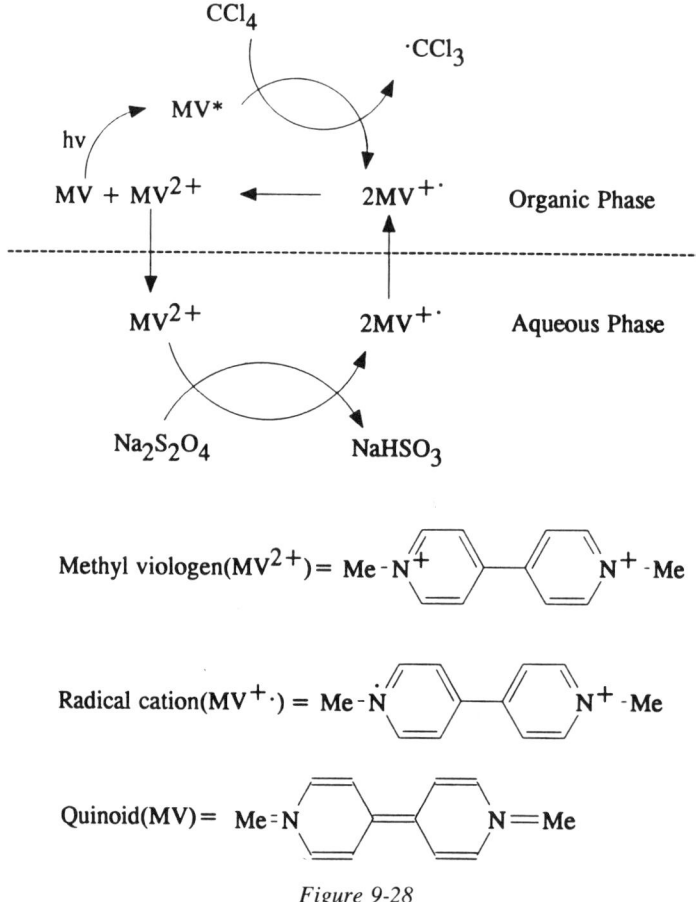

Figure 9-28

tetrachloride to form tetra-*n*-butylammonium chloride, sulfur dioxide, and the trichloromethyl radical which initiated polymerization.

C. Chemical Modification of Polymers

1. Chemical Modification of Polymer Backbone

PTC techniques have proven to be useful vehicles for the chemical modification of polymeric backbones. It has been demonstrated that, under homogeneous conditions, chloride ion may be displaced from polyvinyl chloride (PVC) by azide ion in dimethylformamide (DMF), dimethyl sulfoxide (DMSO), and hexamethylphosphoramide (HMPA) [84].

$$-CH_2-CHCl-CH_2-CHCl- + N_3^- \rightarrow -CH_2-CHCl-CH_2-CHN_3$$

$$2\text{TBA}^+ + \text{S}_2\text{O}_4^{-2} \rightleftharpoons (\text{TBA})_2\text{S}_2\text{O}_4$$

$$(\text{TBA})_2\text{S}_2\text{O}_4 + \text{CCl}_4 \xrightarrow{\text{slow}} \text{TBACl} + (\text{TBA})\text{S}_2\text{O}_4{}^{\cdot} + \text{CCl}_3{}^{\cdot}$$

$$(\text{TBA})\text{S}_2\text{O}_4{}^{\cdot} + \text{CCl}_4 \xrightarrow{\text{fast}} \text{TBACl} + 2\text{SO}_2 + \text{CCl}_3{}^{\cdot}$$

Initiation:

$$\text{CCl}_3{}^{\cdot} + \text{M} \longrightarrow \text{P}_1{}^{\cdot}$$

Propagation:

$$\text{P}_n{}^{\cdot} + \text{M} \longrightarrow \text{P}_{n+1}$$

Termination:

$$\text{P}_n{}^{\cdot} + \text{P}_m{}^{\cdot} \longrightarrow \text{P}_{n+m} \text{ or } \text{P}_n + \text{P}_m$$

Figure 9-29

In solvents in which sodium azide is insoluble, such as tetrahydrofuran (THF), dioxane, and cyclohexane, no reaction takes place. In the presence of laurylpyridinium chloride, however, reaction does take place. It was also reported that a variety of quaternary cations catalyze this reaction, and the order of catalytic activity of the quaternary halides is $Cl > Br > I$. The results are summarized in Figure 9-30; plots of initial rates of reaction vs. initial concentrations of nucleophilic reagent are shown. It is interesting to note that the initial rates of reaction on PVC are greater than the corresponding rates involving isopropyl chloride in spite of the fact that secondary halides are displaced in both cases. In addition, Figure 9-30 shows that the initial rates are not linear with respect to the concentration of nucleophilic species. For instance, in the reaction of pyridinium azide with PVC, a saturation effect was observed. It was suggested that the mechanism involves an initial absorption step followed by a nucleophilic reaction step. Using the Langmuir absorption isotherm as a model, the following equation for initial rates was derived:

$$v_0 = \frac{k_a AB\,[\text{PVC}]_0\,[\text{LPA}]_0}{1 + A\,[\text{LPA}]_0}$$

where $[\text{LPA}]_0$ is the initial concentration of quaternary azide, A is a constant related to the association of the quaternary azide with the PVC polymer, and B represents the maximum association of the quaternary azide with the PVC polymer. The equation predicts that initial rates should be first-order with respect to PVC and that a plot of $1/v_0$ vs. $1/[\text{LPA}]_0$ should be linear. From the slopes and intercepts the terms A and K_aB were evaluated. From this analysis it was reported that (1) the values of A appear to get larger as the length of the alkyl chains of the quaternary cation increases and (2) the values of k_aB increase as the lengths of the alkyl chains decrease. The interpretation of these parameters is consistent with the observation that the rate of reaction of quaternary azide with isopropyl chloride increases linearly over the same concentration ranges as with PVC; association between the quaternary salt and isopropyl chloride appears to be unimportant. Because chloride ion is the product of reaction, competitive association of the quaternary chloride with PVC was also important. The values of A and K_aB were found to vary with the polarity of the solvent. A is very small in DMF and substantially larger in RHF, cyclohexane, and dioxane. Also consistent with the model is the observation that A decreases with increasing degree of polymerization.

It has been reported that 18-crown-6 has been employed to promote the reaction between polyvinyl chloride and potassium acetate in tetrahydrofuran solvent [85].

$$-CH_2-CHCl-CH_2-CHCl-+CH_3CO_2^- \rightarrow -CH_2-CHCl-CH_2-CH(O_2CCH_3)-$$

The extent of acetate incorporation was followed spectrophotometrically by monitoring the increase of carbonyl adsorption at 1740 cm^{-1}. As the concentration of crown increased, the rate of acetate incorporation increased. The kinetic order with respect to crown was found to be 0.7. This order was rationalized in terms of the equilibrium between the potassium acetate tight ion pair and the "free" acetate ion where the potassium ion is complexed with the crown. It was reported that there was considerable darkening and crosslinking of the polymer with long reaction times.

The PTC dehydrochlorination of polyvinyl chloride by aqueous sodium hydroxide solution in the presence of quaternary ammonium or phosphonium halides has been reported.

$$-CH_2-CHCl-CH_2-CHCl-+OH^- \rightarrow -CH=CH-CH=CH-+2Cl^-+2H_2O$$

PVC powder was treated with 16.7% aqueous sodium hydroxide in the absence and in the presence of tetra-n-butylammonium bromide (0.033–0.01 mol/mol polyvinyl chloride). In the absence of quaternary salt the decrease in weight was

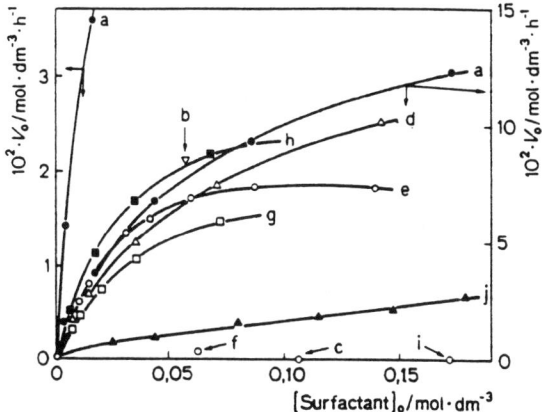

(a) tetra-n-butylammonium chloride
(b) tetra-n-butylammonium bromide
(c) tetra-n-butylammonium iodide
(d) benzyldimethyloctylammonium chloride
(e) laurylpyridinium chloride
(f) laurylpyridinium bromide
(g) benzyldimethylstearylammonium chloride
(h) dimethyldistearylammonium chloride
(i) sodium laurylbenzenesulfonate
(j) laurylpyridinium chloride, reaction of isopropyl chloride

Figure 9-30

negligible at 60°C for a period of 5 h. In the presence of the catalyst, however, reaction was rapid as reflected by the discoloration of the powder (deep brown to black) and the weight loss accompanying reaction. Accompanying the weight loss was the appearance of strong absorption bands at 960 cm^{-1} representing the out-of-plane deformations of C-H bonds. The resulting infrared spectrum was found to be very similar to *trans*-polyacetylene. Immediately after reaction with aqueous base, no substantial amount of oxidation was observed. However, after standing in air, the color faded and hydroxyl and carbonyl bands appeared in the infrared spectrum. The relationship between the activity of the catalyst and its structure was investigated. Tetra-n-butylammonium bromide and tetra-n-butylphosphonium bromide were found to be the most effective catalysts. Triethylbenzyl ammonium bromide and hexadecyltrimethylammonium bromide were ineffective. The effects of tetra-n-butylammonium bromide concentration, aqueous sodium hydroxide concentration, and temperature on percent conversion of PVC were reported. As the concentration of the quaternary salt increased, the conversion first increased and then leveled off to a constant value. It was observed that at a concentration of quaternary salt of 1.5×10^{-3} M, a third "oily" layer was formed, and, at this point, increasing the catalyst concentration had no effect on the percent conversion. Conversions vs. concentration of aqueous sodium hydroxide at 60° and 80°C were reported to exhibit maxima. As in the case with increasing concentration of quaternary salt, a third phase was also observed with increasing concentration of aqueous sodium hydroxide and qualitatively accounts for the observed maxima. The percent conversion also exhibited a maximum with increasing temperature. This was attributed to the decomposition of the quaternary salts at the higher temperatures. It was reported that the percent conversion increased with increasing molecular weight of the polyvinyl chloride. Explanations based upon the number of end groups and/or difference in surface area on rates of diffusion of the quaternary hydroxide were suggested.

The PTC dehydrochlorination reactions of solid copolymers of vinylidene chloride and vinyl chloride with aqueous sodium or potassium hydroxide solutions in the presence of quaternary ammonium and phosphonium salts have been reported [86].

$$-CH_2-CHCl-CH_2-CCl_2- + OH^- \rightarrow -CH=CH-CH=CCl- + 2Cl^- + 2H_2O$$

The effect of structure of the quaternary salts on the effectiveness of the dehydrochlorination was systematically investigated by varying the hydrophilic–organophilic balance of the cation. It was found that tetra-n-propylammonium bromide was the best catalyst. In addition, potassium hydroxide was more effective than sodium hydroxide. As in the case with the dehydrochlorination of polyvinyl chloride, increasing the concentration of quaternary salt or aqueous base produced a third "oily" phase. Unlike the reactions on polyvinyl chloride, a maximum in conversion was not observed with increasing temperature of the reaction. It was suggested that these observations lend serious doubt to the suggestion that the catalyst was decomposing at the higher temperatures. An alternative suggestion based upon the decrease of porosity of polyvinyl chloride above 80°C (the glass transition temperature) was put forth.

The dehydrofluorination of powdered polyvinylidene fluoride with aqueous solutions of sodium hydroxide in the presence of quaternary ammonium and phosphonium halides has been reported at temperatures varying from 25° to 48°C [87].

$$-CH_2-CF_2-CH_2-CF_2- + OH^- \rightarrow -CH=CF-CH=CF- + 2F^- + 2H^2O$$

Tetra-n-butyl ammonium bromide and tetra-n-butylphosphonium bromide were found to be the most active of the catalysts investigated. In comparison, 18-crown-6 had only modest activity. Based upon infrared spectroscopic measurements the structure of the resulting polymer was suggested to be a fluoro-substituted conjugated polyene. The fluoro-polyene was found to be stable to air oxidation. In addition, the wetability of the surface and the electrical conductivity were substantially increased. The dehydrofluorination of polyvinylidene fluoride in dimethylformamide under homogeneous conditions has also been reported [88].

Vinylidene fluoride–trifluoroethylene copolymers have been synthesized from polyvinylidene fluoride films and powders by the following synthetic sequence [89]. The first step involved the PTC dehydrofluorination of polyvinylidene fluoride using 50% aqueous sodium hydroxide in the presence of tetra-n-butylammonium hydrogen sulfate. The resulting unsaturated polymer was then reacted with molecular chlorine or bromine followed by substitution of the halogen on the backbone of the polymer by fluoride. This latter reaction was conducted under PTC conditions using an excess of potassium fluoride or potassium bifluoride in a variety of solvents (water, benzene, acetonitrile, or tetrahydrofuran) in the presence of tetra-n-butylammonium hydrogen sulfate. Displacement reactions were incomplete when tetra-$n$butylammonium hydrogen sulfate and potassium fluoride were used in an organic solvent. In contrast, employing tetra-n-butylammonium hydrogen sulfate, potassium bifluoride, and water resulted in a high degree of displacement on the brominated polymer. 18-Crown-6 with either the fluoride or the bifluoride was found to be ineffective in this displacement process.

The chemical modification of polyvinyl bromide using sodium methylsulfinate, methanethiol/base, and 2-hydroxyethanethiol/base in the presence of tetra-n-butylammonium hydroxide has been reported [90]. Reactions with polyepichlorohydrin and polyvinylidene bromide were also explored but these proved to be less successful.

PTC techniques have been employed in the scission of polysulfide crosslinks in rubber particles [91]. It has been reported that hydroxide ion transferred from water into benzene-swollen rubber particles by quaternary ammonium and phosphonium salts rapidly cleaves the polysulfide crosslinks. In general, quaternary salts having several large alkyl substituents were the most effective. No detectable main chain scission was found during the process. It was discovered, however, that the quaternary salt is consumed during the course of reaction. Thus the process is not truly catalytic. Model studies employing N-methyl-N,N,N-tri-n-alkylammonium chlorides with di-2-cyclohexen-1-yl disulfide revealed that the catalyst decomposition involved a highly selective demethylation of the quaternary ammonium ion by a 2-cyclohexenylthiolate counteranion. The decomposition was inhibited when active alkylating agents, such as benzyl chloride, were present.

2. Chemical Modification of Polymer Terminal Positions

Williamson etherification of α, ω-bis(hydroxyphenyl)polysulfone with a mixture of p- and m-chloromethylstyrenes in the presence of tetra-n-butylammonium hydrogen sulfate has been reported (Figure 9-31) [92,93].

Polyethersulfones containing pendant vinyl or ethynyl groups have been prepared employing a combination of etherification with benzyl chloride, Wittig reactions with formaldehyde, and dehydrobromination of vicinal dibromides all catalyzed by either tetra-n-butylammonium hydrogen sulfate or benzyltrimethy-

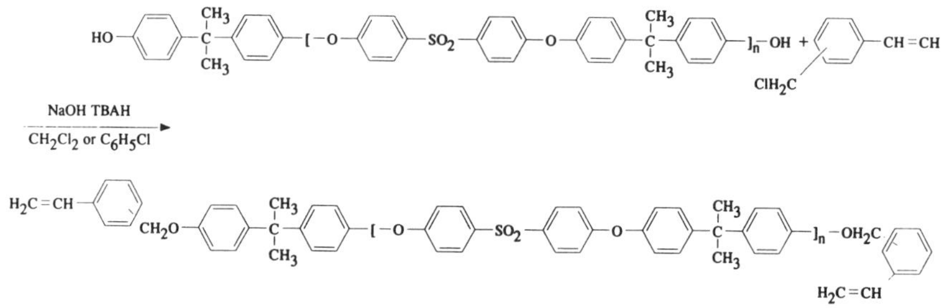

Figure 9-31

lammonium hydroxide under phase-transfer catalytic conditions (Figure 9-32) [94].

Although 2,6-dimethylphenol cannot be etherified to high conversion under normal Williamson reaction conditions, it can be quantitatively etherified in the presence of phase-transfer catalysts [95]. This procedure has been applied to the modification of oligomeric and polymeric species. It has been demonstrated that the 2,6-dimethyl chain end of the poly(2,6-dimethyl-1,4-phenylene oxide) could be quantitatively etherified under PTC conditions [96–98]. This reaction has been carried out with m- and p-chloromethylstyrene to give a vinylbenzylether macromonomer. The etherification has also been carried out with an α,ω-di(electrophilic) oligomer producing an ABA triblock copolymer containing poly(2,6-dimethyl-1,4-phenylene oxide) as the A segment. It was shown that α,ω-bis(vinylbenzyl)aromatic polyether sulfones represent an interesting new class of thermally reactive oligomers [99].

Narrow molecular weight copolymers derived from the oxidative coupling of 2,2-di(4-hydroxy-3,5-dimethylphenyl)propane and 2,6-dimethylphenol have been synthesized. These molecules were characterized by two terminal phenolic hydroxyls. Successful PTC bis-etherification with m- and p-chloromethylstyrene was accomplished using an organic solvent (tetrahydrofuran or benzene)—50% aqueous sodium hydroxide two-phase system in the presence of catalytic quantities of tetra-n-butylammonium hydrogen sulfate [100,101].

Two-phase transfer catalytic approaches have been published for the synthesis of α,ω-di(2-oxazoline)polymers. The first dealt with the etherification of α,ω-di(electrophilic) polymers with 2-(p-hydroxyphenyl)2-oxazoline. The second consisted of either extension of an α,ω-di(phenol) polymer with methylene chloride or the polymerization of bisphenol A with methylene chloride, both in the presence of 2-(p-hydroxyphenyl)-2-oxazoline. In both approaches, stoichiometric quantities of tetra-n-butylammonium hydrogen sulfate were used [102,103].

3. Chemical Modification of Pendant Groups Attached to Polymer Backbone

Early in the development of PTC modification of polymers, it was reported that crown ethers were effective catalysts in solubilizing and activating potassium

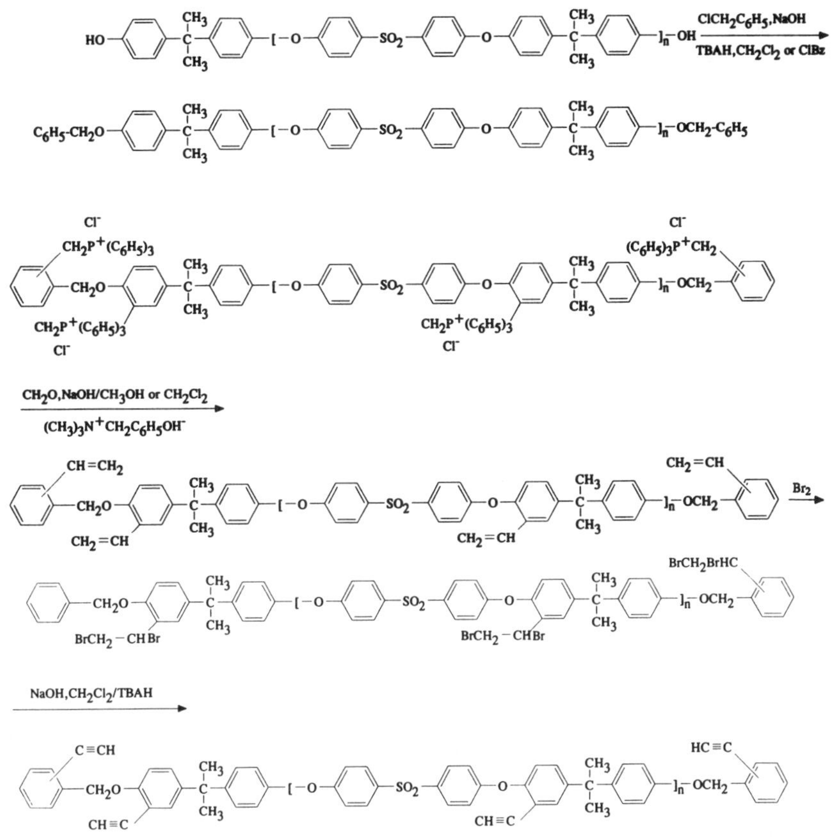

Figure 9-32

acetate in nonpolar solvents for reaction with partly chloromethylated polystyrene to form the corresponding acylated polymer [104]. Since this time, review articles dealing with the applications of phase-transfer catalysts in the modification and functionalization of macromolecules have appeared [105].

The base-promoted PTC reaction of carbazole, 2,3-dimethylindole, and N,N-diethylethanol amine with chloromethylated polymers and copolymers has been reported (Figure 9-33) [106]. The polymers included polyepichlorohydrin, a

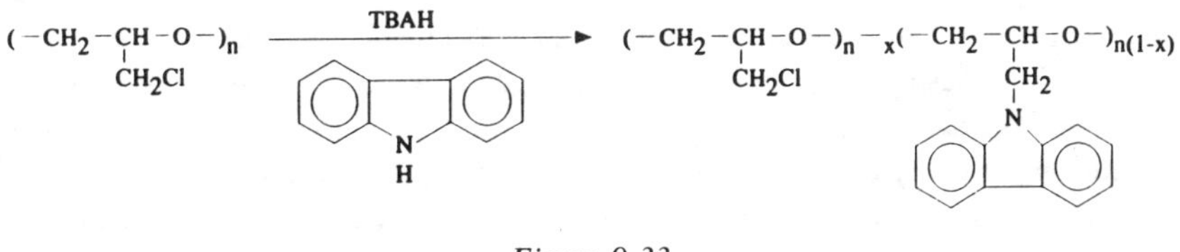

Figure 9-33

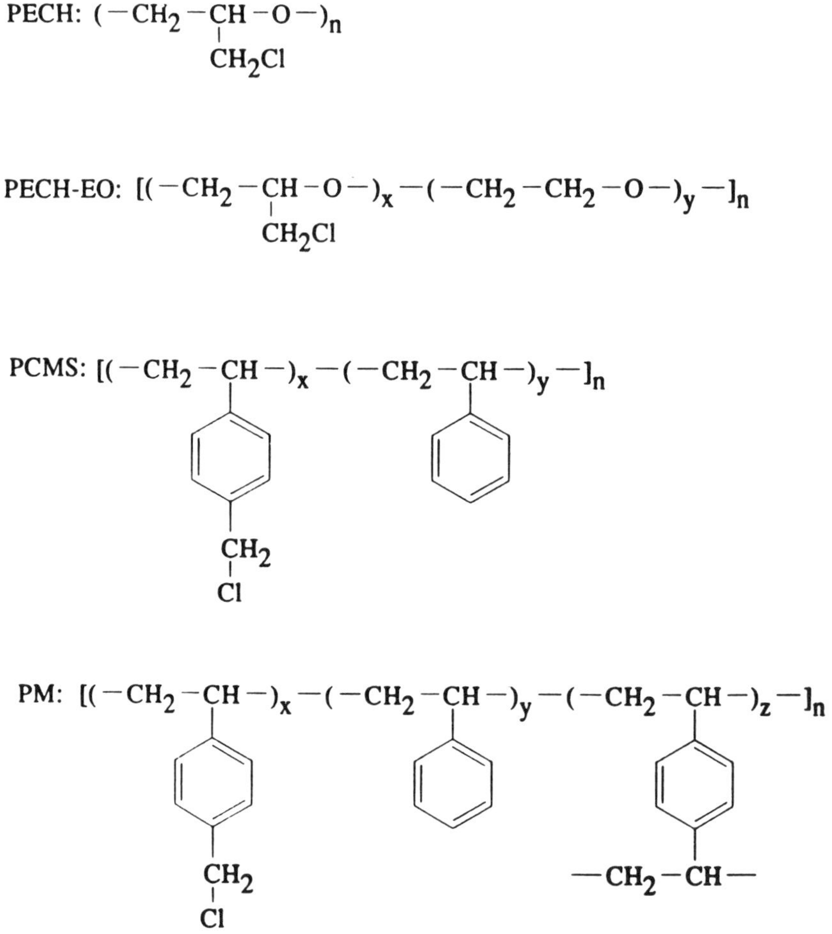

PECH: $(-CH_2-CH-O-)_n$
             |
             CH_2Cl

PECH-EO: $[(-CH_2-CH-O-)_x-(-CH_2-CH_2-O-)_y-]_n$
                     |
                     CH_2Cl

PCMS: $[(-CH_2-CH-)_x-(-CH_2-CH-)_y-]_n$ with p-(CH_2Cl)-phenyl and phenyl substituents

PM: $[(-CH_2-CH-)_x-(-CH_2-CH-)_y-(-CH_2-CH-)_z-]_n$ with p-(CH_2Cl)-phenyl, phenyl, and phenyl crosslinked through $-CH_2-CH-$

Figure 9-34

random copolymer of epichlorohydrin, and ethylene oxide, a soluble chlormethylated polystyrene; and a chloromethylated partially crosslinked polystyrene (Figure 9-34). The phase-transfer catalysts investigated were tetra-*n*-butylammonium hydrogen sulfate, dicyclohexyl-18-crown-6, and the [2.2.2] cryptand. It was found that dimethylformamide was the most efficient solvent for the chemical transformations. The solvents chloroform, nitrobenzene, methyl ethyl ketone, benzene, and benzene-dimethylformamide did not promote the same degree of

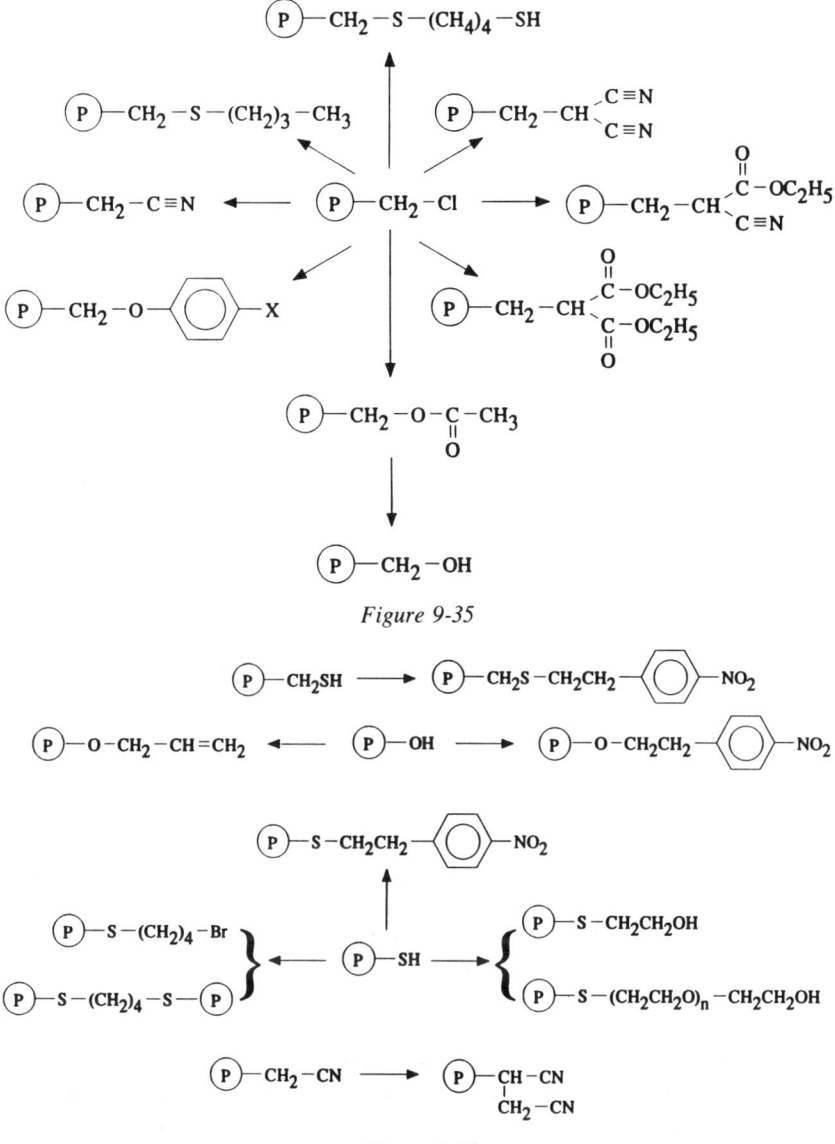

Figure 9-35

Figure 9-36

substitution under similar conditions. The catalyst effectiveness followed the decreasing order: [2.2.2] cryptand > dicyclohexyl-18-crown-6 > tetra-*n*-butylammonium hydrogen sulfate.

The introduction of a wide variety of functionalities in several chloromethylated crosslinked polystyrene resins by PTC techniques has been reported [107]. Dis-

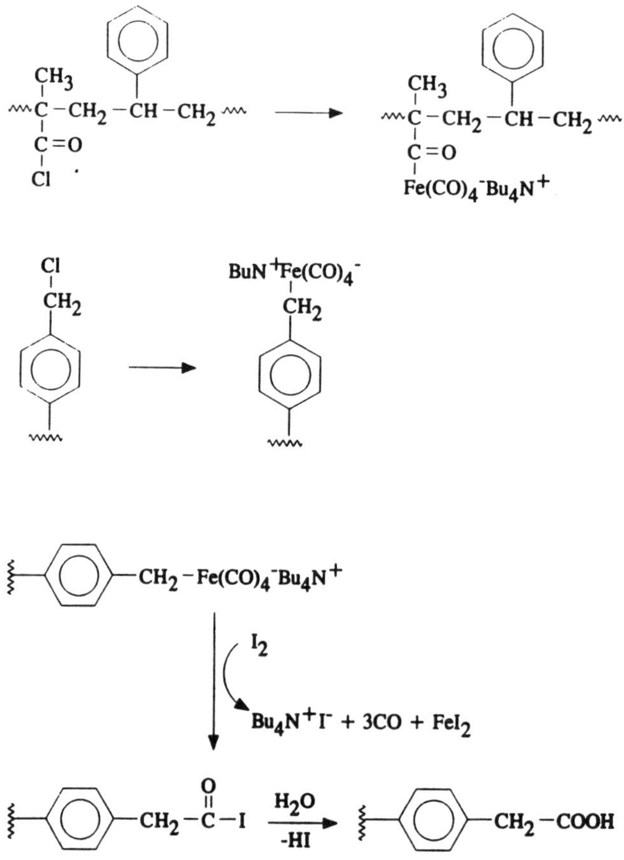

Figure 9-37

placement reactions involving potassium acetate, sodium cyanide, and aqueous base systems with n-1-butane thiol, 1,4-butanedithiol, p-formylphenol, p-nitrophenol, malononitrile, ethyl malonate, and ethyl cyanoacetate were explored (Figure 9-35). Adogen 464 and tetra-n-butylammonium chloride and hydroxide were employed as catalysts. The organic solvents include o-dichlorobenzene, 1,2-dichloroethane, methylene chloride, and benzene. The reported yields of functionalized polymer were excellent. Reaction of cyanide ion with the polystyrene functionalized with p-formylphenol produced the corresponding cyanohydrin which was trapped as the acetate by reaction with acetic anhydride. Other phase-transfer reactions involving polystyrene functionalized with groups that could be made nucleophilic by reaction with base (thiol, hydroxyl, and cyanomethyl) were also reported (Figure 9-36).

Nucleophilic substitution reactions of the potassium salts of iodide, nitrite, bromide, hydrosulfide, and thiocyanate on chlormethylated poly(styrene-co-divenylbenzene) in the presence of 18-crown-6 in acetonitrile, benzene, acetonitrile/ water, and benzene/water have been investigated [108]. Reactions of linear chloromethylated polystyrene with the salts of acetate, phenolate, thioacetate, N,N-diethyldithiocarbamate, and p-toluenesulfinate employing crown ethers and quaternary ammonium and phosphonium salts have also been reported [109].

The effects of phase-transfer catalysts on the alkaline hydrolysis of a sterically hindered polymeric ester such as partially crosslinked polymethyl methacrylate have been reported [110]. A slightly crosslinked copolymer of methyl methacrylate with 0.2% divinylbenzene was employed. 18-Crown-6, the [2.2.2] cryptand, and PEG (mol wt 2000) were employed as the phase-transfer catalysts in a variety of solvents (1,4-dioxane, toluene, dimethylformamide, and combinations of these with water). It was observed that PEG was only slightly less effective than the crown. Quite surprisingly, the [2.2.2] cryptand was found to be less active than the crown.

PTC functionalizations of chloromethylated styrene–divinylbenzene copolymers and styrene–methacryloyl chloride copolymers with tetra-n-butylammonium tetracarbonylferrate dianion have been reported (Figure 9-37) [111]. Carboxylic acid groups were subsequently produced from the styrene–divinylbenzene copolymers by treatment with iodine.

References

1. V. Percec and B.C. Auman, Makromol. Chem., **185,** 617 (1984).
2. P.W. Morgan, Condensation Polymers by Interfacial and Solution Methods, Interscience, New York, 349 (1965).
3. H-B. Tsai and Y-D. Lee, J. Polym. Sci., Part A: Polym. Chem., **25,** 1506 (1987).
4. B. Yamada, Y. Yutaka, T. Matsushita, and T. Otsu, J. Polym. Sci., Part C: **14,** 277 (1976).
5. G.G. Cameron and K.S. Law, Polym. Rep., **22,** 272 (1981).
6. A. Conix and U. Laridon, Angew. Chem., **72,** 116 (1960).
7. Y. Shuto, Kogyo Kagaku Zasshi, **67,** 367 (1964); Chem. Abstr. **61,** 8419d (1964).
8. D.W. Thomson and G.F.L. Ehlers, J. Polym. Sci., Part A, Polym. Chem., **2,** 1051 (1964).
9. Y. Imai, M. Ueda, and M. Ii, Makromol. Chem., **179,** 2085 (1978).
10. Y. Imai, J. Macromol. Sci.-Chem., **A15,** 833 (1981).
11. Y. Imai, N. Sato, and M. Ueda, Makromol. Chem., Rapid Commun., **1,** 419 (1980).
12. M. Sata, Makromol. Chem., Rapid Commun., **5,** 151 (1984).
13. Y. Imai, S. Abe, and M. Ueda, J. Polym. Sci., Part A: Polym. Chem., **19,** 3285 (1981).
14. W. Podkoscielny and D. Wdowicka, J. Polym. Sci., Part A: Polym. Chem., **21,** 2961 (1983).

15. T.D. N'Guyen and S. Boileau, Polym. Bull., **1,** 817 (1979).
16. A.K. Banthia, D. Lunsford, D.C. Webster, and J.E. McGrath, J. Macromol Sci.-Chem., **A15,** 943 (1981).
17. T.D. N'Guyen and S. Boileau, Polym. Bull., **1,** 817 (1979).
18. N. Yamazaki and Y. Imai, Polym. J. **15,** 603 (1983).
19. V. Percec and H. Nava, Makromol. Chem., Rapid Commun., **5,** 319 (1984).
20. V. Percec, B.C. Auman, H. Nava, and J.P. Kennedy, J. Polym. Sci., Part A: Polym. Chem., **26,** 721 (1988).
21. V. Percec, T.D. Shaffer, and H. Nava, J. Polym. Sci., Part C: Polym. Lett., **22,** 637 (1984).
22. T.D. Shaffer and V. Percec, Makromol. Chem., Rapid Commun., **6,** 97 (1985).
23. T.D. Shaffer and V. Percec, J. Polym. Sci., Part C: Polym. Lett., **23,** 185 (1985).
24. V. Percec, H. Nava, and H. Jonsson, J. Polym. Sci., Part A: Polym. Chem., **25,** 1943 (1987).
25. V. Percec and H. Nava, J. Polym. Sci., Part A: Polym. Chem., **25,** 405 (1987).
26. T.D. Shaffer and V. Percec, J. Polym. Sci., Part A: Polym. Chem., **25,** 2755 (1987).
27. T.D. Shaffer, M. Jamaludin, and V. Percec, J. Polym. Sci., Part A: Polym. Chem., **23,** 2913 (1985).
28. T.D. Shaffer and V. Percec, J. Polym. Sci., Part A: Polym. Chem., **24,** 451 (1986).
29. T.D. Shaffer and V. Percec, Makromol. Chem., **187,** 111 (1986).
30. V. Percec and B.C. Auman, Polym. Bull. **12,** 253 (1984).
31. V. Percec and B.C. Auman, Polym. Bull. **10,** 385 (1983).
32. T.D. Shaffer, K. Antolin, and V. Percec, Makromol. Chem., **188,** 1033 (1987).
33. V. Percec, H. Nava, and B.C. Auman, Polym. J., **16,** 681 (1984).
34. M. Ueda, Y. Oishi, N. Sakai, and Y. Imai, Macromolecules, **15,** 248 (1982).
35. Y. Imai, R. Takasawa, and M. Ueda, Polym. Preprints, Japan, **30,** 884 (1981).
36. L.H. Tagle, E.R. Diaz, M.P. De La Maza, and J.C. Vega, J. Polym. Sci., Part A: Polym. Chem., **24,** 495 (1986).
37. M. Ueda, N. Sakai, M. Komatsu, and Y. Imai, Makromol. Chem., **183,** 65 (1982).
38. R. Kellman, R.F. Williams, G. Dimotsis, D.J. Gerbi, and J.C. Williams, Aromatic substitution in condensation polymerization catalyzed by solid–liquid phase transfer, in Phase Transfer Catalysis: New Chemistry, Catalysts, and Applications, ACS Symposium Series 326, ed. C.M. Starks, 128 (1985).
39. R. Kellman, D. Gerbi, R.F. Williams, and J.L. Morgan, Polym. Prepr., Am. Chem. Soc., Div. Polym. Chem., **21,** 164 (1980); **22,** 383, 385, 387 (1981).
40. D.J. Gerbi, G. Dimotsis, J.L. Morgan, and R.F. Williams, J. Polym. Sci., Part C: Polym. Lett., **23,** 551 (1985).
41. K. Soga, S. Hosoda, and S. Ikeda, J. Polym. Sci., Part A: Polym. Chem., **17,** 517 (1979).

42. K. Soga, S. Hosoda, and S. Ikeda, J. Polym. Sci., Part C: Polym. Lett., **15,** 611 (1977).
43. L.H. Tagle and F.R. Diaz, Eur. Polym. J., **23,** 109 (1987).
44. N.S. McPherson, M.L. Clachan, and K.R. Tatchell, Chem. Abstr., **59,** 65926 (1963); Brit. 927178, Bexford Ltd., invs. (1963).
45. E.P. Goldberg and F. Scardiglia, Chem. Abstr., **62,** 5398a; Fr. 1358680, (1965); Borg-Warner Corp., inves. (1964).
46. E.P. Goldberg and F. Scardiglia, Chem. Abstr., **62,** 4182f; Fr. 1358685 (1965); Borg-Warner Corp., inves. (1964).
47. E.P. Goldberg and F. Scardiglia, Chem. Abstr., **66,** 3015t; U.S. 3271368 (1967), Borg-Warner Corp., invs. (1966).
48. Chem. Abstr., **71,** 13509p (1969); Fr. 1538973, Borg-Warner Corp. (1968).
49. G.L. Bronde and T.L. Pickering, Chem. Abstr. **76,** 155741k (1972); U.S. 3640965, Union Carbide Corp., invs. (1972).
50. E.P. Goldberg and F. Scardiglia, Chem. Abstr., **66,** 3019y (1967); U.S. 3227055, Borg-Warner Corp., invs. (1966).
51. Chem. Abstr. **58,** 6949a (1963); Belg. 597208, Gevaert Photo-Producten N.V. (1961).
52. R.J. Schott, F. Scardiglia, E.P. Goldberg, and D.F. Hoeg, Macromol. Synth., **3,** 63 (1972); Chem. Abstr. **83,** 28622s (1975).
53. L.H. Tagle, F.R. Diaz, and A.M. Munoz, Polym. Bull., **11,** 493 (1984).
54. L.H. Tagle, F.R. Diaz, J.C. Vega, and P.F. Alquinta, Makromol. Chem., **186,** 915 (1985).
55. L.H. Tagle, F.R. Diaz, and P.E. Riveros, Polym. J., **18,** 501 (1986).
56. L.H. Tagle, F.R. Diaz, and P. Salas, J. Macromol. Sci.-Chem., **A26,** 1321 (1989).
57. L.H. Tagle, F.R. Diaz, and N. Valdebenito, Polym. Bull. **18,** 479 (1987).
58. L.H. Tagle, F.R. Diaz, and M.P. De La Maza, Polym. Bull., **18,** 485 (1987).
59. L.H. Tagle, F.R. Diaz, and P. Salas, Eur. Polym. J., **25,** 549 (1989).
60. L.H. Tagle, F.R. Diaz, and C. Margozzini, Polym. Bull. **25,** 319 (1991).
61. Y. Imai, T-Q. Nguyen, and M. Ueda, J. Polym. Sci., Part C: Polym. Lett., **19,** 205 (1981).
62. Y. Imai, A. Kameyama, T-Q. Nguyen, and M. Ueda, J. Polym. Sci., Part A: Polym. Chem., **19** (1981).
63. V. Percec and T.D. Shaffer, J. Polym. Sci., Part C: Polym. Lett., **24,** 439 (1986).
64. K. Muhlbach and V. Percec, J. Polym. Sci., Part A: Polym. Chem., **25,** 2605 (1986).
65. V. Percec and J.H. Wang, J. Polym. Sci., Part A: Polym. Chem., **29,** 63 (1991).
66. J.H. Wang and V. Percec, Polym. Bull., **25,** 33 (1991).
67. V. Percec and J.H. Wang, Polym. Bull., **24,** 63 (1990).
68. V. Percec and J.H. Wang, Polym. Bull., **24,** 71 (1990).

69. J.H. Wang and V. Percec, Polym. Bull., **25,** 25 (1991).
70. M.T. Reetz and R. Ostarek, J. Chem. Soc., Chem. Commun., **213** (1988).
71. D.J. Cram and D.Y. Sogah, J. Am. Chem. Soc., **107,** 8301 (1985).
72. T.N. Rakhmatulina, E.N. Baiborodina, A.V. Rzhepka, V.A. Lopyrev, and M. Voronkov, Vysokomol. Soedin., Ser. B, **21,** 229 (1979); Chem Abstr., **90,** 187436v (1979).
73. J.K. Rasmussen and H.K. Smith II, J. Am. Chem. Soc., **103,** 730 (1981).
74. J.K. Rasmussen and H.K. Smith II, Makromol. Chem., **182,** 701 (1981).
75. J.K. Rassmussen, S.M. Heilmann, P.E. Toren, V.P. Alphonsus, and T.A. Kotnour, J. Am. Chem. Soc., (1983).
76. M. Takeishi, H. Ohkawa, and S. Hayama, Makromol. Chem. Rapid Commun., **2,** 457 (1981).
77. N. Kunieda, H. Taguchi, S. Shiode, and M. Kinoshita, Makromol. Chem., Rapid Commun., **3,** 395 (1982), H. Taguchi, N. Kunieda, and M. Kinoshita, ibid. **3,** 495 (1982); H. Taguchi, M. Kunieda, and M. Kinoshita, Makromol. Chem., **149,** 925 (1983).
78. N. Kunieda, S. Shiode, H. Ryoshi, H. Taguchi, and M. Kinoshita, Makromol. Chem., Rapid Commun., **5,** 137 (1984).
79. S. Shimada, Y. Obata, K. Nakagawa, and K. Tabuchi, Polym. J., **22,** 777 (1990).
80. S. Shimada, K. Nakagawa, and K. Tabuchi, Polym. J., **21,** 275 (1989).
81. K. Tabuchi and N. Sakota, Polym. J., **15,** 57 (1983).
82. K. Tabuchi and N. Sakota, Polym. J., **15,** 569 (1983).
83. S. Shimada, Y. Obata, K. Nakagawa, and K. Tabuchi, Polym. J., **23,** 305 (1991).
84. M. Takeishi, R. Kawashima, and M. Okawara, Die Makromolekulare Chemie, **167,** 261 (1973).
85. J. Lewis, M.K. Naqvi, and G.S. Park, Makromol. Chem., Rapid Commun., **1,** 119 (1980).
86. F-F. He and H. Kise, J. Polym. Sci., Part A: Polym. Chem., **21,** 1729 (1983).
87. H. Kise and H. Ogata, J. Polym. Sci., Part A: Polym. Chemi., **21,** 3443 (1983).
88. A.J. Dias and T.J. McCarthy, J. Polym. Sci., Part A: Polym. Chem., **23,** 1057 (1985).
89. B. Hahn and V. Percec, J. Polym. Sci., Part A: Polym. Chem., **25,** 783 (1987).
90. J.M.J. Frechet, J. Macromol. Sci.-Chem., **A15,** 877 (1981).
91. P.P. Nicholas, The scission of polysulfide cross-links in rubber particles through phase-transfer catalysis, in Phase Transfer Catalysis: New Chemistry, Catalysts, and Applications, ed. C.M. Starks, ACS Symposium Series **326,** 155 (1985).
92. V. Percec and B.C. Auman, Makromol. Chem., **185,** 1867 (1984).
93. V. Percec, P.L. Rinaldi, and B.C. Auman, Polym. Bull., **10,** 215 (1983).
94. V. Percec and B.C. Auman, Makromol. Chem., **185,** 2319 (1984).

95. A. McKillop, J.C. Fiaud, and H.P. Hug, Tetrhedron, **30,** 1379 (1974).
96. V. Percec and H. Nava, Makromol. Chem. Rapid Commun., **5,** 319 (1984).
97. V. Percec, P.L. Rinaldi, and B.C. Auman, Polym. Bull., **10,** 397 (1983).
98. V. Percec, P.L. Rinaldi, and B.C. Auman, Polym. Bull., **10,** 397 (1983).
99. V. Percec and B.C. Auman, Makromol. Chem., **185,** 1867 (1984).
100. H. Nava and V. Percec, J. Polym. Sci., Part A: Polym. Chem., **24,** 965 (1986).
101. H. Nava and V. Percec, J. Polym. Sci., Part A: Polym. Chem., **24,** 965 (1986).
102. V. Percec, H. Nava, and J.M. Rodriguez-Parada, Polym. Bull., **12,** 261 (1984).
103. V. Percec, H. Nava, and J.M. Rodriguez-Parada, J. Polym. Sci., Part C: Polym. Lett., **22,** 523 (1984).
104. J.F.L. Rovers, Polymers, **17,** 1107 (1976).
105. L.J. Mathias, J. Macromol. Sci.-Chem., **A15,** 853 (1981).
106. T.D. N'Guyen, A. Deffieux, and S. Boileau, Polymers, **19,** 423 (1978).
107. J.M.J. Frechet, M.D. de Smet, and M.J. Farrall, J. Org. Chem., **44,** 1774 (1979).
108. A.S. Gozdz and A. Rapak, Makromol. Chem., Rapid Commun., **2,** 359 (1981).
109. T. Nishikubo, T. Iizawa, and K. Kobayashi, Makromol. Chem., Rapid Commun., **1,** 765 (1980).
110. A.S. Gozdz, Makromol. Chem., Rapid Commun., **2,** 443 (1981).
111. C. Ungurenasu and C. Cotzur, Polym. Bull., **22,** 151 (1989).

10

Phase-Transfer-Catalyzed Oxidations

A. Introduction

Phase-transfer catalysis (PTC) offers many excellent opportunities for conducting oxidation reactions using inexpensive primary oxidants such as oxygen, sodium hypochlorite, hydrogen peroxide, electrooxidation, permanganate, periodic acid, and others. Reviews on including the use of PTC oxidations have been published [1,2].

PTC oxidations occur by a variety of mechanisms. One of the most important aspects of PTC oxidations is the frequent use of transition metal cocatalysts, often in the form of easily transferrable anions as tungstate, molybdate, chromate, and others, as described in this chapter. This combination of PTC catalyst plus metal cocatalysts allows the chemist to use the low-cost basic oxidants as a source of oxidizing power, but yet to obtain products with high specificity and selectivity.

B. Permanganate Oxidations

1. General Comments

Conventional methods for using permanganate as an oxidant for organic substances are severely limited by the lack of solvents that can dissolve permanganate salts without excessive oxidation of the solvent itself. The PTC technique eliminates the solvent problem, as many oxidations can be run without solvent; and, if one is required, the chemist can choose from a variety of solvents that are attacked slowly or not at all by permanganate.

PTC permanganate oxidations are easy to conduct in the laboratory, but they pose certain problems for commercial operations. Unless certain complexing

Table 10-1. Extraction of quaternary ammonium permanganates into CH_2Cl_2 [20].

Quaternary ammonium cation	Log K_E^a
Tetraethylammonium	1.64
Tetra-n-propylammonium	3.71
Methyl tri-n-butylammonium	4.72
Tetra-n-butylammonium	4.98

$^a K_E$ = [QMnO$_4$] (in CH$_2$Cl$_2$)/[Q$^+$][MnO$_4^-$] (in water).

agents are present, manganese dioxide precipitates as a difficult-to-filter, finely divided solid, sometimes surface-coated with adsorbed product. Separation and recovery of organic products, and handling and disposal of the manganese dioxide, increase the expense of using permanganate. In principle, it is possible to stabilize intermediate manganese species to prevent their disproportionation to manganese dioxide, such as by complexation with polyphosphoric acid, and to electrochemically regenerate permanganate as rapidly as it is used in the organic phase oxidation. In practice, however, this is difficult to do [3].

2. Transfer of Permanganate into Organic Phases

Permanganate anions or salts are among the easiest to transfer from aqueous to organic solutions. Quaternary ammonium, phosphonium, and arsonium salts [4–8], crown ethers [9–12], and some linear polyethers [13,26,28] are all suitable transfer agents. These catalysts have large extraction constants [14,28], particularly when the organic-phase solvent is somewhat polar such as methylene chloride. In acidic solutions even trialkyl amines transfer permanganate to organic solutions, presumably by formation of $R_3NH^+MnO_4^-$, but perhaps also by formation of amine oxides, which can function as phase-transfer catalysts [15].

Preformed quaternary ammonium permanganates can be prepared simply by mixing a quaternary ammonium chloride in organic solution with aqueous potassium permanganate solution. Removal of solvent from the organic solution leaves solid quaternary ammonium permanganates, useful for general oxidation applications [16,30]. However, these products tend to undergo violent thermal decomposition [17–19] and should be used only with the utmost care and precaution.

The exceptional extractability of quaternary ammonium permanganate salts into methylene chloride from water is illustrated in Table 10-1.

The solubilities of several permanganate salts and complexes in four organic solvents are listed in Table 10-2.

Harris and Case [20] examined permanganate transfer into benzene using dialkyl ethers of polyethylene glycol (PEG) and they measured the corresponding rate of oxidation of 1-octene in benzene solution. With the exception of methylene chloride as organic-phase solvent, most simple PEGs are themselves partitioned into the aqueous potassium permanganate solutions [21]. To improve organic-

Table 10-2. Solubilities of quaternary ammonium permanganates in four organic solvents.

Cation	Solubility (moles/liter) in solvent			
	CH_2Cl_2	$CHCl_3$	CCl_4	$C_6H_5CH_3$
Tetra-n-butylammonium	0.417	unst[a]	2.96×10^{-5}	34.4×10^{-5}
Tetra-n-octylammonium	0.713	0.604	59.3×10^{-5}	40.2×10^{-5}
Methyl tri-n-butylammonium	1.83	1.14	5.62×10^{-5}	42.3×10^{-5}
Methyl tri-n-octylammonium	1.38	1.07	0.016	0.798
n-Heptyl triphenylphosphonium	1.36	1.28	insol.	20.2×10^{-5}
Benzyl triphenylphosphonium	0.43	0.093	insol.	insol.
18-Crown-6	0.813	0.257	insol.	16.4×10^{-5}

[a]Compound was unstable.

phase solubility they tested the ability of several dialkyl ethers of PEGs to improve permanganate transfer and to catalyze 1-octene oxidation, as shown by the data in Table 10-3.

Plotting oxidation rate against the ratio of alkyl carbons to PEG molecular weight (an indicator of hydrophobic–hydrophilic balance) illustrates, as in Figure 10-1, the sensitivity of catalysis to partitioning, and possibly also to the surface activity of the catalyst [22]. These data suggest that the oxidation rate is almost entirely dependent on the rate of the permanganate transfer step.

In these studies, the purple coloration of permanganate in benzene could be observed when small amounts of PEG ethers were added to potassium permanganate–water–benzene systems, but not when original PEGs were added.

Solid potassium permanganate is readily transferred into benzene, toluene, or methylene chloride, by using 15-crown-5, 18-crown-6, or tris(2,5-dioxaheptyl)amine (TDA-1) as the phase-transfer agents. These solutions gradually decom-

Table 10-3. Partitioning and catalytic activity of dialkyl ethers of PEG for potassium permanganate into benzene, and corresponding oxidation rates for 1-octene.

PEG Ether[a]	Partitioning (% in C_6H_6 vs. % in water)	Oxidation rate const. k min^{-1}
C_4–PEG1500–C_4	14	0.008
C_6–PEG1500–C_6	84	—
C_{18}–PEG6000–C_{18}	Emulsion	0.04
C_{18}–PEG750–Me	108	0.018
C_{18}–P EG1900–Me	39	0.058
C_{18}–PEG5000–Me	37	0.016
C_8–PEG5000–Me	12	0.008
C_4–PEG5000–Me	13	<0.002
PEG6000	<1	<0.0002

[a]The designation, for example, "C_4–PEG1500–C_4," represents a PEG of mol wt = 1500, capped at both ends by a n-butyl group.

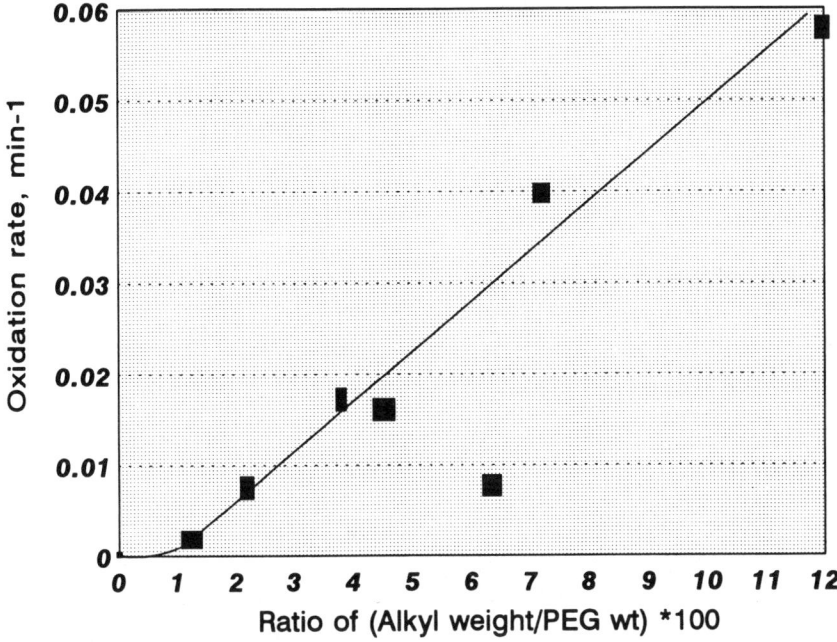

Figure 10-1. Effect of PEG diethers as phase-transfer catalysts for permanganate oxidation of 1-octene.

pose at rates that markedly depend on the transferring agent used, and the extent of solvation [23]. Use of TDA-1, $N[CH_2CH_2OCH_2CH_2OCH_3]_3$, as phase-transfer catalyst offers a good laboratory method for liquid–solid PTC permanganate oxidations [24,25]. TDA-1 addition to a mixture of methylene chloride and powdered potassium permanganate rapidly causes transfer of the oxidizing agent into the organic phase, forming a deep purple color. If the concentration of TDA-1 complex with $KMnO_4$ is too high, fairly rapid decomposition produces manganese dioxide; but dilute solutions are stable for hours or days, with slow decomposition.

3. PTC Permanganate Oxidations

a. Oxidation of Olefins

PTC permanganate oxidations of olefins can be made to yield *cis*-diols, α-hydroxycarbonyls, carboxylic acids, or aldehydes as major products depending on reaction conditions and work-up, as outlined in Figure 10-2, and as discussed in reviews [20,21]. Under liquid–liquid PTC conditions, where permanganate is added as an aqueous solution, oxidation of olefins yields different products depending on pH of the aqueous phase: (1) under basic conditions *cis*-diol is the major product, (2) under neutral conditions the products are primarily α-ketols,

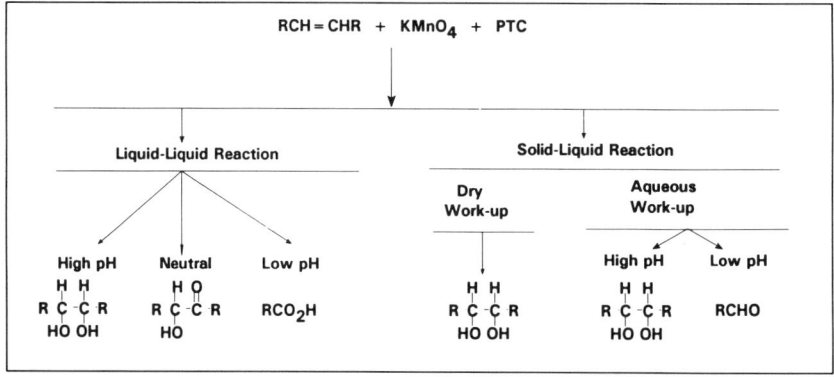

Figure 10-2. Principal products from PTC permanganate oxidation of olefins.

and (3) under acidic conditions the products are mostly carboxylic acids or aldehydes. Under solid–liquid conditions where, for example, solid potassium permanganate is transferred by use of a polyether phase-transfer agent such as TDA-1, and the olefin is dissolved in a nonpolar solvent, the major products depend on how the reaction is worked up. If no water is used during work-up, that is, filtration to remove solids and evaporation of solvent, the reaction gives *cis*-diols and/or aldehydes as major products, the cleavage apparently being more likely if the developing carbonyl is conjugated with an aromatic ring [26,42]. If water is used during work-up, and it is basic, cis-diols are obtained, but with acidic aqueous work-up aldehydes are produced [27].

Use of acetic acid in oxidation of olefins with $KMnO_4$ is an excellent route to carboxylic acids [28,29]. Detailed tabulations for olefin oxidations have been published [21,30].

$$C_7H_{15}CH_2CH=CH_2 + KMnO_4 \xrightarrow[0-25°C]{C_{16}H_{33}N^+(CH_3)_2CH_2C_6H_5\ Cl^-} C_7H_{15}CH_2CO_2H \quad (10\text{-}1)$$
$$80-90\%$$

If the pH of the reaction mixture is not controlled, permanganate oxidation reaction mixtures tend to become alkaline, leading to more complex product mixtures. For example, oxidation of methyl oleate gave the expected nonanoic acid (67%) and azelaic acid (72%), as well as up to 25% methyl 9,10-dihydroxyoctadecanoate with no acid added [31]. However, when the aqueous solution is maintained highly basic from the beginning the principal product (ca. 80%) is *cis*-diol.

The PTC method is a preferred technique for permanganate hydroxylation of olefins to *cis*-diols [32–34]. For example, with cyclooctene [35]:

(10-2)

cyclooctene $+ KMnO_4 \xrightarrow[NaOH]{PhCH_2NEt_3^+\ Cl^-}$ cis-cyclooctane-1,2-diol

With careful pH control dicyclopentadiene yields either a *cis*-diol or a bis-aldehyde [43,36].

(10-3)

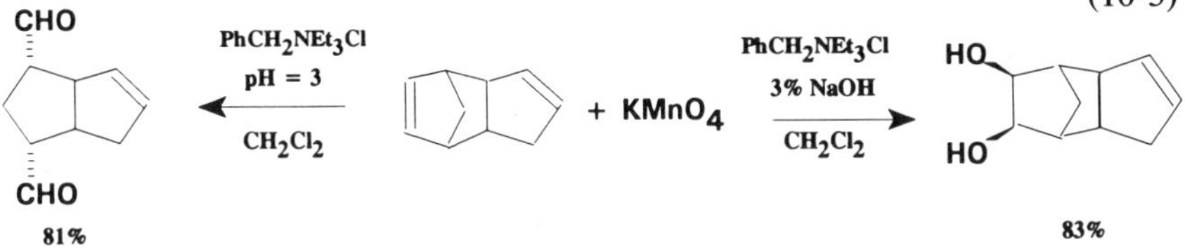

[Reaction scheme: dicyclopentadiene + KMnO$_4$ with PhCH$_2$NEt$_3$Cl, pH = 3, CH$_2$Cl$_2$ gives bis-aldehyde (CHO, CHO) in 81%; with PhCH$_2$NEt$_3$Cl, 3% NaOH, CH$_2$Cl$_2$ gives cis-diol (HO, HO) in 83%]

Hydroxylation of perfluoroalkyl ethylenes to *cis*-diols (80% yield) with aqueous permanganate, using PTC, was found to be superior to other methods tested [37]. Long-chain 1-*S*-alkyl-glycerols have been prepared by a PTC reaction of long-chain primary alkylthiols with allyl bromide in the presence of tetrabutylammonium bromide and aqueous alkali, followed by hydroxylation of the allyl alkyl thioether products to give 1-*S*-alkylglycerols by use of a molar amount of cetyltrimethylammonium permanganate in methylene chloride [38].

Other reports on permanganate oxidation of olefins using PTC techniques are included in the following references: [25,26,29,39–44,51].

b. Oxidation of Alkynes

PTC oxidation of alkynes in liquid–liquid systems, where one of the phases is water, usually produces carboxylic acids in good yields [42,44–47].

$$R\,C\equiv C-R' + KMnO_4 \xrightarrow{PTC} RCO_2H + R'CO_2H \quad (10\text{-}4)$$

However, PTC alkyne oxidations using solid–liquid PTC conditions under near anhydrous conditions, with methylene chloride as the solvent, produce α-diketones as the major product [48,61].

$$R\,C\equiv C-R' + \underset{(dry)}{KMnO_4} \xrightarrow{PTC} R\text{-}CO\text{-}CO\text{-}R' \quad (10\text{-}5)$$

c. Oxidation of Alcohols

Use of TDA-1 as a phase-transfer catalyst for dry permanganate oxidation of alcohols gives excellent yields of the corresponding carbonyl compounds as shown by the data in Table 10-4 [40]. In aqueous systems and with added acid, KMnO$_4$ oxidation of alcohols can give other products [49].

Other examples for PTC permanganate oxidations of alcohols [50–53,65], phenols, [40,54,55], aldehydes [56], and ethers [57] usually involve two-liquid phase processes, and produce carboxylic acids, aldehydes, and ketones. Oxidative ring opening of cyclic acetals with aqueous permanganate produces esters under PTC conditions. For example [58],

$$R-\text{C}_6\text{H}_4-\text{CH}(\text{OCH}_2\text{CH}_2\text{O}) \xrightarrow[\text{KMnO}_4 \text{ (aq)}]{\text{PhCH}_2\text{NEt}_3{}^+ \text{ Cl}^-} R-\text{C}_6\text{H}_4-\text{CO}_2\text{CH}_2\text{CH}_2\text{OH}$$ (10-6)

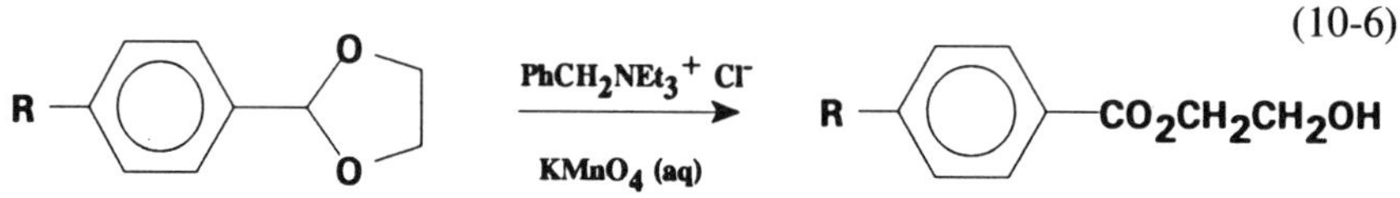

Oxidation of benzaldehyde and nitro-, chloro-, and methyl-substituted benzaldehydes with potassium permanganate under PTC conditions with different catalysts and solvents gives the corresponding carboxylic acids [59].

d. Oxidation of Other Hydrocarbons

Oxidation of aromatic hydrocarbons containing methyl or methylene substituents at 75–90°C gives the corresponding carboxylic acids or ketones, as represented in Table 10-5.

PTC permanganate oxidation of polyethylene gives oxidative degradation products (carbonyl, ether, hydroperoxy, olefin groups) by a free-radical process [66,67].

e. Other Permanganate Oxidations

Oxidation of sulfides to sulfoxides and sulfones, and thiols to sulfonic acids using PTC permanganate oxidations, has been demonstrated [40,68,69]. Chemoselective oxidation of the known naphtho-dithiocin, dithioether,

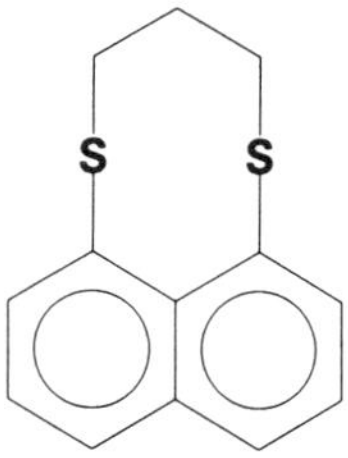

(10-7)

to the corresponding monosulfone can be accomplished with potassium permanganate and a phase-transfer catalyst in a two-phase system in 66% yield [70].

Amines and nitrogen compounds can likewise be oxidized by permanganate under PTC conditions to yield several products [65,71,72]. Triazolines are dehydrogenated [73] and triarylformazans are converted to tetrazolium salts [74]. In fact, oxidation of alkylformazans with permanganate in a methylene chloride–water mixture gives alkyltetrazolium salts which themselves are useful as phase-transfer catalysts in neutral or acidic systems [75].

Phase-transfer-catalyzed permanganate oxidizes coals to soluble and insoluble products which can be analyzed by physical techniques for use as a structural probe for various coals [76].

Table 10-4. TDA-1:$KMnO_4$ oxidation of alcohols.

Substrate	Time (h)	Percent AcOH present	Product (% Yield)
Benzyl alcohol	3	0	Benzaldehyde (82)
2-MeOC$_6$H$_4$CH$_2$OH	3.5	0	2-MeOC$_6$H$_4$CHO (80)
4-MeOC$_6$H$_4$CH$_2$OH	4	0	4-MeOC$_6$H$_4$CHO (77)
3-PhOC$_6$H$_4$CH$_2$OH	3	0	3-PhOC$_6$H$_4$CHO (76)
3,4(OCH$_2$O)C$_6$H$_4$CH$_2$OH	2	0	3,4(OCH$_2$O)C$_6$H$_4$CHO (91)
1-NaphthylCH$_2$OH	8	0	1-Naphthyl-CHO (90)
2-Thienyl-CH$_2$OH	3	0	2-Thienyl-CHO (81)
1-Furyl-CH$_2$OH	15	0	1-Furyl-CHO (74)
2-Indolyl-CH$_2$OH	7	0	2-Indolyl-CHO (54)
2-Pyridyl-CH$_2$OH	2.25	0	2-Pyridyl-CHO (61)
PhCH(OH)CH$_3$	6	20	Acetophenone (90)
PhCH(OH)CH$_2$OH	4	3	PhCOCHO (84)
PhCH(OH)CH$_2$OH	4	20	PhCOCO$_2$H (71)
PhCH(OH)CH(OH)Ph	1.5	0	PhCO-COPh (71)

Table 10-5. Oxidation of alkylaromatic hydrocarbons.

Substrate	Catalyst	Products	Conditions	References
Toluene	C$_{16}$H$_{33}$NMe$_3$Br	Benzoic acid (93%)	85°C, 2.5 h	[60]
p-Xylene	C$_{16}$H$_{33}$NMe$_3$Br	Terephthalic (95%)	75°C, 1.5h	[76]
Mesitylene	C$_{16}$H$_{33}$NMe$_3$Br	Trimesic (90%)	85°C, 1 h	[76]
Durene	C$_{16}$H$_{33}$NMe$_3$Br	Pyromelitic (80%)	90°C, 4.5 h	[76]
Stilbene	C$_{16}$H$_{33}$NMe$_3$Br	Benzoic (84%)	65°C, 0.17 h	[76]
Diphenylmethane	Bu$_4$NBr	Benzophenone	CH$_2$Cl$_2$ Sol	[61]
Anethole	C$_{16}$H$_{33}$NMe$_3$Br	Anisaldehyde (68% yield)		[62]
5-Benzoylacenaph-thene	Bu$_4$NBr or (C$_{10-16}$alkyl) Me$_3$N Cl + CuSO$_4$ co-catalyst	4-Benzoyl-naphthalic anhydride		[63]
Styrene or xylene, stilbene 18-Crown-6	Crowns	Aryl carboxylic acids		[64]
Xylenes	2,3,5-Triphenyltetra-zolium Cl$^-$ or C$_{16}$H$_{33}$NMe$_3$Br	Toluic and phthalic acids	95°C	[65]

C. Oxidations with Hypochlorite and Hypobromite

1. Hypochlorite Compositions in Aqueous Solutions

Lee and Freedman first recognized the value of hypochlorite as an inexpensive and readily available oxidant for use in PTC reactions [77,78]. Hypochlorite anion in aqueous solutions exists in equilibrium with HOCl and chlorine and water, with the major species determined by pH.

$$Cl_2 + H_2O \underset{HCl}{\rightleftarrows} HOCl \underset{}{\overset{OH^-}{\rightleftarrows}} H_2O + OCl^- \quad (10\text{-}8)$$

Hypochlorous acid concentration is maximum in the pH range of 8.5–10.5. At higher pH the primary species is hypochlorite anion, OCl^-, and at low pH, particularly if chloride ion is present, the equilibrium shifts to produce chlorine. Thus, depending on pH, the species transferred to the organic phase may be (1) OCl^-, (2) HOCl or a related form, such as $HOCl_2^-$, or (3) Cl_2 or a related form such as Cl_3^-. In view of these equilibria, PTC oxidations with hypochlorite may be complicated due to the presence of more than one oxidative species.

2. Oxidation of Alcohols and Carbonyl Compounds

Lee and Freedman [77,78] oxidized benzylic and secondary aliphatic alcohols to the corresponding carbonyl compounds,

$$\text{PhCH}_2\text{OH} + OCl^- \text{(aq)} \xrightarrow{R_4N^+ X^-} \text{PhCHO} \quad (10\text{-}8A)$$

as indicated by the data in Table 10-6.

Aliphatic primary alcohols oxidize slowly to aldehydes, and these are rapidly oxidized to carboxylic acids in high yield, in contrast to aromatic aldehydes which are stable under the reaction conditions. Only a low steady-state concentration of aldehyde from primary alcohols is detected.

Abramovici et al. [84] examined the PTC hypochlorite oxidation of benzyl alcohol as a function of pH and alcohol concentration. Both benzaldehyde and benzyl benzoate were produced, the former favored by dilute organic solutions and low pH, and the latter favored by high pH and concentrated organic phase. Dodecyl- or tetradecyltrimethylammonium salts perform well for PTC hypochlorite oxidation of benzyl alcohol to benzaldehyde [85].

In the oxidation of aromatic aldehydes to carboxylic acids with quaternary ammonium salts as catalysts, reaction rates are strongly pH dependent, with maximum rates at pH 9–11. Extraction of OCl^- was maximal at these pH values. The maxima are attributed to coextraction of HOCl together with OCl into the organic phase [100].

Table 10-6. PTC Hypochlorite oxidation of alcohols [79,94].

Alcohol substrate	Solvent	Reaction Time (min)	Product (% yield)
Benzyl	CH_2Cl_2	75	Benzaldehyde (76)
o-Methoxybenzyl	CH_2Cl_2	90	o-Methoxybenzaldehyde (41)
p-Methoxybenzyl	CH_2Cl_2	75	p-Methoxybenzaldehyde (79)
p-Methylbenzyl	CH_2Cl_2	83	p-Methylbenzaldehyde (78)
p-Chlorobenzyl	CH_2Cl_2	60	p-Chlorobenzaldehyde (82)
Benzhydryl	CH_2Cl_2	150	Benzophenone (82)
9-Fluorenyl	CH_2Cl_2	35	Fluorenone (92)
p-Methylbenzyl	EtOAc	30	p-Methylbenzaldehyde (100)
o-Methoxybenzyl	EtOAc	72	o-Methoxybenzaldehyde (94)
p-Methoxybenzyl	EtOAc	28	p-Methoxybenzaldehyde (92)
Cycloheptanol	EtOAc	58	Cycloheptanone (89)
2-Norbornanol	EtOAc	78	2-Norbornanone (36)
4-tert-Butylcyclohexanol	EtOAc	75	4-tert-Butylcyclohexanone (49)
Pyridine-4-methanol	—	60	4-Formylpyridine (93) [80]
$Me_2C=CHCOMe$	$PhCH_2N^+Et_3Cl^-$		$Me_2C=CHCO_2H$ (60) [81]
Aromatic aldehydes	$R_4N^+Cl^-$		Carboxylic acids [82]
$PhCH(OH)CO_2Me$	$Bu_4N^+ Br^-$		$PhCH(O)CO_2Me$ (98) [83]

[a] At room temperature with 25 ml solvent and 25 ml aq. NaOCl with magnetic stirring and 5 mol% tetrabutylammonium bisulfate catalyst, unless marked by new reference.

Lee and Freedman examined the kinetics of hypochlorite oxidation of p-chlorobenzyl alcohol to the aldehyde using 5 mol% tetrabutylammonium hydrogen sulfate as catalyst. They found the rate to follow second-order kinetics in most solvents (methylene chloride, chloroform, benzene, carbon tetrachloride),

$$\frac{-d[ROH]}{dt} = k[ROH][Q^+OCl^-] = k'[ROH]$$

but, since the concentration of Q^+OCl^- is nearly constant during any individual run with excess sodium hypochlorite, the kinetics become pseudo-first-order. However, in ethyl acetate as solvent, the reaction kinetics approximated zero-order between 10% and 90% alcohol conversion.

Rate constants for different solvents are listed in Table 10-7. From the data, Lee and Freedman concluded a normal PTC mechanism for the first four solvents, but something clearly different in ethyl acetate. Consistent with the observed zero-order kinetics they suggested that the rate-determining step occurs prior to the involvement of the substrate, and it is probably the rate of formation of the tetrahedral intermediate anion,

$$CH_3-\underset{\underset{}{\parallel}}{C}-OC_2H_5 \; + \; OCl^- \; \rightleftarrows \; CH_3-\underset{\underset{OCl}{|}}{\overset{\overset{O^-}{|}}{C}}-OC_2H_5 \qquad (10\text{-}9)$$

which should be much more easily transferred from the aqueous phase or the interphase to the organic phase than hypochlorite anion because of its organic groups.

Do and Chou [86], after an extensive investigation of the kinetics of PTC oxidation of benzyl alcohol (in dichloromethane) by aqueous hypochlorite, catalyzed by tetrabutylammonium chloride, concluded that the reaction was rate limited by the organic-phase oxidation step when the stirring rate was faster than 500 rpm. The kinetics were first-order in benzyl alcohol and Q^+OCl^-, with an activation energy of 10.4 kcal/mol. In all respects this kinetic behavior is analogous to PTC kinetics wherein the organic-phase reaction is the slow step of the reaction sequence. In their analysis of the data, they determined relative partitioning of Cl^- and OCl^- between aqueous and organic phase, with the result:

$$K_{OCl/Cl} = \frac{[Cl^-]_{aq}[Q^+OCl^-]_{org}}{[OCl^-]_{aq}[Q^+Cl^-]_{org}} = 2.48$$

showing that the tetrabutylammonium hypochlorite ion pair can be extracted from the aqueous phase into dichloromethane 2.48 times more easily than tetrabutylammonium chloride.

Kinetics for hypochlorite oxidations using resin-bound insoluble catalysts have also been studied. In the oxidation of benzyl alcohol with aqueous hypochlorite, using ion-exchange resins as phase-transfer catalysts, the consecutive first-order reaction rate was higher with ethyl acetate solvent than with other solvents tested (benzene, toluene, xylene, carbon tetrachloride) [87–89]. Decomposition of sodium hypochlorite was also observed in the aqueous and solid phases when ethyl acetate was the solvent, but did not occur at all when toluene was the solvent. Oxidation of benzyl alcohol in toluene with sodium hypochlorite at room tempera-

Table 10-7. Effect of solvent on rate of oxidation of p-chlorobenzyl alcohol at 21°C by NaOCl (aq) in presence of 5 mol% $Bu_4N\ HSO_4$.

Solvent	k (sec^{-1})	k (rel)
Chloroform	1.40×10^{-4}	1.00
Methylene chloride	3.41×10^{-4}	2.44
Carbon tetrachloride	4.36×10^{-4}	3.11
Benzene	4.62×10^{-4}	3.30
Ethyl acetate	23.4×10^{-4a}	16.77[a]

[a] Closest fit to a first-order plot.

ture was catalyzed with two kinds of anion-exchange resins: a gel type (Dowex 1X8) and a macroporous type (AGMP-1). For both resins, decreasing the particle size increased the oxidation rates and decreased selectivity.

a. An Electrochemical Hypochlorite Oxidation of Benzyl Alcohol

Electricity is a particularly interesting oxidant, since it is inexpensive and environmentally clean, although capital costs for commercial electrolysis are frequently high. Electrolysis of sodium chloride solutions to sodium hypochlorite is relatively simple, so the combination of PTC for oxidation of organic substrates with electrochemically generated sodium hypochlorite represents an oxidation route with high potential for commercial utility.

Do and Chou [90] experimentally demonstrated an electrochemical/PTC combination system for oxidation of benzyl alcohol to benzaldehyde, as outlined in the following representation:

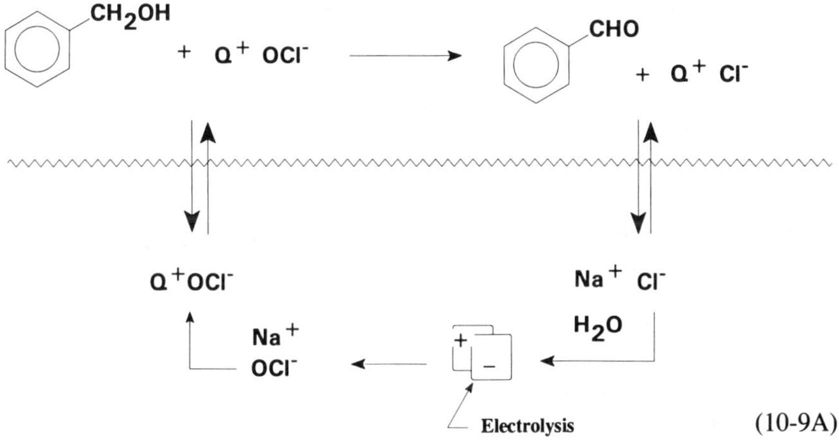

$$PhCH_2OH + Q^+OCl^- \longrightarrow PhCHO + Q^+Cl^- \quad (10\text{-}9A)$$

The results showed that in the presence of $0.05\ M\ Bu_4NHSO_4$ the rate-determining step shifts from the transfer rate of the hypochlorite ions to the rate of anodic oxidation of Cl^- when the agitation rate changes from 400 to 500 rpm. The current efficiency is governed mainly by pH and the nature of the organic solvent as well as the types and the concentration of PTC, and is slightly affected by both the current density and temperature.

In a novel process for the preparation of quinone-amine polymers by reaction of a diamine with p-benzoquinone by an oxidative additive reaction, a phase-transfer catalyst and calcium hypochlorite are used to provide efficient use of the quinone [91]. These polymers have a strong affinity for metal and they can displace water from a wet, rusty steel surface and render it nonwettable, suitable then for painting with an oil-based paint and for providing outstanding protection against corrosion and build-up of bacterial growths on ships or other marine applications.

3. Oxidation of Amines, Amides, Thioamides, and Related Compounds

Oxidation of 2,4-dimethyl-2-amino-pentanonitrile with aqueous sodium hypochlorite under phase-transfer conditions, and at low temperatures provides a low-cost route to azoisoheptanonitrile, useful as a polymerization initiator [92].

$$(CH_3)_2CHCH_2-\underset{\underset{NH_2}{|}}{\overset{\overset{CH_3}{|}}{C}}-CN + NaOCl\ (aq) \xrightarrow[C_{18}H_{37}NMe_3^+\ Cl^-]{C_{12}H_{25}N(Me)_2CH_2C_6H_5^+\ Cl^-\quad 5-10°C} (CH_3)_2CHCH_2-\underset{\underset{CN}{|}}{\overset{\overset{CH_3}{|}}{C}}-N=N-\underset{\underset{CN}{|}}{\overset{\overset{CH_3}{|}}{C}}-CH_2CH(CH_3)_2$$

(10-10)

A most interesting aspect of this oxidation is the use of *two* quaternary ammonium catalysts, one that catalyzes the organic-phase reaction so that it can take place at low temperature, and the other to speed transfer of difficultly transferrable hypochlorite into the organic phase.

PTC hypochlorite oxidation of primary amines, where the nitrogen is attached to a terminal CH_2- group yields nitriles, a synthetically interesting and useful reaction:

(10-11)
$$R\text{-}CH_2\text{-}NH_2 + 2\ NaOCl\ (aq) \xrightarrow{R_4N^+\ Cl^-} R\text{-}CN + 2\ NaCl + H_2O$$

Primary amines, where nitrogen is attached to a secondary $-CH-$ group, give ketones (after hydrolysis) as illustrated by the data in Table 10-8.

Aliphatic amides give nitriles through successive Hofmann rearrangement and oxidation of the resulting amine.

$$R\text{-}CH_2CONH_2 + NaOCl(aq) \xrightarrow{R_4N^+\ X^-} R\text{-}CH_2\text{-}NH_2 \xrightarrow{R_4N^+\ X^-} R\text{-}CN$$
(10-11A)

For example, phenylacetamide gave 61% benzonitrile and 4% benzaldehyde; octamide gave a 34% yield of 1-cyanohexane. Cleanest oxidations of amides were obtained using methylene chloride or carbon tetrachloride as solvent, whereas ethyl acetate gave highest yields in oxidation of amines. Oxidation of aromatic amides stops at the amine stage, after Hofmann rearrangement [93].

Oxidation of benzyl cyanide by sodium hypochlorite under phase-transfer conditions gives products chlorinated at the benzylic carbon [94].

Broda and Dehmlow [95] oxidized thioamides with hypochlorite under PTC conditions to produce synthetically useful materials. For example, thioureas produce carbodimides:

$$\text{RNH-CS-NHR} + \text{NaOCl} \xrightarrow{\text{PhCH}_2\text{NEt}_3^+ \text{Cl}^-} \text{R-N=C=N-R} \quad (10\text{-}12)$$
(in CH$_2$Cl$_2$) (aq) 50-88% Yield

Some ortho-substituted thioureas gave benzoxazole or benzimidazole derivatives, while simple thiamides gave 1,2,4-thiadiazoles:

(10-13)

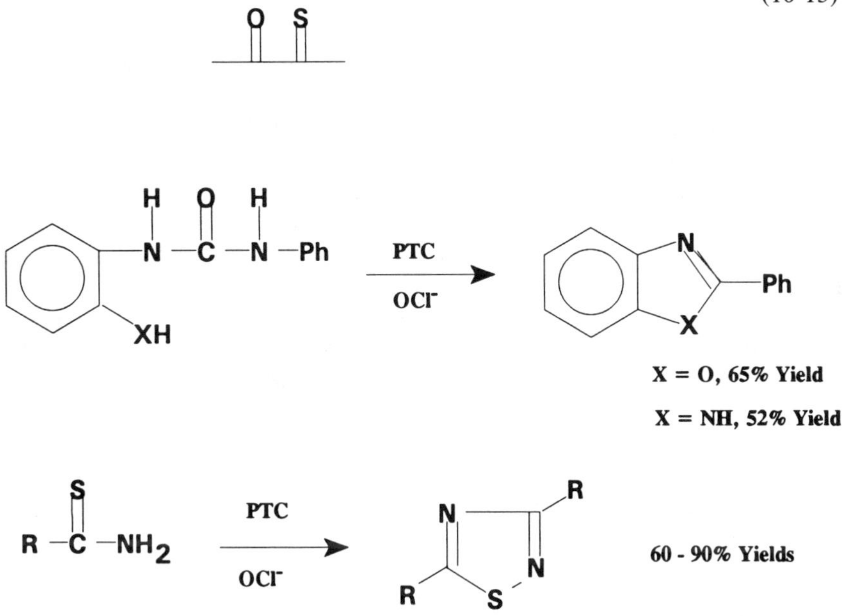

X = O, 65% Yield

X = NH, 52% Yield

60 - 90% Yields

Table 10-8. *PTC hypochlorite oxidation of amines [94].*

Substrate	Reaction time (min)[a]	Product (% yield)
Cyclohexylmethyl	70	Cyclohexanecarbonitrile (76)
1-Octyl	35	1-Cyanoheptane (60)
Cyclohexyl	120	Cyclohexanone (98)
Norbornyl	80	Norbornanone (84)
Benzhydryl	120	Benzophenone (94)
α-Methylbenzyl	140	Acetophenone (98)

[a] At room temperature with 25 ml solvent and 25 ml aq. NaOCl with magnetic stirring and 5 mol% tetrabutylammonium bisulfate catalyst in EtOAc solvent.

4. Oxidation of Sulfides and Related Compounds

PTC hypochlorite oxidation of dialkylsulfides to sulfoxides provides an excellent example of transfer of hypochlorite anion to an organic phase, followed by direct oxygen transfer to substrate [96]:

$$C_4H_9\text{-}S\text{-}C_4H_9 + OCl^- \xrightarrow[CH_2Cl_2/H_2O]{R_4N^+ X^-} C_4H_9\text{-}\overset{O}{\underset{\|}{S}}\text{-}C_4H_9 + Cl^- \quad (10\text{-}14)$$

Nearly complete oxidation occurs in 20 min at room temperature with high selectivity to sulfoxide, compared to 5 h in the absence of Aliquat 336 catalyst. Kinetic analysis of the process shows that the rate-limiting step is anion transfer into the organic phase. Further kinetic investigation of this oxidation by varying pH, hypochlorite concentration, sulfide structure, and the organic phase showed the rate of sulfide oxidation in acetate esters to be accelerated by a pH decrease, which was caused by hypochlorite-catalyzed hydrolysis of the esters [97].

Certain benzoates useful for pesticide and herbicide applications are prepared in a tandem two-step PTC reaction sequence, first by substitution of a MeS-group for a chloro- group, then oxidation of the sulfide to a sulfone, with both steps catalyzed by TDA-1 [98].

In the oxidation of thiophosphinic, thiophosphonic, and thiophosphoric acid esters with hypochlorite to $RR'P(O)OH$ [R,R' = Et, EtO], reaction rates are different depending on R, R' and the use of optimum pH [99].

5. Oxidation of Olefins

a. Direct Epoxidation

Direct epoxidation of olefins with aqueous hypochlorite solutions using simple PTC catalysts such as quaternary ammonium salts as catalysts, without metal cocatalysts, have generally been unsatisfactory because of excessive byproduct formation. Only certain substrates, such as perfluoroolefins and aromatic hydrocarbons such as phenanthrene, give good yields of epoxides. However, special catalysts can be used for epoxidation of ordinary olefins, as outlined in Section 5.b.

Epoxidation of phenanthrene and other aromatics by hypochlorite with ordinary phase-transfer catalysts give the corresponding epoxides in good yield [324]. For example:

$$\text{anthracene} \xrightarrow[\substack{\text{NaOCl (aq)} \\ \text{pH} = 8.5, 40^\circ\text{C}}]{\substack{\text{Bu}_4\text{N}^+ \text{HSO}_4^- \\ \text{CH}_2\text{Cl}_2 \text{ Solution}}} \text{anthracene epoxide} \quad 96\% \text{ Yield}$$

(10-15)

Aliphatic olefins tested in this oxidation system were very reactive, but even at pH 13 gave substantial amounts of products containing both chlorine and oxygen. The epoxide from norbornene was exclusively the *exo*-isomer, suggesting that chlorohydrins were not involved. Chalcones are also converted to the *trans*-epoxides in good yield with NaOCl in methylene chloride using Bu$_4$NBr as the phase-transfer catalyst [101].

An interesting three-phase PTC process for direct epoxidation of perfluoro-olefins with NaOCl has been developed [102]. The system consists of (1) an aqueous solution containing a metal hypohalite, for example, NaOCl; (2) an epoxide product phase; and (3) a reaction phase comprising an organic liquid having a perfluoroolefin–epoxide partition coefficient > 1.3 under reaction conditions. A PTC catalyst that is soluble in phases (1) and (3) and insoluble in the phase (2) is also included. Toluene and hexadecylbenzyl bis(β-hydroxypropyl)ammonium chloride, along with aqueous NaOCl, formed the basic system for epoxidation of perfluoropropylene at -1°C to give hexafluoropropylene epoxide with 53.2% selectivity and 62.7% conversion. Perfluorocyclohexene, perfluorocyclopentene, and perfluoro$-1-$heptene were also converted into their epoxides in > 75% selectivities. A continuous reaction system was developed for these oxidations. A PTC process is also used to clean up wastewater from the perfluoropropylene oxidation process [103]. The perfluoroolefin, F$_2$C=CFCF$_2$OCF(CF$_3$)$_2$ [104], perfluoropropene [105,106], and 1,4-dichloro-2-butene [107] are also directly converted to epoxides by PTC reactions with hypochlorite. Epoxidation of perfluoroolefin sulfides has also been reported, using mild conditions [108].

b. Special PTC Catalysts for Olefin Epoxidation with Hypochlorite

Manganese tetraarylporphyrin catalysts

Much work has been done on development of tetraarylporphyrin compounds as special phase-transfer catalysts for oxidation of olefins to epoxides, using sodium hypochlorite as the primary oxidant [109–113]. Montanari and co-workers [114] have elegantly extended this technology by binding a pyridine or imidazole group to the porphyrin structure in such a way that these groups can axially

Table 10-9. *Olefin epoxidations at pH 9.5, 0°C, with axial pyridine ligand (catalyst A) catalysts or imidazole axial ligand (catalyst B).*

Substrate	Catalyst	Olefin % conversion	Selectivity to epoxide	Reaction time (min)
Cyclooctene	A	95	96.2	4
	B	95	93.5	5
cis-Stilbene	A	95	66.5	1
	A	100[b]	70.9	8
	B	100	81.0	<1
	B	100[b]	71.4	<3
trans-Stilbene	A	26	100	1
	A	94.4	82.6	8
	B	23.1	100	1
Styrene	A	95.5[b]	61.4	6
	B	91.3[b]	62.0	5
2-Methyl-2-heptene	A	97.9[b]	96.2	5
	B	94.4[b]	95.4	10

[a] Carried out in methylene chloride–water, with 0.005 molar equivalent of catalyst and 0.0012 molar equivalent of Aliquat 336.

[b] No quaternary ammonium salt present.

complex the manganese(III), as represented by the two compounds represented by structures A and B:

(10-16)

These catalysts are extremely efficient for two-phase olefin epoxidations using sodium hypochlorite as the primary oxidant, as illustrated by the data in Table 10-9. Axial pyridine or imidazole ligands in these catalysts plays an important role; without them, as in manganese tetraphenylporphyrin, very low epoxidation efficiency is observed [115]. Note the extremely short reaction times required to obtain high olefin conversion.

Reaction rates increase rapidly at lower pH of the aqueous NaOCl solution, and at pH 9.5 catalyst turnovers are in the range of 0.8–3.3 per second at 0°C. At the lower pH values, reaction rates are only slightly affected by the presence of quaternary ammonium phase-transfer catalysts, such as tricaprylylmethylammonium chloride, but at the pH of commercial NaOCl bleach solutions (12.5–13) the presence of quaternary ammonium salts increases the rate significantly. This dependence on pH is due to the presence of sufficient free HOCl at the lower pH to rapidly transfer to the organic phase to keep the catalyst completely oxidized and active. At high pH the oxidant is mostly in the anionic form, OCl$^-$, which is rapidly transferred only by the quaternary ammonium salt.

Propylene epoxidation by manganese tetraphenylporphryin catalyst with NaOCl proceeds in the presence of axial ligands [116]. With 4-benzylpyridine as axial ligand, catalytic activity of Mn(TPP)Cl for propylene epoxidation was increased remarkably compared with 4-methylpyridine as axial ligand. Under best reaction conditions the propylene epoxide selectivity was 100% at 80% propylene conversion. Tetra(2,6-dichlorophenyl)porphyrin-based catalysts are more resistant to oxidative degradation than tetraphenylporphyrin catalysts, yet are still highly active catalysts [117]. Some typical phase-transfer catalysts such as quaternary ammonium salts and organophilic crown ethers or cryptands enhance the rate of Mn(III)-tetrakis(2,6-dichlorophenyl)porphyrin-catalyzed olefin epoxidation with aqueous hypochlorite or hypochlorous acid [118]. The design and use of robust metalloporphyrins as efficient biomimetic catalysts for oxidation reactions has been reviewed [119].

Chiral manganese tetraphenylporphyrins epoxidize terminal or *cis*-disubstituted alkenes in yields above 90% and enantioselectivities ranging from 41 to 76% enantiomeric excess [120]. With the exception of styrene all alkenes were epoxidized from the same face. 4-*tert*-Butylpyridine as an added axial ligand and PhCH$_2^+$NMe$_2$(CH$_2$)$_{13}$MeCl$^-$ as a phase-transfer catalyst were employed.

Nickel Complexes as Catalysts

Certain square planar nickel(II) complexes are active as catalysts for hydrocarbon oxidation reactions including alkene epoxidation, oxidative C=C bond cleavage, and hydroxylation [121]. The reactions are highly dependent upon the structure of the ligand encapsulating Ni(II) and upon the terminal oxidant.

(10-17)

Table 10-10. Alkene oxidation with OCl^-, catalyzed by N(II)salen, and benzyltributylammonium bromide.

Substrate	Percent conversion	Percent epoxide	Percent PhCHO
Styrene	98	44	6
(Z)-β-Methylstyrene	100	84	10
(E)-β-Methylstyrene	100	89	0
(Z)-Stilbene	45	12	12
(E)-Stilbene	80	46	0
Cyclohexene	87	23	—
Norbornene	94	30	—

Quaternary ammonium salts, with Ni(II) salen as catalyst, facilitate the PTC–hypochlorite oxidation of olefins to produce epoxides, and other products, as listed in Table 10-10 [122].

Ruthenium Catalysts

Complexation of ruthenium with 2,2'-bipyridine and pyridine provides a cation, $[(bpy)_2pyRu(O)]^{2+}$, that functions as a PTC catalyst with aqueous sodium hypochlorite to oxidize styrene and *cis*- and *trans*- stilbenes to benzaldehyde (70–99%) as the major product and epoxides (0–22%) as minor products [123]. When ruthenium is added as a cocatalyst in the PTC oxidation of phenanthrene with hypochlorite, further oxidation occurs (compared to phenanthrene epoxide produced in the absence of ruthenium) to yield 2,2'-biphenyldicarboxylic acid in 85% yield [124].

6. *Oxidation of Nonolefinic Hydrocarbons*

a. *Oxidation of Aryl-methyl Groups to Carboxylic Acids*

Phase-transferred hypochlorite with ruthenium oxide as a cocatalyst oxidizes methylbenzenes containing electron-withdrawing substituents to the corresponding carboxylic acids in high yields [125].

$$\text{X-C}_6\text{H}_4\text{-CH}_3 + 4\text{NaOCl} \xrightarrow[\text{RuO}_4]{\text{Bu}_4\text{N}^+ \text{Br}^-} \text{X-C}_6\text{H}_4\text{-CO}_2^- \text{Na}^+ + 3\text{NaCl} + \text{HOCl} + \text{H}_2\text{O} \quad (10\text{-}18)$$

Results from several experiments are listed in Table 10-11.
Regarding oxidation of toluenes containing electron-withdrawing groups, these workers made the observations:

- Ruthenium tetraoxide is formed in situ by the action of sodium hypochlorite, and is selectively dissolved in the organic phase as a necessary

Table 10-11. Oxidation of methylbenzenes, $XC_6H_4CH_3$, with RuO_4–NaOCl in a biphasic transfer-catalyzed system [141].

Substituted toluene	Product	Isolated yield (%)
p-NO$_2$	Carboxylic acid	94
o-NO$_2$	Carboxylic acid	95
m-NO$_2$	Carboxylic acid	93
p-Cl	Carboxylic acid	98
o-Cl	Carboxylic acid	93
p-Br	Carboxylic acid	95
o-Br	Carboxylic acid	95
p-NO	Carboxylic acid	97
H	Carboxylic acid	92
p-OCH$_3$	Ring + α-chlorinated cpds	99[a]
p-OH	Ring + α-chlorinated cpds	98[a]
p-CH$_3$	Unidentified insoluble acid[b]	

[a] Less than 1% carboxylic acid.

[b] No p-toluic or terephthalic acid identified.

component of the reaction system. As long as sodium hypochlorite is present the organic phase retains a yellowish (RuO$_4$) color. Disappearance of the hypochlorite causes immediate precipitation of black RuO$_2$, concluding the reaction.

- Phase-transfer catalyst, Bu$_4$NBr in this case, is also a necessary prerequisite for the reaction. In the absence of quaternary salt no reaction occurs.

- The pH of the sodium hypochlorite solution must be maintained between 8.0 and 10.5. At lower pH the sodium hypochlorite solution decomposes, whereas at higher pH there is no reaction, evidently due to low transfer of hypochlorite into the organic phase.

- Kinetically, the reaction is first order in organic substrate, and zero order in sodium hypochlorite. Catalyst dependency is more complex, tending to first-order dependence for either catalyst when the relative amount of the other catalyst is high, and tending to zero order when the concentration of both catalysts are high.

The authors suggest a mechanism wherein the rate-determining step is slow hydride extraction by ruthenium tetroxide, forming a benzyl cation, which is rapidly oxidized successively to alcohol, aldehyde and acid:

(10-19)

b. Chlorination and Bromination

Sodium hypochlorite in the presence of phase-transfer catalysts has been used also as a chlorination agent [127–131]. Alkylaromatic compounds are selectively α-halogenated [132], as, for example, reaction of ethylbenzene with NaOCl in methylene chloride containing HCl at pH from 12.5 to 8.0, Bu_4NBr as phase-transfer catalyst, and *tert*-butyl alcohol solvent at 25–38°C gave the following products:

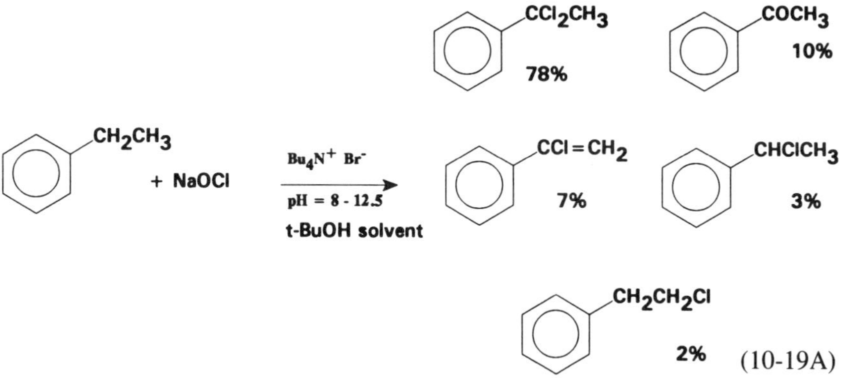

PhCH$_2$CH$_3$ + NaOCl $\xrightarrow[\text{t-BuOH solvent}]{\underset{\text{pH = 8 - 12.5}}{Bu_4N^+\ Br^-}}$ PhCCl$_2$CH$_3$ (78%) + PhCOCH$_3$ (10%) + PhCCl=CH$_2$ (7%) + PhCHClCH$_3$ (3%) + PhCH$_2$CH$_2$Cl (2%) (10-19A)

Ethylbenzene with *tert*-butyl hypochlorite in the absence of water but in the presence of light as initiator gave 93% 1,1-dichloroethylbenzene. Chlorination of toluene with hypochlorite under PTC conditions gives high yields of benzyl chloride [133].

PTC chlorination of vinyl toluene–divinyl benzene resins produces chloromethylated styrene–divinylbenzene resins, useful as intermediate products in production of ion-exchange resins. This route avoids use of highly carcinogenic chloromethyl ethers [129].

(10-20)

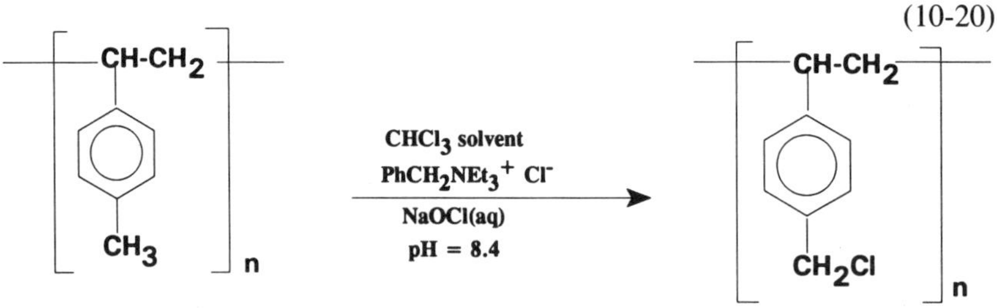

[–(CH-CH$_2$)–(C$_6$H$_4$-CH$_3$)–]$_n$ $\xrightarrow[\text{pH = 8.4}]{\underset{\text{NaOCl(aq)}}{\underset{\text{PhCH}_2\text{NEt}_3^+\ \text{Cl}^-}{\text{CHCl}_3\ \text{solvent}}}}$ [–(CH-CH$_2$)–(C$_6$H$_4$-CH$_2$Cl)–]$_n$

At conversion levels of < 20% of the methyl groups, no detectable formation of dichloromethyl groups occurred, but at 61% methyl group conversion about

4.4% dichloromethyl groups were produced. With NaOBr, bromination proceeded without phase-transfer catalyst.

Replacement of chlorine for bromine in bromobenzenes is effected by using Bu_4NHSO^- as phase-transfer agent [134]. At pH 8.5 substrates bearing electron donating substituents undergo halo-group exchange. Bromination of chloroform with sodium hypobromite under PTC conditions gives bromotrichloromethane in high yield [135]. Cyclopentadiene is brominated to hexabromocyclopentadiene by sodium hypobromite in the presence of trioctylmethylammonium chloride [136]. Addition of carboxylic acid increases the yield, believed to be due to improved transfer of the catalyst to the organic phase. Styrene, with sodium hypobromite, in the presence of PEG as a phase-transfer catalyst, gives 92% β-Br-α-OH-ethylbenzene [137].

Correia [138] found that addition of tetra-n-butylammonium hydrogen sulfate to a solution of NaBr and aqueous NaOCl results in the formation of a third phase consisting largely of organic solvent and $Bu_4N^+Br_3^-$ along with water, hypohalite, and halide ions, and possibly bromine. This material can be used for *trans*-1,2-bromination of olefins, oxidation of 1-alkynes to 1-bromo-1-alkynes, and oxidation of active methylene groups, for example, diphenylmethane to benzophenone.

Bromobenzene derivatives on treatment with aqueous sodium hypochlorite solutions in the presence of tetrabutylammonium salt catalysts are converted to the corresponding chlorobenzene derivatives [139].

D. PTC Oxidations with Hydrogen Peroxide

Barak and Sasson [140] suggest that hydrogen peroxide is an attractive oxidation agent because it is inexpensive, readily available, and its products cause no pollution problems. They note that the performance of hydrogen peroxide as oxidant in two-phase systems is remarkably improved by the addition of quaternary ammonium phase-transfer agents [141], for example in alkene epoxidation [142], and the oxidations of alcohols [143], aldehydes [144], sulfides [145], and ketones [146].

1. Hydrogen Peroxide Transfer into Organic Solutions

Barak and Sasson ascertained that for PTC hydrogen peroxide oxidations, in the presence of certain metal salts, quaternary ammonium salts have several roles, including:

- Transfer of peroxide into the organic phase (either directly to the hydration sphere, or indirectly by oxidation of the metal salt in the aqueous phase)
- Transfer of metal compound into the organic phase, either as the anion, or by complexation with the anion associated with quaternary cation

Table 10-12. *Transfer of H_2O_2 into methylene chloride by various quaternary ammonium salts [161].*

Quaternary salt	Equivalents H_2O_2 transferred per equivalent of quaternary salt
$PhCH_2NEt_3Cl$	0.013
$(Bu_4N)_2SO_4$	0.09
Bu_4NHSO_4	0.1
Bu_4NCl	0.3
Bu_4NBr	0.68
Bu_4N β-naphthalenesulfonate	0.92
Aliquat 336	0.88
Hex_4NBr	0.99
Hep_4NBr	1.00
Oct_4NBr	1.00

- Modifies the nature of the metal by complexation, resulting in improved catalytic activity and in higher resistance of the metal toward reduction.

Quaternary ammonium salts have been widely used as phase-transfer agents for hydrogen peroxide, even in the absence of metal compounds. On shaking quaternary ammonium salt solutions in methylene chloride with aqueous 35% hydrogen peroxide in water, various amounts of hydrogen peroxide are transferred into the organic phase, according to the quaternary salt structure, as indicated in Table 10-12 [46].

Transfer of the hydrogen peroxide may occur by formation of complexes with anions similar to hydration by water. Indeed "hydrogen peroxide of solvation" adducts of hydrogen peroxide with the quaternary salts has been confirmed by their isolation as crystalline complexes [147]. With quaternary ammonium bromide salts oxidation of Br^- to HOBr or its anions can also occur, and in fact the presence of bromide or chloride is essential in some oxidations.

2. Hydrogen Peroxide Oxidations

a. Oxidation of Olefins

Hydrogen peroxide oxidations of cyclohexene using a quaternary ammonium salt and a cocatalyst multivalent metal tend to give different products, depending on the metal cocatalyst used [148].

The metal cocatalysts either react with hydrogen peroxide to form an oxidatively active oxo- or peroxo-metal complex, which tend to be rather selective oxidants [149], or they initiate free radical intermediates, which result in allylic oxidation products. This technique of using catalytic amounts of metal compounds in conjunction with phase-transfer catalysts is attractive for industrial-scale synthetic procedures [157,160].

products favored with:
Mo, W and Os
co-catalysts

+ H_2O_2

(10-21)

products favored by:
Fe,V, Cr,Ce,Ni,Mn,
Co,Pt,Pd,Pb,Se,Ti
co-catalysts

Epoxidation of Olefins

Epoxidation with tungstate and molybdate cocatalysts

Epoxidation of olefins with hydrogen peroxide has been a long-standing goal in commercial oxidation chemistry. Simple transfer of hydrogen peroxide into an organic phase containing olefins is not sufficient for reaction to occur, except with highly activated olefins, such as perfluoro-olefins. Addition of cocatalyst metalate ions, particularly tungstate and molybdate, produces epoxides [150–153] but the reaction stops and conversion is low. Venturello et al. [154] provided an outstanding improvement in these oxidations by addition of forms of phosphoric acid to the PTC systems containing tungstate or molybdate, to obtain excellent yields of epoxides in high selectivity and conversion. Selectivities to epoxide of 70–90% on both hydrogen peroxide and olefin at substantially complete hydrogen peroxide conversion are usually obtained after short reaction times and under mild conditions, as shown by the data in Table 10-13.

Table 10-13. Epoxidation of olefins with H_2O_2, catalyzed by phase-transfer catalysts and tungstate/phosphate[a] [154].

Olefin	Quaternary Salt	pH	Time (min)	% Yield to Epoxide[a]
1-Octene	Aliquat 336	1.6	45	82
	Arquad 2HT	1.6	60	81
	$(C_{16}H_{33})_2NMe_2Cl$	1.6	60	71
1-Dodecene	Aliquat 336	1.6	60	87
Allyl Cl	Aliquat 336	2	150	80
Styrene	Arquad 2HT	3	180	77
α-Methylstyrene	Arquad 2HT	4.5	240	79
Cyclohexene	Arquad 2HT	3	25	88
	Aliquat 336	3	45	81

[a] Based on hydrogen peroxide, at quantitative hydrogen peroxide conversion. Normally the epoxidations were run in a 1,2-dichloroethane (15 ml)/8% aqueous H_2O_2 (120 mmol) mixture with a molar ratio phosphate/tungstate/quaternary salt of 5:2.5:1, made from adding sodium tungstate and phosphoric acid.

The cocatalytic composition between tungstates and phosphoric acid or its salts is formed in situ simply by mixing these two components in approximately 1:2 molar ratio. Acidity is important in that epoxidation effectiveness increases as pH decreases, although too high acidity in the aqueous phase causes hydrolysis of the epoxide formed. Epoxidation yields from long-chain olefins such as 1-octene or 1-dodecene are optimal at about pH 1.6. However, even under much more acidic conditions hydrolytic cleavage of the oxirane ring is largely prevented because of phase separation. Use of an excess of olefin is recommended to obtain the best yields of epoxide and to minimize hydrolysis reactions. Indeed, when the olefin was employed in a stoichiometric amount with hydrogen peroxide, a substantial reduction (3–80%) in yield was observed. The reaction is stereospecific giving only *trans*-2-hexene oxide when *trans*-2-hexene was epoxidized, and *cis*-2-hexene gave only *cis*-2-hexene oxide.

Berentsveig and co-workers [155,156] developed a mathematical model for 1-octene epoxidation with hydrogen peroxide, catalyzed by benzyltrioctylammonium chloride and sodium tungstate, using 1,2-dichloroethane as organic-phase solvent. The proposed mechanism involved rapid hydrogen–peroxide oxidation of a hetropoly anion, formed from the $Na_2WO_4 \cdot 2H_2O$, and slow bi- and trimolecular subsequent oxidation steps in the organic phase to produce epoxide. The inorganic species formed from molybdate and tungstate salts reaction with hydrogen peroxide in the presence of phosphoric acid or arsenic acids, and which are readily transferred into organic phases, have been the subject of several investigations [157–159].

Molybdenum-containing dihydroxyboryl polymers, stabilized by picolinato- and pyridine-2-dicarboxylato ligands, have been evaluated as catalysts for epoxidation of cyclohexene with hydrogen peroxide [160].

Olefin epoxidations are summarized in Table 10-14. *cis*-1,2-Bis(formylmethyl) cyclopentane derivatives, useful as intermediates for synthesis of certain prostaglandins,

(10-22)

are prepared by treatment of bicyclononenes with H_2O_2 in a two-phase system containing nonpolar solvents in the presence of sodium tungstate, phosphoric acid, and phase-transfer catalysts, followed by treatment of the resulting epoxybicyclononanes with HIO_4 in two-phase systems containing nonpolar solvents [173].

Epoxidation of allyl methacrylate with hydrogen peroxide using an Aliquat

Table 10-14. *Summary of some epoxidations of olefins with hydrogen peroxide in the presence of phase-transfer catalysts and molybdate or tungstate cocatalysts.*

Substrate	Catalysts	Product (Yield)	Reference
1-Octene	$C_{16}H_{33}N(CH)_{33}Br/ Na_2WO_4$	Epoxide	[161]
1-Octene	Quats + Na_2WO_4	Epoxide	[162]
1-, 2-Octene	$PhCH_2NMe_3Cl/WO_4 + H_3PO_4$	Epoxides (96%)	[163]
Allyl-Cl, -Br	$PhCH_2NMe_3Cl/WO_4 + H_3PO_4$	Epoxides	[157]
Eugenol, isosafrole	$PhCH_2NMe_3Cl/WO_4 + H_3PO_4$	No epoxidation	[157]
Cyclohexene	$PhCH_2NMe_3Cl/WO_4 + H_3PO_4$	1,2-Diol	[157]
Cyclohexene (mechanism studied)	Quats + molybdate or tungstate + P or As acid	Epoxide	[164, 165]
Cyclohexene	Amines/quats (Quat 336) Hexavalent anions + phosphates	Epoxide ("good" %)	[166]
α-Methylstyrene	$Me(CH_2)_{15}N^+Me_3 Br^-$ + Na_2WO_4/H_3PO_4	$PhCMe(OH)CH_2OH$	[167]
Dicyclopentadiene derivatives	Cetylpyridinium chloride + tungstate + phosphate	Epoxides (90%)	[168, 169]
Olefinic alcohols	Amines or quats (Adogen 464) + molybdate or tungstate + H_3PO_4	Epoxides (81%)	[170]
Allyl esters	$Me(n-C_8H_{17})_3NCl$ + /Na_2WO_4 + 85% H_3PO_4	Glycidyl esters (94%)	[171]
CF_2=CCFCF$_3$	$Bu_4N^+OH^-$	Epoxide (86%)	[172]

336, sodium tungstate, and phosphoric acid PTC catalyst system was selective for epoxidation of allyl group, producing glycidyl methacrylate in 94% selectivity at 74% conversion [174,175]. Rigorous control of temperature and of the initial pH prevents polymerization. In a similar way, epoxidation of the unsaturated-ester addition product of dicyclopentadiene to methacrylic acid could be selectively epoxidized at ring unsaturation and not on the methacrylate group [176,177]. Likewise, diallyl glutarate was converted to diglycidyl glutarate [178].

Polyoxotungstates, for example, tetrakis(cetylpyrdinium) decatungstate, have been prepared and used as phase-transfer photooxidation catalysts with hydrogen peroxide with hydrocarbons [179]. For example, photooxidation of cyclohexane produced dicyclohexyl (31%) and dicyclohexyl ether (36%) as major products. Photooxidation of adamantane with hydrogen peroxide gave 39% 1- and 61% 2-adamantol.

1,1-Dioxomolybdatranic acids,

$$H[MoO_2(OCH_2)_3N] \text{ (A) or } H[MoO_2(CH(CH_3)CH_2)_3N] \text{ (B)},$$

have been used as cocatalysts in the presence of imidazole and quaternary salts such as tetrapentylammonium bromide or cetylpyridinium bromide as phase-transfer catalysts for epoxidation of cyclohexene with 30% hydrogen peroxide

[180]. The molar ratio of imidazole to molybdenum affected the reaction greatly, and the reaction gave the highest epoxide yield when the ratio was 2:1. A red molybdenum(VI) peroxy complex intermediate formed in the reaction was the active species for epoxidation.

Epoxidation with manganese catalysts

The complex of manganese(III) with a bis(imine) of (R,R)- or (S,S)-1,2-diamino-1,2-diphenylethane with 3-*tert*-butylsalicylaldehyde, known as the *Jacobson catalyst*, enables the two-phase oxidation of olefins to enantiomeric glycols with aqueous sodium hydroxide. This technology has been licensed for applications to make enantiomeric versions of antihistamines, antihypertensive drugs, nonsteroidal antiinflammatory drugs (such as ibupropfen and ketoprofen), and semisynthetic taxol, an anticancer drug [181].

Epoxidation of olefins with nickel catalysts

The macrocycles represented by the structure below (R = CH_2Ph, CH_2CHMe_2, $CHMe_2$), prepared by condensation of phenylalanine, leucine, and valine with 1,3-propanediamine followed by hydride reduction and cyclocondensation with dimethyl malonate, are phase-transfer catalysts for the epoxidation of (E)-PhCH=CHMe, norbornene, and cyclohexene with aqueous hydrogen peroxide [182].

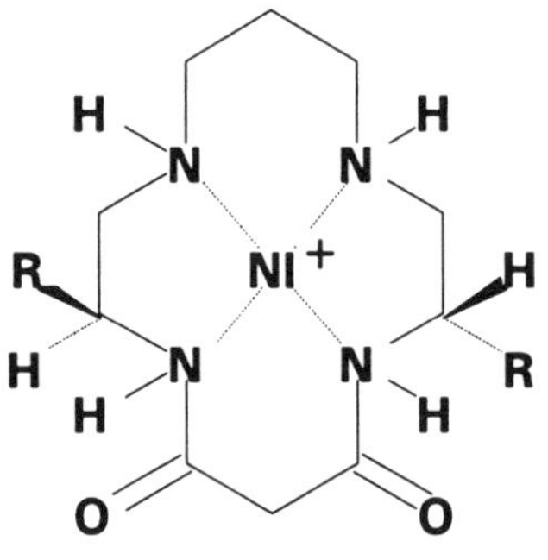

(10-23)

Other Olefin Oxidations with Hydrogen Peroxide

B4arak and Sasson [140] demonstrated that ruthenium trichloride cocatalyst with phase-transfer catalysts oxidize styrene with aqueous H_2O_2 to benzaldehyde and formaldehyde.

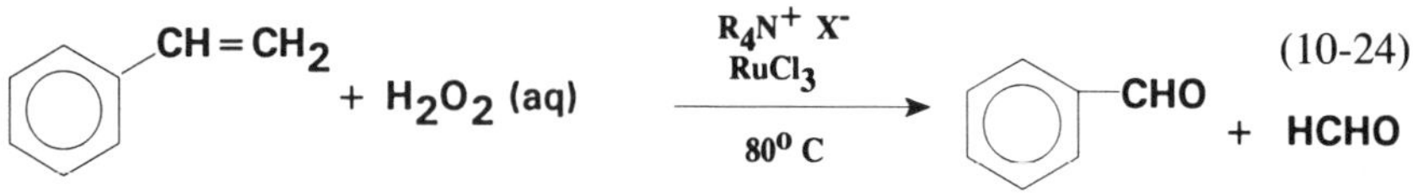

(10-24)

Using didecyldimethylammonium bromide as the PTC catalyst with ruthenium chloride cocatalyst, hydrogen peroxide oxidation gave 64% benzaldehyde, 6% benzoic acid, and 4% styrene oxide, together with some side products. In the absence of quaternary ammonium salt, no reaction took place, except some decomposition of the peroxide. More organophilic quaternary salts (tetraheptylammonium bromide and Aliquat 336) led to lower styrene conversion and increased the peroxide decomposition. Less organophilic tetrabutylammonium bromide gave complete styrene conversion, but less selectivity to benzaldehyde.

When palladium chloride was substituted for ruthenium in the above oxidation, the primary product was acetophenone:

$$PhCH=CH_2 + H_2O_2 \text{ (aq)} \xrightarrow[\substack{PdCl_2 \\ 80°C}]{R_4N^+ X^-} \underset{56\%}{PhCOCH_3} + \underset{12\%}{PhCHO} + \underset{14\%}{PhCO_2H} \quad (10\text{-}25)$$

Salts of rhodium, iridium, and cobalt showed much lower selectivity than ruthenium and palladium. Tests with ruthenium or palladium on carbon black were found to yield inferior results, compared with soluble catalysts.

b. Oxidation of Alkynes

Oxidation of terminal alkynes by customary oxidants such as per-acids, potassium permanganate, or osmium or ruthenium tetraoxide usually leads to the formation only of carboxylic acids. Ballistreri and co-workers [183] found that hydrogen peroxide oxidized terminal acetylenes in the presence of catalytic quantities of a quaternary ammonium PTC agent and sodium molybdate to give keto aldehydes:

$$C_4H_9-C\equiv CH + H_2O_2 \xrightarrow[\substack{Na_2MoO_4 \\ pH = 3.5}]{(C_8H_{17})_3NCH_3^+ \text{ Cl}^-} \underset{\substack{43\% \text{ Yield} \\ 100\% \text{ Selectivity}}}{C_4H_9-\overset{O}{\overset{\|}{C}}-CHO} \quad (10\text{-}26)$$

When sodium tungstate was used instead of sodium molybdate, conversions of the acetylene were much higher, sometimes 100%, but the primary product was carboxylic acid, that is, pentanoic acid, in ca. 90% yield, with about 10% ketoaldehyde. However, Ishii and Sakata oxidized alkynes to epoxy ketones and unsaturated ketones with 35% aqueous hydrogen peroxide in the presence of quaternary ammonium peroxotungstophosphate catalyst [184]. Aromatic alkynes gave diketones as the major product.

c. Oxidation of Aromatic Hydrocarbons

Aqueous hydrogen peroxide oxidizes polynuclear aromatic hydrocarbons such as phenanthrene, pyrene, benzo[*a*]pyrene, chrysene, *o*-phenanthroline, etc, in the presence of a phase-transfer catalyst plus a tungstate cocatalyst and phosphoric acid to give aromatic dicarboxylic acids in high selectivity and conversion [185]. For example, with phenanthrene:

$$\text{phenanthrene} + H_2O_2 \text{(aq)} \xrightarrow[\substack{H_2WO_4 \\ 80°\text{ C, 6 Hr.}}]{(C_8H_{17})_3NCH_3^+ \ Cl^-} \text{biphenyl-2,2'-dicarboxylic acid} \quad (10\text{-}27)$$

In a similar experiment without phosphoric acid, the selectivity was still high, 97%, but the conversion had fallen to 39%. Oxidation of pyrene using phosphoric acid gave phenanthrene-4,5-dicarboxylic acid in 81% selectivity with 99.2% conversion of pyrene.

Barak and Sasson [186] observed that when catalyzed by quaternary ammonium salts and ruthenium salts, hydrogen peroxide oxidized cumene at 67% conversion to a mixture of 2-phenylpropanol-2 and acetophenone:

$$\text{cumene} + H_2O_2 \text{ (30\% Aq)} \xrightarrow[\substack{RuCl_3 \\ (C_{10}H_{21})_2NMe_2^+ \ Br^- \\ 80°C, 4 \ Hr.}]{} \text{2-phenylpropanol-2} + C_6H_5COCH_3 \quad (10\text{-}28)$$

70%   24%

Further conversion could be realized by addition of more hydrogen peroxide. During reaction the ruthenium chloride complex with didecyldimethylammonium bromide stayed in the organic phase while the aqueous phase remained clear and colorless. However, in the absence of quaternary salt, ruthenium metal precipitated with simultaneous decomposition of hydrogen peroxide. Symmetrical quaternary salts, such as tetrabutylammonium bromide and tetrahexylammonium bromide, failed to stabilize the ruthenium salt and performance was poor due to precipitation of ruthenium metal. Aliquat 336 performed much like didecyldimethylammonium bromide. Other transition metal salts tested, rhodium chloride, iridium chloride, and iron chloride, were all significantly inferior to ruthenium.

Two-phase hydrogen peroxide oxidation of ethylbenzene to acetophenone, and similar oxidations of other aromatic hydrocarbons, was accomplished using PEG plus a cocatalyst of ferric nitrate [187]. Higher molecular weight PEG gave higher yields, as did the use of a water–acetonitrile homogeneous reaction mixture in place of the two-phase mixture.

Catalytic hydroxylation of anisole by hydrogen peroxide in the presence of

Table 10-15. PTC oxidation of alcohols with hydrogen peroxide [194][a]

Alcohol	Product	Co-Catalyst	Time (min)	Percent Yield
Cyclohexanol	Ketone	Mo	150	88
Cyclohexanol	Ketone	W	49	97
Menthol	Ketone	Mo	210	97
Menthol	Ketone	W	120	89
Borneol	Ketone	Mo	120	100
Borneol	Ketone	W	45	96
2-Octanol	Ketone	W	75	97
Benzyl alcohol	Aldehyde	W	30	85

[a] Catalyzed by Aliquat 336 and molybdenum or tungsten complexes, 75°C, pH = 3.0 (Mo) or 1.4 (W).

quaternary salts or dibenzo-18-crown-6 co-catalyzed with Fe(III)/Cu(II) salts gave up to 40% conversion to o- and p-hydroquinone [188].

d. Oxidation of Alcohols

Aqueous hydrogen peroxide oxidizes primary and secondary alcohols in the presence of trioctyl-methylammonium chloride plus molybdenum- or tungsten-co-catalyst, to give high yields of the corresponding aldehydes or ketones, as illustrated by the data in Table 10-15 [158].

These oxidations are strongly dependent on pH, with best yields and fastest reactions at pH 3.0 when sodium molybdate was used as co-catalyst, at pH 1.4 when sodium tungstate was used. Sulfuric acid was added to increase acidity. At higher acidities oxidations are slower, and lower acidities reduce product selectivities because of concomitant decomposition of hydrogen peroxide.

Solid tungsten oxide has been used as a support and co-catalyst with cetyltrimethylammonium bromide for the hydrogen peroxide oxidation of benzyl alcohol [321].

e. Oxidation of Aniline

Barak and Sasson [189] investigated hydrogen peroxide oxidations of aniline, finding that in the absence of catalyst, in the presence of ruthenium trichloride, or in the presence of didecyldimethylammonium chloride, the only product was azoxybenzene. However, if the bromide form of the quaternary salt, or if a binary mixture of ruthenium trichloride and quaternary ammonium chloride, were used as catalyst, then oxidation produced mixtures of nitro- and azoxybenzene. When ruthenium chloride and quaternary ammonium bromide are used as catalysts, nitrobenzene becomes the main product:

$$\text{PhNH}_2 + \text{H}_2\text{O}_2\text{(aq.)} \xrightarrow[\text{or } R_4N^+ \, Br^-]{RuCl_2 + R_4N^+Cl^-} \text{PhN(O)=NPh} + \text{PhNO}_2$$

$$\text{PhNH}_2 + \text{H}_2\text{O}_2\text{(aq.)} \xrightarrow{RuCl_2 + R_4N^+ \, Br^-} \text{PhNO}_2 \qquad (10\text{-}29)$$

The structure and the concentration of quaternary ammonium bromide also has a significant effect on the composition of the product mixture, as shown in Table 10-16.

The effect of catalyst concentration on product distribution is illustrated in Figure 10-3. Nitrobenzene is produced by a reaction parallel to that which produces azoxybenzene, since when azoxybenzene is exposed to the same oxidation conditions no reaction occurs. It appears that simple reaction of aniline with hydrogen peroxide produces azoxybenzene at rates depending on how well hydrogen peroxide is transferred into the organic phase, assuming that in the absence of quaternary catalyst none of the ruthenium chloride is transferred to the organic phase. Also, it appears that introduction of free-radical species into the reaction is responsible for a concurrent reaction that produces nitrobenzene. Thus, when quaternary ammonium chloride is present and $RuCl_3$ can be transferred into the organic phase as an anionic complex, $RuCl_4^-$, ruthenium can catalyze decomposition of hydrogen peroxide into free radical species, or when bromide (easily convertible to a bromo-free radical) is present, nitrobenzene is formed. When both RuCl, and quaternary ammonium bromide are present, nitrobenzene is formed in even greater yields. In fact when no catalysts are added, but sodium bromide is added, the same amount of nitrobenzene (60%) is produced as when a combination of the two systems (didecyldimethyl ammonium

Table 10-16. Effect of size and structure of the quaternary cation on product distribution in hydrogen peroxide oxidation of aniline[a] in presence of $RuCl_3$ [195].

Quaternary cation	Total No. C-Atoms	Percent Nitrobenzene	Percent Azoxybenzene
Tetraethyl	8	24	60
Tetrapropyl	12	25	62
Tetrabutyl	16	28	55
Tetrapentyl	20	48	33
Tetrahexyl	24	44	35
Tetraheptyl	28	45	32
Dodecyltrimethyl	15	42	41
Didecyldimethyl	22	60	15
Tricaprylylmethyl	25	55	25

[a] Aniline 54 mmol, $RuCl_3$ 0.077 mmol, R_4NBr 3 mmol, H_2O_2 640 mmol, dichloroethane 10 ml, 90°C, 24 h. The mass balance is made up by a constant 5% azobenzene together with bromo- and hydroxylation products.

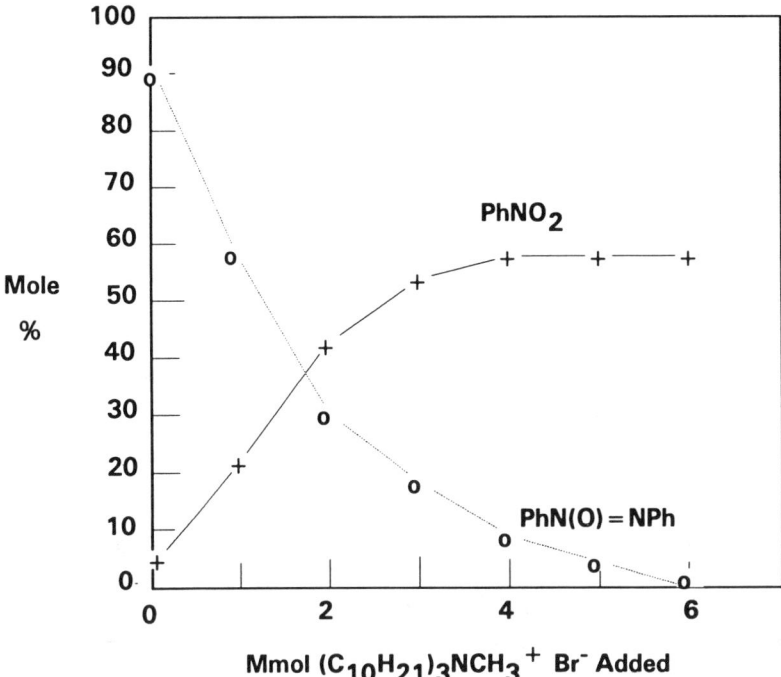

Figure 10-3. Effect of catalyst concentration on product distribution in PTC-H_2O_2 oxidation of aniline.

bromide and $RuCl_3$) is employed. Large amounts of didecyldimethyl ammonium bromide present in the reaction mixture caused bromination and hydroxylation products in > 25% yields.

It was also shown that the amount of hydrogen peroxide available in the organic phase affected product distributions, since more oxidant is required to produce nitrobenzene than azoxybenzene. Other metal chlorides were tested in place of ruthenium; only cobalt chloride, which is also known for its tendency to participate in free radical reactions, showed a tendency to produce nitrobenzene in addition to azoxybenzene. In the presence of $PdCl_2$ and $RhCl_2$ only azoxybenzene is produced.

f. Halogenations with Hydrogen Peroxide and HCl or HBr

Bromine, aqueous hydrogen bromide, or hydrogen chloride, in the presence of hydrogen peroxide and a quaternary ammonium salt, are highly effective ring-halogenating agents for aromatic hydrocarbons such as benzene, toluene, ethylbenzene, and chlorobenzene [190]. The quaternary salt in these two-phase systems appears to perform both as a transfer agent for bromine and as a Lewis acid catalyst for bromination of the aromatic hydrocarbon.

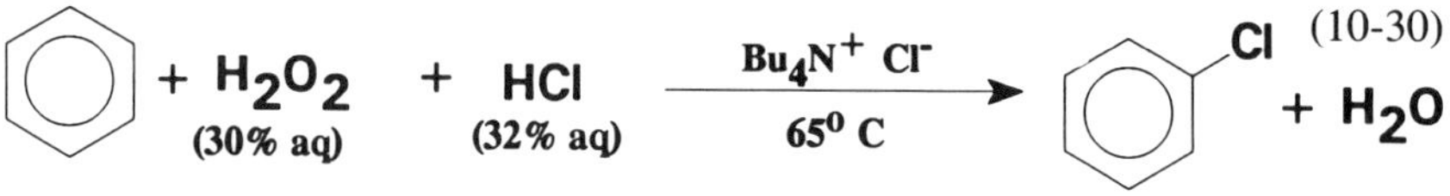

$$C_6H_6 + H_2O_2 \text{ (30\% aq)} + HCl \text{ (32\% aq)} \xrightarrow[65^\circ C]{Bu_4N^+ Cl^-} C_6H_5Cl + H_2O \quad (10\text{-}30)$$

Halogenation of alkylbenzenes produced similar yields of aryl halides with ortho para isomer distributions much the same as in ferric chloride-catalyzed halogenations.

Dakka and Sasson noted:

- Quaternary ammonium salts catalyze the reaction of elementary bromide with benzene (about 50 to 70-fold rate increase) relative to the rate in the absence of quaternary salt.
- Catalyst structure effect:

$$Et_4N^+ > Pr_4N^+ > Bu_4N^+ > Hex_4N^+ > Oct_4N^+ > Me_4N^+ \quad (10\text{-}31)$$

- Catalytic activity of quaternary salts was enhanced by hydrobromic acid.
- Oxidation of aqueous HBr to bromine was twice as fast in the presence of an inert organic phase (methylene chloride) when 5 mol% of $Bu_4N^+Br^-$ was added.

These workers suggested that activation and transfer of bromine and chlorine was achieved by formation of complex anions, Q^+Br_{n-} ($n = 3$ or 5) wherein the bromine or chlorine atoms are partially polarized. They concluded that the polarization would be stronger with the smaller quaternary cations due to the proximity of the bromine molecule to the charged nitrogen, a species known to be a mild brominating agent [191].

Use of a chiral PTC catalyst for chlorination of olefins with HCl plus hydrogen peroxide gave dichloro-products with some asymmetric induction [192].

Use of an organophilic quaternary ammonium bromide can significantly affect the orientation of bromination of aniline. Bromination of anilines normally produces a para/ortho ratio near 10:1. But the use of a hydrophobic quaternary ammonium bromide in water gives up to 70% ortho-substitution [193]. The effect is due to attraction for the attacking species (or its counterion) by the amino-group of the aniline. Preference for ortho-substitution is in the order $PhNMe_2 > PhNHMe > PhNH_2$.

g. Other PTC–Hydrogen Peroxide Oxidations and Applications

In the presence of iron salts, hydrogen peroxide gives products via free-radical intermediates. Thus, hydroxylation of anisole with hydrogen peroxide in the presence of catalysts (dibenzo-18-crown-6, CTAB) and Fe(III)/Cu(II) salts proceeded with up to 40% conversion and afforded *o*- and *p*-hydroxyphenol [194].

Trialkylboranes, generated in situ by hydroboration of alkenes, with chloroform and aqueous sodium hydroxide, followed by PTC oxidation with hydrogen peroxide, provides a convenient method for the synthesis of trialkylmethanols [195]. For example:

$$C_4H_9CH=CH_2 + BH_3 \longrightarrow (C_6H_{13})_3B \xrightarrow[\text{NaOH (aq)}]{\text{PTC, CHCl}_3}$$

$$(C_6H_{13})_3COH \xleftarrow{H_2O_2} (C_6H_{13})_3C\text{-BCl}_2 \qquad (10\text{-}32)$$

h. Use of Resin-Bound Catalysts for "Stoichiometric PTC"

One of the simplest approaches for oxidations using hydrogen peroxide is a two-step "stoichiometric PTC" process where an insoluble resin having carboxylic acid groups attached is (1) first converted to the percarboxylic acid by reaction with hydrogen peroxide, catalyzed by methanesulfonic acid and (2) in a second step an equivalent amount of organic substrate is passed through the resin for oxidation. Thus, reaction of olefins with peracid groups attached to a resin produce epoxides in 40–90% yields [196,197].

Step 1: (10-33)

$$\text{Resin-CO}_2\text{H} + \text{H}_2\text{O}_2\text{(aq)} \xrightarrow{\text{CH}_3\text{SO}_3\text{H}} \text{Resin-CO}_3\text{H} + \text{H}_2\text{O}$$

Step 2:

$$\text{Resin-CO}_3\text{H} + \text{RCH=CH}_2 \longrightarrow \text{Resin-CO}_2\text{H} + \text{R-CH}\overset{O}{-}\text{CH}_2$$

Although strong acids are required to catalyze equilibration of hydrogen peroxide with carboxylic acids to make peracids, use of arsonic acids or seleninic acids obviates this problem. For example, aqueous hydrogen peroxide catalyzed by arsonic acid groups bound to an insoluble resin is efficient for epoxidation of alkenes to epoxides without hydrolysis, and for Beyer Villiger oxidation of ketones [198,199]:

(10-34) 63%

Resin-bound seleninic acid groups also catalyze hydrogen peroxide oxidation of olefins and ketones, for example producing *trans*-1,2-cyclohexanediol from cyclohexene [200].

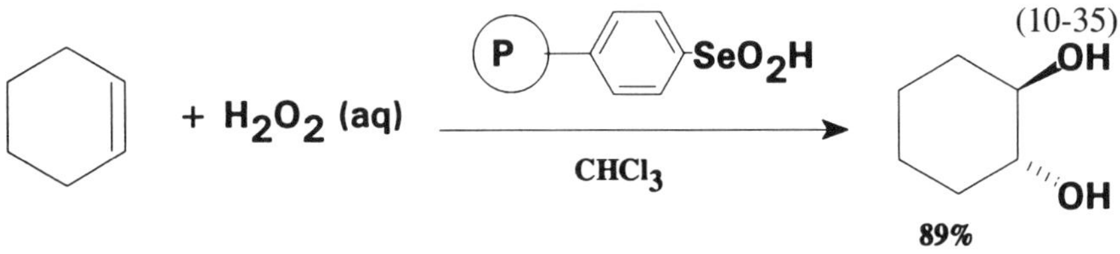

$$\text{cyclohexene} + H_2O_2 \text{ (aq)} \xrightarrow[CHCl_3]{\text{P}-\text{C}_6H_4-SeO_2H} \text{trans-cyclohexane-1,2-diol} \quad 89\% \qquad (10\text{-}35)$$

E. PTC Air or Oxygen Oxidations

PTC has been used in a surprisingly diverse range of air or oxygen oxidations. These include reactions involving carbanions, free-radicals, "activated" oxygen, singlet-oxygen, and transition-metal-mediated oxidation systems. As might be expected, the kind of assistance provided by the phase-transfer catalyst differs considerably for each reaction type, and therefore the optimal catalyst chemistry and structural requirements are rather different for the different reaction types.

1. Carbanion Oxidations

Oxidation of compounds with an acidic C−H bond proceed by formation of a carbanion using sodium hydroxide, followed by addition of oxygen, then rearrangement or decomposition of the intermediate peroxide anion [70,201–203]. These reactions are normally conducted with oxygen under pressure and at temperatures of 40–50°C. For example, oxidation of phenyl isopropyl ketone [204]:

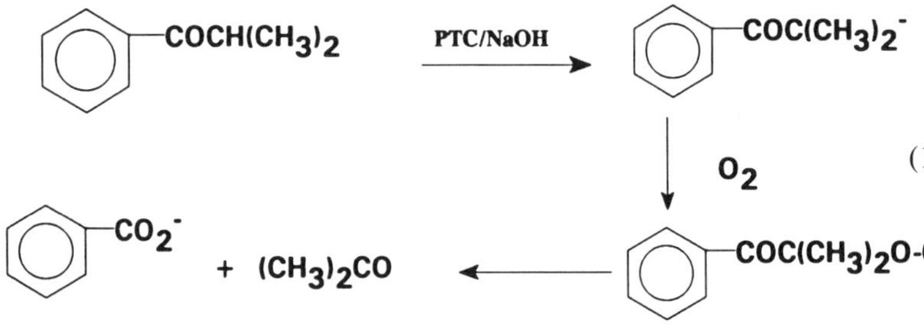

$$Ph\text{-}COCH(CH_3)_2 \xrightarrow{PTC/NaOH} Ph\text{-}COC(CH_3)_2^{-} \xrightarrow{O_2} Ph\text{-}COC(CH_3)_2O\text{-}O^{-} \longrightarrow Ph\text{-}CO_2^{-} + (CH_3)_2CO \qquad (10\text{-}36)$$

The important role played by the PTC catalyst is to assist in the formation of carbanion by transfer of hydroxide into the organic phase in a highly active form where it can react with acidic C−H bond of the substrate.

Sasson and co-workers have examined PTC carbanion oxidations in detail [206]. Kinetics for oxidation of diphenylmethane to benzophenone in the presence of 50% NaOH and tetrabutylammonium bisulfate were first-order in quaternary salt, zero-order in diphenylmethane, and something between zero- and first-order in oxygen concentration.

Prior to Sasson's work, it was known [205] that polyethylene glycol (PEG 6000) gave good results for carbanion oxidations, but that $(Bu)_4N^+Br^-$ was practically useless. Sasson demonstrated that the oxidations do have a high sensitivity to catalyst structure. Quaternary ammonium salts are efficient catalysts provided that (1) the quaternary cation has substantial organophilicity, enough to be partitioned into the organic phase when coupled with hydroxide anion, and (b) that the counteranion of the quaternary salt charged to the reaction mixture not have a high affinity for association with the quaternary salt cation, compared to hydroxide. With regard to the latter point:

> A necessary condition for a successful process is the right choice of catalyst counterion. The only catalysts that gave high yields when diphenylmethane (DPM) was used as the standard substrate consisted of the lipophilic tetrabutyl- and tetrahexyl-ammonium cations and hydrogen sulfate (HSO_4^-) anion. In the presence of a wide range of catalysts, e.g. $(Hex)_4N$ Br; $Bu_4)N$ Br, Cl, I; Bu_4P Br, Et_4NHSO_4, Me_4NHSO_4, $BzEt_3NCl$, and 15-crown-5, we could obtain only negligible yields of oxidation products (less than 5%). The importance of the counterion is demonstrated by the reactivity of a series of similar ammonium cations with different counterions. X = I, Br, Cl, HSO_4, . . . It can be seen that only those catalysts in which X = HOS_4^- were reactive. . . . inorganic anions such as bromide or iodide are not catalysts. Even the chloride anion presents a problem as the extraction constants of the hydroxide anion are lower by a factor of 10,000 as compared with the chloride anion . . ."

Halpern and Lysenko [207] examined the PTC–NaOH catalyzed oxidation of dihalodiphenylmethanes to the corresponding dihalobenzoquinones with oxygen, with findings similar to those of Sasson.

(10-37)

Cl–C₆H₄–CH₂–C₆H₄–Cl + O_2 $\xrightarrow[\text{NaOH}]{R_4N^+ X^-}$ Cl–C₆H₄–C(=O)–C₆H₄–Cl

The strong effect of catalyst structure and associated anion is shown in Table 10-17. Use of NaOH pellets induced a 50% higher initial rate than when 50% aqueous NaOH was used for oxidation of 4,4′-dichlorodiphenylmethane.

In PTC carbanionic oxidations with O_2 and NaOH, the loss of color as carbanion is consumed acts as an indicator to signal completion of reaction. Highly colored carbanions from fluorene, indene, and a para-substituted phenylacetonitriles are good examples [208].

Masui et al. [209] demonstrated that achiral cyclic ketones could be oxidized in the presence of base and Dolling's chiral phase-transfer catalysts to produce chiral α-hydroxyketones in 90% yield and up to 79% enantiomeric excess.

Several other chiral catalysts, including one attached to a crosslinked insoluble polymer, were tested. None had as good overall performance as Dolling's catalyst. Triethylphosphite was added to the reaction mixture to reduce in situ any labile hydroperoxides that may form.

In some situations the hydroperoxide anion intermediate that is formed in the

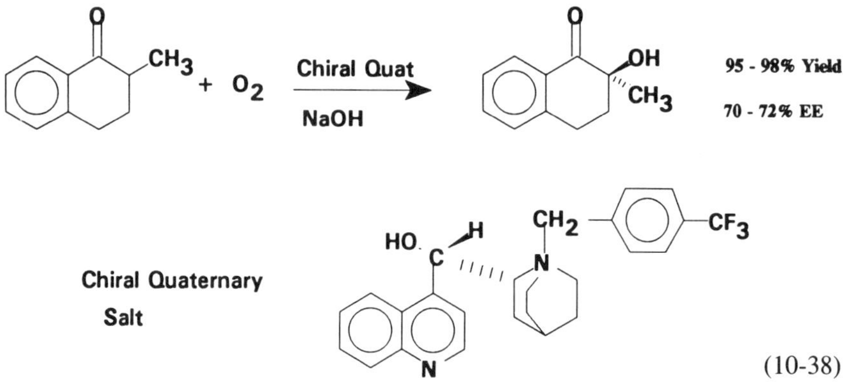

536 / Phase-Transfer Catalysis

$$\text{(10-38)}$$

presence of O_2 and strong NaOH can be intercepted and itself used as an oxidant, as for example in epoxidation of cyclohexeneones using 9-hexylfluorene as the intermediate peroxide anion substrate [210]. Further, use of 9-hexylfluorene with quaternary salts derived from (+)-chinchonine, in the presence of aqueous sodium hydroxide, give chiral epoxides [211].

Aryl pyridinyl ketones were prepared in a two-step PTC procedure involving first alkylation of phenylacetonitrile, then oxidation of the resulting nitrile with air [212]:

$$\text{(10-39)}$$

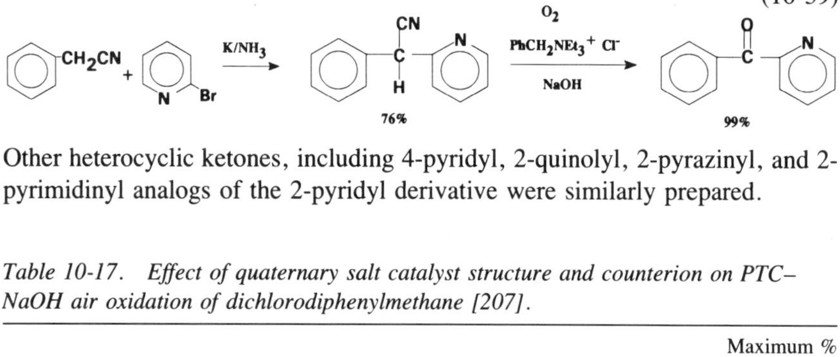

Other heterocyclic ketones, including 4-pyridyl, 2-quinolyl, 2-pyrazinyl, and 2-pyrimidinyl analogs of the 2-pyridyl derivative were similarly prepared.

Table 10-17. *Effect of quaternary salt catalyst structure and counterion on PTC–NaOH air oxidation of dichlorodiphenylmethane [207].*

Quaternary cation	Counterion	Maximum % conversion[a]
$(Oct)_3NCH_3^+$	Cl^-	98
$(Oct)_4N^+$	Br^-	93
Bu_4N^+	HSO_4^-	90
Bu_4N^+	Br^-	67
$PhCH_2NEt_3^+$	Cl^-	44
None	—	0

[a] At 200 psi oxygen pressure, 18 h, 50% NaOH, with 20 mol% catalyst, dichlorodiphenylmethane is 61% 4,4'-isomer and 39% 2,4'-isomer.

Table 10-18. PTC oxidation with oxygen in the presence of sodium hydroxide.

Substrate	Catalyst	Product (% yield)	References
Fluorene	DHDMAC[a]	Fluorenone (100)	[213]
Xanethene	DHDMAC	Xanthone (63)	[213]
Thioxanthene	DHDMAC	Thioxanthone (99)	[213]
Acridane	DHDMAC	Acridine (97)	[213]
9,10-Dihydroanthracene	DHDMAC	Anthraquinone (75) and anthracene (9)	[213]
1,4,4a,9a-Tetrahydroanthraquinone	DHDMAC	Anthraquinone (100)	[213]
Alkyl-phenylacetonitriles	Et$_4$NCl	R-ArCOCN	[214]
Aniline	18-Crown-6	Azobenzene (35–40)	[215]

[a] DHDMAC = dihexadecyldimethylammonium chloride.

Some additional examples of carbanionic oxidation for several compounds containing acidic C-H bonds are listed in Table 10-18. (See also reference [213].)

Neumann and Sasson found that oxidation of *p*-nitrotoluene with oxygen in a PTC system with sodium hydroxide, and with mechanical agitation, gave only dimeric products, whereas use of ultrasonic agitation gave significant yields of *p*-nitrobenzoic acid [216]. (See also reference [217].)

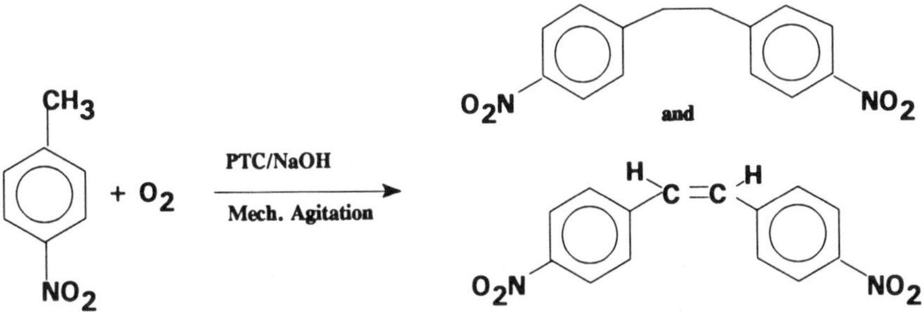

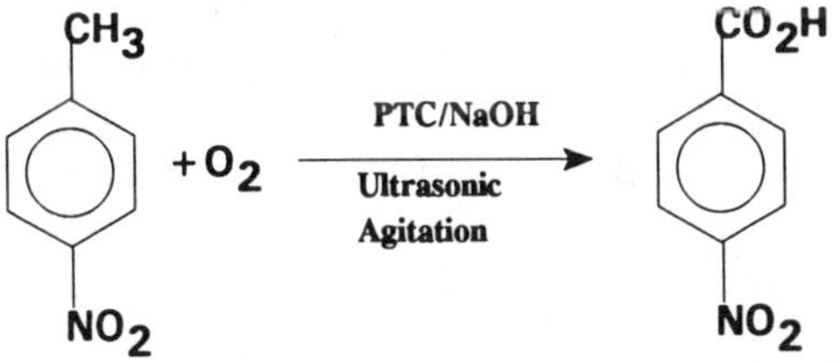

(10-40)

Heterogeneous reaction of acylhydrazones with aqueous NaCN in an inert organic solvent, and acetic acid in the presence of air and a catalytic amount of a quaternary ammonium salt produces α-cyanoketones [218].

Table 10-19. Autoxidations catalyzed by PTC.

Substrate	Catalyst	Conditions	Products	[Ref.]
Toluene	$(C_{10}H_{21})_2N^+Me_2 Br^-$ + $CoCl_2$	Air, 12–15 atm. 135–160°C 3 h	Benzoic acid 92% yield 99.5% purity	[225]
p-Xylene	$RNMe_3Br + MnO_4^-$		Toluic acid (20%)	[226]
n-Nonane	$Bu_4N^+Br^- + CoBr_2$	Air	Carboxylic acids	[227]
Cyclohexane	$Bu_4NBR + CoBr_2$	Air	Acids	[233]
n-Alkanes	PTC catalysts	160°C 0.1–1MPa	Fatty Acids	[228]
$Me(CH_2)_{11}SH$	$PhCH_2N^+Et_3Cl^-$ + $CuSO_4$	20–22°C, 1 h	R-S-S-R (100%) (20% in absence of catalyst)	[229]

2. PTC Involvement in Free-Radical Oxidations

Sasson and co-workers [219] have studied the oxidation of hydrocarbons with air in the presence of quaternary ammonium bromides, with and without the presence of cobalt bromide.

In the absence of cobalt, the kinetics of high-temperature (120–140°C) oxidation of p-xylene by oxygen in a two-phase system indicate that the quaternary ammonium cation is simply a phase-transfer agent for bromide ion, which functions as a source of initiator free radicals by reaction with hydroperoxide. Kinetic data support this conclusion in that the reaction follows a half-order dependence on quaternary cation and bromide ion concentrations [221].

$$\frac{dO_2}{dt} = \frac{k\,[p\text{-xylene}]\,[Q]^{0.5}[Br]^{0.5}}{[H_2O]^{0.2}}$$

In the presence of added cobalt bromide or chloride, the reaction becomes a high-conversion process for production of terephthalic acid [222]. Thus, oxidation of p-xylene with air at 20 atm pressure in the presence of didecyldimethylammonium bromide and cobalt chloride as catalysts at 135–170°C gave 95% terephthalic acid and 5% p-toluic acid. The use of heterocyclic quaternary cations has been recommended for oxidation of p-xylene by oxygen [223].

Air oxidation of styrene, α-methylstyrene, cyclohexene, and C_6–C_{14} α-olefins at 120°C in the presence of cetyltrimethylammonium bromide and aqueous solutions of $NiBr_2$, $MnBr_2$, $Co(py)_2Br_2$, $PbBr_2$, $CoBr_2$, and $CrBr_3$, gave products derived from both abstraction and addition mechanisms with Br^- promoting the former [224].

Other PTC autooxidations are listed in Table 10-19.

3. PTC with "Activated" Oxygen Carriers

"Activated" oxygen requires a special type of phase-transfer catalyst be used, one which can complex with gaseous oxygen to transfer it into the desired phase, and also activate the oxygen so that it can be used at the reaction site. Indeed, perhaps the most common of all phase-transfer reactions is the transfer and activation of oxygen from the air by hemoglobin into the blood and throughout the bodies of animals to the cells where energy production is necessary.

A simple chemistry for "activated" oxygen carriers is the reversible complex formation of cobalt "salcomine" with oxygen:

(10-41)

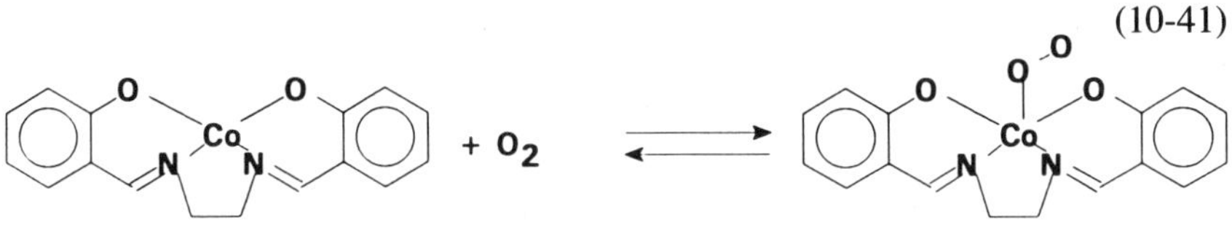

Salcomine   Salcomine: O_2 Complex

X-Ray data indicate the O–O bond in this complex is longer than in ordinary oxygen gas, suggesting that the oxygen is indeed activated.

Salcomine is a phase-transfer catalyst for oxygen to oxidize methylated phenols to the corresponding quinones (and/or diphenoquinones) [230]. An especially important example is oxidation of 2,3,6-trimethylphenol to the corresponding quinone:

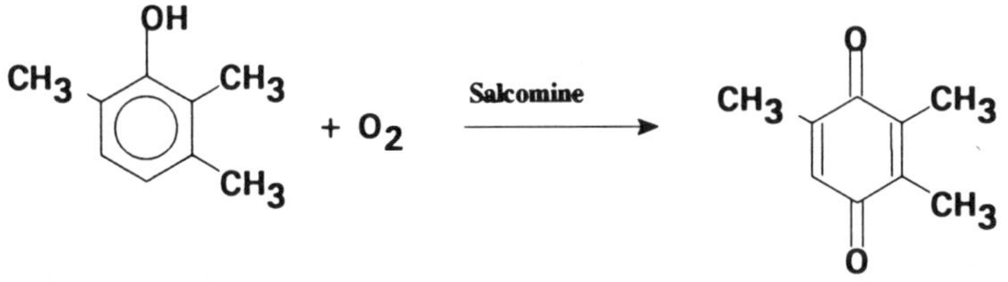

(10-42)

Hydrogenation of the quinone produces 2,3,6-trimethylhydroquinone, which is used on a multi-million pounds per year scale for manufacture of synthetic vitamin E.

4. PTC with Singlet Oxygen Generation

Use of rose bengal [3,4,5,6-tetrachloro-2-(2,4,5,7-tetraiodo-6-hydroxy-3-oxo-3H-xanthen-9-yl)benzoic acid, disodium salt] as a photosensitizer to convert triplet oxygen to singlet oxygen it is not normally useful for oxidation of organic compounds because of the insolubility of rose bengal (a disodium salt) in organic solutions [231,232]. However, suitable quaternary ammonium cations transfer the rose bengal dianion into virtually any organic solution where photochemical generation of singlet oxygen can occur, producing selective oxidation of many organic compounds [233]. For example:

$$\text{benzene} + O_2 \xrightarrow{(R_4N^+)_2 \text{(Rose Bengal)}^-} \text{endoperoxide} \qquad (10\text{-}43)$$

5. Transition-Metal-Mediated Oxidations Involving PTC

The Wacker reaction for oxidation of ethylene using an aqueous solution of palladium and cupric chlorides as catalysts to give acetaldehyde is extensively practiced on an industrial scale throughout the world. However, its use with higher olefins is strongly curtailed because of difficulty of solubilizing the higher olefins in the aqueous oxidizing solution. With an inverse PTC system, for example where the PTC catalyst is β-cyclodextrin, the β-cyclodextrin one can transfer olefins into an aqueous phase containing $CuCl_2$ and $PdCl_2$. Air oxidation then produces ketones [234].

$$RCH=CH_2 + O_2 \xrightarrow[\substack{CuCl_2 \text{ (aq)} \\ PdCl_2 \text{ (aq)}}]{\text{cyclodextrin}} R\text{-}CO\text{-}CH_3 \qquad (10\text{-}43A)$$

This reaction proceeds in the absence of cyclodextrin, but in the case of higher olefins, for example, decene, β-cyclodextrin causes a 23-fold increase in reaction rate, indicating that the process is transfer limited. The PTC process is useful for oxidation of a variety of terminal and internal olefins, and has the useful characteristic that the catalyst remains in the aqueous phase, avoiding the need for its separation from the organic products.

In ordinary PTC reactions rhodium and ruthenium cocatalysts with quaternary ammonium salts also catalyze air oxidation of terminal olefins to methyl ketones, but with concurrent production of internal olefins [235]. Use of palladium chloride as the transition metal gives much better results.

α-Chloroketones are produced from the parent carbonyl compounds by oxidation in a two-phase mixture with $EtNH_3NO_3$-H_2O-$CuCl_2 \cdot 2H_2O$-O_2-Q^+Cl^- (Q^+Cl^- = quaternary ammonium chloride) [236]. In the absence of water oxidative cleavage of the ketones to carboxylic acids occurs, and gives only low yields of α-chloroketones.

Oxidation of myrcene with a catalyst system consisting of cetyltrimethylammonium bromide, cupric chloride, palladium chloride, and lithium molybdate gave citral in 66% yield, after 15 min reaction time [237]. Cobalt catalysts suspended on colloids, useful for autooxidations, have been reviewed [238].

F. Oxidation by Persulfates

Persulfuric acids exist in two forms:

$$\text{H-O-}\underset{\underset{\text{O}}{\|}}{\overset{\overset{\text{O}}{\|}}{\text{S}}}\text{-O-O-H} \qquad\qquad \text{H-O-}\underset{\underset{\text{O}}{\|}}{\overset{\overset{\text{O}}{\|}}{\text{S}}}\text{-O-O-}\underset{\underset{\text{O}}{\|}}{\overset{\overset{\text{O}}{\|}}{\text{S}}}\text{-O-H}$$

  Peroxomonosulfuric Acid   Peroxydisulfuric Acid

   (Caro's Acid)   (10-44)

These acids are unstable in aqueous solutions, hydrolyzing to sulfuric acid and hydrogen peroxide. Their salts are generally stable, commercially available, being produced by electrolysis of, for example, potassium hydrogen sulfate. They are powerful oxidants, being able to convert Mn^{2+} to MnO_4^- when catalyzed by silver salts.

Some oxidations with persulfates do not require use of a phase-transfer catalyst. Thus, oxidation of alkenes with aqueous potassium peroxymonosulfate in the absence of solvent or catalyst oxidizes some alkenes to diols and epoxides at room temperature [239]. Acidic (pH $\leq$ 1.7) aqueous solutions of $2KHSO_5 \cdot KHSO_4 \cdot K_2SO_4$ produced epoxide from cyclooctene and diols from all other reactive alkenes studied. Adjustment of initial pH to $\geq$ 6.7 with $NaHCO_3$ enabled selective epoxidations of tetramethylethylene, 1-methylcyclohexene, cyclohexene, styrene, and 1-phenylpropene. 1-Octene and tetrachloroethylene did not react. Phase-transfer catalysts have little effect on the reaction.

Peroxysulfate anions can be used with phase-transfer systems in several ways. They may be transferred directly to an organic phase for oxidations. Cocatalysts, such as cerium, osmium, or ruthenium in anionic form, may be used as transferred oxidants, while the persulfate acts as a primary oxidant to keep the cocatalyst anion in its oxidized form. Another important application persulfates is their use as free radical initiators, wherein the radical anion, $\cdot OSO_3^-$, formed by peroxide fission in the aqueous phase, is transferred to an organic phase [240].

1. Direct PTC Oxidations with Persulfates

Both persulfates derived from sulfuric acid can be effectively used when combined with a PTC technique. For example, peroxymonosulfate (caroate) epoxidized water-insoluble olefins in benzene/water–acetone–phosphate buffer (pH 7.5) at 60°C in the presence of 18-crown-6 or Bu_4NHSO_4 [241]. Oxidation of aromatic methyl groups by potassium peroxydisulfate with benzyltriethylammonium chloride in benzene gives the corresponding aldehyde [242].

Potassium persulfate, under PTC conditions, oxidizes sulfides to sulfoxides and sulfones, at rates highly dependent on the electronic nature of attached groups, as illustrated by the data in Table 10-20 [243].

$$\text{X}\underset{Y}{\overset{Z}{\bigcirc}}\text{-S-}\underset{Z}{\overset{Y}{\bigcirc}}\text{-X} + \text{KHS}_2\text{O}_8 \xrightarrow[25°, 18\text{h}]{R_4N^+ X^-} \text{Ar-SO-Ar and ArSO}_2\text{Ar} \qquad (10\text{-}45)$$

Little difference was found among the quaternary salts tested, but polyethers were generally less active catalysts, as in Table 10-21.

Increasing the quantity of catalyst or increasing the ratio of methylene chloride to water led to more sulfone being produced. Polar solvents, such as ethyl acetate, dioxane, or methanol resulted in quantitative yields of sulfone from 4,4'-bis-(N-methylphthalimido)sulfide and 4,4'-bis(methoxyphenyl)sulfide, while use of o-dichlorobenzene as solvent resulted in very slow reaction and significant quantities of unreacted sulfide after 18 h.

Table 10-20. PTC persulfate oxidation of sulfides [243].

Substituents			Percent in product mixture[a]		
X	Y	Z	Ar-S-Ar	Ar-SO-Ar	Ar-SO$_2$Ar
−NO$_2$	H	H	99.9	—	—
−H	H	−CN	95.0[b]	5.0	—
−CN	H	H	—	83.0	17.0
−H	H	H	—	30.0	70.0
−CH$_3$	H	H	—	66.0	34.0
−OCH$_3$	H	H	—	92.8	7.2
N-methyl phthalimido			—	92.1	7.9
PhCH$_2$-S-CH$_2$Ph			—	70.0	30.0
Ph-S-C$_2$H$_5$			—	—	99.9

[a] Yields of diaryl sulfides were determined after 18 h at room temperature, using 5 mol% Bu$_4$NBr in methylene chloride solution.

[b] Reaction proceeded slowly; after 13 days, 69% sulfoxide and 31% sulfide was detected.

Table 10-21. Effect of PTC catalyst structure on persulfate oxidation of bis-(4-methoxyphhenyl) sulfide, 5 mol% catalyst.

	Percent in product mixture		
Catalyst	Sulfide	Sulfoxide	Sulfone
Bu$_4$NBr	—	92.8	7.2
Bu$_4$NHSO$_4$	—	93.5	6.5
Aliquat 336	—	92.8	7.4
18-Crown-6	50	50	0
Dicyclohexyl-18-crown-6	8.6	79.0	12.4
Polyethylene glycol (PEGM-550)[a]	95.0	5.0	—

[a] With 4,4'-bis (N-methylphthalimido) sulfide as substrate.

Table 10-22. Conversion of carboxylic acids to peracids with potassium persulfate.

Carboxylic acid	Percent yield peracid in presence of Quat or PEG	Percent yield in absence of catalyst
$PhCO_2H$	79.5	50.3
$m\text{-}NO_2C_6H_4CO_2H$	93	65
$p\text{-}NO_2C_6H_4CO_2H$	88	63.5
$m\text{-}ClC_6H_4CO_2H$	87.5	57
$3,5\text{-}(NO_2)_2C_6H_3CO_2H$	90	50
Phthalic	79.5	48.2
$PhCH_2CO_2H$	70	43.5
$3,5\text{-}(MeO)_2C_6H_3CH_2CO_2H$	85	58

Aqueous solutions of potassium persulfate convert water-insoluble carboxylic acids in ether or methylene dichloride solution to peracids, both in the presence and absence of phase-transfer catalysts [244]. Peracid yields are 80–90% in the presence of benzyltriethyl ammonium chloride. Alternatively this liquid–liquid conversion can be catalyzed by use of an acidic (sulfonic) ion-exchange resin. Peracids are also obtained in 80–90% yield with polyethylene glycol (PEG 400) as catalyst in liquid–solid PTC systems. Comparisons showing the effect of PEG as phase-transfer catalyst are shown in Table 10-22.

2. PTC Oxidations with Persulfates Using a Cocatalyst

a. Ruthenium Cocatalyst

Catalytic amounts of ruthenate anion, generated by an aqueous phase reaction of $RuCl_2$ with potassium peroxydisulfate, are readily transferred into organic solutions by quaternary ammonium salts for selective oxidation of benzylic and allylic alcohols to the corresponding aldehydes and ketones [245]. For example, diols containing an allylic or benzylic hydroxyl group are oxidized only at the activated group.
Results from several oxidations are listed in Table 10-23.

$$PhCH(OH)CH_2CH_2OH + K_2S_2O_8 \xrightarrow[CH_2Cl_2/10\% \text{ NaOH}]{\substack{25^\circ C \\ (C_8H_{17})_3NCH_3^+ \, Cl^-}} PhCOCH_2CH_2OH \quad (10\text{-}46)$$

95% Yield

b. Cerium as Cocatalyst

Cerium(IV) can be extracted into an organic solution as $Q_2^+[Ce(NO_3)_6]$, or as the sulfate, which when combined with $S_2O_8^-$ as an aqueous-phase primary oxidant, is a useful oxidizing system for organic reactions [246]. For example,

Table 10-23. Oxidation of allylic and benzylic alcohols with potassium peroxydisulfate, catalyzed by Adogen 464 and potassium ruthenate [245].

Substrate alcohol	Product	Yield (%)
Benzyl	Benzaldehyde	92
2-Chlorobenzyl	2-Chlorobenzaldehyde	88
4-Methoxybenzyl	4-Methoxybenzaldehyde	90
4-Nitrobenzyl	4-Nitrobenzaldehyde	95
1-Phenylethanol	Acetophenone	92
Benzhydryl	Benzophenone	90
Cinnamyl	Cinnamylaldehyde	91
2-Buten-1-ol	Crotonaldehyde	99
1,3-Diphenyl-2-propen-1-ol	1,3-Diphenyl-2-oxy-2-propene	86
3-Phenyl-1,3-propandiol	3-Phenyl-3-oxy-1-propanol	95
5-Phenyl-1,3-dihydroxy-4-pentene	5-Phenyl-1-hydroxy-4-pentene-3-one	90

Table 10-24. PTC oxidation of sulfides to sulfoxides with cerric ammonium sulfate.

Sulfide	Time (min)	Percent RSOR
Bu-S-Bu	10	100
Ph-S-CH_3	15	98
$PhCH_2CH_2$-S-Ph	20	99
Ph-S-Ph	360	96
$PhCH_2$-S-CH_2Ph	15	97
Tetrahydrothiophene	15	100
$PhCH_2$SPh	15	90
$PhCH_2$SPh (no catalyst)	23 h	41

[a] In water/methylene chloride in presence of 5 mol% Bu_4NBr, with ceric ammonium nitrate substrate ratio = 2:1, at room temperature.

PTC–cerric ammonium nitrate oxidation of sulfides provides a selective route to sulfoxides, as shown in Table 10-24 [247].

This system is further activated by addition of a surfactant, for example, sodium dodecylsulfate, and by addition of silver as $AgNO_3$. This composite oxidized naphthalene to p-naphthoquinone.

$$\text{naphthalene} \xrightarrow[\text{(NH}_4)_2S_2O_8 \text{ (aq)}]{\substack{Bu_4N^+ \, HSO_4^- \\ (NH_4)_2Ce(NO_3)_6}} \text{p-naphthoquinone} \quad (10\text{-}47)$$

Use of the surfactant, sodium dodecyl sulfate, boosts reactivity, possibly by providing organic-phase solubilization for the silver ion. Ceric cations with surfactants alone, for example, sodium dodecyl sulfate in the presence of 2,6-pyridine-dicarboxylic acid, are also powerful but selective oxidants, as shown by their ability to convert a variety of toluenes (*o*-methyl, *m*-methyl, *p*-isopropyl, *p*-fluoro, *p*-chloro, *p*-bromo, and *m*-methoxy) to the corresponding benzaldehydes [248].

c. Osmium Cocatalyst

Asymmetric PTC dihydroxylations by persulfate, mediated by osmium tetraoxide, have been reported for several alkenes [249]. For example, treatment of *trans*-stilbene with disodium monopersulfate in water at 0–4°C, using osmium tetroxide with *N*-methylmorpholine-*N*-oxide as phase-transfer catalyst, gave (*R,R*)-PhCH(OH)CH(OH)Ph in 55% yield.

3. Manganese Porphyrin as a Special Oxidation Phase-Transfer Catalyst

Potassium monopersulfate, $KHSO_5$, is an efficient single-oxygen-atom donor to metalloporphyrins. It can be used in a dichloromethane/water–phosphate buffer, pH 6–7, as a primary oxidizing agent in olefin epoxidation and alkane hydroxylation with Mn porphyrin complexes as catalysts [250]. Saturated hydrocarbons as well as methoxybenzene derivatives can be oxidized by $KHSO_5$/Mn(porphyrin)Cl at room temperature with a conversion rate in the range of 0.1–4 turnovers per minute.

G. PTC Oxidations with Chromium Compounds

Chromate ion is easily transferred from acidic aqueous solution into organic media, but not from basic or neutral aqueous media [251]. Chromate has been used as a mild and selective oxidant for reactions in a "stoichiometric PTC" technique wherein the chromate is first attached to a cationic anion-exchange resin, and then organic material to be oxidized is passed through the insoluble solid, using a full molar quantity of oxidant [252].

Dichromate dianion is more difficult to transfer than chromate, although it has been used in a solid–liquid PTC technique with trioctylmethylammonium chloride in organic solvents for oxidation of benzyl and secondary alcohols to the corresponding carbonyl compounds [253].

Dichromate can be extracted into organic solutions with Adogen 464 and these solutions oxidize allylic and benzylic alcohols to aldehydes or ketones in a rather slow reaction (15–18 h at 55°C) [259]. Primary alcohols are almost inert, and secondary alcohols oxidize very slowly. If the aqueous dichromate phase is acidified with sulfuric acid, a species can be extracted with tetrabutyl ammonium

Table 10-25. PTC oxidations with chromium oxidants.

Substrate	Catalyst and oxidant	Product	References
α-Amyrin acetate	$C_{16}H_{33}NMe_3Br + CrO_3$	11-Oxo-derivatives	[256]
β-Amyrin acetate		> 80% yields	
Isoborneol	$C_{16}H_{33}N^+Me_3$	Ketone	[257]
Camphene	CrO_{x-} + electrochem		
$PhCH_2OH$	PTC	PhCHO (high yield)	[258]
Benzylic primary and secondary alcohols	$PhCH_2N^+Et_3\ ClCrO_3^-$	Carbonyl compounds	[259]
$4\text{-MeOC}_6H_4CH_2OH$	$PhCH_2N^+Et_3\ ClCrO_3^-$	4-MeOC_6H_4CHO	[265]
$ArCH_2Cl$ (or Br)	$PEG + K_2CrO_4$	ArCHO (81–95%)	[260]
$n\text{-}C_7H_{15}CH_2\text{-Br}$ (or Cl)	$PEG + K_2CrO_4$	$n\text{-}C_7H_{15}CHO$ (21%)	[266]

hydrogen sulfate into methylene chloride which oxidized primary alcohols to aldehydes within minutes [254,255].

Several chromate oxidations are listed in Table 10-25.

Kinetics of PTC benzyl alcohol oxidation by chromic acid showed that high yields of aldehyde were obtained because the two-phase system prevented further oxidation of aldehyde to acid. The reaction rates were similar in both aqueous solution and in organic medium containing a phase-transfer agent [261]. Nonisothermal catalytic kinetics of phase-transfer oxidation of xylene with tetrabutylammonium bisulfate as phase-transfer agent and potassium dichromate–sulfuric acid as oxidizing agent have also studied [262].

A method for the production of N-ethoxycarbonylnorcodeinone, an intermediate for manufacture of strong analgesics and antagonists of morphine-type narcotics, for example, butorphanol, naloxone, and naltrexone, involves PTC oxidation of the allylic secondary hydroxyl group of N-ethoxycarbonylnorcodeine with alkali metal or ammonium dichromate in the presence of quaternary ammonium salts, as phase-transfer catalysts [263]. For example, tetrabutylammonium yields 85–88% carbonyl compounds.

A commercial electrochemical two-phase process using graphice electrodes with dichromate as a regenrateable oxidant, and tetrabutylammonium salts as phase-transfer catalyst, has been patented for oxidation of anthracene to anthraquinone [322].

1. Chromate Salts as Cocatalysts with Phase-Transfer Agents

Oxidation by sodium perborate of selected alcohols and unsaturated compounds to ketones and acids can be achieved at 60–80°C in the presence of catalytic amounts of chromium(VI) oxide and methyltridecylammonium chloride [264]. For example, oxidation of fluoren-9-ol, benzyl alcohol, cyclohexene, and benzil gave fluorenone, benzaldehyde, adipic acid, and benzoic acid, respectively.

Chromate esters are produced in about 80% yield by reaction of allylic or benzylic halides with chromate/dicyclohexano-18-crown-6 in hexamethylphosphodtriamide (HMPT) solvent, at 100°C [265]. The corresponding alcohols were inert; n-alkyl bromides reacted only very slowly (20% yield). Formation of aldehydes and ketones from benzyl and allyl halides was achieved by reaction with a dichromate salt, $(Bu_4N)_2Cr_2O_7$, obtained by extraction from aqueous K_2CrO_4/Bu_4NHSO_4 [266]. Chromium trioxide in the presence of 18-crown-6 in methylene chloride solution is an efficient and selective oxidant conversion of thiophenols to disulfides, and hydroquinones to benzoquinones in quantitative yields [267].

2. Electrochemical Regeneration of Chromium Oxidants in Combination with PTC Systems

Electrochemical oxidation of primary and secondary alcohols by Cr(VI)/Cr(III) systems under phase-transfer catalysis show good selectivity for oxidation of benzyl alcohols [268]. Benzhydryl alcohol was electrochemically oxidized for 6 h in the presence of Bu_4NBr in chloroform containing $KCr(SO_4)_2$ and sulfuric acid to give 87% benzophenone at a current efficiency of 28%. Other alcohols, including 1-phenylethyl alcohol, benzyl alcohol, cinnamyl alcohol, 1-pentanol, and 2-hexanol were similarly oxidized to 90% or higher yields of the corresponding aldehydes or ketones. A review on indirect electrosynthesis using phase-transfer catalysis includes an example for the oxidation of anthracene to anthraquinone using a dichromate cocatalyst, and oxidation of p-xylene using a cerium catalyst [269].

Cetyltrimethylammonium chromate in its oxidized form is soluble in the nonpolar organic phase of a two-phase mixture and the reduced form is soluble in the aqueous solution. Using an electrochemical system to keep the chromate in the oxidized form provides a technique wherein isoborneol and camphene are selectively oxidized [270]. In electroorganic syntheses using indirect-electrode reactions, chromium ion itself can be a phase-transfer catalyst. That is, chromium ions are oxidized at the electrode to a higher oxidation state, and these move into an organic phase where oxidation of an organic substrate occurs [271]. Silver catalyzes the anodic oxidation of chromium(III) to dichromate [272].

H. PTC Oxidations with Nitric Acid

Nitric acid oxidation of thioethers, mediated by gold salts, $AuCl_4^-$, in combination with quaternary cations, gives sulfoxides [273,274].

$$R\text{-}S\text{-}R + HNO_3 \text{ (aq)} \xrightarrow{(Bu_4N)^+ (AuCl_4)^-} R\overset{\overset{\displaystyle O}{\|}}{-S}-R \qquad (10\text{-}48)$$

Attack on the thio group is selective even in the presence of other functional groups, and is regioselective in the case of disulfides. Nitric acid, in conjunction with cerium salts, can also be used in an electrochemical system using a phase-transfer catalyst [275].

I. PTC Carbon Tetrachloride/Sodium Hydroxide Oxidations

Carbon tetrachloride in the presence of sodium hydroxide functions as an oxidant, as, for example, in conversion of aromatic ketones to the acyloin derivative, useful as photopolymerization initiators [276].

$$Ph-\overset{O}{\underset{}{C}}-CH(CH_3)_2 + NaOH + CCl_4 \xrightarrow[15°C]{(C_8H_{17})_3NCH_3^+ Cl^-} Ph-\overset{O}{\underset{}{C}}-\overset{OH}{\underset{CH_3}{C}}-CH_3 \qquad (10\text{-}49)$$

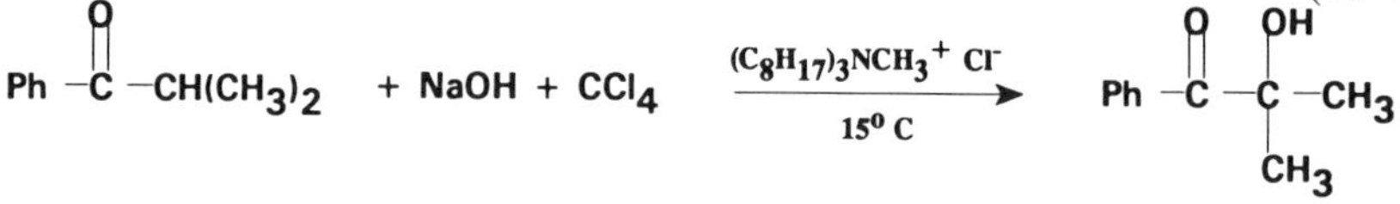

Other aromatic ketones gave different products when treated with carbon tetrachloride and aqueous sodium hydroxide under phase-transfer conditions [277]. Thus, acetophenone and phenylacetone gave halogenated epoxides:

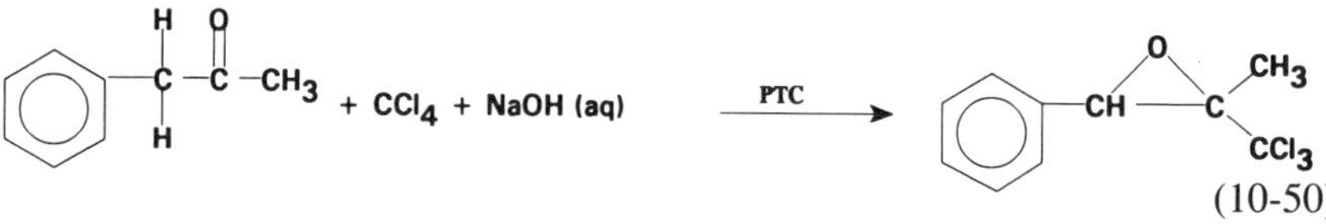

$$\text{PhCH(H)}-\overset{O}{\underset{}{C}}-CH_3 + CCl_4 + NaOH\ (aq) \xrightarrow{PTC} \text{Ph-CH}\overset{O}{\underset{}{\diagdown}}\overset{CH_3}{\underset{CCl_3}{C}} \qquad (10\text{-}50)$$

while α-phenylacetophenone gave coupling products and fragments:

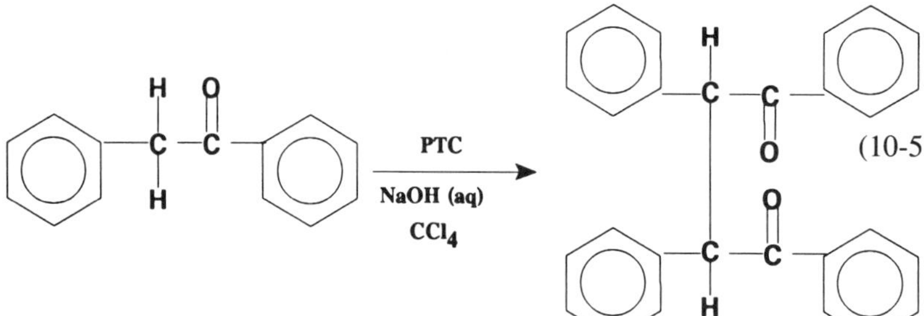

$$\text{PhCH}_2-\overset{O}{\underset{}{C}}-\text{Ph} \xrightarrow[CCl_4]{PTC, NaOH\ (aq)} \text{coupled product} \qquad (10\text{-}51)$$

Chupp and co-workers [278] found that PTC reactions of sodium hydroxide with either carbon tetrachloride or hexachloroethane to chlorinate benzyl chlorides and benzal chlorides having ortho-nitro substituents give the corresponding benzotrichlorides in 70–90% yields. The PTC method was claimed to be superior to other available methods.

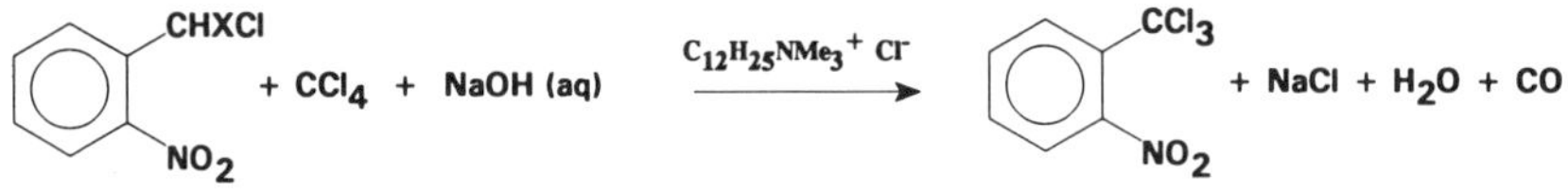

$$\underset{NO_2}{\text{o-}}\text{C}_6\text{H}_4\text{CHXCl} + CCl_4 + NaOH\ (aq) \xrightarrow{C_{12}H_{25}NMe_3^+ Cl^-} \underset{NO_2}{\text{o-}}\text{C}_6\text{H}_4\text{CCl}_3 + NaCl + H_2O + CO$$

X = Cl or H $\qquad (10\text{-}52)$

Carboxylic acids containing a hydrogen and at least one electron-withdrawing group attached to the same carbon can be chlorinated by use of carbon tetrachloride and potassium carbonate, in the presence of a phase-transfer catalyst [323].

Benzyl alcohols are oxidized by CCl_4 and sodium carbonate selectively to benzaldehydes in the presence of quaternary ammonium catalysts and $RuCl_3$ as cocatalyst [279].

J. PTC Oxidations with Periodate and Related Oxidizing Anions

1,2-Diols are cleaved with periodate under phase-transfer conditions to produce the expected aldehydes and ketones in 85–95% yields [280].

Sodium periodate, under PTC conditions, oxidizes certain unsaturated compounds to the allylic oxygenated products [281,282]. For example, α- and β-amyrin acetates, and oleanolic acid, and ursolic acid derivatives gave allylic oxygenated products. However, oleanenols gave epoxides in 24–37% yields [283].

With polystyrene-bound phosphonium or ammonium ions, oxidation of alcohols with preexchanged periodate ion is successful [284,285].

Tetraalkylammonium periodates in organic solvents may be prepared by mixing of an organic solution of a quaternary salt with sodium periodate in water, and these homogeneous ion-pair extraction type oxidations have been used to oxidize sulfides, 2-hydroxyacids, 2-bromoketones, and aryl acetic acids [286,287]. The PTC version of these reactions also works well:

$$R-\underset{\underset{R'}{|}}{\overset{\overset{OH}{|}}{C}}-CO_2H \;+\; NaIO_4 \;\xrightarrow{PTC}\; R\text{-}CO\text{-}R' \quad (80\text{-}90\%)$$

$$Ph\text{-}CHBr\text{-}CO_2H \;+\; NaIO_4 \;\xrightarrow{PTC}\; PhCHO \quad (80\%)$$

$$R\text{-}CH(CO_2H)_2 \;+\; NaIO_4 \;\xrightarrow{PTC}\; RCHO \;+\; RCO_2H \quad (80\%)$$

$$Ph\text{-}S\text{-}R \;+\; NaIO_4 \;\xrightarrow{PTC}\; R\text{-}SO\text{-}R \quad (75\text{-}85\%)$$

(10-53)

Sodium bromate plus a phase-transfer catalyst with a ruthenium complex cocatalyst has been used in place of *tert*-butyl hydroperoxide for oxidation of α-hydroxy esters and nitriles into the corresponding α-ketoesters or nitriles [288].

1. Osmium as Cocatalyst

PTC oxidation of olefins with periodic acid in the presence of catalytic amounts of osmium tetraoxide and a quaternary ammonium salt or amine produces carboxylic acids [20]. Aldehydes are intermediates in this reaction, and may be isolated in good yield by stopping the reaction at an appropriate conversion level. In this system the periodic acid can be regenerated electrochemically [289].

2. Ruthenium as Cocatalyst

Ruthenium tetroxide is a strong oxidant, capable of converting alcohols to aldehydes or ketones, aldehydes to carboxylic acids, ethers into esters or lactones, amides to imides, and olefins into aldehydes and ketones [290]. Ruthenium tetraoxide may also be used in a catalytic quantity with periodic acid as the primary oxidant (or with electrochemical regeneration of the periodic acid), in the presence of quaternary ammonium as phase-transfer catalysts, for oxidation of olefins to carboxylic acids in quantitative yields [20]. Aqueous hypochlorite may also be used as the primary oxidant to regenerate ruthenium tetroxide, with addition of sodium hydroxide to suppress formation of chlorinated products [57].

Although periodate salts are rather expensive for most large-scale chemical reactions, IO_4^- can be used on a catalytic scale by continuous cyclical reoxidation of its reduction products with another primary oxidant. Thus, periodate may be made in situ by oxidation with chlorine, persulfate, or electrochemically [291]. For example, Johnson et al. [289] patented a process for electrolytic oxidation of olefins to carbonyl compounds using a Group VII metal or compound (e.g., OsO_4, RuO_2) in conjunction with a phase-transfer catalyst. The metal compound was maintained in the high oxidation state by reoxidation in the aqueous phase with periodic acid, or periodate anion, which in turn was maintained by electrochemical oxidation of iodate and/or its lower oxidation states.

K. PTC Oxidations with Perborate

In the presence of phase-transfer catalysts, perborate is extracted from aqueous $NaBO_3$ into methylene chloride, where α,β-unsaturated ketones can be oxidized to epoxides:

(10-54)

$(CH_3)_2CH=CHCOCH_3 + NaBO_3$ (aq) $\xrightarrow[\text{75 min}]{(C_6H_{13})_4N^+ \, HSO_4^-}$

$\begin{array}{c} CH_3 \\ CH_3 \end{array} \!\!>\!\! C \overset{O}{-\!\!\triangle\!\!-} C \!\!<\!\! \begin{array}{c} H \\ COCH_3 \end{array}$

78%

L. PTC Oxidations with Ferrate and Ferricyanide

Potassium ferrate, under phase-transfer conditions, has been found to be a highly selective oxidant for conversion of unsaturated alcohols to unsaturated aldehydes and ketones [292]. Yields for unsaturated aldehydes ranged from 80 to 96 percent.

A unique kind of PTC action involves aqueous phase generation of relatively stable triaryl phenoxy radicals in a ferricyanide aqueous phase or at the interface, which then react in the organic phase with hydrazines to produce azo-compounds [293].

M. PTC Oxidations with Superoxide

Potassium superoxide reacts very rapidly with water to produce oxygen and potassium hydroxide. However, in the absence of water potassium superoxide can be used in PTC systems, for example in a solid–liquid system catalyzed by 18-crown-6, as a powerful oxidizing reagent. Thus, KO_2 oxidizes α-disubstituted compounds to the corresponding carboxylic acids [294]:

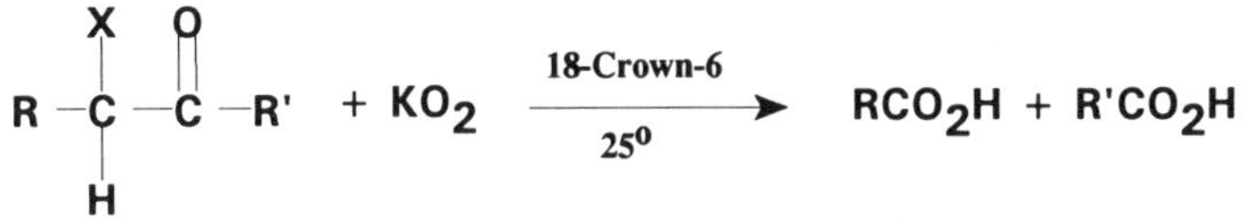

$$R-\underset{H}{\overset{X}{\underset{|}{C}}}-\overset{O}{\overset{\|}{C}}-R' + KO_2 \xrightarrow[25^\circ]{\text{18-Crown-6}} RCO_2H + R'CO_2H \quad (10\text{-}55)$$

X = OH, Cl, Br

Potassium superoxide in benzene, with 18-crown-6 as catalyst, can also behave as a peroxidizing agent, via an SN2 reaction to produce peroxides from alkyl halides [295]. However, if dimethyl sulfoxide is also present, alcohols are produced instead of peroxides [296–298]. For a review of these reactions see [46].

Copolymers of chloromethylated styrene with styrene and divinylbenzene react with KO_2 in the presence of 18-crown-6 in organic solvents (dimethyl sulfoxide and DMF, benzene) to give corresponding hydroxymethyl derivatives [299]. Tetrabutylammonium iodide was a better phase-transfer catalyst, giving 85% conversion with 80% yield of hydroxyl groups, and 85% conversion.

N. PTC Electrochemical Oxidations

Electrochemical regeneration of primary oxidants has been mentioned in the preceding sections as a technique for maintaining a steady-state concentration of chemical oxidants. For example, electrochemical PTC oxidation of benzyl alcohol, mediated by electrochemically generated hypochlorate oxidant, was discussed in Section C.2. Kinetics of that process have been correlated with a theoretical model [300].

In addition to use of electrochemical techniques to maintain a steady-state concentration of an oxidant, *direct electrochemical oxidation using PTC systems* also can be achieved. For example, acetoxylation of the three dimethoxybenzenes in two-phase water/methylene chloride can be accomplished in an electrochemical cell in the presence of phase-transfer catalysts, such as $Bu_4N^+HSO_4^-$, $Ph_3BuP^+Br^-$, and $C_{16}H_{33N}Me_3^+Br^-$ [301].

$$\underset{\text{OCH}_3}{\underset{|}{\text{C}_6\text{H}_3}}\text{-OCH}_3 + \text{CH}_3\text{CO}_2\text{H} \xrightarrow[\text{NaOAc}]{\text{PTC} \atop \text{electrolysis}} \underset{\text{OCH}_3}{\underset{|}{\text{C}_6\text{H}_2}}(\text{OCH}_3)(\text{O}_2\text{CCH}_3) \quad (10\text{-}56)$$

The chemical and current yield for acetoxylated products were much higher in PTC systems than in homogeneous systems [302]. An electrochemical version of a thin film contactor, based on a vertical rotating disk electrode, has been described for use with two-phase electrolytic acetoxylations [303].

Electrochemical cyanation of 1,2- and 1,3-dimethoxybenzenes in water/methylene chloride containing a phase-transfer catalyst are most successful when maximal coverage of the anode by organic phase occurs [304]. Correspondingly, Aliquat-336, a very hydrophobic catalyst [composition is equivalent to $(C_8H_{17})_3NCH_3^+Cl^-$] gives the best chemical (81%) and current (77%) yields of cyano-1,2-dimethoxybenzenes. The anodic cyanation of naphthalene has also been demonstrated [305].

Anodic chlorination of naphthalene in an aqueous NaCl/methylene chloride system containing tetrabutylammonium chloride as PTC gave 1-chloronaphthalene in 56% yield (33% current efficiency). Addition of $ZnCl_2$ to the electrolysis increases the yield of 1-chloronaphthalene up to 92% (49% current efficiency), apparently due to the higher oxidizing selectivity of $[Bu_4N]_2\ ZnCl_4$ [306–310].

Production of γ-butyrolactones, useful for making Nylon-4, is accomplished in high current efficiency by electrochemical reduction of a mixture of an acrylic acid ester and an aldehyde in an aqueous emulsion and in the presence of a phase-transfer catalyst in an electrolytic cell [311–313]. A PTC electrochemical oxidation method is used to oxidize the surface of pitch-based carbon fibers, allowing better adhesive and coating properties through changes in surface tension of the fiber surface [314].

Electrochemical PTC dichromate and cerium oxidations have been reviewed [315]. As an example, the oxidation of anthracene to anthraquinone via dichromate is described. The Ce^{2+}/Ce^{3+} redox system can be also used as a catalyst for the *p*-xylene oxidation.

Other reviews and mechanistic studies on electrochemical PTC studies are given in references [316–319].

O. PTC Oxidations with Other Oxidants

Sodium tellurite with a phase-transfer catalyst has been used as a mild and highly selective oxidizing agent for conversion of thiols to unsymmetrical disulfides [320].

References

1. R.A. Sheldon, Stud. Surf Sci Catal. **66** (Dioxygen Act. Homogeneous Catal Oxid.), 573–94 (1991).
2. R.A. Sheldon, Chemtech, **21,** 566 (1991).
3. P.H. Washecheck, C.M. Starks, K. Yang, and M.L. Johnson, unpublished data.
4. D.G. Lee, E.J. Lee, and K.C. Brown, in Phase-Transfer Catalysis: New Chemistry, Catalysts, and Applications, C.M. Starks, ed., ACS Symp. Ser., **326,** 82 (1987).
5. C.M. Starks and C. Liotta, Phase Transfer Catalysis, Academic Press, New York, 1978.
6. W.A. Gibson and R.A. White, Anal. Chim. Acta, **12,** 413 (1955).
7. A.W. Herriott and D. Picker, Tetrahedron Lett., 1511 (1974).
8. N.A. Gibson and J.W. Hosking, Aust. J. Chem., **18,** 123 (1065).
9. D.J. Sam and H.E. Simmons, J. Am. Chem. Soc., **94,** 4024 (1972).
10. D.G. Lee and V.S. Chang, J. Org. Chem., **43,** 1532 (1978).
11. C.J. Pederson, J. Am. Chem. Soc., **89,** 7017 (1967).
12. D.G. Lee and H. Karaman, Can. J. Chem., **60,** 2456 (1982).
13. M.J. Harris and M.G. Case, J. Org. Chem., **48,** 5390 (1984).
14. H. Karaman, R.J. Barton, B.E. Robertson, and D.G. Lee, J. Org. Chem., **49,** 4509 (1984).
15. C.M. Starks and P.H. Washecheck, to Conoco Inc., U.S. Patent 3,547,962 (1970).
16. D.G. Lee and K.C. Brown, J. Am. Chem. Soc., **104,** 5076 (1982).
17. J.A. Morris and D.C. Mills, Chem. Ind. (Lond.), 446 (1978).
18. H.P. Jager, J. Lutolf, and M.W. Meyer, Angew. Chem., Int. Ed. Engl. **18,** 786 (1979).
19. H.J. Schmidt and H.J. Shafer, Angew. Chem., Int. Ed. Engl., **18,** 787 (1979).
20. D.G. Lee, E.J. Lee, and K.C. Brown, in Phase-Transfer Catalysis; New Chemistry, Catalysts, and Applications, C.M. Starks, ed., ACS Symp. Ser. **326,** 82 (1987).
21. J.M. Harris, N.H. Hundley, T.G. Shannon, and E.C. Struck, J. Org. Chem., **47,** 4789 (1982).
22. D. Mason, S. Magdassi, and Y. Sasson, J. Org. Chem., **55,** 2714 (1990).
23. V. Holba and J. Muchova, React. Kinet. Catal. Lett., **32,** 365–9 (1986).
24. A. McKillop and L.S. Mills, Synth. Commun., **17,** 647 (1989).
25. G. Soula, J. Org. Chem., **50,** 3717 (1985).
26. V. Bhushan, R. Rathore, and S. Chandrasekaran, Synthesis, 431 (1984).
27. T. Ogino and K. Mochizuki, Chem. Lett., 443 (1979).
28. C.M. Starks, J. Am. Chem. Soc., **93,** 195 (1971).
29. A.P. Krapcho, J.R. Larson, and J.M. Eldridge, J. Org. Chem., **42,** 3749 (1977).

30. E.V. Dehmlow and S.S. Dehmlow, Phase Transfer Catalysis, 2d Ed., Verlag Chemie, Deerfield Beach, Florida (1983).
31. T. Okimota and D. Swern, J. Am. Oil Chem. Soc., **54,** 867A (1977).
32. A. Chollet, C. Mahaim, C. Foetisch, M. Hardy, and P. Vogel, Helv. Chim. Acta, **60,** 59 (1977).
33. K. Inoue, H. Noguchi, H. Massanobu, and Y. Uchida, Yukagaku, **29,** 397 (1980) [CA 94.102469].
34. M.M. Campbell, R.C. Craig, A.C. Boyd, I.M. Gilbert, R.T. Logan, J. Redpath, R.G. Roy, D.S. Savage, and T. Sleight, J. Chem. Soc. Perkin Trans. I, 2235 (1979).
35. W.P. Weber and J.P. Shepherd, Tetrahedron Lett., 4907 (1972).
36. T. Ogino, Tetrahedron Lett., 177 (1980).
37. A. Ayari, S. Szonyi, E. Rouvier, and A. Cambon, J. Fluorine Chem., **50,** 67–75 (1990) [CA: 114.080974].
38. B.V.S.K. Rao and R. Subbarao, J. Am. Oil Chem. Soc., **68,** 646 (1991).
39. T. Sala and M.V. Sargent, J. Chem. Soc., Chem. Commun., 253 (1978).
40. T. Okimoto and D. Swern, J. Am. Oil Chem. Soc., **54,** 862A (1977).
41. T.A. Foglia, P.A. Barr, and A.J. Malloy, J. Am. Oil Chem. Soc., **54,** 858A (1977).
42. G.M. Lampman and S.D. Sharpe, J. Chem. Ed., **60,** 503 (1983).
43. S. Marmor, Laboratory Methods in Organic Chemistry, Burgess, Minneapolis, 1983, pp. 352–356.
44. D.G. Lee, S.E. Lamb, and V.S. Chang, Org. Synth., **60,** 11 (1981).
45. D.G. Lee and V.S. Chang, J. Org. Chem., **44,** 2726 (1979).
46. D.G. Lee, E.J. Lee, and W.D. Chandler, J. Org. Chem., **50,** 4306 (1985).
47. D.G. Lee and V.S. Chang, Synthesis, 462 (1978).
48. T. Fang, Polym. Prepr. (Am. Chem. Soc., Div. Polym. Chem.), 32 (10 371-2 (1991).
49. H.J. Schmidt, and H.J. Schafer, Angew. Chem. Int. Ed. Engl., **20,** 104, 109 (1981).
50. Takasago Perfumery Co., Ltd., Japan Kokai Tokkyo Koho JP 61/33184 (1986) [CA 105.134193].
51. Otsuka Pharmaceutical Co., Ltd., Japan Kokai Tokkyo Koho, JP 59/141572 (1984) [CA 102.185437].
52. U.R. Nayak, V.S. Dalavoy, and V.B. Deodhar, Indian J. Chem., Sect. B., **28B,** 1055 (1989) [CA 113.059574].
53. Yu. A. Zhdanov, Yu E. Alekseev, and S.S. Doroshenko, Dokl. Akad. Nauk SSSR, **262,** 1147 (1982).
54. G.W. Gokel and H.D. Durst, Synthesis, 168 (1976).
55. H. Bock and D. Jaculi, Angew. Chem., Int. Ed. Engl., **23,** 305 (1984).
56. F.M. Menger, J.U. Rhee, and H.K. Rhee, J. Org. Chem., **40,** 3803 (1975).
57. H.J. Schmidt and H.J. Schafer, Angew. Chem. Int. Ed., Engl., **18,** 68, 69 (1979).

58. N. Huang and L. Xu, Synth. Commun., **20**, 1563 (1990).
59. Rao, K.H. and Rao, M.B., J. Indian Chem. Soc., **68**, 132–4 (1991) [CA: 115.231794].
60. G.A. Artamkina, A.A. Grinfel'd, and I.P. Beletskaya, Zhur. Org. Khim., **16**, 698 (1980). [Engl. Transl. p. 612].
61. S.M. Gannon and J.G. Krause, Synth. Commun., 915 (1987).
62. W. Huang and Q. Jiang, Huaxue Shijie, **29**, 253–6 (1988); [CA 109.189937].
63. V.N. Vostrova, V.A. Shigalevskii, S.S. Tkachenko, and N.L. Nikulina, USSR Patent SU 1,643,522 (1991), Appl 4,631,050, (1991) [CA 116,083528].
64. Y.S. Lin and J.S. Shih, J. Chin. Chem. Soc. (Taipei), **38**, 451 (1991) [CA: 115.279242].
65. J. Kulic, M. Adamek, A.B. Zhivich, G.I. Koldobskii, and Yu E. Myznikov, Zh. Obsch. Khim., **60**, 2370 (1990).
66. J. Konar and R. Ghosh, Polym. Degrad. Stab., **21**, 263–75 (1988) [CA 109.129798].
67. J. Konar, R. Ghosh, and S.K. Ghosh, Polym. Degrad. Stab., **22**, 43–52 (1988) [CA 109.231678].
68. D. Scholz, Monatsh. Chem., **112**, 241 (1981).
69. D.G. Lee and N.S. Srinivasan, Sulfur Lett., **1**, 1 (1981).
70. R.S. Glass and J.L. Broeker, Tetrahedron, **47**, 5477 (1991).
71. L.M. Rossi and P. Trimarco, Synthesis, 743 (1978).
72. W. Skupinski, L. Pichnej, R. Pakula, W. Jahn-Andrychowska, Z. Trojanowaks, and K. Butkiewicz, Prezem. Chem., **65**, 545 (1986) [CA 107.217195].
73. P.K. Kadaba, J. Prakt. Chem., **324**, 857 (1982).
74. T.F. Osipova, G.I. Koldobskii, V.A. Ostrovskii, and Yu. E. Myznikov, Khim. Geterostskl. Soedin., 841–5 (1985) [CA 103.178201].
75. I.V. Nikonova, G.I. Koldobskii, A.B. Zhivich, and V.A. Ostrovskii, Zh. Obshch. Khim., **61**, 2104 (1991).
76. S. Burke, A.W.P. Jarvie, and A.F. Gaines, Fuel, **69**, 1370 (1990) [CA: 114.009370].
77. G.A. Lee and H.H. Freedman, Tetrahedron Lett., **20**, 1641 (1976).
78. G.A. Lee and H.H. Freedman, Israel J. Chem., **26**, 229 (1985).
79. T. Ido, H. Tariki, and S. Goto, Kagaku Kogaku Ronbunshu, **11**, 424 (1985) [CA 104.185779].
80. T. Kakimoto and A. Katsura, Jpn. Kokai Tokkyo Koho, JP 01305062 (1989) [CA 112.178698].
81. K. Gu, Faming Zhuanli Shenqing Gongkai Shuomingshu CN 85100500 (1986) [CA 107.078110].
82. S. Abramovici, R. Neumann, and Y. Sasson, J. Mol. Catal., **29**, 291 (1985).
83. M.E. Brokke and W.L. Magee, Jr., to Stauffer Chemical Co., Eur. Pat. Appl., E.P. 140,454 (1985) [CA 103.123041].

84. S. Abramovici, R. Neumann, and Y. Sasson, J. Mol. Catal., **29**, 299 (1985).
85. F.E. Friedli, T.L. Vetter, and M.J. Bursik, J. Am. Oil Chem. Soc., **62**, 1058 (1985).
86. J.-S. Do and T.-C. Chou, Ind. Eng. Chem., **29**, 1095 (1990).
87. T. Ido, N. Ohyama, S. Goto, and H. Teshima, Kagaku Kogaku Ronbunshu, **9**, 58 (1983) [CA 98.125171].
88. T. Ido, H. Tariki, and S. Goto, Kagaku Kogaku Ronbunshu, **11**, 424 (1985) [CA 104.185779].
89. T. Ido, H. Tariki, K. Sakurai, and S. Goto, Kagaku Kogaku Ronbunshu, **10**, 287 (1984) [CA 101.090171].
90. J.S. Do. and T.C. Chou, J. Appl. Electrochem., **19**, 922 (1989) [CA 112.044272].
91. S. Erhan, F. Chertok, K. Kaleem, V.S. Nithianandam, and T.A. Reddy, as reported in Chem. Eng. News, July 20, 1991, p. 20.
92. G. Qin, et al., Faming Zhuanli Shenqing Gongkai Shuomingshu CN 1,039,413 (1990) [CA: 113.230791].
93. M. Takayama and T. Suzuki, to Ihara Chemical Industry Co., Ltd., Jpn. Kokai Tokkyo Koho JP 61/271255 (1986) [CA 106.213563].
94. S.N. Mathur, S.N. Rao, and U.T. Bhalerao, Indian J. Chem., Sect. B, **27B**, 666 (1988) [CA 110.114065].
95. W. Broda and E.V. Dehmlow, Israel J. Chem., **26**, 219 (1985).
96. J.H. Ramsden, R.S. Dragoo, and R. Riley, J. Am. Chem. Soc., **111**, 3958 (1989).
97. L. Horner and J. Gerhard, Phosphorus Sulfur, **22**, 5 (1985).
98. P.K. Wehrenberg, to Stauffer Chemical Co., U.S. Patent 4,704,467 (1987) [CA 108.150060].
99. L. Horner and J. Gerhard, Phosphorus Sulfur, **22**, 13–21 (1985) [CA 103.071387].
100. H. Mitsui, T. Hayashi, and I. Maeda, Japan Kokai Tokkyo Koho JP 61/109784 (1986) [CA 106.020291].
101. D.R. Gupta and S. Naitani, Pol. J. Chem., **64**, 747 (1990) [CA: 115.091767].
102. J.R. Lawson, (to Dupont) Eur. Pat. Appl., EP 414,569 (1991) [CA:115.008555].
103. M. Ikeda, Y. Suzuki, and A. Aoshima, (to Asahi Chemical Industry Co., Ltd.,) Jpn. Kokai, Tokkyo Koho JP 03,154,683 (1991) [CA: 115.165863].
104. M. Ikeda, M. Miura, and A. Aoshima, to Asahi Chemical Industry Co., Ltd. Eur. Pat. Appl. 100,488 (1983) [CA 101.072593].
105. Daikin Kogyo Co., Ltd., Jpn. Kokai Tokkyo Koho JP 58/131976 (1983) [CA 100.005309].
106. J.R. Lawson, to Dupont, Eur. Pat. Appl. EP 473,398 (1992) [CA: 116.235424].
107. G.S. Girgoryan, I.L. Karoyan, A.Ts. Malkhasyan, G.T. Martirosyan, G.A. Artamkina, and I.P. Beletskaya, Zh. Org. Khim., **23**, 1149–53 (1987).
108. R. Winter and G.L. Gard, J. Fluorine Chem., **50**, 141 (1990).

109. B. Meunier, E. Guilmet, M.E. DeCarvalho, and R. Poiblanc, J. Am. Chem. Soc., **106,** 6668 (1984), and references therein.
110. O. Bartolini, M. Momenteau, and B. Meunier, J. Chem. Soc., Perkin Trans., 1967 (1984).
111. B. DePoorter and B. Meunier, Tetrahedron Lett., **25,** 1895 (1984).
112. O. Bartolina, M. Momenteau, and B. Meunier, Tetrahedron Lett., **25,** 5773 (1984).
113. J.P. Collmann, J.L. Brauman, B. Meunier, T. Hayashi, T. Kodadek, and S.A. Raybuck, J. Am. Chem. Soc. **107,** 2000 (1985), and references contained therein.
114. F. Montanari, M. Penso, S. Quici, and P. Vigano, J. Org. Chem., **50,** 4888 (1985).
115. B. Meunier, E. Guilmet, M.E. Carvlho, and R. Poiblanc, J. Am. Chem. Soc., **106,** 6668 (1984).
116. J. Suo, S. Li, and H. Wang, Fenzi Cuihua, 5, 103 (1991) [CA: 115.207311].
117. S. Banfi, F. Montanari, and S. Quici, Gazz. Chim. Ital., **120,** 435 (1990) [CA: 114.034766].
118. S. Banfi, F. Montanari, S. Quici, and G. Torosyan, Inclusion Phenom. Mol. Recognit. Chem., **12,** 159 (1992).
119. F. Montanari, S. Banfi, G. Pozzi, and S. Quici, Rev. Heteroat. Chem., **6,** 94 (1992).
120. R.L. Halterman and S.T. Jan, J. Org. Chem., **56,** 5253 (1991).
121. C.J. Burrows, Inclusion Phenom. Mol. Recognit. [Proc. Int. Symp.] 5th., 1988, (Pub. 1990) 199–207 [CA: 115.048614].
122. H. Yoon and C.J. Burrows, J. Am. Chem. Soc., **110,** 4087 (1988).
123. J.C. Dobson, W.K. Seok, and T.J. Meyer, Inorg. Chem., **25,** 1513 (1986).
124. M. Yao and Y. Au, Faming Zhuanil Shenqing Gongkai Shuomingshu CN 1,046,326 (1990) [CA: 115.008315].
125. Y. Sasson, G.D. Zappi, and R. Neuman, J. Org. Chem., **51,** 2880 (1986).
126. A.B. Sorokin and A.M. Khenkin, Kinet. Katal., **30,** 368 (1989).
127. S. Mohanraj and W.T. Ford, Polym. Prepr. (Am. Chem. Soc., Div. Polym. Chem.), **27,** 5–6 (1986).
128. A.E. Qureshi and W.T. Ford, React. Polym., **10,** 279–85 (1989).
129. S. Mohanraj and W.T. Ford, U.S. Patent 4,713,423 (1987).
130. S. Mohanraj and W.T. Ford, Macromolecules, **19,** 2470 (1986) [CA 105.173171].
131. D.K. Parker, to Goodyear Tire and Rubber Co., Eur. Pat. Appl. 281,495 (1988) [CA 110.115586].
132. M.E. Waleter, G.M. St. George, W.F. Ritchey, to Dow Chemical Co., U.S. Patent 4,992,151 (1991).
133. T.V.N. Bui and P.N.S. Chu, Tap Chi Hoa Hoc, **23,** 6–7 (1985) [CA 105.193306].
134. J.T. Arnold, T.O. Bayraktaroglu, R.G. Brown, C.R. Heiermann, W.W. Magnus, A.B. Ohman, and R.G. Landolt, J. Org. Chem., **57,** 391 (1992).

135. I.N. Prokhorova, F.S. Sirovskii, L.M. Kartashov, and T.V. Chernysheva, Khim. Prom.st. (Moscow) 879 (1989) [CA 112.121059].

136. F.S. Sirovskii, A.V. Karmyshaneva, L.I. Virin, and R.M. Kecher, Khim Promst. (Moscow) 649 (1989) [CA 112.038657].

137. S. Yan and S. Song, Yingyong Huaxue, **7,** 69 (1990) [CA 113.061680].

138. J. Correia, J. Org. Chem., **57,** 4555 (1992).

139. J.T. Arnold, T.O. Bayraktaroglu, R.G. Brown, C.R. Heiermann, W.W. Magnus, A.B. Ohman, and R.G. Landolt, J. Org. Chem., **57,** 391 (1992).

140. G. Barak and Y. Sasson, J. Chem. Soc., Chem. Commun., 1266 (1987).

141. E.V. Dehmlow and S.S. Dehmlow, "Phase Transfer Catalysis," 2d.ed., Verlag Chemie, Weinheim (1983).

142. C. Venturello, E. Alneri, and M. Ricci, J. Org. Chem., **48,** 3831 (1983).

143. O. Bortolini, V. Conte, F. diFuria, and G. Modena, J. Org. Chem., **51,** 2661 (1986).

144. B.M. Trost and Y. Masuy, Tetrahedron Lett., **25,** 173 (1984).

145. O. Bortolini, F. diFuria, and G. Modena, J. Org. Chem., **50,** 2689 (1985).

146. S.E. Jacobsen, F. Mares, and P.M. Zambri, J. Am. Chem. Soc., **101,** 6938 (1979).

147. N.A. Sokolov, G.Y. Perchugov, and O.S. Morozov, Zh. Obsch. Khim., **49,** 1856 (1979), Engl. translation, p. 1633.

148. C.M. Starks and D.R. Napier, S. African Pat. 7,101,495 (1971); Brit. Pat. 1,324,763 (1971).

149. R.A. Sheldon and J.K. Kochi, Metal Catalyzed Oxidation of Organic Compounds, Academic Press, New York, 1981.

150. G.B. Payne and P.H. Williams, J. Org. Chem., **24,** 54 (1959).

151. H.J. Reich, F. Chow, and S.L. Peake, Syntheses, 299 (1978).

152. T. Hori and K.B. Sharpless, J. Org. Chem., **43,** 1689 (1978).

153. S.E. Jacobson, F. Mares, and P.M. Zambri, J. Am. Chem. Soc., **101,** 6946 (1978).

154. C. Venturello, E. Alneri, and G. Lana, Ger. Offen, 3,027,349 (1981) [CA 95.042876].

155. T. Dovganyuk, V.V. Berentsveig, E.A. Karakhanov, and T.A. Chemielva, Neftekhimiya, **28,** 232–7 (1988) [CA 109.092672].

156. V.V. Dovganyuk, E.A. Karakhanov, and I.G. Syschikova, Vestn. Mosk, Univ., Ser. 2, Khim., **29,** 515 (1988). [CA 110.211892].

157. L.J. Csanyi and K. Jaky, J. Mol. Catal., **61,** 75 (1990).

158. C. Aubry, G. Chottard, N. Platzer, J.M. Bregeault, R. Thouvenot, R. Chauveau, C. Huet, and H. Ledon, Inorg. Chem., **30,** 4409 (1991).

159. J. Ma, X. Ye, and Y. Wu, Cuihua Xuebao, **12,** 451 (1991) [CA: 116.068351].

160. E. Tempesti, L. Fiuffre, C. Mazzocchia, F. DiRenzo, and P. Gronchi, Stud. Surf. Sci. Catal., **41** (Heterog. Catal. Fine Chcm.) 403 (1988) [CA 110.010046].

161. T. Dovganyuk, V.V. Berentsveig, E.A. Karakhanov, and T.A. Chemielva, Neftekhimiya, **28**, 232–7 (1988) [CA 109.092672].

162. M.L.O. Penido, D.L. Nelson, and D. Pilo-Veloso, J. Braz. Chem. Soc., **1**, 35 (1990) [CA: 114.023371].

163. T.V. Dovganyuk, V.V. Berentsveig, E.A. Krakhanov, and A.A. Borisenko, Neftekhimiya, **30**, 602 (1990) [CA: 114.061852].

164. L.J. Csanyi, K. Jaky; J. Catal. **127**, 42 (1991).

165. J.L. Csanyi and K. Jaky, Stud. Surf. Sci. Catal., **66** (Dioxgyen Act. Homogeneous Catal. Oxid.) 437–43 (1991) [CA: 115.279196].

166. FMC Corp., Jpn. Kokai Tokkyo Koho JP 62/230778 (1987) [CA 108.096629].

167. L. Cerveny, K. Truxova, and V. Ruzicka, Chem.-Ztg., **110**, 303 (1986) [CA 106.175859].

168. S. Teshigahara, T. Ishigaki, and Y. Kano, to Shin-Daikyowa Petrochemical Co., Ltd., Jpn. Kokai Tokkyo Koho JP 03,240,780 (1991) [CA: 116.107004].

169. S. Teshigahara and Y. Kano, to Tosoh Corp., Japan, Toso Kenkyu Hokoku, **35**, 47 (1991) [CA: 116.084740].

170. L.T. McElliott, Union Camp Corp., Eur. Pat. Appl. EP 151941 A2, Aug., 1985. [CA 104.149198].

171. A.T. Au, to Dow Chemical Co., U.S. Patent US 5,036,154 (1991).

172. G. Bornengo, M. Filippo, M. Pontevivo, and G. Bottaccio, to Montedison S.p.A., Eur. Pat. Appl. 143,655, A1, June, 1985. [CA 103.215141].

173. T. Shimoju and K. Shimako, to Sumitomo Pharmaceuticals Co., Ltd., Jpn. Kokai Tokkyo Koho JP 03,236,343 (1991) [CA: 116.105696].

174. P. Caubere, Y. Fort, and A. Ortar Agnes, to Atochem, Eur. Pat. Appl. EP 468,840 [CA: 116.215092].

175. Y. Fort, A. Olszewski-Otar, and P. Caubere, Tetrahedron, **48**, 5099 (1992).

176. S. Teshigahara and Y. Kano, to Tosoh Corp., Toso Kenkyu Hokoku, **35**, 47 (1991) [CA: 116.084740].

177. S. Teshigahara, T. Ishigaki, and Y. Kano, to Shin-Daikyowa Petrochemical Co., Ltd., Jpn. Kokai Tokkyo Koho JP 03,240,780 (1991) [CA: 116.107004].

178. A.T. Au, Andrew, to Dow Chemical Co., U.S. Patent US 5,036,154 (1991).

179. E.N. Karaulova, E.I. Bagrii, A.I. Mikaya, and I.R. Barykina, Neftekhimiya, **32**, 12 (1992) [CA: 117.025727].

180. Z. Lu, Q. Guo, Y. Yin, and D. Jin, Cuihua Xuebao, **13**, 198 (1992) [CA: 117.111395].

181. As reported in Chem. Eng. News, September 21, 1992, p. 8; Apr. 2, 1990, p. 20.

182. C. Burrows, H. Yoon, and T.R. Wagler, to State University of New York Research Foundation, U.S. Patent, US 4,987,227 (1991).

183. F.P. Ballistreri, S. Falla, and G.A. Tomaselli, J. Org. Chem., **53**, 830 (1988).

184. Y. Ishii and Y. Sakata, J. Org. Chem., **55**, 5545 (1990).

185. Y. Salto, S. Araki, Y. Sugita, and N. Kurata, to Nippon Shokubai Kagaku Kogyo Co., Ltd., Eur. Pat. Appl. EP 193,368 (1986) [CA 105.190703].
186. G. Barak and Y. Sasson, J. Chem. Soc., Chem. Commun., 637 (1988).
187. E.A. Karahhanov, E.A. Ivanova, S. Yu. Narin, and A.G. Dedov, Vestn. Mosk. Univ., Ser. 2, Khim., **30**, 510 (1989) [CA 112.234604].
188. P.C. Reteesh, T. Yu. Filippova, A.G. Dedov, and E.A. Karakhanov, Vestn. Mosk. Univ., Ser. 2: Khim., **32**, 418 (1991) [CA: 116.058559].
189. G. Barak and Y. Sasson, J. Org. Chem., **54**, 3484 (1989).
190. J. Dakka and Y. Sasson, J. Chem. Soc., Chem. Commun., 1421 (1987).
191. M. Fournier, F. Frounier, and J. Berthelot, Bull. Soc. Chim. Belg., **93**, 157 (1983).
192. S. Julia and A. Ginebreda, Tetrahedron Lett., 2171 (1979).
193. G. Cerichelli, L. Luchetti, and G. Mancini, Tetrahedron Lett., **30**, 6209 (1989).
194. P. Reteesh Pulippuracceril, T. Yu. Filippova, A.G. Dedov, and E.A. Karakhanov, Vestn. Mosk. Univ., Ser. 2: Khim., **32**, 418 (1991) [CA: 116.058559].
195. J. Bai and H. Ding, J. Chem. Soc., Chem. Commun., 323 (1990).
196. C.R. Harrison and P. Hodge, J. Chem. Soc., Perkin Trans 1, 605 (1976).
197. J.M.J. Frechet and K.E. Hague, Macromolecules, **8**, 130 (1975).
198. S.E. Jacobson, F. Mares, and P.M. Zambri, J. Am. Chem. Soc., **101**, 6938 (1979).
199. S.E. Jacobson, F. Mares, and P.M. Zambri, J. Am. Chem. Soc., **101**, 6946 (197).
200. R.T. Taylor and L.A. Flood, J. Org. Chem., **48**, 5160 (1983).
201. B. Dietrich and J.M. Lehn, Tetrahedron Lett., 1225 (1973).
202. E. Alneri, G. Bottaccio, and V. Carletti, Tetrahedron Lett., 2117 (1977).
203. J. Yamashita, S. Ishikawa, and H. Hashimoto, Bull. Chem. Soc., Japan, **53**, 736 (1980).
204. Z. Stec and Z. Kulicki, Chem. Stosow., **32**, 515 (1988) [CA 112.181828].
205. D. Feldman and M. Rabinovitz, J. Org. Chem., **53**, 3779 (1988).
206. R. Neumann and Y. Sasson, J. Org. Chem., **49**, 1282 (1984).
207. M. Halpern and Z. Lysenko, J. Org. Chem., **54**, 1201 (1989).
208. H.W. Hill, L.J. Enloe, and R.R. Doyle, J. Chem. Educ., **62**, 608 (1985).
209. M. Masui, A. Ando, and T. Shioiri, Tetrahedron Lett., **29**, 2838 (1988).
210. N. Baba, J. Oda, S. Kawahara, and M. Hamada, Bull. Inst. Chem. Res., Kyoto Univ., **57**, 121 (1989) [CA 113.023546].
211. N. Baba, S. Kawahara, M. Hamada, and J. Oda, Bull. Inst. Chem. Res., Kyoto Univ., **65**, 144 (1987) [CA 109.170140].
212. C.K.F. Hermann, Y.P. Sachdeva, and J.F. Wolfe, J. Heterocyclic Chem., **24**, 1061 (1987).
213. G.A. Artamkina and I.P. Beletskaya, Zh. Vses, Khim. O-va. im. D.I. Mendeleeva, **31**, 196 (1986) [CA 105.171422].
214. J. Jarrouse and J.C. Raulin, C.r. Acad. Sci. Paris, **284**, 503 (1977).

215. B.J. Garcia, G.W. Gokel, and P.W. Tudor, unpublished results, 1973.
216. R. Neumann and Y. Sasson, J. Chem. Soc., Chem. Commun, 616 (1985).
217. L. Guglielmetti, to Ciba-Geigy A.-G., Eur. Pat. Appl. EP 305,648 (1989) [CA 111.098965].
218. T. Chiba and M. Okimoto, J. Org. Chem., **56,** 6163 (1991).
219. J. Dakka, A. Zoran, and Y. Sasson, to Gadot Petrochemical Industries Ltd., Eur. Pat. Appl. EP 318,399 (1989) [CA 111.134924].
220. M. Hronec, M. Harustiak, and J. Ilavaky, React. Kinet. Catal. Lett., **27,** 231 (1985).
221. M. Harustiak, M. Hronec, and J. Ilavsky, J. Mol. Catal., **48,** 335 (1988).
222. M. Harustiak, M. Hronec, and J. Ilavsky, J. Mol. Catal., **53,** 209 (1989).
223. M. Harustiak, M. Hronec, J. Ilavsky, and S. Witek, Stud. Org. Chem. (Amsterdam), **35** (Chem. Hetrocyclic Compounds) **308** (1988) [CA 110.172760].
224. M. Harustiak, M. Hronec, J. Ilavsky, and M. Ciffrova, Ropa Uhlie, **28,** 315 (1986) [CA 107.006617].
225. J. Dakka, A. Zoran, and Y. Sasson, to Gadot Petrochemical Industries Ltd., Eur. Pat. Appl., 300,921 (1989) [CA 111.023218].
226. C. Iditoiu, A. Ghenciu, and C. Andrei, Bul. Stiint, Teh. Inst. Politeh. "Traian Vuia" Timisora, Ser. Chim., **34,** 65–70 (1989) [CA: 114.101274].
227. M. Harustiak, M. Hronec, and J. Ilavaky, Ropa Uhlie, **29,** 41 (1987) [CA 107.175323].
228. H. Bednarska, et al., Ger. Offen. DE 3624417 (1986) [CA 106.121842].
229. E.M.G.A. Van Kruchten and C.M.J. Leenaars, to Shell Internationale Research, Eur. Pat. Appl. EP 288,104 (1988), [CA 110.094511].
230. H.M. Van Dort and H.J. Geursen, Recl. Trav. Chim. Pays-Bas, **86,** 520 (1967).
231. D.R. Kearns, Chem. Revs., **71,** 395 (1971).
232. A.A. Frimer, Singlet Oxygen, CRC Press, Boca Raton, FL, 1984.
233. A. Guarini and P. Tundo, J. Org. Chem., **52,** 3501 (1987).
234. H.A. Zahalka, K. Januszkiewicz, and H. Alper, J. Mol. Catal., **35,** 249 (1986).
235. K. Januszkiewicz and H. Alper, Tetrahedron Lett., **24,** 5163 (1983).
236. A. Atlamsani and J.M. Bregeault, New J. Chem., **15,** 671 (1991).
237. J.B. Woell, to Union Camp Corp., Eur. Pat. Appl. EP 439,368 (1991) [CA: 115.136438].
238. W.T. Ford, R. Chandran, and H. Turk, Pure Appl. Chem., **60,** 395 (1988).
239. W. Zhu and W.T. Ford, J. Org. Chem., **56,** 7022 (1991).
240. J.K. Rasmussen, S.M. Heilmann, L.R. Krepski, and H.K. Smith II, ACS Symp. Ser., **326,** 115 (1987), and references contained therein.
241. R. Curci, M. Fiorentino, L. Troisi, J.O. Edwards, and R.H. Pater, J. Org. Chem., **45,** 4758 (1980).

242. Seitetsu Kagaku Co., Ltd, Jap. Kokai, 81/20539 (1981) [CA 94.023557].
243. E.L. Evans and M.M. Grade, Synth. Commun., **16**, 1207 (1986).
244. C.S. Pande and N. Jain, Synth. Commun., **18**, 2123 (1988).
245. K.S. Kim, S.J. Kim, Y.H. Song, and C.S. Hahn, Syntheses Commun., 1017 (1987).
246. E.V. Dehmlow and J.K. Makrandi, J. Chem. Res., Synop., 32 (1986).
247. E. Baciocchi, A. Piermatte, and R. Ruzziconi, Synth. Commun., **18**, 2167 (1988).
248. J. Skarzewski, Pr. Nauk. Inst. Chem. Org. Fiz. Politech. Wroclaw, (No. 30) 3-66 (1986) [CA 109.092399].
249. I.E. Marko and K.B. Sharpless, WO 8906225 (1989) [CA 112.075971].
250. A. Robert and B. Meunier, New J. Chem., **12**, 885 (1988).
251. J. Hala, O. Navatil, and V. Neuchta, J. Inorg. Nucl. Chem., **28**, 553 (1966) and references contained therein.
252. G. Cainelli, G. Cardillo, M. Orena, and S. Sandri, J. Am. Chem. Soc., **98**, 6737 (1976).
253. R.D. Hutchins, N.R. Natale, W.J. Cook, and J. Ohr, Tetrahedron Lett., 4167 (1977).
254. D. Pletcher and S.J.D. Tait, Tetrahedron Lett., 1601 (1978); J. Chem. Soc. Perkin Trans. 2, 788 (1979).
255. D. Landini, F. Montanari, and F. Rolla, Synthesis, 134 (1979).
256. C. Singh, Indian J. Chem., Sect. B, **24B**, 300-1 (1985) [CA 104.034208].
257. D.C. Trivedi, Bull. Electrochem., **2**, 285 (1986) [CA 105.160706].
258. F.C. Thyrion, Bull. Soc. Chim. Belg., **93**, 281 (1984) [CA 101.110096].
259. C.S. Rao, A.A. Deshmukh, M.R. Thakor, and P.S. Srinivasan, Indian J. Chem., Sect. B, **25B**, 324–5 (1986) [CA 106.66828].
260. Z. Huang and X. Huang, Huaxue Shiji, **8**, 18–19 (1986) [CA 105.152626].
261. F.C. Thyrion, Bull. Soc. Chim. Belg., **93**, 281 (1984) [CA 101.110096].
262. S. Li, Q. Chen, and K. Pang, Shenyang Yaoxueyuan Xuebao, **8**, 87 (1991) [CA: 115.135295].
263. J. Hodkova, Z. Vesely, and J. Trojanek, Czech Patent CS 266,942 (1990) [CA: 115.183648].
264. J. Muzart and A.A. N'ait, Synth. Commun., **21**, 575 (1991).
265. G. Cardillo, M. Orena, and S. Sandri, J. Chem. Soc., Chem. Commun., 190 (1976).
266. D. Landini and F. Rolla, Chem. Ind. (Lond.) 213 (1979).
267. M. Juaristi, J.M. Aizpurua, B. Lecea, and C. Palomo, Can. J. Chem., **62**, 2941 (1984).
268. Y. Huang, Z. Li, G. Wang, and K. Cai, Gaodeng Xuexiao Huaxue Xuebao, **12**, 351–2 (1991) [CA: 115.255406].

269. D. Pletcher, Electroorg. Synth [Manuel M. Baizer Meml. Symp.] 1990 (Pub. 1991), 255–62 ed R.D. Little, N.L. Winberg, Dekker: New York, N.Y.
270. D.C. Trivedi, Bull. Electrochem., **2**, 285 (1986) [CA 105.160706].
271. T. Shono, Kagaku (Kyoto), **4**, 756 (1986) [CA 106.092327].
272. J. Lin-Cai and D. Pletcher, J. Electroanal. Chem. Interfacial Electrochem., **152**, 157 (1983).
273. F. Gasparrini, M. Givovannoli, G. Natile, and G. Palmieri, Congr. Naz. Chim. Inorg. [Atti] 15 th., 178–80 (1982) [CA 101.006358].
274. F. Gasparrini, M. Giovannoli, D. Misiti, G. Natile, and G. Palmieri, Congr. Naz. Chim. Inorg. [Atti], 16th., 215 (1983) [CA 101.022765].
275. D. Pletcher and E.M. Valdes, Electrochim. Acta, **33**, 509 (1988) [CA 109.081925].
276. M. Koehler, M. Roemer, and C.P. Herz, to Merck Patent G.m.b.H., Ger. Offen. DE 3,512,541 (1986) [CA 106.032575].
277. W.P. Reeves, M.W. Creswell, D.S. Glass, and G.M. Scheide, Israel J. Chem., **26**, 225 (1985).
278. J.P. Chupp, R.C. Grabiak, K.L. Leschinsky, and T.L. Neumann, Synthesis Commun., 224 (1986).
279. Y. Sasson, H. Wiener, and S. Bashir, J. Chem. Soc., Chem. Commun., 1574 (1987).
280. P.S. Kalsi, P.P. Kaur, J. Singh, and B. Chhabra, Chem. Ind. (Lond.), 394 (1987).
281. C. Singh, Indian J. Chem., Sect. B, **24B,** 859 (1985) [CA 105.227063].
282. C. Singh, Indian J. Chem., Sect. B, **24B,** 857 (1985) [CA 105.172775].
283. R. Mehrotra, C. Singh, and S.P. Popli, Indian J. Chem., Sect. B, **23B,** 695 (1984) [CA 101.230800].
284. M. Schneider, J.V. Weber, and P. Faller, J. Org. Chem., **47,** 364 (1982).
285. P. Hodge, E. Koshdel, and J. Waterhouse, J. Macromol. Chem., **185,** 489 (1984).
286. E. Santaniello, A. Manzochi, and C. Farachi, Synthesis, 563 (1980).
287. E. Dantaniello, F. Ponti, and A. Manzocchi, Tetrahedron Lett., **21,** 2655 (1980).
288. F. Abe, T. Kobayashi, T. Sakakura, and M. Tanaka, Masato, Kagaku Gijutsu Kenkyusho Hokoku, **86,** 151 (1991). [CA: 116.058337].
289. M.A. Johnson, P.H. Washecheck, K. Yang, and C.M. Starks, (to Conoco Inc.) U.S. Patent 3,650,918 (1972).
290. L.M. Berkowitz and P.N. Rylander, J. Am. Chem. Soc., **80,** 6682 (1958).
291. C.L. Mantell, Ind. Eng. Chem., Process Des. Dev., **1,** 144 (1962); Chem. Eng. Prog. 129–35 (1977).
292. K.S. Kim, Y.K. Chang, S.K. Bae, and C.S. Hahn, Synth. Commun., 866 (1984).
293. K. Dimroth and W. Tuncher, Synthesis, 339 (1977).
294. J. San Filippo, C. Chern, and J.S. Valentine, J. Org. Chem., **41,** 1077 (1976).
295. R.A. Johnson and E.G. Nidy, J. Org. Chem., **40,** 1680 (1975).
296. J. San Filippo, Jr., C.-I. Chern, J.S. Valentine, J. Org. Chem., **40,** 1678 (1975).

297. M. Shibasaki and S. Ikeyami, Tetrahedron Lett., 4037 (1977).
298. E.J. Corey, K.C. Nicolaou, and M. Shibasaki, J. Chem. Soc., Chem. Commun., 658 (1975).
299. B.N. Kolarz and A. Rapak, Makromol. Chem., **185**, 2511 (1984).
300. J.S. Do and T.C. Chou, J. Appl. Electrochem., **20**, 978 (1990).
301. C.-B.L. Laurent, G. Rauniyar, and M. Thomalla, C.R. Seances Acad. Sci., Ser. 2, **295**, 339 (1982).
302. E. Laurent, G. Rauniyar, and M. Thomalla, J. Appl. Electrochem., **15**, 121 (1985).
303. M. Fleischmann, C.L.K. Tennakoon, H.A. Bampfield, and P.J. Williams, J. Appl. Electrochem., **13**, 593 (1983).
304. E. Laurent, G. Rauniyar, and M. Thomalia, J. Appl Electrochem., **14**, 741 (1984).
305. S.R. Ellis, D. Pletcher, P. Gough, S.R. Korn, J. Appl. Electrochem., **12**, 687 (1982).
306. S.R. Ellis, D. Pletcher, W.N. Brooks, and K.P. Healy, J. Appl. Electrochem., **13**, 735 (1983).
307. S.R. Forsyth, D. Pletcher and K.P. Healy, J. Appl. Electrochem., **17**, 905 (1987).
308. Z. Ibrisagic, D. Pletcher, W. Brooks, and K.P. Healy, J. Appl. Electrochem., **15**, 719 (1985).
309. S.R. Forsyth, D. Pletcher, and K.P. Healy, J. Appl. Electrochem., **17**, 905 (1987).
310. J. Koryta, Chem. Listy, **8**, 897 (1987) [CA 107.207237].
311. Asahi Chemical Industry Co., Ltd. Japan Kokai Tokkyo Koho JP 8358289 (1983) [CA 99.095855].
312. Asahi Chemical Industry Co., Ltd. Japan Kokai Tokkyo Koho JP 58/207383 (1983) [CA 100.164348].
313. Asahi Chemical Industry Co., Ltd. Japan Kokai Tokkyo Koho JP 58/207382 (1983) [CA 100.164349].
314. A.H. Gilbert, B. Goldstein, and G. Marom, Composites (Guildford, U.K.), **21**, 408 (1990) [CA: 115.137309].
315. D. Pletcher, Electroorganic Synthesis [Manuel M. Baizer Meml. Symp.] 1990 (Pub. 1991), 255–62. Ed. Little, R.D.; Weinberg, N.L.; Dekker: New York, N.Y. [CA: 116.115434].
316. L. Jiang, Huaxue Tongbao (10)1–4 (1985) [CA 104.070631].
317. J. Koryta, Chem. Listy, **81**, 897 (1987) [CA 107.207237].
318. D.C. Trivedi, J. Electrochem., **2**, 285 (1986) [CA 105.160706].
319. M. Fleischmann, C.L.K. Tennakoon, P. Gough, and J.H. Steven, J. Appl. Electrochem., **13**, 603 (1983).
320. H. Suzuki, S. Kawato, and A. Nasu, Bull. Chem. Soc. Japan, **65**, 626 (1992).
321. J.Y. Ma, Y. Wang, Z. Yu, S.-W. Zhang, Catal. Lett. **15**, 275 (1992); [CA: 117.211838]
322. H. Aragao dos Santos, L. Espindola, V.F. Ferreira, T. Nakamura, Brazil Patent 91 02,011 (1991) [CA: 118.134747]
323. Y. Sasson (to du Pont) U.S. Patent 5,138,107 (1992).
324. H. Fonouni, S. Krishnan, D.G. Kuhn, and G.A. Hamilton, J. Am. Chem. Soc., **105**, 7672 (1983).

11

Phase-Transfer-Catalyzed Reductions

Phase-transfer catalysis (PTC) procedures that have been developed for use with sodium borohydride, lithium aluminum hydride, and several other reducing agents involving anion transfer to organic media are described in this chapter. Reductions and hydrogenations employing both metal catalysts and phase-transfer catalysts are described in Chapter 13. PTC reductions have not been as intensively studied as other kinds of PTC reactions, evidently due to an excellent array of low-cost hydrogenation techniques and reduction systems already available to chemists, eliminating a strong need for development of PTC systems.

A. Sodium Borohydride Reductions

1. Reduction of Carbonyl Compounds

Under phase-transfer catalysis (PTC) conditions with simple tetraalklammonium salts as catalysts, aqueous sodium borohydride rapidly reduces aliphatic, aromatic, and unsaturated aldehydes to give high yields of the corresponding alcohols [1]. However, under the same conditions dialkl, diaryl, araalkl, and cycloaliphatic ketones are practically inert. For example:

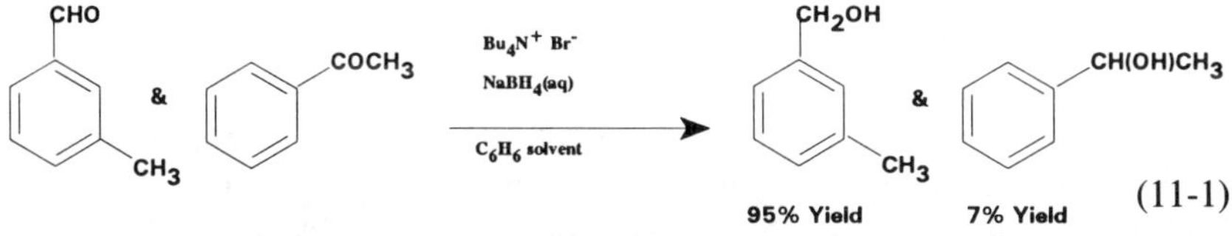

$$\text{(m-CH}_3\text{C}_6\text{H}_4\text{CHO)} \& \text{(C}_6\text{H}_5\text{COCH}_3\text{)} \xrightarrow[\text{C}_6\text{H}_6 \text{ solvent}]{\substack{\text{Bu}_4\text{N}^+\text{Br}^- \\ \text{NaBH}_4\text{(aq)}}} \text{(m-CH}_3\text{C}_6\text{H}_4\text{CH}_2\text{OH)} \underset{95\% \text{ Yield}}{} \& \text{(C}_6\text{H}_5\text{CH(OH)CH}_3\text{)} \underset{7\% \text{ Yield}}{} \quad (11\text{-}1)$$

PTC borohydride reduction of acetophenone in water/octane or water/1,2-dicloroethane, catalyzed by $(Bu_4N)_2SO_4$, occurs only slowly [2]. Use of crown ethers as phase-transfer catalysts for a liquid-solid borohydride system is slow to reduce ketones, and it is complicated by base-catalyzed side reactions to produce unsaturated ketones and their reduction products [3]. Optimization of a PTC process for borohydride reduction of 2-[(dimethylamino)-methyl]cyclohexanone to the alcohol has been described [4].

Although simple tetraalklammonium salts do reduce ketones slowly, use of quaternary ammonium salts containing a β-hydroxy group significantly accelerates the reduction, as first noted by Colonna and Fornasier [5], as shown by the data in Table 11-1.

$$CH_3(CH_2)_5COCH_3 + NaBH_4 \text{ (aq)} \xrightarrow{\text{Catalyst}} CH_3(CH_2)_5CH(OH)CH_3 \qquad (11\text{-}2)$$

Other workers have confirmed the great superiority of β-hydroxyquaternary salts as phase-transfer catalysts for borohydride reductions [6–10]. The ephedrine-based catalyst (middle entry in Table 11-1) can also be made into an insoluble PTC catalyst by supporting it on a montmorillonite clay, such that it is still quite active for reduction of ketones to alcohols, yet easily separated by filtration [11].

Polyenthylene glycol(PEG) [12], crown ethers [13], and various polymers have also been reported to be effective phase-transfer catalysts for borohydride reduction of ketones to alcohols. The polymers include soluble copolymers of vinyl pyridine [14], soluble polymers prepared from 2,6-dichloropyridine and $HS(CH_2)_nSH$ ($n = 6$ or 10) [15], and soluble and insoluble polymers from reaction of 2,4-dichloro-6-phenyl-1,3,5-triazine with $HS(CH_2)_nSH$ [16].

An interesting route involving two tandem PTC reactions has been described for preparation of α,α-dimethyl alcohols [17]:

$$\text{PhCH}_2\text{Cl} + H-\underset{\underset{CH_3}{|}}{\overset{\overset{CH_3}{|}}{C}}-CHO \xrightarrow[\text{NaOH}]{R_4N^+} C_6H_5CH_2-\underset{\underset{CH_3}{|}}{\overset{\overset{CH_3}{|}}{C}}-CHO \xrightarrow[\text{NaBH}_4]{R_4N^+} C_6H_5CH_2-\underset{\underset{CH_3}{|}}{\overset{\overset{CH_3}{|}}{C}}-CH_2OH \qquad (11\text{-}3)$$

Calcium chloride, added to aqueous sodium borohydride solutions, behaves as an inverse phase-transfer catalyst for certain corticosteroidal ketones by forming bidentate complexes with the side chains of 17-α-hydroxy steroids [18]. These complexes not only greatly enhance the solubility of the steriods in the aqueous borohydride phase, but also fix them in an orientation favorable to 20α-reduction, resulting in asymmetric reduction.

2. Azide Reductions

Most azides, when treated with sodium borohydride in homogeneous media, are usually inert or give only poor yields of the corresponding amines [19]. However, Rolla [20] has demonstrated that aqueous sodium borohydride and a PTC system using hexadecyltributyl phosphonium bromide catalyst gives amines in excellent yields, as shown by the data in Table 11-2. Moreover, since the alkl azides are easily made in 80–95% yield by PTC azide displacement on alkyl halides or alkyl methanesulfonates, a simple one-pot method is provided for production of

Table 11-1. Effect of PTC catalyst structure on borohydride reduction of 2-octanone [5].

Catalyst	Reaction Time (Hr.)	% Yield
$(C_8H_{17})_3NMe^+$	6.5	80
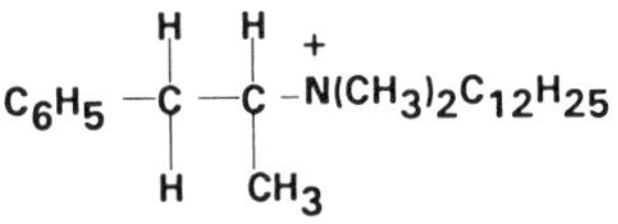 $C_6H_5-CH_2-CH(CH_3)-N(CH_3)_2C_{12}H_{25}^+$	7.0	100
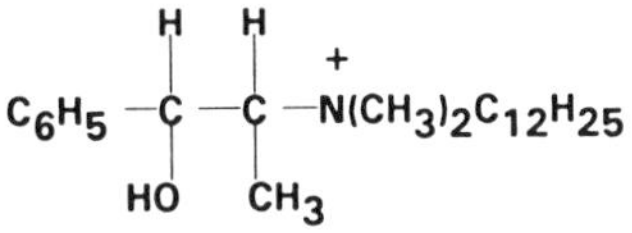 $C_6H_5-CH(OH)-CH(CH_3)-N(CH_3)_2C_{12}H_{25}^+$	0.5	100

Table 11-2. PTC borohydride reduction of azides [20].

Azide	Product	Percent Yield
n-Octyl azide	n-Octylamine	92
n-Hexadecyl azide	n-Hexadecylamine	88
sec-Octyl azide	sec-Octylamine	79
Benzyl azide	Benzylamine	89
Phenyl azide	Aniline	92
1-Naphthyl azide	1-Naphthylamine	89
Toluenesulfonyl azide	Toluenesulfonamide	72
tert-Butyl 2-azido-2-phenylacetate	Phenylglycine	72

amines from alkyl halides or alcohols. A similar two-step PTC process for preparing 9-aminoacridines has also been reported [21].

3. Other PTC Borohydride Reductions

a. Halo-compounds

PTC reduction of substituted benzoyl chlorides with aqueous sodium borohydride produces the corresponding alcohols in good yields at room temperature [22,23]:

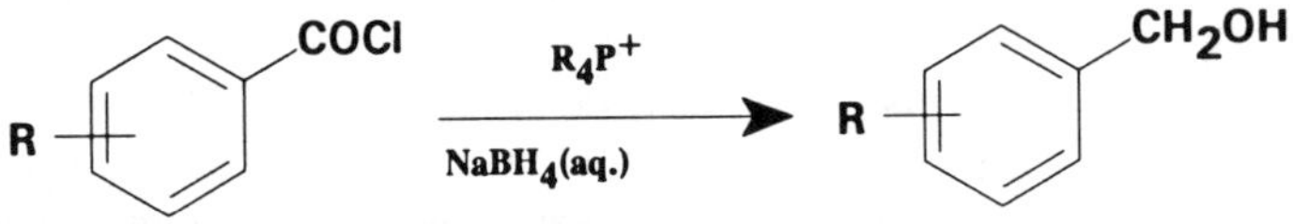

$$R\text{-}C_6H_4\text{-}COCl \xrightarrow[NaBH_4(aq.)]{R_4P^+} R\text{-}C_6H_4\text{-}CH_2OH \qquad (11\text{-}4)$$

Borohyride reduction of alkyl halides using PEG [24] or with a crown ether or PEG combined with tri-n-butyltin chloride [25] as catalysts replaces the halo-group by hydrogen.

b. Mercurial compounds

β-Alkoxy-and β-hydroxyalkyl radicals are formed by PTC-borohydride reduction of the corresponding mercurial compounds [26]. These radicals add to electron deficient olefins to give addition products. For example,

$$CH_3OCH(Ph)CH_2\text{-}HgCl + CH_2=CHCN \xrightarrow[\text{NaBH}_4\text{(aq.)}]{\text{PTC}} CH_3OCH(Ph)CH_2CH_2CH_2CN \quad (11\text{-}5)$$

B. Lithium Aluminum Hydride Reductions

Lithium aluminum hydride is solubilized into nonpolar organic solvents by 15-crown-5, allowing this powerful reducing agent to be used in solid-liquid PTC systems [27]. Results from several experiments are summarized in Table 11-3.

In contrast to 15-crown-5, addition of [2.1.1]-cryptand, which easily solubilizes LiAlH$_4$ into organic solutions, retards or completely inhibits reduction of carbonyl compounds unless an excess of LiI or NaI or LiClO$_4$ is added [28,29]. Tetrahexylammonium bromide, crown ethers, and tris(2,-5-dioxahexyl)amine have all been reported to function as phase-transfer catalysts for LiAlH$_4$ in the solid-liquid reduction of cyclohexanone to give 100% yields of cyclohexanol in 30-60 min at 25°C [30]. Solid-liquid PTC reductions by LiAlH$_4$ have also been reported for conversion of potassium acetate or propionate to the corresponding alcohols [31,32], for conversion of thioesters of alkylphosphinic acids to alkylphosphines [33], and for reduction of alkylsilicon or germanium halides or alkoxides to the corresponding alkylsilanes or alkylgermanes [34,35].

C. Reductions with Sodium Formate

Aromatic aldehydes are reduced by aqueous sodium formate in the presence of RuCl$_2$(Ph$_3$P)$_3$ as cocatalyst with a quaternary ammonium salt at 20–80°C [36]. Rates of the process depend on the structure and concentration of aldehyde, quaternary ammonium salt, polarity of the organic solvent, stirring rate, and are linearly related to the concentration of sodium formate in the aqueous phase.

Replacement of halogen by hydrogen for the bromo-groups of 2-,3- or 4-bromo-4-methoxybenzene with aqueous sodium formate in the presence of cyclodextrin as catalyst produced anisole via an inverse PTC process [37]. Cyclodextrin dissolves in the aqueous phase and functions to transfer the organic substrate to the aqueous phase for reduction. Use of an organic-soluble palladium-phosphine

Table 11-3. Solid-liquid PTC reduction of various compounds with lithium aluminum hydride, catalyzed by crown or quat, in aromatic hydrocarbon solvents [27].

Substrate	Product	Catalyst	Temp. (°C)	Time (h)	Yield (%)
Cyclopentanone	Cyclopentanol	—	25	6	3
		—	80	1	16
		15-C-5	25	3	15
		15-C-5	25	6	>95
		15-C-5	80	1	>95
		18-C-6	25	3	>95
		18-C-6	80	1	>95
		$PhCH_2NEt_3Cl$	25	3	47
		$PhCH_2NEt_3Cl$	80	1	84
		$PhCH_2NEt_3Cl$	80	6	>95
		$(Oct)_4NBr$	25	3	7
		$(Oct)_4NBr$	25	6	12
		$(Oct)_4NBr$	80	1	82
$n\text{-}C_6H_{13}Br$	N-Hexane	—	80	6	trace
		15-Crown-5	80	3	22
		15-Crown-5	80	6	83
		$PhCH_2NEt_3Cl$	80	3	41
		$PhCH_2NEt_3Cl$	80	6	92
C_6H_5Br	Benzene	—	80	6	0
		15-Crown-5	80	3	23
		15-Crown-5	80	6	47
		$PhCH_2NEt_3Cl$	80	3	27
		$PhCH_2NEt_3Cl$	80	6	62
C_6H_5CN	$C_6H_5CH_2NH_2$	—	80	1	16
		15-Crown-5	80	1	>95
		$PhCH_2NEt_3Cl$	80	1	86
CH_3CONEt_2	Et_3N	—	80	1	12
		15-Crown-5	80	1	78
		$PhCH_2NEt_3Cl$	80	1	86

catalyst with aqueous sodium formate provides another route for hydrogenolysis of aryl bromides [38].

D. Reductions with Sulfur-Containing Anions

PTC reduction with sodium dithionate provides selective conversion for 2,4-alkadienoates to the corresponding monounsaturated 3-alkenoates in 30–70% yields [39,40].

$$R''RC=\underset{R}{\overset{R}{C}}-\underset{R}{\overset{}{C}}=\underset{}{\overset{}{C}}-CO_2R' \xrightarrow[Na_2S_2O_4 \text{ (aq)}]{(C_8H_{17})_3NMe^+ \ Cl^-} R''RCH-\underset{R}{\overset{R}{C}}=\underset{R}{\overset{H}{C}}-\underset{}{\overset{}{C}}-CO_2R' \quad (11\text{-}6)$$

The products were mixtures of E and Z isomers, the ratios of which were independent of the starting configurations. Similar results were obtained for reduction of chalcones [41]. Sodium dithionate with trioctylmethylammonium chloride ("Adogen" 464) as catalyst has also been used to reduce the double bond of trimethylcyclohexenone and related compounds, but the procedure is unsatisfactory for hydrophilic ketones which gave mostly water-soluble sulfur derivatives [42].

In the presence of sodium dithionite, under PTC conditions, perfluoroalkyl iodides add to olefins, the products from which undergo further dehydrohalogenation and reduction to yield perfluoroalkyl-alkenes and -alkanes, respectively [43].

Steroidal ketones are reduced by sodium dithionite in remarkable selectivity and high yields under phase-transfer conditions to yield alcohols, and enones are reduced ketones [44]. An application of these selectivities was demonstrated by reducing an important steroidal intermediate, 16-dehydropregnenolone acetate, to pregnenolone acetate in very high yields. Reductive debromination of 1,2-dibromo-1,2-diarylethanes with sodium dithionite in a two-phase system using dioctyl-4,4'-dibyridinium as a catalyst produces diarylethenes through a free-radical process [45].

Aqueous sodium di- or trithiocarbonate dehalogenates vic-dihaloalkanes under PTC conditions to yield the corresponding alkenes in high yields and under mild conditions [46].

$$R-\underset{Br}{\underset{|}{\overset{H}{\overset{|}{C}}}}-\underset{Br}{\underset{|}{\overset{H}{\overset{|}{C}}}}-R' + Na_2CS_3(aq) \xrightarrow{PTC} RCH=CHR' \quad (11\text{-}7)$$

Sodium sulfide under PTC conditions reduces nitrobenzenes substituted by hydrogen or electron-withdrawing groups to the corresponding anilines [47]. For example:

$$\text{IN:} \quad \underset{NO_2}{\underset{|}{C_6H_3}}-NO_2 + Na_2S(aq) \xrightarrow{Bu_4N^+ Cl^-} \underset{NO_2}{\underset{|}{C_6H_3}}-NH_2 \quad (11\text{-}8)$$

Reduction of some organic sulfur compounds by formamiinesulfinic acid can also be accomplished under phase-transfer conditions [48].

E. Hydrogenation

Phase-transfer-catalyzed hydrogenations have mostly been conducted using metal cocatalysts, as discussed in Chapter 13.

β-Cyclodextrin and PEGs are useful phase-transfer catalysts for hydrogenation of conjugated dienes, using hydridopentacyanocobaltate, which is generated in situ from hydrogen and aqueous cobalt cyanide [49]. Cerium or lanthanum chloride promotes the reaction, which proceeds by 1,2-addition of hydrogen to the diene.

A mixture of trioctylmenthylammonium chloride ("Aliquat" 336) and rhodium trichloride forms a complex, $[(C_8H_{17})_3N+Me]^+$ $[RhCl_4]^-$, which selectively catalyzes hydrogenation of the double bond of α,β-unsaturaturated ketones and esters at low hydrogen partial pressure [50].

Hydrogenation of nitrobenzene to p-aminophenol is accomplished with a platinum catalyst and in the presence of a surfactant that can act as a phase-transfer agent [51]. If a palladium salt is used instead of platinum, then aniline is the major product.

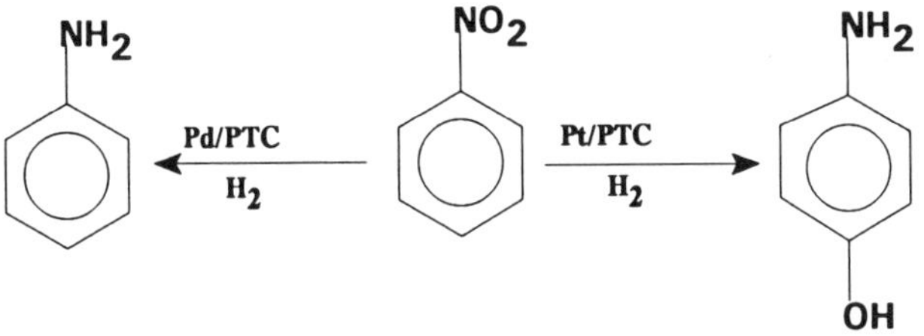

$$\text{PhNH}_2 \xleftarrow[\text{H}_2]{\text{Pd/PTC}} \text{PhNO}_2 \xrightarrow[\text{H}_2]{\text{Pt/PTC}} p\text{-HOC}_6\text{H}_4\text{NH}_2 \quad (11\text{-}9)$$

When an aqueous phase is present hydrogenations can also be accomplished using synthesis gas instead of hydrogen [52]. Iron carbonyls have been used with PTC systems to catalyze the reduction of ketones using alcohols as a hydrogen source [53,54].

F. Reductions with Formaldehyde

Phase-transfer-catalyzed reductions by formaldehyde give alcohols from ketones [55], and 2-phenyltriazoles from o-nitroazobenzene in high yield [56]. A proposal has been made to use a PTC formaldehyde reduction process to convert viscous bitumen deposits into low-viscosity oils in underground formations, allowing for easier recovery [57].

G. Electrochemical Reduction

γ-Substituted butyrolactones are produced by electrochemical reduction of a mixture of acrylic acid ester and aldehyde, using quaternary ammonium salts as phase-transfer catalysts, with lead electrodes and phosphate as the supporting electrolyte [58].

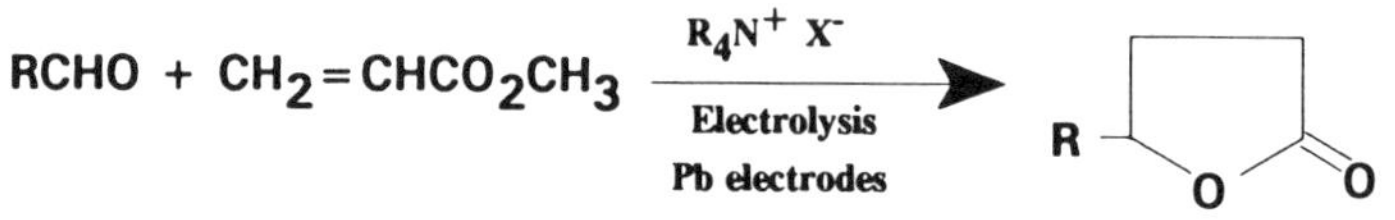

$$\text{RCHO} + \text{CH}_2=\text{CHCO}_2\text{CH}_3 \xrightarrow[\substack{\text{Electrolysis}\\ \text{Pb electrodes}}]{R_4N^+ \ X^-} \quad \underset{\substack{\\ }}{R-\!\!\!\overset{\displaystyle\frown}{\underset{O}{}\!\!\!\!\underset{}{}\!\!\!\!\underset{O}{}}} \quad (11\text{-}10)$$

H. Photochemical Reduction

Oil-soluble N,N'-dioctyl-4,4′-bipyridinium dibromide, a "viologen," can be used in a photochemical process as an electron carrier between water and oil phases to reduce nitrobenzenes in the oil phase to the corresponding oxime or quinone [59].

I. Wolff-Kishner Reduction

Wolff-Kishner reduction of aldehydes and ketones gave higher yields of reduced products than the conventional procedure when phase-transfer-catalyzed by 18-crown-6, [60,61], a polypod ether [62], or a polymeric crown ether from cyclization of poly(β-chloromethylglycidyl ether) with [63].

J. Reduction by Dodecacarbonyltriiron and Related Species

$Fe_3(CO)_{12}$ reduces nitrobenzene to aniline under triphase conditions [64]. Reduction occurs under mild conditions and isolation of the product is easy. Treating benzal bromides with KOH and 18-crown-6 in benzene in the presence of $Fe(CO)_5$ gives the corresponding stilbenes [65]. Other transition metal complexes tried were less effective than $Fe(CO)_5$

References

1. C. Rao, D. Someswara, A.A. Deshmukh, and B.J. Patel, Indian J. Chem., Sect. B, **25B,** 626 (1986) [CA 106.101813].
2. G. Lamaty, M.H. Riviere, and J.P. Roque, Bull. Soc., Chim. Fr., 33 (1983) [CA 99.021730].
3. T. Matsuda and K. Koida, Bull. Chem Soc., Japan, **46,** 2259 (1973).
4. H. Goncalves, M.T. Maurette, E. Oliveros, E. Puch-Costes, E. Mathieu, and R. Phan Tan Luu, New. J. Chem., **11,** 43 (1987).
5. S. Colonna and R. Fornasier, Synthesis, 531(1976).
6. J. Balcells, S. Colonna, and R. Fornasier, Synthesis, 266 (1976).
7. J.P. Masse and E.A. Parayre, J. Chem. Soc., 438 (1976).
8. R. Franklin, to AKAZO America, Inc. U.S. Patent 4,912,265 (1990).

9. T.C. Pochapsky, P.M. Stone, and S.S. Pochapsky, J. Am. Chem. Soc., **113**, 1460 (1991).
10. C. Passarotti, G.L. Bandi, A. Fossati, M. Valenti, and L. Dal Bo, Bull. Chim. Farm, **129**, 195-8 (1990).
11. A. Sakar and B.R. Rao, Tetrahedron Lett., 32, 1247-50 (1991).
12. Y. Jiang, Z. Xu, and Q. Guo, Xiamen Daxue Xuebao, Ziran Kexueban, **28**, 519–22 (1989) [CA: 114.024312].
13. Y.S. Lin, and J.S. Shih, J. Chin. Chem. Soc. ((Taipei), **38**, 451-9 (1991) [CA: 115.279242].
14. S. Kondo, M. Nakanishi, K. Yamane, K. Miyagawa, and K. Tsuda, J. Macromol. Sci., Chem., **A27**, 391 (1990).
15. S. Kondo, M. Nakanishi, and K. Tsuda, J. Polym. Sci., Chem. Ed., **23**, 581 (1985).
16. S. Kondo, T. Yamamoto, H. Kunisada, and Y. Yuki, J. Macromol. Sci. Chem., **A27**, 1515-30 (1990).
17. V.A. Krivoruchko, A.V. Gurevich, L.A. Kheifits, and S.A. Voitkevich, Maslo-Zhir. Prom.-st. (8) 25 (1986) [CA 107.133979].
18. C.A. Han and C. Monder, Steroids, **42**, 619 (1983) [CA 102.149599].
19. T. Sheradsky, The Chemistry of the Azide Group, S. Patai, Ed., Interscience, New York, 1971, Chapter 6, and references contained therein.
20. F. Rolla, J. Org. Chem., **47**, 4327 (1982).
21. M. Vlassa and M. Kezdi, J. Prakt. Chem., **327**, 521 (1985).
22. A.T. Costello and D.J. Milner (to Imperial Chemical Industries), Brit. UK Pat. Appl. 2,155,464 (1985) [CA 104.168107].
23. A.T. Costello and D.J. Milner, to ICI PLC, Brit. Appl. 2,155,464 (1985) [CA 104.168107].
24. W. Hou and X. Jiang, Hangzhou Daxue Xuebao, Ziran Kexueban, **12**, 275 (1985) [CA 104.014071].
25. D.E. Bergbreiter and J.R. Blanton, J. Org. Chem., **52**, 472 (1987).
26. J. Barluenga, J. Lopez-Prado, P.J. Campos, and G. Asensio, Tetrahedron, **39**, 2863 (1983).
27. V. Gevorgyan and E. Lukevics, J. Chem. Soc., Chem. Commun., 1234 (1985).
28. J.L. Pierre and H. Handel, Tetrahedron Lett., 2317 (1974).
29. A. Loupy, J. Seyden-Penne, and B. Tchoubar, Tetrahedron Lett. 1677 (1976).
30. E.V. Dehmlow and R. Cyrankiewicz, J. Chem. Res., Synop., 24 (1990).
31. S. Szakacs, S. Gobolos, and J. Szammer, Magy. Kem. Foly, **86**, 273 (1980) [CA 93.185442].
32. S. Szakacs, S. Gobolos, and J. Szammer, Monatsh. Chem., **112**, 883 (1981) [CA 95.168439].
33. O.G. Sinyashin, I. Yu. Gorshunov, E.S. Batyeva, and A.N. Pudovik, Zh. Obshch. Khim., **54**, 1917 (1984).

34. V.N. Gevorgyan, L.M. Ignatovich, and E. Lukevics, J. Organomet. Chem., **28,** C31 (1985).

35. J. Zech and H. Schmidbaur, Chem. Ber., **123,** 2087-91 (1990).

36. R. Bar, L.K. Bar, Y. Sasson, and J. Blum, J. Mol. Catal., **33,** 161 (1985).

37. S. Shimizu, Y. Sadaki, and C. Hirai, Bull. Chem. Soc. Japan, **63,** 176 (1990).

38. R. Bar, Y. Sasson, and J. Blum, J. Mol. Catal., **16,** 175 (1982).

39. F. Camps, J. Coll, and J. Guitart, Tetrahedron, **43,** 2329 (1987).

40. F. Camps, J. Coll, A. Guerrero, J. Guitart, and M. Riba, Chem. Lett., 715 (1982).

41. J.K. Makrandi and V. Kumari, Synth. Commun., **20,** 1885-8 (1990).

42. F. Camps, J. Coll and J. Guitart, Tetrahedron, **42,** 4603 (1986).

43. G. Rong and R. Keese, Tetrahedron, Lett., **31,** 5615-16 (1990).

44. K.G. Akamanchi, H.C. Patel, and R. Meenakshi, Synth. Commun., **22,** 1655 (1992).

45. Z. Goren and I. Wilner, J. Am. Chem. Soc., **105,** 7764 (1983).

46. A. Sugawara, A. Nakamura, A. Araki, and R. Sato, Bull. Chem. Soc. Jpn., **62,** 2739 (1989).

47. V.F. Shner, N.N. Aramonova, T.I. Petrunina, V.F. Seregina, and B.V. Salov, Zh. Org. Khim., **25,** 879 (1989).

48. G. Borgogno, S. Colona, and R. Fornasier, Synthesis, 529 (1975).

49. J.T. Lee and H. Alper, J. Org. Chem., **55,** 1854 (1990).

50. J. Azran, O. Buchman, I. Amer, and J. Blum, J. Mol. Catal., **34,** 229 (1986).

51. T.M. Juang, J.C. Hwang, H.O. Ho, and C.Y. Chen, J. Chin. Chem. Soc. (Taipei), **35,** 135 (1988) [CA 110.172490].

52. K. Januszkiewicz and H. Alper, J. Mol. Catal., **19,** 139 (1983).

53. K. Hothimony and S. Vancheesan, J. Mol. Catal., **52,** 301 (1989).

54. K. Jothimony, S. Vancheesan, and J.C. Kuriacose, J. Mol. Catal., **32,** 11 (1985).

55. B.G. Zupancic and M. Kokalj, Synth. Commun., **12,** 881 (1982).

56. T. Kanechika, H. Okamura, and C. Ebina, to Sumitomo Chemical Co., Ltd., Jpn. Kokai Tokkyo Koho JP 61/161269 (1986) [CA 106.033067].

57. G.D. Derdall, Can. Pat. 1172159 (1984) [CA 101.213839].

58. Asahi Chemical Industry Co., Ltd., Japan Kokai Tokkyo Koho JP 58/207382 (1983) [CA 100.164349]; Jpn. Kokai Tokkyo Koho JP 58/207383 (1983) [CA 100.164348].

59. H. Tomioka, K. Ueda, H. Ohi, and Y. Izawa, Chem. Lett., 1359 (1986).

60. Q. Huang, K. Yu, S. Tang, W. Gan, H. Wang, B. Liang, and W. Huang, Gaodeng Xuexiao Huaxue Xuebao, **6,** 336 (1985) [CA 104.088218].

61. D. Feng, Huaxue Shijie, **29,** 489 (1988) [CA 110.175470].

62. Y. Xu, X. Zhang, F. Wang, and H. Wang, Wuhan Daxue Xuebao, Ziran Kexyeban, (4) 71 (1985) [CA 105.098046].

63. Y. Xu, S. Dong, X. Bai, and H. Wu, Gaofenzi Tongxun, 266 (1983) [CA 100.121716].
64. K. Jothimony, S. Vancheesan, and J.C. Kuriacose, J. Mol Catal., **52,** 297 (1989).
65. S.C. Shim, C.H. Doh, W.H. Park, and H.S. Lee, Bull. Korean Chem. Soc., **10,** 475 (1989) [CA 112.197704].

12

Phase-Transfer Catalysis: Chiral Phase-Transfer-Catalyzed Formation of Carbon-Carbon Bonds

A. Introduction

It is only within the last two decades that heterogeneous enantioselective syntheses involving chiral phase-transfer catalysts (quaternary ammonium and phosphonium salts, macrocyclic miltidentate ligands, polypeptides, polymer-bound quaternary salts and macrocyclic multidentate ligands, etc.) were initiated. The technique has been applied to a wide variety of organic reactions. These include reductions of carbonyls with sodium borohydride [1], epoxidations of α, β-unsaturated carbonyls [2], oxidation of ketones to α-hydroxyketone [3], alkylations [4] involving carbon and nitrogen acids, Michael reactions on α, β-unsatuarated carbonyls [5], Darzens reactions [6], Cory-Chaykovshy reactions [6], chloroform addition reactions to carbonyls [6], and dichlorocarbene additions to alkenes [6]. The asymmetric induction accompanying each of the above reactions has been reported in terms of optical rotation, optical purity, and enantiomeric excess. In some cases only chemical yields were reported with no indication of the degree of asymmetric induction. The reported enantiomeric excesses have varied from just a few percent to the mid to upper nineties. In several cases, the reported enantiomeric excesses were found to be incorrect due to contamination by degradation products of the chiral catalyst [40]. Since each of the above reaction types has been extensively reviewed [6], this chapter will not attempt to cover the broad topic of chiral phase transfer catalysis (PTC). Instead, only a few examples dealing with the *formation of carbon–carbon bonds* will be discussed. These examples involve several detailed studies in which high enantioselectivities have been achieved. It is anticipated that these examples will provide a fundamental understanding of some of the variables that must be addressed in the successful design of phase-transfer-catalyzed enantioselective reactions.

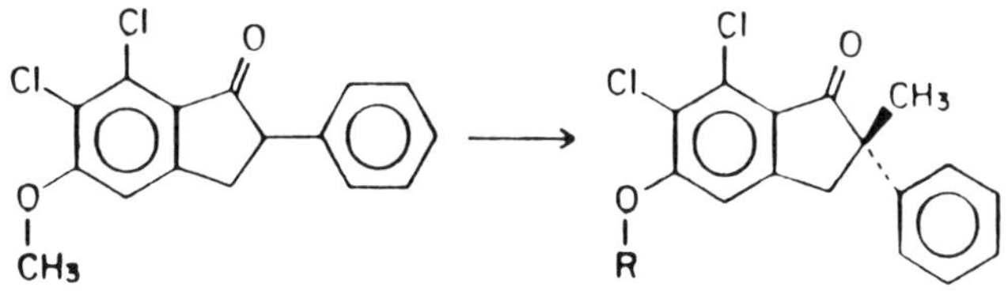

Figure 12-1

B. Alkylation Reactions

1. Methylation of 6,7-Dichloro-5-methoxy-2-phenyl-1-indanone [4g,4h,4i]

The PTC reaction of 6,7-dichloro-5-methyoxy-2-phenyl-1-indanone with methyl chloride in a 50% sodium hydroxide/toluene two-phase system in the presence of catalytic quantities of a variety of substituted N-benzylcinchoninium halides has been reported (Fig 12-1). The chiral methylation product was obtained in 98% chemical yield and up to 94% enantiomeric excess (ee). The high stereoselectivity was attributed to the presence of a tight ion pair between the enolate and the chiral catalyst. The effects of solvent, alkylating agent, temperature, catalyst, and catalyst concentration, and base concentration on the kinetics and stereoselectivity were the subjects of a thorough and scholarly investigation. The details of this pioneering work will now be presented and discussed.

a. The Nature of the Tight Ion Pair

Molecular models and the single-crystal X-ray structure suggest that the dominant conformation of the n-benzylcinchonium cation is one in which the N-benzylcinchonium cation is one in which the N-benzyl, the quinoline, and the carbon-oxygen bond associated with the alcohol functionality lie very close to a single plane. The assumed planar nature of the enolate ion led to the suggestion that the picture of the tight ion pair should be based upon a three-point interaction model as shown in Figure 12-2. In this model, two of the interaction points are provided by the N-benzyl and the quinoline groups of the catalyst interacting with the pi-electrons of the phenyl and the dichloromethoxy aromatic ring of the enolate, respectively. The hydroxyl group at the 9-position of the catalyst cation provides the third interaction point by hydrogen bonding with the negatively charged enolate oxygen. As a consequence of this model, the methylating agent can approach the tight ion pair only from one side.

b. Kinetics and Mechanism

Detailed studies were reported for each of the following steps in the overall reaction sequence:(a) deprotonation of the indanone by the aqueous sodium

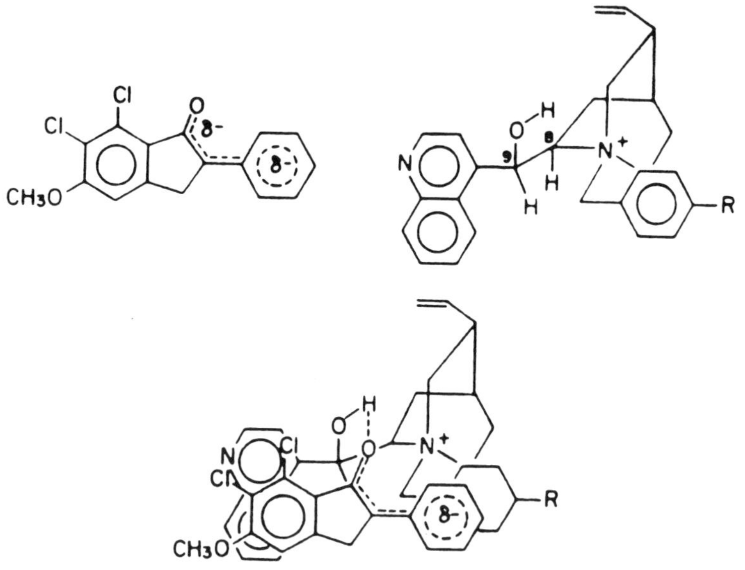

Figure 12-2

hydroxide (30%, 40%, and 50% by weight) to produce the enolate: (b) the transfer of the enolate ion by the quaternary ammonium cation into the toluene phase; and (c) the organic phase reaction of the methylating agent with the transferred cation-enolate ion pair. Since major differences were observed when the concentration of aqueous base was changed from 50% to 30%, the two conditions will be discussed separately.

Formation of the Indanone Enolate in the Presence of 50% Aqueous Sodium Hydroxide

The indanone has been determined to have a relatively high solubility in toluene (16mM). Its concentration drops to approximately 10^{-4} M, however, when brought in contact with a 50% aqueous sodium hydroxide solution. The reason for this observation is that the indanone reacts almost completely to form the sodium enolate which exists as a separate solid phase suspended between the aqueous and organic phases (Figure 12-3). The deprotonation was found to take place in the absence of quaternary ammonium cation. It was suggested that this step represents an interfacial process.

Transfer of the Enolate into the Toluene Phase in the Presence of 50% Aqueous Sodium Hydroxide

While the solubilities of the N-benzylcinchoninium halides are $<10^{-5}$ M in toluene, the quenched reaction mixtures showed the presence of the catalysts in

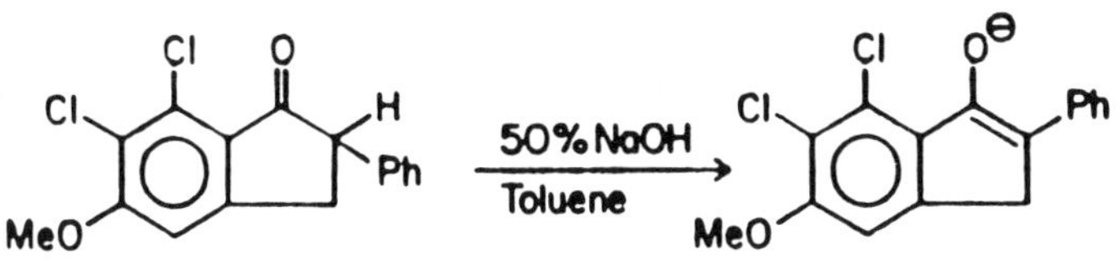

Figure 12-3

millimolar quantities. Experimental evidence was presented that indicated that in the presence of aqueous sodium hydroxide, the catalyst exists as a "dimer" in the toluene phase (Figure 12-4). It is believed that the base reacts with the quaternary ion by deprotonating the hydroxyl functionality located at the 9-position and it is this zwitterion that complexes with another quaternary salt to produce a "dimeric" species that has a reasonably high solubility in the toluene phase. The transfer of the enolate into the toluene phase is then promoted by the catalyst "dimer."

Three mechanistic transfer modes were suggested. These include (1) dissolution of a small quantity of indanone enolate in the aqueous phase followed by liquid-liquid extraction, (2) solid-liquid extraction of the solid indanone enolate, and (3) deprotonation of the neutral indanone in the toluene phase by the zwitterionic portion of the solubilized catalyst. In support of mechanism (3) it was experimentally observed that upon addition of the indanone to a toluene solution of the catalyst "dimer" the orange color of the enolate developed. Nevertheless, these observations do not eliminate the pathway involving deprotonation by hydroxide ion followed by complexation of the chiral catalyst with the enolate on the surface of the solid enolate and subsequent transfer into the organic phase. Aside from the mechanistic details of the phase-transfer step(s) it was proposed that a catalyst-indanone enolate complex once formed resides in the toluene phase where it undergoes alkyation.

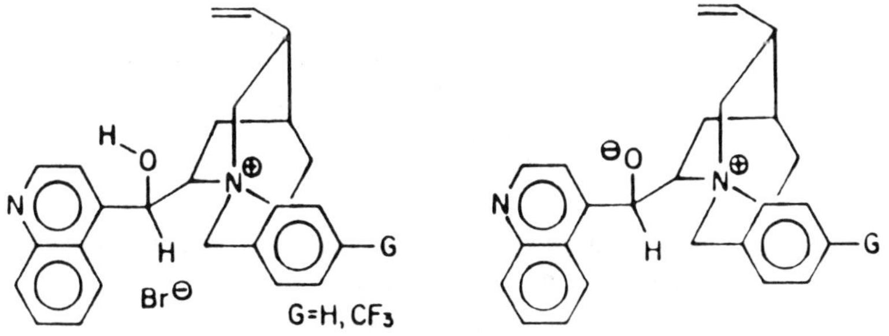

Figure 12-4

Table 12-1. Chiral methylation of 6,7-dichloro-5-methoxy-2-phenyl-1-indanone in the presence of 50% aqueous sodium hydroxide.

Initial catalyst (mmol)	Initial indanone (mmol)	$(Cat)_{tol}$ (mM)	$(CH_3Cl)_{tol}$ (M)	$k \times 10^6$ $M\ sec^{-1}$
1,0	10	6	0.56	3.7
1.0	20	4.5	0.56	4.1
0.5	10	2.5	0.56	3.2
0.5	5.0	2.9	0.56	2.8
0.25	10	0.65	0.56	1.9
0.25	2.5	1.5	0.56	1.5
0.125	1.25	0.9	0.56	1.2
1.0	10	5	0.68	7.5
1.0	10	6	1.12	6.7
1.0	10	6	0.22	2.3
1.0	10	6	0.16	1.5
2.0	20	11	0.56	5.4
1.0	10	6	0.56	3.9
1.0	10	5	1.68	8.1

[a]Catalyst: N-[p-(trifluoromethyl)benzyl]cinchoniunium bromide.

[b]Reaction conditions: 125 mL of toluene, 25 mL of 50% aqueous NaOH, and indanone were stirred together until the level of indanone in toluene dropped to approx. $10^{-4}\ M$ (1–5 h). The specified quantities of catalyst and methyl chloride wre then added to begin the reaction.

[c]The isomer ration remained unchanged (96/4 [t]).

D.L. Hughes, U.-H. Dolling, K.M. Ryan, E.F. Schoenewaldt, and E.J.J. Grabowski, J. Org. Chem., **52**, 4745 (1987).

Methylation of the Enolate in the Toluene Phase in the Presence of 50% Aqueous Sodium Hydroxide

It was reported that the kinetics of methylation are complex. When the chiral catalyst is present at the beginning of the reaction an induction period is observed for the disappearance of the indanone. This induction period essentially disappears when the catalyst is added to the reaction mixture after the enolate is formed. In an attempt to simplify this rather complex system and to eliminate complications that would interfere with mechanistic interpretations, detailed kinetic studies were reported on reaction systems in which the sodium enolate was preformed in the absence of phase-transfer catalyst. The results are summaried in Table 12-1. The data indicate that:

1. The rate of methylation is nearly independent of the quantity of charged sodium enolate at constant catalyst concentration.

2. As the quantity of enolate increases, the level of catalyst in the toluene phase decreases. This observation may be consistent with the formation

of an omega phase which may provide the vehicle for the formation of the complex between the chiral catalyst and the enolate (see Chapter 3).

3. The quantity of indanone (neutral or enolate) in the toluene is initially 10^{-4} M. As the reaction progresses the quantity increases.

4. The enantiomer excess remains essentially unchanged (96/4) as the indanone/catalyst ratio is changed, as the methyl chloride concentration is changed from 0.16 to 1.68 M, and as the stirring rate is varied from 500 to 1200 rpm.

5. A bilogarithmic plot of rate constant for methylation vs. the amount of charged catalyst was reported to have a solpe of 0.56±0.05. This observation is consistent with the catalyst being present predominately in the "dimeric" form with a small equilibrium concentration of the monomer (see Chapter 3).

6. A bilogarithmic plot of rate constant vs. concentration of methyl chloride was reported to have a slope of 0.67±0.06. It was suggested that this observation was consistent with a step prior to the methylation step being rate competitive with methylation. The order in methyl chloride is near 1.0 when Aliquat 336 is used.

7. It was observed that in the methylation of the indanone in the presence of Aliquat 336 (methyltri-*n*-octylammonium chloride), the orders with respect to quaternary ammonium catalyst and methyl chloride were both 1.0.

Formation of Indanone Enolate in the Presence of 30% Aqueous Sodium Hydroxide

In the presence of 30% (by weight) aqueous sodium hydroxide, the solid sodium enolate did *not* separate from the two liquid phases and its rate of formation is substantially slower. This behavior was attributed to the solvation associated with OH$^-$ in 30% sodium hydroxide as opposed to 50%. It is believed that rapid deprotonation still occurs by means of an interfacial process but the less basic nature of the 30% solution results in a smaller amount of enolate in equilibrium with the ketone.

Methylation in the Toluene Phase in the Presence of 30% Aqueous Sodium Hydroxide.

The results of kinetic studies dealing with the chiral methylation of the phenylindanone using 30% aqueous sodium hydroxide are summarized in Table 12-2. The following observations are pertinent:

Table 12-2. Chiral methylation of 6,7 dichloro-5-methoxy-2-phenyl-1-indanone in the presence of 30% aqueous sodium hydroxide.

Initial catalyst (mmol)	Initial indanone (mmol)	$(Cat)_{tol}$ (mM)	$(CH_3Cl)_{tol}$ (M)	$k \times 10^6$ $M\ sec^{-1}$	ee (%)
2.0	10	13	0.56	2.2	55
1.0	10	6.5	0.56	1.6	64
0.5	10	3.3	0.56	0.90	70
0.25	10	1.8	0.56	0.54	78
0.20	10	1.1	0.56		78
0.25	2		0.56	0.52	
0.50	5	3.3	1.68	2.15	69
0.50	5	3.3	0.56	0.82	72
0.50	5	3.3	0.28	0.43	72
0.50	5	2.9	0.135	0.20	75
0.50	5	3.3	0.56	0.77	69

[a]Catalyst: N-[p-(trifluoromethyl)benzyl]cinchoninium bromide.

[b]Reaction conditions: 125 mL toluene, 25 mL of 30% aqueous NaOH, 1200 rpm.

D.L. Hughes, U.-H. Dolling, K.M. Ryan, E.F. Schoenewaldt, and E.J.J. Grabowski, J. Org. Chem., 52, 4745 (1987).

1. The rate of alkylation is zero-order in indanone.

2. Changing the stirring rate from 500 to 1200 rpm had no effect on the rate of reaction.

3. As the catalyst concentration increased, the enantiomeric excess decreased.

4. Bilogarithmic plots of rate constants for (+)-methylation vs. catalyst concentration and rate constants for racemic methylation vs. catalyst concentration resulted in slopes of 0.5 and 1.0, respectively, suggesting that the mechanisms for formation of chiral and racemic products are different.

5. A bilogarithmic plot of rate constant for methylation vs. the concentration of methyl chloride had a slope close to unity. In contrast to the results using 50% aqueous sodium hydroxide, this suggests that the rate-controlling step is solely the alkylation step.

The contrast between the results in 50% and 30% sodium hydroxide is striking. It was pointed out that when dealing with 30% sodium hydroxide, the amount of water present in the system is greater than in the corresponding system involving 50% sodium hydroxide. Thus, the hydroxyl functionality located at the 9-position of the quaternary ammonium cation can be more adequately hydrated and diminish the extent of catalyst "dimer" formation.

Table 12-3. Chiral methylation of 6,7-dichloro-t-methoxy-2-phenyl-1-indanone: dependence of enantioselectivity on catalyst structure.

G	X⁻	ee (%)	G	X⁻	ee (%)
H	Cl	79	m-Cl	Br	89
H	Br	83	p-Cf$_3$	Br	94
p-MeO	Cl	70	m-Cf$_3$	Cl	87
p-Me	Cl	75	m-Cf$_3$	Br	94
p-F	Cl	87	p-NO$_2$	Cl	30
p-Cl	Cl	88	3,4-Cl$_2$	Cl	94
m-Cl	Cl	81	3,4-Cl$_2$	Br	92

$^a \rho = 0.21 \pm 0.02$.
^{b}THe p-CF$_3$ catalyst required 7 h while the H catalyst erequired 36 h.
^{c}Only small differences were observed for Cl⁻ and Br⁻.

D.L. Hughes, U.-H. Dolling, K.M. Ryan, E.F. Schoenewaldt, and E.J.J. Grabowski, J. Org. Chem., **52**, 4745 (1987).

Effect of Catalyst Structure on Enantiomeric Excess in the Presence of 50% Aqueous Sodium Hydroxide

A series of meta- and para-substituted N-benzylcinchonium structures were employed as catalysts in the methylation of the phenylindanone using 50% aqueous sodium hydroxide. The results are summarized in Table 12-3. A plot of the log of the enantiomeric excesses vs. the Hammett substituent parameters resulted in a correlation with a ρ-value of $+0.21$. Electron-withdrawing groups increased the enantiomeric excess of the reaction—an observation consistent with the three-point ion pair model suggested for the origin of the asymmetric induction. Only small differences were observed between chloride and bromide counteranions. The catalysts n-benzylquinidinium; N-methylcinchoninium; and N-methylcinchoninium; and N-methyl-, N-benzyl-, and N-dodecylephedrinium bromides were not particularly successful catalysts. All gave ee's of <25%.

Effects of Solvent, Methylating Agent, and Temperature on Enantiomeric Excess in the Presence of 50% Aqueous Sodium Hydroxide

The following observations were reported for the effects of solvent, methylating agent, and temperature on enantiomeric excess:

1. Nonpolar, polarizable solvents such as toluene and tetralin gave enantiomeric excesses of 92%. These values decreased to 82% with o-dichlorobenzene, and to still lower values with nonpolar, non polarizable solvents such as hexane and cyclohexane.

2. The following order for enantiomeric excess was obtained for various methylating agents: methyl chloride (92%) > methyl bromide (68%) > methyl

iodide (36%) > methyl sulfate (36%). No reaction was observed with methyl phosphate.

3. From 0 to 40° C using methyl chloride in toluene, the enantiomeric excess changed from 92 to 94%.

Catalyst Degradation

It is well known that quaternary ammonium salts degrade in the presence of strong base [40]. Since the chiral catalyst could decompose during the course of reaction and since the products of decomposition could be chiral and could contaminate the product, a careful study was undertaken to determine the products of decomposition of the catalyst. The degradation pathway is shown in Figure 12-5. Oxiranes have been shown to be degradation products from the reaction of β-hydroxy quaternary ammonium salts with base. Indeed, oxirane derivatives were identified from the degradation of the cinchonidium catalyst. The dominant pathway for catalyst degradation, however, was oxygen methylation of the zwitterion followed by a Hofmann elimination process.

2. *Alkylation of 2,3-Dichloro-5-methoxy-2-*n-*propyl-1-indanone with 1,3-Dichloro-2-butene in Toluene-50% Aqueous Sodium Hydroxide [4j].*

The enantioselective alkylation of 2,3-dichloro-5-methoxy-2-*n*-propyl-1-indanone with 1,3-dichloro-2-butene has been reported. In a toluene-50% aqueous sodium hyroxide two-phase system in the presence of catalytic quantities of *N*-(*p*-trifluoromethylbenzyl) cinchoninium bromide the (*S*)-(+) enantiomer was obtained with a chemical yield of 99% and an enantiomeric excess of 92%. In contrast, when the alkylation is carried out in the presence of catalytic quantities of the diasteriomeric *N*-(*p*-trifluoromethylbenzyl) cinchonidinium bromide the (*R*)-enantiomer was obtained with a chemical yield of 99% and an enantiomeric excess of 78%. The enolate–chiral catalyst ion pairs illustrated in Figure 12-6 were believed to be the crucial intermediates in the alkylation step.

It is interesting to note that one of the π-π interactions proposed for the methylation of 2,3-dichloro-5-methoxy-2phenyl-1-indanone is missing in the above ion pairs. In spite of this, there is no loss in enantiomeric excess for the formation of the (*S*)-enantiomer. It appears that the remaining π-π and hydrogen bond interactions are sufficient for achieving a high enantiomeric excess.

The effect of substituents located at the para-position of the benzyl group on enantiomeric excess was investigated. A plot of the log of the enantiomeric excesses vs. the Hammett substituent parameters resulted in a correlation with a ρ-value of 0.67. This value is substantially greater than the ρ-value for the chiral methylation of 2,3-dichloro-5-methoxy-2-phenyl-1-indanone (+0.21), indicating that the 2-*n*-propyl analogue is more susceptible to substituent effects.

The decrease in enantioselectivity when employing *N*-(*p*-trifluoromethylben-

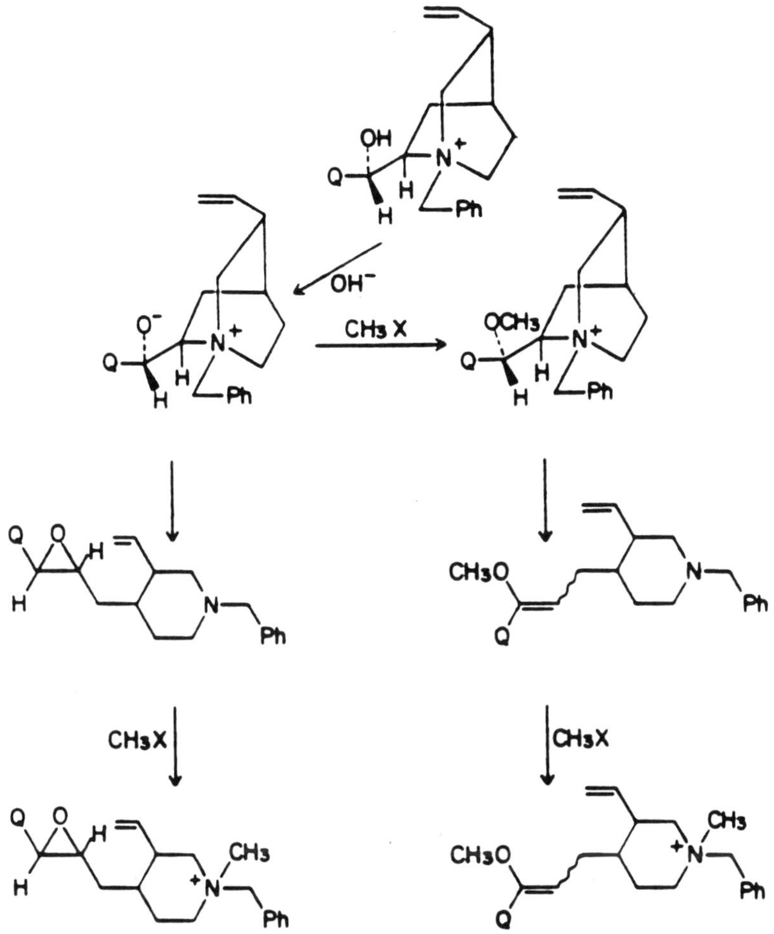

Figure 12-5

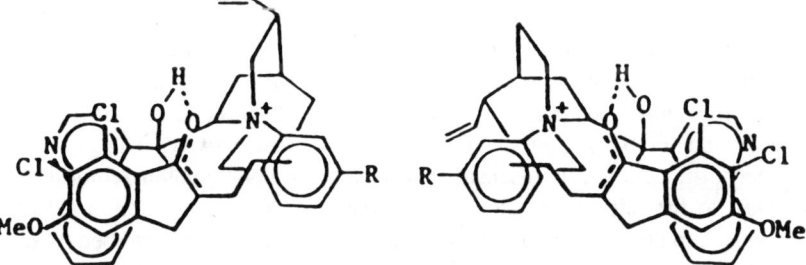

Figure 12-6

zyl) cinchonidinium bromide was attributed to the relative position of the vinyl group located on the bicyclic portion of the molecule. Based upon space-filling models it was conjectured that the vinyl group in N-(p-trifluoromethylbenzyl) cinchoninium bromide "buttresses" the hydroxyl group of the catalyst enforcing the hydrogen bonding association with the oxygen of the enolate. This vinyl group interaction is absent in the cinchonidinium catalyst.

The reaction of 6,7-dichloro-2,3-dihydro-5-methoxy-2-propyl-1H-inden-1-one and 1,3-dichloro-2-butene in a toluene–50% aqueous sodium hydroxide two-phase system in the presence of N-(3,4-dichlorobenzyl) cinchonidinium chloride and Triton X-405 has been reported. A quantitative yield of product was realized. Both the E- and Z-isomeric products were obtained with an overall enantiomeric excess of approximately 60%. The presence of the achiral, nonionic surfactant as cocatalyst not only allowed a reduction in the amount of expensive chiral catalyst required, but also reduced reaction time and increased the enantiomeric excess without substantially affecting the yield of product. The alkylation of 2-n-propylindanone with 1,3-dichloro-2-butene in the presence of N-(p-trifluoromethylbenzyl) cinchonidinium bromide and Triton X-405 resulted in a 99% chemical yield with 80% enantiomeric excess. The observations reported in this work are extremely important. They may provide the foundations for the development of a wide variety of PTC procedures where more than one catalyst is used. Recalling the mechanistic models described in Chapter 3 it is easy to visualize the use of one catalyst to enhance the phase-transfer step of the process and a second catalyst to promote the reaction step. Both rates and selectivities may thus be enhanced.

3. Asymmetric Alkylation of Oxindoles [41]

The reaction of chloroacetonitrile with 5-methoxy-1,3-dimethyloxindole in a toluene-50% aqueous sodium hydroxide two-phase system in the presence of N-(3,4-dichlorobenzyl) cinchoninium chloride produces the (S)-enantiomer with a chemical yield of 83% with an enantiomeric excess of 73% (Figure 12-7). The results are summarized in Table 12-4. The highest enantioselectivity was achieved with the slow addition of the chloride to the oxindole. Rapid addition of chloroacetonitrile or decreasing the concentration of base from 50% to 25% lowered the enantiomeric excess to 66% and 61%, respectively. Progressively substituting the electron-withdrawing subsituents on the 3- and/or 4-positions of the benzyl group of the catalyst increases the enantioselectivity. In contrast, substitutions at the 2- and/ or 6-positions resulted in no enantioselectivity. The degree of asymmetric induction was insensitive to change in counteranion (chloride vs. bromide), to the rate of agitation, and to the temperature of reaction (15–25° C). The asymmetric alkylation using a variety of alkylating agents (benzyl bromide, 36% ee; benzyl chloride, 58% ee; allyl chloride, 78% ee; 4-chloro-2-methyl-2-butene, 73% ee) was also reported.

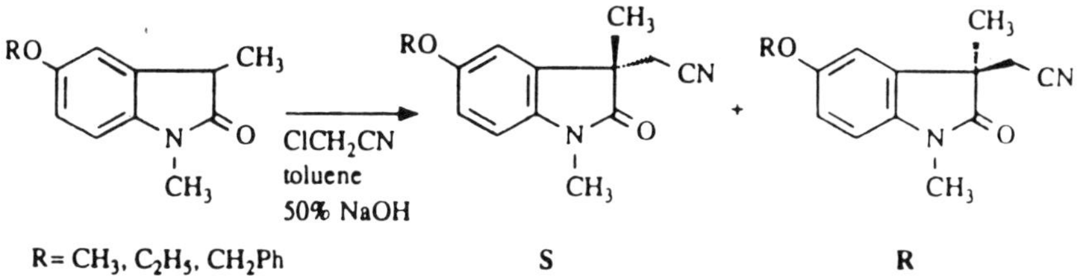

Figure 12-7

4. Synthesis of Chiral Amino Acids

The asymmetric synthesis of α-amino acids by alkylation of a prochical diphenylketimine protected glycine derivative in a methylene chloride–17% aqueous sodium hydroxide two-phase system in the presence of a cinchonine or cinchonidine chiral phase-transfer catalyst has been reported (Figure 12-8) [4k]. Although there is only a distant structural relationship between the protected glycine derivative and the 2-phenylindanone discussed above, several additional complications are encountered with the reaction of this particular structure which were not encountered in the alkylation of the 2-phenylindanone. First of all, the glycine derivative is acyclic and, as a consequence, has at least two reactive conformations that could lead to structurally different carbanions. Second, it is necessary to selectively monoalkylate the carbanion, and, once monoalkylation is achieved, the product must not undergo base-catalyzed racemization. A systematic study dealing with the effects of substrate, catalyst, electrophile, aqueous base concentration, organic solvent, and reaction conditions on the chemical yield and accompanying enantiomeric excess were reported.

Table 12-5 summarizes the percent enantiomeric excess (R-enantiomer) as a function of the alcohol portion of the ester for the reaction of allyl bromide with

Table 12-4. Asymmetric alkylation of oxindole.

R_1	R_2	X	(S)-5, % ee
H	H	Br	10
H	2-CF_2	Br	4
H	3-CF_3	Br	69
H	4-CF_3	Br	72
H	3-Br	Br	48
H	4-Br	Br	68
H	4-Cl	Br	69
H	3,4-Cl_2	Br	77
H	3,4-Cl_2	Cl	78
H	2,6-Cl_2	Br	0
OCH_3	H	Br	39
OCH_3	3,4-Cl_2	Br	77

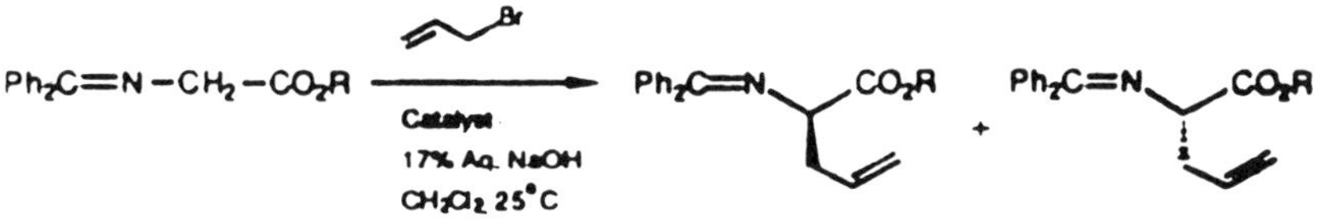

$$Ph_2C=N-CH_2-CO_2R \xrightarrow[\substack{\text{Catalyst} \\ 17\% \text{ Aq. NaOH} \\ CH_2Cl_2\ 25°C}]{\includegraphics{Br}} Ph_2C=N\diagdown\diagup^{CO_2R}_{\diagup\!\!\!\diagup} + Ph_2C=N\diagdown\diagup^{CO_2R}_{\diagup\!\!\!\diagup}$$

Figure 12-8

the glycine derivative in a methylene chloride–17% aqueous sodium hydroxide two-phase system at 25°C in the presence of the chiral catalyst N-(p-trifluoromethylbenzyl) cinchoninium bromide. The highest percent enantiomeric excess was obtained for the *tert*-butyl ester. Replacing the p-trifluoromethyl groups with hydrogen resulted in essentially the same enantioselectivity. When toluene was used as the organic solvent only a small asymmetric induction was realized (approximately 5% enantiomeric excess). Changing the catalyst to the cinchonidine series resulted in the production of the S-enantiomer in 20% ee (R' = OCH_3) to 56% ee (R' = H).

Increasing the concentration of aqueous base decreased the reaction time and increased the % ee (17% NaOH, 56% ee, >16h; 33% NaOH, 59% ee, 8h; 50% NaOH, 65% ee, 4 h). The leaving group associated with the alkylating reagent affected % ee and reaction times (Cl, 34% ee, 12 h; Br, 65% ee, 4 h; I, 44% ee, 5 h). Increasing the concentration of the protected glycine from 0.04 to 0.64 M changed the reaction time from 24 to 1 h respectively. One-tenth an equivalent of catalyst and 20 equivalents of base gave the best results.

Table 12-6 summarises the effects of the structure of the alkylating reagent. "Pseudoenantiomeric" chiral phase-transfer catalysts in the cinchonine and the cinchonidine series were compared in many of the experiments. In all cases, the *tert*-butyl ester of the protected glycine was used, and the leaving group associated with the alkylating agent was bromide. Although catalysts in both series do not have an enantiomeric relationship, the percent enantiomeric excess, the percent chemical yield, and the reaction times were quite similar.

Studies related to the PTC enantioselective synthesis of α-methyl amino acid

Table 12-5. Synthesis of chiral amino acids: effect of ester substituent on % enantiomeric excess.

R	% ee (R)	R	% ee (R)
$PhCH_2$	28	Et	36
4-NO_2PhCH_2	22	Me_2CH	42
4-$MeOPhCH_2$	30	Me_3C	56
Ph_2CH	14	Me_3CCH_2	44
1-NaphthylCH_2	28	Et_3C	40
Me	30		

[a]CH_2CL_2/17% aqueous NaOH.

Table 12-6. Synthesis of chiral amino acids: effect of alkyl malide structure on asymmetric alkylation.

R−X	Equivalent	Major enantiumer	% ee	Yield (%)	Time (h)
Allyl-Br	5	R	66	75	5
	5	S	62	78	5
PhCH$_2$-Br	1.2	R	66	75	9
	1.2	S	64	85	9
Me-Br	5	R	42	60	24
n-Bu-Br	5	R	52	61	14
4-ClphCH$_2$-Br	1.2	R	66	81	12
	1.2	S	62	82	12
2-NaphthylCH$_2$-Br	1.2	R	54	82	18
	1.2	S	48	81	18

M.J. O'Donnell, W.D. Bennett, anfd S. Wu, J. Am. Chem. Soc., **111**, 2353 (1989).

derivatives have been reported (Figure 12-9) [4p]. Schiff base derivatives of aromatic aldehydes and alanine *tert*-butyl ester were alkylated under solid-liquid phase-transfer catalytic conditions in the presence of optically active cinchonium and cinchonidium-derived catalysts. The results are summarized in Table 12-7. In all cases reported in Table 12-7 *N*-benzylcinchonidium chloride was used as the phase-transfer catalyst, producing the *R*-enantiomer as the major product. It was discovered that the base system used in the formation of the carbanion had a major effect on chemical yield while having a very small effect on the percent enantiomeric excess. The mixed base system KOH-K$_2$CO$_3$ used under solid-liquid phase-transfer catalytic conditions was superior. Depending on which chiral catalysts was used (cinchonine or cinchonidine series), either enantiomer could be obtained. Unlike the alkyation of the protected glycine derivatives reported above, substantially different percent enantiomeric excesses were realized with these "pseudoenantiomeric catalysts." It was pointed out that the percent enantiomeric excess was slightly higher for the Schiff base *tert*-butyl ester of L-alanine compared to the corresponding derivative of D-alanine and D,L-alanine (48% ee, 44% ee, 42% ee, respectively). The alkylation of L-alanine with an achiral phase-transfer catalyst resulted in racemic product. The imine protecting group and alkylating agent employed in the reaction had a major influence on the percent realized. Liquid-liquid PTC conditions were reported to give comparable enantioselectivity; however, the chemical yields were lower.

5. Michael Addition Reactions [5p]

The Michael addition of the 6,7-dichloro-5-methoxy-2-*n*-propyl-1-indanone to methyl vinyl ketone in a toluene–50% aqueous sodium hydroxide two-phase system in the presence of catalytic quantities of *N*-(*p*-trifluoromethylbenzyl) cinchoninium bromide has been reported (Figure 12-10). Chemical yields of

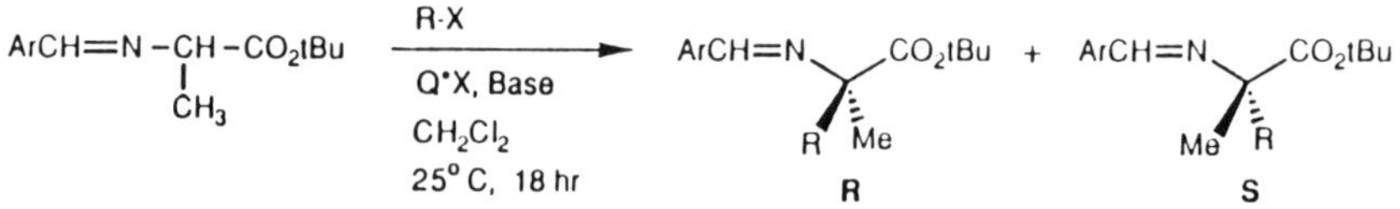

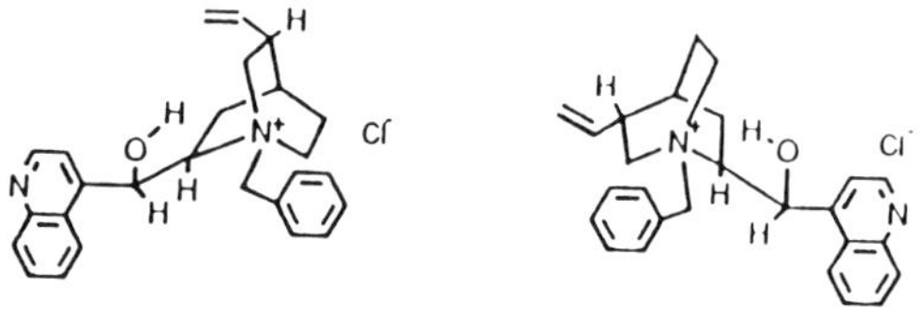

ArCH=N−CH(CH₃)−CO₂tBu →[R-X, Q*X, Base, CH₂Cl₂, 25°C, 18 hr] ArCH=N−C(R)(Me)−CO₂tBu (R) + ArCH=N−C(Me)(R)−CO₂tBu (S)

Figure 12-9

Table 12-7. Asymmetric PTC alkylation of Schiff bases with alkyl halides under various conditions.

Variable	Chemical yield (%)	% ee
Base		
NaOH(50% aq.)	72	49
NaOH	73	41
KOH	70	46
KOH:K₂CO₃(1:1)	87	48
KOH(melted):K₂CO₃(1:1)	78	46
Aryl group of Aldimine		
4−ClC₆H₄−	78	46
4−MeOC₆H₄−	80	42
Ph−	78	22
1-Naphthyl−	82	16
2-Naphthyl−	79	42
2,4,6−Cl₃C₆H₂−	68	24
RBr		
4−FC₆H₄CH₂Br	84	50
4−ClC₆H₄CH₂Br	87	48
4−BrC₆H₄CH₂Br	80	44
PhCH₂Br	80	44
2−NaphthylCH₂Br	87	42
CH₂=CHCH₂Br	78	36

95% and enantiomeric excesses up to 80% were realized. In accordance with the "ion pair model," the S-enantiomer was the predominant product.

In an attempt to synthesize the corresponding R-enantiomer, N-(3,4-dichlorobenzyl) cinchonidinium chloride was used as the phase-transfer catalyst. Although an excellent chemical yield was obtained, only a 20% enantiomeric excess was

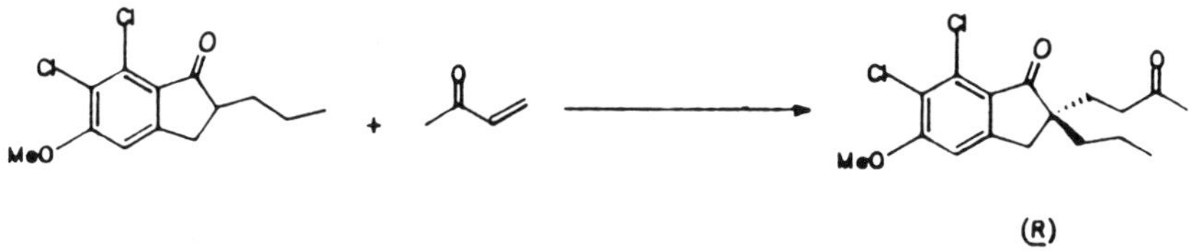

Figure 12-10

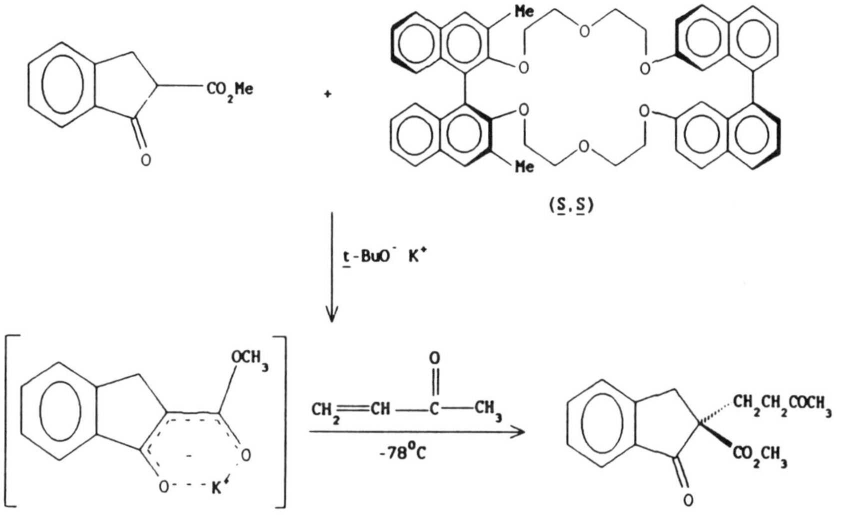

Figure 12-11

achieved. The percent enantiomeric excess increased to 40% when N-(p-trifluoromethylbenzyl) cinchonidinium bromide was used as the phase-transfer catalyst. The reaction was relatively insensitive to changes in organic solvent, base, base strength, temperature, stirring rate, indanone concentration, and quantity of catalyst. Hydrogenation of the vinyl group of N-(p-trifluoromethylbenzyl) cinchonidinium bromide produced a catalyst that resulted in excellent chemical yields of the R-enantiomer in 52% enantiomeric excess.

Chiral crown ethers have been employed as catalysts in the Michael addition 2-carbomethyoxy-1-indanone to methyl vinyl ketone in the presence of potassium *tert*-butoxide (Figure 12-11) [50]. The reactions were conducted at −78°C. An enantiomeric excess of 99% was realized. When the reaction was carried out at 25°C, the enantiomeric excess decreased to 67%. The (S,S)-crown produced the R-enantiomer whereas the (R, R)-crown produced the corresponding S-enantiomer. The identical reaction using quinine as the catalyst afforded the S-enantiomer with an enantiomeric excess of 76% [5a].

References

1. (a) S. Colonna, and R. Fornasier, Synthesis, 531, (1975); (b) C.A. Bunton, L. Robinson, and M.F. Stam, Tetrahedon Lett., 121, (1971; (c) J. Balcells, S. Colonna, and R. Fornasier, Synthesis, 266, (1976); (d) J.P. Masse, E.R. Parayre, J. Chem. Soc., Chem Commun., 438, (1976); (e) S. Colonna, R. Fornasier, J. Chem. Soc., Perkin Trans. 1, 371, 1978; (f) E.V. Dehmlow, P. Singh, and J. Heider J. Chem. Res. (S), 292, (1981); (g) S. Julia; A. Ginebreda, J. Guixer; J. Masana; A. Tomas; S. Colonna, J. Chem. Soc., Perkin Trans. 1, 8 (1978); (h) R. Kinishi; Y. Nalkajima; J. Oda; Y. Inouye, Agric. Biol. Chem., **42,** 869 (1978); (i) Y. Shida; N. Ando; Y. Yamamoto; J. Oda; Y. Inouye, Agric. Biol. Chem., **43,** 1797 (1979); (j) R. Kinishi; N. Uchida; J. Yamamoto; J. Oda; Y. Inouye, Agric. Biol. Chem., **44,** 643 (1980); (k) J.M. McIntosh, Tetrahedron Lett., 403 (1979); (l) C. Innis; G. Lamaty, Nouv. J. Chim., **1,** 503 (1977); (m) J.P. Mazaleyrat, Tetrahedron Lett., 1243 (1983); (n) I.A. Yesikova; E.P. Serebryakov, Izv. Akad. Nauk SSSR. Ser. Khim., 1836 (1989); (o) E. Chiellini; R. Solaro; S. D'Antone, Crown Ethers and Phase Transfer Catalysis in Polymer Science, Ed. I.J. Mathias and C.E. Carraher, Jr., New York and London, Plenum Press, pp. 227–248 (1984).

2. (a) R. Helder; J.C. Hummelen; R.W.P.M.Laane; J.S. Wiering; H. Wynberg, Tetrahedron Lett., 1831 (1976); (b) H. Wynberg, Chemica, **30,** 445 (1976); (c) H. Wynberg; B. Greijdanus, J. Chem. Soc., Chem. Commun., 427 (1978); (d) J.C. Hummelen; H. Wynberg, Tetrahedron Lett., 1089 (1978); (e) H. Wynberg; B. Marsman, J. Org. Chem., **45,** 158 (1980); (f) Y. Harigaya; H. Yamaguchi; M. Ondo, Heterocycles, **15,** 183 (1981); (g) N. Baba; J. Oda; M. Kawaguchi, Agric. Ciol. Chem., **50,** 3113 (1986); (h) E.V. Dehmlow; C. Sauerbier, Liebigs Ann. Chem., 181 (1989); (i) S. Julia; J. Masana; J.C. Vega, Angew, Chem., Int. Ed. Engl., **19,** 929 (1980); (j) S. Julia; J. Guixer; J. Masana; J. Rocas; Sl. Colonna, R. Annunziata; H. Molinari, Chem. Soc., Perkin Trans. 1, 1317 (1982); (k) S. Colonna; H. Molinari; S. Banfi; S. Julia; J. Masana; A. Alvarez, Tetrahedron, **39,** 1635 (1983); (l) S. Banfi; S. Colonna; H. Molinari; S. Julia; J. Guixer, Tetrahedon, **40,** 5207 (1984); (m) B. Marsman; H. Wynberg, J. Org. Chem., **44,** 2312 (1979); (n) R. Curci; M. Fiorentino; L. Troisi; J.O. Edwards; R.H. Parter, J. Org. Chem., **45,** 4758 (1980); (o) R. Curci; M. Fiorentino; M.R. Serio, J. Chem. Soc., Chem. Commun., 155 (1984); (p) J.P. Collman; J.I. Brauman; J.P. Fitzgerald; P.D. Hampton; Y. Naruta; T. Michida, Bull. Chem. Soc. Jpn., **61,** 47 (1988).

3. M. Masui; A. Ando; T. Shioiri, Tetrahedron Lett., **29,** 2838 (1988).

4. (a) J.-C. Fiaud, Tetrahedron Lett., 3495 (1975); (b) E. Chiellini, R. Solaro, J. Chem. Coc., Chem. Commun., 231 (1971); (c) R. Chiellini; R. Solaro; S. D'Antone, Makromol. Chem., **178,** 3165 (1977); (d) K. Saigo; H. Koda; H. Nohira, Bull. Chem. Soc. Jpn., 52, (1979; (e) S. Julia; J. Guixer, J. Chem. Soc., Chem. Commun., 742 (1978); (f) S. Julia; A. Ginebreda; J. Guixer; A. Tomas, Tetrahedron Lett., **21,** 3709 (1980); (g) U.-H. Dolling; P. Davis, E.J.J. Grabowski; S.H. Pines, Eur. Pat. Appl. 121, 872, (1984) (Chem. Abstr., 102, 113080uz0 (1985); (i) D. L. Hughes; U.-H. Dolling; K. M. Ryan; E. F. Schoenewaldt; E.J.J. Grabowski, J. Org. Chem., **52,** 4745 (1987); (j) A. Bhattacharya; U.-H. Dolling; E.J.J.Grabowski; S. Karady; K. M. Rayan; L. M. Weinstodk, Angew, Chem., Int. Ed. Engl., **25,** 476 (1986);

(k) M.J. O'Donnell; W.D. Bennett; S. Wu, J. Am. Chem. Soc., **111**, 2353 (1989); (l) T.B.K. Lee; G.S.K. Wong, J. Org. Chem., 56, 872 (1991); (m) N. Belokon'Yu; V.I. Maleev; J.T. Savel'yeva; N.S. Garbalinskaya; M.B. Saporovskaya; V.I. Bakhmutov; V.M. Belikov, Izv. Akad. Nauk SSSR. Ser. Khim., 631 (1989); (n) J.W., Verbicky, Jr.; E.A.O'Neil, J. Org. Chem., **50**, 1786 (1985); (o) E.V. Dehmlow; A. Sleegers, J. Org. Chem., **53**, 3875 (1988); (p) M.J. O'Donnell; S. Wu, Tetrahedron: Asymmetry, **3**, 591 (1992)

5. (a) H. Wynberg; R. Helder, Tetrahedron Lett., 4057 (1975); (b) H. Pluim; H. Wynberg, Tetrahedron Lett., 1251 (1971); (c) K. Hermann; H. Wynberg, J. Am. Chem. Soc., **103**, 417 (1981); (d) N. Kobayashi; K. Iwai, Polymer J. (Tokyo), **13**, 263 (1981); (e) S. Colonna; H. Hiemstra; H. Wynberg, J. Chem. Soc., Chem. Commun., 238 (1978); (f) R. Annunziata; M. Cinquini; S. Colonna, Chem. Ind. (Lond.), 238 (1980; (g) S. Colonna; A. Re; H. Wynberg, J. Chem. Soc., Perkin Trans., **1**, 547 (1981); (h) A. Loupy; J. Sansoulet; A. Zaparucha; C. Merienne, Tetrahedron Lett., 333 (1989); (i) N. Kobayashi; K. Iwai, J. Org. Chem., **46**, 1823 (1981); (j) P. Hodge; E. Khoshdel; J. Waterhouse, J. Chem. Soc., Perkin Trans. **1**, 2205 (1983); (k) K. Hermann; H. Wynberg, Helv. Chim. Acta, **60**, 2208 (1977); (l) P. Hodge; E. Khoshdel; J. Waterhouse; J.M.J. Grechet, J. Chem. Soc., Perkin Trans. 1, 2327 (1985); (m) S. Banfi; M. Cinquini; S. Colonna, Bull. Chem. Soc. Jpn., **54**, 1841 (1981); (n) R. S. E. Conn; A.V. Lovell; S. Karady; L.M. Weinstock, J. Org. Chem., **51**, 4710 (1986); (o) D.J. Cram; G.D.Y. Sogah, J. Chem. Soc., Chem. Commun., 624 (1981)

6. Y. Goldberg, Phase Transfer Catalysis, Selected Problems and Applications, Gordon and Breach Science Publishers, Philadelphia, 1992.

13

Phase-Transfer Catalysis–Transition Metal Cocatalyzed Reactions

A. Introduction

Transition metals are active catalysts for a variety of organic reactions. Sometimes they are also useful in conjunction with phase-transfer catalysts, particularly when hydroxide anions and other inorganic species are also required to complete the entire reaction sequence. Use of phase-transfer catalysts as cocatalysts with transition metal compounds offers some additional options and advantages compared to conventional reactions. For example, catalytic production of phenylacetic acid from benzyl bromide, carbon monoxide, and sodium hydroxide, catalyzed by cobalt carbonyls and phase-transfer agents, such as tetrabutylammonium salts, is believed to occur by the following sequence [1]:

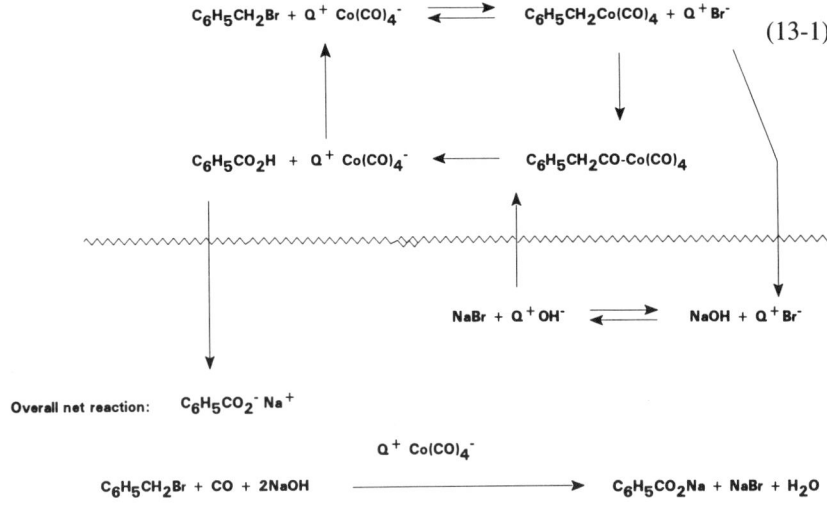

(13-1)

Table 13-1. Some acidity constants of metal carbonyl hydrides [7].

Hydride	pK^a
$HCo(CO)_4$	<2
$HCo(CO)_3P(OC_6H_5)_3$	10^{-3}
$HCo(CO)_3P(C_6H_5)_3$	10^{-7}
$HCo(CO)_2[P(PC_6H_5)_3]_2$	extremely weak
$HCo(PF_3)_4$	strong acid
$HMn(CO)_5$	8×10^{-8}
$HRe(CO)_5$	extremely weak
$H_2Fe(CO)_4$	3.6×10^{-5} (K_1)
	1×10^{-14} (K_2)

Although the quaternary cation participates in the transfer of the $Co(CO)_4^-$ anion from the aqueous to the organic phase, the rate-determining step is cleavage of the acylcobaltcarbonyl species by $R_4N^+OH^-$. Thus, the quaternary salt is a key element in facilitating the whole reaction sequence.

Carbonylation reactions using transition metal–phase-transfer cocatalysts have been the most widely studied of this area in phase-transfer catalysis (PTC), but hydrogenation, coupling, and other reactions have also been described. Des Abbayes and co-workers [2,3] have provided an excellent review of the role of phase-transfer catalysis in metal carbonyl chemistry. Alper [4] and others [5,6] have also contributed heavily and reviewed this area.

B. Carbonylation and Reactions with Carbon Monoxide

1. Formation of Metal Carbonyl Anions

Conversion of metal carbonyls to anions by hydroxide is an important step in many carbonylation reactions, and one of the most important properties to consider for their use in PTC. This reaction, which presumably takes place in the aqueous hydroxide phase (or on the surface of, e.g., KOH), is a nucleophilic attack on coordinated CO:

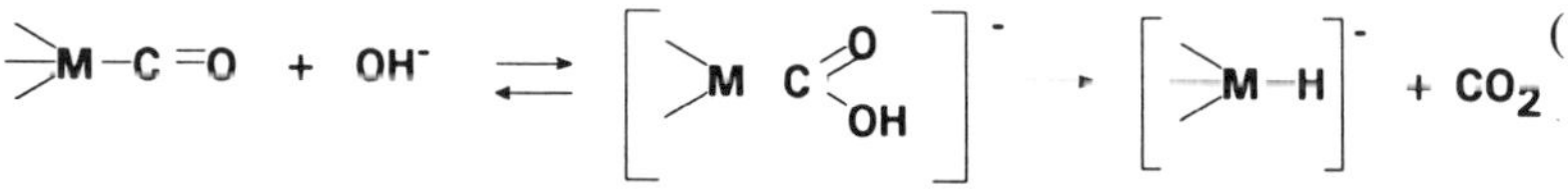

$$\mathrm{M-C=O} + OH^- \rightleftharpoons \left[\mathrm{M-C}\begin{smallmatrix}O\\OH\end{smallmatrix} \right]^- \rightarrow \left[\mathrm{M-H} \right]^- + CO_2 \quad (13\text{-}2)$$

The acidities of several metal carbonyl hydrides, which allow this conversion to proceed, are given in Table 13-1.

2. Carbonylation of Alkyl Halides and Aryl Halides [8–10]

Conversion of halides to carboxylic acids by metal carbonyls and sodium hydroxide, as represented by equation 13-1, proceeds in the absence of phase-transfer catalysts and is believed to involve three steps [11]:

Displacement (13-3)

$$R\text{-}X + M(CO)_n^- \longrightarrow R\text{-}M\text{-}(CO)_n + X^-$$

CO Insertion

$$R\text{-}M(CO)_n + CO \longrightarrow R\text{-}CO\text{-}M(CO)_n$$

Scission

$$R\text{-}CO\text{-}M(CO)_n + OH^- \longrightarrow RCO_2H + M(CO)_n^-$$

This sequence is also followed in the phase-transfer-catalyzed system, but the PTC reaction is faster and yields carboxylic acids in higher yields [12,13]. Addition of phase-transfer catalyst enhances this reaction sequence because:

1. Phase-transfer generation of metal carbonyl anions is a convenient method for conversion of metal carbonyls to anions, as in equation 13-2 above.

2. The metal carbonyl anions are easily transferred from the aqueous phase to the organic phase by phase-transfer catalysts, and are activated by bulky quaternary salts so that displacement of $Co(CO)_4$ on alkyl halide is speeded.

3. PTC transfer of OH^-, and its reaction with the acyl compound, normally a slow step, is greatly enhanced by the presence of appropriate phase-transfer catalysts [14].

4. The two-phase system limits or avoids side reactions of the organic halide with hydroxide.

5. CO is approximately 10 times more soluble in many organic solvents than in water, and consequently side reactions that tend to give formate are limited.

Reaction rates for carboxylation of benzyl chloride, studied by des Abbayes and co-workers [1], depend strongly on stirring speed up to 900 rpm, but level off at higher stirring rates. Kinetics showed the rate-limiting step to be cleavage of the acylcobaltcarbonyl species by the ion pair $Bu_4N^+OH^-$. Under optimum conditions a 85% yield of phenylacetic acid could be obtained, but several byproducts were also present in yields depending rather strongly on carbon monoxide pressure and solvent. Yields of some of the products formed by carbonylation of benzyl chloride at nonoptimal conditions for carboxylic acid formation are illustrated by the data in Table 13-2. In one report, carbonylation of benzyl chloride was said to give phenylacetic acid in 97% yield when reaction was run with *tert*-amyl alcohol as solvent and polyethylene glycol (PEG) as the phase-transfer catalyst, in the presence of dicobalt octacarbonyl and NaOH [15]. Benzyl phenylacetate is produced by simple PTC reaction between benzyl bromide and $Bu_4N^+C_6H_5CH_2CO_2^-$. Dibenzyl ketone is believed to result from forma-

Table 13-2. Product distributions from PTC catalytic carbonylation of benzyl bromide with cobalt carbonyl at low CO pressure, and various levels of agitation [14].

	Percent yield in solvent/stirring rate		
	C_6H_6/H_2O		CH_2Cl_2/H_2O
Product	650 rpm	1100 rpm	1100 rpm
$C_6H_5CH_2CO_2H$	34	36	8
$C_6H_5CH_2CO_2CH_2C_6H_5$	11	8	3
$[C_6H_5CH_2]_2CO$	15	14	12
$C_6H_5CH_3$	22	14	20
$C_6H_5CH_2$-$CH_2C_6H_5$	17	21	50
CO_2	37	45	70

"Yields based on unrecovered benzyl bromide. Experimental conditions: 10 mL H_2O; 10 mL organic solvent; 20°C; 1 atm CO; 20 mmol NaOH; 2mmol $C_6H_5CH_2Br$; 0.1 mmol $NaCo(CO)_4$; 0.01 mmol Bu_4NBr.

tion and decomposition of an alkyl-acylcobalt carbonyl, while toluene and bibenzyl arise from benzyl radicals.

$$C_6H_5CH_2Br + C_6H_5CH_2CO\text{-}Co(CO)_4 \xrightarrow{-CO} \xrightarrow{\text{Several steps}} C_6H_5CH_2CO\text{-}Co(CO)_3CH_2C_6H_5 \quad (13\text{-}4)$$

$$C_6H_5CH_3 \xleftarrow{RH} C_6H_5CH_2\cdot \longrightarrow [C_6H_5CH_2]_2 \qquad [C_6H_5CH_2]_2CO + Co(CO)_4^-$$

$$\downarrow CO$$

In some instances with benzyl halides, further reactions occur to produce other products. For example, with *p*-methylbenzyl chloride [16]:

$$\text{(p-CH}_3\text{-C}_6\text{H}_4\text{-CH}_2\text{X)} \xrightarrow[\text{NaOH, TEBA}^+\text{Cl}^-]{C_6H_6,\ Co_2(CO)_8} \begin{array}{l} p\text{-CH}_3C_6H_4CH_2CO_2H\ (24\%) \\ p\text{-CH}_3C_6H_4CH_2COCO_2H\ (25\%) \\ p\text{-CH}_3C_6H_4\text{-CHCOCO}_2H\ (26\%) \\ \qquad\qquad |\ C_6H_4\text{-}CH_3\text{-}p \end{array} \quad (13\text{-}5)$$

Further development of this chemistry has led to a new route to make α-ketocarboxylic acids by bis-carbonylation of nonactivated alkyl halides [17]. The reaction is carried out at 30–100°C and a CO pressure of 50 atm with *tert*-butyl alcohol as solution mediator. HCl or HBr formed during the reaction is neutralized by slow addition of alkali.

Using iron carbonyl as catalyst, des Abbayes and co-workers [18] showed that various substituted benzyl halides are readily converted to either 1,3-diaryl-

2-propanones or arylacetic acids in a liquid–liquid phase-transfer system. These products arise by first formation of an acyltetracarbonyl anion, which then reacts in the organic phase with benzyl chloride to yield ketone, or with hydroxide to yield carboxylic acid. The product composition is complicated by a PTC reaction between produced carboxylic acid anion and benzyl chloride to produce ester.

$$
\begin{array}{c}
\text{Fe(CO)}_5 + \text{OH}^- \\
\downarrow \\
\text{R-C}_6\text{H}_4\text{-CH}_2\text{X} \xrightarrow{\text{Fe(CO)}_4^{-2},\ \text{PTC}} \text{R-C}_6\text{H}_4\text{-CH}_2\text{CO-Fe(CO)}_4^- \xrightarrow{\text{OH}^-,\ \text{ArCH}_2\text{X}} (\text{R-C}_6\text{H}_4\text{CH}_2)_2\text{CO} \\
\downarrow \text{OH}^- \\
\text{R-C}_6\text{H}_4\text{-CH}_2\text{CO}_2\text{H} \xrightarrow{\text{OH}^-,\ \text{ArCH}_2\text{X}} \text{R-C}_6\text{H}_4\text{-CH}_2\text{CO}_2\text{CH}_2\text{C}_6\text{H}_4\text{-R}
\end{array}
\quad (13\text{-}6)
$$

Maximum production of ketone (78%) is favored in part by factors that increase the concentration of iron acyltetracarbonyl anion in the organic phase, and by factors that reduce the rate of transfer of OH^- to the organic phase, providing more opportunity for the acyltetracarbonyl anion to react with alkyl halide rather than hydroxide. Thus, experimental factors that maximize ketone formation include:

- Keep a low concentration of hydroxide in the aqueous solution (obtained by use of calcium hydroxide instead of sodium hydroxide).
- Low stirring speed (inhibits OH^- transfer to organic phase).
- Increase temperature (e.g., from 20 to 35°C), and maintain very low partial pressure, or use of a molar amount of $Fe(CO)_5$.
- Use an inert atmosphere (introduction of CO inhibits ketone formation.)

With less reactive alkyl halides the reaction with $Fe(CO)_5$ and NaOH under PTC conditions tends to stop at the acyltetracarbonyl anion stage, although ketones can be produced from this intermediate product by addition of a reactive alkyl iodide. For example, using 1-bromobutane for carbonylation, treatment with methyl iodide gives 2-hexanone:

$$
CH_3(CH_2)_3\text{-Br} \xrightarrow[\substack{2M\ \text{NaOH} \\ CO\ (1\ \text{atm}) \\ Fe(CO)_5}]{(Bu_4^+)_2\ SO_4^{-2}} CH_3(CH_2)_3\text{-CO-Fe(CO)}_4^- \xrightarrow{CH_3I} CH_3(CH_2)_3\text{-CO-CH}_3 \quad (13\text{-}7)
$$

Under the same reaction conditions 1,4-dibromobutane gave cyclopentanone, although α,α'-dibromo-o-xylene gave only disubstituted iron tricarbonyl derivative:

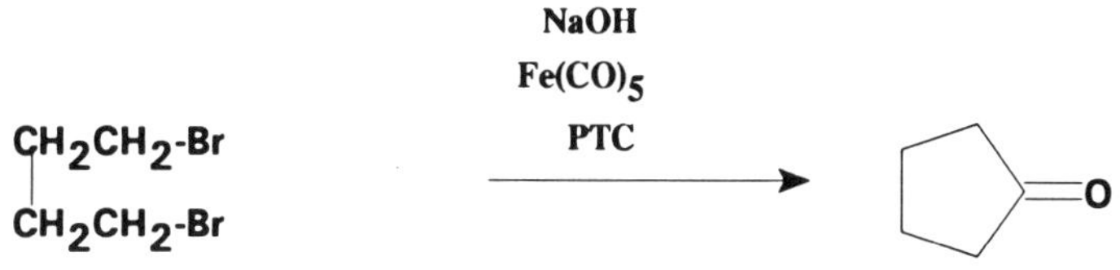

$$\begin{array}{c} CH_2CH_2\text{-Br} \\ | \\ CH_2CH_2\text{-Br} \end{array} \xrightarrow[\text{PTC}]{\substack{\text{NaOH} \\ \text{Fe(CO)}_5}} \text{cyclopentanone}$$

(13-8)

$$o\text{-C}_6\text{H}_4(\text{CH}_2\text{Br})_2 \longrightarrow o\text{-xylylene-Fe(CO)}_3$$

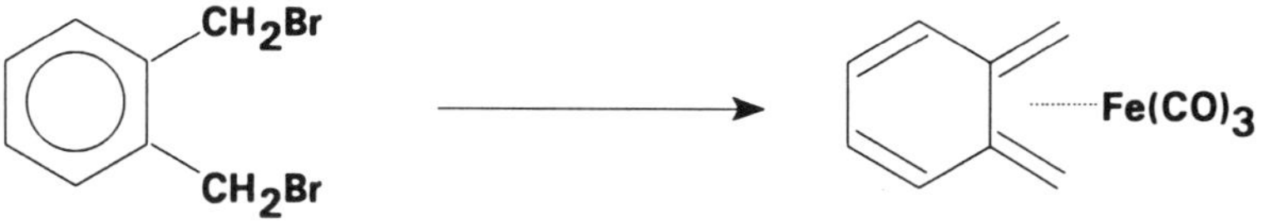

Ungurenasu and Cotzur [19] treated macroporous chloromethylated crosslinked polystyrene with $Fe(CO)_5$ in the presence of Bu_4NBr with slow addition of aqueous NaOH, followed by subsequent oxidation with iodine, to produce resin-bound carboxylic acid groups. These steps are represented as follows:

$$\text{P-C}_6\text{H}_4\text{-CH}_2\text{Cl} \xrightarrow[\text{NaOH(aq)}]{\substack{Bu_4N^+Br^- \\ Fe(CO)_5}} \text{P-C}_6\text{H}_4\text{-CH}_2Fe(CO)_4^-\ Bu_4N^+$$

$$\downarrow I_2$$

$$\text{P-C}_6\text{H}_4\text{-CH}_2CO_2H \xleftarrow[-\text{HI}]{H_2O} \text{P-C}_6\text{H}_4\text{-CH}_2COI$$

$$+ Bu_4NI,\ CO,\ FeI_2$$

(13-9)

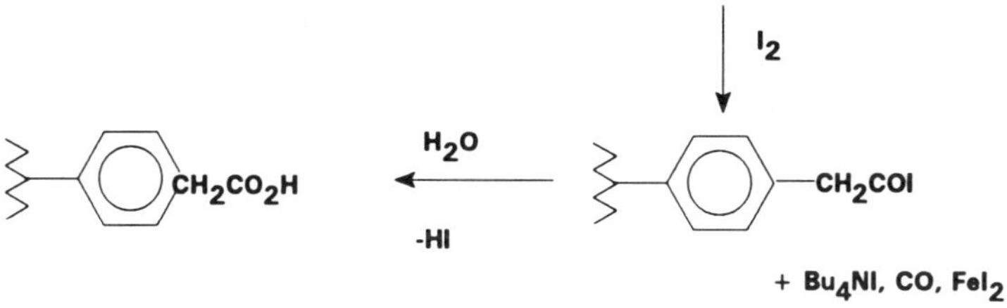

p-Xylylene dichloride on treatment with CO, $Fe(CO)_5$, and a phase-transfer catalyst gave p-phenylenediacetic acid and $4\text{-MeC}_6\text{H}_4\text{CH}_2\text{CO}_2\text{H}$ [20]. PTC mono carbonylation of xylylene dihalides could be accomplished in the presence of Pd(0) complexes to give (halomethyl)phenylacetic acids in 32–51% yields. [21]

Carboxylation of aryl chlorides and bromides aided by photoexcitation, gives the corresponding benzoic acids.

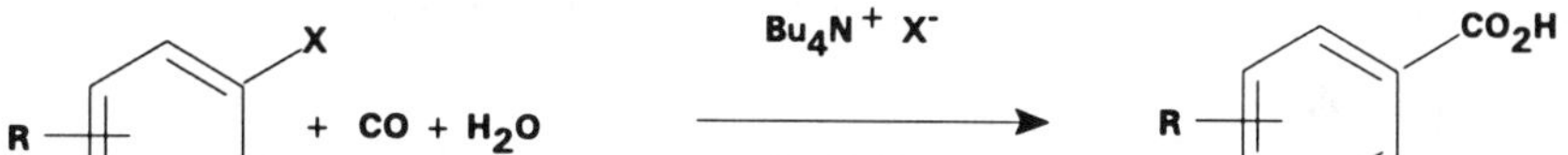

$$R\text{-C}_6\text{H}_4\text{-X} + CO + H_2O \xrightarrow[Co_2(CO)_8]{Bu_4N^+X^-} R\text{-C}_6\text{H}_4\text{-CO}_2H$$

+ KOH   to 95% Yields

(13-10)

Table 13-3. PTC carbonylation/carboxylation studies of halo-compounds.

R–X type	Metal carbonyl	PTC catalyst	References
Benzylic halides	$Co_2(CO)_8$	TEBA	[26–30]
	$Fe(CO)_5$	Bu_4NHSO_4	[31–33]
	$Pd(L)_n$[a]	$Hex_4N^+HSO_4^-$	[34–36]
Aryl halides	$Co_2(CO)_8$	Bu_4NBr	[37–40]
Allyl chloride	$Ni(CO)_4$	R_4NX	[41,42]
Alkyl bromides	$M(CO)_n$	R_4NX	[23,43]
Vinyl dibromides	Pd complexes	PTC	[24]

[a]Ligands used with palladium include $(PPh_3)_4$, dibenzylideneacetone, and 1,2-bis(diphenylphosphino)ethane.

Aryl bromides are carbonylated under 1 atm CO pressure in DMF in the presence of potassium fluoride and phase transfer catalysts to give aroyl fluorides in excellent yields [22].

A commercially useful synthesis of malonic esters involves carbonylation of ethyl chloroacetate with cobalt-carbonyl and phase-transfer catalyst under 1 atm CO pressure at 25°C [23]. No carbonylation occurs in the absence of phase transfer agent. 1-Bromoalkenes, including β-bromostyrenes are converted into carboxylic acids in reasonable yields, but 2-bromoalkenes give poor yields [24].

Aryl vinyl dibromides with zero-valent Pd compounds and a phase transfer agent in the presence of 5N NaOH and benzene give diynes in modest yields, but when the solvent is t-amyl alcohol, di-carbonylation to diacids occurs [25].

$$ArCH=CBr_2 + CO + NaOH \xrightarrow[PhCH_2NEt_3Cl]{C_6H_6, PdL_n} Ar\,C\equiv C-C\equiv C-Ar \quad (40\text{-}75\%)$$

$$\xrightarrow[PhCH_2NEt_3Cl, \, t\text{-}AmOH]{Pd(diphos)_2} ArCH=C\begin{matrix}CO_2H\\CO_2H\end{matrix} \quad (89\text{-}93\%) \quad (13\text{-}11)$$

Vinylic dibromides derived from aliphatic aldehydes or ketones give only the alkylidene malonic acid product:

$$\underset{R}{\overset{R'}{>}}C=CBr_2 + CO + NaOH(aq) \xrightarrow[PhCH_2NEt_3Cl]{PdL_n} RR'C=CHCO_2H \,\, \& \,\, RR'C=C(CO_2H)_2 \quad (13\text{-}12)$$

80% if C_6H_6 solv.   80% if t-AmOH solv.

A summary of alkyl and aryl halide carboxylation studies is given in Table 13-3.

3. Carbonylation of Olefins

Carboxylation of α-olefins occurs with carbon monoxide and water, catalyzed by cobalt carbonyl, phase-transfer agents, and surfactants (anionic, nonionic, and cationic surfactants). The yield and selectivity of desired carboxylic acids are markedly improved by the presence of surfactants acting as cocatalysts [44].

The intermediate acyl cobalt carbonyl formed in reactions of alkyl iodides with cobalt carbonyl anions can be trapped with high-reactivity olefins to yield interesting products, as represented in the following equations [45].

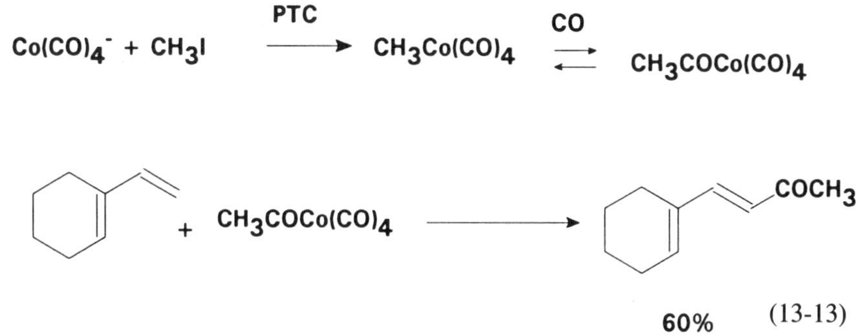

$$Co(CO)_4^- + CH_3I \xrightarrow{PTC} CH_3Co(CO)_4 \xrightleftharpoons{CO} CH_3COCo(CO)_4$$

$$\text{cyclohexenyl-CH=CH}_2 + CH_3COCo(CO)_4 \longrightarrow \text{cyclohexenyl-CH=CH-COCH}_3$$

60%   (13-13)

Similar work has been reported for other olefins [46–48], and for manganese carbonyl and phase-transfer-catalyzed stereospecific hydroacylation of allenes to α,β-unsaturated ketones [49].

By use of a water-soluble triarylphosphine complex with rhodium, inverse phase carbonylations of olefins occur to yield aldehydes [50]. Tris(sulfonated or carboxylated) triarylphosphine complexes with rhodium are charged as catalysts, and these complexing agents keep the rhodium in the aqueous phase.

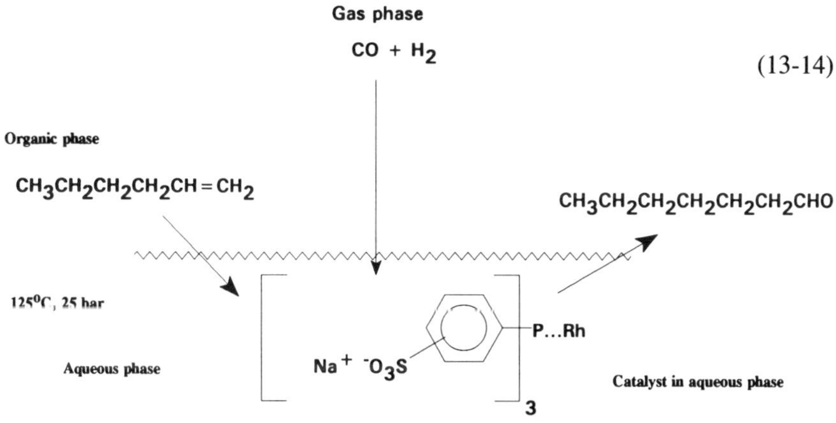

(13-14)

Gas phase: CO + H$_2$

Organic phase: CH$_3$CH$_2$CH$_2$CH$_2$CH=CH$_2$ → CH$_3$CH$_2$CH$_2$CH$_2$CH$_2$CH$_2$CHO

125°C, 25 bar

Aqueous phase: $[Na^+ \ ^-O_3S\text{-C}_6H_4\text{-P}...Rh]_3$ Catalyst in aqueous phase

In the presence of Me(CH$_2$)$_{13}$N$^+$Me$_3$ MeSO$_4^-$ as a phase-transfer catalyst, a 95:5 ratio of n-/iso C$_7$ aldehydes was obtained in 41% conversion, while in the absence of PTC, but otherwise under the same reaction conditions, the ratio was 98/2, and product aldehydes were obtained in 22% conversion.

Satyanarayana et al. [51] found that allenes are carboxylated in an aqueous base–toluene two-phase system, with cetyltrimethylammonium bromide as the

Table 13-4. PTC-Nickel carbonylation of allenes [51].

R	R'	Product	Percent Yield
CH_3	CH_3	$(CH_3)_2C = CHCH_2CO_2H$	52
CH_3	C_2H_5	$C_2H_5(CH_3)C = CHCH_2CO_2H$	58
H	C_2H_5	$C_2H_5CH = CHCH_2CO_2H$	48
CH_3	$(CH_3)_2CHCH_2$	$i\text{-}C_4H_9(CH_3)C = CHCH_2CO_2H$	66
$-(CH_2)_5-$		$cyc\text{-}(CH_2)_5C = C(CO_2H)CH_2CO_2H$	72

phase-transfer agent and nickel cyanide as the metal catalyst, to give β, γ-unsaturated acids in 48–66% yields.

$$\underset{R'}{\overset{R}{>}}C=C=CH_2 + CO + NaOH \xrightarrow[\text{Toluene}]{\underset{Ni(CN)_2}{C_{16}H_{33}NMe_3^+ Br^-}} \underset{R'}{\overset{R}{>}}C=C\underset{H}{\overset{CH_2\text{-}CO_2H}{<}} \quad (13\text{-}15)$$

Some experimental results are listed in Table 13-4.

4. Carbonylation of Acetylenes

Acetylenes react with carbon monoxide, aqueous NaOH, toluene, catalytic amounts of nickel cyanide cocatalyst, and cetyltrimethylammonium bromide as the phase-transfer catalyst, to afford unsaturated acids in reasonable yields [44,52–54].

$$R\text{-}C\equiv CH + CO \xrightarrow[\underset{90°C, 1\,atm}{C_{16}H_{33}NMe_3^+ Br^-}]{Ni(CN)_2,\, PhCH_3} R\text{-}\underset{CO_2H}{\overset{|}{C}}=CH_2 \quad (13\text{-}16)$$

The nickel–cyanide-catalyzed carboxylation reaction of Amer and Alper was applied to a variety of terminal alkynes forming acrylic acid derivatives in fair to good yields (Table 13-5). The reaction is also useful for diynes including 1,7-octadiyne and 1,8-nonadiyne. No carboxylic acids were formed when internal alkynes were used as reactants.

(13-17)

$$HC\equiv C(CH_2)_nC\equiv CH + CO \xrightarrow[\underset{C_{16}H_{33}NMe_3^+ Br^-}{NaOH}]{Ni(CN)_2,\, PhCH_3} H_2C=\underset{CO_2H}{\overset{|}{C}}\text{-}(CH_2)_n\text{-}\underset{CO_2H}{\overset{|}{C}}=CH_2$$

The preferred phase-transfer catalysts for carbonylation of acetylenes, like carbonylation of benzyl chlorides, are quaternary ammonium salts having one long-chain alkyl group such as hexadecyl- or tetradecyl-trimethylammonium bromide, as indicated in Table 13-6.

Phase-transfer-catalyzed carbonylation of alkynes in the presence of cobalt chloride, potassium cyanide, and nickel cyanide affords saturated carboxylic

Table 13-5. Carbonylation of alkynes catalyzed by nickel cyanide under phase-transfer conditions [53].

Alkyne	NaOH (N)	Reaction Time (h)	Product yield (%)
Phenylacetylene	7.5	1	95
	6.25	3	93
	5	3	95
	3.25	4	90
	1	4	68
	1	18	68
4-Phenyl-1-butyne	3	5	50
3-Methyl-1-pentyne	6.25	6	95
1-Hexyne	5	5	62
1-Heptyne	5	3.5	70
1,7-Octadiyne	6.25	6	73
1,8-Nonadiyne	5	8	69

[a]Reaction conditions: 10 mmol alkyne, 1.0 mmol $Ni(CN)_2$ tetrahydrate, 0.165 mmol CTAB, 20 ml toluene, 20 ml NaOH, CO (1 atm), 90°C.

Table 13-6. Effect of phase-transfer catalyst on the hydrocarbonylation of phenylacetylene to atropic acid [a] [53].

Phase-transfer agent	Reaction time (h)	Atropic acid % yield
$C_{16}H_{33}N(CH_3)_3^+Br^-$	3	62
$C_{14}H_{29}N(CH_3)_3^+Br^-$	4	59
$(C_8H_{17})_4N^+Br^-$	2	34
$(C_6H_{13})_4N^+Br^-$	3	20
$(C_4H_9)_4N^+HSO_4^-$	33	41

[a] Reaction conditions: 10 mmol phenylacetylene, 1.0 mmol $Ni(CN)_2$ tetrahydrate, 0.165 mmol catalyst, 20 ml toluene, 20 ml 6.25N NaOH, CO (1 atm), 90°C.

acids $RCH(CO_2H)Me$ and $RCH_2CH_2CO_2H$ with good selectivity observed for the branched-chain isomer [55].

α-Haloalkynes also react with CO to give either allenic monoacids or unsaturated diacids in 70–96% overall yields, when catalyzed by nickel cyanide under phase-transfer conditions [56]. Carboxylation of α-hydroxyalkynes, catalyzed by nickel cyanide under phase-transfer conditions, gives either allenic monoacids or unsaturated diacids in 70–96% overall yields [57].

$$C_2H_5-\underset{\underset{OH}{|}}{\overset{\overset{CH_3}{|}}{C}}-C\equiv CH + CO + 5\,NaOH(aq) \xrightarrow[Ni(CN)_2]{PTC} C_2H_5-\underset{\underset{CO_2H}{|}}{\overset{\overset{CH_3}{|}}{C}}=C-CH_2CO_2H \quad (13\text{-}18)$$

The reaction occurs stepwise: first to the allenic monoacid via a nucleophilic substitution, then carbonylation of this latter to yield diacid [58]. Stereoselectivity is determined by the phase-transfer catalyst.

Bimetallic PTC has been demonstrated in carboxylation of alkynes with methyl iodide with carbon monoxide in an aqueous–organic two-phase system, using dodecyltrimethylammonium chloride as the phase-transfer agent, and cobalt and ruthenium carbonyls as metal catalysts, affording γ-keto acids [44,59].

Alkynes also undergo regiospecific cobalt–carbonyl-catalyzed conversion to butenolides [60].

5. Carbonylation of Aziridines and Azobenzenes

Alper and Mahatantila [61] found that PTC-Pd(0) carbonylation of 2-phenylaziridine and related compounds gave modest yields of indoles, along with substituted acetophenones:

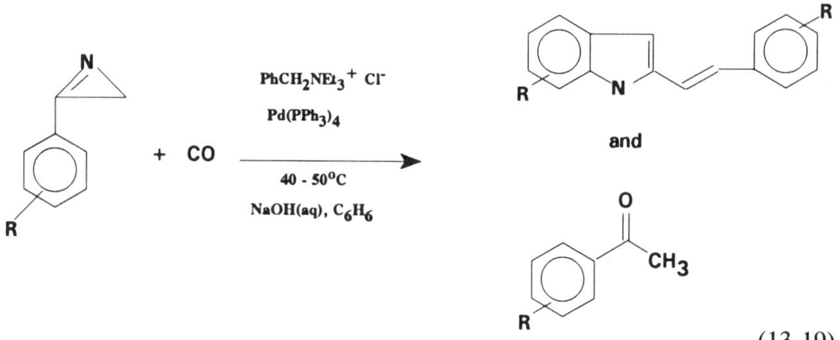

(13-19)

Azobenzenes undergo di- or mono-acylation with cobalt carbonyl and methyl iodide under phase-transfer conditions [62].

$$Ar_1N=NAr_2 + Co_2(CO)_8 + CH_3I \xrightarrow[C_6H_6/H_2O]{PhCH_2NEt_3{}^+ \, Cl^-}$$

$Ar_1N(COCH_3)NHAr_2$ (major)

$Ar_1N(COCH_3)\text{-}NAr_2(COCH_3)$

$Ar_1NHCOCH_3$

$Ar_2NHCOCH_3$ (13-20)

Maximum yield of monoacylated product is obtained when p-toluenesulfonic acid is also added to the reaction mixture.

6. Carbonylation of Thiiranes

Calet and Alper [65] found that treatment of 2-phenylthiirane and other thiiranes with carbon monoxide, methyl iodide, aqueous KOH, benzene solvent, and catalyzed by a small amount of cobalt carbonyl and polyethylene glycol (PEG-400), gave thiol acid in 78% yield. For example:

$$\text{Ph}\underset{S}{\triangle} + CO + KOH \xrightarrow[\substack{25°/1\ \text{atm} \\ Co_2(CO)_8}]{\substack{MeI \\ PEG\text{-}400}} \underset{\substack{| \\ CO_2H}}{\overset{\substack{Ph \\ |}}{H-C-CH_2SH}} \quad (13\text{-}21)$$

Reaction does not proceed in the absence of a phase-transfer catalyst, but the yield is only 17% with Bu_4NBr as catalyst, and 39% with cetyltrimethylammonium bromide as catalyst. The slow step in the reaction sequence is transfer of hydroxide from aqueous to organic phase.

7. Carbonylation Reaction with Phenol

Hallgren and Lucas [64] found palladium bromide to be a good catalyst for the phase-transfer-catalyzed synthesis of diphenyl carbonate from phenol, carbon monoxide, and oxygen:

$$C_6H_5OH + CO + 0.5\ O_2 \xrightarrow[\substack{Mn(acac)_2 \\ Mol.Sieves}]{\substack{Bu_4NBr \\ PdBr_2,\ NaOH}} (C_6H_5)_2CO_3 + H_2O \quad (13\text{-}22)$$

B. PTC Reduction and Hydrogenation with Metal Cocatalysts

1. Hydrogenolysis of Aryl and Alkyl Halides

In reductive hydrogenolysis of alkyl and aryl halides the PTC technique allows for efficient removal of hydrogen halides by neutralization in the aqueous phase, and for use of reducing agents such as sodium formate. Bar et al. [65] demonstrated the reductive dehydrohalogenation of aryl bromides with aqueous sodium formate using palladium complexes with tetrahexylammonium hydrogen sulfate. With D_2O instead of H_2O in the aqueous phase, selective formation of the aryl monodeuterated products was observed. Aryl bromides and iodides reacted cleanly under the reaction conditions but aryl chlorides reacted either very slowly or not at all.

The effect of catalyst structure is illustrated in Figure 13-1. Of the several

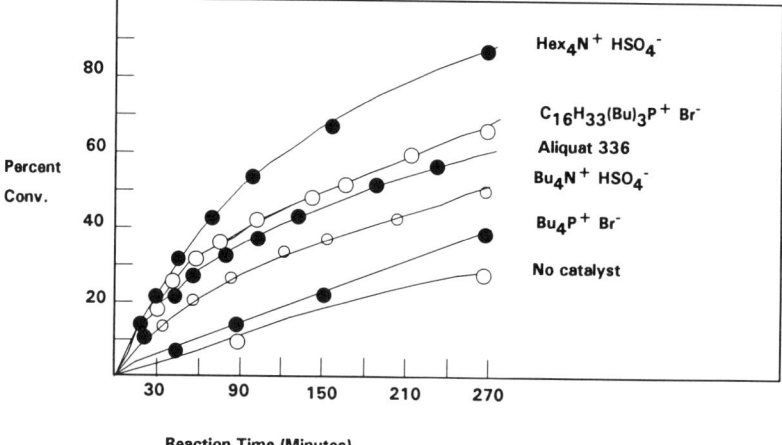

Figure 13-1. Concentration time profiles for formation of naphthalene from 2-bromonaphthalene, aq. sodium formate, and $PdCl_2(PPh_3)_2$ at 120°C.

metal cocatalysts tested, $PdCl_2(PPh_3)_2$ was the best, as shown by the rate data in Table 13-7.

The palladium complex suffered, however, from the disadvantage of being easily reduced to metal by aqueous sodium formate. Addition of excess triphenylphosphine was necessary to avoid catalyst deterioration. Metallic palladium in the absence of base promoted rapid formate decomposition.

In related work, Zoran et al. [66] found that hydrogenolysis of aryl bromides could also be accomplished using benzyl alcohol and aqueous sodium hydroxide as a reducing system.

Table 13-7. Rates for hydrodebromination of bromobenzene by aqueous sodium formate under PTC conditions at 100°C in the presence of Aliquat 336 and various metal catalysts [63].

Metal catalyst	Initial rate (mmol/L/min)
$IrCl(CO)(PPh_3)_2$	0.00
$RhCl(PPh_3)_3$	0.31
$RhCl(CO)(PPh_3)_2$	0.42
$RuCl_2(PPh_3)_3$	0.42
$Pd(PPh_3)_4$	2.12
$PdCl_2(PPh_3)_2$	2.63

Table 13-8. Effect of quaternary salt structure on $PTC-PdCl_2(PPh_3)_2^-$ catalyzed transfer hydrogenolysis of 4-bromotoluene by benzyl alcohol and NaOH [64].

Quaternary salt catalyst	Maximum rate (mol/L/min/mol Pd)
Aliquat 336	1.37
Cetyltrimethylammonium bromide	0.97
Tetraoctylammonium bromide	0.71
Benzyltriethylammonium chloride	0.63
Tetrahexylammonium bromide	0.55
Tetrabutylphosphonium bromide	0.49
Tetraethylammonium bromide	0.19

Table 13-9. PTC–vanadium dehalogenation of halides [65].

Halide	Product	Percent Yield
2-(Bromomethyl)-naphthalene	2-Methylnaphthalene	76
1-(Chloromethyl)-naphthalene	1-Methylnaphthalene	92
Bromodiphenylmethane	Diphenylmethane	62
p-Chlorobenzyl chloride	p-Chlorotoluene	44
2-Chloroacetophenone	Acetophenone	39
3-Bromostyrene	Styrene	38
2-Bromonaphthalene	Naphthalene	72
Bromocyclohexane	Cyclohexane	66
1-Bromooctane	Octane	60

$$\text{Ar-Br} + \text{PhCH}_2\text{OH} + \text{NaOH} \xrightarrow[\text{PPh}_3,\text{ PTC}]{\text{PdCl}_2(\text{PPh}_3)_2} \text{Ar-H} + \text{PhCHO} + \text{NaBr} + \text{H}_2\text{O} \quad (13\text{-}23)$$

The structure of the quaternary salt catalyst was found to have a substantial effect on catalytic activity, as shown by the data in Table 13-8. Falicki and Alper [67] described the generation of (η^5-cyclopentadienyl)-tricarbonylhydridovanadate anion by reaction of (η^5-cyclopentadienyl)-vanadium tetracarbonyl with 5 N NaOH, benzene, and tetrabutylammonium hydrogen sulfate, as an excellent and experimentally simple method for generation of the vanadium hydride anion.

$$\text{CpV(CO)}_4 + \text{NaOH(aq)} \xrightarrow{R_4N^+\ X^-} Q^+\ \text{HV(CO)}_3\text{Cp}^- \quad (13\text{-}24)$$

This anion could then be used for reductive dehalogenation of alkyl or aryl halides as shown in Table 13-9. Shimizu, et al. [68] examined an inverse PTC–metal hydrogenolysis of bromoanisoles with sodium formate to produce both dehydrohalogenated product as well as an aryl coupling product under mild conditions using palladium on charcoal and cyclodextrin as the inverse phase-transfer catalyst.

(13-25)

Br—C6H4—OCH3 (org) + HCO2⁻Na⁺ (aq) →[Pd/C, 60°C, NaOH, H2O]→ Cyclodextrin: CH3O—C6H5 (50 - 84%) & CH3O—C6H4—C6H4—OCH3 (15 - 50%)

These reactions proceed in the absence of phase-transfer catalyst, but the β-cyclodextrin accelerates the rate by up to fourfold.

Okano and co-workers [69] demonstrated inverse PTC hydrogenolysis of allyl chlorides with sodium formate using palladium complexed with a phosphine containing ethoxylate chains, $P[CH_2CH_2O)_3CH_3]$. This material is a liquid that is freely soluble in water at 30°C, and its complex with palladium, $PdCl_2L_2$, is also freely soluble in water at 20°C, but is also soluble in toluene, cyclohexane, and heptane. When this complex is used as a catalyst for PTC reaction of aqueous sodium formate with 1-chloro-2-octene, quantitative hydrogenolysis results:

C_6H_{13}—CH=CH—CH2—Cl (org) + $HCO_2^-Na^+$ (aq) →[$PdCl_2L_2$, Heptane, water]→ $C_7H_5CH=CH_2$ + $C_6H_{13}CH=CHCH_3$ (13-26)

For $PdCl_2[P(n-Bu_3)]_2$, Conv = 26%   20%   3%

For $PdCl_2[P\{(CH_2CH_2O)_3CH_3\}_3]_2$, Conv. = 100%   82%   17%

By use of the apparatus represented in Figure 13-2 for hydrogenolysis of allyl acetate with sodium formate, the amount of reaction taking place in each phase could be determined.

Results with two catalysts and two different organic solvents are listed in Table 13-10.

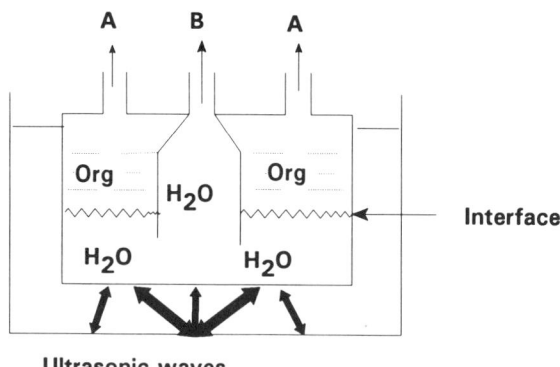

Figure 13-2. Apparatus to determine percent of reaction taking place in each phase.

Table 13-10. Reduction of allyl acetate to propene in a two-phase system. Determination of the phase where reaction occurs [67].

Catalyst	Organic layer	Percent Yield of Propene	Aqueous	Organic + interface
$PdCl_2[P(n-Bu)_3]_2$	Heptane	38	12	88
	Toluene	31	15	85
$Pd[P\{(CH_2CH_2O)_3CH_3\}_3]_2$	Heptane	55	98	2
	Toluene	41	23	77

2. Reduction of Acid Chloride Groups to Aldehydes

Ungurenasu and Cotzur [70] showed that acid chloride groups attached to a styrene–methacrylic acid(chloride)–divinylbenzene insoluble polymer could be converted to aldehyde in 70–96% yields by treatment with iron pentacarbonyl and NaOH under PTC conditions, followed by treatment with acetic acid.

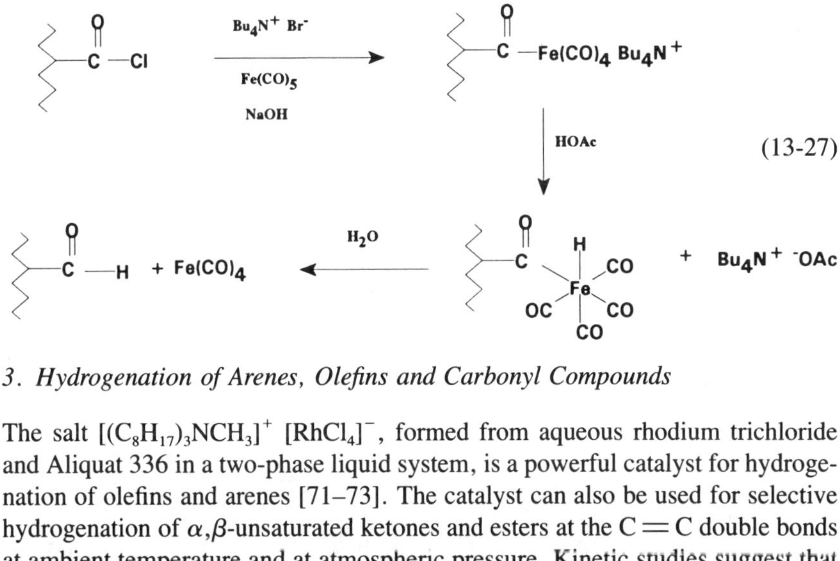

$$\text{(13-27)}$$

3. Hydrogenation of Arenes, Olefins and Carbonyl Compounds

The salt $[(C_8H_{17})_3NCH_3]^+ [RhCl_4]^-$, formed from aqueous rhodium trichloride and Aliquat 336 in a two-phase liquid system, is a powerful catalyst for hydrogenation of olefins and arenes [71–73]. The catalyst can also be used for selective hydrogenation of α,β-unsaturated ketones and esters at the $C=C$ double bonds at ambient temperature and at atmospheric pressure. Kinetic studies suggest that the process is both reaction and diffusion controlled.

$$C_6H_5COCH=CHC_6H_5 + H_2 \xrightarrow[ClCH_2CH_2Cl]{\substack{Q^+ [RhCl_4]^- \\ H_2O}} C_6H_5COCH_2CH_2C_6H_5 \quad (13\text{-}28)$$

Results with several substrates are given in Table 13-11. Long-chain tertiary amines, such as tri-n-dodecylamine, could be used instead of Aliquat 336 without significant loss of yield; however, tertiary phosphines were found to act as potent inhibitors. The solvent was also important: high yields (93%) were obtained in

Table 13-11. Hydrogenation of some α,β-unsaturated carbonyl compounds by rhodium trichloride and Aliquat 336[a] [71].

Substrate	Products	Yield (%)
$C_6H_5COCH = CHC_6H_5$	$C_6H_5COCH_2CH_2C_6H_5$	93
$4\text{-}CH_3C_6H_4COCH = CHC_6H_5$	$4\text{-}CH_3C_6H_4COCH_2CH_2C_6H_5$	96
$4\text{-}ClC_6H_4COCH = CHC_6H_5$	$4\text{-}ClC_6H_4COCH_2 CH_2C_6H_5$	96
$C_6H_5CH = CHCO_2C_2H_5$	$C_6H_5CH_2CH_2CO_2C_2H_5$	96
$C_6H_5CH = CHCO_2H$	$C_6H_5CH_2CH_2CO_2H$	87
	and $C_6H_5CH_2CH_2CH_2OH$	13

[a]Reaction conditions: 1 mmol substsrate, 0.0053 mmol $RhCl_3$ in 1 ml water, 0.055 mmol Aliquat 336 in 1 ml 1,2-dichloroethane; 1 atm H_2, stirring rate 280 rpm at 30°C, 4 h for ketones, 3 h for acid and ester.

1,2-dichloroethane, 74% in ethyl acetate, 45% in benzene, and 38% in cyclohexane.

Inverse PTC complexing agents, particularly the highly water-soluble tris(sulfonated) compounds, R = $(CH_2)_2OH$ (polar), Me (nonpolar), or C_5H_{11} (organophilic), have been used to stabilize a catalytically active colloidal suspension of polyhydroxylated rhodium particles allowing the hydrogenation of liquid alkenes in biphasic medium [74]. High concentration of the hydrotropic trisulfonates along with quaternary salts produces efficient hydrogenation. Hydrogenations under D_2 or H_2 give rise to a high level of isotopic exchanges and also to deuterium scrambling, in agreement with previous results obtained with polyhydroxylated rhodium colloids electrostatically stabilized by other triphenylsulfonates. These are efficient catalytic systems that provide an easy route for recovery and recycle of the aqueous catalyst phase.

Januszklewicz and Alper [75] found the dimer of chloro(1,5-hexadiene)rhodium and a quaternary ammonium salt combination in a two-phase system to be an exceedingly mild, selective, and stereospecific method for hydrogenation of aromatics and heterocyclic compounds, as illustrated by the data in Table 13-12. The stereospecificity was demonstrated with naphthalene and p-methylanisole.

A rather comprehensive patent by Henkle claims hydrogenation of aromatics, aldehydes, olefins, nitroalkanes, nitriles, and chlorohydrocarbons using a metal salt catalyst in phase-transfer systems [76]. As an example, methyl benzoate was reduced to 99.8% methyl cyclohexanecarboxylate by hydrogenation with a catalyst of $RhCl_3$, $[CH_3(CH_2)_7]_3NMeCl$ in a mixture of water, ether, and triethylamine.

4. Reduction of Nitrogen Compounds

Reduction of nitrobenzene to aniline by $Fe_3(CO)_{12}$ under triphase conditions has been investigated and the efficiency of various polymer-supported phase-transfer

Table 13-12. Hydrogenation of arenes by PTC-[1,5-HDRhCl]$_2$. Room temperature hydrogen pressure at 1 atm [73].

Reactant	Catalyst[a]	Buffer pH	Solvent	Products (% yields)		
PhCOCH$_3$				PhCH(OH)CH$_3$	cy-C$_6$H$_{11}$COCH$_3$	cy-C$_6$H$_{11}$CH(OH)CH$_3$
	CTAB	7.6	C$_6$H$_6$	31	53	9
	CTAB	7.6	C$_6$H$_{14}$	10	39	51
	THS	7.6	C$_6$H$_{14}$	23	77	—
	CTAB	9.2	CH$_3$OH/ C$_6$H$_6$	84	8	8
PhCH$_2$COCH$_3$					cy-C$_6$H$_{11}$COCH$_3$	cy-C$_6$H$_{11}$CH(OH)CH$_3$
	CTAB	7.6	C$_6$H$_6$	54	14	
	THS	7.6	C$_6$H$_{14}$	100	0	
PhCH$_2$CH$_2$COCH$_3$					cy-C$_6$H$_{11}$CH$_2$CH$_2$COCH$_3$	cy-C$_6$H$_{11}$CH$_2$CH$_2$CH(OH)CH$_3$
	THS	7.6	C$_6$H$_{14}$	97		
	DTAC	7.6	C$_6$H$_{14}$	80		
	CTAB	7.6	C$_6$H$_{14}$	82	10	
	CTAB	7.6	C$_6$H$_6$	30		
	CTAB	9.2	C$_6$H$_6$	12		
	CTAB	H$_2$O	C$_6$H$_{14}$	4		
	CTAB	6N/ NaOH		20 + 80% PhCH$_2$CH$_2$CH(OH)CH$_3$		
p-CH$_3$C$_6$H$_4$OCH$_3$	THS	THS	C$_6$H$_{14}$	cis-4-Methylcyclohexyl methyl ether (92%)		
PhC$_6$H$_5$	THS	7.6	C$_6$H$_{14}$	n-Butylcyclohexane (100%)		
PhCO$_2$CH$_3$	THS	7.6	C$_6$H$_{14}$	C$_6$H$_{11}$CO$_2$CH$_3$ (69%)		
PhCONH$_2$	THS	7.6	C$_6$H$_{14}$	C6H$_{11}$CONH$_2$ (79%)		
CH$_3$CO$_2$Ph	THS	7.6	C$_6$H$_{14}$	CH$_3$CO$_2$C$_6$H$_{11}$ (31%)		
C$_6$H$_6$	THS	7.6	C$_{10}$H$_{22}$	Cyclohexane (100%)		
Naphthalene	CTAB	7.6	C$_6$H$_6$	Tetralin (73%), decalin (trace)		
	CTAB	7.6	C$_6$H$_{14}$	Tetralin (20%), cis-decalin (80%)		
2-Methylpyridine	THS	7.6	C$_6$H$_{14}$	2-Methylpiperidine (39%)		
Quinoline	THS	7.6	C$_6$H$_{14}$	1,2,3,4-Tetrahydroquinoline (100%) at 75°C		
Isoquinoline	THS	7.6	C$_6$H$_{14}$	No reaction at r.t. or 75°C		
2-Ethylfuran	THS	7.6	C$_6$H$_{14}$	2-Ethyltetrahydrofuran (51%)		
Phenol	THS	7.6	C$_6$H$_{14}$	Cyclohexanol (73%)		

[a]CTAB = cetyltrimethylammonium bromide; THS = tetrabutylammonium hydrogen sulfate.

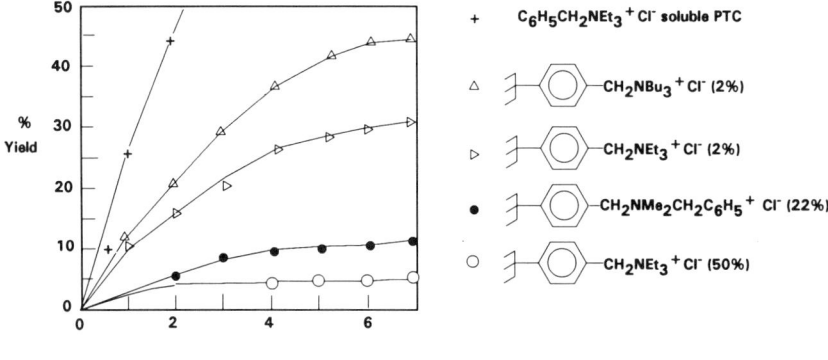

Figure 13-3. PTC-$Fe_3(CO)_{12}$ nitrobenzene reduction. Comparison of triphase and soluble catalysts.

Table 13-13. Reduction of nitroaromatics by PTC–vanadium carbonyl–sodium hydroxide [75].

Substrate	Product	Percent yield
Nitrobenzene	Aniline	78
o-Nitrotoluene	o-Toluidine	83
p-Nitrotoluene	p-Toluidine	88
o-Nitroanisole	o-Anisidine	85
m-nitroanisole	m-Anisidine	77
2,6-Dimethylnitrobenzene	2,6-Dimethylaniline	98
4-Nitrostilbene	4-Aminostilbene	76
4-$O2NC_6H_4CH = CHCOPh$	4-$H2NC_6H_4CH_2CH_2COPh$	66

agents compared [77]. The behavior of the different catalysts tested is shown in Figure 13-3.

Falicki and Alper [65] used a PTC–vanadium carbonyl–sodium hydroxide system to reduce nitroaromatics to anilines:

$$ArNO_2 + CpV(CO)_4 + NaOH \xrightarrow{Bu_4N^+ \ HSO_4^-} Ar\text{-}NH_2 \quad (13\text{-}29)$$

Results for these experiments are listed in Table 13-13. Four metal cocatalysts—platinum, palladium, rhodium, and ruthenium—were compared in a phase-transfer-catalyzed nitrobenzene to p-aminophenol [78]. Quaternary ammonium salt catalyst concentration, sulfuric acid concentration, hydrogen pressure, and temperature affected the reaction, for which the rate-determining step is the catalytic process by which the hydrogen atom is absorbed on the metal surface. Platinum favors p-aminophenol production, while palladium favors formation of aniline.

Electrochemical reduction of 4-amino-3,4'-dinitrodiphenyl sulfide to 3,4,4'-triaminodiphenyl sulfide in an acidic medium was accomplished by conducting reaction in a sulfuric acid medium in the presence of vanadyl sulfate and tetraalkylammonium salts at 50–60° [79].

Bimetallic cocatalysts with PTC, i.e., cobalt carbonyl and chloro(1,5-hexadiene)rhodium(I) chloride, with dodecyltrimethylammonium chloride gave smooth reduction of nitroaromatics to the corresponding anilines in good yields [80]:

$$\text{Ar-NO}_2 + \text{CO} + \text{NaOH} \xrightarrow[\text{Co}_2(\text{CO})_8]{\substack{\text{C}_{12}\text{H}_{25}\text{HMe}^+ \text{ Cl}^- \\ [1,5\text{-HDRhCl}]_2 \\ \text{r.t., 1 atm, 2.5h}}} \text{Ar-NH}_2 \qquad (13\text{-}30)$$

5. Other Reductions and Hydrogenations

Model coal constituents are reduced by iron carbonyl-catalyzed reductions under water gas shift conditions [81].

Alper and Sibtain [82] found that the two-phase reaction of benzylic mercaptans with $Fe_3(CO)_{12}$ and tetrafluoroboric acid gave desulfurized products with product distributions depending on the presence or absence of an acidic phase transfer catalyst. For example:

			% Yield with	
	RSO_3H^*		RSO_3H	no catal.
	HBF_4	p-Cl-C$_6$H$_4$CH$_3$	6	44
p-Cl-C$_6$H$_4$CH$_2$SH + Fe$_3$(CO)$_{12}$	$\longrightarrow$	(p-Cl-C$_6$H$_4$CH$_2$)$_2$S	11	19
		(p-C$_6$H$_4$CH$_2$S)$_2$	40	6

* $RSO_3H = C_{12}H_{25}C_6H_4SO_3H$

(13-31)

Rhodium complexes are excellent catalysts in conjunction with quaternary salts for hydrogenolysis of chloroarenes [83] under biphasic conditions [40% NaOH, toluene] using benzyltriethylammonium chloride as PTC. The reaction occurs under mild conditions, and many functional groups, e.g. R, OR, CF_3, COAr, CO_2H, NH_2, survive the carbon–chlorine bond cleavage.

C. Coupling Reactions of Alkenes, Alkynes, and Alkyl Halides

1. Acetylene and Olefin Reactions with Halo-compounds

A multiphase Heck reaction forms the basis of a process for preparation of substituted olefins and acetylenes, and consists of treatment of organic halocompounds with olefins or acetylenes in the presence of a Pd-phosphane complex and a base [84]. For example:

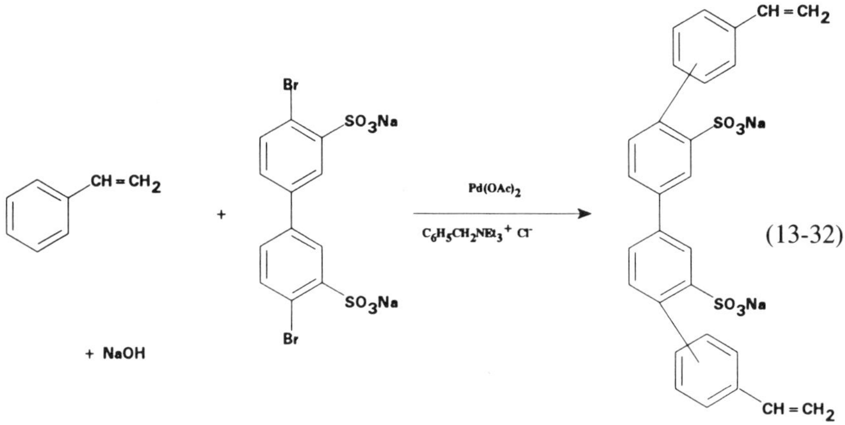

614 / Phase-Transfer Catalysis

[Reaction scheme 13-32: styrene + dibrominated biphenyl disulfonate with Pd(OAc)₂, C₆H₅CH₂NEt₃⁺ Cl⁻, NaOH → terphenyl-type coupling product] (13-32)

The reaction is carried out in a multiphase system such that the solubilities of the complex compound and the product lead to their existence in different phases at the end of the reaction. The products thus obtained are patented as optical brighteners, dyes, or intermediates for brighteners or dyes.

Palladium-catalyzed vinylation of vinylic iodides can be stereocontrolled under solid–liquid phase transfer conditions, providing an efficient way to stereodefine dienoic esters [85]. Vinyl iodides couple with methyl acrylate, catalyzed by PTC–palladium, to give a new carbon–carbon bonds [86], for example, in a one-pot synthesis to give pellitorine in an overall yield of 48%:

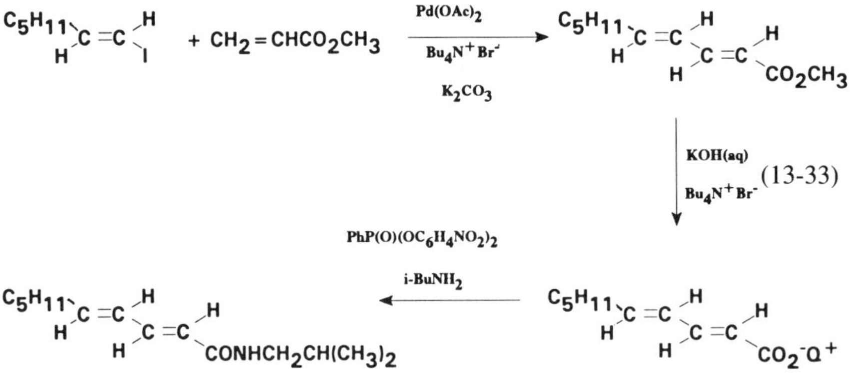

[Reaction scheme 13-33: C₅H₁₁-vinyl iodide + CH₂=CHCO₂CH₃ with Pd(OAc)₂, Bu₄N⁺Br⁻, K₂CO₃ → dienoic methyl ester; then KOH(aq)/Bu₄N⁺Br⁻ → carboxylate salt; then i-BuNH₂, PhP(O)(OC₆H₄NO₂)₂ → pellitorine amide CONHCH₂CH(CH₃)₂] (13-33)

The starting vinylic iodide was prepared by addition of diisobutylaluminum hydride to 1-heptyne, followed by iodination [87].

Similarly, acrylic acid esters react with 1-iodo-1-alkynes to give coupling products in 50–70% yields [88]:

$$R-C\equiv C-I \quad + \quad \underset{\underset{R'}{|}}{H_2C}\overset{H}{\underset{}{-}}\overset{}{C}\overset{}{-}\overset{}{C}\overset{\nearrow O}{\diagdown} \quad \xrightarrow[\text{DMF/rt}]{\underset{\underset{Na_2CO_3}{}}{\overset{Bu_4N^+\ Cl^-}{\overset{Pd(OAc)_2}{}}}} \quad R-C\equiv C\overset{H}{\diagdown}C=C\overset{R'}{\diagup}\overset{}{\diagdown}\overset{C=O}{\underset{H}{}}$$

(40 - 60%) (13-34)

Palladium(0)-induced conversion of vinylic dibromides to diynes [89] can also be done under PTC conditions.

The known reaction of aryl iodides with copper acetylides and with terminal acetylenes in the presence of potassium carbonate, cuprous iodide, or powdered copper which occurs under rather severe conditions can be carried out under mild conditions (20°C, 0.3–3 h) with quantitative yields in the presence of PTC catalysts and halopalladium complexes [90]. Similarly, cross-coupling of copper acetylides with aryl iodides could be accomplished. Coupling of terminal acetylenes and aryl vinyl, and heterocyclic halides can also carried out [13,91] in the interfacial systems of benzene–aqueous alkali–$PhCH_2NEt_3^+Cl^-$ in the presence of Pd complexes and cuprous iodide. Addition of 10 mol% 18-crown-6 produced high yields of substitution products.

$$PhC\equiv CH\ +\ p\text{-}O_2NC_6H_4\text{-}I \quad \xrightarrow[\underset{THF,\ 20°\ C,\ 3\ h}{NaOH\ (aq)}]{\overset{18\text{-crown-}6}{PhPdI(Ph_3P)}} \quad p\text{-}O_2NC_6H_4C\equiv CPh$$

(13-35)

(100%)

Aryl iodides containing less electron-accepting substituents lowered reaction rates significantly. Best yields of products were obtained when potassium carbonate was used instead of NaOH, and DMF was used instead of THF. Some results are given in Table 13-14.

Terminal alkynes react with allylic bromides and cuprous chloride, under phase-transfer conditions (CH_2Cl_2, 30–50% NaOH, quaternary ammonium salt), to give enynes in 38–90% yields [92]. The coupling reaction occurs without allylic or allene rearrangement in most cases. A copper (III) intermediate may participate in the reaction:

$$C_6H_5(CH_2)_3-C\equiv C-H\ +\ CH_2=CHCH_2Br \quad \xrightarrow[NaOH(aq)]{\overset{R_4N^+}{CuCl}} \quad C_6H_5(CH_2)_3-C\equiv C-CH_2CH=CH_2$$

(13-36)

Phase-transfer-catalyzed coupling of 3-iodo-4-methoxy-2(1H)-quinoline with alkynes gave coupling products and/or substituted furoquinolines, such as the dictamnine derivative shown [93].

Arenesulfonyl chlorides react with acrylate esters in the presence of a catalytic amount of $PdCl_2(PhCN)_2$ under solid–liquid PTC conditions to give the corresponding (E)-3-arylprop-2-enoates in good yield. For example, from 1-naphthalenesulfonyl chloride and butyl acrylate [94].

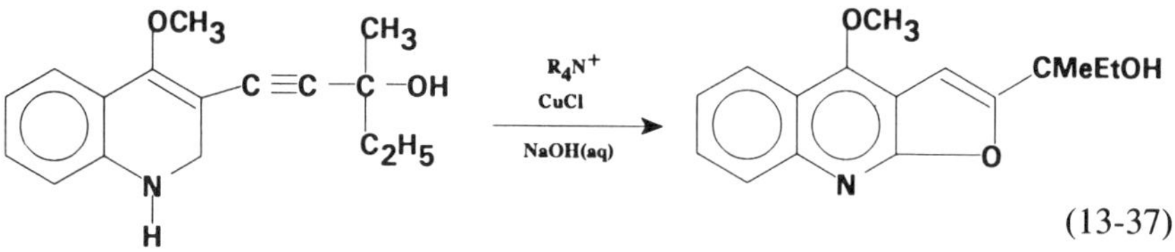

$$\text{(Structure: 4-methoxy-3-(1-hydroxy-1-ethylpropyl-ethynyl)-dihydroquinoline)} \xrightarrow[\text{NaOH(aq)}]{\substack{R_4N^+ \\ CuCl}} \text{(Structure: methoxy-furoquinoline-CMeEtOH)} \tag{13-37}$$

$$CH_2=CHCO_2Bu + \text{(1-naphthyl-SO}_2\text{Cl)} \xrightarrow[\substack{PdCl_2(C_6H_5CN)_2 \\ 140°\ C,\ K_2CO_3}]{(C_8H_{17})_3NCH_2C_6H_5{}^+\ Cl^-} \text{(1-naphthyl-CH=CHCH}_2\text{CO}_2\text{Bu)} \tag{13-38}$$

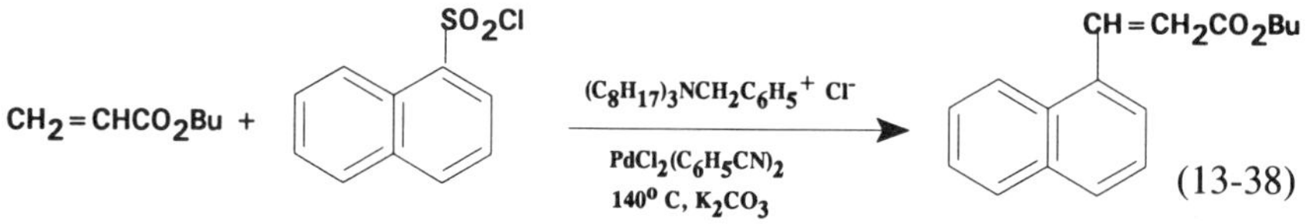

2. Acetylene and Olefin Coupling Reactions

Cyclooligmoerization of alkynes, both terminal and internal, can be accomplished under phase-transfer conditions by a $RhCl_3$-Aliquat 336 catalyst combination [95].

$$CH_3-C\equiv C-CH_3 \xrightarrow[RhCl_3]{(C_8H_{17})_3NCH_3{}^+\ Cl^-} \text{hexamethylbenzene} \tag{13-39}$$

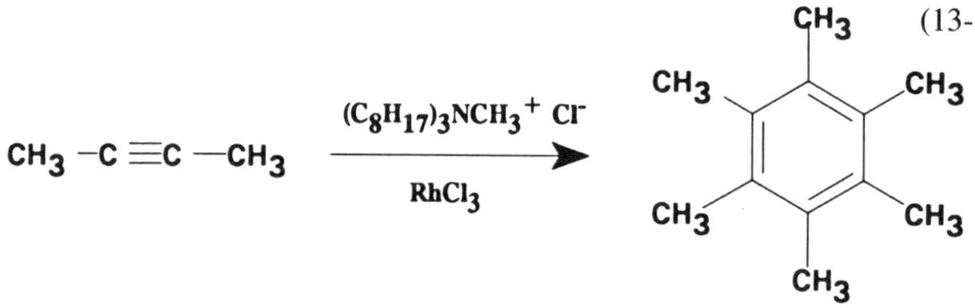

Cyclotrimerization of 1-heptyne followed second-order kinetics, being dependent on acetylene concentration and $RhCl_3$ concentrations. Cyclooligomerization of phenylated diynes by quaternary ammonium complexes with rhodium chloride and with chloroplatinic acid gives a mixture of products, depending on the metal catalyst and the reaction conditions [96].

Table 13-14. PTC–palladium reaction of phenylacetylene with aryl iodides [89].

Ar-I	Time-(h)	Percent Ar-I Conv.	Product	Percent yield
p-$O_2NC_6H_4$-I	1.5	100	p-$O_2NC_6H_4C\equiv CPh$	100
p-NCC_6H_4-I	0.5	84	p-$NCC_6H_4C\equiv CPh$	80
p-ClC_6H_4-I	0.5	—	p-$ClC_6H_4C\equiv CPh$	60
Ph-I	1.5	—	$PhC\equiv CPh$	60
p-$MeOC_6H_4$-I	1.5	37	p-$MeOC_6H_4C\equiv CPh$	40

Table 13-15. PTC–Pd-catalyzed hydrolysis of vinylic dibromides to carboxylic acids [98].

Carbonyl compound	Product	Percent yield (isolated)
C_6H_5CHO	$C_6H_5CH_2CO_2H$	95
$o\text{-}CH_3C_6H_4CHO$	$o\text{-}CH_3C_6H_4CH_2CO_2H$	51
$p\text{-}CH_3C_6H_4CHO$	$p\text{-}CH_3C_6H_4CH_2CO_2H$	68
$p\text{-}ClC_6H_4CHO$	$p\text{-}ClC_6H_4CH_2CO_2H$	35
$p\text{-}CH_{30}C_6H_4CHO$	$p\text{-}CH_3OC_6H_4CH_2CO_2H$	85
$C_6H_5CH=CHCHO$	$C_6H_5CH=CHCH_2CO_2H$	58
$C_6H_5COCH_3$	$C_6H_5CH(CH_3)CO_2H$	23

D. Other Reactions

Many examples of the use of metal compounds, particularly anionic metallates, with quaternary ammonium salts and other phase-transfer catalysts for oxidation reactions are described in some detail in Chapter 10.

Wang and Alper [97] found that pentacarbonylmanganese bromide in the presence of PTC catalyzes the hydrolysis of nitriles to amides in 30–70% yields. For example, treatment of benzonitrile with a stoichiometric amount of the manganese compound in the presence of 0.5 M NaOH and benzyltriethylammonium chloride at 40°C for 32 h gave benzamide in up to 58% yield. Reaction does proceed in the absence of the PTC catalyst, but the yield is only 32%. In the absence of pentacarbonylmanganese bromide the yield of benzamide was only 6%. Similar chemistry by Trogler and Jensen [98], using platinum, palladium, or nickel complexes in conjunction with PTC has been patented for hydration of nitriles to amides and for hydration of olefins to alcohols.

Ethoxylated alcohols and phenols, $RO(CH_2CH_2O)_nH$, used as phase-transfer catalysts in conjunction with copper or copper salts, catalyze the condensation of aminobromoanthraquinone sulfonic acids with anilines in the presence of sodium bicarbonate to yield anthraquinone dyes [99].

Vinylic dibromides, readily prepared by treatment with carbon tetrabromide and triphenyl phosphine, can be hydrolyzed to carboxylic acids in 20–95% yields by treatment with NaOH, catalyzed by PTC-Pd(diphos)$_2$ [100]:

$$\text{R-CO-R'} \xrightarrow[\text{CBr}_4]{\text{Ph}_3\text{P}} \text{R(R')CH=CBr}_2 \xrightarrow[\substack{\text{Pd(diphos)}_2 \\ \text{PEG-400}}]{\text{NaOH}} \text{RCHCO}_2\text{H} \quad (13\text{-}40)$$
$$\overset{\text{R'}}{\underset{|}{}}$$

This route provides a convenient method for a one-carbon addition to a carbonyl group, ending with a carboxyl group. Some representative results are shown in Table 13-15.

References

1. H. des Abbayes, A. Buloup, and G. Tanguy, Organometallics, **2**, 1730 (1983).
2. H. des Abbayes, Israel J. Chem., **26**, 249 (1985).
3. H. des Abbayes, New Journal of Chemistry, **11**, 535 (1987).
4. H. Alper, J. Organomet. Chem., **300**, 1 (1986).
5. S. Vancheesan, J. Sci. Ind. Res., **44**, 75 (1985).
6. H. Liu, Huaxue Tongbao (4), 9 (1985) [CA 103.196133].
7. J.P. Collman and L.S. Hegedus, Principles and Applications of Organo-Transition Metal Chemistry, University Science Book, Mill Valley, CA, 1980.
8. H. Alper, Adv. Chem., Ser., **326**, 8 (1987).
9. H. des Abbayes, Israel J. Chem., **26**, 249 (1985).
10. B.M. Choudary, N. Reddy, and A.B. Prabhakar, Appl. Catal., **32**, 357 (1987) [CA 108.186270].
11. I. Wender and P. Pino, eds., Organic Syntheses via Metal Carbonyls, John Wiley & Sons, (a) Vol. 1, (1986), (b) Vol. 2 (1977).
12. H. des Abbayes and A. Buloup, Tetrahedron Lett., **21**, 4343 (1980).
13. H. Alper, J.K. Currie, and H. des Abbayes, J. Chem. Soc., Chem. Comm., 311 (1978).
14. Z. Jin, Y. Wang, Z. Xia, and H. Zuo, Fenzi Chihua, **2**, 79 (1988) [CA 110.134648].
15. G. Palyi, S.E. Sampar, V. Galamb, J. Palagyi, and L. Marko, Hung. Patent 49,843 (1989) [CA 113.005940].
16. H. des Abbayes and A. Buloup, J. Chem. Soc., Chem. Commun., 1090 (1978).
17. B. Fell, H. Chrobaczek, and W. Kohl, Chem.-Ztg., **109**, 167 (1985) [CA 104.109143].
18. H. des Abbayes, J.-C. Clemént, P. Laurent, G. Tanguy, and N. Thilmont, Organometallics, **7**, 2293 (1988).
19. C. Ungurenasu and C. Cotzur, Polymer Bull., **22**, 151 (1989).
20. S.C. Shim, W.H. Park, C.H. Doh, and H.K. Lee, Bull. Korean chem. Soc., **9**, 61 (1986) [CA 110.007793].
21. S.C. Shim, W.H. Park, C.H. Doh, and J.O. Baeg, Bull. Korean Chem. Soc., **9**, 185 (1988) [CA 110.074976]
22. T. Okano, N. Harada, and J. Kiji, Bull. Chem. Soc., Japan, **65**, 1741 (1992).
23. M.L. Kantam, N.P. Reddy, and B.M. Choudary, Synth. Commun., **20**, 2631-40 (1990).
24. V. Galamb and H. Alper, Transition Met. Chem. (Weinheim, Ger), **8**, 271 (1983).
25. V. Galamb, M. Gopal, and H. Alper, Organometallics, **2**, 801 (1983).
26. H. Alper and H. des Abbayes, J. Organometal. Chem., **134**, C11 (1977).
27. L. Cassar and M. Foa, J. Organomet. Chem., **134**, C15 (1977).

28. S. Gambarotta and H. Alper, J. Organomet. Chem., **212**, C23 (1981).
29. F. Francalanci and M. Foa, J. Organomet. Chem., **232**, 59 (1982).
30. I. Amer and H. Alper, J. Am. Chem. Soc., **111**, 927 (1989).
31. G. Tanguy, B. Weinberger, and H. des Abbayes, Tetrahedron Lett., **24**, 4005 (1983).
32. C.B. Ou and K.L. Liu, J. Chin. Chem. Soc.(Taipei); **32** (1), 23 (1985) [CA 104.050497].
33. L.K. Liu, Y.H. Lin, and B.C. Ou, Bull. Inst. Chem., Acad. Sin., **33**, 41 (1986) [CA 106.119402].
34. H. Alper, K. Hashem, and J. Haeveling, Organometallics, **1**, 775 (1982).
35. S. Zhang, S. Xiao, M. Ran, and H. Dai, Fenzi Cuihua **1**, 115 (1987) [CA 109.075653].
36. S. Zhang, S. Xiao, H. Lai, and J. Fan, Shiyou Huagong, **18**, 263 (1989) [CA 111.216257].
37. J.J. Brunet, C. Sidot, and P. Caubere, J. Org. Chem., **48**, 1166, 1919 (1983).
38. J.J. Brunet, C. Sidot, and P. Caubere, Tetrahedron Lett., **22**, 1013 (1981).
39. I. Amer and H. Alper, J. Org. Chem., **53**, 5147 (1988).
40. J.T. Lee and H. Alper, Organometallics, **9**, 3064 (1990).
41. M. Foa and L. Cassar, Gaz. Chim. Ital., **109**, 619 (1979).
42. F. Joo and H. Alper, Organometallics, **4**, 1775 (1985).
43. M.L. Kantam, N.P. Reddy, and B.M. Choudary, Synth. Commun., **20**, 2631–40 (1990).
44. J. Hagen and P. Bauermann, Chem.-Ztg., **110**, 151 (1986) [CA 107.006745].
45. H. Alper and J.F. Petrignani, J. Chem. Soc., Chem. Comm., 1154 (1983).
46. H. Alper and J.K. Currie, Tetrahedron Lett., **29**, 2665 (1979).
47. H. Alper and D.E. Laycock, Tetrahedron Lett., **22**, 33 (1981).
48. S. Gambarotta and H. Alper, J. Org. Chem., **46**, 2142 (1981).
49. N. Satynarayana and H. Alper, J. Chem., Soc., Chem. Commun., (1) 8–9 (1991).
50. H. Bahrmann, B. Cornils, W. Konkol, and W. Lipps, to Ruhrchemie A.-G., Ger. Offen., DE 3,412,335 (1985) [CA 104.185989].
51. N. Satyanarayana, H. Alper, and I. Amer., Organometallics, **9**, 284 (1990).
52. H. Arzoumanian and J.F. Petriganani, Tetrahedron Lett., **27**, 5979 (1986).
53. J.X. Wang and H. Alper, J. Org. Chem., **51**, 275 (1986).
54. I. Amer and H. Alper, J. Organomet. Chem., **383**, 573 (1990).
55. J.T. Lee and H. Alper, Tetrahedron Lett., **32**, 1769–70 (1991).
56. H. Arzoumanian, F. Cochini, D. Nuel, J.F. Petrignani, and N. Rogag, Organometallics, **11**, 493 (1992).
57. H. Arzoumanian, F. Cochini, D. Nuel, J.F. Petrignani and N. Rogag, Organometallics, **11**, 493 (1992).

58. N. Satyanarayana and H. Alper, Organometallics, **10**, 804-7 (1991).
59. K.E. Hashem, J.-F. Petriganani, and H. Alper, J. Mol. Catal., **26**, 285 (1984).
60. H. Alper, Adv. Organometal. Chem., **19**, 183 (1981).
61. H. Alper and C.P. Mahatantila, Hetrocycles, **20**, 2025 (1983).
62. D. Roberto and H. Alper, Organometallics, **9**, 1245 (1990).
63. S. Calet and H. Alper, Organometallics, **6**, 1625 (1987).
64. J.E. Hallgren and G.M. Lucas, J. Organometal. Chem., **212**, 135 (1981).
65. R. Bar, Y. Sasson, and J. Blum, J. Mol. Catalysis, **16**, 175 (1982).
66. A. Zoran, Y. Sasson, and J. Blum, J. Mol. Catal., **27**, 349 (1984).
67. S. Falicki and H. Alper, Organometallics, **7**, 2548 (1988).
68. S. Shimizu, Y. Sasaki, and C. Hirai, Bull. Chem. Soc., Japan, **63**, 176 (1990).
69. T. Okano, Y. Moriyama, H. Konishi, and J. Kiji, Chem. Lett., 1463 (1986).
70. C. Ungurenasu and C. Cotzur, Polymer Bull., **22**, 151 (1989).
71. J. Azran, O. Buchman, I. Amer, and J. Blum, J. Mol. Catal., **34**, 229 (1986).
72. J. Blum, I. Amer, A. Zoran, and Y. Sasson, Tetrahedron Lett., **24**, 4139 (1983).
73. Y. Sasson, A. Zoran, and J. Blum, J. Mol. Catal., **11**, 293 (1981).
74. C. Larpent, F. Brisse-LeMenn, and H. Patin, New J. Chem., **15**, 351-6 (1991).
75. G.R. Januszklewicz and H. Alper, Organometallics, **2**, 1055 (1983).
76. A. Laufenberg, A. Behr, and W. Keim, to Henkel, Ger. Pat DE 3,841,698 (1988) [CA 112.0977024].
77. K. Jothimony, S. Vancheesan, and J.C. Kuriacose, J. Mol. Catal., **52**, 297 (1989).
78. T.M. Juang, J.C. Hwang, H.O. Ho, and C.Y. Chen, J. Chin. Chem. Soc. (Taipei), **35**, 135 (1988) [CA 110.172490].
79. Yu.B. Khokhraykov, I.A. Avrutskaya, S.S. Kucherov, G.V. Itov, A.P. Tomilov, V.S. Pilyugin, and T.V. Chernyshev, USSR Patent SU 1,565,838 (1900) [CA: 113.233715].
80. K.E. Hashem, J.-F. Petriganani, and H. Alper, J. Mol. Catal., **26**, 285 (1984).
81. T.J. Lynch, M. Banah, H.D. Kaesz, and C.R. Porter, Prepr. Pap.-Am. Chem. Soc., Div. Fuel Chem., **28**, 172 (1983).
82. H. Alper and F. Sibtain, J. Organometallic Chem., **285**, 225 (1985).
83. V.V. Grushin and H. Alper; Organometallics, **10**, 1620 (1991).
84. A. Bader, D. Arlt, and F. Seng, to Bayer A.-G., Eur. Pat Appl. EP 459,258 (1991) [CA:116.083372].
85. T. Jeffery, Tetrahedron Lett., **26**, 2667 (1985).
86. T. Jeffery, Synth. Commun., **18**, 77 (1988).
87. G. Zweifel and C.C. Whitney, J. Am. Chem. Soc., **89**, 2753 (1967).
88. T. Jeffery, Synth. Commun., 70 (1987).

89. H. Alper, J.K. Currie, and H. des Abbayes, J. Chem. Soc., Chem. Commun., 311 (1978).
90. N.A. Bumagin, A.B. Ponomarev, A.N. Ryabtsev, and I.P. Beletskaya, Izvest. Akad. Nauk, Ser. Khim., (3) 604 (1988).
91. A. Carpita, A. Lessi and R. Rossi, Synthesis, 571 (1984).
92. V.V. Grushin and H. Alper, J. Org. Chem. **57**, 2188 (1992).
93. J. Reisch and P. Nordhaus, J. Hetrocycl. Chem. **28**, 167-71 (1992).
94. H. Hashimoto, K. Itoh, and M. Nomura, J. Chem. Soc., Perkin Trans. I, 2207-11 (1990).
95. I. Amer, T. Bernstien, M. Eisen, J. Blum, and K.P.C. Volhardt, J. Mol. Catal., **60**, 313 (1990).
96. Y. Badrieh, J. Blum, I. Amer, and K.P.C. Volhardt, J. Mol. Catal., **66**, 295 (1991).
97. J.-X. Wang and H. Alper, J. Chem. Research, (S) 456 (1986).
98. W.C. Trogler and C.M. Jensen, U.S. Patent 4,684,751 (1987) assigned to Univ. California.
99. Yu. Shlykov, W. Hepp, G. Knoechel, E. Schick, H.J. Riedel, U. Benndorf, and H. Weise, VEB Chemiekombinat Bitterfeld, Ger.(East) DD 226,294 (1985) [CA 105.210411].
100. P. Li and H. Alper, J. Org. Chem., **51**, 4354 (1986).

14

Phase-Transfer Catalysis in Analytical Chemistry

A. The Phase-Transfer Catalysis/Analytical Chemistry Match

In analytical chemistry, analytes must sometimes be reacted with derivatizing reagents to facilitate analysis. Derivatization is needed when the analytes, without modification, cannot be easily separated using conventional chromatography or when elution times are too long. Successful derivatization methods require rapid, convenient, and complete conversion of the analytes to ensure high reproducibility, accuracy, minimum operations, and rapid turnaround. Rapid, complete, and convenient conversions happen to be among the significant advantages that phase-transfer catalysis (PTC) generally offers. In addition, the wide applicability of PTC in the most common organic reactions renders PTC a particularly good fit for analytical chemistry derivatization methods. Indeed, the most common derivatization techniques involve nucleophilic substitutions/additions, such as esterification and etherification. Naturally, PTC is becoming more prominent in these procedures.

B. Esterification, Etherification, and Other Nucleophilic Derivatizations

Pentachlorophenol can be determined (in human and horse serum) by liquid chromatography after derivatization with dansyl chloride, 5-(dimethylamino)-1-naphthalenesulfonyl chloride, in the presence of methylene chloride and tetrabutylammonium bromide [1]. The reaction was complete within 2 min at room temperature. The relative standard deviation of the total procedure for a human serum sample containing 4 ng/mL pentachlorophenol was 4.5%. Dansyl chloride is commonly used as a fluorescent reagent for labeling of proteins and amino acids. In this case fluorescent detection was employed. Dansylation was also used for the detection of phenolic steroids, such as estradiol, in urine samples [2]. Using commercially available automated equipment for precolumn derivatization, a rate of 3–8 analyses per hour was achieved with detection limits of 3–5 ng/mL. Other PTC dansylations for HPLC have been reported [3,4].

PTC was also used for fluorescence labeling of carboxylic acids on a microanalytical scale by esterification using N-(9-acridinyl)-bromoacetamide [5] and N-(1-pyrenyl) bromoacetamide [6]. Fatty acids were derivatized, online, within 5 min at 60°C with the fluorophore 6-bromomethylacridine in the presence of a surfactant and tetrakis-(decyl) ammonium bromide [7,8]. HPLC was used and the detection limit was 300 nM with precision better than 3% using an internal standard. More than 100 samples could be injected on a single precolumn.

Alcohols [9,10] and carboxylic acids [11,12] have been alkylated by PTC in preparation for analytical determination by gas chromatography. Fairly reactive alkylating agents were used such as benzyl chloride, alkyl iodide, and propyl bromide in the presence of symmetrical quaternary ammonium cations such as tetrabutylammonium, tetrapentylammonium, and even tetramethylammonium.

The N-alkylation of sulfonamide diuretics was reported using extractive alkylation [13]. Sixteen sulfonamides could be alkylated and rapidly analyzed after the PTC derivatization, using gas chromatography/mass spectrometry (GC/MS). It was found that tetrahexylammonium allowed the alkylation and identification of all 16 sulfonamide diuretics whereas tetrapentylammonium and tetrabutylammonium successfully alkylated 10 and 3 sulfonamides, respectively. The alkylating agent was methyl iodide and an organophilic quaternary ammonium salt was, therefore, preferable (see guidelines in Chapter 6). The materials analyzed represented a wide range of illegal diuretics taken by olympic athletes and one diuretic masking agent. This test provided a rapid and accurate method for urine screening of these athletes.

A variety of inorganic nucleophilic anions can be determined simultaneously by GC after PTC derivatization [14]. Cyanide, iodide, nitrite, sulfide, and thiocyanate were converted to their volatile pentafluorobenzyl derivatives in the presence of cryptand (Kryptofix 222) and methylene chloride. Flame ionization detection was sufficient to simultaneously determine these ions at the submicromole level. Polymer-bound PTC was used in the derivatization of dialkyl phosphates by pentafluorobenzyl bromide for analysis of pesticide degradation products [15]. Pentafluorobenzyl halide was also used for the PTC derivatization of carboxylic acids and anilines for analysis by capillary GC [16]. Pentafluoropropionic anhydride was used for the two-phase derivatization of di-(4-amino-2-chlorophenyl) methane for GC/MS analysis of urine of workers in polyurethane plants [di(aminophenyl)methane is the large-scale raw material used] [17]. A method for the analysis of ethanol in beer was developed, by using PTC insitu derivatization of the ethanol to ethyl dithiocarbonate [18].

3. Non-Nucleophilic PTC Reactions Used in Analytical Chemistry

The PTC carbene addition was used to develop a method for the identification of the location of double bonds in olefins and fatty acids [19]. Using GC/MS, it was routinely possible to locate a single double bond in a molecule at the microgram level.

Transition metal PTC was used in the HPLC analysis of oxidation products from lignite [20]. Using 1-methylnaphthalene as a model compound, oxidation was performed using Adogen 464 (Sherex) and RuO_2 as the catalysts and $NaIO_4$ as the oxidant to yield phthalic acid and 3-methylphthalic acid. HPLC characterization was performed on the *p*-bromophenacyl esters.

Porphyrin carboxylic acids were determined directly, without derivatization, using HPLC in the presence of phase-transfer catalysts [21]. Separation was simply achieved by associating the carboxylic acids with tetrabutylammonium (($H_2PO_4^{-2}$) as the counteranion) in alcohol/water solvents.

References

1. C. De Ruiter, J. Brinkman, R. Frei, H. Lingman, and P. Van Zoonen, Analyst, **115**, 1033 (1990) [CA 113: 167005u].

2. C. De Ruiter, J. Tsoi, U. Brinkman, and R. Frei, Chromatographia, **26**, 267 (1988) [CA 110: 225604q].

3. J. Halvax, G. Wiese, W. Van Bennekom, and A. Bult, J. Pharma. Biomed. Anal., **10**, 335 (1992) [CA 117: 178411g].

4. Y. Wen, S. Wu, S. Lin, and H. Wu, Zhonghua Yaoxue Zazhi, **44**, 221 (1992) [CA 117: 184038g].

5. S. Allenmark, M. Chelminska-Bertilsson, and R. Thompson, Anal. Biochem., **185**, 279 (1990) [CA 112: 194679c].

6. S. Allenmark and M. Chelminska-Bertilsson, Chromatographia, **28**, 367 (1989) [CA 112: 191042d].

7. F. Van der Horst, M. Post, J. Holthius, and U. Brinkman, J. Chromatogr., **500**, 443 (1990) [CA 112: 154591m].

8. F. Van der Horst, M. Post, J. Holthius, and U. Brinkman, Chromatographia, **28**, 267 (1989) [CA 112: 54466f].

9. H. Brink, Acta Pharm. Seuc., **17**, 233 (1980) [CA 94: 64772b].

10. H. Brink, R. Modin and J. Vessman, Acta Pharm. Seuc., **16**, 247 (1979) [CA 92: 75438g].

11. A. Arbin, H. Brink, and J. Vessman, J. Chromatogr., **196**, 255 (1980) [CA 93: 220084r].

12. A. Arbin, H. Brink, and J. Vessman, J. Chromatogr., **170**, 25 (1979) [CA 90: 174747x].

13. A. Lisi, G. Trout and R. Kazlauskas, J. Chromatogr., **563**, 257 (1991).

14. S. Chen, H. Wu, M. Tanaka, T. Shono, and K. Funazo, J. Chromatogr., **502**, 257 (1990) [CA 113: 17112n].

15. A. Miki, H. Tsuchihashi, K. Ueda, and M. Yamashita, Jpn. J. Toxicol. Environ. Health, **38**, 168 (1992) [CA 117: 84766z].
16. E. Goosens, M. Broekman, M. Wolters, R. Strijker, D. De Jong, G. De Jong, and U. Brinkman, J. High Resolut. Chromatogr., **15**, 242 (1992) [CA 117: 177974f].
17. K. Jedrzejczak and V. Gaind, Analyst, **117**, 1417 (1992) [CA 117:186134j].
18. W.Chan and A. Lee, Analyst, **117**, 1509 (1992) [CA 117:190407t].
19. V. Purghart, C. Hocart, and U. Schlunegger, Rapid Commun. Mass Spectrum., **5**, 596 (1991) [CA 116: 105558e].
20. E. Olson and B. Farnum, Prepr. Pap.-Am. Chem. Soc. Div. Fuel. Chem., **26**, 60 (1981) [CA 97: 147361e].
21. R. Bonnet, A. Charalambides, K. Jones, I. Magnus, and R. Ridge, Biochen. J., **173**, 693 (1978) [CA 90: 83047c].

15

Phase-Transfer Catalysis: Industrial Perspective

A. Industrial Background

In the early 1990s, it was estimated that approximately 500 commercial phase-transfer catalysis (PTC) processes were being performed using at least 25 million pounds per year of catalyst. It was also estimated that sales of products manufactured by processes consisting of at least one major PTC step were at least $10 billion/year (approximately $5 billion in polymers; $3 billion in pharmaceuticals; $2 billion in agricultural chemicals; $1 billion in monomers; and unestimated sales in general chemicals, flavors and fragrances, dyes, surfactants, explosives, and others).

Since its "birth" in the late 1960s, PTC technology experienced rapid growth, especially in the 1980s and continuing into the 1990s (see Figure 15-1 and Figure 15-2). A sampling of patents shows that many of the common industrial reactions have been patented using PTC technology (see Figure 15-3). More than 50 companies, including all of the major global chemical companies, have patented PTC applications.

B. Evaluation of PTC as a Commercial Manufacturing Process Technology

1. General Considerations

Sometime during the development of each new manufacturing process, a decision is made as to which specific process steps and technologies will be implemented. Choosing the specific process technology for a commercial manufacturing process takes into account many factors. Among these are cost and availability of raw materials, utilization of raw materials (yield), cost and availability of unit operations, safety, environmental aspects, waste treatment, recycle, plant capacity,

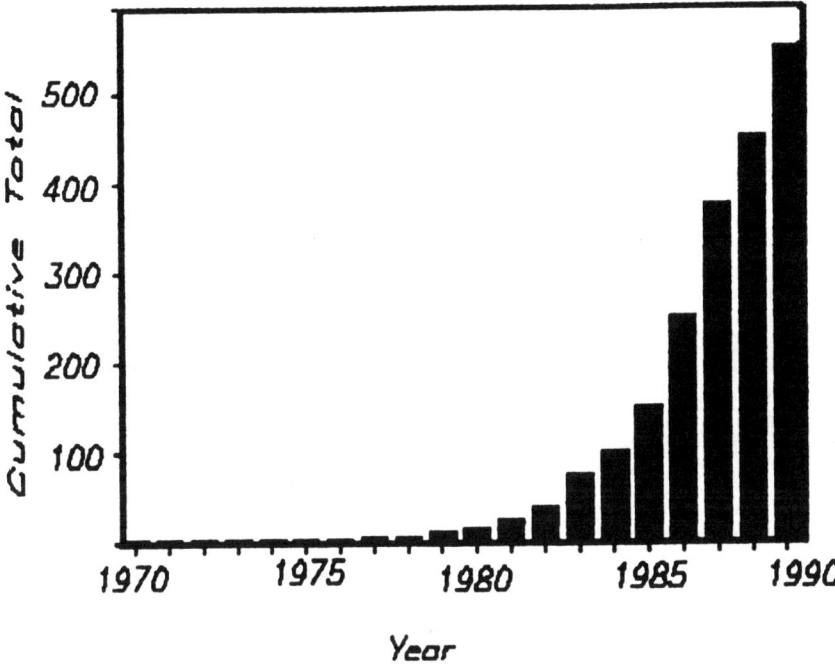

Figure 15-1. Estimated worldwide PTC applications.

flexibility, predictability, understanding of the process, many other plant operability issues, and even psychological factors. This section addresses PTC as a manufacturing process technology as industrial chemists and engineers may see it.

PTC offers many advantages as a process technology, though it also has some drawbacks. The primary reasons that chemists and engineers consider PTC for their processes are yield, environmental aspects, safety, and choice of alternate raw materials. A less cited though important reason for choosing PTC is selectivity in applicable cases. Secondary reasons for choosing PTC include lower energy costs (lower reaction temperature), alleviating the need for special conditions such as reactor dryness and others. The key drawbacks of commercial PTC relate to the catalyst: the need to separate the catalyst (and its decomposition products) from the product, treatment or recycle of recovered catalyst, catalyst decomposition, cost, and toxicity.

2. *Advantages of PTC—Industrial Viewpoint*

a. *High Yield and Increased Reaction Rate*

High yield in industry translates into effective utilization of raw materials from a cost standpoint. High yield and increased reaction rate both contribute to

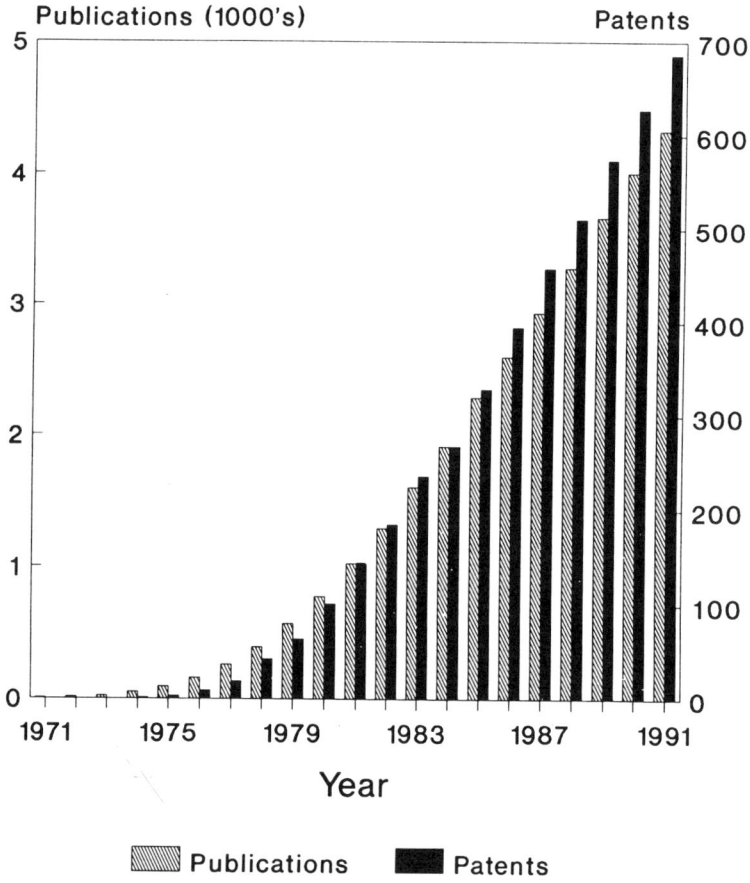

Figure 15-2. PTC literature.

increased plant capacity for a given capital investment. Increasing yield may alleviate the need for plant expansion during market growth. Also, high yield translates into pollution prevention. In fact, the "Non-Product Output" of a process may be significantly reduced from ($100\% - x$), where x is the percent yield, to ($100\% - x + y$), where y is the yield increase using PTC. Often, the ratio ($100 - x$)/($100 - x + y$) is high. Increased reaction rate of the desired reaction also affects selectivity, especially in cases where the desired reaction is phase-transfer-catalyzed and the side reactions are noncatalyzed. Selectivity (i.e., product purity) is important for all applications, but is particularly important in pharmaceuticals.

b. Mild Reaction Conditions

Mild reaction conditions are usually the result of anion activation due the solubility of the anion in the organic reaction phase and its reduced state of

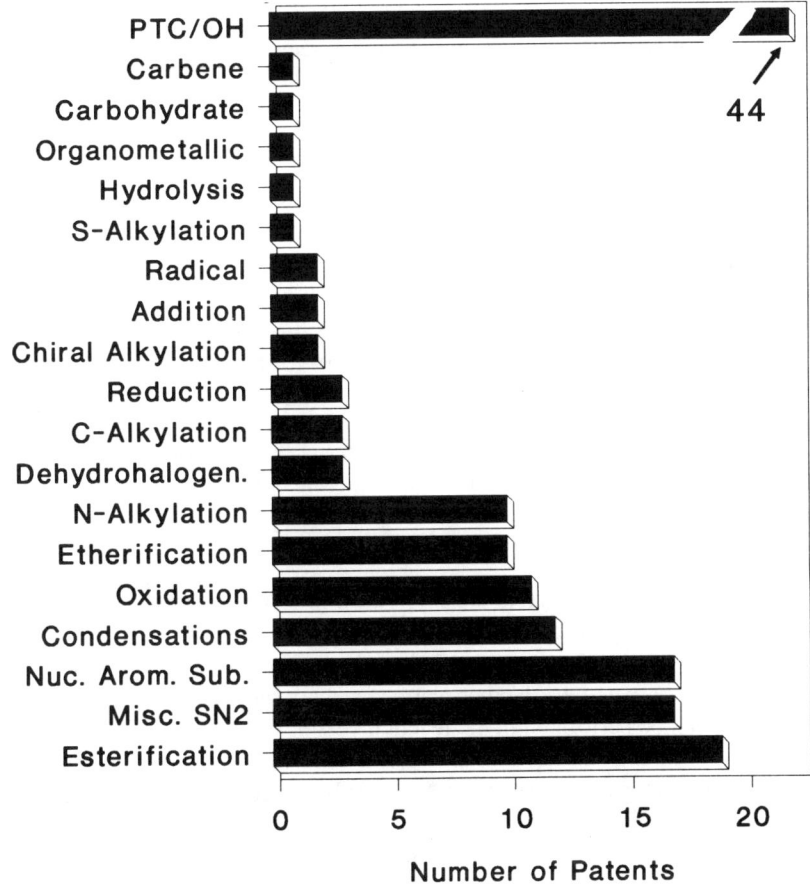

Figure 15-3. PTC patents by reaction type.

hydration. The key benefit of mild reaction conditions is safety and the controlled thermal properties of the reaction. Mild reaction conditions usually also result in greater reliability and reproducibility of a process. Lower utilities cost is another obvious ramification of mild reaction conditions.

c. Opportunities: Choice of Solvent

The flexibility in choice of solvent is a key advantage of PTC. Due to the nature of PTC systems, solvent can be chosen based on desirable safety, health, environmental, cost, and handling considerations. For example, solvent may be chosen based on containment, flammability, toxicity, recoverability, cost, or boiling point (for separation from product). The reduction or elimination of solvent provides less environmental problems, larger batch size for a given reactor size, reduced energy consumption (no solvent recovery) and debottlenecking of solvent purification facilities.

d. Opportunities: Alternate Raw Materials

Developing new processes using existing PTC technology includes the use of alternate raw materials. Among the major advantageous alternate raw materials are inorganic bases; solid–liquid PTC conditions for enhanced selectivity and reactivity; and the use of oxygen, hypochlorite, and hydrogen peroxide for oxidations. The most widely used alternate raw material is NaOH (and K_2CO_3). PTC using strong base, as compared to classic organic bases, usually offers greater safety, lower cost of the base, a more acceptable inorganic waste stream, easier handling (nonflammable, nonanhydrous, and flowable liquid), and better availability. An additional alternate raw material that can sometimes be used is aqueous HCl instead of anhydrous HCl (which is used extensively in industry).

e. Opportunities: Simplicity and Versatility of Process and Equipment

Simple stirred tank reactors are often used in PTC reactions. These reactors allow easy scale-up of laboratory runs (compared to continuous systems); heat evolution is easy to control with stirring speed; one can use excess reagent in one (reused) phase with fresh other-phase reagent in subsequent batches; washing/extraction can be performed in the reaction vessel; sequential PTC reactions can be run by draining the aqueous phase and replacing the aqueous phase for a second reaction (example: benzyl chloride to phenylacetonitrile to 2-phenylbutyronitrile, by cyanide displacement then alkylation); and equilibrium-controlled reactions may be easy to drive to completion by removal/replacement of one phase.

The nature of the two-phase reactions not requiring strictly anhydrous conditions is an advantage in itself and additionally allows easy turnover of the reactor to another process in a multipurpose facility. In such a facility, PTC also allows the easy manufacture of development quantities of products and use toll manufacture. PTC often provides a path to debottleneck existing facilities.

The two-phase nature of PTC reactions provides the ability to conduct sequential PTC reactions by simple removal of the aqueous or solid phase and replacing it with another reagent to conduct the second reaction.

Sometimes PTC can be used in continuous reactions and in the use of countercurrent extraction equipment.

f. Human Aspects of Industrial Process Development

An often overlooked aspect of new process development is the human factor. In order for a process technology to be implemented it first must be demonstrated by a process chemist in the lab. In order for the process chemist to try the technology, he/she must be familiar and comfortable with the technology and be able to easily evaluate the option within the time management constraints of industry. PTC reactions are generally easy to set up in the lab and analyze, lowering the barrier to evaluation. PTC is a "culturally desirable" technology in many companies (especially the larger companies) due to the advantages cited above. This is a particularly important issue when the process chemist needs to "sell the idea" to management and engineering.

It is actually worthwhile for process chemists to keep up to date with PTC technology since PTC is applicable to such a wide range of common reactions (J. March's 3rd edition of Advanced Organic Chemistry has 45 references to PTC syntheses). In addition, PTC information is readily available to process chemists. Chemical Abstracts indexes phase-transfer catalysis and publishes CA Selects for phase-transfer catalysis. The common journals read by process chemists (J. Org. Chem., Synthesis, Tetrahedron Lett., etc.) carry most of the PTC synthetic and mechanistic publications and index them under "phase-transfer catalysis" in the yearly index. Courses for industrial chemists are available (such as "Phase-Transfer Catalysis in Industry" presented by the authors of this book) which provide in-depth training on the fundamentals, applications, and guidelines for evaluating and optimizing new potential applications of phase-transfer catalysis in an industrial setting. Finally, even though the actual optimization of PTC processes is not trivial, the act of learning the basic Starks extraction mechanism is simple, thereby reducing the barrier to introduction to PTC. In summary, the more comfortable the process chemist feels with a technology and the more the technology provides real advantage, the more the technology will have the opportunity to be properly evaluated.

3. Limitations of PTC and Barriers to Commercialization

The limitations of PTC in industrial processes and the barriers to evaluating and implementing PTC in commercial processes relate to the catalyst, the human aspect of the development effort, and other noncatalyst process issues.

a. Catalyst

The key component of any PTC system is the catalyst. Therefore, the major opportunities and pitfalls of PTC processes may be associated with the catalyst.

Catalyst Separation

Once a candidate PTC reaction is shown to work, the separation of the catalyst from the product (or reaction matrix) is essential in order to render it a process. As shown in Chapter 6, catalyst separation is generally performed by extraction or distillation. This usually adds an extra unit operation to the process. Depending on the nature of the catalyst and reaction matrix, the separation may not be easy to develop or implement (and may not be considered as interesting to the process chemist as to the process engineer). Often, the catalyst can be chosen to facilitate separation from the product, even if reactivity is not optimal with that catalyst. For example, it is often advantageous to attenuate reactivity by adding more of a catalyst that is more easily recoverable and is less expensive than having less catalyst that is difficult to recover and more expensive. This is one reason that catalysts such as tetrabutylammonium bromide, methyltributylammonium chloride, and polyethylene glycols are desirable catalysts. Catalyst separation is greatly simplified by the use of insoluble phase-bound catalysts, but these catalysts often suffer from reactivity deficiencies (see Chapter 5).

Catalyst Recycle

Once the catalyst is separated from the product it must be recycled and/or waste treated. This may require another unit operation. Catalyst recycle is often integrated with the unit operation of catalyst separation (see, for example, the separation/recycle of tetrabutylammonium bromide in Chapter 6). This is particularly beneficial when the catalyst separates as the third phase upon settling of the reaction mixture or as a second phase during catalyst recovery. Occasionally, it is necessary to periodically purge the system from recycled phase-transfer catalyst due to the accumulation of anionic byproducts that may poison the catalyst.

Catalyst Waste Treatment

The Material Safety Data Sheets usually recommend waste treatment by incineration. In some cases, the catalysts may be destroyed in bioponds. In some cases, special dedicated bioponds are used for catalyst destruction, since some catalysts (especially benzyl-containing quaternary ammonium cations) interfere with the microorganisms required to treat the other waste generated by the plant. Catalysts may also be treated by oxidative chlorination, but this will usually generate chlorinated hydrocarbons which are subject to emissions regulations. [The information provided here does not constitute recommendations for operation of a plant or waste treatment. Any evaluation and/or implementation of an operation or waste treatment must be fully and comprehensively evaluated and must comply with good manufacturing procedure, safety, and environmental responsibility and must comply with federal, state, local, and any other governmental or legal regulations.]

Catalyst Decomposition

Catalyst decomposition is a process issue that almost always needs to be dealt with, especially with quaternary onium salts. Since the most common PTC systems involve hydroxide and quaternary ammonium salts, Hofmann decomposition is often encountered. The ramifications of catalyst decomposition include: reduced reaction rates, sometimes leading to the inability to complete the reaction; miscalculation of plant capacity and other engineering parameters when catalyst decomposition is not accounted for in the reaction kinetics; separation of the catalyst decomposition products from the desired reaction products and overall reaction matrix; and the extent and ability to recycle the catalyst. When the catalyst is separated from the product by distillation, further decomposition may occur and the catalyst byproducts may contaminate the product. It is sometimes possible to choose the catalyst so that the catalyst decomposition products do not distill with the product fraction (for example, one can design a quaternary ammonium cation with appropriate alkyl chains). For more information on catalyst decomposition see Chapter 6.

Catalyst Toxicity

The Material Safety Data Sheet of each catalyst and material must be consulted before use. The handling and additional storage of the catalyst or its residual presence in the product may constitute another drawback in the use of PTC. Toxicities of catalyst range from highly toxic to innocuous (see extensive data in Chapter 6). Several polyethylene glycols have an LD_{50}'s of $>$ 10,000 mg/kg (oral-rat; check Material Safety Data Sheet of each material before use). Most of the commercial quaternary ammonium salts have LD_{50}'s in range of 100–1000 mg/kg (oral-rat; check Material Safety Data Sheet of each material before use).

Catalyst Cost and Availability

Catalyst cost and/or availability may also be a consideration in choosing the catalyst or even in determining the feasibility of the entire PTC process. As shown in Chapter 6, catalyst cost, not including recycle, may range from less than $1/lb to several $100/lb. Most polyethylene glycols and quaternary ammonium salts are and will continue to be readily available since they are used in may other non-PTC applications. Other compounds such as specialized crown ethers and cryptands are expensive and may be difficult to obtain in large quantities. The technology exists for the semi-large-scale preparation of simple crown ethers and their availability and price fluctuate with demand.

b. *Human Aspects of Industrial Process Development*

Just as familiarity with PTC technology is a major advantage of PTC, lack of familiarity is a major disadvantage (although if you are reading this, you are probably not suffering from lack of familiarity with PTC). Specialized knowledge of PTC chemistry, such as that in this book, is needed to develop the best process.

634 / Phase-Transfer Catalysis

One or two exploratory experiments may give negative results and discourage further efforts. This is a particularly potent pitfall when choosing catalyst. As seen in Chapter 6, strong base PTC reactions may have optimal reactivity with either small accessible quaternary ammonium salts or large organophilic quaternary ammonium salts, exclusively. This may lead to statements of the type "I once tried A phase-transfer catalyst and THEY don't work." This is usually easily overcome by the knowledge of the principles of PTC. It should be noted that many of the commercial manufacturers of organic chemicals (e.g., pharmaceuticals, agricultural chemicals, etc.) have much valuable empirical experience with PTC. We recommend caution in being complacent with the empirical PTC knowledge only, since opportunities may be missed by not being up to date on the recent fundamental and application developments of PTC. Some companies ensure keeping up to date by bringing in consultants and in-house courses such as "Phase-Transfer Catalysis in Industry."

Sometimes phase-transfer catalysts are not readily on hand in departments where PTC is not widely practiced. This may be a barrier to trying it out for the first time, and is a particular problem if the reaction requires a special catalyst. Some applications are not considered as PTC candidates for fear of water being a major detriment. Many surprising applications are missed because of the fear of hydrolysis such as carbene applications and reactions requiring the use of acyl or sulfonyl chlorides or dimethyl sulfate. For example, phenols may be removed from large aqueous waste streams in over 99% by the extraction of phenol into an organic phase and esterification with 1.0 (!) equivalents of benzoyl chloride [1]. In this case the benzoyl chloride is "protected" from hydrolysis by its solubility in the organic phase and the immiscibility of the organic and basic aqueous phases. Many other such examples exist.

The development of a PTC process is often more complicated than conventional processes, usually due to the catalyst factors cited above. Anticipation of added effort may constitute a barrier to beginning to develop the PTC process. Finally, clear incentive for the development of a PTC process may not be readily apparent to the inexperienced chemist. Again, the human barriers to evaluating PTC as a process option can usually be overcome by knowledge of the technology or obtaining the knowledge from sources such as this book.

c. Noncatalyst Process Issues

In general new process development (not necessarily PTC), the two most difficult process parameters to scale up from lab to plant are usually agitation and heat transfer. As seen above, both agitation and heat transfer play particularly significant roles in multiphase systems in general and in PTC systems in particular. Additional consideration must be given to handling solids in solid–liquid PTC systems (which is often required for selectivity and/or reactivity). As seen in Chapter 3, the kinetics of PTC are usually complicated and often misleadingly simple (e.g., assuming integral order kinetics when it is not the case). Since reactor design should be based on kinetic considerations, complicated kinetics and improper kinetic conclusions can result in improper reactor design and misestimation of heat transfer requirements and nameplate plant capacity. The approach to addressing scalability of agitation and heat transfer in PTC should be the same as for general process development, usually preliminary calculations and design, pilot plant–lab iterations, and plant "EvOp."

4. Identifying Future Opportunities for Making Economic Impact Using PTC

Large and small companies alike can derive significant economic benefit from using PTC. The implementation of PTC in industry takes many forms: improvement of existing non-PTC processes by retrofitting with PTC; improvement of current PTC processes by optimization through better understanding; incorporating commonly known PTC technology into new processes being developed which include common organic and polymer reactions; and finally, developing new PTC technology. Following are some process characteristics that should signal to chemists and engineers that PTC technology is likely to be advantageously applied.

a. Improvement of Existing Non-PTC or PTC Manufacturing Processes

Following are some process characteristic triggers for considering PTC alternatives for improving existing non-PTC manufacturing processes:

- Reaction of anions with organic substrates
  Use of an alkoxide, amide, or hydride as a base
  Nucleophilic displacements
- Use of DMSO, DMF, NMP, or acetonitrile as a solvent
- Erratic process behavior (exothermicity, yield, purity)
- Environmental, emissions, emulsion, or separation problems

The guidelines presented in Chapter 6 should also be considered for improving existing PTC processes. Cases are known where PTC processes have been improved several years after commercialization. Improvements of existing PTC processes usually deal with catalyst recycle, choice of catalyst, choice or elimination of solvent, and yield/selectivity improvement. The criteria for performing solvent-free PTC are described in Chapter 6.

b. Incorporating Known PTC Technology into New Process/Product Development

Often, in pharmaceutical and agricultural chemical process development, some of the "mundane" reaction steps of a multistep reaction sequence are the common esterification, alkylation, or other standard reactions. These reactions are the ones in which PTC excels in relation to yield, selectivity, purity, choice of solvent, separation, etc. PTC should be part of the process option evaluation program particularly if the reactions include:

- Nucleophilic displacements
  Esterification
  Etherification
  C-, N-, or S-alkylation
  Using inorganic nucleophiles such as cyanide, azide, halide, etc.
- Oxidation
  Using oxygen, hypochlorite, permanganate, or dichromate
- Dehydrohalogenation

It is worthwhile to evaluate PTC technology for these reactions alongside the "classical" synthetic procedures.

c. Developing New PTC Technology

In contrast to using known PTC technology to improve existing processes or developing new processes, developing totally new PTC technology usually requires significant resources. Examples of *new* PTC technologies that were developed with significant investment are the Merck chiral alkylation/addition reactions using optically active catalysts [2], the high-temperature General Electric catalysts [3], and TDA-1 complexing catalyst by Rhone Poulenc [4].

Fertile areas for development include the coupling of engineering or waste treatment technology with PTC. Examples of this are the membrane reactor systems developed by Sepracor [5,6] and PTC grafted on a capsule membrane [7]; the fixed-bed systems using ion-exchange resins by Ragaini [8]; the segmented flow assembly by Tomlinson [9]; the use of cascade reactor trains with mid-stream recycle loops for the large-scale manufacture of chloroprene by DuPont [10]; the PTC treatment of wastewater from the PTC manufacture of hexafluoropropylene oxide (with recycle) [11]; the combination of clays with PTC [12,13]; and the combination of organometallic chemistry [14]; electrochemistry, and photochemistry with PTC.

Other applications of PTC with industrially interesting aspects include the unique ability to control condensation polymerization (e.g., high molecular weight and extremely high selectivity for electrophilic end groups for the preparation of block copolymers see Chapter 9); the ability to chemically modify the surface of polymeric materials (see Chapter 9); the removal of ppm levels of phenols from waste streams by esterification using reactor-settlers [15,16]; the destruction of PCBs from transformer oils by Brunelle [17]; the removal of organic compounds from drinking water by PTC oxidation [18]; the preparation of aziridines with enhanced safety [19]; the PTC synthesis of materials used for ore separation [20]; the ability to sometimes alkylate aldehydes with little aldol condensation; and the ability to simulate high dilution in the preparation of macrocycles. A review of industrial perspectives prior to 1986 may be found in reference [21].

5. Conclusion

PTC has an outstanding track record in profitable and environmentally acceptable commercial manufacturing processes for a very wide range of products and industries. Because of the many advantages PTC has to offer for so many common applications, it is worthwhile for companies to encourage their process chemists to be familiar with PTC technology. Companies should invest the effort to properly evaluate PTC for reactions that involve the reaction of anions with organic substrates and other selected reactions. The companies should properly weigh the potential advantages and disadvantages of each individual PTC process based on the chemical and processing criteria delineated above. Even the companies that successfully employ PTC in commercial processes should not be complacent with the empirical knowledge of PTC, since the fundamentals have major impact on the optimization and evaluation of new potential applications. It is anticipated that the fundamentals will continue to evolve to provide enhanced insights into the optimization and evaluation of new potential PTC applications. The general and commercial PTC knowledge is in the rapid growth stage. The state of the art of PTC is anticipated to continue to grow much beyond what is known at the time of writing of this book, as the creativity of chemists and engineers finds new applications of PTC, integrates other technologies with PTC, applies the known fundamentals of PTC, and expands the fundamentals of PTC.

References

1. V. Krishnakumar and M. Sharma, Ind. Eng. Chem. Proc. Des. Dev., 23, 410 (1984).
2. U. Dolling, E. Grabowski, and S. Pines (Merck, Sharpe and Dohme), U.S. Patent 4,605,761 (1986).
3. D. Brunelle, Stable Catalysts for Phase-Transfer at Elevated Temperatures, in "Phase-Transfer Catalysis, ACS Symposium 326, C. Starks, ed., Washington D.C., pp. 38–53 (1987).
4. G. Soula, et al. (Rhone-Poulenc), Eur. Patents 76,718 (1981); 60,171 (1981); 43,303 (1980); 21,868 (1980) 22,387 (1979); 21,927 (1979); 16,673 (1979).
5. S. Matson and T. Stanley, Sepracor, U.S. Patent Appl. 938,230 (1986).
6. T. Stanley and J. Quinn, Chem. Eng. Sci., 42, 2313 (1987).
7. Y. Okahat and K. Ariga, J. Org. Chem., 51, 5064 (1986).

8. V. Ragaini, G. Verzella, A. Ghignone, and G. Colombo, Ind. Eng. Chem. Process Des. Dev., **25**, 878 (1986).
9. J. Kinkel and E. Tomlinson, Sep. Sci. Tech., **18**, 857 (1983).
10. L. Maurin, DuPont, U.S. Patent 4,418,232 (1983).
11. M. Ikeda, Y. Suzuki, and A. Aoshima (Asahi Chemical), JP 03,154,683 (1991) [CA 115: 165683q].
12. T. Pinnavaia and C. Lin, Michigan State University, U.S. Patent 5,099,054 (1992).
13. A. Sarkar and B. Rao, Tetrahedron Lett., **32**, 1247 (1991).
14. H. Alper, Adv. Organomet. Chem., **19**, 183 (1981).
15. V. Krishnakumar and M. Sharma, Ind. Eng. Chem. Proc. Des. Dev., **23**, 410 (1984).
16. N. Dutta, S. Borthakur, and G. Patil, Sep. Sci. Technol., **27**, 1435 (1992).
17. D. Brunelle, General Electric, U.S. Patent 4,410,422 (1983).
18. Environ. Sci. Technol., **23**, 1030 (Sept 1989).
19. R. Kostyanovski, G. Kadorkina, and A. Mkhitaryan, USSR Patent SU 1,668,359 (1989) [CA 116:128640q].
20. E. Saphores, A. Gleisner, J. Vega, and W. Mardones, U.S. Patent 5,011,965 (1991).
21. H. Freedman, Pure Appl. Chem., **58**, 857 (1986).

Index

Accessible PTC catalysts, 125, 128, 275, 280
   charge density view of, 267
   effect on PTC reactions, 274
   on thiophenolate displacements, 272
   Q# as a quantitative description of, 277
Acetals, permanganate oxidation of, 505
Acetanilide, N-alkylation of, 405
Acetate displacement reactions, 355
   displacement on poly(vinyl chloride), 486
Acetone, condensation with aldehydes, 432
Acetonitrile, C-alkylation of, 397
Acetophenone, C-alkylation of, 390
Acetoxylation, aromatics by electrochemical oxidation, 552
Acetylacetone
   alkylation of, 385
   vapor phase alkylation of, 390
Acetylenes
   carbonylation of, 602
   oxidation with hydrogen peroxide, 527
   PTC dehydrohalogenation, 423
   PTC Michael additions to, 430
   trimerization of to aromatics, 616
   oxidation with permanganate, 505
Acid chlorides
   azide reactions with, 359
   borohydride reduction of, 567
   hydrogenolysis to aldehydes, 609
   magnesium peroxide reaction, 369
   thiocyanate reaction, 366
   sulfide displacement, 364
Acid strength, effect on air oxidations, 535
Acidity of anions, effect on PTC, 328–332
Acidity of C–H bonds in alkylations, 384
Acidity effect on deuterium exchange, 440
Acids, transfer from aqueous phase, 183
Acrylic acid, polymerization to polyesters, 454
Activated carbon, catalyst separation by, 296
Active hydrogen compounds, oxidation of, 535
Active site density, triphase catalysts, 225
Active site locus in triphase catalysts, 218
Active sites of insoluble PTC catalysts
   kind of PTC groups attached, 232
   crown ether and cryptand groups, 235
Activity of insoluble catalysts
   active site density effect on, 225
   agitation effect on, 222
   crosslinking effect on, 227
   effect of support composition on, 228
   particle size effect on, 223
Activity of three-liquid phase systems, 255

Acyl azides from acid chlorides, 359
Acyl cyanides from acid chlorides, 346
Adenine, N-alkylation of, 403
Adsorption, for PTC catalyst separation, 295
Advantages for PTC in industry, 627
Aggregation of cation–anion pairs, 87
Aggregation of anions, kinetic effect, 79
Agitation in PTC reactions, 14, 102–104
  choice of, by PTC rate matrix, 322
  effect on displacement reactions, 340
  variable in PTC reactions, 319
  effect on anion transfer rates, 321
  effect on side reactions, 321
  with insoluble PTC catalysts, 222
  in C-alkylation, 320
  in cyanide displacement reaction, 320
  ultrasonic, effect on products, 537
Air oxidation using PTC, 534
  of diphenylmethanes to ketones, 334
  free radical type, PTC in, 538
  olefins to ketones, 540
Alcohols
  derivatization for analytical use, 626
  from alkyl halides, 356
  alkylation of sugars, co-catalyst, 178
  borohydride reduction co-catalyst, 565
  co-catalysts for hydroxide transfer, 37–40, 177
  cocatalysts for hydroxide reactions, 326
  dehydrohalogenation cocatalyst, 178, 422
  dihalocarbene generation cocatalyst, 426
  effect on PTC polymerization, 139
  HCl or HBr reactions with, 183
  hydroxide ion displacement on RX, 368
  ketones by lithium aluminum hydride, 568
  oxidation, bromate and Ru cocatalyst, 549
  oxidation with carbon tetrachloride, 548
  oxidation, chromate and dichromate, 546
  oxidation with hydrogen peroxide, 529
  oxidation with hypochlorite, 508
  oxidation with permanganate, 505
  oxidation with persulfate, 543
  O-alkylation of, 410
  periodate oxidation, Ru co-catalyst, 550
  derivatives for analytical use, 626
  by reduction of aldehydes, 568
  by reduction of carboxylate salts, 568
Aldehydes
  by alcohol oxidation, 505
  aldol condensation of, 431
  alkylation of, 391
  borohydride reductions of, 565
  by carbonylation of olefins, 601
  by chromate oxidations, 546
  in Darzens condensation reaction, 435
  by decarboxylation of halo-acids, 549
  by diol cleavage with periodate, 549
  by hydrogenolysis of acid chlorides, 609
  from olefins with hydrogen peroxide, 526
  by olefin oxidation with periodate, 549
  by oxidation of alcohols, 529
  by oxidation of olefins, 503
  by oxidations with hypochlorite, 508
  by oxidations with permanganate, 505
  by oxidations with persulfate, 541
  by periodate oxidation, Ru co-catalyst, 550
  by persulfate oxidation of alcohols, 543
  reaction, dichlorocarbene, ammonia, 429
  reduction of, with sodium formate, 568
  in Wittig reaction, 433
  Wolff–Kishner reduction of, 572
Aldimines, PTC alkylation of, 393
Aldol condensation reactions, 431
Alkanes, from alkyl halides, 568
Alkoxides, by alcohols with hydroxide, 37–40
Alkyl structure, effect on displacement, 340
Alkyl halides
  borohydride reduction to alkanes, 567
  carbonylation of, 595

coupling with olefins and alkynes, 613
hydrogenolysis of, 605
olefin carbonylation to ketones, 601
oxidation with chromate, dichromate, 546
reduction, lithium aluminum hydride, 568
Alkyl phosphines from alkylphosphinates, 568
Alkyl silanes from alkylsilicon halides, 568
Alkyl silicon halides, reduction of, 568
Alkylation reactions with PTC
  of acetophenone, 390
  of acetylacetone in vapor phase, 390
  acidic hydrocarbons, 399
  acidity effect on, 328–332
  of aldehydes, 391
  catalyst comparison for, 236
  chiral alkylations, 577, 583, 586
  correlation by Q# of catalyst, 282
  C–C chain polymers by, 474
  C- vs O-alkylation, 139, 385, 387
  of diethyl malonate in vapor phase, 390
  of esters and carboxylic acids, 391
  of ethyl acetoacetate, vapor phase, 390
  hydroxide ion concentration effect, 106
  of imines, amino acid synthesis, 392
  insoluble PTC catalysts for, 397
  mono- vs di-alkylation, 397
  of nitriles, 395
  of phenylacetonitriles, 99
  of polymer-bound phenylacetonitrile, 493
  polysulfoxides, selective catalysts, 173
  PTC variables involved in, 277
  Q# use to correlate yields, 282
  rate correlation by "Q#", 281
  solvent-effect on catalyst choice, 284
  of sulfones, 397
C- vs O-alkylation in PTC reactions, 385
  in alkylation of 2-pyridone, 405
  catalyst structure effect on, 387
  in phenol alkylations, 416
  solvent effect on, 387
  in vapor-phase PTC reactions, 390

C-Alkylation using PTC, 383–388
  C–H acidity and, 384
  chiral products from, 576–591
  choice of bases for, 325
  of ketones, 384
  polyethylene glycols PTC catalysts, 168
N-Alkylation using PTC, 400–410
  of amides, 405, 408
  choice of bases for, 325
  of heterocycles, 400
  polyethylene glycols catalysts for, 166
  of sulfonamides, 407
O-Alkylation using PTC, 410
  choice of bases for, 325
  polyethylene glycols PTC catalysts, 417
  of sugars, alcohols as co-catalysts, 178
S-Alkylation, 418
  thiols and dialkyl sulfides from, 362
S- vs O-Alkylation in PTC reactions, 419
Alkynes
  bromination of, 256
  carbonylation of, 602
  carbonylation of vinyl bromides, 600
  coupling olefins and alkyl halides, 613
  oxidation with hydrogen peroxide, 527
  oxidation with permanganate, 505
  trimerization of, to aromatics, 616
Alkylurethanes by cyanate displacements, 371
Allylanisole isomerization, three-phase, 257
Allylbenzene, deuterium exchange of, 439
Allylbenzene, isomerization of, 439
Alternating block copolymers by PTC, 459
Alumina
  as co-catalyst for displacements, 363
  as PTC catalyst, 249
  as support for PTC catalysts, 250
Aluminum chloride
  as PTC co-catalyst, 179
  transfer to organic phases, 186
Ambident ions in PTC reactions, 389
  alkylation of desoxybenzoin, 385
  beta-diketones as, 388

catalyst structure effect on, 139, 387
effect of alkylating agent on, 387
hydration effect on, 386
phenoxide and napthoxide as, 416
PTC alkylation of 2-pyridone, 405
vapor phase reaction effects, 390
water effect on, 386
Amides
N-alkylation of, 401, 405
oxidation by hypochlorite, 512
periodate oxidation, Ru co-catalyst, 550
reduction of, 568
Amines
by borohydride reduction of azides, 566
oxidation of, by hypochlorite, 512
oxidation of, by permanganate, 506
oxidation with hydrogen peroxide, 529
as PTC catalysts, 140
by reduction of nitroaromatics, 612
by reduction of dialkylamides, 568
Amino acid polymers as PTC catalysts, 243
Amino acids
chiral, by PTC alkylations, 587
dichlorocarbene, aldehydes, ammonia, 429
synthesis of, general, 392
4-Aminopyridinium salts
high-stability PTC catalysts, 142, 291
PTC for divalent anions, 148
separation of, by extraction, 293
Ammonia transfer to organic phases, 188
Ammonium or phosphonium groups on resin, 220
Analysis for trace anions using PTC, 351, 622
Analysis of PTC reaction kinetics, 51–71
Anhydrous solid-liquid PTC reactions, 110
Anilines
air oxidation of in basic PTC system, 535
halogenation with HX + hydrogen per, 532

hydrogenation of nitrobenzene, 571
N-alkylation of, 408
oxidation with hydrogen peroxide, 529
Anion activation by quaternary salts, 128
catalyst structure effect on, 11
in polyethylene glycol with salts, 162
Anion poisoning of PTC reactions, 134
Anion reaction on chloromethyl-polymers, 495
Anion valence effect on transfer of, 26
Anion-activating catalyst types, 125
Anion–cation distance, solvent polarity, 267
Anion–radical coupling of phenols, 477
Anionic polymerization by PTC, 479
Anions, analysis of traces of, PTC in, 626
Anions, associated with quaternary salt, 135
Anions, character of, displacements, 343
Anions, concentration effect on PTC, 42
Anions, hydration effect on reactivity, 340
Anions, lipophilicity of, 324
Anions, nucleophilicity effect of, 340
Anions, structure and aggregation of, 79
Anions, transfer of to organic phases, 15, 23, 322
charge density view of, 269
effect of anion hydration level, 40–46
effect of nature of anion on, 310
effect of stirring on, 102
effect on PTC displacement reaction, 340
phase transfer catalyst role in, 2, 5
with polyethylene glycol, 161
solvent effect on, 306, 307
Applications of PTC in industry, 626–630
variables in design of, 266
Aqueous phase volume effect, 246
Aromatics compounds
acetoxylation, cyanation of, 552
hydrogenation to saturated compound, 611
hydroxylation by hydrogen peroxide, 528

Aromatic hydrocarbons
　from aryl halides by reduction, 568
　halogenation by HX + hydrogen peroxide, 532
　oxidation with hydrogen peroxide, 527
　oxidation with permanganate, 506
Aromatic ketones, air oxidation of, 535
Aromatic methyl groups, oxidation of, 541
Aromatic with activated halides
　etherification of, 412
　fluoride exchange with, 353
　polychlorinated biphenyls reaction, 419
　substitution on aryl chlorides, 352
　S-alkylation with thiols, 418
Arsonium quaternary salts
　as phase transfer catalysts, 140
　polymer-bound for oxidations with H, 533
Aryl halides
　carbonylation of, 595
　coupling with olefins and alkynes, 614
　hydrogenolysis of, 605
　reduction of, 568
Azapropellane-based salts as PTC, 146
Azide displacement reaction, 358
　on acid chlorides, 359
　on alkyl halides, 359
　on poly(vinyl chloride), 359, 484
　on sulfonyl chlorides, 359
Azides, alkyl, reduction to amines, 566
Aziridines, carbonylation of, 604
Azobenzenes, carbonylation of, 604
Azo-nitriles, from of amino-nitriles, 512

Barium hydroxide as PTC catalyst, 249
Bases in PTC reaction, selection of, 325
Benzaldehydes, oxidation of, 548
Benzamide, N-alkylation of, 406
Benzanilide, N-alkylation of, 406
Benztriazole, N-alkylation of, 401
Benzyl alcohols, oxidation with NaOCl, 509
Benzyl chloride, carbonylation products, 596
Benzylcinchonium salts as chiral PTC, 577

Benzylic chlorination with hypochlorite, 520
Benzylic groups, oxidation of
　air, 535
　hydrogen peroxide, 527
　hypochlorite, 518
　permanganate, 506
　persulfate, 541
Benzyltriethylammonium salt separation, 292
Betaines, as phase transfer catalysts, 152
Beyer–Villiger oxidation of ketones, 533
Bicarbonate ion displacement, 368
Bio-oxidation of PTC catalysts, 293
Bisphenol-A, O-alkylations of, 413, 452–470
Bisulfide displacements to thiols, 362
Bis-binaphthyl crown ethers, as PTC, 156
Bis-phenolates, transfer of, 148
Bis-quaternary salts as PTC catalysts, 147
Block copolymers by PTC, 459
Bonding of PTC catalysts to silica, 251
Boranes, PTC hydrogen peroxide oxidation, 533
Borohydride reductions, 565–569
　of acid chlorides to alcohols, 567
　of alkyl halides to alkanes, 567
　alkyl mercury compounds to radicals, 568
　reduction of azides to amines, 566
Bromate, oxidations with, 549
Bromide displacement reactions, 347
　catalyst activity comparison in, 238, 251
　third liquid phase system of, 254
Bromide-chloride exchange reactions, 349
Bromination
　of aromatics with aq. HCl or HBr, 531
　of benzylic hydrogens with NaOCl, 520
　of olefins, three-phase systems, 256
Bromochloropropane, selective reaction, 137
2-Bromoethylbenzene, dehydrobromination, 420
2-Bromooctane, dehydrobromination of, 421

Bromo-groups on aromatics, replacement, 521
4-Bromo-2,6-dimethylphenol polymers, 415
Butyraldehyde, C-alkylation of, 391

Cadmium iodide as a PTC catalyst, 249
Calixarenes
  O-methylation of, 415
  as phase transfer catalysts, 185
Caprolactam, N-alkylation of, 401
Carbanions, air oxidation of, 535
Carbazole, N-alkylation of, 401
Carbene reactions in PTC systems, 423
Carbodimides, by oxidation of thioureas, 513
Carbohydrates
  O-alkylation of, 411
  transfer to organic phases, 185
Carbon monoxide in carbonylations, 595
Carbon monoxide as reducing agent, 612
Carbon number, "C#," correlation of PTC, 270, 276
Carbon tetrachloride and NaOH as oxidant, 547
Carbonate ion displacements, 368, 409
Carbonyls, borohydride reduction, 565
Carbonylation reactions, 595–605
Carbonylation
  of alkynes, 602
  of aziridines, 604
  of azobenzenes, 604
  of chloromethyl groups on polymers, 495
  of olefins, 600
  mechanism, involving PTC, 596
  of phenol, 605
  of thiiranes, 605
Carbon–carbon bond formation using PTC
  alkylation reactions, 383
  chiral phase transfer catalysts for, 576–591
  Darzen's condensation, 435
  Michael addition reactions, 429
  Wittig reactions, 433

Carbon–carbon chain polymers by PTC, 474
Carboxylate anion displacements, 355–360
Carboxylates, reduction to alcohols, 568
Carboxylic acids
  alkylation of, 391
  by carbonylation of alkyl and aryl, 596
  by carbonylation of alkynes, 602
  derivatives for analytical use, 626
  by hydrolysis of vinylic dibromides, 617
  by olefin oxidation with periodate, 549
  by oxidation of alkynes, 505
  by oxidation of benzylic methyls, 518
  by oxidation of olefins, 503
  peracids from, 543
Catalysts for PTC (see PTC catalysts)
Catalyst concentration and kinetic order, 104
Catalyst degradation, insoluble PTC, 246
Catalyst selection, criteria for, 266
Catalyst separation and recycle, 291
Catalyst structure
  and catalyst efficiency, 13, 127, 236, 284
  C vs O-alkylation selectivity, 387
  hydroxyethyl-, for alkylations, 391
Catalysts for inverse PTC, 180
Catalysts for PTC (see Phase transfer catalysts)
Catalysts for PTC, 128–180
  commercial sources of, 304
  containing hydroxyethyl-, hydroxide, 286
  C# description of, 270
  methyltributylammonium salts as, 285
  phenylacetonitrile alkylations, 236
  quantitative correlations of, 270
  Q# description of, 277
  selection of, 284
  separation and recycle of, 291
  solvent-effect on choice of, 284
  stability of quaternary salts in, 288
  tetrabutylammonium salts as, 284
  trioctylmethylammonium salts as, 285

Catalysts for transfer of other species, 179
Cation effect, with polyethylene glycol, 161
Cations, transfer to organic phases, 186
Cation–anion interaction energies, 85
Cation–anion pairs, aggregation, 87
Cation–anion separation in PTC, 85
Cation–anion–solvent polarity relations, 267
Cellulose, O-alkylation of, 411
Cerium co-catalyst, with persulfate, 543, 544
Cesium fluoride as co-catalyst for PTC, 179
Chain length of quaternary salts, effect on transfer, 28, 29
Charge density of cations and anions, 267
Chemical modification of polymers, 484–495
  of pendant groups on polymers, 491
  of terminal positions of polymers, 489
Chiral phase transfer catalysis, 576
Chiral alkylation, 576–591
  in amino acid synthesis, 586
  of ketimines, 587
  of oxindoles, 586
  of Schiff bases, 589
Chiral epoxidation, Jacobson catalyst, 526
Chiral halogenation of olefins, 532
Chiral induction with PTC
  catalyst structure and, 583
  effect of reaction conditions on, 583
  mechanism of, 578
  Michael addition reactions, 589
Chiral insoluble PTC catalysts, 243
  amino-acid polymers as, 244
Chiral catalysts for olefin epoxidation 517, 526
Chiral PTC catalysts
  cinchonium and cinchionidine salts, 587
  Darzens reaction, 436
  Michael addition reactions, 430
  oxidations with air, 535

Chiral quaternary salts as PTC catalyst, 146
Chloride displacement reactions, 347–350
Chlorination
  of aromatics with aqueous HCl, 531
  of aromatics, electrochemically, 552
  of benzylic hydrogens with NaOCl, 520
  with carbon tetrachloride and NaOH, 547
Chloroacetonitrile, Darzen condensation, 436
Chloroacetylene, from dichloroethylenes, 423
Chlorobromopropane, selective reaction, 137
2-Chloroethyl vinyl ether, alkylation, 414
Chloroform and NaOH to dichlorocarbene, 423–429
Chloromethyl-polymers, displacements, 495
Chloromethylstyrene, in preparation of, 216
Chloroprene, by dehydrohalogenation of, 422
Choice of PTC catalysts, variables in, 266
Choice of solvents for industrial PTC, 315
Chromate and dichromate, oxidants, 546
  electrochemical regeneration of, 546, 547
  transfer to organic phases, 546
Chromate as a co-catalyst in perborate, 546
Chromium trioxide, transfer by crowns, 547
Cinchonium salts, as chiral PTC, 577, 583
  for chiral amino-acid synthesis, 587
Clay, oleophilic, as support for PTC, 249
Cluster-type quaternary salts as PTC, 149
Coal, PTC oxidation with permanganate, 506
Cobalt, as co-catalyst for PTC hydrogen, 571

646 / Index

Cobalt carbonyl, co-catalyst, 595, 601
Commercial applications of PTC, 626–630
Commercial process for monomers, 414
Commercial PTC processes, solvent choice, 315
Commercial sources of PTC catalysts, 304
Comparison soluble and insoluble PTC, 238, 239
Complexation, 111
  of potassium salts and crown ethers, 111
  of inorganics, by polyethylene glycol, 159
  metal salts as PTC co-catalysts, 178
  with nickel compounds, as catalysts, 517
Condensations under PTC conditions, 429, 431
Condensation polymerization by PTC, 452
Condensation reactions
  aldol reactions, 431
  Darzen's reaction, 435
  polyethylene glycols as PTC for, 167
Copper co-catalyst for hydroxylation of, 528
Copper-palladium co-catalysts, 540
Cost of common commercial PTC catalysts, 304
Counterion effects
  ambident ion selectivities, 388
  on hydroxide ion transfer, 35
  on catalyst choice, 323
Coupling alkene, alkyne, alkyl halide, 613
Co-catalysts for PTC reactions, 17, 125, 175
  alcohols, for cyanide displacement, 178
  alcohols, for dehydrohalogenation, 178, 422
  alcohols, dihalocarbene generation, 426
  alcohols, for hydroxide reactions, 37–40, 177, 326
  alcohols, for sugar alkylation, 178
  alumina as, in sulfide displacement, 363
  aluminum chloride as, 179
  cerium, in persulfate oxidations, 543, 544
  cesium fluoride as, 179
  chromate and dichromate as, 546
  cobalt carbonyl, for carbonylations, 595, 601
  cobalt as, for hydrogenations, 571
  cuprous salts as, 178
  effect on PTC polymerization, 139
  formate reductions, ruthenium as, 568
  gold, for nitric acid oxidations, 547
  iodide as, for displacements, 179
  iodide as, in aldehyde alkylations, 391
  iron carbonyl as, for carbonylation, 597
  Fe or Cu for hydroxylations, 527, 528
  metal salts and complexes as, 178
  molybdate for alkyne oxidation with, 527
  osmium, for persulfate oxidation of, 545
  osmium for periodate oxidations, 549
  palladium as, for oxidations, 540
  palladium as, cyanide displacement, 346
  palladium as, for coupling reaction, 614
  palladium as, for hydrogenolysis, 607
  ruthenium, in persulfate oxidations, 543
  ruthenium as, hypochlorite oxidn., 518
  ruthenium as, for hydrogen peroxide, 526, 527
  ruthenium as, in aniline oxidation, 530
  ruthenium as, for bromate oxidation, 549
  transition metal and PTC, 594
  transition metal-hydrodebromination, 606
  transition metals, for oxidations, 522
  tungstate, molybdate, epoxiation, 523–529

Criteria for PTC catalyst selection, 266
Criteria for solvent selection for PTC, 305
Cross-link effect on triphase catalysts, 217, 227
Crown ethers as PTC catalysts, 153–158
  bound to insoluble supports as PTC, 220, 235
  as catalysts for solid-liquid PTC, 113
  for dichromate oxidations, 547
  as high-temperature PTC catalysts, 291
  by oligomerization of epoxides, 156
  preparation of, 153
  on quaternary salts, as PTC, 150
  separation of, by extraction, 293
  solubilization of potassium salts, 111
  transfer, lithium aluminum hydride, 568
  for transfer of permanganate, 502
12-Crown-4, preparation of, 154
18-Crown-6, preparation of, 154
Cryptand groups, on insoluble resins, 220, 235
Cryptands as PTC, 153
  as high-temperature PTC catalysts, 291
  preparation of, 155
Cuprous salts as co-catalysts for PTC, 178
Cyanamide, N-alkylation of, 408
Cyanate ion, in displacement reactions, 370
Cyanation, electrochemical, aromatics, 552
Cyanide displacement reactions, 343–348
  acid chlorides to acyl nitriles, 346
  alcohols as solvents, co-catalysts, 178
  catalyst comparison for, 164, 238, 239
  catalyzed by PTC on clay supports, 249
  solid–liquid PTC systems, 114, 234
Cyanide, reaction with epoxides, 346
Cyanoacetate alkylation to polymers, 474
Cyclodextrins as inverse PTC catalysts, 180, 608
  for hydrogenation, 571
  for olefin oxidation, 540
  for polymerizations, 482
  for reduction of aryl halides, 568
Cyclohexadiene oxidation by singlet oxygen, 539
Cyclopentadiene, alkylation of, 399
Cyclophosphazenes, as PTC catalysts, 169
C- vs O-alkylation in PTC reactions, 385
  in alkylation of 2-pyridone, 405
  catalyst structure effect on, 387
  in phenol alkylations, 416
  solvent effect on, 387
  in vapor-phase PTC reactions, 390
C-Alkylation using PTC, 383–388
  C–H acidity and, 384
  chiral products from, 576, 591
  choice of bases for, 325
  of ketones, 384
  polyethylene glycol PTC catalysts, 168
N-Alkylation using PTC, 400–410
  of amides, 405, 408
  choice of bases for, 325
  of heterocycles, 400
  polyethylene glycols catalysts for, 166
  of sulfonamides, 407
O-Alkylation using PTC, 410
  choice of bases for, 325
  polyethylene glycol PTC catalysts, 417
  of sugars, alcohols as co-catalysts, 178

Darzens condensations, chiral PTC, 435, 436
Decarboxylation of hydroxy-acids, 549
Decomposition of quaternary salts, 131
  during distillation steps, 295
  by Hofmann reaction, 288
  of insoluble PTC catalysts, 246
  by internal displacement, 288
  of phosphonium salts, 132
  temperature effect on, 290
Dehydrochlorination
  of poly(vinyl chloride), 486
  of vinylidene chloride copolymers, 488
Dehydrofluorination
  of poly(vinylidene fluoride), 488

Dehydrohalogenation
  alcohols as co-catalysts for, 178
  catalysts for, 422
  in manufacture of chloroprene, 422
  mechanisms of, 107
  polyethylene glycol catalysts for, 163, 166
  three-phase PTC system, 255, 258
Density of active sites on insoluble, catalysts, effect on activity of, 225
Derivatization of compounds by PTC for, 622
Design of PTC reactions, variables for, 266
Desoxybenzoin, alkylation of, 385
Desulfuriation with iron carbonyls, 613
Deuterium exchange using PTC, 438
  of allylbenzene, 440
  of chloroform, 101
  of fluorene, 99, 440
  of heterocycles, 440
  of imidazoles, 440
  of indenes, 440
  of ketones, 440
  as a measure of acidity, weak acids, 440
  of phenylacetonitrile, 101
  of pyridines, 440
  of thiazoles, 440
  of thiophenes, 440
  of weak acids, 331
Diacyl disulfides, from acid chlorides, 364
Diacyl peroxides from acid chlorides, 369
Diacyl sulfides, from acid chlorides, 364
Dialkyl carbonates, from carbonate salts, 368
Dialkyl peroxides from alkyl halides, 369
Dialkyl sulfides
  from alkyl halides, 362
  permanganate oxidation of, 506
  hypochlorite oxidation of, 514
Diarylimines, in Michael reactions, 430
Diarylmethanes, air oxidation of, 535
Diaryl sulfides, persulfate oxidation, 541
Dibenzazepine, N-alkylation of, 402
Dibenzo-18-crown-6, preparation of, 153

Dibromoalkanes
  dehydrohalogenation to alkynes, 423
  polythioetherfication with, 462
  reaction with potassium carbonate, 465
Dibromocarbene
  addition to olefins, 426
  generation of from bromoform, 426
Dichlorobenzene, in displacements, 418
1,4-Dichlorobutene, in displacements, 358
Dichlorocarbene, other dihalocarbenes, 423–428
  addition to olefins, 423
  in amino acid synthesis, 429
  analytical use of, 623
  catalyst structure effect on, 138, 139
  catalysts for preparation of, 423
  chloroform, 423
  correlation by Q# of catalyst, 282
  hydroxyethyl quaternary catalysts, 423
  in imidazopyridines synthesis, 428
  polyethylene glycols as catalysts, 167, 423
  reaction with alcohols, 427
  reaction with sulfur compounds, 428
  from sodium trichloroacetate, 429
gem-Dichlorocyclopropanes, 412
  in alkylation of phenylacetonitrile, 397
Dichromate and chromate
  oxidations with, 546
  transfer to organic phases, 546
Diene reduction to mono-ene, 569
Diethyl malonate, vapor phase alkylation, 390
Diffusion effects with insoluble PTC, 245
Diffusion of reactants in insoluble PTC, 222
Dihaloalkanes, reduction of, 570
Diketones, by oxidation of alkynes, 505
Dimer by-products in cyanide with acid, 346
Diols
  by cleavage with periodate, 549
  cocatalysts, for hydroxide transfer, 37–40, 177
  by olefin oxidation, 533, 545
Diphenic acid by phenanthrene oxidation, 527

Diphenylmethane, alkylation of, 399
Dipolar groups, insoluble PTC supports, 241
Dipolar polymers, as PTC, 171
Disadvantages for commercial use of PTC, 631
Displacement reactions using PTC, 339–370
  with acetate, 355
  advantages of use of PTC for, 339
  with azide, 358
  with bromide, 352
  with carbonate and bicarbonate, 368
  with carboxylate salts, 355
  on chloromethylated polymers, 495
  with cyanate anions, 370
  with cyanide, 343
  with fluoride, 353
  with formate, 355
  with hydroxide, 367
  with iodide, 352
  iodide as co-catalyst for, 179
  with nitrite ion, PTC in, 366, 367
  on pendant groups on polymers, 491
  with peroxide anions, 369
  phosphide and phosphinite anions in, 370
  polyethylene glycol catalysts for, 166
  polysulfoxides and polysulfides as, 174
  on poly(vinyl chloride), kinetics, 486
  reaction variables involved in PTC, 340
  solid-liquid, crown-ether catalysts, 113
  with sulfides, 363
  with sulfite, to alkanesulfonates, 366
  with superoxide anions, 369
  temperature effect on, 341
  with thiocyanate, 364
Disposition of waste material from PTC, 293
Distillation for catalyst separation, 294
Distribution of quaternary salts, 132
Disulfides, from oxidation thiols, 547, 552
Disulfide anion, in displacements, 364
Dithionite reductions, 569
  of dihaloalkanes to alkenes, 570

  of ketones to alcohols, 570
  of perfluoroalkyl iodides, 570
Divalent anions, transfer of, 31, 148
Doublet anions in organic phases, 79
Dual PTC, two catalysts simultaneously, 125, 175
  in vapor-phase PTC alkylations, 390

Electrochemical PTC reactions, 551
  acetoxylation of aromatics, 552
  benzyl alcohol oxidation, 511
  cerium oxidations, 552
  chlorination of aromatics, 552
  chromate regeneration, 546, 547
  cyanation of aromatics, 552
  dichromate oxidations, 552
  hypochlorite regeneration, 511
  periodate regeneration, 550
  permanganate regeneration, 501
  reductions with, 571
Elimination reactions with PTC, 419
Ephedrine salts, selective catalysts, 152
Epichlorohydrin
  of phenols, 414
  O-alkylations with, 411
  polyetherification with, 457
Epoxidation
  with hydrogen peroxide, 523
  by hypochlorite oxidation of olefin, 514
  and Jacobson manganese catalyst, 526
  and nickel catalysts, 526
  NaOCl in three-phase PTC system, 256
  of olefins, 514, 533, 541
  perfluoroolefins, 515
  of unsaturated ketones, perborate, 550
Equivalent conductance quat. solutions, 80
Equilibrium features of halide exchanges, 347
Ester interchange with PTC, 357
Esterification, betaines as PTC for, 152
Esterification by PTC, analytical use, 622
Esterification (see carboxylate salts)
Esters, alkylation of, 391
Ethers
  for analytical uses, 622

650 / Index

by PTC reactions, 368, 410
oxidation of, with permanganate, 505
polyethers, 457
polyethylene glycol as catalysts, 166
at polymer terminal positions, 490
Ethoxylated quaternary salts as PTC, 150
Ethyl acetate, solvent for hypochlorite, 509
Ethyl acetoacetate, vapor phase alkylation, 390
Ethylene oxide oligomerization to, 156
Extraction into organic phases
  anion effect on, 310
  hydroxide, 35, 271
  picrate, 273
  quaternary bromides, 272
  quaternary picrates, 273
  quaternary salts, 26
  solvent effect on, 309
Extraction processes for separation of, 291
Extraction of anions (see transfer)
Extraction mechanism of PTC, 89
  kinetics of, 91
  mechanism of, 49
  early criteria for distinguishing, 97

Factors in choice of PTC catalysts, 284
Favorski reaction, 391
Ferrate, oxidations with, 550
Ferricyanide, oxidations with, 550
Fluoalkylation of, 399
  deuterium exchange of, 438
Fluoride displacements with PTC, 347–354
  aromatic halides, 353
  sulfonyl chlorides, 354
Fluorinated alcohols, O-alkylation of, 412
Fluorobenzene, from cyclopentadiene, 428
Fluorouracil, N-alkylation of, 403
Formaldehyde
  reductions with, 571
  transfer to organic phases, 186
Formate
  displacement reactions with, 355

insoluble PTC catalysts for, 237
reduction with, 568
Free energy of halide transfer, 24
Free radical oxidation, use of PTC in, 538
Free radicals from alkylmercury cpds., 568
Functionalization of insoluble PTC, 216
Fundamentals of PTC, 23–120

Generation of quaternary salts *in situ*, 140
Glycerol, transfer to organic phases, 185
Glycidyl ethers by PTC etherifications, 411
Glycols, co-catalysts for hydroxide ion, 37–40
Gold, co-catalyst, 547
Guanidinium bis-quaternary salts, PTC, 144, 148
Guidelines
  exploring new PTC reactions, 326
  choice of insoluble PTC catalysts, 299
  PTC catalyst choice, 284
  screening new PTC reactions, 332

Halide displacement and exchange, 347
Halide ions
  charge to volume ratio, 24
  free energy of transfer to organics, 24
  ionic radius effect on transfer, 24
Haloacetylenes, by dehydrohalogenation, 423
Halogenation of aromatic hydrocarbons, 531
Halo-acids, decarboxylation of, 549
Hard–soft cations and anions, effect of, 267
Heck reaction, 613
Hectorite clay, support of PTC catalyst, 249
Heterocycles, deuterium exchange on, 440
Heterocyclic quaternary salts as PTC, 142, 145
Hexaalkyldisilanes by PTC reactions, 165

Hexaalkylguanadinium salts as PTC, 144
Hexamethylenetetraamine, N-alkylation
  of, 410
Higher-temperature PTC catalysts
  4-aminopyridinium salts, 291
  crown ethers, 291
  cryptands, 291
  polyethylene glycols and ethers, 291
  TDA-1, 287, 291
Hofmann degradation of quaternaries,
  288
Hydration of anions, during PTC, 41
  ambident ions, effect on, 386
  anion transfer effect, 40–46
  catalyst activity effect, 270
  hydroxide, kinetic factor, 106
  hydroxide ion transfer, 37
Hydrazinobetaines, as PTC, 153
Hydrideopentacyanocobaltate co-catalyst,
  571
Hydrides, transfer to organic phases, 186
Hydrocarbons, acidic, alkylation of, 399
Hydrogen bonding, effect on transfer, 29
Hydrogen bromide, aqueous
  brominations with, 531
  transfer to organics, 183
Hydrogen chloride, aqueous
  chlorinations with, 531
  transfer to organics, 183
Hydrogen chloride gas, exchange on
  RBr, 350
Hydrogen peroxide
  alkyne oxidations with, 527
  epoxidation of olefins, 533
  epoxidation with Jacobson catalyst,
    526
  epoxidation with nickel catalysts, 526
  olefin oxidations with, 526
  oxidation of alcohols, 529
  oxidation of aromatic amines, 529
  oxidation of aromatic hydrocarbons,
    527
  oxidation of benzylic groups, 527
  oxidation of ketones to lactones, 533
  oxidation of olefins, 522
  oxidation of olefins to diols, 533

oxidations with, 521
transfer to organic phases, 521
use with W, Ru, Fe co-catalysts, 527
Hydrogenation using PTC systems, 570,
  605
  aromatics to saturated compounds, 611
  olefins, selective, 609
Hydrogenative dehalogenation of PCBs,
  258
Hydrogenolysis of alkyl and aryl halide,
  605
Hydrolysis of alkyl halides, 368
  during malonic ester alkylation, 392
  of polymers, 495
  polyesters, 465
  polyethylene glycols PTC catalysts,
    166
  vinylic dibromide to carboxylates, 617
Hydroquinone, oxidation of, 512, 539,
  547
Hydroxide ion in PTC systems, 282, 383
  alcohols as co-catalysts for, 326
  betaines as catalysts for, 152
  counterion effect on, 36
  dehydrohalogenation mechanisms with,
    107
  displacements reactions of, 367
  early criteria for mechanisms of, 97
  effect of PTC structure on transfer, 35
  effect of stirring on, 102
  effect of water concentration on, 36
  enhancement by alcohol co-catalysts,
    37–40
  factors affecting, 32
  hydration as a kinetic factor in, 106
  hydration effect on ambident ions, 387
  hydration of, reactivity, transfer, 319
  hydroxyethyl-quaternary salts PTC,
    286
  inverse PTC mechanism for, 108
  mechanistic types with, 89–108
  polyethylene glycols as catalysts, 287
  isomerization of olefins, 439
  quaternary structure, stability and, 289
  Q# to optimize catalysts for, 278
  reverse, PTC mechanism for, 106

selection of base for reactions of, 325
selectivity values for, 35
transfer by hydroxyalkyl salts, 151
transfer to organic phases, 92, 271
Hydroxyls on polymers, activity rise, 493
Hydroxyacids, decarboxylation of, 549
Hydralkylation reactions, 391
2-Hydroxyethyl groups, in catalysts for, 423
  for hydroxide transfer enhancement, 151, 286
  ephedrine-based as PTC, 152
Hydroxyl groups on crown ethers, effect, 241
Hydroxylation, of olefins and aromatics, 523, 528, 532
Hypobromite oxidations using PTC, 508
Hypochlorite ion, transfer of, 510
Hypochlorite oxidations using PTC, 508
  alcohols, 508
  aldehydes, 508
  amides, 512
  amines, 512
  benzylic hydrogens, Ru co-catalyst, 518
  chlorination of benzylic hydrogens, 520
  dialkyl sulfides, 514
  nickel catalysts for, 517
  nitriles, 512
  olefins, 514
  olefins to epoxides, 515
  oxidation of quinones, 512
  perfluoroolefins, 515
  phenanthrene, 515
  polynuclear aromatic hydrocarbons, 518
  thioamides, 512
  thioureas to carboximides, 513
Hypochlorous acid, species present in, 508

Ibuprofen, alkylation of phenylacetates, 392

Imidazoles
  N-alkylation of, 401
  deuterium-hydrogen exchange, 440
Imidazole-based quaternary salts as PTC, 146
Indene, alkylation of, 399
  deuterium-hydrogen exchange, 438
Indoles
  N-alkylation of, 401
  carbonylation of aziridines, 604
Industrial use of PTC, variables in, 266
Industrial PTC, solvent choices, 315
Industrial use of PTC techniques, 626
Inorganic reagent concentration effects, 15, 51–71
Inorganic polar surface, aid to PTC, 252
Inorganic solids as supports for PTC, 228, 248, 250, 251
Insoluble PTC catalysts, 20, 207–300
  acetate displacement reactions, 357
  active site density effect, 225
  activity of liquid-third-phase, 253
  alkylation of phenylacetonitrile, 236
  amino acid polymers as, 243
  arsonium groups on, for oxidations, 533
  choice of, for PTC reactions, 299
  comparison with soluble catalysts, 238, 239
  crosslinking effect on, 227
  crown ethers and cryptand groups on, 235
  cyanide displacement reactions with, 234
  degradation of, during reactions, 246
  dipolar functions on, as PTC, 241
  effect of agitation on, 222
  effect of catalyst quantity on rate, 231
  effect of particle size of, 223
  effect of porosity on activity, 229
  effect of ratio of reactant volumes, 246
  effect reagent size on diffusion in, 245
  factors affecting reactivity of, 300
  hypochlorite oxidations, kinetics, 510
  inorganic solid supports, 248
  kind of PTC groups attached, 232

kinetic equations for, 247
polyethylene glycol as active group, 240
polyethylene glycol in formation of, 256
quaternary salt groups on, 232
quaternary salts on clays as, 249
selenium, for oxidations, 533
silica and alumina as, 250
solvent effect on, 230
spacer chain effect on activity, 230
thiocyanate displacement reactions, 365
three-liquid phase systems, 255
use of, to facilitate separation, 296
Insoluble PTC catalysts (see Triphase and polymer-bound PTC)
Intercalated salts on clay as PTC, 250
Interfacial area in PTC, rate effect, 98
Interfacial catalyst-rich layers in PTC, 99
Interfacial concentration of PTC agents, 99
Interfacial mechanism of PTC
  deuterium exchange in chloroform, 101
  early criteria for distinguishing, 97
  kinetics for Makosza mechanism, 94, 95
  mechanistic details of, 92
  modified PTC mechanism for, 89
Interfacial tension, effect on PTC rate, 98
Intrinsic rate, organic-phase reactions, 2
  vs. mass-transfer rates, 51–71
Inverse micelles, mechanism of transfer, 9
Inverse PTC
  cyclodextrin as catalyst, for olefins, 540
  cyclodextrins as, 180
  hydrogenolysis of alkyl halides, 608
  hydroxide ion in, 108
  metal salts and complexes as, 181
  pyridine-N-oxide as, 182
  reduction of aryl halides, 568
  tetramethylammonium salts as, 182
  transfer of organics to water, 179
Iodide co-catalyst for PTC displacement, 65, 179, 358, 391

Iodide anion PTC reactions
  displacement by bromide, 352
  displacement reactions, 347
  catalyst comparison for, 238, 251
Iodide ion "poisoning" effect, 323
  kinetics of, 64
  phenylacetonitrile alkylation, 396
Ion pairs in chiral PTC, 81, 577
  correlation with Q#, 283
Ion-effect on kinetics of PTC reactions, 87
Ion-exchange resins, as PTC catalysts, 295
Iron carbonyl
  co-catalyst for carbonylation, 597
  reduction of nitroaromatics, 611
  reductions with, using PTC, 572
Iron co-catalysts, for oxidation, 527, 528
Isobutyraldehyde, C-alkylation of, 391
Isocyanates, from cyanate displacement, 370
Isocyanates, from acyl azides, 359
Isomerization of olefins
  allybenzene, 271
  catalyst structure effect on, 274
  hydroxide ion concentration effect, 106
  three-phase PTC system, 257
Isotactic anionic polymerization by PTC, 479

Jacobson catalyst, stereospecific epoxidation, 526

Ketimines, alkylation, amino acid synthesis, 393, 588
Ketocarboxylic acids, carbonylation of, 597
Ketols by olefin oxidation, 503
Ketone alkylation
  alcohol oxidation with permanganate, 505
  alkylation of, 384
  Beyer–Villiger oxidation, 533
  borohydride reduction, 565
  chiral alkylation of, 577

chromate oxidation of, 546
condensation with aldehydes, 432
decarboxylation of hydroxy-acids to, 549
deuterium-hydrogen exchange in, 438
olefin carbonylation of, with R-I, 601
by olefin oxidation, 527
by oxidation of alcohols, 529
oxidation of, carbon tetrachloride, 547
oxidation of with hydrogen peroxide, 533
oxidation of, with hypochlorite, 508
by persulfate oxidation of alcohols, 543
reduction of with dithionite, 570
reduction, lithium aluminum hydride, 568
Wolff–Kishner reduction, 572
Keto-aldehydes by alkyne oxidations, 527
Kinetics, of solid-liquid PTC reactions, cyanide displacement reactions, 114–119
Kinetic order of PTC reactions, 51, 105
in solid–liquid PTC, 119
Kinetics of PTC reactions, 51–119
air oxidations, 538
analysis of, 71
bromide–chloride exchanges, 351
carbonylation of benzyl chloride, 596
catalyst concentration kinetic order, 104
of chiral alkylations, 577, 581
condensation polymerization, 454
dependence on ion-pairing, 82
deuterium exchanges, 99, 440
dichromate oxidation of xylene, 546
displacement reactions, 357
displacements, poly(vinyl chloride), 485
effect of interfacial tension on, 98
effect of ion-pair aggregation on, 87
effect of stirring and agitation, 102–104
extraction mechanism for, 51
extraction mechanism kinetics for, 91
fractional kinetic orders in, 51–71, 87
hypochlorite oxidations, 509, 510

insoluble catalysts, effect on, 247
intrinsic rate-limited steps in, 57, 68
Langmuir effects in, 486
limiting cases of, 54
Makosza interfacial mechanism, 94
mass-transfer limited reactions, 51–71
non-steady state, 74
olefins isomerization, 99
phenol-O-alkylation, 417
phenylacetonitrile alkylation, 99, 396
poisoning, co-catalysis by iodide, 65
2,4,6-tribromophenol alkylation, 413
solid–liquid displacement reactions, 113
thiocyanate displacements, 365

Lactams, N-alkylation of, 401
Lactones, Beyer–Villiger oxidation of, 533
Langmuir kinetics with PVC, 486
Leaving group choice in displacements, 322, 340
Lignin, O-alkylation, epichlorohydrin, 411
Liquid crystalline polyethers by PTC, 460
Liquid–liquid phase transfer catalysis, 49
Liquid–solid PTC reactions, 108–119
Lithium aluminum hydride
reductions with, 568
transfer with crown ethers, 568
LiI to aid in transfer of, 568
Low-temperature oxidation with PTC, 512

Macrocyclic and macrobicyclic compounds, 153
Macrocyclic polyethers, s-l PTC catalysts, 109
Macroporous resins, for insoluble PTC, 216, 229
Magnesium peroxide, with acid chlorides, 369
Makosza interfacial mechanism, 89, 92, 94, 99, 100
Malonate esters
carbonylation ethyl chloroacetate, 600
minimization of hydrolysis in, 392

Michael additions, 430
  vapor phase alkylation of, 390
Malonic acids, conversion to aldehydes, 549
Malononitrile, alkylation of, 397
Manganese carbonyls, co-catalyst, 515, 617
Manganese catalyst, for epoxidation, 526
Manganese dioxide, complexation of, 501
Manufacture of dinitriles by PTC, 344
Mass transfer to organic phases, 23
  effect on kinetics of PTC, 51–71
  kinetics of, 65, 68
  relationship to agitation, 102
Material safety data, PTC catalysts, 293, 303
Mechanisms of PTC reactions, 48–119
  Brandstrom–Montonari mechanism, 49
  chiral induction, 577
  early criteria for distinguishing, 97
  extraction mechanism, 89
  extraction mechanisms of, 49
  interfacial mechanism details, 89, 92
  involving hydroxide reactions, 89–108
  kinetic order to distinguish, 105
  liquid–liquid reactions, 49
  Makosza interfacial mechanism, 89, 92, 99
  polymerization by PTC, 453
  reverse PTC mechanism, 106
  sequence of steps in, 48
  solid–liquid PTC, 109
  Starks extraction mechanism, 49
Membrane reactors, for PTC, 302
Mercaptans, by NaSH displacements, 362
Mercury alkyls, borohydride reduction, 568
Metal carbonyl anions, transfer, 595
Metal co-catalysts for oxidation, 522
Metal oxides supports for insoluble PTC, 250
Metal salts
  complexation with polyethylene glycol, 159
  as co-catalysts, 178
  as inverse PTC, 181
Metals and metal hydrides, transfer of, 186

Methacrylate polymers, as insoluble-PTC, 228
Methyl acrylate, dihalocarbene addition to, effect of catalyst structure on, 139
Methylation of deoxybenzoin
  catalyst structure in, 277
  correlation by Q#, 281
Methylene chloride, S-alkylation with, 419
Methyltributylammonium salts
  separation of, by extraction, 292
  as PTC catalysts, 285
Michael additions using PTC, 429
  acetylenes, 430
  allylphenylsulfones as addends, 431
  chiral phase transfer catalysts in, 589
  chiral PTC, 430
  malonate esters as addends, 430
  phenylacetonitrile as addend, 430
  thiophenols as addends, 430
  trichloromethylate anion in PTC, 139
Microwave heating in PTC reactions, 17, 412
Modification of polymers by PTC, 452–495
Modified interfacial mechanism, 89
  deuterium exchange in chloroform, 101
  kinetics of, 96
  mechanistic details of, 95
Molecular weight variation in polymers, 454
Molybdate co-catalyst, for oxidations, 523, 527, 529
Monomers from O-alkylations, 414
Morpholine-based salts as PTC, 145
Multisite phase transfer catalysts, 148

Naphthalene, persulfate oxidation of, 544
Naproxen by PTC alkylation, 392
Naphthoquinone by persulfate oxidation, 544
Napthoxide anion, as ambident ion, 416
Neopentyl alcohol, O-alkylation of, 412
Nernst layer on insoluble catalysts, 222
Nickel compounds as co-catalysts, 517, 526

Nickel cyanide co-catalyst, 602
Nitric acid, oxidations with, 547
Nitriles
    hypochlorite oxidation of, 512
    alkylation of, 395
    reduction of, 568
Nitrite
    analysis for traces of with PTC, 367
    displacement reactions, 366
    oxidation with, 367
Nitrobenzene, hydrogenation of, 571
Nitro-compounds, reduction of, 570, 612
Nitrogen heterocycles, N-alkylation of, 401
Non-steady-state kinetics in PTC, 74
Nylon-66, modified, as a PTC catalyst, 365

Olefins
    bromination of, 256
    carbonylation of, 600, 601
    catalysts for epoxidation of, 515
    coupling, alkynes and alkyl halides, 613
    dehydrohalogenation of, 422
    dihalocarbene addition to, catalyst, 139
    dihalocarbene additions, 423
    epoxidation with hydrogen peroxide, 523
    epoxidation with hypochlorite, 515
    halogenation with HX and oxidant, 532
    isomerization of, 167, 331
    isomerization, kinetics of, 99
    isomerization in three-phase PTC, 257
    oxidation to diols and epoxides, 541, 545
    oxidation to ketones, 540
    oxidation, chromate or dichromate, 546
    oxidation with hydrogen peroxide, 522, 533
    oxidation with hypochlorite, 514, 518
    oxidation with periodate, 549, 550
    oxidation with permanganate, 503
    from Wittig reactions, 434
Olefins, fluoro-, oxidation of, 256, 515
Oligomeric ethers by PTC O-alkylation, 415

Omega phase, definition, in s-1 PTC, 116
Optically active catalysts (see Chiral)
Organic phase polarity, effect on PTC, 29, 31
Organic phase volume, insoluble PTC, 246
Organic reactant size, in insoluble PTC, 245
Organic solvents for PTC reactions, 131
    ion-aggregation effect on, 81
    polarity effects on PTC, 267
    transfer of PTC catalysts into, 24
Organophilicity of PTC catalysts, 270
Osmium co-catalyst for oxidations, 545, 549
Oxidations using PTC, 500
    of acetals with permanganate, 506
    by air or oxygen in PTC systems, 537
    of alcohols, benzylic and allylic, 543
    of alcohols with bromate, 549
    of alcohols with hydrogen peroxide, 529
    of alcohols with hypochlorite, 508
    of alcohols with permanganate, 505
    of aldehydes with hypochlorite, 508
    of aldehydes with permanganate, 506
    of alkynes with hydrogen peroxide, 527
    of alkynes with permanganate, 505
    of amides with hypochlorite, 512
    of amines with hypochlorite, 512
    of amines with permanganate, 506
    of aromatic amines, 529
    of aromatic hydrocarbons, 527
    of benzylic groups, 506, 518, 541
    of carbanions by air, 535
    with carbon tetrachloride + NaOH, 547
    with chromate and dichromate, 546
    of coal with permanganate, 506
    of dialkyl sulfides, 514
    electrochemical, with PTC systems, 551
    epoxidation of olefins, 533
    with ferrate and ferricyanide, 550
    free radical type, PTC in, 538
    halo-acids to aldehydes, 549

hydrogen peroxide for, 521
hydroxy-acids to aldehydes, 549
hypochlorite, nickel catalysts for, 517
hypochlorite, polynuclear aromatics, 518
hypochlorite and hypobromite, 508
hypochlorite of perfluoroolefins, 515
of ketones with hypochlorite, 508
of amlonic acids to aldehydes, 549
naphthalene, 544
with nitric acid, 547
of nitriles with hypochlorite, 512
of olefins, 503
of olefins, with hydrogen peroxide, 243, 522, 526, 533
of olefins, with air, Ru, Pd, Rh, Cu, 540
of olefins, with amino-acid polymer, 243
of olefins to epoxides, 514, 515
of olefins to ketones, with air, 540
of olefins with periodate with Os, 549
of olefins to epoxides, diols, 541
oxidation of ketones to lactones, 533
with perborate, 550
with periodate, 549, 550
with persulfates, 540
pH dependence in hypochlorite, 508
of phenanthrene with hypochlorite, 515
of phenols with permanganate, 506
with polyethylene glycol catalysts, 167
of polyethylene with permanganate, 506
of quinones by hypochlorite, 512
with singlet oxygen, using PTC, 539
with sodium nitrite, 367
of sulfides with permanganate, 506
of sulfides with persulfate, 542, 544
with superoxide, 551
of thioamides with hypochlorite, 512
of thioureas with hypochlorite, 513
of triarylformazans, permanganate, 506
of triazolines with permanganate, 506
using "oxygen carriers" as PTC, 538
of weak acids, with KOH and air, 331
Oxidative coupling of phenols, 476
Oxidative replacement of Br by Cl, 352

Oxygen, singlet, generation of, PTC, 539
Oxygen carriers, use as PTC for oxidation, 538
Oxygen, oxidations with, using PTC, 534–540
Oxygen transfer to organic phases, 188

Palladium compounds
　with copper, air oxidation co-catalyst, 540
　co-catalyst for coupling reactions, 614
　co-catalyst for hydrogenolysis, 607
　co-catalyst, cyanide on vinyl halides, 346
Parameters, quantitative, PTC catalysts, 270
　Q-number to define "accessibility", 277
Particle size effect, solid-liquid PTC, 114, 223
Patents for commercial uses of PTC, 627
PCB destruction by thiol S-alkylations, 419
Pentaerythritol, etherification of, 256
Peptides, for transfer of inorganics, 172
　as insoluble PTC, 243
Perfluoralkyl iodides
　sulfite displacement on, 366
　reduction with dithionite, 570
Perfluoroolefins, epoxidation of, 515
Periodate, oxidations using PTC, 549
　for decarboxylation of halo-acids, 549
　for decarboxylation of malonic acid, 549
　electrical regeneration of, 550
　in oxidation of alcohols, 550
　in oxidation of aldehydes, 550
　in oxidation of amides, 550
　in oxidation of hydroxy acids, 549
　in oxidation of olefins, 550
Permanganate anion
　transfer to organic phases, 501
　transfer by polyethylene glycols, 160
　electrochemical regeneration of, 501
　solubility of in organic solvents, 502
Permanganate oxidations using PTC, 500
　of acetals, 505

658 / Index

of alcohols, 505
of aldehydes, 505
of alkynes, 505
of amines, 506
of benzylic groups by, 506
of coal, 506
of ethers, 505
of olefins, 503
of phenols, 505
of polyethylene, 506
of sulfides, 506
of triarylformazans, 506
of triazolines, 506
Peroxide ion displacements using PTC 369
Per-carboxylic acids by persulfate, 543
Persulfate, 540
   conversion of acids to peracids, 543
   olefins to diols and epoxides, 541
   oxidation of alcohols, 543
   oxidation of benzylic hydrogens, 541
   naphthalene oxidation, Ce-cocatalyst, 544
   sulfides oxidation, Ce cocatalyst, 544
   polymerization initiation with, 481
   sulfides to sulfoxides and sulfones, 542
Pharmaceuticals by PTC O-alkylations, 412
Pharmaceuticals by phenylacetonitriles, 396
Pharmaceuticals by PTC N-alkylation, 402
Phase transfer catalysis (see also PTC)
   analytical chemistry use of, 622
   anion activation in, 11
   basic concepts of, 1
   displacement reactions using, 339
   fundamentals of, 23, 48
   general steps in, 23, 48
   industrial perspectives in, 626
   kinetics for extraction mechanism, 51
   mechanisms for, 23, 48
   organic solvents in, 16
   physical chemistry of, 23, 48
   rate matrix for, 5
   reaction types with, 3
   reactive type catalysts for, 11
   reductions with, 565
   temperature effects on, 17
Phase transfer catalysts, 123–258
   accessibility character of, 274–277
   accessible type, 128
   acetate displacement reactions, 356
   agitation effect on decomposition o, 134
   4-aminopyridinium salts, stable PTC, 143
   amino-acid polymers as, 243
   anion activation by, 127
   arsonium salts as, 140
   azapropellane-based, 146
   betaines as, 152
   bis-quaternary salts for, 147
   calixarenes as, 185
   catalyst comparison for cyanide PTC, 164
   characteristics of common catalysts, 304
   characteristics which affect use, 123
   chiral, 576–591
   chloride-bromide exchange reactions, 350
   choice of, 266
   clay and inorgnaic supports for, 249
   cluster-type quaternary salts for, 149
   commercial sources of, 124, 304
   cost and availability, 124
   co-catalysts for, 177
   crown ethers as, 153
   crown-ether substituted quaternary, 150
   cryptands as, 153
   C# description of, 270
   for deuterium-hydrogen exchange, 440
   in dichlorcarbene reactions, 423
   dipolar polymers as, 171
   displacement reactions, 342, 344
   distribution into organic phases, 24, 126
   dual PTC catalysts, 176
   effect of size of, 276
   ephedrine-based quaternary salts as, 152
   ephedrine-based salts as, 430
   ethoxylated cyclophosphazenes, 169

ethoxylated quaternary salts as, 150
ethoxylated triazines as, 169
fractional kinetic orders in, 87
generation of, *in situ,* 140
guanidinium salts, high stability, 144
heterocyclic quaternary-onium salts, 142
high-temperature, 287
hydrazinobetaines as, 153
hydroxyl-substituted, as PTC, 151
imidazole-based, 146
insoluble catalysts for PTC, 207–258
inverse, cyclodextrins for, 180
kinetic order of PTC reactions in, 104
macrocyclic and macrobicyclic ethers, 153
morpholine-based, 145
multisite catalysts, 148
organophilicity of, 270
phosphonium salts as, 140
piperidine-based, 145
in polyetherification reactions, 463
polyesterification, 465
polyethylene glycols and ethers as, 157, 162
in polymerization reactions, 455
polypode ethers as, 168
polysiloxanes as, 175
polysulfides as, 174
polysulfones as, 175
polysulfoxides as, 173
poly(amides) as, 171
poly(sulfoxides) as, 171
poly(vinylpyrrolidone) as, 172
quantitative correlations of, 270
quaternary salts as, 125–153
quaternary salts on clays, 249
quinine and quininium salts as, 430
quinuclidine-based quaternary salts, 146
Q# description of, 277
recovery and recycle of, 124, 188, 291
selection of, 284
selenonium salts as, 140
separation and recycle of, 124, 188, 291
siloxane-substituted salts as, 153
soluble polymers with polar groups, 171
special types of, 20
stability of, 124, 132
structure effect on efficiency, 127
structure effect on selectivity, 137
structure of, effect on selectivity, 135
structure-activity relationships of, 267
substituted quaternary salts as, 151
sulfides as, via sulfonium salt for, 362
sulfonium salts as, 140
tetraalkylammonium salts as, 125
tetrazolium salts as, 147
toxicity of, 124
tris(3,6-dioxaheptyl)amine as, 157, 165
types of, 18
volume to charge ratio effect in, 23
Phase transfer catalyst structure, effects on counter ion of, selection of, 323
air oxidations using PTC, 536
aniline halogenation, 532
aromatics halogenation, 532
carbonylation, 603
C vs O-alkylation, 387
dehydrohalogenation, 487
hydrogenolysis of bromides, 606, 607
poly(vinyl chloride) reactions, 485
product distributions, 530, 542
reductions with iron carbonyl, 612
PTC, improvement commercial processes, 626
PTC functional groups, on supports, 220
PTC in industrial processes, 293
PTC process design, 319
PTC rate matrix, 5–8
anion placements in, 325
and choice of agitation, 322
Phase transfer reactions
acidity of anions effect, 328–332
advantages for industrial uses, 627
agitation effect on, 319
condensation, commercial utility of, 432
condensations, 429
C-alkylations, 383
dehydrohalogenation, 419
displacement reactions with, 432

effect of solvent on, 305
guidelines, of new reactions, 327
kinetic orders in, 105
kinetics of, 49
optimal water concentrations in, 318
oxidations using PTC, 500–552
oxidative coupling of phenols, 476
polymerization, polymer chemistry, 452–495
polymerizations by, 452–495
poly(vinyl bromide), 489
poly(vinyl chloride), 486
poly(vinylidene fluoride), 488
PTC dihalocarbene generation, 427
quick screening of, 332
reductions using PTC, 565–572
using strong bases, 383
variables in design of, 266
Wittig reaction, 433
Phenanthrene
oxidation to diphenic acid, 518, 527
oxidation to phenanthrene epoxide, 515
Phenetoles by PTC O-alkylation, 412
Phenols
carbonylation of, 605
alkylation of, 143
O-alkylation of, 411, 412
C-vs O-alkylation, catalyst, 139
oxidation to quinones with oxygen, 539
oxidation with permanganate, 505
oxidative coupling by PTC, 476
nitrite reaction on aryl halides, 367
Phenolphthalein polythiocarbonates, 471
Phenothiazine, N-alkylation of, 402
Phenoxide anion, as ambient ion, 416
Phenylacetic acid, C- and O-alkylation, 392
Phenylacetone, alkylation of, 384
Phenylacetonitrile alkylation, 395
comparison of catalysts for, 236
hydroxide concentration effect on, 106
in Michael addition to acetylenes, 430
Phenylethyl bromide dehydrobromination, 255
Phosgene, in PTC polymerizations, 465

Phosphide ion displacements using PTC, 370
Phosphinite ion displacements using PTC, 370
Phosphonium quaternary salts as PTC, 140
bound to insoluble supports, 220
decomposition from hydroxide, 289
in Wittig reactions, 434
Phosphate co-catalyst for epoxidation, 523
Photochemistry in PTC
oxidations with singlet oxygen, 572
polymerization, 483
reductions using PTC, 572
thiocyanate reactions with, 366
Phthalimide, N-alkylation, 401, 408
Pinacol, co-catalyst with hydroxide, 38
Piperazines, N-alkylation of, 402
Piperidine-based salts as PTC, 145
Poisoning effect of iodide ion, 64, 323
Polarity of organic solvents, effect on, 131, 270
Polyamides by PTC polymerization, 454
Polycarbonate formation using PTC, 464
Polycarbonates from PTC displacement, 369
Polychloro-aromatics, dehalogenation of, 258
Polyesters by PTC polymerization, 454
catalyst effect on molecular weight, 465
from acrylic acid, 454
Polyethers, by PTC polymerizations, 457
from bis-phenol-A, 413
Polythioethers by PTC polymerization, 457
Polyethoxylated compounds as PTC, 168
Polyethylene oxidation with permanganate, 506
Polyethylene glycols as PTC catalysts, 157–168
anion effect on, 161
bound to insoluble supports, 220
catalysts for O-alkylation of, 411
catalyst for dihalocarbene reaction, 167
catalysts for condensation reaction, 167
catalysts for C-alkylations, 167

catalysts for dehydrohalogenations, 166
catalysts for displacement reaction, 166
catalysts for hydrolysis reactions, 166
catalysts for N-alkylation reaction, 166
catalysts for olefin isomerization, 167
catalysts for oxidation reactions, 167
catalysts for O-alkylations, 416
catalysts Williamson ether synthesis, 166
cation effect on, 162
comparison with other catalysts, 164
complexes with metal salts, 159
effect of molecular weight, 164
formation of three-phase systems, 256, 420
high-temperature PTC catalysts, 291
in hydroxide transfer reactions, 287
on insoluble PTC supports, 240
preparation of, 157
quaternary-groups on, as PTC, 150
separation, third-layer formation, 293
solvent effects on, 162
third-phase formation with, 162
transfer of inorganic salts, 159
Polymerization, of phosgene, diphenols, 139
Polymer backbone modification using PTC, 484
Polymers, as supports for insoluble PTC 218–230
  active site density effect, 225
  active site location on, 218
  agitation effect on activity, 221
  characteristics affecting activity, 220
  chloromethylstyrene in preparation, 216
  comparison with soluble PTC, 208, 210
  components of, 211
  cross-linking effect on, 227
  crown ethers on, 220
  cryptands on, 220
  diffusion effect in, 209
  effect of catalyst quantity on rate, 231
  effect of particle size, 223
  effect on ambident ion selectivity, 416
  functionalization of, 216
  functionalization of resins for, 216
  macroporous resins for support of, 216
  microporous resins for support of, 216
  molar scale, 210
  polyethylene glycol groups on, 220
  polymer support, preparation, 211
  pore size of insoluble resins for, 216
  porosity effect on activity of, 229
  preparation of polymer-bound, 210
  quaternary onium salts on, 220
  solvent effect on, 230
  spacer chain effect on activity, 230
  spacer chains in, 219
  styrene–divinylbenzene supports for, 211
  suspension polymerization procedure, 216
Polymers, chemical modification, 484–495
Polymerization by PTC reactions, 452–495
  acrylic acid to polyesters, 454
  anionic polymerizations, 479
  carbon–carbon chains by, 474
  catalyst effect on molecular weight, 465
  condensation polymerization, 452
  dibromoalkanes in, 465
  free-radical polymerizations, 481
  mechanisms for, 453
  phosgene in, 465
  polyamides, 454
  polycarbonates from, 464
  polyesters, 454
  polyethers, 457
  polyphenylene oxides, 477
  polyphosphonates, 454
  polysulfonates, 454
  polysulfones, 454
  polythiocarbonates from, 464
  polythioesters, 454
  polythioethers, 457
  Poly-C-alkylation, 474
  radical anions in, 187
  redox initiators for, 482
  solvent effect on condensations, 468
  thiophosgene in, 467

662 / Index

Polymer-end functional groups, reaction, 489
Polymer-modification by PTC reactions, 452–495
Polynuclear aromatics oxidation, 527, 535, 544
Polyphenylene oxide, end-capping, 490
Polyphenylene oxides, oxidative coupling, 476
Polyphosphonates by PTC polymerization, 454
Polypode polyethers as PTC catalysts, 168
Polysiloxanes as PTC catalysts, 175
Polysulfide-links in rubber, scission of, 489
Polysulfides as PTC catalysts, 174
Polysulfonates by PTC polymerization, 454
  as PTC catalysts, 175
Polysulfoxides as PTC catalysts, 173
Polythiocarbonates, PTC polymerization, 464
Polythioesters by PTC polymerization, 454
Polythioethers, by PTC polymerization, 461
Poly(2,6-dimethylphenol), capping of, 490
Poly(amides), by PTC polymerization, 171
Poly(methyl methacrylate), hydrolysis, 495
Poly(sulfoxides), as PTC catalysts, 171
Poly(vinyl bromide), reactions of, 489
Poly(vinyl chloride), in displacements, 485
  acetate displacement, 486
  azide displacement on, 486
  azide displacement reactions with, 359
  dehydrochlorination of, 486
  kinetics of, 486
Poly(vinylpyrrolidone), as PTC catalyst, 172
Pore size, of resins for support of PTC, 216, 229

Potassium carbonate as base in PTC, 325
Potassium hydroxide as a base in PTC, 325
Potassium metal, transfer with TDA-1, 165
Potassium permanganate, transfer of, 502
Potassium salts, solubilization, crowns, 111
Product selectivity, effect of catalyst, 135, 139
Propionaldehyde, C-alkylation of, 391
Protein fragments, as insoluble PTC, 243
Purines, N-alkylation of, 403
Pyrazole, N-alkylation of, 401
Pyridines, deuterium-hydrogen exchange, 440
Pyridine-N-oxide as inverse PTC, 182
Pyridones, ambident ions in alkylation, 416
Pyridones, selective O-alkylation of, 416
2-Pyridone, N- and O-alkylation of, 405
Pyrrole, N-alkylation of, 401

Quadrupole ions, effect on PTC kinetics, 87
Quantitative parameters PTC catalysts, 270, 277
Quaternary ammonium salts, 124–153
  accessible type, 128
  adsorbed on clays, 249
  aggregation of, in organic phases, 80
  anion activation by, 128
  anion, effect on catalyst activity, 135
  arsonium salts, as PTC, 140
  bases for dehydrohalogenations, 107
  bound to insoluble supports, 220
  carbonylation, 603
  cation–anion–solvent effects, 267
  concentrations at interfaces, 99
  containing hydroxyethyl groups, 286
  crown-ether substituted, as PTC, 150
  decomposition of, temperature on, 290
  distribution into organic phases, 126
  ethoxylate-substituted, as PTC, 150
  extractability, of, 28, 272, 309, 501
  for dehydrohalogenation, 107
  heterocyclic, as PTC catalysts, 142

hydroxy-substituted as PTC, 151
in situ generation of, 140
insoluble supports, 232
ionization constants of, 83
permanganate salts, extractability, 501
phosphonium salts, as PTC, 140
Q-number correlations with, 283
stability of during PTC reactions, 131, 288
structure effect on air oxidations, 536
structure, allylbenzene isomerization, 274
structure effect on decomposition of, 290
structure effect on displacements, 485
structure effect on hydrogenolysis, 606, 607
structure effect, picrate extraction, 309
structure effect on rates, 13
structure effect on transfer, 521
structure effects on distribution, 26, 29
structure, effect on halogenation, 532
structure, on hydroxide transfer, 35
structure, on product distributions, 530
structure-activity relationships of, 267
third liquid phase formation with, 255
sulfonium salts, as PTC, 140
Quinuclidine-based salts for PTC, 146
Q-number, to describe catalysts, 277–282
Q-number, on ambident ion alkylation, 386

Radical initiated polymerizations using, 481
Radical-anions, transfer of, 187
Rates (see Kinetics)
Rate enhancement in three-phase PTC, 258
Rate increase by use of dual catalysts, 176
Rate matrix for PTC reactions, 5
Rates of PTC reactions
 effect of anion hydration on, 42
 ionic strength effects on, 43
 using insoluble PTC catalysts, 248
Reactants, size effect, insoluble PTC, 245

Reaction variables in PTC reactions, 12
Reactions catalyzed by phase transfer, 3
Reactions catalyzed by insoluble PTC, 212
Reactive-inverse PTC catalysts, 183
Recovery of phase transfer catalysts, 188, 291
Recycle of insoluble PTC catalysts, 124, 246
Redox polymerization by PTC, 482
Reductions using PTC, 565–572
 borohydride for, 565
 of dihaloalkanes to alkenes, 570
 dithionate anion for, 569
 electrochemical, 552
 formate salts for, 568
 hydrogen, using PTC, 605
 lithium aluminum hydride for, 568
 nitroaromatics to anilines, 570
 of nitrobenzene, effect of catalyst, 611
 nitrobenzene, variable products by, 571
 of nitrogen compounds, using PTC, 610
 of perfluoroalkyl iodides, 570
 photochemical, using PTC, 572
 polyethylene glycols as PTC, 167
 with sulfur-containing anions, 569
Removal of phase transfer catalysts, 124, 188
Resin supports for insoluble PTC, 219
Resin-bound phase transfer catalysts (see Polymer-bound catalysts)
Rhodium, as co-catalyst for carbonylation, 601
Rhodium, as co-catalyst for oxidation, 540
Ring substitution degree, insoluble PTC, 225
Robinson, annelations, chiral, 430
Ruthenium compounds, as co-catalysts, 540
 aromatic hydrocarbons, 527
 bipyridine complex, 518
 bromate oxidations, 549
 hydrogen peroxide oxidations, 526, 530
 hypochlorite oxidations, 518

664 / Index

periodate oxidations, 550
persulfate oxidation, 543

Salcomine, as an oxygen transfer agent, 539
Schiff bases, chiral alkylation of, 589
Screening of new PTC applications, 332
Sebacic acid, in polymerizations, 455
Selectivity in PTC reactions, 137
  in acetate on bromochloral, 356
  bis-quaternary salts for alkylation, 147
  catalyst structure effect on, 135, 139
  hydrogenation unsaturated ketones, 609
  hydroxy-substituted salts, hydroxide, 151
  polysulfoxides for alkylation, 173
  structure effect of catalyst on, 135, 139
  in transfer of anions, 24, 27
Selenonium salts as PTC catalysts, 140
Separation and recovery of PTC catalyst, 124, 188
  adsorption methods for, 295
  of aminopyridinium catalysts, 293
  of benzyltriethylammonium catalysts, 292
  of crown ethers, 293
  distillation processes for, 294
  extraction processes for, 291
  of methyltributylammonium catalysts, 292
  of polyethylene glycols, 293, 295
  for recycle of PTC catalysts, 291
  of TDA-1 as PTC catalyst, 293
  with tetrabutylammonium catalysts, 292
  using three-liquid-phase PTC system, 255
Silica, as support for PTC catalysts, 250
  adsorption of PTC catalysts on, 295
  with chemically bound PTC groups, 251
Silochrome as support for insoluble PTC, 250
Siloxane-substituted quaternary salts, 153
Singlet oxygen generation, 539

Size of quaternary cations on transfer, 126
Sodium borohydride (see Borohydride)
Sodium formate (see Formate salts)
Sodium hydroxide, as a strong base, 325
Sodium metal, transfer of, with TDA-1, 165
Solid-liquid PTC reactions, 108–119
  kinetics of, 113
  omega phase formation in, 116
  polyethers as catalysts for, 109
  water effect in, 110
Solubilization of catalyst plus anions, 307
Solubilization (see Transfer), 568
Soluble polymers containing polar group, 171
Solvents, and their effects in PTC, 16, 131
  alkylation of deoxybenzoin, 312
  anion solvation, 306
  anion transfer effect of, 29, 31, 132, 307
  bromide displacements, effect of, 312
  cation-anion distance effect on, 267
  chiral alkylations, effect of, 583
  criteria for selection of, 305
  C vs O-alkylation selectivity, 387
  dielectric constant and, 307
  diffusion of, through insoluble PTC, 231
  displacement reactions, effect on, 340
  distribution of quaternary salts, 132
  effect on insoluble-PTC activity, 230
  examples for choice of, 311
  hydrogen bonding effect of, 30
  in hypochlorite oxidations, 509
  industrial process considerations, 315
  ion-pair separation effect, 306
  leveling effect in, 29
  polarity of, 307
  in polycondensations, effect of, 468
  quaternary salt transfer effect, 309
  reactivity effect of, 305
  screening of, 318
  solubilization of PTC ion pairs, 307

stabilization of transition state, 306
transfer of salts with PEGs, effect, 162
trifluoroethoxide displacements, 312
use of no solvents in PTC, 314, 316
Solvent-swelling of insoluble PTC, 230
Solvent-free PTC, 314, 316, 401
Spacer chains in insoluble-PTC catalysts
  effect, triphase catalyst activity, 219, 230
  synthesis of, 219
  effect on activity, 230
Special quaternary ammonium salts as PTC, 142
Stability of PTC catalysts
Stability of salts in PTC
  guanidinium salts, 144
  insoluble PTC catalysts, 246
  4-aminopyridinium salts, 143
  simple quaternary salts, 131, 288
  phosphonium salts, 132
Stereoselective PTC (see Chiral)
Sterically hindered phenols, O-alkylation, 413
Sterically hindered thiols, S-alkylation, 418
Stirring effect in PTC reactions, 102–104
Stoichiometric phase transfer reactions, 533
Strong base reactions using PTC (see Hydroxide)
Structure of phase transfer catalysts, and effect of
  accessibility relationship effects, 281
  choice of, effect of, 284
  distribution into organic phases, 26
  effect, C vs O-alkylation select., 387
  effect on transfer, 24
  effect on picrate extraction, 309
  effect on PVC displacements, 487
  hydroxide ion transfer, 35
  quaternary salt PTC catalysts, 267
  selectivity effect of, 137, 138
  size relationship effects, 276
Suberonitrile manufacture, PTC process, 344

Substitution reactions (see also Displacement reactions)
  cyanide displacements, 234
  iodide as co-catalyst for, 179
  phase transfer catalysts for, 151
  polyethylene glycols as catalysts, 166
  polysulfoxides and polysulfides as, 174
Sulfide displacement reactions, 362
  with acid chlorides, 364
  dialkyl sulfides from, 362
  PTC systems for, 362
  PCB removal with, 364
Sulfide reduction of nitroaromatics, 570
Sulfides, dialkyl- or diaryl-
  nitric acid oxidation to sulfoxides, 547
  oxidation by permanganate, 506
  oxidation with hypochlorite, 514
  persulfate oxidations of, 541, 544
  as phase transfer catalysts, 362
Sulfinate salts, bis-, in polymerization, 456
Sulfite displacements, alkanesulfonates, 366
Sulfonamides, N-alkylation of, 407
Sulfones in PTC reactions
  alkylation of, 397
  by permanganate on sulfides, 506
  by persulfate oxidation of sulfides, 541
Sulfonium salts as PTC catalysts, 140
  generation *in situ*, as PTC catalyst, 141
Sulfonyl azides from sulfonyl chlorides, 359
Sulfonyl chlorides, in polymerizations, 455
Sulfonyl fluorides, sulfonyl chlorides, 354
Sulfoxides
  by hypochlorite oxidation sulfides, 514
  by nitric acid oxidation of sulfide, 547
  by permanganate oxidation sulfides, 506
  by persulfate oxidation of sulfides, 541, 544
Superoxide anion, oxidations with, 369, 551
Suppliers of common PTC catalysts, 304

Surface activity of PTC catalysts, 98
Surface layers of salts, ion transfer, 115, 116
Swelling of insoluble PTC catalysts, 230
Synergy from use of two PTC catalysts, 176

TDA-1, [see tris(3,6-dioxaheptyl)amine], 165
Teflon solid support for insoluble PTC, 250
Tellurite, in oxidation of thiols, 552
Temperature effects on PTC reactions, 17
  on displacement reactions, 341
  quaternary salt decomposition, 290
Terminal polymer positions, reactions, 489
Tertiary alcohols from boranes, 533
Tertiary amides, reduction of, 568
Tert-butyl chloride, PTC displacements, 365
Tetraalkylammonium or phosphonium salts (see Quaternary salts)
Tetraarylporphyrins with Mn as PTC, 515
Tetrabutylammonium salts, 284
  in third-liquid phase PTC systems, 253
  separation of, 291
Tetramethylammonium salts as inverse PT, 182
Tetrazolium salts as PTC catalysts, 147
Theobromine, N-alkylation of, 403
Theophylline, N-alkylation of, 403
Thiadiazoles, from thioamides, 513
Thiazoles, deuterium exchange in, 440
Thia-crown compounds as phase transfer, 362
Thiiranes, carbonylation of, 605
Thioamides, oxidation of, 512
Thiocarbenes, generation of, 429
Thiocresylate displacement reactions, 238
Thiocyanate in PTC displacements, 364
  comparison of catalysts for bidentate ion, selectivity of, 364
  kinetics of, 365
Thioesters, alkylphosphinate, reduction, 568

Thioetherification by PTC reactions, 418
Thiol groups on polymers, reactions of, 494
Thiolates, in Michael reactions, 431
Thiols
  oxidation to disulfides, 552
  from NaSH displacements, 362
Thiol-carboxylic acids from thiiranes, 605
Thiophenes, deuterium exchange in, 440
Thiophenol, Michael additions in, 430
Thiophenols, oxidation of, 547
Thiophosgene, in polymerization, 467
Thiophosphinic, oxidation, 514
Thioureas, oxidation of, 513
Third-layer (liquid) systems in PTC, 10, 252
  for dehydrohalogenation, 258, 420
  interfacial systems in s-l PTC, 99
  omega phase in s-l PTC reactions, 116
  polyethylene glycol catalyst for, 162, 417
  for separation of PTC catalysts, 293
Toluenes, nitro-substituted, oxidation, 537
Toxicity of common PTC catalysts, 304
Trace water effects on solid–liquid PTC, 114
Transfer to organic phases
  of acids, 183
  agitation and stirring effect on, 102
  aluminum chloride, 186
  ammonia, 188
  anion hydration effect on, 40–46
  anion lipophilicity and, 324
  anion structure effect on, 25, 310
  anion valence effect on, 26
  of anions, 322
  of carbohydrates, 185
  catalysts for, 23
  of cations, 186
  charge density view of, 269
  of chromate and dichromate, 546
  crown ethers for, 111
  divalent anions in, 31, 148
  effect on PTC displacements, 340
  extraction constants for, 26
  of formaldehyde, 186

of glycerol, 185
of HCl and HBr from aqueous phases, 183
of hydrogen peroxide, 521
of hypochlorite anion, 510
of ion pairs, 23–46
of metals and metal hydrides, 186
omega phase formation in, 116
of oxygen, 188, 539
permanganate by polyethylene glycol, 501
of permanganate by quaternary salts, 501
of permanganate with crown ethers, 502
polarity of organic phase, effect, 29
of polyethylene glycol, 160
of quaternary ammonium bromides, 272
of quaternary ammonium picrate, 273
of radical anions, 187
of rose bengal, singlet oxygen with, 539
selectivity constants for, 24
in solid–liquid PTC reactions, 116
soluble polar polymers for, 171
solvent effect on, 306, 307
of species other than anions, 21
stirring effect on, 102
TDA-1 as PTC catalyst for, 165, 287
of water, 185
Transfer step, mechanisms for, 2, 7
Transfer-rate limited reactions, 276
Transition metal, on hydrodebromination, 606
Transition metal-PTC co-catalysis, 594
Trialkylmethanols, from boranes, 533
Triarylformazans, oxidation of, 506
Triazines, alkylated, 370
Triazines, ethoxylated, as PTC catalyst, 169
Triazolines, oxidation of, 506
Tribromide ion, formation of in PTC, 256, 532
Trichloroacetate displacement, products, 358

Trichloroacetate to dichlorocarbene, 429
Trihaloacetamides, N-alkylation of, 408
Trioctylmethylammonium PTC, separation, 294
Trioctylmethylammonium salts as PTC, 285
Triphase catalysis (see also Insoluble PTC catalysts)
 active site density effect on, 225
 active site location on, 218
 agitation effect on reactivity with, 222
 catalyst parameters effect, 221
 catalyst quantity effect, 231
 chloromethylstyrene in preparation, 216
 comparison with soluble PTC, 208, 210
 crosslinking effect on, 227
 crown ethers on, 220
 cryptands on, 220
 diffusion effect in, 209
 functionalization of resins for, 216
 macroporous resins for support of, 216
 microporous resins for support of, 216
 molar scale, 210
 particle size effect on activity, 223
 polyethylene glycol groups on, 220
 polymerization procedure effect on, 218
 pore size of insoluble resins for, 216
 porosity effect on activity, 229
 preparation and types of, 210, 211
 quaternary onium salts on, 220
 solvent effect on, 230
 spacer chain effect on activity, 219, 230
 styrene-divinylbenzene supports for, 211
 suspension polymerization procedure, 216
 tabular survey of, 212
Triphenyl methane, C-alkylation of, 400
Tris(2,5-dioxaheptyl)amine [TDA-1], 157, 165
 as high-temperature PTC catalyst, 287, 291

permanganate oxidations with, 502
separation of, by extraction, 293
Trithiocarbonate as a reducing agent, 570
Tungstate co-catalyst
  in alkyne oxidations, 527
  in epoxidation of olefins, 523
  in oxidation of aromatics, 527
  in hydrogen peroxide oxidation, 529

Ultrasonic agitation in PTC reactions, 222, 537
Urea derivatives, N-alkylation of, 408
Ureas, alkylated, from cyanate reaction, 370

Vanadium carbonyl anions, in PTC, 607
Vapor phase PTC reactions, 21
  alcohols with HCl or HBr, 185
  alkylation of acetylacetone, 390
  alkylation of diethyl malonate, 390
  alkylation of ethyl acetoacetate, 390
  ambident ion selectivities in, 390
Variables in design of PTC systems, 266
Vinyl benzyl chloride for insoluble PTC, 493
Vinyl bromides, carbonylation of, 600
Vinyl dibromides, hydrolysis to acids, 617
Vinyl halides, in cyanide displacements, 346

Vinyl monomers, PTC-polymerization of, 481
Vinylation of polymer terminal groups, 489

Wacker process, PTC modification of, 540
Wastewater clean-up with hypochlorite, 515
Waste-treatment questions in PTC, 293
Water
  C vs O-alkylation, effect on, 386
  fluoride displacements, effect on, 353
  nitrite displacements, effect on, 367
  optimal concentrations of, 318
  solid–liquid PTC reactions, 110–119
  trace level effects on s-l PTC, 114
  transfer to organic phase, 185
Weak- and strong acid-anions in PTC, 328–332
Williamson ether synthesis, 410–433
  polyethylene glycol catalysts for, 166
  in polyetherification, 459
Wolff–Kishner reductions using PTC, 572

Xanthine, N-alkylation of, 403
Xylylene dihalides in polymerizations, 455, 457